FRUIT FLIES OF ECONOMIC SIGNIFICANCE:

THEIR IDENTIFICATION AND BIONOMICS

D1732996

FRUIT FLIES OF ECONOMIC SIGNIFICANCE:

THEIR IDENTIFICATION AND BIONOMICS

Ian M. White
International Institute of Entomology
London, UK

and

Marlene M. Elson-Harris
Department of Primary Industries
Queensland, Australia

C·A·B International

In association with

ACIAR
(The Australian Centre for International Agricultural Research)

C·A·B International
Wallingford
Oxon OX10 8DE
UK

Tel: Wallingford (0491) 32111
Telex: 847964 (COMAGG G)
Telecom Gold/Dialcom: 84: CAU001
Fax: (0491) 33508

© C·A·B International 1992. All rights reserved. No part of this publication may be reproduced in any form or by any means, electronically, mechanically, by photocopying, recording or otherwise, without the prior permission of the copyright owners.

Published on behalf of
International Institute of Entomology
(an Institute of C·A·B International)
56 Queen's Gate
London SW7 5JR
UK
Tel: 071 584 0067/8
Telex: 265871 (MONREF G)
Telecom Gold/Dialcom: 84: CAU006
Fax: 071 581 1676

For full details of services available from the International Institute of Entomology please contact the Institute Director at the address given

In association with
ACIAR
(The Australian Centre for International Agricultural Research)
GPO Box 1571
Canberra
ACT 2601
Australia
Tel: 062 48 8588
Telex: AA 62419
Telecom Gold/Dialcom: 6007: IAR001
Fax: 062 573051

A catalogue entry for this book is available from the British Library

ISBN 0 85198 790 7

Printed and bound in the UK by Redwood Press Ltd, Melksham

Contents

Preface

In 1988 the authors were assisting Dr R.A.I. Drew (Department of Primary Industries, Queensland Government, Australia), Professor D.E. Hardy (University of Hawaii) and A.J. Allwood (then of the Department of Primary Production, Northern Territory, Australia) to present a training course in fruit fly identification, held at the Universiti Pertanian Malaysia. Each course tutor provided the students with course notes that were bound together to form a training manual. Inevitably, those notes lacked consistency of approach, style, and worst of all, terminology. One evening a group of us discussed the need for a general textbook on fruit fly pest identification that could be used for future training courses and be of value to plant quarantine services and, as a result, the present work was born. Initially however, the two of us each planned separate publications, one on adults (IMW) and the other on larvae (MMEH). Dr P. Ferrar of the Australian Centre for International Agricultural Research suggested and encouraged the unification of these projects and we are grateful to him for that suggestion.

No text book is ever the *last word* on its subject. This work is being produced at a time when several important complexes of pest species are being researched and major field programmes are being carried out in Asia, the Pacific and South America, which will improve on the sometimes fragmentary host data presented in this work. In addition the larval stages of most species of economic importance still need to be described in terms of the many new characters visible with the scanning electron microscope. Although subsequent editions of this work cannot be promised at this stage, the authors would be pleased to receive specimens and data which in any way add to the information presented here, so that any future edition may be an improvement on the first.

Ian M. White
Marlene M. Elson-Harris

Foreword

Tropical fruits are attractive commodities, being both appealing to look at and delicious and healthy to eat. Temperate countries and large population centres in the tropics that cannot produce their own supplies will pay high prices for exotic imports, and nations of the tropical Americas, South-east Asia and the South Pacific see good sources of export income from tropical produce, and are rapidly increasing their horticultural plantings.

Unfortunately for all concerned, fruit flies find most tropical fruits equally attractive, which produces a two-fold problem. The high prices are only paid for fruit imports of top quality; cases unpacked to reveal fruit fly larval feeding and ensuing decay are not well received, and orders quickly switch to more reliable competitors. Even more seriously, markets such as the United States, Japan, and Australia, which are warm enough for fruit fly establishment, impose quarantine restrictions against produce likely to carry new fruit flies into their countries.

Fumigation with ethylene dibromide, an effective answer for many years past, is now virtually unusable because of residue restrictions, and other methods are being researched. These require much more detailed knowledge of the fruit fly problems concerned; also expansion into new areas of production and new species of tropical fruit is leading to fruit fly problems not previously encountered. For these reasons a world reference on fruit flies, their identification and their hosts has become greatly needed.

The present work provides just such a reference, compiled by two experts in the field. Dr Ian White has travelled widely in tropical countries of the world, and has studied collections of adult fruit flies from all regions; and Ms Marlene Elson-Harris has pioneered the use of larval characters in fruit fly taxonomy - which can often help to distinguish taxa whose adults are virtually inseparable. It is particularly pleasing to see this concentration on young stages as well as adults, and on host fruit relationships as well as morphological characteristics; after all, it is a maggot in a fruit that is the economic problem.

This comprehensive work should stimulate further research in an already exciting field, to the benefit of all concerned. A good number of farmers in developing and developed countries will earn better incomes from enhanced fruit exports as a result, and consumers in many countries will gain access to a better and more varied supply of some very healthy foods.

Paul Ferrar
Research Program Coordinator (Crop Sciences)
Australian Centre for International Agricultural Research, Canberra.

Acknowledgements

The authors wish to thank CAB International (CABI) for covering development costs incurred by the senior author (IMW), and the Australian Centre for International Agricultural Research (ACIAR) and the Department of Primary Industries, Queensland Government (QDPI), for funding the larval taxonomy studies of the second author (MMEH). We are also grateful to ACIAR for funding the publication of this work. We also wish to thank Dr K.M. Harris (Director, International Institute of Entomology; IIE [an Institute of CABI]), Dr J.M. Ritchie (Assistant Director, IIE), Dr M. Bengston (Director, Entomology Branch, QDPI) and Dr R.A.I. Drew (Head of the Fruit Fly Research Team, QDPI) for their support and encouragement in producing this work. We are also indebted to The Natural History Museum, London (NHM), for allowing IMW access to their extensive collections and library facilities, CAB International Library Services for carrying out monthly literature searches, and the Royal Society for funding a visit to QDPI by the senior author.

The following specialists kindly reviewed early drafts of sections of the manuscript: Professor S.H. Berlocher (University of Illinois), Professor G.L. Bush and Mr J. Jenkins (Michigan State University), Mr J.E. Chainey and Dr B.L. Pitkin (NHM), Dr R.A.I. Drew and Dr D.L. Hancock (QDPI), Dr P. Ferrar (ACIAR), Dr A.L. Norrbom (Systematic Entomology Laboratory, United States Department of Agriculture; USDA-SEL) and Dr D.K. Yeates (Department of Agriculture, Western Australia; WADA).

We would like to thank the following for kindly providing unpublished data of value to this project: Mr A.J. Allwood (South Pacific Commission Fruit Fly Programme, Fiji; SPC), Dr F.L. Banham (Agriculture Canada); Dr C.O. Calkins (USDA Behavior and Biological Control Research Laboratory, Florida); Dr A.C. Courtice (Tasmania); Mrs S. Facknath (University of Mauritius); Dr C.S.K. Lau (Deptartment of Agriculture, Hong Kong; HKDA); Ms L.E. McComie (Ministry of Food Production, Trinidad and Tobago); Dr N.J. Mills and Dr J.K. Waage (International Institute of Biological Control; IIBC); Mrs S. Niedbala (IIE); Mr B. Merz (Eidgenössiche Technische Hochschule, Zurich); Dr J. Payne (USDA, Byron, Georgia), Dr C.Y.L. Schotman (Caribbean Plant Protection Commission, and Food and Agriculture Organisation; CPPC; FAO); Dr F.C. Thompson (USDA-SEL); Professor

X.-j. Wang (Academia Sinica, Beijing); Dr D. Waterhouse (SPC); and Mr A.M.Wood (CABI Information Services). We would also like to thank the following colleagues for their helpful comments, support and encouragement in producing this work: Dr P.S. Baker and Dr D.J. Greathead (IIBC); Mrs L. Carroll (USDA-SEL); Mr J.C. Deeming (National Museum of Wales); Dr B.S. Fletcher (Commonwealth Scientific and Industrial Research Organisation, Canberra; CSIRO); Dr A. Freidberg (Tel Aviv University); Professor D.E. Hardy (University of Hawaii); Dr J. LaSalle, Dr R. Madge, Dr A. Polaszek, Ms A.K. Walker and Dr M.R. Wilson (IIE); Dr A.C. Lloyd (QDPI); Mr A.C. Pont (formerly NHM); Dr G.L. Steck (Florida Dept. of Agriculture); Dr R.A. Wharton (Texas A&M University); and Mr N.P. Wyatt (NHM). We are also grateful to the late Dr E.M. Hering (Berlin) for compiling a card catalogue of tephritid host data (now in the NHM).

The following entomologists kindly lent or donated specimens for this study, or permitted us to extract data from collections in their charge: Dr M. Bigger (Natural Resources Institute, UK; NRI), Dr R. Contreras-Lichtenberg (Naturhistorisches Museum, Vienna; NHMV), Dr P.S. Cranston (Australian National Insect Collection, CSIRO; ANIC), Dr R.A.I. Drew (QDPI), Dr M.W. Mansell (National Collection of Insects, Pretoria; NCIP), Dr D.K. McAlpine (Australian Museum, Sydney; AMS), Dr W. Mathis (National Museum of Natural History, Washington DC), Dr B.L. Pitkin (NHM) and Dr R.A. Wharton (Texas A&M University). We would also like to thank the following for assisting in the field collection of larval material or for supplying larvae from laboratory cultures: Dr R.A.I. Drew, Mr E.L. Hamacek, Dr D.L. Hancock, Miss J. Grimshaw, Mr B. McCulloch, Mr R. Piper and Ms B. Waterhouse (QDPI); Dr P. Ferrar (ACIAR); Dr C. Lauzon (University of Vermont, USA); Dr C.S. Ooi (University of Malaya, Kuala Lumpur); Dr A. Postle (WADA); Mr E.S.C. Smith (Department of Agriculture and Fisheries, Northern Territory, Australia); plus staff of the Atomic Energy Agencies of the Philippines and Thailand, and the USDA laboratories in Homestead (Florida) and Honolulu (Hawaii).

Many of the illustrations used in this work were produced by artists working for IIE and QDPI. These were Mr G. Kibby and Mr G. duHeaume at IIE (cover; figs 97, 128-130, 163-174, 177, 179, 186, 194, 196-198, 201, 204, 205, 208-213, 217-218, 230-248, 250-254), and Mrs M.C. Romig and Mrs S.A. Sands at QDPI (figs 1, 175, 176, 178, 180-185, 187-193, 195, 199, 200, 202, 206, 219-221). Dr D.L. Hancock (formerly of the Natural History Museum of Zimbabwe) kindly provided illustrations of many African species drawn by Miss J. Duff (figs 214-216, 222-229, 249); Dr M. Bigger (NRI) generously allowed us to reproduce his excellent illustrations of fruit flies that he found in Bhutan (figs 203, 207); and International Pheromone Systems Ltd. kindly provided fig. 2.

The authors are also grateful to Mr D.J. de Courcy Henshaw for producing the camera ready copy and advising on chemical nomenclature, and Mrs A. Greathead for her careful reading of the manuscript. We are also grateful to our spouses, Joy and Harry, for their tolerance during compilation of this work, and for Harry's considerable help with field work which resulted in many new larval and host plant data.

Introduction

The family Tephritidae (=Trypetidae), the true fruit flies, includes about 4000 species arranged in 500 genera. As such, it is amongst the largest families of Diptera (true flies), and one of the most economically important. The larvae of most species develop in the seed-bearing organs of plants, and about 35% of species attack soft fruits, including many commercial fruits. Hence the name fruit flies, although geneticists and some entomologists use that term for the Drosophilidae. Besides attacking soft fruit, the larvae of about 40% of species develop in the flowers of Asteraceae (=Compositae) and most of the remaining species are associated with the flowers of other families, or their larvae are miners in leaf, stem or root tissue. Very few species of known biology are non-phytophagous. Most of the species described in this work are pests or potential pests of soft fruit, but some leaf, stem and flower pests are included, and the use of tephritids in the biological control of noxious weeds is briefly reviewed.

The Fruit Fly Problem

There are fruit pest tephritids in almost all fruit growing areas of the world and their economic importance can be summarized as follows:

- They attack commercially produced fruit;
- Some species have become pests in regions far removed from their native range;
- Quarantine restrictions have to be imposed to limit further spread of fruit fly pests;
- Quarantine regulations imposed by an importing country can either deny a producing country a potential export market, or force the producer to carry out expensive disinfestation treatment.

Monetary estimates of fruit production and fruit fly damage are not available for most countries. However, Australia may be taken as an example, with annual fruit production running at over A$850 million, and potential losses if fruit flies were not controlled are believed to exceed A$100 million (Anon., 1986). The cost of a fruit fly free area being invaded are even greater, e.g. Dowell & Wange (1986) listed eight species that are a major threat to California, and estimated that the statewide establishment of those species would cause crop losses of US$910 million and cost US$290 million to control. The cost of eradicating a fruit fly from even a small island is very large, e.g. it cost Japan Y5 billion (about US$32 million) and 200,000 man days work to eradicate the Oriental fruit fly, *Bactrocera dorsalis* (Hendel), from its south-western islands using the sterile insect release method (Anon., 1986). It is therefore very important that quarantine entomologists can make rapid identifications of fruit flies intercepted with imported fruit produce, so that measures can be taken quickly to try to prevent the establishment of new fruit fly pests. Recent regional reviews of the fruit fly problem are as follows: Australia and the South Pacific, Anon. (1986) and Hooper & Drew (1989); Central and South America, Enkerlin *et al.* (1989) and Schwarz *et al.* (1989b); Europe and temperate Asia, Fimiani (1989) and Fischer-Colbrie & Busch-Petersen (1989); Hawaiian Islands and North America, E.J. Harris (1989); Southern Africa, Hancock (1989); and tropical Asia, Anon. (1986), Kapoor (1989) and Koyama (1989a). The role of taxonomic services in helping to combat the fruit fly problem was discussed by Hardy (1991) and White (1989b; 1991a; 1991b).

Details of import restrictions are too numerous and rapidly changing to be included in the present work, and publications produced by national agricultural departments, the Food and Agriculture Organization of the United Nations (FAO) and journals such as *Citrograph*, should be consulted. Private travellers usually fail to understand the need to abide by regulations banning the import of fruit. The incidence of infested fruit in aircraft baggage has been discussed by Satoh *et al.* (1985).

Another little studied aspect of the fruit fly problem is the effect of ingested larvae on human health. The only available information refers to *Anastrepha* spp. larvae causing abdominal pain and diarrhoea, particularly in children (Jirón & Zeledón, 1979).

Aims and Arrangement of the Book

Previously available identification guides for fruit flies did not allow recognition of flies from outside a small area of coverage, e.g. if an American fruit pest had been transported to the South Pacific region it could not have been correctly identified using the South Pacific literature. Furthermore the last host catalogue with a world coverage is now very out of date (Phillips, 1946). The dangers inherent in the lack of even rudimentary host data for some potential pest species were recently noted by Liquido & Cunningham (1991); those authors cited the example of melon fly, *Bactrocera cucurbitae* (Coquillett), a native of tropical Asia which only became known to science after its establishment in Hawaii. The present work aims to be an identification guide of use to quarantine and pest control entomologists in all regions of the world; a host catalogue listing fruits and other crops attacked by tephritids; a manual for teaching courses; a general introduction to the family; and a review and bibliography of the specialist literature.

About 250 species of Tephritidae have been known to attack fruits that are either grown commercially, or harvested from the wild, and these are all mentioned in the text of the present work. However, many of these species are very rare, and a selection of about 100 species was included in the keys and descriptions. In selecting species for inclusion in this book, the aim was to form a balance between describing only about twenty major pests and covering several hundred fruit associated species, any of which may one day be reared from a commercial host. There was little value in covering only a few major pests as the staff of plant quarantine services already have datasheets on those species. Conversely, to cover several hundred species would have made for a large book that would have been of value only to specialists in tephritid taxonomy. Instead, we decided to include all species known to have been found damaging an economically important plant, our definition of which was any plant listed by Terrell *et al.* (1986). Most of the tephritids injurious to those plants are discussed in detail. However, some tephritids recorded from economic plants are only mentioned briefly for one or more of the following reasons: they are both rare and difficult to identify; they are normally associated with a wild plant and only once reared from an economic host; their only economic host is one of low value; or the part of the plant attacked is not of economic value. Although some tephritids not included in the present work will in the future be reared from economically important plants, it is hoped that this approach has selected the vast majority of species likely to be intercepted in plant quarantine, found as pests or otherwise encountered by economic entomologists.

Each species account includes the following sections:

— **Taxonomic notes**, giving other names (synonyms) by which the species has been known. Further details may be obtained from regional catalogues (Cogan & Munro, 1980; Foote, 1965; 1967; 1984; Hardy, 1977; Hardy & Foote, 1989).
— **Commercial hosts**, notes the pest status of the species and lists those host plants which were listed as being of economic importance by Terrell *et al.* (1986), plus a few other plant species which we regarded as being worthy of mention. The same data are presented in reverse form (listed by plant genus and species) in a host catalogue at the end of the book (p. 433).
— **Wild hosts**, summarizes the natural host range of the species, if known. Specific names were validated using *Index Kewensis*, plus local floras when appropriate.
— **Adult identification**, gives diagnostic notes to supplement the keys to adults. The diagnostic notes are intended to assist in confirming identifications made using the keys and do not give the technical details needed to separate the named species from all others.
— **Description of third instar larva**, gives a description of the final larval instar. Some of these descriptions were based on previously published data and the sources of material studied or published data are indicated.
— **Distribution**, lists the regions and countries from which the fruit fly is known.
— **Other references**, lists selected additional references, with an emphasis on post-1980 publications.

Important note: At the time of going to press the continuing division of the former USSR into independant states has made it impractical to revise its use, in this text, with the new republics which were formerly included in the Soviet Union.

Distribution and Host Relationships

The family Tephritidae is represented in all world regions, except Antarctica. The major pest genera each have a limited natural distribution, as follows:

- *Anastrepha* spp. attack a wide range of fruits in South and Central America and the West Indies, with a few species occurring in the extreme south of the USA. No *Anastrepha* species have become established outside those areas.
- *Bactrocera* spp. (formerly included in *Dacus*) are native to tropical Asia, Australia and the South Pacific regions, with a few species found in Africa and warm-temperate areas of Europe and Asia. One section of the genus, typified by subgenus *B. (Zeugodacus)*, is almost exclusively associated with the flowers and fruits of Cucurbitaceae, and the rest of the genus is associated with a wide range of fruits predominantly of tropical wet forest origin. Some *Bactrocera* spp. have become established in Hawaii, French Guiana and Suriname as a result of modern fruit movement.
- *Ceratitis* spp. attack a wide range of fruits and are native to tropical Africa. *Ceratitis capitata* (Wiedemann) has been established in all other world regions except Asia, whilst several outbreaks in North America have been eradicated.
- *Dacus* spp. are almost all associated with the flowers and fruits of Cucurbitaceae, or with the pods of Asclepiadaceae, and most species are found in Africa. *Dacus ciliatus* Loew has become established in the Indian subcontinent and in the Indian Ocean Islands.
- *Rhagoletis* spp. are found in South and Central America, mostly on Solanaceae, and in the temperate areas of Europe and North America, where most species are associated with the fruits of a single family of plants, and often a single genus. The most important pest species are associated with the Rosaceae and some of these have the potential to become established in new areas.

A summary of the tephritid fauna of each region, and the major identification literature available, is as follows:

- Africa south of the Sahara is known as the **Afrotropical Region** (=Ethiopian Region of earlier workers). About 140 genera are known from that region, including 14 *Bactrocera* spp., 65 *Ceratitis* spp. and about 170 *Dacus* spp. Munro (1984) revised the Dacini and Munro (1964a) produced a useful guide to some of the fruit pest Tephritidae found in the region. Most *Ceratitis* spp. can be separated using papers by Hancock (1984; 1985a; 1987). Unfortunately the only general work on the region is now very out of date (Bezzi, 1924a; 1924b). Other papers by H.K. Munro cover individual genera and full references were given in the regional catalogue by Cogan & Munro (1980); see also Hancock (1985b; 1985c; 1986a; 1986b; 1990; in press, a; in press, b).
- Tropical Asia, including Indonesia to the west of Irian Jaya, the Ryukyu Islands

of Japan and China south of the Yangtze River, forms the **Oriental Region**. About 160 genera are known from that region, including about 180 *Bactrocera* spp. and about 30 *Dacus* spp. Kapoor *et al.* (1980) presented a key to the Indian genera, and monographic works cover all the species known from Thailand and the Philippines (Hardy, 1973; 1974). Recently, Hardy (1982b; 1983a; 1983b; 1985; 1986a; 1986b; 1987; 1988a; 1988b) completed a series of papers describing the Indonesian tephritid fauna; some keys by Hardy (1973; 1974; 1986a; 1987) were recently reprinted in a work by Ibrahim & Ibrahim (1990). The work of Zia (1937) is still of value for southern Chinese species and other references for the Oriental tephritid fauna were given in the regional catalogue by Hardy (1977).

- Australia and the New Guinea area form the **Australasian Region**, and New Zealand plus the Pacific Islands form the **Oceanic Region**. About 130 genera are found in those regions, including about 270 *Bactrocera* spp., *Ceratitis capitata* and 27 *Dacus* spp. Drew (1989a) revised the Dacini and Drew *et al.* (1982) is a useful guide for economic entomologists wanting both taxonomic and other details; a key by Drew (1982a) was recently reprinted in a work by Ibrahim & Ibrahim (1990). Unfortunately there is no general work on other groups from the Australasian Region, but Malloch (1939a) and Hardy (1951) are helpful. The only general work for any parts of the Oceanic Region is Hardy & Adachi (1956) for Micronesia and Hardy & Delfinado (1980) for the Hawaiian Islands. Hardy & Delfinado (1980) discussed the large number of *Trupanea* spp. that have evolved in the Hawaiian Islands, and gave good diagnostic details of several species introduced for weed biological control. Further references can be found in the regional catalogue by Hardy & Foote (1989).
- Europe, temperate Asia, the Middle East and North Africa form the **Palaearctic Region**. About 140 genera are known from that region, including 13 *Bactrocera* spp., *Ceratitis capitata*, 5 *Dacus* spp. and 22 *Rhagoletis* spp. Rohdendorf (1961) provided a key to most of the *Rhagoletis* spp. and many of the *Bactrocera* and *Dacus* spp. were included by Ito (1983-5). Key works covering individual countries include Freidberg & Kugler (1989) for Israel; Ito (1983-5) for Japan; Richter (1970) for European USSR; Shiraki (1933) and Munro (1935c) for Taiwan; Zia & Chen (1938) for northern China; and White (1988) for the United Kingdom. The only regional work is now very old (Hendel, 1927), and White (1988) and the regional catalogue of R.H. Foote (1984) should be consulted for other references.
- Canada, the USA and the northern mountains of Mexico form the **Nearctic Region**. About 60 genera are found in that region, including 20 *Anastrepha* spp. and 24 *Rhagoletis* spp. Most *Rhagoletis* spp. can be identified using a key by Bush (1966), and keys to other genera were listed by Foote & Steyskal (1987), who also provided a generic key; a monograph keying all North American species will soon be available (R.H. Foote, F.L. Blanc & A.L. Norrbom, in prep.). The species were catalogued by Foote (1965) and host records by Wasbauer (1972).
- The remaining areas of the Americas form the **Neotropical Region**. About 90 genera are found in that region, including about 180 *Anastrepha* spp., a member of the *Bactrocera dorsalis* species complex in Suriname and French Guiana, *Ceratitis capitata* and 21 *Rhagoletis* spp. *Anastrepha* and *Rhagoletis* spp. can be

identified using Steyskal (1977) and Foote (1981) respectively. Foote (1980) provided a key to the genera, and the species were catalogued by Foote (1967).

Each major pest genus has a typical pattern of host relationships. Most *Rhagoletis* spp. are restricted to a single plant genus, or a few closely related plant genera, and that pattern of host relationships is described as stenophagous (Fletcher, 1989b). That pattern of host relationship is also typical of the non-pest genera throughout the family. Most *Dacus* and *Bactrocera (Zeugodacus)* spp. show a strong preference for attacking species of a single plant family, typically Asclepiadaceae or Cucurbitaceae, but the pest species will attack other hosts. True polyphagy, that is attacking plants belonging to a wide range of families, is exhibited by the pest species of *Anastrepha*, *B. (Bactrocera)* and *Ceratitis*. *B. (Daculus) oleae* (Gmelin) is the only major pest species of *Bactrocera* which is associated with a single host species. Many non-pest species of *Anastrepha*, *B. (Bactrocera)* and *Ceratitis* have a very narrow range of known hosts, but in many cases that may have more to do with lack of data than fact. However, there is evidence that some pest species have a narrow range of wild hosts, and in some cases unrelated commercial hosts are more likely to be attacked when preferred wild hosts are in short supply (Fitt, 1986). Fitt (1986) found differences in behaviour and physiology between *B. tryoni* (Froggatt), a generalist, and some more specialist species, namely *B. cacuminata* (Hering), *B. cucumis* (French) and *B. jarvisi* (Tryon). He showed that when deprived of any fruit for four days, females of *B. tryoni* readily oviposited into the fruits of host species which were previously unacceptable, but the more specialist species did not so readily accept alternative fruits.

Table 1 summarizes the host and geographic range of each major group within the family Tephritidae, including the non-pest groups. This table also draws attention to the fact that the genera which include polyphagous pests tend to have their origins in tropical wet forest habitats, while those with a narrower range of hosts are either temperate (*Rhagoletis*) or associated with tropical dry area plants such as Asclepiadaceae and Cucurbitaceae (see Drew, 1989a; Hancock, 1989).

The host catalogue presented at the end of this book (p. 433) lists the plants of economic importance and the tephritids recorded from them. Ideally there should be a world catalogue of host data for the Tephritidae. However, construction of such a catalogue will involve scanning thousands of publications and perhaps millions of data labels on specimens in collections scattered through many countries. Even then, the data will be inadequate by comparison to critically gathered survey data. In the present work we have tried to find as many host data as was practical in a limited period of time.

Recently published host lists collated by specialists studying a narrow taxonomic group were the primary source of host data, namely *Anastrepha* spp. (Norrbom & Kim, 1988b), African Dacini (Munro, 1984), Australian and South Pacific Dacini (Drew, 1989a), and North and South American *Rhagoletis* spp. (Bush, 1966; Foote, 1981). However, there are no reliable host catalogues for the Dacini of tropical Asia or the Ceratitini. For species falling into those groups some regional host surveys proved to be of considerable value, for example papers by Syed (1970; 1971) and Syed *et al.* (1970a; 1970b) for the Dacini of Pakistan. Papers surveying the hosts of Asian species adventive in the Hawaiian Islands were assumed to be reliable sources of data

Table 1. Major taxa arranged by habitat, host range and distribution, and their size expressed as a percentage of the family.

Habitat	Host range	Geographic range	Taxa	%
Hosts mostly derived from temperate, or dry/moist tropical habitats; most spp. attack plants of one family or preferred family	Asteraceae; flowers, rarely stems & roots	all regions	Tephritinae, not Tephrellini	40
	Acanthaceae, Lamiaceae & Verbenaceae; flowers & stems	Old World, mostly Africa	Tephrellini	4
	Cucurbitaceae flowers and fruit	Old World, not Africa	*Bactrocera (Zeugodacus)*	2
		Old World, mostly Africa	*Dacus (Didacus)*	3
	Passifloraceae fruit, or pods of Apocynaceae & Asclepiadaceae			
	fruit feeders, or leaf/stem miners, each on a single family of plants	all regions	Adramini & Trypetini, e.g. *Rhagoletis*	19
	fruit feeders, some polyphagous	mostly Africa	Ceratitini (Ceratitina)	4
	stem miners of asparagus	Europe & Africa	Zaceratini	0.1
	cycads	Asia & Africa	Rivelliomimini	0.1
	Poaceae stems	Old World tropics	Ceratitini (Gastrozonina)	3
Hosts mostly derived from tropical wet forest; most pest spp. attack plants of several families	fruits and flowers, mainly of Cucurbitaceae	Africa and Asia	*Dacus (Callantra), D. (Dacus)*	3
	fruits of many families, often polyphagous	Old World, few from Africa	*Bactrocera*, except *B. (Zeugodacus)*	9
		New World tropics	*Anastrepha*	5
		Old World tropics	Acanthonevrini & Phytalmiini	8
	wood, usually decaying			

(Harris & Lee, 1986; Harris *et al.*, 1986b; Vargas & Nishida, 1985a; Vargas *et al.*, 1983b), as were surveys carried out in Guatemala by Eskafi & Cunningham (1987) and Tonga by Litsinger *et al.* (1991).

Some host lists do not give full details of their data sources, for example Kapoor (1970) and Kapoor & Agarwal (1983) for India, Le Pelley (1959) for East Africa and Yunus & Ho (1980) for Malaysia. However, those publications did provide many records that are noted in the text as requiring confirmation. Similarly, a card index compiled by E.M. Hering (now NHM property), from a search of over 150 publications between 1868 and 1948, provided further records that need to be confirmed. Those old records may be incorrect for one or more of the following reasons: an error in the identification of the tephritid; an error in the identification of the host plant; or due to a casual observation of a tephritid adult on a plant being published and then misquoted as a host record in subsequent publications. A good example of that derives from Bezzi (1916) who mentions a specimen of *B. caudata* (Fabricius) that was collected "on" sorghum. Kapoor & Agarwal (1983) included that in a host catalogue without indicating that it was not based on rearing. Anyone with a little knowledge of tephritids will recognise that record as being implausible, but there is no way of knowing how many plausible records were also based on casual observations of resting adults. In collections, long series of specimens labelled with the name of a plant sometimes comprise only males, indicating that they were almost certainly bait trapped, and the plant name presumably refers to the tree in which the trap was placed. The problem there is that flies caught in a trap may be derived from the fruits of several plant species and not just from the species in which the trap was placed.

With the exception of *Anastrepha* host records, all doubtful or questionable records were included in the host data presented in this work. Had we simply omitted those records it might have appeared that we had missed them in error and some future workers might then uncritically add them back into their host lists. Norrbom & Kim (1988b) have already listed *Anastrepha* records that should be disregarded and they are not repeated here. In the host catalogue questionable (requiring confirmation) and doubtful (presumed erroneous) records are respectively marked, ? and ??. However, the choice of category was sometimes subjective and that is another reason why it was desirable to list even those records that we believed to be erroneous.

Simple lists of tephritids and their recorded host plants also obscure the fact that some hosts are preferred to others. Host choice may also be dependent on fruit variety or the stage in development of a fruit or flower, and possibly interactions with other species competing for the same hosts. Sometimes there may even be host races of a tephritid species, each of which attacks a subset of the total host list for the species. Taken together, those factors can result in a tephritid species which will attack particular plants in one part of its range but not in another, for example *Anastrepha ludens* (Loew) does not attack citrus (*Citrus* spp.) in Costa Rica although it does in other areas of its range (Jiron *et al.*, 1988). Similarly complex relationships are known amongst species associated with the flowers and stems of Asteraceae (Anderson *et al.*, 1989; White & Clement, 1987; White *et al.*, 1990). The possibility of using host plant resistance as a control strategy was discussed for fruit associated species by Greany

(1989) and for the safflower pest, *Acanthiophilus helianthi* (Rossi), by Jakhmola & Yadav (1980).

Sibling Species Complexes

The genus *Rhagoletis*, notably the apple maggot fly complex (*R. pomonella* (Walsh) complex), has been extensively studied by evolutionary and population geneticists, and its host races and sibling species form a classical example of presumed sympatric speciation (Bush, 1966; 1969; and other references on p. 369). Since the discovery of the *Rhagoletis* species complexes, others have been discovered in all major sections of the family. Examples which are further discussed in the present work are the *Anastrepha fraterculus* (Wiedemann) (p. 133), *Bactrocera dorsalis* (p. 186), *R. cingulata* (Loew) (p. 356) and *R. tabellaria* (Fitch) (p. 384) complexes. Other genera known or believed to include species complexes are *Chaetorellia* (White & Marquardt, 1989), *Tephritis* (Seitz & Komma, 1984) and *Urophora* (White & Korneyev, 1989).

Biology

The following account of fruit fly biology is brief and readers requiring more detail should consult the recent two volume book edited by Robinson & Hooper (1989), which included detailed papers on many aspects of tephritid biology, ecology and control. Other important review papers include Bateman (1972), Boller & Prokopy (1976), Christenson & Foote (1960), Fletcher (1987; 1989b), Prokopy (1977) and Zwölfer (1983). Reports of fruit fly symposia are also useful compendiums of literature (Cavalloro, 1983; 1986; 1989; Economopoulos, 1987; IAEA, 1990; Mangel *et al.*, 1986; Vijaysegaran & Ibrahim, 1991).

One of the characteristic features of the Tephritidae is the long extendible ovipositor of the females (e.g. figs 10, 167, 248). The fruit associated species use this to deposit eggs within the host fruit, and flower associated species usually place their eggs between parts of the host flower, for example between the bracts of a thistle (Asteraceae). Although the fruit associated species normally attack intact fruit, it has been shown that *Ceratitis capitata* selectively attacks oranges (*Citrus* sp.) that are already damaged, and will oviposit into the wound (Papaj *et al.*, 1989a). It is likely that many of the species which are normally associated with thin skinned fruits, only attack thick skinned fruits, such as citrus, when the skin of those fruits has been damaged.

Some tephritids have been shown to use a pheromone to mark fruit in which they have oviposited, as a signal to other members of the same species that the fruit is already attacked. These oviposition deterrent or host marking pheromones have been recorded for *C. capitata*, and several *Anastrepha* and *Rhagoletis* spp. (Averill & Prokopy, 1989a). A wide range of other fruit and flower associated species drag their ovipositor around the oviposition area after depositing eggs, suggesting that production of deterrent pheromones is very widespread in the family. Evidence of a deterrent pheromone has only been found for two Asteraceae associated species so far (Pittara

& Katsoyannos, 1990; Straw, 1989b). Oviposition deterrent pheromones are not known for any species of Dacini (Averill & Prokopy, 1989a), but following oviposition the olive fly, *Bactrocera oleae*, spreads the juice of the host fruit over the fruit surface using the labellum, and that acts as an oviposition deterrent chemical (Averill & Prokopy, 1989a).

There are three larval instars, although some flower associated species complete the first instar before emerging from the egg (White & Clement, 1987). Fruit tissue is a poor source of protein and there is evidence that bacteria provide essential nutrients (Fletcher, 1987). Artificial larval diets have been defined for many species and these are reviewed by Boller (1989b), Fay (1989), Leppla (1989) and Tzanakakis (1989). Some fruit feeding species are actually seed feeders, e.g. *Toxotrypana curvicauda* Gerstaecker and probably most of the *Anastrepha* spp. that have long ovipositors (Norrbom & Kim, 1988b). However, about half of all species of Tephritidae are flower associated (Table 1), including almost all species of Tephritinae. Some of the Cucurbitaceae associated *Bactrocera (Zeugodacus)* and *Dacus* spp., are either exclusively or occasionally associated with flowers rather than fruit, and in some cases it is specifically the male flowers which have been recorded as the larval host. As most field surveys involve only trapping and fruit collections, it is likely that the hosts of many *B. (Zeugodacus)* and *Dacus* spp. have been overlooked by a failure to collect cucurbit flowers. Many Tephritinae induce galls, including root galls, stem galls and galls hidden within the flower heads of Asteraceae, and gall formation was reviewed by Freidberg (1984). A few of the Cucurbitaceae associated *B. (Zeugodacus)* and *Dacus* spp. have been associated with galls, but in the only properly investigated cases (Bhatia & Mahto, 1968; Sugimoto *et al.*, 1988; Syed, 1971) these were galls induced by other insects and then attacked by the fruit fly. The melon fly, *B. (Z.) cucurbitae*, has also been known to develop in the stems of both cucurbits and tomato (Carey & Dowell, 1989; Syed, 1971), and *Bactrocera eximia* Drew has only been reared from a plant stem (Drew, 1989a). The larvae of some Trypetinae develop in leaf mines, e.g. the celery fly, *Euleia heraclei* (Linnaeus), and many Ceratitini (subtribe Gastrozonina) develop in bamboo shoots. Some species have more unusual larval habitats, namely some Acanthonevrini and Phytalmiini in dead wood, and the larvae of *Euphranta toxoneura* (Loew) develop in the leaf galls of a sawfly (p. 409).

Most fruit feeders drop to the ground and move into the soil where they form a puparium, but most flower feeding Tephritinae pupariate within the host tissue. The larvae of many of the fruit feeders can jump along the ground to find suitable pupariation sites and this is probably a common feature of the Dacinae (e.g. *Bactrocera* and *Ceratitis* spp.), but it is unknown in *Anastrepha* spp. (Christenson & Foote, 1960). A summary of the number of eggs laid and duration of each life-cycle stage is presented in Table 2 (data mostly from Christenson & Foote, 1960).

Following adult emergence, most species require a protein source to permit egg maturation, and recent studies have shown that plant surface bacteria are a very important source of nutrients (Drew, 1989a; Drew & Lloyd, 1989; Lloyd, 1991), at least for some *Bactrocera* spp. Such bacteria are probably spread by mature females

Table 2. Duration of each life-cycle stage for selected species of fruit associated Tephritidae.

Species	Egg (days)/ No.laid	Larva (days)	Pupa (days)	Adult (months)
	Species attacking many host families			
Anastrepha fraterculus	3-6/ 200-400	15-25	15-25	8
Anastrepha ludens	6-12/ 1500	15-30	12-20	11
Bactrocera dorsalis	1-20/ 1200-1500	9-35	10-30	1-3
Bactrocera tryoni	2-3/ >1000	10-30	7+	>1
Ceratitis capitata	2-4/ 300	6-11	6-11	2-3
	Species usually attacking a single host family (Cucurbitaceae)			
Bactrocera cucurbitae	1/300-1000	4-17	7-13	1-5
	Species only attacking a single host family			
Bactrocera oleae	2-4/ 200-250	10-14	10	1-2
Epochra canadensis	6-8/ ?	15-25	?	?
Myiopardalis pardalina	2-7/ 100+	8-18	13-20	?
Rhagoletis cerasi	6-12/ 50-60	30	winter	1-2
Rhagoletis completa	5/ 200-400	28-37	winter	1-2
Rhagoletis pomonella	2-8/ 280	14-22 or overwinter	winter	1-2

as they feed on the fruit surface, and the same range of bacteria species has been found in the gut contents and stung fruit (Drew & Lloyd, 1989). The role of these bacteria is complex and not yet fully understood, and many authors regard their role as symbiotic although that is doubted by others (Drew & Lloyd, 1989; Girolami, 1983; Howard, 1989).

Most fruit associated tephritids are attracted to substances which give out ammonia, for example hydrolysed or autolysed protein. The males of many Dacinae, including *Bactrocera*, *Ceratitis* and *Dacus* spp., are attracted to chemicals known as male lures or parapheromones (p. 17). No male lures have been identified for any species of Tephritinae or Trypetinae, including *Anastrepha* and *Rhagoletis* spp. However, males of *Anomoia purmunda* (Harris), a European trypetine, have been observed congregating on fresh paint and on lady's bedstraw (*Galium verum* L.) (Smith, in press), suggesting some male lure activity. The exact role of male lures is unknown (Cunningham, 1989a). However, Drew (1989a) has drawn attention to bacterial odours as attractants, and as sources of naturally occurring substances which may be mimicked by male lures, and lures based on bacteria are now being manufactured (Sivinski & Calkins, 1986). Pheromones are involved in the mating of fruit associated species (Jones, 1989; Katsoyannos, 1989a; Koyama, 1989b; Mazomenos, 1989; Nation, 1989a), but odours associated with the host plant also play an important role in bringing the sexes of *Bactrocera* and *Rhagoletis* spp. together (Boller & Prokopy, 1976; Drew, 1989a), and the host is an important rendezvous for flower associated Tephritinae (Zwölfer, 1974a). *Rhagoletis* spp. mate on or near their host fruit, and *Bactrocera* and *Ceratitis* spp. usually mate on or near their host tree, but *Anastrepha*, some *Dacus* spp. and the leaf miner *Euleia heraclei* do not (Boller & Prokopy, 1976; Fletcher, 1987; 1989b; 1989c; Leroi, 1975b). Many tephritids have a pattern of lek behaviour, in which congregating males defend territories in which they wait for receptive females. Mating displays involve the use of wing pattern and movement, and sometimes also acoustic signals, the presentation of a nuptial gift of dried secretion from the mouthparts, or kissing behaviour (Freidberg, 1981; 1982; Pritchard, 1967; Sivinski, 1988; Sivinski & Burk, 1989; Zwölfer, 1974b). *Bactrocera* spp. usually mate at dusk, but species belonging to other genera do not show such a uniform preference (P.H. Smith, 1989). There is some evidence that competition for fertilization takes place by sperm-loading (Tsubaki & Sokei, 1988).

Temperate species with a narrow host range, such as *Rhagoletis* spp., are usually univoltine, i.e. they only have one generation per year. However, tropical pest species of *Anastrepha*, *Bactrocera*, *Ceratitis* and *Dacus* are typically multivoltine, i.e. they have several generations per year. These different life history strategies were discussed in review papers by Fletcher (1989b) and Zwölfer (1983).

Natural Enemies

No comprehensive parasitoid-host catalogue has been compiled, but recorded associations of parasitoids with major pest species were listed by Narayanan & Chawla (1962) and more comprehensive lists may be extracted from parasite catalogues, as follows: 1913-1937 published records in Thompson (1943); 1938-1962 records in Herting & Simmonds (1978); 1987 records in Fry (1990). Due to difficulties in verifying the identifications on which those records were based, no attempt was made to list parasitoids in the present work, but some sources of parasitoid data are given after each species account.

The larvae and puparia of fruit associated tephritids are attacked by a variety of parasitic Hymenoptera, particularly by species of Opiinae (Braconidae) (Christenson & Foote, 1960; Wharton & Gilstrap, 1983), but Chalcidoidea and other groups are also important. The adults of some *Bactrocera* spp. are parasitized by a stylops (Drew & Allwood, 1985) and tephritid puparia in soil are subject to attack by a variety of predators.

The role of parasitoids in the natural regulation of tephritid populations was reviewed by Debouzie (1989), and the parasitoids of *Bactrocera* and *Rhagoletis* spp. by Fletcher (1987) and Boller & Prokopy (1976), respectively. Rates of parasitism were high (90%) in some of the examples discussed by those authors, but Fletcher (1987) noted that low (0-30%) levels of parasitism are more typical. There is a long history of attempts to use parasitoids for the biological control of tephritid pests, beginning with the survey work carried out by Silvestri (1913) for parasitoids of the olive fly, *B. oleae*. A catalogue listing many of the releases of biological control agents was produced by Clausen (1978), and the history and effectiveness of biological control was reviewed by Wharton (1989a; 1989b).

Drew (1987b) showed that birds and rodents ate sufficient attacked fruit to account for a far higher level of larval mortality than invertebrate predators and parasitoids, e.g. rodents consumed larvae in 78% of fallen *Planchonella australis* (R. Br.) Pierre (Sapotaceae) fruits. Puparia in soil are very vulnerable to predators as well as parasitoids. Ants are of particular importance and 38% mortality has been attributed to them (Wong *et al.*, 1984a), although the ants found in some regions are unable to detect or crack puparia (Boller & Prokopy, 1976). Bateman (1972) also listed ground dwelling Coleoptera (Carabidae and Staphylinidae), Neuroptera (Chrysopidae) and Hemiptera (Pentatomidae) as predators. Some Trypetini, namely *Rhagoletis zephyria* Snow and *Zonosemata vittigera* (Coquillett), appear to have evolved wing markings that enable them to mimic jumping spiders (Salticidae) and thereby reduce predation by those spiders (Mather & Roitberg, 1987; Whitman *et al.*, 1988). It is possible that the resemblance of *Toxotrypana* (fig. 248) to an ichneumon wasp and Dacini spp., especially *B. (Tetradacus)* (figs 203, 204) and *D. (Callantra)* spp. (figs 219-221), to aculeate Hymenoptera, may also be some sort of protective mimicry, but that has not been investigated.

Other Diptera Sometimes Associated with Fruit

Some other families of Diptera are sometimes found in association with fruit. Many of these records are probably based on the observation of flies attacking fruit already damaged by some other agent and some examples follow.

DROSOPHILIDAE

Drosophila spp. are the "fruit flies" of the genetics laboratory, although they are primarily microfungi feeders. There is no evidence of members of that family attacking undamaged fruit, and when *Drosophila* spp. are found in association with fruit they are probably attracted to fermentation products. However, it has been suggested that *Drosophila* spp. may help spread fungal infection among packed fruit (Louis *et al.*, 1989).

LONCHAEIDAE

Silba spp. (Lonchaeidae) have often been recorded from fruit (Ferrar, 1987), but Souza *et al.* (1983) found that eggs of *Silba* spp. were always found in oviposition holes made by tephritids. Ferrar (1987) also noted that members of the closely related genus *Neosilba* have been recorded from fruit, but in many cases as a secondary invasion following attack by tephritids; however *N. perezi* (Romero & Ruppel) is truly phytophagous but it attacks the shoots of cassava (*Manihot esculenta*). Ferrar (1987) also listed species of *Dasiops*, *Lamprolonchaea* and *Lonchaea* reared from fruit, and *Earomyia* spp. develop in the seeds of coniferous trees.

MUSCIDAE

The genus *Atherigona* is divided into two groups; subgenus *Atherigona (Atherigona)* are truly phytophagous and develop in the stems of Poaceae (Gramineae), and this group includes many pest species; subgenus *A. (Acritochaeta)* spp. are often found in abundance in fruit crops, but when found in fruit they are usually thought to be only secondary invaders. That assumption is based on the fact that *Atherigona (Acritochaeta)* spp. are often found in decaying plant material and sometimes they are even facultatively predacious on other larvae (see Ferrar, 1987). However, there have been recent reports of these flies causing primary damage to fruit in Australia, Hong Kong, India and Nigeria (Chughtai *et al.*, 1985; Ogbalu, 1989; HKDA, unpublished data, 1990; QDPI, unpublished data, 1990).

NERIIDAE

Larvae of this family develop in decaying vegetable substances, particularly fruit (Ferrar, 1987).

OTHER FAMILIES

A *Ptecticus* sp. (Stratiomyidae) has been reared from fallen mango (*Mangifera indica*) fruits, and a few individuals were also reared from unripe fruit that was attacked while still on the tree (Cordero-Jenkins *et al.*, 1990). Many species of Agromyzidae and a few species of Cecidomyiidae develop in the pods of legumes (Fabaceae); however, very few tephritids regularly attack legume pods and the larvae of those families are unlikely to be confused with those of the Tephritidae.

Pest Management

The following account of methods used for detection, control and eradication is intended as a general introduction to the major techniques used. Comprehensive details of control and eradication techniques are beyond the scope of the present work and further details can be obtained from the references given here and at the end of many of the species accounts.

Prevention of Fruit Fly Attack

Preventing fruit flies from becoming established in fruit fly free areas such as New Zealand is largely achieved by strictly enforced quarantine regulations. For example, forbidding the import of fruit fly susceptible crops from infected areas without post-harvest disinfestation treatment, and forbidding travellers to carry fruit in their baggage.

Preventing crops from being attacked in fruit fly infested areas is seldom practical. However, in areas where labour costs are low, large fruits may be individually wrapped in paper or cloth before they reach a suitable stage for fruit fly attack. For example, in Taiwan crop yields of bitter gourd (*Momordica charantia*) and angled luffa (*Luffa acutangula*) could be increased by about 45 % when the fruits were wrapped with two layers of paper bags, replaced every two to three days (Fang, 1989). Repellency has also be tried; the oviposition deterrent pheromone of *Rhagoletis pomonella* was found to deter oviposition for up to three weeks, providing it was not washed away by rain (Averill & Prokopy, 1987).

Detection

Fruit flies may be detected as eggs or larvae in fruit imports, as attacked growing fruit, or as adults caught in detection and monitoring traps. Detection of larval infested fruit in cargo is usually carried out by hand sampling crates of fruit, although acoustic detection systems are now being developed to listen for larvae in individual fruits (Webb *et al.*, 1988). Detection of potentially infested fruit in the baggage of airline passengers poses a major problem and many countries susceptible to fruit fly infestation impose heavy penalties against anybody found carrying fruit of any sort into their country. Many fruit fly outbreaks may be attributable to undetected illegal imports of a few fruits in an airline passenger's baggage. There are few figures to indicate the likely scale of this problem, but in a survey carried out over a one year period in Japan, over 1000 individuals of *Bactrocera dorsalis* were collected (Satoh *et al.*, 1985). In New Zealand, Baker & Cowley (1991) recorded 7 to 33 interceptions of fruit flies per year in cargo, and 10 to 28 per year in passenger baggage, with up to 750 larvae in a single fruit (*B. xanthodes* (Broun) in breadfruit). Carey & Dowell (1989) suggested that even if only 1 passenger in a thousand carried fresh fruit into California, that would be 7000 illegal imports per year, any one of which could give rise to a fruit fly outbreak.

Collecting Attacked Fruit

Attacked fruit will often have puncture marks made by the entry of the female's ovipositor. Sometimes there may be some tissue decay around these 'stings' and some fruits with a very high sugar content, e.g. peach (*Prunus persica*), exude globules of sugar. However, it is not always possible to recognise infested fruit and large samples of a potential host should be collected even when there are no obvious signs of oviposition. Field samples should be transported back to the laboratory in heavy paper bags as plastic bags or sealed plastic containers cause heavy mortality of larvae and eggs in fruit. When sampling wild hosts it is important to collect leaves, flowers and bark, as well as the fruit, to aid in accurate plant identification. In hot tropical conditions, fruit samples must be stored in large insulated (e.g. expanded polystyrene or styrofoam) boxes to reduce 'heat stress' during transportation.

Upon return to the laboratory, collected fruit should be placed in a container that has a gauze or muslin top and a dry medium at its base, such as sterilized sawdust or sand, in which emerging larvae can pupariate. In the tropics, fruit holding containers should be covered in a non-drying adhesive (e.g. *Bird Stop* from Rentokil) which acts as an ant barrier. Samples should be checked every two days for puparia, and fruit from which larvae have emerged should be discarded. When all the larvae have emerged from the fruit, or when any sign of mould growth appears, the sawdust should be sieved and the puparia collected. Puparia can then be transferred to petri dishes, covered with a thin layer of moist heat-sterilized sawdust and then placed in a small muslin emergence cage; this cage must also be protected by ant barriers. It is vital to provide some sugar solution as food for the emerging adults, and to keep the adults alive for at least four days after emergence. During that time the flies will develop their full body coloration and normal shape. Failure to feed the flies will result in specimens that have shrivelled abdomens and dull colours. The key to *Bactrocera* and *Dacus* spp. depends on colour characters, and misidentifications will be made if reared specimens are not fed for a few days prior to their being killed and mounted. For further details about rearing from fruit see Aluja *et al.* (1987a) and Drew (1982b).

Trapping and Baits

Newly adventive populations of fruit flies are sometimes first detected in monitoring traps set in areas susceptible to fruit fly attack. For example, New Zealand has no fruit associated species of Tephritidae, but susceptible areas of the country are covered by a grid of monitoring traps designed to detect any arriving fruit flies. In 1990 those traps caught a few specimens of *Bactrocera passiflorae* (Froggatt) (R.A.I. Drew, pers. comm., 1990), which had probably emerged from illegally imported fruit that was dumped because of its maggot infestation; luckily this species did not become established from that introduction. Trapping programmes of that sort have also detected *B. cucurbitae*, *B. dorsalis* and *C. capitata* in California (Dowell & Wange, 1986), and *B. tryoni* in South Australia (Anon., 1991). Numerous trap designs have been published and these were reviewed by Drew (1982b).

The males of most *Bactrocera*, *Ceratitis* and *Dacus* spp. can be collected in traps which have been baited with special chemicals, sometimes called parapheromones. The physiological basis of this male only attraction is still the subject of debate and the topic was reviewed by Cunningham (1989a). The most important male lures are as follows:

1. Cue Lure - This attracts males of many *Bactrocera* and *Dacus* spp. Its chemical description is 4-(*p*-acetoxyphenyl)-2-butanone although chemical companies may list it as 4-(3-oxobutyl)-phenylacetate. Very similar but less effective chemicals attracting the same range of species are anisylacetone and Willison's lure (Drew, 1982b). It is also possible to use 2-butanone, but it is highly volatile and must be kept in a slow release dispenser rather than being placed on a cotton wick.

2. Methyl Eugenol - This attracts males of many *Bactrocera* spp., but not members of subgenus *B. (Zeugodacus)*, and it attracts some species of subgenus *Ceratitis (Pardalaspis)*. Under recent chemical nomenclature this should be called methoxy eugenol and it has been known as eugenol methyl ether, although it is not an ether in the strictest sense; it can be chemically described as 4-allyl-1,2-dimethoxybenzene or 3,3,dimethoxy (1) 2 propenyl benzene. Before its discovery many of the same range of species were bait trapped using citronella oil and huon pine oil (Drew, 1982b). Some plants also attract at least part of this range of species, namely tulsi plant or holy basil (*Ocimum sanctum* L.; Lamiaceae), the flowers of canon ball tree (*Couroupita guianensis* Aublet; Lecythidaceae) and the flower spike of *Spathiphyllum candicum* Poepp. & Endl. (Araceae) (Lewis *et al.*, 1988; Shah & Patel, 1976; Yong, 1990a). See also Mitchell *et al.* (1985) and Wong *et al.* (1989).

3. Trimedlure - This attracts males of many species belonging to the subgenera *Ceratitis (Ceratitis)* and *C. (Pterandrus)*. It is t-butyl 4, (or 5), -chloro-2-methyl cyclohexane carboxylate. Chemicals known as siglure and med-lure attract the same species, but they are less effective and now rarely used. See also Hancock (1985a), Hill (1987), Leonhardt *et al.* (1984; 1987), McGovern & Cunningham (1988), McGovern *et al.* (1987) and Rice *et al.* (1984).

4. Terpinyl Acetate - This attracts males of a wide range of *Ceratitis* spp., including those attracted to either methyl eugenol or trimedlure, plus at least some species of subgenus *C. (Ceratalaspis)*, which are not attracted to either of those baits. The ester terpinyl acetate probably consists of mixed isomers, in dynamic equilibrium, containing C_{10} groups, but differing from the eugenols in that the basic C_6 ring structure is not derived from benzene (phenyl) but from two isoprene C_5H_{10} molecules, which is then esterified. The ability to attract those tephritids which are also attracted by either methyl eugenol or trimedlure may be related to the oxidation of terpinyl acetate when exposed to air (D.J. de Courcy Henshaw, pers. comm., 1991).

5. Vert Lure - This is only known to attract *D. vertebratus* Bezzi and it is methyl-4-hydroxybenzoate (Hancock, 1985b). The discovery of this specialist lure suggests that other species which do not respond to known lures may also have as yet undiscovered lures. Propyl-4-hydroxybenzoate also attracts *D. vertebratus* (Hancock, 1985b).

Commercial suppliers of cue lure, methyl eugenol, trimedlure and suitable traps, include the following:

Agri Sense - BCS Ltd.,
Treforest Industrial Estate,
Pontypridd,
Mid Glamorgan,
CF37 5SU,
United Kingdom.

International Pheromone Systems Ltd.,
Units 12/13 Meadow Lane,
Meadow Lane Industrial Estate,
Ellesmere Port,
South Wirral,
L65 4EH,
United Kingdom.

Traps that use male lures are usually based on the Steiner trap design. This is a horizontal cylinder with a large opening at each end (fig. 1) and chemical lure is impregnated into a cotton wick suspended within the trap (Drew, 1982b); see Cowley *et al.* (1990) for an alternative design. To avoid flies escaping, and to avoid predation of captured flies, an insecticide is normally mixed with the lure, or a strip of filter paper impregnated with insecticide, is also placed in the trap. The organophosphorus insecticides malathion and dichlorvos are the preferred options; some other insecticides may alter the structure of the lure and render the bait ineffective. Fly paper strips, such as *Vapona*, are also suitable.

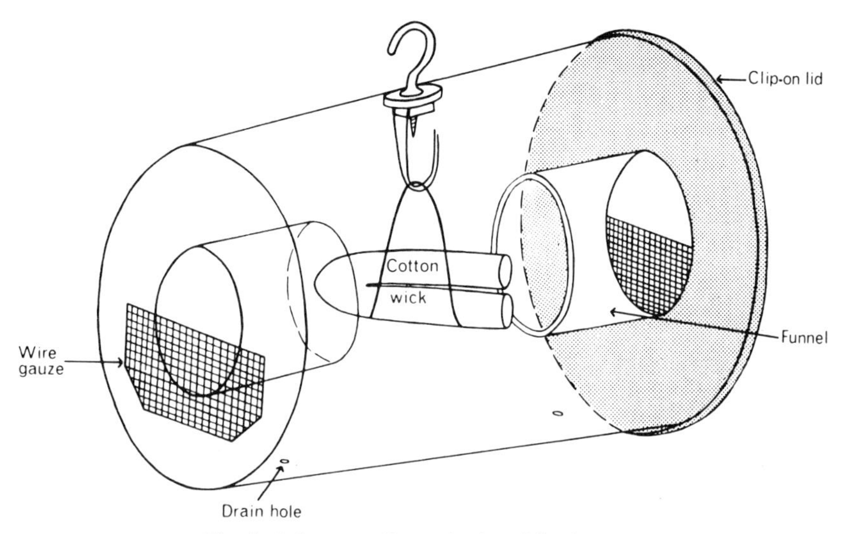

Fig. 1. Steiner trap (Queensland modification).

If the lure which attracts a male fly is to be used as an aid to identification, then it is essential that the lure chemicals do not become cross contaminated with each other. Methyl eugenol and cue lure are such potent lures that even allowing a person to handle containers of both lures, or traps baited with each lure, could contaminate those lures sufficiently for them to appear to attract the 'wrong' species of fly. Drew (1982b) suggested that collectors should wash their hands with alcohol between servicing traps containing each of these lures. Ideally, a survey team should always carry one of the lures inside their vehicle and the other in a box fixed to the outside of their vehicle so that the supply bottles do not become cross contaminated; the team should also allocate different personnel to the handling of each lure so that traps do not get contaminated. Plastic traps tend to become impregnated with lure and once a trap has been used for one lure it should never be used for any other (Drew, 1982b). Mixtures of methyl eugenol and cue lure used to be marketed for the capture of *B. dorsalis*, although work by Hooper (1978b), working with Australian *Bactrocera* spp., showed that traps baited with a mixture caught fewer flies than traps baited with each lure separately. However, that does not always appear to be the case; Ramsamy *et al.* (1987) showed that a mixture of 3 parts methyl eugenol to 7 parts cue lure caught more *B. cucurbitae* than pure cue lure or any other proportional mixture.

Both females and, to a lesser extent, males, of any fruit associated species may be collected in traps baited with a substance that emits ammonia fumes (e.g. yeast autolysate, hydrolysed protein or ammonium carbonate). However, male lures are preferred whenever possible because of their high specificity and efficiency in attracting flies over a very wide area (up to 0.8km for methyl eugenol; less for cue lure and much less for trimedlure).

No male lures are known for species of *Anastrepha*, *Bactrocera oleae* (olive fly) and *Rhagoletis* spp. These may be collected in a trap that relies solely on an ammonia source as bait, or with a trap that combines visual and olfactory attraction.

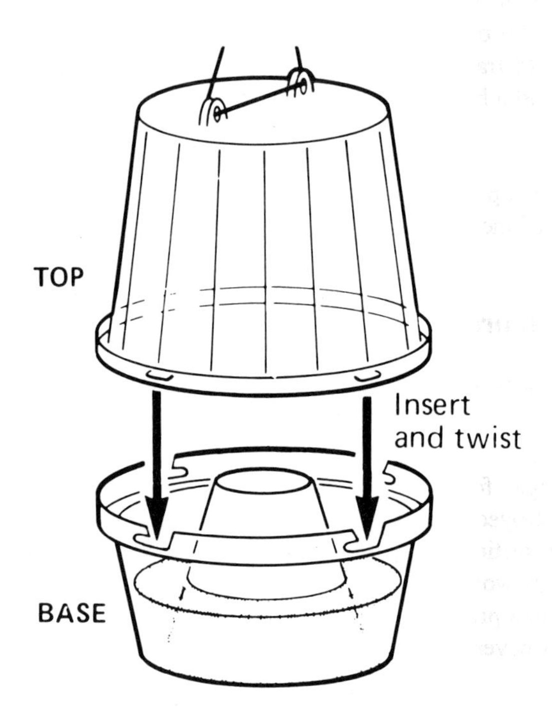

Fig. 2. A modern version of the McPhail trap (*Liquibaitor* trap).

Traps that rely on the ammonia bait are usually based on the McPhail trap; that type of trap has an entrance hole in its base and a trough around the base to hold a liquid bait, such as a protein solution (Drew, 1982b). A plastic trap based on that principle is available from International Pheromone Systems Ltd (fig. 2). A combination of yellow sticky trap and a slow release dispenser for ammonium acetate is one of many forms of visual trap that has been used. A review of trap design involving colour and shape was presented by Economopoulos (1989); see also Burditt (1982; 1988), Calkins *et al.* (1984), Cunningham (1989b), Economopoulos *et al.* (1986), Hedstrom & Jimenez (1988), Hedstrom & Jiron (1985), Jones *et al.* (1983), Katsoyannos (1987; 1989b), Reissig *et al.* (1985), Rhode & Sanchez (1982), Riedl & Hislop (1985), Sharp (1987), Sharp & Chambers (1983), Souza *et al.* (1986) and Witherell (1982). Traps emitting ammonia may also attract insects other than tephritids, e.g. members of the New World family Richardiidae (Ferrar, 1987) which are both closely related to Tephritidae and sometimes superficially similar in appearance.

When planning a fruit fly survey using traps it is necessary to set traps at a rate of only one per km^2 if methyl eugenol or cue lure is used (Drew, 1982b), but considerably closer for other baits, in particular for ammonia based baits; e.g. Calkins *et al.* (1984) used 45 traps per hectare for *Anastrepha suspensa* (Loew). Trap height is also important and a height of about 2m is normal for orchard trapping, but in rain forest the traps should be placed as high as possible within the forest canopy (Drew, 1982b; Hooper & Drew, 1979). Traps baited with methyl eugenol or cue lure may remain effective for up to two weeks, but they should be emptied ever few days to minimize the risk of mould or other damage to the specimens.

Control and Suppression

When an infestation is detected, it is important to gather all fallen and infected host fruits, and destroy them. Those species whose males are attracted to lures should be continually monitored using bait traps (see Bateman, 1982). In the case of species with a narrow host range, for example *Rhagoletis* spp., wild and abandoned host trees should also be destroyed whenever practical.

Insecticidal protection is possible by using a cover spray or a bait spray (Roessler, 1989b). Bait sprays work on the principle that both male and female tephritids are strongly attracted to a protein source from which ammonia emanates. Bait sprays have the advantage over cover sprays that they can be applied as a spot treatment so that the flies are attracted to the insecticide and there is minimal impact on natural enemies. Bait and cover sprays may be used for either *control* or *suppression*; the distinction being that control refers to procedures designed to protect a single orchard while suppression refers to procedures covering a large area (Bateman, 1982).

The practical details of bait spraying are summarized here because bait spraying is likely to be the first measure taken after a fruit fly outbreak is detected; for further details see Bateman (1982), from which the following details were taken:

- The spray includes 20g protein 'solids' plus 10g (active ingredient; a.i.) malathion (also known as maldison) per litre of solution.
- Spray a grid of bait spots at 15m intervals in orchard or forest areas; double that if the technique is being used for eradication; in urban areas spray bait spots at 6m intervals along each side of a street.
- Each spot is ideally 100ml of solution delivered as a solid jet (rather than a dispersed spray); that is best carried out using a power sprayer with a hand trigger release.
- Treatment should be carried out at regular intervals throughout the active season, typically every week at time of peak activity; rain forest species (e.g. *Bactrocera tryoni*) will be naturally stressed in dry areas, but in warm wet areas the spray concentration or the spraying frequency may need to be increased.

Hydrolysed protein is the usual choice of bait but some supplies of that material are acid hydrolysed, highly phytotoxic to some crops and have a high salt content

(Smith & Nannan, 1988; R.A.I. Drew, pers. comm., 1990). Researchers at the Queensland Department of Primary Industries (QDPI) have recently developed the use of autolysed yeast as a protein bait. Smith & Nannan (1988) described trials of bait sprays using a yeast preparation that had a low salt content, and in which the yeast had been killed by subjecting it to low temperature heating; no temperature was specified but 40°C will kill yeast. They found that the most effective insecticide was chlorpyrifos (2g a.i. per litre), but malathion (5g a.i. per litre) was almost as good with the autolysed yeast (10g a.i. per litre). Those mixtures caused no phytotoxicity to passion vines (*Passiflora* spp.), although the malathion was more toxic to beneficial insects in that crop than the chlorpyrifos.

A cheap source of unwanted yeast is available in most countries in the form of brewery waste. Sometimes that may be very acidic because brewers waste oxidizes to produce acetic acid (it will also esterify to produce ethyl acetate). The acidity can be removed by adding sodium hydroxide (NaOH) until the pH is between 5 and 7 (A.J. Allwood, pers. comm., 1988). Successful trials of a bait spray consisting of brewery waste (pH raised from 4 to almost 7), malathion and a sticking agent have been carried out in Malaysia by QDPI. Plant surface bacteria are an important food source for fruit associated tephritids (Drew *et al.*, 1983), and Smith & Nannan (1988) suggested that low pH and high salt content could damage the leaf surface bacterial flora. Drew & Fay (1988) showed that the addition of bacteria increased the attraction of a bait spray.

Bait spraying is primarily a control technique. However, it may be used for sampling as well if a knapsack sprayer is used to spray a single tree branch and a sheet is spread across the ground under the branch; the sheet will catch the flies attracted to, and killed by, the bait plus insecticide mixture.

Some other methods of chemical control have been used against *Rhagoletis* spp. Boller & Prokopy (1976) noted that systemic organophosphates, such as dimethoate, are highly effective against most species, killing eggs, larvae and adults; they also discussed soil application of insecticide to destroy pupae and juvenile hormone analogs to prevent development into adults.

Biological control has been tried against some fruit fly species, namely *Anastrepha ludens*, *Bactrocera dorsalis*, *B. cucurbitae*, *B. tryoni* and *Ceratitis capitata*, but introduced parasitoids have had little impact (Wharton, 1989a; 1989b).

Eradication

Bateman (1982) listed three main eradication procedures, namely bait spraying, male annihilation and sterile insect release. Bait spraying can be used at short notice against any species and has been used successfully for incipient outbreaks of *B. tryoni* in South Australia and large invasions of *C. capitata* in Florida (Bateman, 1982).

Male annihilation utilizes the attraction of males of many species to chemical lures (methyl eugenol and cue lure for *Bactrocera* spp.) and was used to eradicate *B. dorsalis* from the northern Ryukyu Islands, Japan (Cunningham, 1989c). That technique has also been used against *B. cucurbitae* and *C. capitata* in Hawaii, where it did have some impact on population size (Cunningham, 1989c). However, Bateman (1982) noted that male lures could only be used for eradication when traps (or fibre board impregnated with bait plus insecticide) were set at a very high density over the entire range of the target population.

Sterile insect release has been used to eradicate some populations of fruit flies. The sterile insect technique (SIT) requires the release of millions of sterile flies into the wild population so that there is a strong likelihood of wild females mating with sterile males (Gilmore, 1989). SIT was used to eradicate *B. dorsalis* from the Ogasawara Islands and *B. cucurbitae* from Kume Island, Japan (Shiga, 1989). SIT has been used against *C. capitata* in California, Costa Rica, Hawaii, Italy, Mexico, Nicaragua, Peru, Spain and Tunisia (Gilmore, 1989). The largest of those programmes (Programa Moscamed) is being carried out in southern Mexico and is designed to stop the fly spreading north, and ultimately, to eradicate it from Central America (Schwarz *et al.*, 1989a). SIT has also been tried against *A. ludens* (Gilmore, 1989) but no major control programme has been carried out. SIT depends on the ability to mass rear millions of sterile flies and Vargas (1989) reviewed the required procedures.

Combinations of the above techniques may also be used. A combination of bait spraying and male annihilation was used to eradicate two successive outbreaks of *B. tryoni* in Easter Island (Bateman, 1982). Similarly, it is current policy to use a combination of bait spraying and SIT against *C. capitata* outbreaks in the USA (Mitchell & Saul, 1990).

Post-Harvest Disinfestation

There is a large body of literature covering the subject of post-harvest disinfestation of fruit and that was reviewed by Armstrong & Couey (1989). The main techniques available are fumigation, usually with methyl bromide, heat treatment, either with hot vapour or hot water, cold treatments, insecticidal dipping and irradiation. Many specific references are given in the species accounts which follow later in this work.

Methods of Collection and Preparation

Fruit associated tephritids are usually collected by trapping the adults or by rearing them from infested fruit using the techniques described in the Pest Management section (p. 16). However, special care is needed when using the keys in this work to identify trapped adults. The keys include all of the species likely to be found by rearing adults from cultivated fruit hosts, but trapping could potentially collect species that are not included in these keys.

Leaf mining species are also best collected by rearing as their adults tend not to spend long periods associated with their host plants. Tephritids associated with flower heads are easily collected by sweeping a net (a white fine-gauze butterfly net) across the tops of potential host plants, or by rearing (see White, 1988 for details).

Preservation of Larvae

The procedure for preserving and processing the immature stages requires the following series of steps:

A. Preservation:

1. Wash larvae thoroughly in clean cold water (preferably distilled);
2. Kill by immersing in hot water (just off the boil);
3. Allow to cool to room temperature before removing larvae;
4. Transfer larvae to 30% ethanol for 30 minutes;
5. Transfer larvae to 50% ethanol for 30 minutes;
6. Transfer and preserve larvae in 70% ethanol.

B. Processing larvae for scanning electron microscopy:

1. Transfer larvae from 70% to 80% ethanol (at least 30 minutes);
2. Transfer from 80% to 90% ethanol (at least 30 minutes);
3. Transfer from 90% to 100% ethanol overnight;
4. Transfer to diethyl ether for at least 24 hours (longer if time permits);

5. Remove from ether, allow to dry thoroughly;
6. Mount on SEM stubs;
7. Coat with platinum;
8. Store stubs in a dry air-tight container with a layer of silica gel on the bottom to act as a moisture indicator.

C. Slide preparation of larvae:
1. Place larva in 70% alcohol in a small excavated block;
2. The anterior and posterior ends of the body should be removed and prepared separately; with fine forceps, detach head and first thoracic segment from remainder of body; and similarly detach posterior region (for examination of the spiracles and anal lobes) by severing the body between sixth and seventh abdominal segments;
3. Hold head by first thoracic segment; or hold posterior region taking care not to damage the spiracles or anal lobes;
4. Head only - with ventral surface uppermost, carefully make a median incision on the ventral surface of the labium and the remainder of first thoracic segment using small dissecting needles;
5. Using fine forceps gently pull muscle and other tissue from the head or posterior region;
6. If tissue is difficult to remove, soak in cold potassium hydroxide (KOH) for 10 minutes, place in water for 5 minutes, then transfer to 70% alcohol;
7. On a clean glass slide, place a small droplet of a gum arabic based mountant (p. 29);
8. Spread out dissected part with dorsal surface uppermost;
9. Place a small glass coverslip over the droplet of mounting medium and gently apply pressure until the section is flattened;
10. Oven or air dry until the mounting medium is firm;
11. Ring coverslip with a sealant.

Preparation of Adults

The keys in this work are primarily designed for use with dry mounted specimens. If specimens are preserved in spirit (70% ethanol) some colour characters will be difficult to interpret correctly and key couplets which ask about the tomentum will be impossible to answer. Specimens to be dry mounted should be killed with ethyl acetate vapour, taking care not to allow the specimens to come in contact with any form of moisture, as this will also damage the tomentum.

Dry mounted specimens may be pinned through the side of the thorax with a micropin; large micropins, such as D2 size, will be needed for most of the fruit associated tephritids, but B2 or even A1 should be used for small species. The micropin should be inserted just behind the wing base (fig. 3) at such an angle that it emerges just in front of the wing base on the other side of the specimen. Pinning

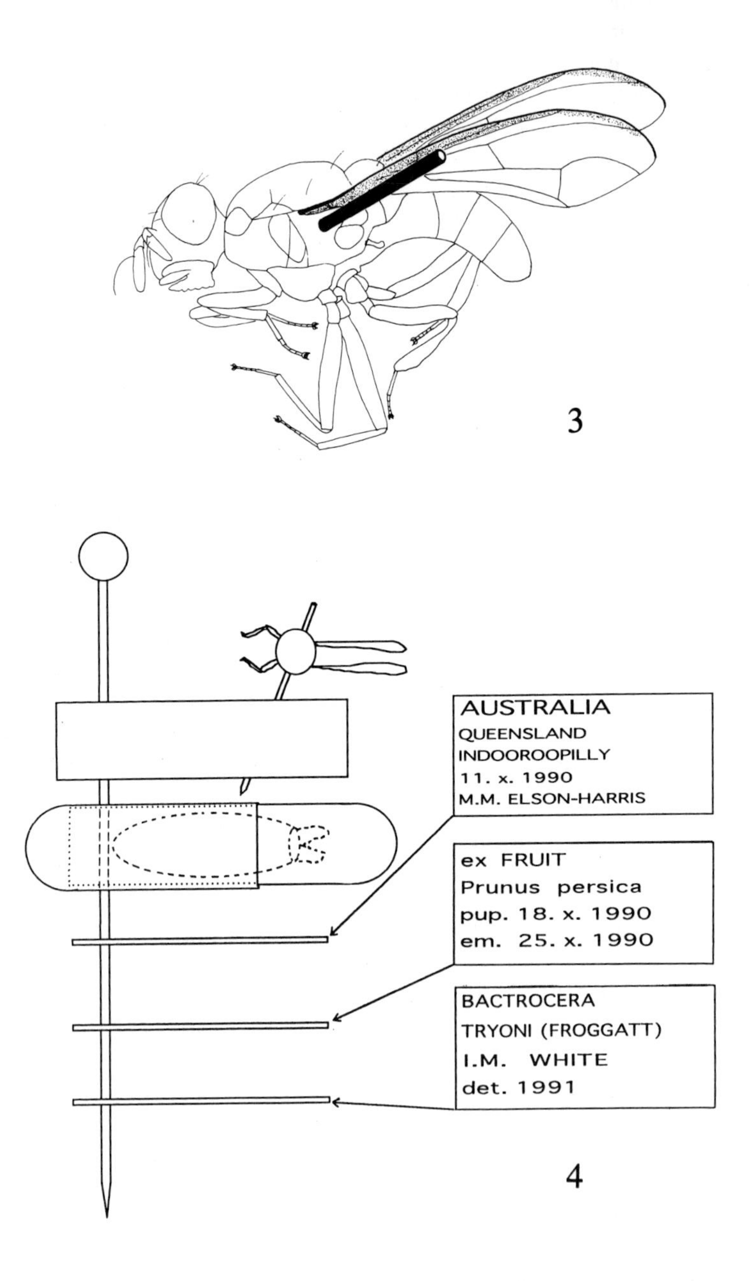

Figs 3-4. Setting and staging; 3, setting a fruit fly, with a pin placed through the side of the thorax; 4, staging and labelling a pinned specimen.

through the top of the thorax is not recommended as it often obscures useful characters and it takes longer to prepare specimens that way. The micropinned specimen should then be temporarily mounted on a sheet of closed cell foam (e.g. *Plastazote*) (9mm or 12mm thickness) with the wings raised and, if necessary, held in place with another micropin; cork may be used if closed cell foam is not available. When pinning females it is often worthwhile to extrude the aculeus by gently squeezing abdominal segment 5 with a pair of fine forceps; the aculeus can then be removed for detailed examination without the need to remove the whole abdomen. After a few days the wing positioning pin, if used, may be removed and the specimen 'staged' (fig. 4) by mounting it on a strip of closed cell foam measuring 4x4x12mm (cut from a 12mm thick sheet); if closed cell foam is not available the specimen may be staged onto a strip of thin card.

The closed cell foam or card stage is then mounted on an entomological pin ('continental' size 5 is ideal) together with a locality label, a host label if the specimen was reared, and a determination label (fig. 4). Note that the collection and emergence dates are given separately. Other details such as the separate dates of pupariation and emergence from the puparium should be given if known, as should the temperature if a constant temperature rearing room was used. The possible host of non-reared specimens should be clearly differentiated from that of reared specimens; for example, "ON PSIDIUM GUAJAVA", would refer to a specimen swept or found resting on guava, as opposed to "EX PSIDIUM GUAJAVA FRUIT" for a specimen reared from guava fruit. Other items which may have to be mounted on the same pin are a gelatine capsule containing the empty puparium and a terminalia preparation, for example a dissected ovipositor.

All of the external characters used in the following keys can be observed with a good stereo or dissection microscope equipped with top or incident light. To see wing pattern detail a white background is necessary, and this can be supplied either by using a white microscope base plate, a piece of white paper or by using a transmitted light base. A summary of the most important identification features of each species is given in the "Adult identification" subsection of each species account, and those summaries should be consulted after using the keys. Unfortunately, it would be almost impossible to give sufficient information to distinguish reliably each of the 100 species included here, from all of the remaining 1400 (estimated) fruit associated species.

Dissection of Adult Females

To identify *Anastrepha* spp. it is almost always essential to dissect the ovipositor of a female specimen and examine it using a high power (x400) compound microscope. An illustration of the aculeus tip shape, and its length measurement, is also given for most species belonging to other genera, as those details may help confirm identifications made using the keys. To measure the aculeus length a calibrated eyepiece micrometer will be needed, and the length should be measured from the base, where it joins the eversible ovipositor membrane, to the apex. The resulting value may be compared to the aculeus length measurements given in the keys. Male terminalia details are not given here, partly because further research is required to determine which features are useful, and also because the characters are very difficult for the non-specialist to interpret. The procedure for dissection of the ovipositor is as follows:

1. Break off the abdomen (removal of only the tip of the abdomen may result in the loss of the base of the aculeus) and place it in 10% KOH (potassium hydroxide); if the aculeus was completely extruded when the specimen was pinned, it is usually safe to remove only the aculeus from the dry mounted specimen and leave the rest of the abdomen intact;
2. Leave the abdomen overnight in a 10% solution of potassium hydroxide (KOH) at room temperature (or 20 minutes at 95°C if a hot plate is available on which to stand the preparation); an alternative quick method of cold clearing was developed by Henshaw & Howse (1982) using hydrogen peroxide;
3. Transfer the abdomen to glacial acetic acid at room temperature;
4. After a period of at least 15 minutes, the abdomen is removed from glacial acetic acid and transferred to a spot of the mountant on a glass microscope slide via the intermediate steps appropriate to the chosen mountant; details are given at the end of this section;
5. The dissection should be done in that drop of mountant and two fine dissection needles are required, each of which can be made by fixing a micropin to the end of a thin wooden stick with an epoxy resin glue;
6. Break the oviscape (fig. 10) from the rest of the abdomen;
7. By gently squeezing the oviscape with one pin it is usually possible to telescope the aculeus out of the oviscape; it will be necessary to finish removal of the aculeus by holding the oviscape with one pin and pulling the aculeus out with the other, taking great care not to damage the tip of the aculeus; if that method fails the oviscape will have to be torn open to remove the aculeus.

If a terminalia preparation is to be kept with the dry mounted specimen a number of methods are available. The simplest technique is to use glycerol as a temporary mountant for examination, and then store the dissected abdomen in a drop of glycerol, in a microvial that has a cap through which the mounting pin may be passed. It is possible to either use glass microvials which have cork tops, like miniature collecting tubes (Irwin, 1978) or to make microvials from fine plastic tubing which can be plugged with plastic rod equal in diameter to the bore of the tube. Some specialists mount terminalia in a permanent mountant between small coverslips placed each side of a punctured slip of card. To do this a piece of card the size of a normal data label is perforated with a stationery punch and a 10mm diameter coverslip is glued across the underside of the hole. A spot of mountant, e.g. Canada balsam is added. When the dissection is completed a second coverslip may be added on top; if a second coverslip is not added reorientation of the dissected parts is facilitated by softening the balsam with xylene (seldom necessary for female terminalia). Uncovered mounts become pitted and wrinkled, but this can be cured by adding a drop of xylene to the mount when it is being re-examined.

If the dissection is to be kept as a permanent slide it is essential to cross reference the slide to the individual that was dissected, e.g. with a code number. Examples of suitable mountants for larval and adult dissections, and the processes needed before their use, are as follows:

1. Berlese's fluid - This is one of the many gum arabic based water soluble mountants and it is very easy to use; however, its permanence is in doubt and it should not be used to make slides of valuable specimens. Specimens should be transferred directly from glacial acetic acid to the Berlese's fluid. When the dissection is complete, leave the slide to set for at least a day at room temperature before adding a further drop of Berlese's fluid and the coverslip. That period of initial drying minimizes the tendency for parts of the preparation to move and roll when the coverslip is added. The slide must then be kept flat to dry before being ringed with *Glyceel* (which is acetone soluble). The completed slide should then be kept flat for several weeks to dry at room temperature, or placed in a hot air cabinet at 35-40°C for at least two weeks. Some microscopists ring Berlese's mounts with *Euparal*, but it is essential that the Berlese's mountant is absolutely dry before doing so, otherwise a chemical reaction occurs which damages the slide. Before ringing, excess mountant must be chipped away from around the coverslip. Methods for making Berlese's fluid were given by K.M. Harris (in litt., in Freeman & Lane, 1985) and Henshaw (1980). Other gum arabic based water soluble mountants are Hoyer's, Faure's and Andre's mountants (Walker & Crosby, 1979). Unlike *Euparal* and Canada balsam, those media will clear very small insects, but it is still essential to clear the comparatively large abdomens of tephritids in KOH before dissection.

2. Euparal - This mountant is soluble in absolute alcohol and *Euparal essence*; it is believed to be permanent and it does not need ringing. Specimens should be transferred from glacial acetic acid to absolute alcohol or, clove oil; after a few minutes they may be placed in the *Euparal*. The completed slide should then be kept flat for several weeks to dry at room temperature, or placed in a hot air cabinet at 35-40°C for at least two weeks. There is no need to ring a slide made with *Euparal*. Both *Euparal* and *Eukitt*, which is an equivalent medium, are trade names.

3. Canada balsam - This is a permanent mountant with a long established usage; it is soluble in xylene and clove oil. Specimens should be transferred from glacial acetic acid to absolute alcohol for several minutes; they should then be placed in either xylene or clove oil for several minutes before being transferred to the Canada balsam. When the dissection is complete a coverslip is added and the slide should be kept flat for at least six weeks in a hot air cabinet at 40-45°C, longer for thick mounts. It is not advisable to store Canada balsam slides vertically unless they have been dried for a long period and ringed with *Glyceel* or a similar ringing compound.

The term 'absolute alcohol' may refer either to 100% ethanol (=ethyl alcohol) or propan-2-ol (=iso-propyl alcohol or iso-propanol). Ethanol is very expensive in some countries because it is heavily taxed; conversely, propan-2-ol is very cheap, has no tax on it, and has the advantage that it is considerably less hygroscopic than ethanol; ethanol absorbs water from the air and becomes diluted very quickly so that it is no longer of any use for slide preparation work.

Terminology

The present work uses the system of Diptera terminology proposed in the *Manual of Nearctic Diptera* by McAlpine (1981) for the adults and Teskey (1981) for the larvae. Some subsequent authors found it necessary to make slight modifications to fit that set of unified terms to the Tephritidae, notably Freidberg & Mathis (1986), Foote & Steyskal (1987) and Norrbom & Kim (1988a).

Adult

The morphological terms used in this guide are all explained in the glossary (p. 500), together with many of the terms used by other authors. Foote & Steyskal (1987) should also be consulted for a detailed account of tephritid adult morphology.

Tephritids vary in wing length from about 2mm to 25mm and most species have patterned wings. They may be distinguished from other picture-winged Acalyptratae by the right-angled bend near the end of vein Sc (fig. 34), just before Sc becomes faint and joins C (the costa), and by the presence of frontal setae. Other important features are: vibrissae absent (not to be confused with genal setae); wing with both H and Sc breaks; vein R_1 with dorsal setulae; vein R_{4+5} often with dorsal and/or ventral setulae; cell cu*p* usually with a pointed extension (figs 39-43); ovipositor telescopic (fig. 10). A more comprehensive description of family level characters was provided by Drew (1989b).

The major features of the head (fig. 5) are supplemented by illustrations of the typical head shape found in each genus associated with commercial fruit crops (figs 18-32). The following abbreviations are used:

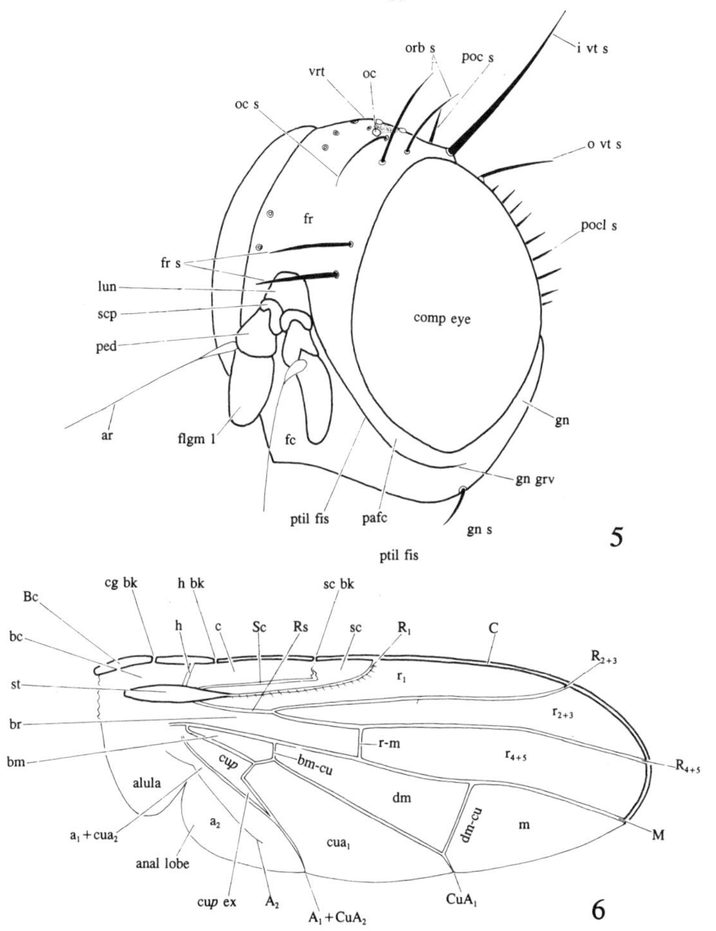

Figs 5-6. Adult morphology; 5, head; 6, wing. Abbreviations are listed on p. 31.

ar - arista;
comp eye - compound eye;
fc - face;
flgm 1 - 1st flagellomere;
fr - frons;
fr s - frontal setae;
gn - gena (plural: genae);
gn grv - genal groove;
gn s - genal seta;
i vt s - inner vertical seta;
lun - lunule;
oc - ocellus;
oc s - ocellar seta;
o vt s - outer vertical seta;
orb s - orbital setae;
pafc - parafacial area;
ped - pedicel;
poc s - postocellar seta;
pocl s - postocular setae;
ptil fis - ptilinal fissure;
scp - scape;
vrt - vertex.

Features of the wing venation (fig. 6) are supplemented by illustrations of the typical wing patterns (figs 35-38) and cell cu*p* shapes (figs 39-43) found in the major fruit pest genera. Full names of wing veins and cells are not used in the following keys and are therefore not listed here (see McAlpine, 1981). The following abbreviations are used for the three costal break positions and to mark the extension to cell cu*p*:

cg bk - costagial break; cu*p* ex - extension to cell cu*p*;
h bk - humeral break;
sc bk - subcostal break;
st - stem vein.

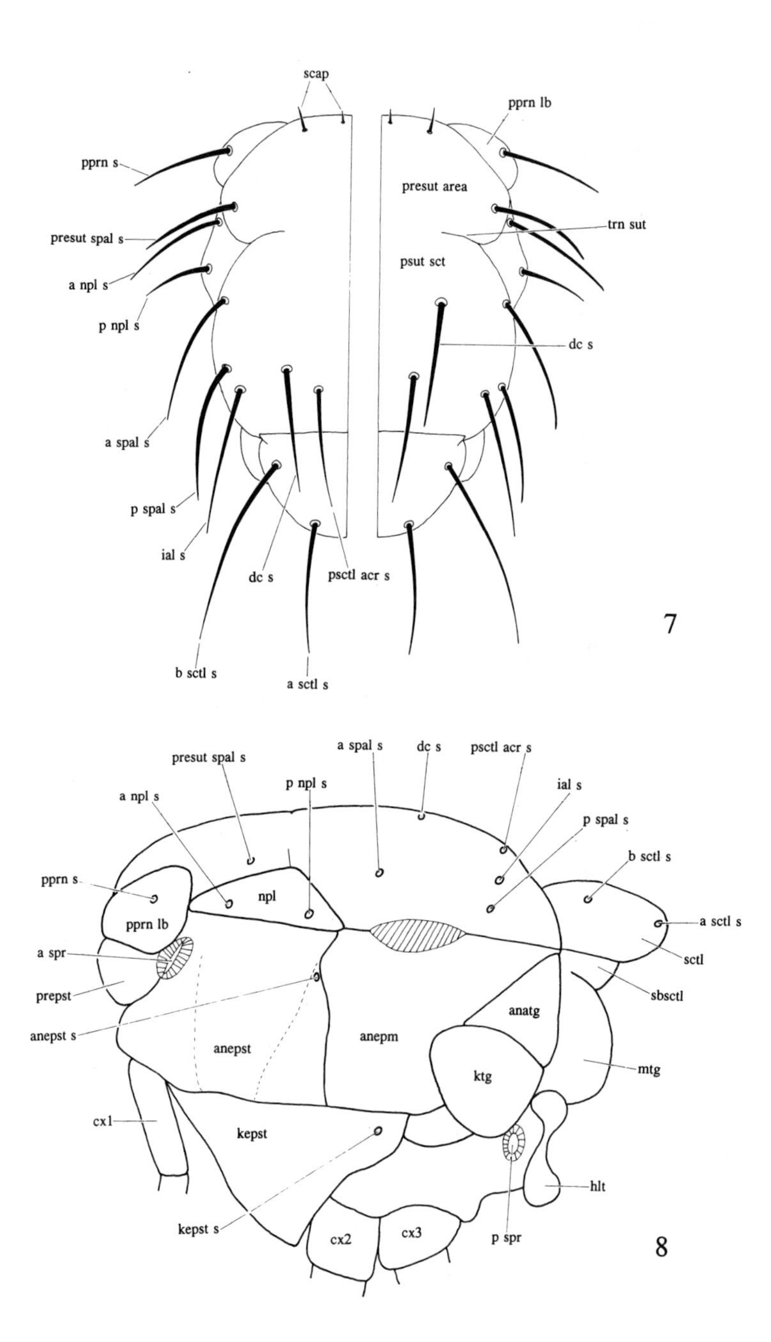

Figs 7-8. Adult morphology, thorax; 7, dorsal features; 8, lateral features. Abbreviations are listed on p. 33.

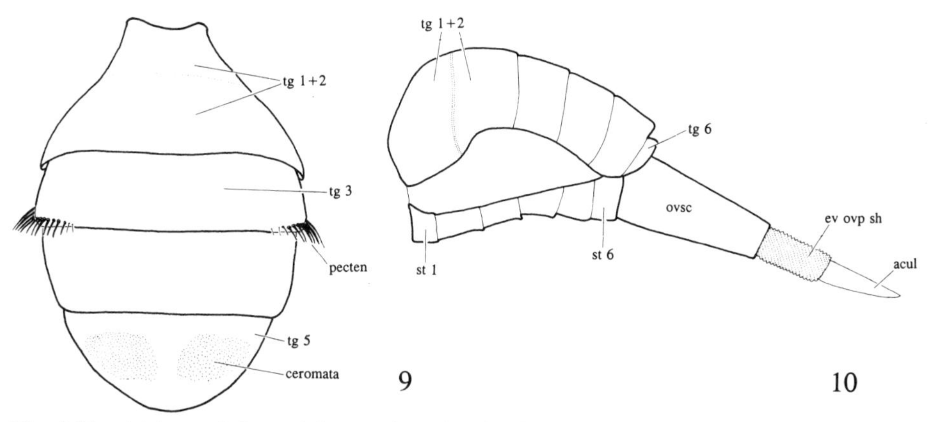

Figs 9-10. Adult morphology, abdomen; 9, male, with features of typical of Dacini ; 10, female, with extended ovipositor. Abbreviations are listed on p. 33.

Features of the abdomen and female terminalia are illustrated (figs. 10), with the following abbreviations:

acul - aculeus;
ev ovp sh - eversible ovipositor sheath;
ovsc - oviscape;
st - sternites numbered 1-5 in the male (fig. 9) and 1-6 in the female (fig. 10);
tg - tergites where 1+2 are fused to form syntergosternite 1+2, followed by tergites 3-5 in the male (fig. 9) and 3-6 in the female (fig. 10).

The features of the thorax are illustrated (figs 7, 8), with the following abbreviations:

a npl s - anterior notopleural seta;
a sctl s - apical scutellar seta;
a spal s - anterior supra-alar seta;
a spr - anterior spiracle;
anatg - anatergite;
anepm - anepimeron;
anepst - anepisternum;
anepst s - upper anepisternal seta;
b sctl s - basal scutellar seta;
cx - coxa;
dc s - dorsocentral seta;
hlt - halter;
ial s - intra-alar seta;
kepst - katepisternum;
kepst s - katepisternal seta;
ktg - katatergite;
mtg - mediotergite;
npl - notopleuron;
p npl s - posterior notopleural seta;
p spal s - posterior supra-alar seta;
p spr - posterior spiracle;
pprn lb - postpronotal lobe;
pprn s - postpronotal seta;
prepst - proepisternum;
presut area - presutural area;
presut spal s - presutural supra-alar seta;
psctl acr s - prescutellar acrostichal seta;
psut sct - postsutural scutum;
sbsctl - subscutellum;
scap - scapula setae;
sctl - scutellum;
trn sut - transverse suture.

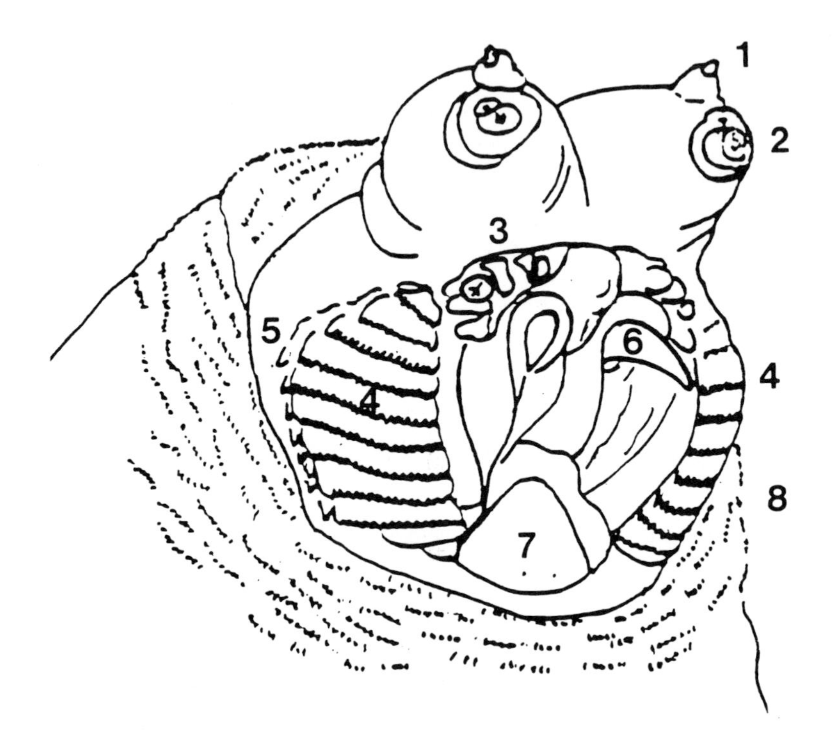

Fig. 11. *Bactrocera*, 3rd instar larva, ventral view of head. *1*, antennal sensory organ. *2*, maxillary sensory organ. *3*, stomal sensory organ. *4*, oral ridges. *5*, accessory plates. *6*, mouthhooks. *7*, labium. *8*, T1 spinules.

Immature Stages

Puparium

Diptera have either *obtect* pupae, in which the head appendages, wings and legs are visible and lie in sheaths attached to the surface of the body, or *exarate* pupae, in which those appendages are free. However, exarate pupae are almost always encapsulated within a *puparium* which is the hardened skin of the last larval instar. Exarate pupae, and puparia, are a feature of a suborder of the Diptera called the Cyclorrhapha, which includes the Acalyptratae, which in turn includes the Tephritidae. The formation of the visible structure that follows the last larval instar is often erroneously called *pupation*; it is actually *pupariation* as true pupation takes place unseen within the puparium.

Tephritid puparia vary in colour from white, through brown to black, although a black puparium in a species that normally has pale coloured puparia, is generally

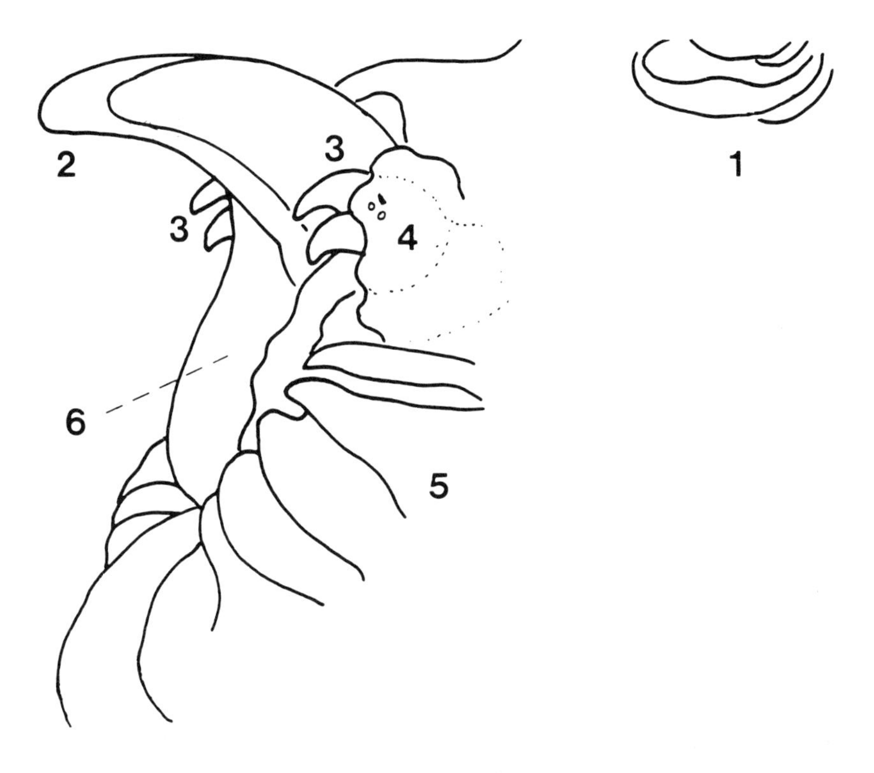

Fig. 12. *Rhagoletis*, 3rd instar larva, head in lateral view. *1*, maxillary sensory organ. *2*, mouthhooks. *3*. preoral teeth. *4*, stomal sensory organ. *5*, oral ridges. *6*, mouth opening.

symptomatic of parasitism by Hymenoptera. They tend to be rounded at the anterior end, have slightly out-curved lateral, dorsal and ventral surfaces, sometimes with distinct segmentation, and the posterior end may be rounded or flat. The number of tubules in each anterior spiracle and the shape of the posterior spiracles of the third instar larva can sometimes be determined by examination of a puparium. The *cephalopharyngeal skeleton* of the third instar larva may also be removed from the puparium for examination. However, separation of the many species that may be associated with a particular fruit is seldom possible using the limited set of larval characters obtainable from a puparium; as most frugivorous species spend only a few days as a puparium it is better to wait for the emergence of adults before attempting identification. Flower associated species can sometimes be determined from puparia, because colour and spiracle form is often enough to separate the small number of species that may attack a given host plant, and some keys for that purpose were provided by White (1988). Although the number of larval characters that can be determined from a puparium is limited, it is still worth preserving it along with the reared adult, as it confirms that the adult was reared and a future worker may find some value in its study.

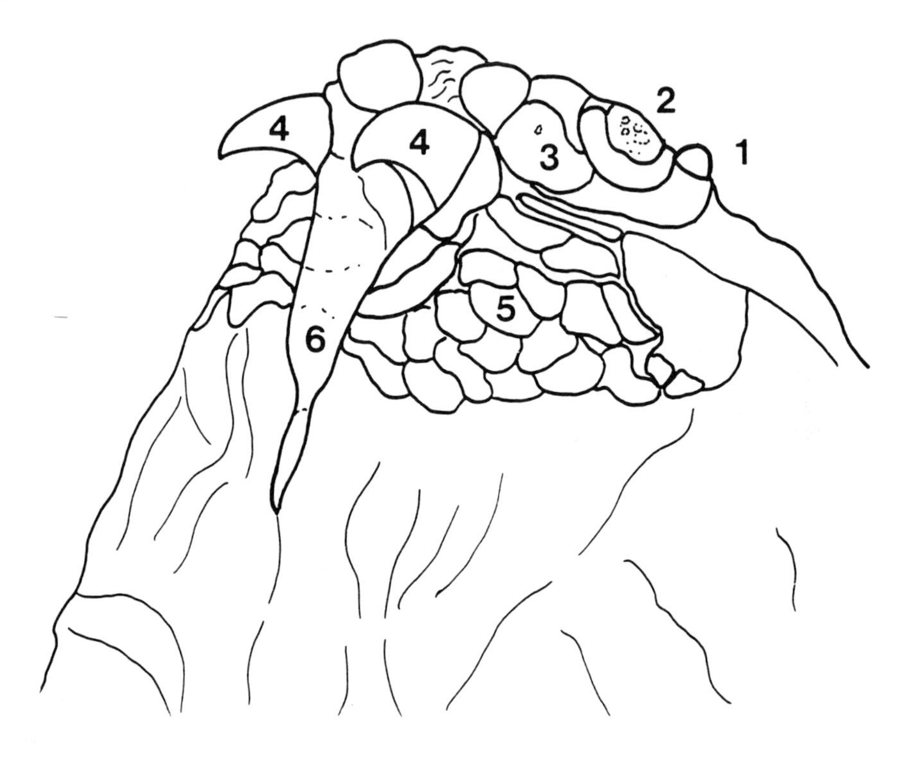

Fig. 13. *Anomoia*, 3rd instar larva, head in ventro-lateral view showing 'face mask'. *1*, antennal sensory organ. *2*, maxillary sensory organ. *3*, stomal sensory organ. *4*, mouthhooks. *5*, oral ridge area. *6*, mouth opening.

Third Instar Larva

Tephritid larvae are variable in shape and size depending on species and availability of essential nutrients in the breeding media. Larvae of Dacinae and Trypetinae that develop in soft fruits, are usually maggot-like, with abdominal segment 8 truncate and the rest of the body tapered to the anterior end. Larvae of Tephritinae tend to be cylindrical and rounded or almost truncate at both ends of the body. Mature larvae are usually creamy-white although some may appear darker due to the gut contents showing through the cuticle. The cuticle is almost translucent, without pigment or sclerotisation and the surface is frequently ornamented with numerous rounded projections or small sharply pointed spinules. A tephritid larva has a small tapered head with 2 distinct black dots (the heavily sclerotised mouthhooks), 3 thoracic and 8 abdominal segments referred to in the text as T1-T3 and A1-A8 respectively.

The head segment (fig. 11) is bilobate anteriorly with 2 pairs of small, but well defined, sensory structures on lobes previously assigned a variety of names including

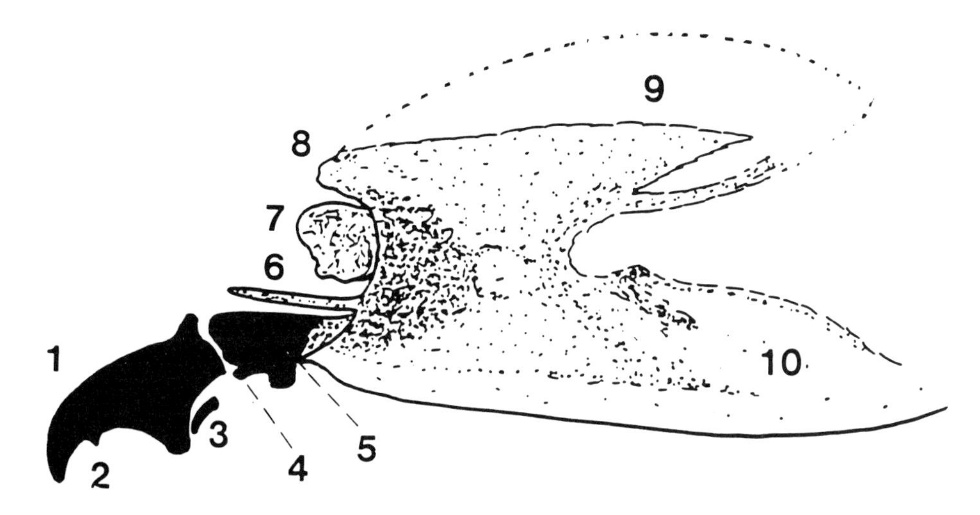

Fig. 14. *Bactrocera (Zeugodacus) cucurbitae*, cephalopharyngeal skeleton of 3rd instar larva. *1*, mouthhook. *2*, preapical tooth. *3*, dental sclerite. *4*, labial sclerite. *5*, hypopharyngeal sclerite. *6*, parastomal bar. *7*, anterior sclerite. *8*, dorsal arch. *9*, dorsal cornu. *10*, ventral cornu.

anterior sense organs and posterior sense organs by Snodgrass (1924), Phillips (1946), and Exley (1955). In this text they are called *antennal sensory organs* and *maxillary sensory organs*. Each antennal sensory organ has 1 to 3 segments, sometimes lightly sclerotised with a large basal segment and a cone-shaped distal segment. The maxillary sensory organs lie just below the antennal sensory organs. Each consists of a broadly flattened segment usually with 2 well defined groups of sensilla surrounded by folds of cuticle.

The mouth opening is ventral with small projecting *mouthhooks* (Teskey, 1981, calls them *mandibles*). Anterolaterally of the mouthhooks lie the *stomal sensory organs* with several small sensilla and surrounded by a series of *preoral lobes* (new term) which may be entire (unserrated) or toothed on the lower edges/posterior margins. In some genera e.g. *Rhagoletis* and *Carpomya* (Kandybina, 1977), additional, often heavily sclerotised, *preoral teeth* (fig. 12; pl. 37.b), or finger-like processes, occur at the base of the stomal sensory organ. At each side of the mouth opening there is a series of transverse radiating furrows or *oral ridges* (fig. 11; pl. 4.b) which may be entire (unserrated) or toothed on their posterior margins. Frequently in fruit feeders, an additional series of small *accessory plates* (new term) is present along the outer edge of the oral ridges. A cellular or reticular 'face mask' (fig. 13) incorporating the area of the head around the antennal and maxillary sensory organs and part of the mouth occurs in Trypetini, e.g. some *Anomoia*, *Acidiella* and *Myoleja* spp. (Kandybina, 1977). This cellular-like structure is absent in the major fruit pest genera.

The *cephalopharyngeal skeleton* has stout, heavily sclerotised mouthhooks which, in *Bactrocera* spp. (fig. 14), are usually strongly curved apically but lack preapical teeth. In a few species, e.g. *Bactrocera cucumis* (fig. 139) and *B. cucurbitae* (fig. 156), a small vestigial preapical tooth is visible. In *Dacus* spp., the Tephritinae and

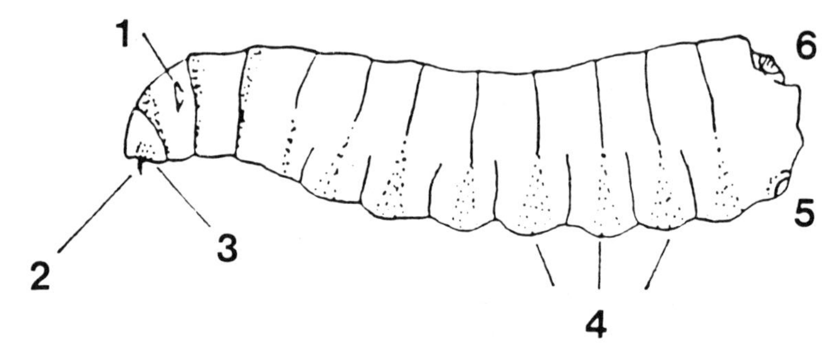

Fig. 15. *Bactrocera*, 3rd instar larva, lateral view. *1*, anterior spiracle. *2*, mouthhook. *3*, oral ridge. *4*, creeping welts. *5*, anal lobes. *6*, posterior spiracles.

Trypetinae, one, or sometimes two, preapical teeth may be present, e.g. the leaf miner *Euleia heraclei* has two preapical teeth as well as some additional cuticular teeth around the mouth (Ferrar, 1987). Posteriorly, the mouthhooks articulate with the *hypopharyngeal sclerite*, otherwise known as the hypostome, hypostomium, hypostomal piece or intermediate sclerite in descriptions by Efflatoun (1927), Phillips (1946) and Ferrar (1987). *Parastomal bars* (in the form of long, rod-shaped sclerites lying dorsally parallel to the hypopharyngeal sclerite) are common in fruit-infesting species, e.g. *B. tryoni* and *R. pomonella*. *Labial sclerites* form a V-shape in the floor of the mouth between the hypopharyngeal sclerites and the mouthhooks, and are otherwise known as subhypostomal and ligulate sclerites (Exley, 1955; Ferrar, 1987). The labial sclerites are small in Dacini but much larger in Tephritinae. Another pair of small sclerites lying close to the mouthhooks are the *dental sclerites* (sometimes called dentite sclerites) which are common in the Dacini but absent or inconspicuous in other groups. The side walls of the pharynx are supported by sclerotised structures referred to as *dorsal* and *ventral cornua* (singular: *cornu*). The dorsal cornua, also known as dorsal wing plates (Exley, 1955), are frequently cleft distally and shorter and thinner than the ventral cornua in Dacini, e.g. *B. tryoni*. Dorsal cornua are joined anteriorly by the *dorsal bridge*, sometimes referred to as the dorsal arch (Exley, 1955). In non-fruit feeding species the dorsal cornua are equal to or longer than the ventral cornua. The dorsal bridge may be present in some species and absent in others. In the Dacini, and some *Anastrepha* spp., an extra sclerite, the *anterior sclerite* (first mentioned by Exley, 1955), occurs on either side of the pharyngeal sclerite, projecting anteriorly from just below the dorsal bridge.

On the body, surface sculpturing or spination is very variable in tephritid larvae. Fruit infesting larvae, e.g. *B. tryoni*, frequently have encircling anterior bands of sharply pointed, posteriorly directed spinules on each thoracic segment. Although in some *Anastrepha* and *Rhagoletis* spp. there are dorsal spinules on the abdomen, most species have only transverse rows of spinules ventrally. These form creeping welts (fig. 15; pl. 4.f) with the first few rows anteriorly directed and the remainder posteriorly directed. Tephritines with gall inducing larvae usually lack long, sharply pointed spinules although some have small, knob-like structures often forming a surface

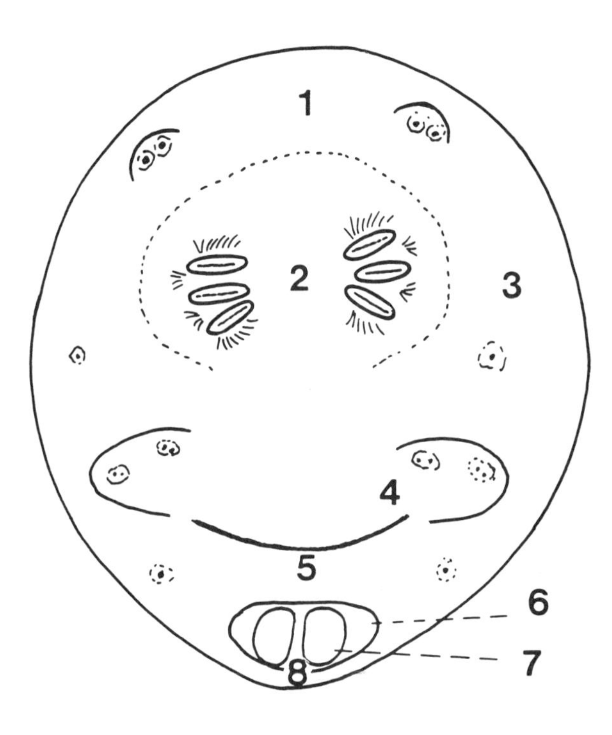

Fig. 16. A8 of 3rd instar larva, posterior view. *1*, dorsal area. *2*, posterior spiracles. *3*, lateral area. *4*, intermediate areas. *5*, ventral area. *6*, anal elevation. *7*, anal lobes. *8*, anal opening.

covering, giving rise to a stippled appearance, or forming reduced patches in particular areas.

Anterior spiracles (fig. 15; pl. 4.c) project laterally on each side of T1. The number of tubules in each anterior spiracle ranges from 2 to over 50. However it is usually of limited use taxonomically, as the number of tubules may vary widely within a species; an exception being the separation of *R. cingulata* (Loew) from *R. indifferens* Curran, in which that provides a better separation than any adult character (p. 357). Generally, gall inducing larvae have very low numbers of tubules, often set in fan-shaped arrangements. In other species, rows of tubules will often bifurcate and in some root-, leaf- and stem-mining species there are multiple rows (White, 1988).

The *caudal segment* (A8) (fig. 16) is the last abdominal segment and represents fused segments A8-A10. It bears the posterior spiracles and the anus. Detailed terminology of the caudal segment follows Phillips (1946) and Heppner (1984). The caudal segment can be divided into several areas for ease of reference: the *anal elevation* surrounds the anal lobes; the area immediately dorsal to that is the *ventral area*. Above and to the side of the posterior spiracles are the *dorsal* and *lateral areas*.

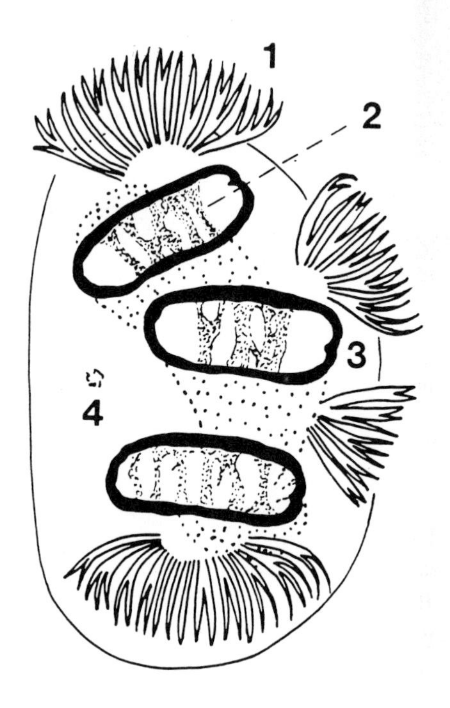

Fig. 17. *Bactrocera*, posterior spiracles of 3rd instar larva. *1*, spiracular hairs. *2*, spiracular slit. *3*, rima. *4*, ecdysial scar.

Each of those areas may bear *tubercles* with small *sensilla* which may be useful in distinguishing species. Berg (1979) and Greene (1929) produced keys to tephritid larvae heavily dependent on the use of the tubercle patterns.

Posterior spiracles (fig. 17; pl. 4.d) occur from the midline to high up on the dorsal edge of the caudal segment and are very useful taxonomic characters, previously well described by Butt (1937), Efflatoun (1927), Exley (1955), Phillips (1946) and Varley (1937). Each spiracle usually has 3 *spiracular openings* or *slits*, with the exception of *Myopites* spp. which only have 2 slits (Freidberg, 1980). The slits are frequently almost parallel to each other in fruit infesting species but arranged at greater angles to each other in non-fruit infesting species. Spiracles can be flush with the surface or the slits may be set on separate protuberant lobes as in some Terelliini larvae. The outer edge of each slit has a supporting sclerotization called a *rima* and attached to the cuticle at the outer ends of the slits are the *spiracular hairs*, also known as interspiracular processes (Phillips, 1946; Exley, 1955). There are usually 4 bundles of varying numbers of hairs associated with each spiracle. There are more branches to these spiracular hairs in fruit infesting species than in flowerhead feeding or gall

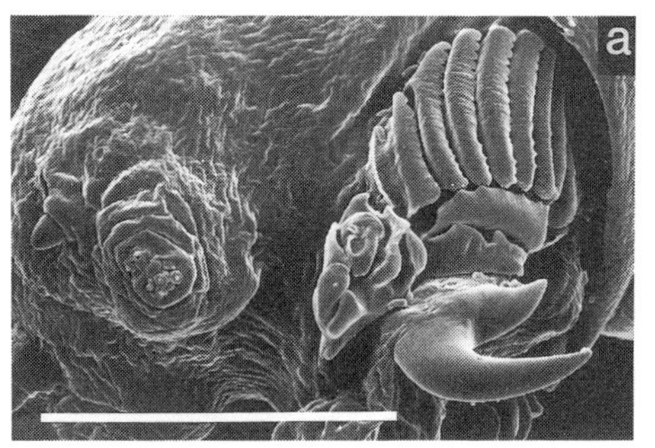

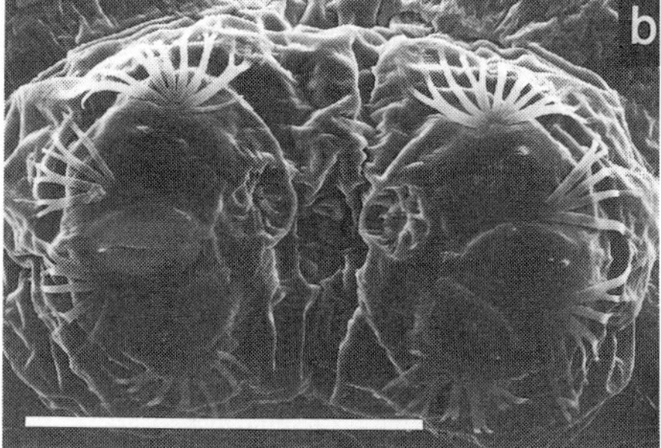

Plate 1. *Bactrocera tryoni*, SEMs of 2nd instar larva; a, head, lateral view (note oral ridges and mouthhooks); b, posterior spiracles. Scales=0.1mm.

inducing larvae. The *ecdysial scar* or button (Exley, 1955) marks the position of the external openings of the spiracles of the previous instar.

The *anal elevation* (fig. 16) surrounds the anal opening, which in fruit infesting larvae is often flanked by 2 large *anal lobes*. Each lobe may be entire, grooved or bilobed and is usually surrounded by several discontinuous rows of spinules whichoften concentrate into a small patch of slightly larger spinules just below the anal opening. In known gall inducing larvae the anal lobes are absent but the anal opening may be surrounded by small knob-like projections.

Second Instar Larva

Larvae are generally creamy-white (although some are discoloured due to gut contents) and very similar though smaller than the third instar of the same species.

Antennal and maxillary sensory organs are similar to those of the third instar (pl. 1.a). Usually the stomal sensory organs are slightly smaller and there are fewer oral ridges. Accessory plates are absent and the 'face mask' is reduced. The cephalopharyngeal skeleton (fig. 131) is very similar to that of the third instar but the mouthhooks have 1 or more preapical teeth and the hypopharyngeal sclerite is comparatively much longer. The dorsal cornua may be cleft distally and in some species the ventral cornua have large 'windows' (large, unsclerotised areas).

In comparison with the third instar, body surface sculpturing and spination is reduced although individual spinules may appear larger. The anterior spiracles are well developed and very similar to those of the third instar; the number of tubules usually corresponds to those of the third instar. The posterior spiracles (pl. 1.b) are similar to those of the third instar with 3 spiracular slits (except *Myopites* spp. which have 2) surrounded by sclerotised rimae. Four spiracular hair bundles are present but there are fewer hairs per bundle. In fruit feeders the anal area has obvious anal lobes with reduced numbers of spinules surrounding the lobes or the anal opening.

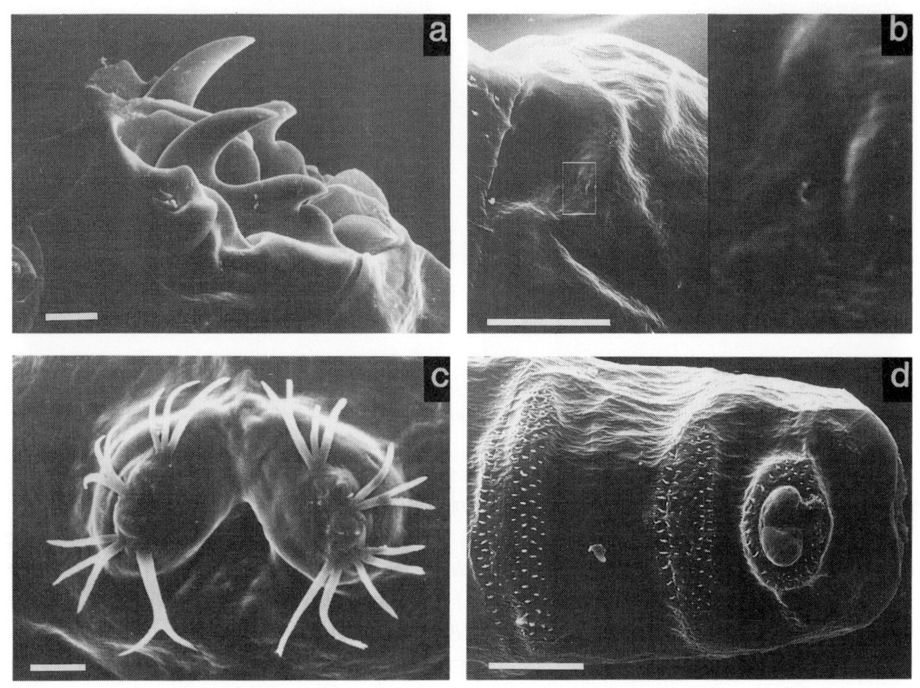

Plate 2. *Bactrocera tryoni*, SEMs of 1st instar larva; a, head, lateral view (scale = 10μm); b, T1, lateral view showing simple pit structure of anterior spiracle (scale = 50μm); c, posterior spiracle (scale = 10μm); d, creeping welts and anal lobes (scale = 0.1mm).

First Instar Larva

First instar larvae are extremely small, almost translucent, with little surface sculpturing. Some *Urophora* spp. complete the first instar within the egg and emerge as second instar larvae (see White & Korneyev, 1989).

Antennal and maxillary sensory organs are well developed and clearly defined. Stomal sensory organs are small; oral ridges and accessory plates are absent (pl. 2.a). The cephalopharyngeal skeleton (fig. 132) is only weakly sclerotised, with mouthhooks yellow to pale amber and with the remainder of the skeleton slightly darker. The mouthhooks usually have 1 or more large preapical teeth. The hypopharyngeal sclerite is fused to the pharyngeal sclerite and the cornua are only weakly sclerotised.

Body surface sculpturing and spination is very reduced. Individual spinules may appear larger by comparison with subsequent instars. The anterior spiracles appear as a minute pore which is only discernible under high magnification using a scanning electron microscope (pl. 2.b). There are 2 spiracular openings which may be rounded or slit-like on the posterior spiracles. Four spiracular hair bundles each with a few relatively long hairs are present (pl. 2.c). Anal area has prominent anal lobes, surrounded by large, stout spinules in the fruit feeders (pl. 2.d). Some non-fruit feeders have well developed spinules or small rounded projections around the anal opening.

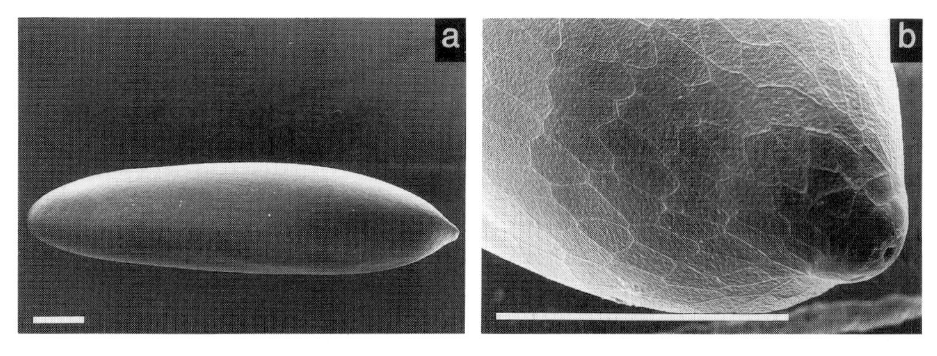

Plate 3. *Bactrocera tryoni*, SEMs of egg; a, whole egg; b, anterior end showing polygonal patterning and micropyle. Scales=0.1mm.

Egg

Egg colour is usually glistening white to creamy-yellow, becoming slightly darker towards the time of hatching. The shape and size vary according to species; e.g. the eggs of *B. tryoni* (pl. 3.a) and *Ceratitis capitata* are elongate and gently tapering whereas those of *Urophora solstitialis* (Linnaeus) are rounded at the anterior end, and the posterior end is long and pointed (Ferrar, 1987). At the anterior end of each egg is a small micropyle which is obvious in most species and very pronounced in some species of Terelliini; in some *Chaetorellia* spp. the micropyle is at the end of a tubular extension of the egg, which may be up to twice as long as the main body of the egg (White & Marquardt, 1989). The eggs of some *Chaetorellia* spp. are also unusual in that the first instar larva turns within the chorion to emerge head first from the posterior end of the egg (White & Marquardt, 1989). The surface or chorion of the egg usually appears smooth under the light microscope. However, at high magnification, or using the SEM, a polygonal pattern is usually visible (pl. 3.b), and this probably represents the boundaries of the follicle cells (see Margaritis, 1985).

The Classification of Tephritid Fruit Flies

Problems of Forming an Agreed Classification

There is no generally accepted classification of the Tephritidae and the system used here follows a suggestion made by Norrbom (1987), that the flower and stem infesting, and gall forming tribes should all be included within the subfamily Tephritinae, which he characterized by the scapular setae being poorly differentiated or lacking. That character is often difficult to interpret in species which have small scapular setae, or in specimens that have the head pressed against the antero-dorsal surface of the scutum. Subsequently, the tribal identity of almost all of the 500 genera was determined as part of a world cataloguing project (F.C. Thompson, A.L. Norrbom, A. Freidberg & I.M. White, in prep.); the data presented here on the numbers of genera in each group were derived from that study. Recently, Hancock (1986a; 1986b; 1990) discussed the classification of some groups, with particular reference to the African fauna and Kitto (1983) analyzed relationships using immunological data, and the partial classifications proposed by those authors concord fairly well with the system used here. All previous classifications were biased in favour of relationships suggested by examination of material from a single region, and works which provided keys to some tribes were Foote (1980), Freidberg & Kugler (1989), Hardy (1973; 1974) and White (1988). The classification presented here uses a number of characters that are difficult to interpret, or involve the male terminalia, which are not discussed to great extent in this work. It is therefore impractical to attempt to produce a reliable key following this system. Instead, some notes on characters which are typically present in each group are given in the following discussions.

Subfamily Dacinae

The larvae of most species belonging to this subfamily develop in fruit. Members of this group are all Old World in origin, although a few species have been accidentally introduced into the New World, notably the Mediterranean fruit fly (*Ceratitis capitata*). All of the Old World fruit pest species that have a broad host range belong to this subfamily which includes 39 genera. The close relationship between the Ceratitini and Dacini was noted by Hancock (1986b) and supported by the work of Kitto (1983); placing them within a single subfamily was suggested by A.L. Norrbom & A. Freidberg (pers. comm., 1989).

Identification Postocular setae all slender and usually all black (figs 19, 22). Scutum with scapular setae. Posterior area of anepisternum clearly separated from rest of anepisternum by a vertical suture. Wing usually with setulae along dorsal surface of vein R_{4+5} at least as far as r-m crossvein. Two spermathecae.

Tribe Ceratitini

This is an Old World group of 36 genera which is divided into two subtribes which are distinct in their biology, but hardly distinct morphologically.

Identification Scutum with dorsocentral setae, which are usually placed well forward, typically in line with the anterior supra-alar setae. Katepisternum usually with a seta. Scutellum usually swollen upwards (especially in the Ceratitina). Wing with cell bm about as deep as cell cu*p*; vein CuA_2 along front of cu*p* extension usually sinuous, so that the extension is at least as broad half way along its length as at base (particularly well developed in the genera of Ceratitina whose larvae develop in fruit, but not *Capparimyia*), and usually only reaching about a quarter of the way to the wing margin from the basal part of cell cu*p* (fig. 42).

Subtribe Ceratitina

This is a predominantly African group of 17 genera, including *Capparimyia* spp. (p. 279) in the buds of Capparidaceae, and *Ceratitella* (p. 417), *Ceratitis* (p. 286), *Neoceratitis* (p. 402), *Trirhithromyia* (p. 393) and *Trirhithrum* (p. 395) spp. whose larvae develop in fruit. The following works should be used for identification: Indonesia and New Guinea, Hardy (1987); Africa, Hancock (1984; 1985a; 1987) and Munro (1934).

Identification Arista usually not plumose (except some *Trirhithrum* spp.; fig. 31). 1st flagellomere evenly rounded at apex. Ocellar setae well developed (fig. 23) (except *Capparimyia*; fig. 20). Scutellum convex in profile and marked with polished black (or dark brown) areas, or entirely polished black (or dark brown).

Subtribe Gastrozonina

This is a group of 19 genera found in the Old World tropics, including *Acroceratitis* (p. 408), *Clinotaenia* (p. 409) and *Gastrozona* (p. 408). Species of known biology have been reared from the shoots of Poaceae (=Gramineae). The following works should be used for identification: Indonesia and New Guinea, Hardy (1988a); Thailand and adjoining countries, Hardy (1973); Philippines, Hardy (1974); Africa, Hancock (1985c).

Identification Arista plumose. 1st flagellomere often with a dorso-apical point. Ocellar setae usually well developed. Scutellum often convex in profile; if so, then shiny and usually marked with yellow and black (as in Ceratitina).

Tribe Dacini

This is an Old World group of three genera, namely *Bactrocera* (p. 165), *Dacus* (p. 313) and *Monacrostichus* (p. 347). The larvae of most species of known biology develop in fruit; however, some *Bactrocera* and *Dacus* spp. develop in the flowers of Cucurbitaceae. The following works should be used for identification: Thailand and adjoining countries, Hardy (1973); Philippines, Hardy (1974); Japan (Ito, 1983-5); China (Zia, 1937); Indonesia, Hardy (1982b; 1983a); Africa, Munro (1984); Australia and the South Pacific, Drew (1989a).

Identification Ocellar, dorsocentral and katepisternal setae absent. Wing with cell bm about two times as deep as cell cu*p*; extension of cell cu*p* very long, usually reaching at least half way to wing margin from basal part of cell cu*p* extension (fig. 40). Tergite 5 usually with a pair of flattened areas (ceromata) (fig. 9) (except *Monacrostichus*; fig. 232).

Subfamily Trypetinae

The larvae of most species with a known biology develop in fruit; however, some species are known to be leaf or stem miners and it is likely that many of the species of unknown biology also fit that category. This subfamily includes 235 genera, most of which have been placed within the seven named tribes. The unplaced genera include *Callistomyia* (p. 401), *Macrotrypeta* (p. 415) and *Nitrariomyia* (p. 402). The classification used here is largely that of Hancock (1986b). However, we have accepted the view that *Anastrepha* and *Toxotrypana* are very closely allied (Kitto, 1983; Norrbom & Foote, 1989), and we have followed the suggestion of A.L. Norrbom (pers. comm., 1989) that they form a tribe within the Trypetinae.

Identification Postocular setae all slender and usually all black (fig. 18). Scutum with scapular setae (very reduced in some species, especially in Phytalmiini). Posterior area of anepisternum clearly separated from rest of anepisternum by a vertical suture. Wing with cell bm about as deep as cell cu*p* (figs 39, 41, 43). Usually with three spermathecae (exceptions include some *Rhagoletis* spp.).

Tribe Acanthonevrini

This is a group of 77 genera, most of which are found in the Old World tropics, with a few representative species in temperate Asia. The larvae of some species are known to develop in fruit, for example some species in Australia and Papua New Guinea have been reared from wild figs (*Ficus*, Moraceae) (Hardy, 1986a) and *Dirioxa pornia* (Walker) (p. 340) may attack damaged cultivated fruit. There is also a record of *Rioxa sexmaculata* (Wulp) from the seeds of rubber tree (*Hevea brasiliensis*) (Yunus & Ho, 1980), but the source of that record is unknown. However, one species has been reared from bamboo shoots and many other species have been collected in bamboo thickets, suggesting that bamboos are important hosts (Hancock, in press, a; Hardy, 1986a). Some *Acanthonevra*, *Dacopsis*, *Diarrhegma* and *Diarrhegmoides* spp. have been reared from decomposing tree trunks (Brimblecomb, 1945; Dodson & Daniels, 1988; Hardy, 1986a), and some *Afrocneros* and *Ocnerioxa* spp. (p. 409) have been reared from under the bark of living trees (Munro, 1967). Remarkably, *Termitorioxa termitoxena* (Bezzi) has been reared from the galleries of a termite in tree trunks in northern Australia (Hill, 1921). The following works should be used for identification: Indonesia and New Guinea, Hardy (1982a; 1986a); Africa, Hancock (1986b) and Munro (1967).

Identification Ocellar setae usually reduced or absent. Arista sometimes plumose (fig. 25). Usually with more than two pairs of scutellar setae, typically three pairs (fig. 230) (only two in *Afrocneros* and *Ocnerioxa*). Anatergite without long hairs. Wing usually with setulae along dorsal surface of vein R_{4+5} at least as far as r-m crossvein; cell cu*p* with a short acute extension. Aculeus broad, usually with several pairs of long preapical sensilla and apical section appearing to be a distinct segment.

Tribes Adramini and Euphrantini

These tribes are doubtfully distinct (A.L. Norrbom, pers. comm., 1989). They include 28 genera, all of which are Old World with the exception of *Epochra* (p. 344) from North America. The larvae of *Epochra* and many *Euphranta* (p. 401) spp. develop in fruit, and some other species develop in seeds, namely *Adrama* (p. 401) and *Munromyia* (p. 402) spp. However, many develop in other parts of plants, e.g. *Coelotrypes vittatus* Bezzi in the buds of *Ipomoea* spp. (p. 414; Convolvulaceae) (Munro, 1953) and *Coelopacidia* spp. of known biology are stem borers, one species having been reared from *Senecio* (Asteraceae) and another from *Polemannia* (Apiaceae) (Munro, 1935b). Remarkably, *Euphranta toxoneura* (p. 409) from Europe is only known from the galls of sawflies and another species develops in the stems of a broomrape (Orobanchaceae) (p. 423). The following works should be used for identification: World genera, plus spp. from Indonesia and New Guinea, Hardy (1983b; 1986b); Japan, Ito (1983-5); Thailand and adjoining countries, Hardy (1973); Philippines, Hardy (1974); Taiwan, Shiraki (1933) and Munro (1935c); African genera and some species, Bezzi (1924a), Hancock (1986b) and Munro (1967).

Identification Arista sometimes pubescent or short plumose. Anatergite with long thin hairs (fig. 44). Ocellar, postocellar, katepisternal, dorsocentral and postpronotal setae absent in Adramini (*Adrama, Coelopacidia, Coelotrypes, Munromyia*). Wing usually with setulae along dorsal surface of vein R_{4+5}, at least near base; cell cu*p* with a short acute extension.

Tribe Phytalmiini

This group of 16 genera is restricted to tropical Asia and the New Guinea area. Some *Phytalmia* spp. have been reared from decomposing tree trunks (Dodson & Daniels, 1988) and a review of the genera whose males have antler-like structures (antler flies) was presented by McAlpine & Schneider (1978). The antlers are used to settle disputes and an illustrated account of that behaviour was given by Attenborough (1990).

Identification Arista usually plumose. Ocellar setae reduced or absent. Males of some genera with antler-like outgrowths of the genae. Anatergite without long hairs. Postocellar, katepisternal, dorsocentral and postpronotal setae absent. Wing vein R_{4+5} without dorsal setulae, except sometimes at base; cell cu*p* usually without an extension.

Tribe Rivelliomimini

This is a group of four genera found in tropical Asia and Africa. The only species with any host data is *Cycasia oculata* Malloch, which has been reared from a cycad (*Cycas* sp.; Cycadaceae) in Guam (Malloch, 1942); Hardy (1973) reported it from queen sago (*Cycas circinalis*) but he did not give the source of that data. [Hancock (in press, a) placed *Cycasia* in synonymy with *Ornithoschema*, making *C. oculata* Malloch a junior homonym of *O. oculata* de Meijere; further work is required to resolve that problem].

Identification Arista not plumose. Ocellar setae reduced or absent. Dorsocentral setae usually close to a line through the anterior supra-alar setae. Wing vein R_{4+5} without dorsal setulae, except sometimes at base; cell cu*p* closed by an oblique crossvein CuA_2, but without any bend in that crossvein to form an extension to the cell. Anatergite without long hairs. Both sexes with a pair of shiny black marks (bullae) on tergite 5. Tergite 6 of female vestigial.

Tribe Toxotrypanini

This is a group of two genera found in the tropical Americas, namely *Anastrepha* (p. 128) and *Toxotrypana* (p. 390). The larvae of most species develop in fruit and like members of the Dacinae, some species have a very broad host range and are serious crop pests. Steyskal (1977) may be used for identification of *Anastrepha* spp. and the genera were included in a key by Foote (1980).

Identification Arista not plumose. Ocellar setae reduced or absent (figs 18, 30). Wing with vein M upcurved before joining wing margin (fig. 35) (markedly so in *Anastrepha*). Dorsocentral setae, if present (reduced or absent in *Toxotrypana*), placed close to line of intra-alar setae. Anatergite without long hairs. Wing with setulae along dorsal surface of vein R_{4+5}, to beyond r-m crossvein; cell cu*p* with a long acute extension, usually reaching about half-way to wing margin from basal part of cell cu*p* (fig. 39). Eversible ovipositor sheath expanded basally and dorsal scales on this part very well developed (figs 45, 46).

Tribe Trypetini

This is a group of 60 genera, many of which have been placed in named subtribes by F.C. Thompson, A.L. Norrbom, A. Freidberg & I.M. White (in prep.). The unplaced genera include species whose larvae develop in fruit, such as *Acidiella* (p. 401), *Anomoia* (p. 401), *Oedicarena* (p. 402), *Taomyia* (p. 402) and some *Myoleja* (p. 402) spp.; leaf miners, such as *Hemilea* (p. 410) and some *Myoleja* (p. 410) spp.; and *Acidoxantha* spp. in large flowers (p. 415).

Identification Arista bare or pubescent, not plumose (figs 29, 32). Ocellar, postocellar, katepisternal, dorsocentral and postpronotal setae usually well developed. Anatergite without long hairs. Wing cell cu*p* with a short acute extension (except *Zacerata* which lacks any extension).

Subtribe Carpomyina

This is a group of 11 genera, most species of which are north temperate or South American (Norrbom, 1989). The larvae of species of known biology develop within the fruits of a narrow range of plants, typically a single genus and at most several genera within a single family. Some *Rhagoletis* spp. (p. 352, 422) are serious pests, and the genera *Carpomya* (p. 281), *Myiopardalis* (p. 349), *Rhagoletotrypeta* (p. 402) and *Zonosemata* (p. 399) also include pest or potential pest species.

Identification Antenna with 1st flagellomere often dorso-apically pointed. Dorsocentral setae usually close to a line through the anterior supra-alar setae (except *Zonosemata* in which they are placed much further back). Vein R_{4+5} usually without dorsal setulae, except for some aberrant specimens. Larva with preoral teeth (pl. 37.b).

Subtribe Trypetina

This is a group of ten genera, most species of which are north temperate. The larvae of species of known biology are leaf or stem miners, e.g. *Euleia* (p. 403), *Strauzia* (p. 410) and *Trypeta* (p. 410, 423) spp.

Identification Dorsocentral setae usually close to a line through the anterior supra-alar setae. Wing usually with setulae along dorsal surface of vein R_{4+5}, at least as far as r-m crossvein.

Subtribe Zaceratina

This is a group of two genera with representatives in Europe and Africa. The larvae of both *Pliorecepta* (p. 408) and *Zacerata* (p. 410) spp. mine the shoots of *Asparagus* spp. (Liliaceae). The relationship between these two genera was first noted by A.L. Norrbom & A. Freidberg (pers. comm., 1989).

Identification Wing with vein M between bm-cu and r-m crossvein downcurved (markedly so in *Zacerata*); vein R_{4+5} without dorsal setulae, except sometimes at base. Larva with a prominence placed just below the posterior spiracles, which terminates in a pair of sharp processes.

Subfamily Tephritinae

The larvae of most species belonging to this subfamily develop in flowers, usually of the family Asteraceae (=Compositae), and members of this group are found in all world regions. Few members of this group are pests, but some are of value for the classical biological control of some noxious weeds. This subfamily includes 211 genera and the classification used here follows Hancock (1990).

Identification Scutum without scapular setae; dorsocentral setae usually placed well forward, typically close to, or in front of, a line through the anterior supra-alar setae. Posterior area of anepisternum usually not clearly separated from rest of anepisternum by a vertical suture because it is obscured by tomentum, especially in the Tephritini and Terelliini. Wing with cell bm about as deep as cell cup.

Tribe Myopitini

This is a group of five genera and the larvae of most species of known biology induce either capitula (flower head) or stem galls on species of Asteraceae, e.g. *Urophora* spp. (p. 420, 423). The following works should be used for identification: Europe, Dirlbek (1973; 1974), Freidberg (1980), Steyskal (1979), and White & Korneyev (1989); North and South America, Steyskal (1979); Middle East, Freidberg & Kugler (1989).

Identification Head with only one pair of orbital setae, which are not convergent. Postocular setae all slender and black. Thorax usually with a fine tomentum. Wing cell cu*p* without a pointed extension; pattern usually of black or brown bands, but entirely hyaline in some spp.

Tribe Tephrellini

This is an Old World group of 24 genera and most authors have called the group the Aciurini. Most species are found in Africa, and all species of known biology attack plants belonging to the families Acanthaceae, Lamiaceae (=Labiatae) and Verbenaceae (Hancock, 1990). The larvae of most species develop in their host's flowers but some are found in seed capsules and a few induce galls (Freidberg, 1984; Munro, 1947). *Aciura coryli* (Rossi) has been reported as a pest (p. 414), and some aciurines may be of interest to future biocontrol programmes against *Lantana camara* L. (Verbenaceae) (see Munro, 1947 for host list). The following works should be used for identification: Africa, Munro (1947); Madagascar, Hancock (in press, b); Europe, Hendel (1927); Thailand and adjacent countries, Hardy (1973); Philippines, Hardy (1974); Indonesia and New Guinea, Hardy (1987); Middle East, Freidberg & Kugler (1989); key to genera from all regions, Hancock (1990).

Identification Posterior pair of orbital setae not convergent. Postocular setae all slender and usually all black. Thorax with a reduced tomentum. Wing usually largely black, often cut by hyaline V-shaped areas; cell c without one or two isolated dark marks; cell cu*p* with a pointed extension.

Tribe Terelliini

This is a group of six genera, most of which are found in north temperate areas. The larvae of most species develop in the capitula of Asteraceae, e.g. *Chaetorellia* (p. 413), *Chaetostomella* (p. 409, 415), *Craspedoxantha* (p. 415) and *Terellia* (p. 413, 414, 420, 423) spp. However, the larvae of some *Orellia* (p. 410) spp. develop in the stems or roots of Asteraceae. The following works should be used for identification: Europe, Korneyev (1985), White (1989a) and White & Marquardt (1989); Middle East, Freidberg & Kugler (1989); North America, Freidberg & Mathis (1986); Africa and Asia, Freidberg (1985).

Identification Posterior pair of orbital setae convergent. At least some of the postocular setae white and scale-like. Thorax covered by a dense tomentum; scutum pattern lyre-shaped. Wing cell cu*p* with a pointed extension; patterned with bands.

Tribe Tephritini

We have followed the suggestion of Hancock (1990) and placed all remaining Tephritinae in this ill defined tribe, although some authors have used the following additional groups: Cecidocharini, Dithrycini, Oedaspidini, Platensinini, Rhabdochaetini, Schistopterini and Spathulinini. Most species of known biology develop in species of Asteraceae, usually in the capitulum (flower head), e.g. *Acanthiophilus* (p. 411), *Acinia* (p. 417), *Campiglossa* (p. 415), *Dioxyna* (p. 414, 415, 422), *Ensina* (p. 415, 422), *Euaresta* (p. 418), *Euarestoides* (p. 418), *Gymnocarena* (p. 413), *Jamesomyia* (p. 414), *Neotephritis* (p. 413), *Oxyna* (p. 422), *Paracantha* (p. 413, 414), *Paroxyna* (p. 415), *Sphenella* (p. 422), *Tephritis* (p. 414, 419), *Tetreuaresta* (p. 420), *Tomoplagia* (p. 402), *Trupanea* (p. 413, 414, 415, 416, 423) and *Xanthaciura* (p. 416, 421) spp. Some species are known to induce stem or root galls, e.g. *Aciurina* (p. 415), *Eurosta* (p. 422), *Procecidochares* (p. 415, 418, 419, 422), some *Spathulina* (p. 416) and *Valentibulla* spp. (p. 415). Stem galls are also induced by *Eutreta* spp. (p. 413, 422, 423), including South and Central American species that induce stem galls on Verbenaceae, and one of those has been introduced into Hawaii as a weed biocontrol agent (p. 419). Another example of not attacking Asteraceae is found amongst Australian species (various genera close to *Oedaspis*) associated with Goodeniaceae. The following works should be used for identification: Europe, Hendel (1927) and White (1988); Indonesia and New Guinea, Hardy (1985); Philippines, Hardy (1974); Middle East, Freidberg & Kugler (1989); North American genera, Foote & Steyskal (1987); South American genera, Foote (1980); Thailand, Hardy (1973).

Identification Head usually with two pairs of orbital setae, neither of which is convergent; at least some of the postocular setae white and scale-like. Wing cell cu*p* with a pointed extension; usually with a reticulate pattern that includes one or two dark marks across cell c.

Keys to Fruit Pest Tephritidae; 1 - Adults

Two keys to genera are provided. The first key only separates the four major groups of fruit pest tephritids, namely the genera *Anastrepha*, *Ceratitis* and *Rhagoletis*, and the tribe Dacini (*Bactrocera* and *Dacus*). The second key (p. 55) also includes 10 other genera that may sometimes be found infesting cultivated fruits.

Keys to economically important species of the following groups are also presented: *Anastrepha* (p. 65); *Bactrocera* and *Dacus* (p. 72); *Carpomya* spp. associated with *Ziziphus* (p. 92); *Ceratitis* and *Trirhithromyia* (p. 92); *Rhagoletis* (p. 99); and *Trirhithrum* spp. associated with coffee (p. 105). The final key facilitates identification of males of *Bactrocera*, *Ceratitis* and *Dacus* spp. that have been bait trapped using the lures known as cue lure, methyl eugenol, trimedlure and vert lure (p. 106).

WARNING: The present work was primarily designed for the identification of fruit flies that have been reared from cultivated fruits. Although some species not included in the following keys will sometimes be found associated with cultivated fruit, these keys should be sufficient for the identification of most fruit flies reared from cultivated fruits. Attempting to use these keys for the identification of fruit flies collected by other means is likely to result in misidentifications.

Simplified Key to Major Fruit Pest Genera

1 Vein Sc abruptly bent forward at nearly 90°, weakened beyond the bend and ending at subcostal break; dorsal side of vein R_1 with setulae (fig. 34). Wing usually patterned by coloured bands (figs 35-38). Wing cell cu*p* with an acute extension (figs 39-43).
. . . . TEPHRITIDAE 2

\- Vein Sc not abruptly bent forward (fig. 33), except in the Psilidae, which lack both dorsal setulae on vein R_1 and frontal setae. Species associated with fruit very rarely

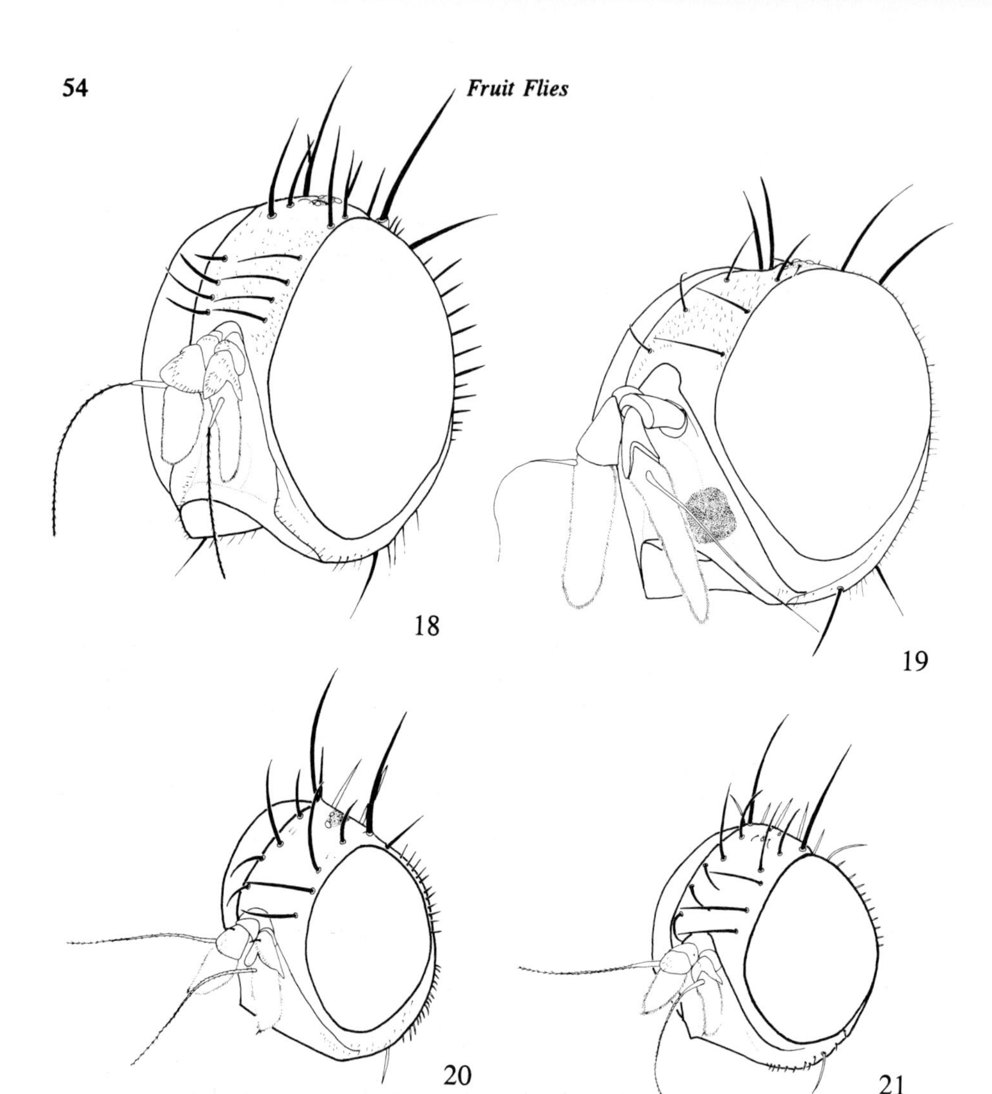

Figs 18-21. Heads, showing features typical of each genus; 18, *Anastrepha fraterculus*; 19, *Bactrocera (B.) dorsalis* (most *Dacus* spp. are similar, except *D. (Callantra)* spp.); 20, *Capparimyia savastani*; 21, *Carpomya vesuviana*.

have any wing patterning. Wing cell cu*p* usually without an acute extension (exceptions include some Otitidae and Pyrgotidae).
. . . . Families other than Tephritidae (p. 13)

2 Cell cu*p* very narrow and extension of cell cu*p* very long (fig. 40). 1st flagellomere (3rd segment of antenna) at least 3 times as long as broad (figs 19, 24). Wing pattern usually confined to a costal band and an anal streak (fig. 36) (for exceptions see figs 177, 180, 181, 184, 195, 206). [Tropical and warm temperate Old World; adventive species in Hawaii and northern South America.]
. . . . ***BACTROCERA*** and ***DACUS*** (p. 72)

\- Cell cu*p* broader and the extension shorter (figs 39, 42, 43). 1st flagellomere shorter (figs 18, 23, 29). Wing pattern usually includes some coloured crossbands (figs 35, 37, 38).

. . . . 3

3 The wing vein that terminates just behind the wing apex (vein M) is curved forwards before merging into the wing edge (fig. 35). Wing pattern usually similar to fig. 35 (for exceptions see figs 164, 167, 170, 172). [South America, West Indies and southern USA.]

. . . . ***ANASTREPHA*** (p. 65)

\- The wing vein that terminates just behind the wing apex (vein M) meets the wing edge at approximately a right angle (figs 37, 38). Wing pattern usually similar to fig. 37 or fig. 38.

. . . . 4

4 Cell cu*p*, including its extension, shaped as fig. 42. Basal cells of wing usually with spot- and fleck-shaped marks, giving a reticulate appearance (fig. 37). Scutellum convex and shiny. [*Ceratitis capitata* is found in most tropical and warm temperate areas; other spp. are African.]

. . . . ***CERATITIS*** (Keys 1 [males], 2 [females]; p. 92)

\- Cell cu*p*, including its extension, shaped as fig. 43. Basal area of wing not reticulate (fig. 38). Scutellum fairly flat and not shiny. [Larvae develop in the fruits of Berberidaceae, Caprifoliaceae, Cornaceae, Cupressaceae, Elaeagnaceae, Ericaceae, Grossulariaceae, Juglandaceae, Oleaceae, Rosaceae and Solanaceae. North temperate regions and South America.]

. . . . ***RHAGOLETIS*** (p. 99)

Key to Genera Associated with Fruit of Economic Importance

The major fruit pest genera most likely to be encountered by quarantine entomologists (those included in the simplified key) are named in ***BOLD CAPITALS***.

1 Vein Sc abruptly bent forward at nearly 90°, weakened beyond the bend and ending at subcostal break; dorsal side of vein R_1 with setulae (fig. 34). Wing usually patterned by coloured bands (figs 35-38). Wing cell cu*p* with an acute extension (figs 39-43) (except in a few genera which are not covered by the present work).

. . . . TEPHRITIDAE 2

\- Vein Sc not abruptly bent forward (fig. 33), except in the Psilidae, which lack both dorsal setulae on vein R_1 and frontal setae. Species associated with fruit very rarely have any wing patterning. Wing cell cu*p* usually without an acute extension (exceptions include some Otitidae and Pyrgotidae).

. . . . Families other than Tephritidae (p. 13)

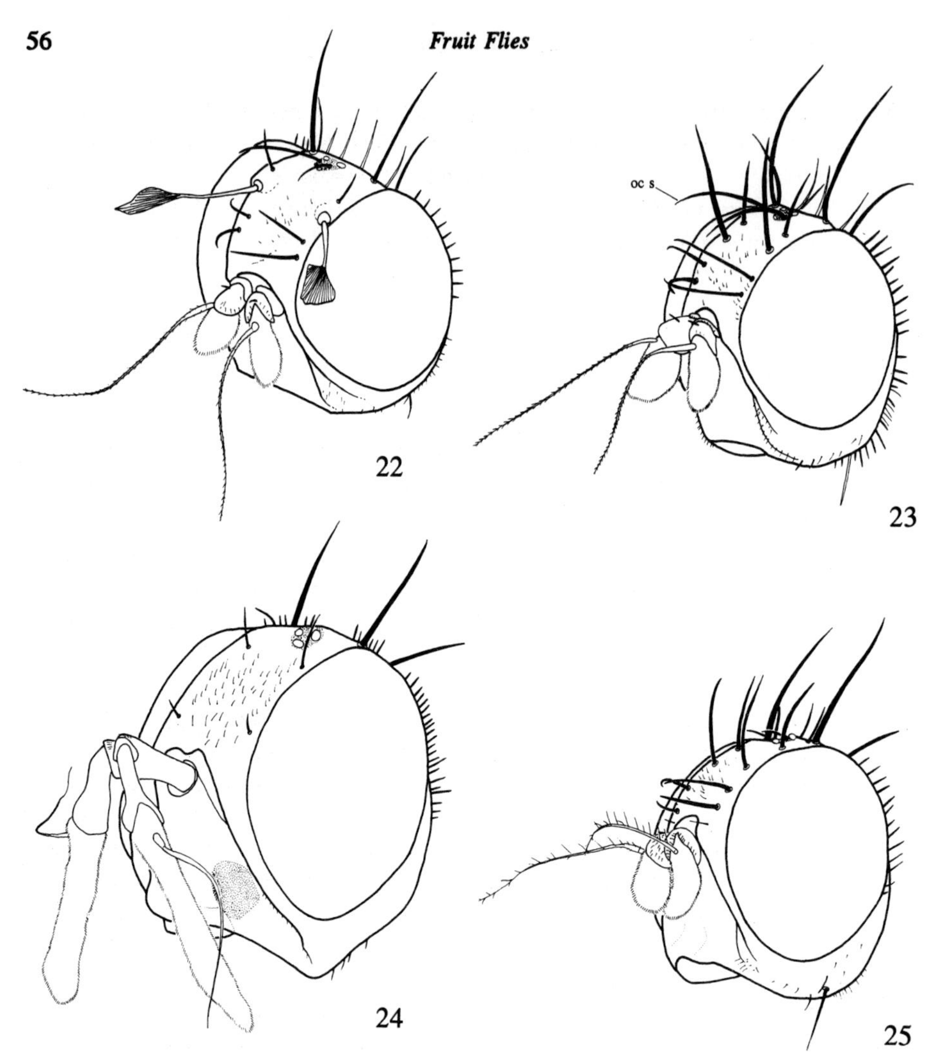

Figs 22-25. Heads, showing features typical of each genus or subgenus; 22, *Ceratitis (Ceratitis) capitata*, male; 23, *C. (C.) capitata*, female; 24, *Dacus (Callantra) smieroides*; 25, *Dirioxa pornia*. *oc s*, ocellar setae.

2 Cell cu*p* very narrow, usually about half depth of cell bm (fig. 40), except *Monacrostichus citricola* in which it is about two-thirds depth of bm (fig. 232).

[1st flagellomere at least 3 times as long as broad (figs 19, 24, 27). Cu*p* extension very long, equal or longer than length of vein $A_1 + CuA_2$ (fig. 40). Head and thorax with reduced chaetotaxy; lacking ocellar, postocellar, dorsocentral and katepisternal setae.]

. . . . 3

\- Cell cu*p* broader, always considerably broader than half depth of cell bm, and usually about as deep as cell bm (figs 39, 41-43).

. . . . 5

3 Vein M curved forward in apical quarter of cell dm, so that cell dm is about twice as wide at apex as in basal three-quarters (fig. 232). Abdominal tergite 5 without a pair of large slightly depressed areas (ceromata). [1st flagellomere very long (fig. 27). Cell r_1 very narrow. Abdominal tergites 1+2 longer than broad, giving a strongly wasp-waisted appearance. Wing length 6.3-8.8mm. Larvae develop in citrus fruits. Malaysia and Philippines.]
. . . . *Monacrostichus citricola* (p. 347)

\- Vein M not curved forward in apical quarter of cell dm (fig. 36). Abdominal tergite 5 with a pair of slightly depressed areas (ceromata) (fig. 9).
. . . . 4

[The following couplet may be ignored; *Bactrocera* and *Dacus* spp. are included in a single set of keys (p. 72) which avoid the use of the difficult tergite fusion character as much as possible.]

4 Abdomen with all tergites fused into a single plate, at most with smooth transverse lines marking the boundaries of each segment (view from side to check that no sclerites overlap the next). [Associated with cucurbit and sometimes other fruit crops. Old World tropics.]
. . . . *DACUS* (p. 72)

\- Abdomen with all tergites separate (view from side to see overlapping sclerites). [Associated with a wide range of fruits. Tropical and warm temperate areas of the Old World; adventive spp. are found in Hawaii and northern South America.]
. . . . *BACTROCERA* (p. 72)

5 Scutellum with 6 marginal setae (basal, subbasal and apical pairs) (fig. 230). [Cell *cup* extension about one-third length of vein A_1+CuA_2. Larvae develop in fruit that was already damaged or over-ripe at time of attack. Australia.]
. . . . *Dirioxa* (p. 340)

\- Scutellum with 4 marginal setae (a basal and an apical pair) (except aberrant individuals of other spp., notably *Myiopardalis pardalina*, fig. 233).
. . . . 6

6 Scutum without dorsocentral setae. Female with an exceptionally long oviscape which is longer than the wing length, and curved when seen in profile (fig. 248). Wing without a pattern of coloured bands, but yellow-brown infuscate in cells bc, c, sc, r_1 and r_{2+3}. Vein R_{2+3} sinuate, sometimes almost touching vein R_1 near the end of cell r_1. Cell r_1 usually with one or more accessory crossveins. [Head with very small hair-like, frontal, orbital and ocellar setae (fig. 30). Scutellar setae small and hair-like. Cell *cup* extension very long, about equal in length to vein A_1+CuA_2. Wing length 10.3-11.5mm. Larvae develop in papaya (*Carica papaya*). New World tropics.]
. . . . *Toxotrypana curvicauda* (p. 390)

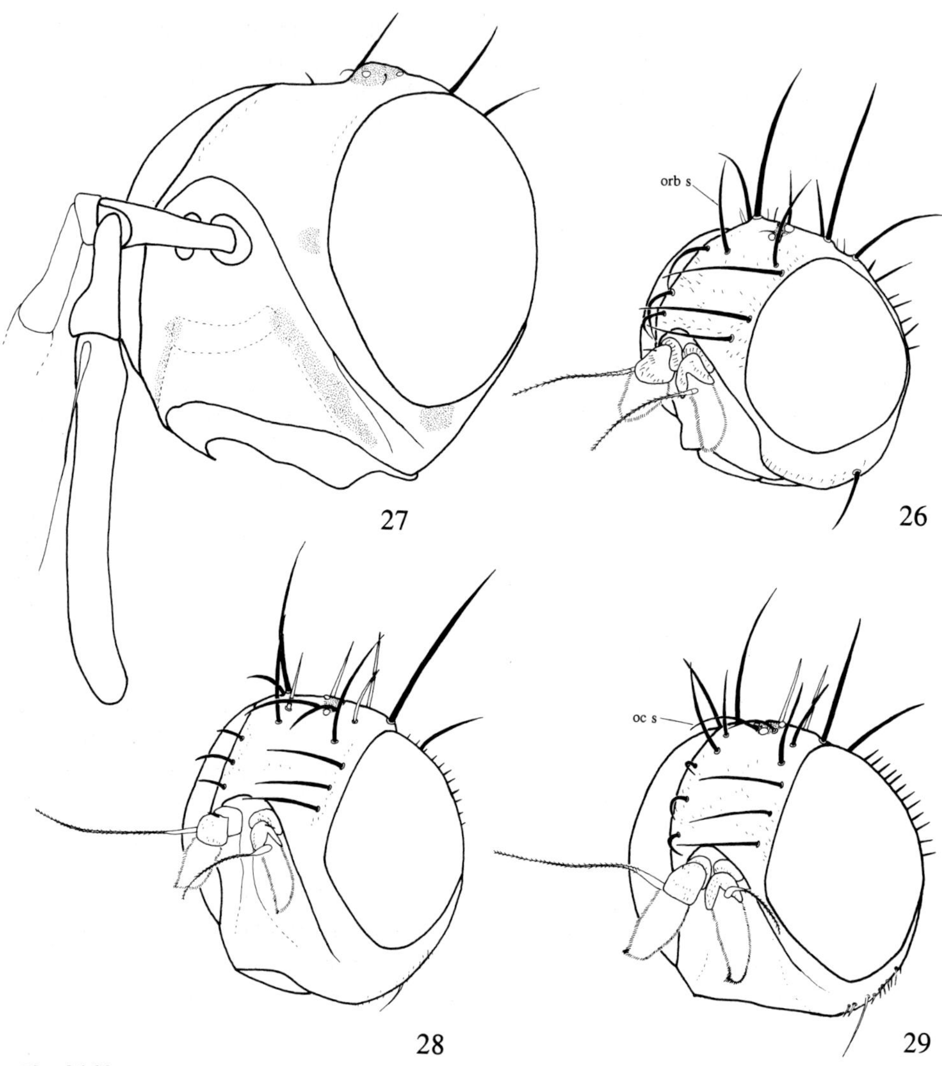

Figs 26-29. Heads, showing features typical of each genus; 26, *Epochra canadensis*; 27, *Monacrostichus citricola*; 28, *Myiopardalis pardalina*; 29, *Rhagoletis pomonella*. *oc s*, ocellar setae. *orb s*, orbital setae.

- Scutum with dorsocentral setae. Female with an oviscape that is straight and usually shorter than the wing length. Wing usually with a pattern of yellow or brown crossbands (except some *Trirhithrum* spp.; figs 128-130, 250). Vein R_{2+3} not sinuate. Cell r_1 without any accessory crossveins (except in some aberrant individuals).

. . . . 7

7 Apex of vein M turned anteriorly to merge with C without a distinct angle (fig. 35). [Wing pattern usually similar to fig. 35. Cell *cup* extension long, similar in length to vein A_1+CuA_2 (fig. 39). Scutum with dorsocentral setae placed closer to a line between the posterior supra-alar setae, than to a line between the anterior supra-alar setae (as in left half of fig. 7). South America, West Indies and southern USA.]
. . . . ***ANASTREPHA*** (p. 65)

\- Apex of vein M meeting C with a distinct angle (figs 37, 38).
. . . . 8

8 Cell *cup* extension long, at least one-third as long as vein A_1+CuA_2; vein CuA_2 usually curved forwards along anterior edge of *cup* extension (fig. 42), except *Capparimyia* (fig. 41). Scutellum usually convex and patterned with yellow and black areas; dorsal surface of vein R_{4+5} with setulae at least as far as r-m crossvein. Head with 2 pairs of frontal setae (figs 20, 22, 23, 31).
. . . . 9

\- Cell *cup* extension short, never more than one-fifth as long as vein A_1+CuA_2, and vein CuA_2 straight along anterior edge of *cup* extension (fig. 43). Scutellum usually flat; if convex and patterned with yellow and black areas (some *Carpomya* spp.) then vein R_{4+5} without dorsal setulae, except at base. Head with 3-4 pairs of frontal setae (figs 21, 26, 28, 29, 32).
. . . . 11

9 Head without ocellar setae, or ocellar setae very small, only about as long as distance between anterior and posterior ocelli (fig. 20). Cell *cup* extension about one-third as long as vein A_1+CuA_2 and vein CuA_2 straight along anterior edge of *cup* extension (fig. 41). Larvae develop in the buds of caper (*Capparis spinosa*). [Wing length 3.1-4.0mm. Aculeus length 0.8mm. Southern Europe, North Africa, the Middle East and Pakistan.]
. . . . *Capparimyia savastani* (p. 279)

\- Ocellar setae longer, usually similar in length and strength to orbital setae (figs 22, 23, 31). Cell *cup* extension longer, usually about half as long as vein A_1+CuA_2 and vein CuA_2 curved forwards along anterior edge of *cup* extension (fig. 42). Larvae develop in fruit.
. . . . 10

10 Scutellum with yellow areas. Wing with a preapical crossband (band on crossvein dm-cu) which is isolated from the rest of the wing pattern (fig. 37), except *Ceratitis catoirii* (fig. 214). Basal cells of wing (c, br, bm, *cup*) usually with spot- and fleck-shaped marks, giving a reticulate appearance (fig. 37). [*C. capitata* is found in most tropical and warm temperate areas; other spp. are African.]
. . . . ***CERATITIS*** and
Trirhithromyia cyanescens (Keys 1 [males], 2 [females]; p. 92)

[The recent removal of *T. cyanescens* from *Ceratitis* (Hancock, 1984) makes the separation of these genera difficult and in the present work *T. cyanescens* is included in the key to *Ceratitis* spp. of economic importance.]

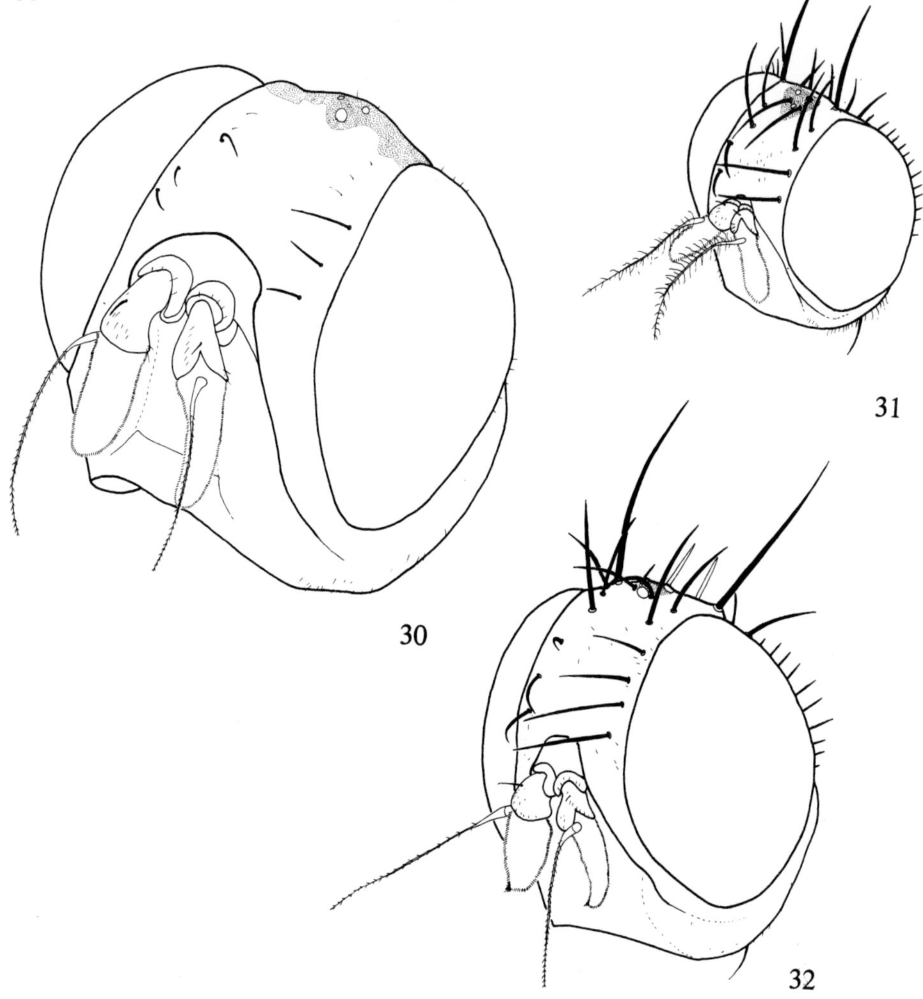

Figs 30-32. Heads, showing features typical of each genus; 30, *Toxotrypana curvicauda*; 31, *Trirhithrum coffeae*; 32, *Zonosemata electa* (this species has 3 or 4 frontal setae and may differ on each side, as shown here).

- Scutellum almost entirely dark brown to black, at most with small yellow spots adjacent to the scutellar setae. If wing has a preapical crossband (band on crossvein dm-cu), then it joins the rest of the wing pattern (fig. 128). Basal cells not reticulate (in the included spp.). [Spp. included in the present work have a plumose arista (fig. 31) and their larvae develop in coffee fruits. Africa.]
  . . . . *Trirhithrum* (p. 105)

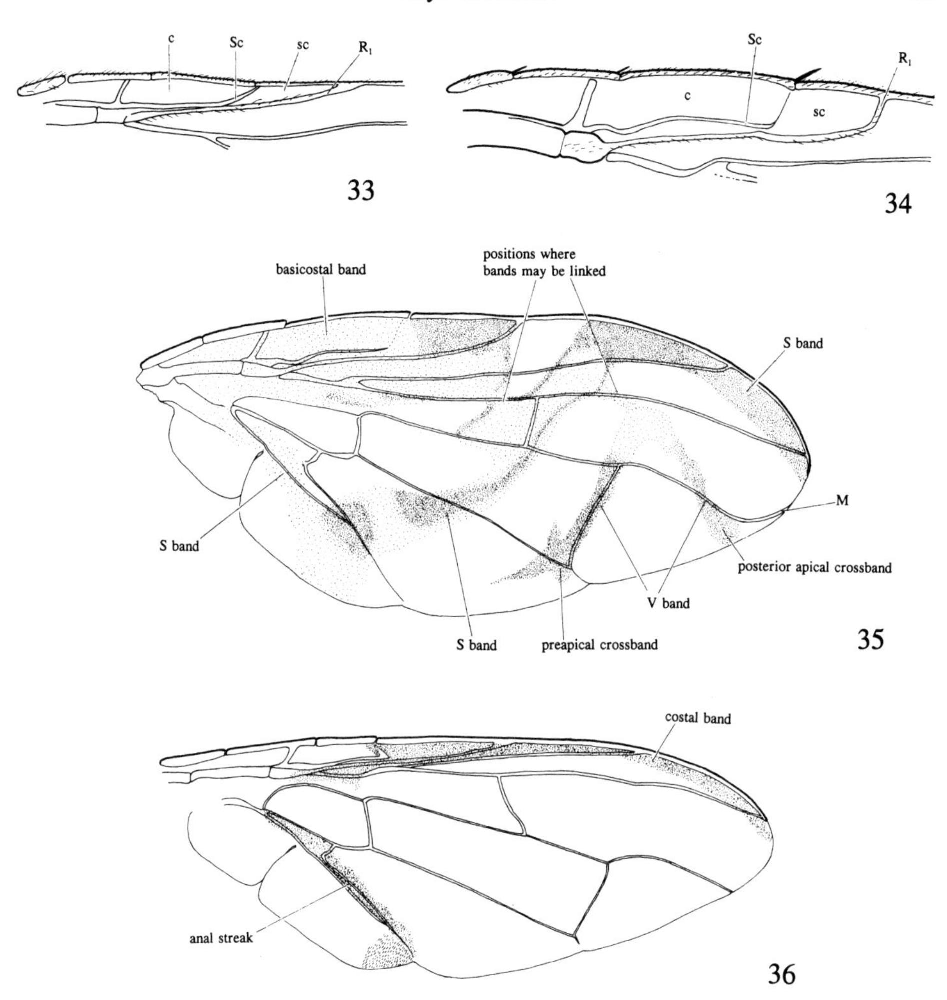

Figs 33-36. Wings; 33-34, cells c and sc, showing shape of vein Sc; 33, Platystomatidae; 34, Tephritidae; 35-36, typical patterns found in major pest genera; 35, *Anastrepha* spp.; 36, *Bactrocera* and *Dacus* spp.

11 Head with only one pair of orbital setae (fig. 26). Scutum without presutural supra-alar setae (fig. 231). Anatergite with long pale hairs which are distinct from the general pubescence (fig. 44). 1st flagellomere rounded at apex (fig. 26). [Dorsocentral setae placed about half-way between anterior and posterior supra-alar setae. Vein R_{4+5} with setulae on the dorsal surface as far as the r-m crossvein. Larvae develop in currants (*Ribes*). North America.]
. . . . *Epochra canadensis* (p. 344)

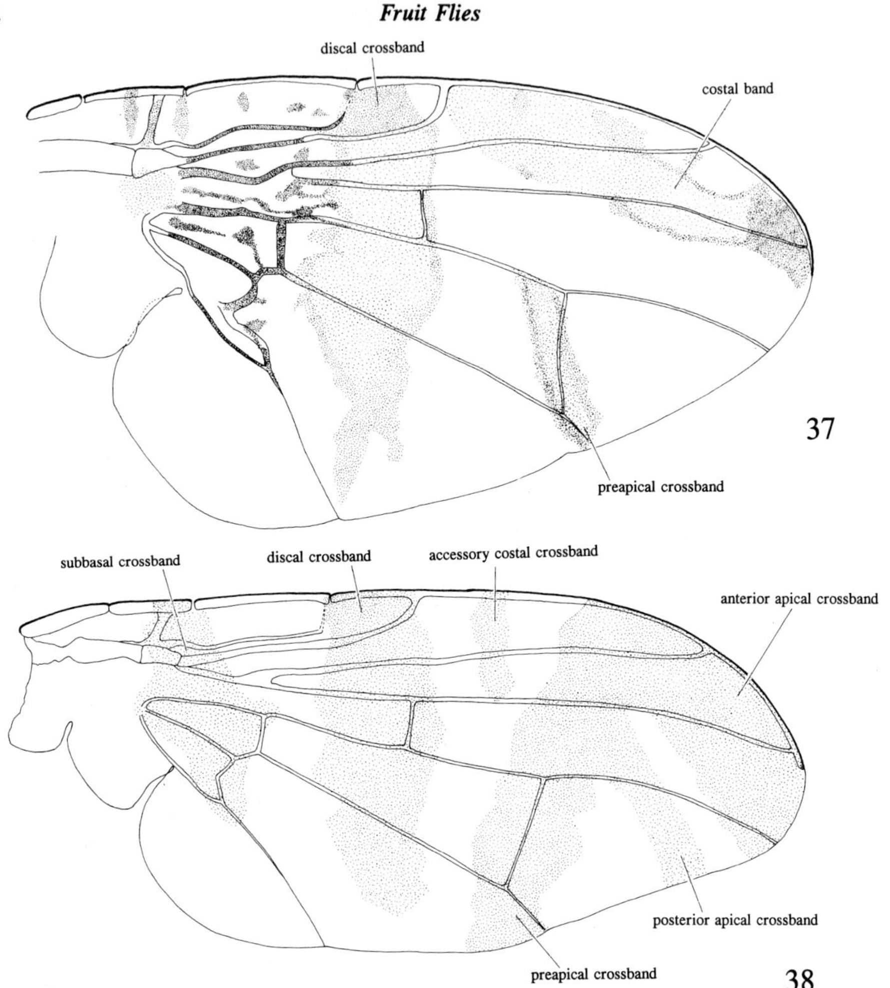

Figs 37-38. Wings, typical patterns found in major pest genera; 37, *Ceratitis* spp.; 38, *Rhagoletis* spp.

- Head with two pairs of orbital setae (figs 21, 28, 29, 32). Scutum with presutural supra-alar setae (e.g. figs 235-247). Anatergite without long pale hairs, at most with a fine pubescence. 1st flagellomere usually with a small antero-apical point (figs 21, 28, 29, 32), except some *Rhagoletis* spp. 12

12 Scutum with dorsocentral setae based very close to a line between the posterior supra-alar setae (as in the left side of fig. 7). Vein R_{4+5} usually with dorsal setulae at least as far as the r-m crossvein. [Wing length 5.8-6.2mm. Larvae develop in peppers (*Capsicum*). Southern North America.] . . . . *Zonosemata electa* (p. 399)

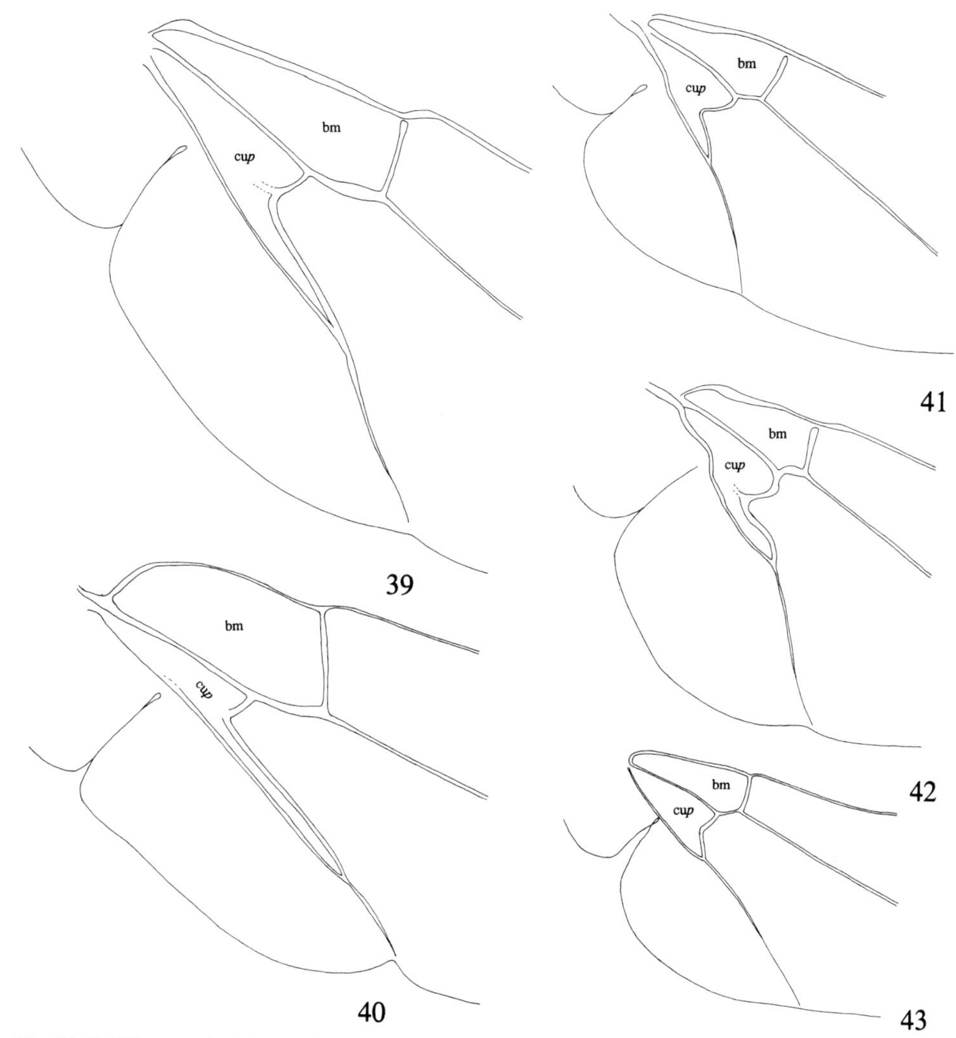

Figs 39-43. Wings, typical shapes of cells bm and cu*p*; 39, *Anastrepha* spp.; 40, *Bactrocera* and *Dacus* spp.; 41, *Capparimyia savastani*; 42, *Ceratitis* spp.; 43, *Rhagoletis* spp.

- Scutum with dorsocentral setae based close to a line between the anterior supra-alar setae (as in the right side of fig. 7). Vein R_{4+5} usually without dorsal setulae, except sometimes at the base of the vein (except in some aberrant individuals). 13

13 Ocellar setae very small, only about as long as distance between anterior and posterior ocelli (fig. 21). [Larvae of spp. included in the present work develop in the fruits of jujube (*Ziziphus*). Southern Europe, Middle East and India; adventive in Mauritius.] . . . . *Carpomya* (p. 92)

- Ocellar setae longer, usually similar in length and strength to orbital setae (figs 28, 29).

. . . . 14

14 Scutellum yellow, with large black patches at the base of each seta and near the centre of the dorsal area (fig. 233). Lateral and posterior areas of scutum yellow with large black patches, giving a *Ceratitis*-like appearance. Genae very deep, about one-third eye height (fig. 28). Larvae develop in the fruits of Cucurbitaceae. [Wing length 3.9-5.3mm. Southern USSR, Middle East and the northern part of the Indian subcontinent.]

. . . . *Myiopardalis pardalina* (p. 349)

- Scutellum entirely cream to yellow (figs 236, 239, 245); if marked with black, the black areas are confined to the base and lateral areas (figs 116-120). Scutum not marked with yellow and black patches (figs 234-247). Genae usually less than one-quarter eye height (fig. 29). Larvae develop in the fruits of Berberidaceae, Caprifoliaceae, Cornaceae, Cupressaceae, Elaeagnaceae, Ericaceae, Grossulariaceae, Juglandaceae, Oleaceae, Rosaceae and Solanaceae. [North temperate regions and South America.]

. . . . ***RHAGOLETIS*** (p. 99)

Key to *Anastrepha* Species of Economic Importance

The following key includes the 15 *Anastrepha* spp. that are likely to be found in association with commercial fruit crops. Some minor pest or potential pest species which are easily identified are also included. Positive identification necessitates dissection of the female's aculeus (p. 27) and many of the major pest species can only be reliably determined by a specialist. Aculeus length values were obtained by examination of specimens, supplemented with values given by Stone (1942).

1 Wing with a posterior apical crossband (figs 163, 166, 168, 169, 171, 173, 174); if wing without a well developed posterior apical crossband (some *A. leptozona*; fig. 165) then scutum orange brown, without any dark markings.
. . . . 2

\- Wing without a posterior apical crossband or any other markings crossing the apical section of vein M (figs 164, 167, 170, 172); scutum largely dark coloured or at least with dark markings. WARNING - teneral specimens of any spp. could be erroneously assumed to lack some wing and scutum markings.
. . . . 12

2 Scutum with dark brown dorsocentral stripes, united on posterior margin to form a U-shaped marking (fig. 173); part of each dark stripe bare of microtrichia (best seen when viewed from in front, from which microtrichose areas appear white frosted, especially in *A. striata*), but with dark brown to black setulae, which are much darker than the white or yellow setulae in the central area of the scutum. Aculeus tip without any distinct serrations, blunt, and at least 0.18mm wide (figs 50, 52).
. . . . 3

\- Scutum without dark brown marks, except sometimes posteriorly; scutum evenly microtrichose; setulae all of one colour or gradually darkened laterally. Aculeus tip serrate and less than 0.18mm wide (figs 54-61, 63). NOTE - female specimens are essential for the identification of these spp.
. . . . 4

3 Aculeus long, 3.3-3.7mm (fig. 50). Scutum with brown longitudinal stripes in line with the dorsocentral setae, which are covered in fine dark setulae; at most narrowed by white microtrichia (view from in front) in the area of the notopleural suture; area lateral to each stripe with at least scattered setulae. [Wing with basicostal band joined to the S band by a broad mark on vein R_{4+5}; preapical crossband not joined to the posterior apical crossband to form a complete V band; V band not joined to S band (similar to fig. 173). Tapered apical section of aculeus 0.3mm long; tip 0.19mm wide. Wing length 7.9-8.4mm. Larvae develop in guava fruit (*Psidium*). Southern Brazil.]
. . . . *A. bistrigata* (p. 131)

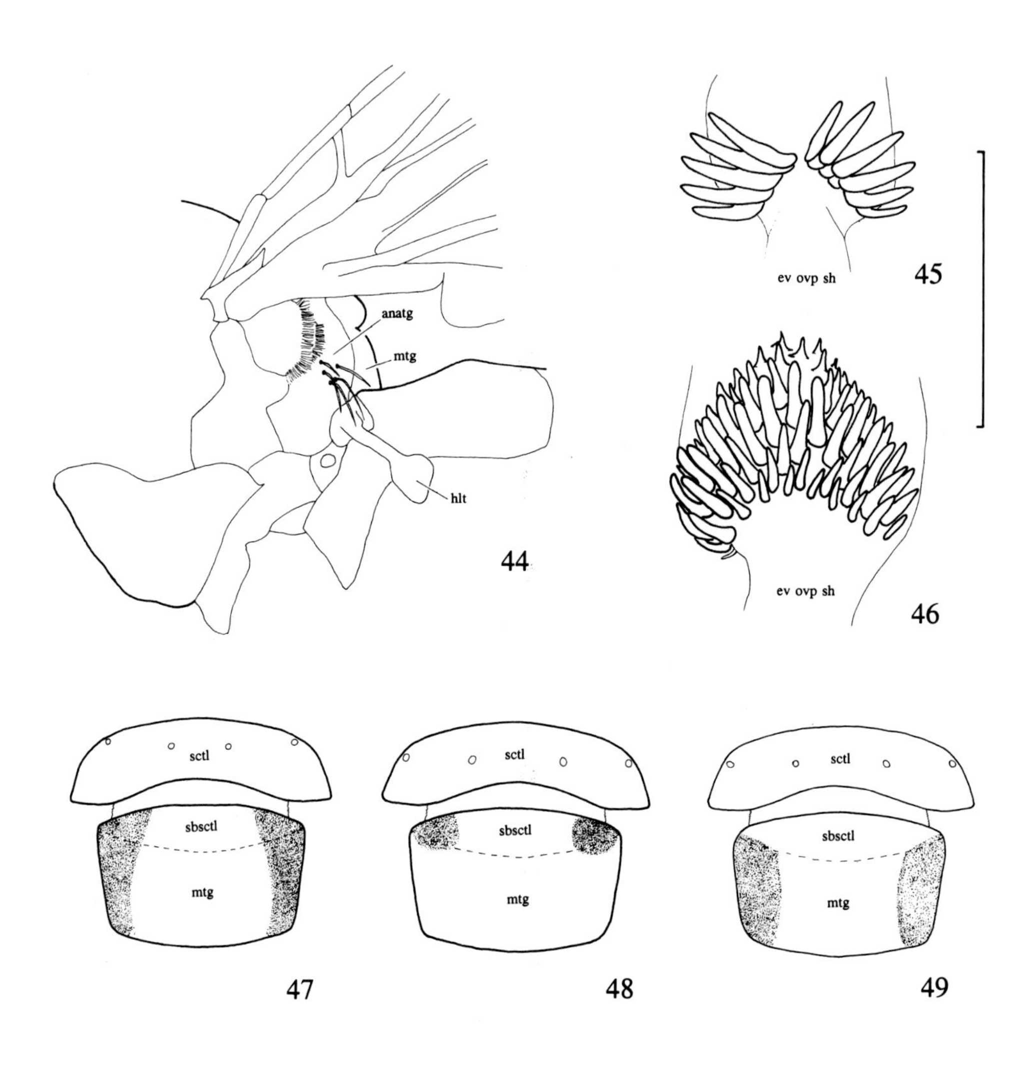

Figs 44-49. Thoracic and abdominal features; 44, *Epochra canadensis*, area behind wing base showing anatergite hairs; 45-46, dorsal views of large 'teeth' in basal area of eversible ovipositor sheath (scale=0.5mm); 45, *Anastrepha macrura*; 46, *A. serpentina* (most other spp. are similar); 47-49, *Anastrepha*, posterior views of area below scutellum showing subscutellum and mediotergite markings; 47, *A. fraterculus*; 48, *A. ludens*; 49, *A. obliqua*. *anatg*, anatergite. *ev ovp sh*, eversible ovipositor sheath. *hlt*, haltere. *mtg*, mediotergite. *sctl*, scutellum. *sbsctl*, subscutellum.

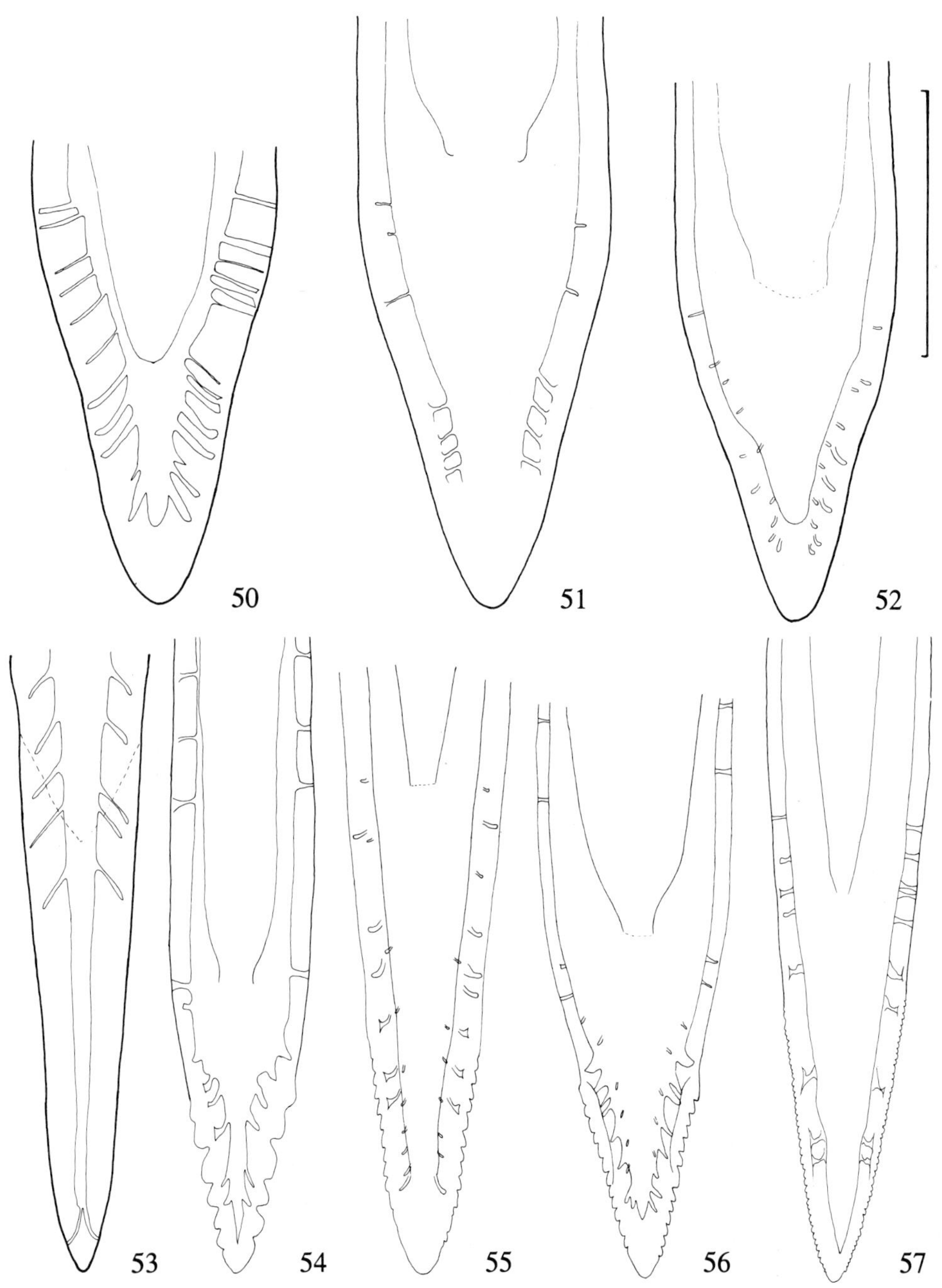

Figs 50-57. *Anastrepha* aculei, dorsal views (optical sections) of apices; 50, *A. bistrigata*; 51, *A. ornata*; 52, *A. striata*; 53, *A. grandis* (dashed lines show dorsal V-shaped ridge); 54, *A. antunesi*; 55, *A. distincta*; 56, *A. fraterculus*; 57, *A. leptozona*. Scale=0.2mm.

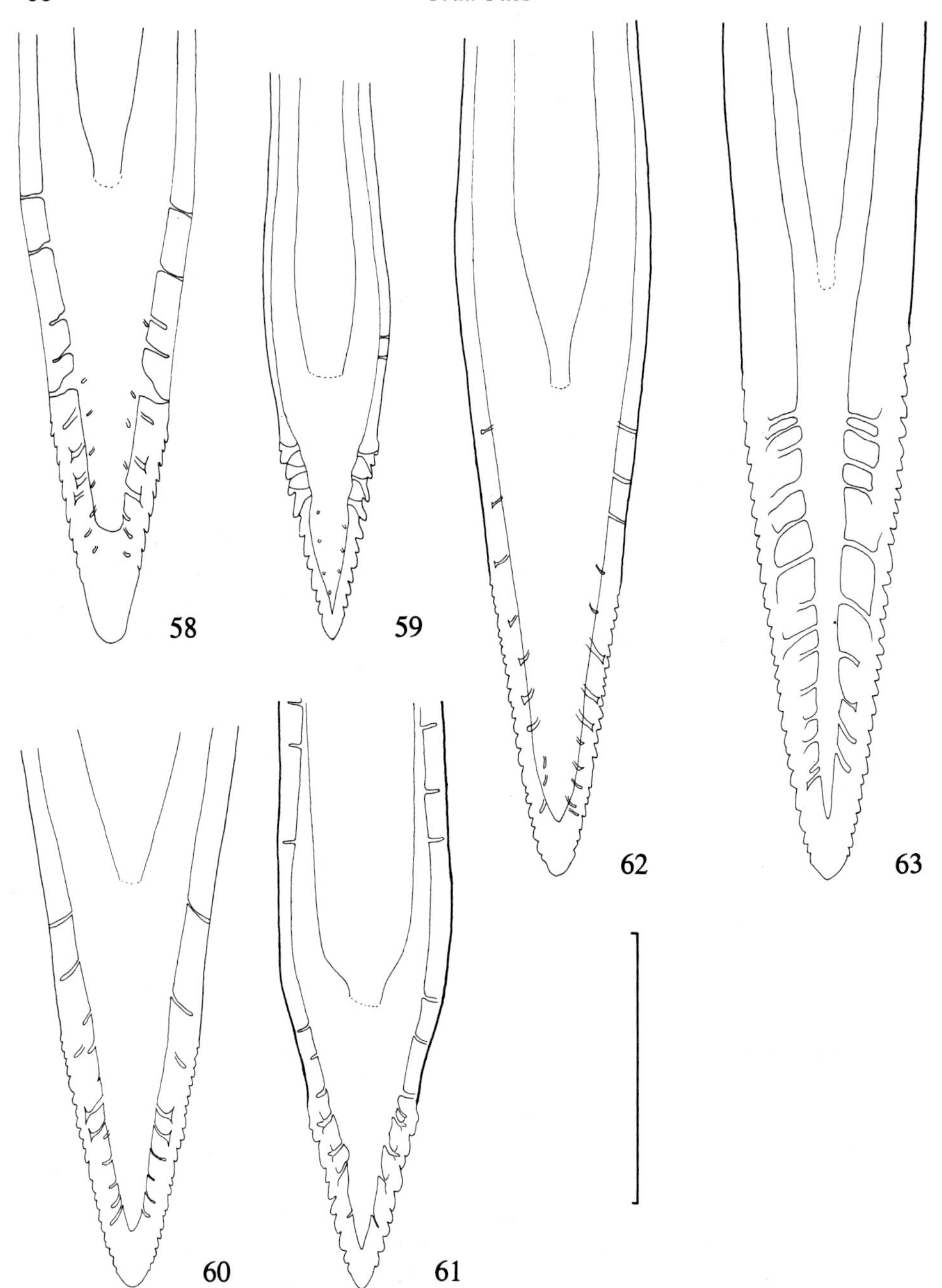

Figs 58-63. *Anastrepha* aculei, dorsal views (optical sections) of apices; 58, *A. ludens*; 59, *A. obliqua*; 60, *A. ocresia*; 61, *A. suspensa*; 62, *A. serpentina*; 63, *A. pseudoparallela*. Scale=0.2mm.

- Aculeus shorter, 2.0-2.3mm long (fig. 52). Scutum with dark brown to black setulae on the dark stripes, broadly separated into two areas by the notopleural suture; lateral (outer) part of each stripe on postsutural scutum without any setulae. [Wing with basicostal band joined to the S band by a narrow mark on vein R_{4+5} and often also on vein R_{2+3}; preapical and posterior apical crossbands joined to form the V band; V band not joined to S band (fig. 173). Tapered apical section of aculeus 0.3mm long; tip 0.18mm wide. Wing length 5.7-7.7mm. Brazil to southern USA.] . . . . *A. striata* (p. 155)

4 Aculeus at most 1.9mm long (usually less than 1.6mm). 5

- Aculeus at least 2.1mm long. 7

5 Apical section of S band touching, or almost touching, vein M. Thorax usually with a distinct black spot across the centre of the suture of the scutum and scutellum (fig. 174). [Wing with basicostal band usually joined to the S band by a narrow mark on vein R_{4+5}; preapical and posterior apical crossbands joined to form the V band; V band usually narrowly joined to S band. Aculeus length 1.4-1.6mm; serrated portion 0.14mm long and occupying about half of the distance between the 'apparent genital opening' and apex; tip 0.13mm wide; tapered apical section 0.2mm long (fig. 61). Wing length 4.9-6.7mm. West Indies and southern USA.] . . . . *A. suspensa* (p. 158)

- Apical section of S band ending well separated from vein M. Thorax usually without a black spot across the suture of the scutum and scutellum (fig. 168) (some *A. fraterculus* have this spot). 6

6 Subscutellum entirely orange; mediotergite usually dark laterally (fig. 49). Aculeus short relative to body size, 0.59-0.67 times as long as wing cell dm measured along vein CuA_1. [Wing with basicostal band joined to S band; preapical and posterior apical crossbands joined to form the V band which is usually joined to the S band, often broadly (fig. 168). Aculeus length 1.3-1.6mm; tip 0.10-0.12mm wide; tapered apical section 0.2mm long; serrated portion 0.13-0.15mm long (fig. 59). Wing length 5.7-7.5mm. Argentina to southern USA.] . . . . *A. obliqua* (p. 144)

- Subscutellum and mediotergite brown laterally (fig. 47). Aculeus longer relative to body size, 0.70-0.86 times as long as wing cell dm measured along vein CuA_1; except in Argentina and southern Brazil where it is only 0.64-0.69 times as long. [Wing with basicostal band usually joined to S band by a narrow mark on vein R_{4+5}, particularly in South American populations; preapical and posterior apical crossbands usually joined to form the V band; V band usually not joined to S band in South American populations, but usually joined in Central American populations (similar to fig. 168). Aculeus length 1.4-1.7(-1.9)mm; tip 0.12-0.14mm wide; tapered apical section 0.2-0.3mm long; serrated portion 0.15-0.18mm long (fig. 56). Wing length 4.4-7.2mm. Argentina to southern USA.] . . . . *A. fraterculus* complex (p. 133)

7 Wing pattern uniformly dark brown, except for the area at the base of cells r_1 and r_{2+3}, which is usually yellow. Basicostal and S bands broadly joined (fig. 169). Abdomen with dark transverse bands across the base of at least tergites 2-3 (unless teneral). Aculeus length 2.7-3.3mm; tip 0.16mm wide; tapered apical section 0.4mm long; serrated portion 0.2mm long (fig. 60). Wing length 6.6-7.8mm. [West Indies and southern USA.]
. . . . *A. ocresia* (p. 148)

- Wing pattern pale yellow-brown, except sometimes at edges of bands where they may be darkened. Basicostal and S bands narrowly joined or separate (figs 163, 165, 166, 171). Abdomen entirely yellow or orange brown (may be discoloured by internal tissues).
. . . . 8

8 Aculeus tip with a few (about 5) large teeth on each side (fig. 54). Subscutellum and mediotergite entirely yellow to orange, not darkened to brown laterally; wing length at most 7.0mm. [Aculeus length 2.4mm; tip 0.10mm wide; tapered apical section 0.2mm long; serrated portion 0.15mm long. Wing length 5.5-7.0mm (fig. 163). Brazil to Panama.]
. . . . *A. antunesi* (p. 129)

- Aculeus tip with many smaller teeth (figs 55, 57, 58, 63). If subscutellum and mediotergite entirely yellow to orange, then wing length usually more than 7.0mm (this permutation of characters is given to enable users to attempt tentative identification of males and should be used cautiously).
. . . . 9

9 Subscutellum and mediotergite entirely yellow to orange, not darkened to brown laterally (some preserved specimens may be translucent in the centre of those areas and orange laterally due to differences in the amount of tissue remaining below the integument). Aculeus tip with numerous very fine teeth in at least apical three-quarters of tapering apical portion (figs 57, 63).
. . . . 10

- Subscutellum and/or mediotergite darkened to brown laterally (figs 47-48). Aculeus tip with fine teeth in at most apical half of tapering apical portion (figs 55, 58).
. . . . 11

10 Wing with apical curve of vein M slight; vein M not touching end of S band; basicostal and S bands narrowly joined (fig. 171). Preapical and posterior apical crossbands joined to form a V-band. [Aculeus length 2.6-3.2mm; tip 0.15mm wide; tapered apical section has no distinct base, but serrated portion 0.4mm long (fig. 63). Wing length 8.0-9.2mm. Larvae usually develop in passion fruits (*Passiflora*). South America.]
. . . . *A. pseudoparallela* (p. 150)

- Wing with apical curve of vein M very strong; vein M touching end of S band; basicostal and S bands separate (fig. 165). Preapical and posterior apical crossbands not joined (posterior apical crossband sometimes very reduced). [aculeus length 2.3-3.1mm; tip 0.13mm wide; tapered apical section has no distinct base, but serrated portion 0.2mm long (fig. 57). Wing length 6.9-9.1mm. Larvae usually develop in the fruits of Sapotaceae. Brazil to Mexico.]
. . . . *A. leptozona* (p. 138)

11 Aculeus usually less than 3.3mm long (aculeus length 2.1-3.4mm); oviscape usually shorter than combined length of scutum and scutellum (except in Andean specimens which may be up to 1.1 times as long). Subscutellum usually entirely orange; mediotergite orange medially and darkened to brown laterally (similar to fig. 47). [aculeus tip 0.13mm wide; tapered apical section has no distinct base, but serrated portion 0.2mm long (fig. 55). Wing length 5.5-7.5mm (similar to fig. 166). Larvae usually develop in the fruits of *Inga* spp. Brazil to southern USA.] . . . . *A. distincta* (p. 132)

- Aculeus usually more than 3.3mm long (aculeus length 3.3-4.7mm); oviscape at least 1.1 times as long as combined length of scutum and scutellum (fig. 166). Subscutellum darkened to brown laterally; mediotergite entirely orange (fig. 48) (may appear darkened laterally if internal tissues only remain adhered to integument in lateral areas). [aculeus tip 0.14mm wide; tapered apical section 0.3mm long; serrated portion 0.2mm long (fig. 58). Wing length 6.5-9.0mm. Costa Rica to southern USA.] . . . . *A. ludens* (p. 140)

12 Wing with a hyaline triangle beyond apex of vein R_1, which crosses cell r_1 and at least part of r_{2+3}; apical half of cell r_{2+3} partly hyaline (figs 170, 172). Head with 2 pairs of orbital setae. [Most wing bands very dark brown.] 13

- Wing without a hyaline triangle beyond apex of vein R_1, so that there is a costal band running the entire length of the wing; apical half of cell r_{2+3} entirely coloured (figs 164, 167). Head often with only 1 pair of orbital setae (the anterior pair). 14

13 Wing with basicostal and S bands joined (fig. 172). Abdomen brown, with a T-shaped yellow medial area. Aculeus tip serrate (fig. 62). [Wing length 6.6-9.0mm. Aculeus length 2.8-3.7mm; tip 0.14mm wide; tapered apical portion 0.4mm long; serrated portion 0.2mm long. Argentina to southern USA.] . . . . *A. serpentina* (p. 152)

- Wing with basicostal and S bands separate (fig. 170). Abdomen mostly yellow, with transverse brown bands. Aculeus tip not serrate (fig. 51). [Wing length 8-9mm. aculeus length 3.8mm. Ecuador.] . . . . *A. ornata* (p. 148)

14 Wing with a reduced pattern; costal band brown (fig. 167). Scutum dark brown to almost black with medial and lateral yellow stripes; scutellum dark brown dorsally and laterally, in basal third to half (fig. 167). Base of eversible ovipositor sheath with a single row of large spines (fig. 45). [Wing length 8mm. aculeus very thin; tip pointed and not serrate (similar to fig 53, but almost parallel sided before the point); aculeus length 5.5-6.5mm. Larvae develop in the fruits of Sapotaceae. South America.] . . . . *A. macrura* (p. 142)

- Wing pattern more extensive; bands yellow (fig. 164). Scutum with yellow medial and lateral stripes, and brown stripes along the dorsocentral lines. Base of eversible ovipositor sheath with a multiple row of large spines (similar to fig. 46). [Wing length 9-11mm. Aculeus tip not serrate, with a dorsal V-shaped ridge close to apex (fig. 53), and a similar ventral V-shaped ridge about twice as far from apex as dorsal ridge; aculeus length 5.2-6.2mm. Larvae usually develop in the fruits of cucurbits. Argentina to Panama.]
. . . . *A. grandis* (p. 136)

Keys to *Bactrocera* and *Dacus* Species of Economic Importance

The following set of nine keys separates the 32 *Bactrocera* and *Dacus* spp. most likely to be found in association with commercial fruit crops. Key 1 separates the major subgenera of both *Bactrocera* and *Dacus*, but not the *B. (Zeugodacus)* group of subgenera as they can only be separated using male characters; see Key 5 (p. 84). Aculeus length values were obtained by examination of specimens; however, values for most *Bactrocera* and *Dacus* spp. were based on the examination of only one or two specimens and users of the keys should expect to find a few specimens that do not exactly fit the values given. The only major exception within these genera is the *B. dorsalis* species complex, in which numerous measurements were taken of most member species.

Key 1: Key to Subgenera of *Bactrocera* and *Dacus*

This key separates the major subgenera. It also enables species level identification for those subgenera that only include a single pest species.

1 Scutellum bilobed (fig. 199). Postpronotal lobes each with a single seta (fig. 199). [Scutum with both anterior supra-alar setae and prescutellar acrostichal setae. Scutum with lateral yellow stripes (vittae) that run from the postpronotal lobe to the scutellum (fig. 199), sometimes with a break adjacent to the notopleura. Scutum sometimes with a short medial yellow stripe (vitta). Scutellum with 2 marginal setae (the apical pair). Wing length 6.3-6.5mm. Aculeus length 1.7mm (fig. 74). Males attracted to methyl eugenol. South Pacific.]
. . . . *B. (Notodacus) xanthodes* (p. 247)

- Scutellum not bilobed (figs 175-198, 200-207, 219-229). Postpronotal lobes without well developed setae (aberrant individuals with a fine seta on each lobe do occur, notably some *B. depressa* and *B. tau*).
. . . . 2

2 Scutellum with 4 marginal setae (e.g. figs 205, 207). [Scutum with a medial yellow or orange stripe (vitta) (e.g. figs 205-207). Abdomen with all tergites separate (view from side to see overlapping sclerites). Larvae usually develop in the flowers or fruits of Cucurbitaceae, but the pest spp. will sometimes attack other hosts, particularly tomatoes.]
. . . *B. (Zeugodacus)* group of subgenera (Keys 5 [males], 6 [females]; p. 84)

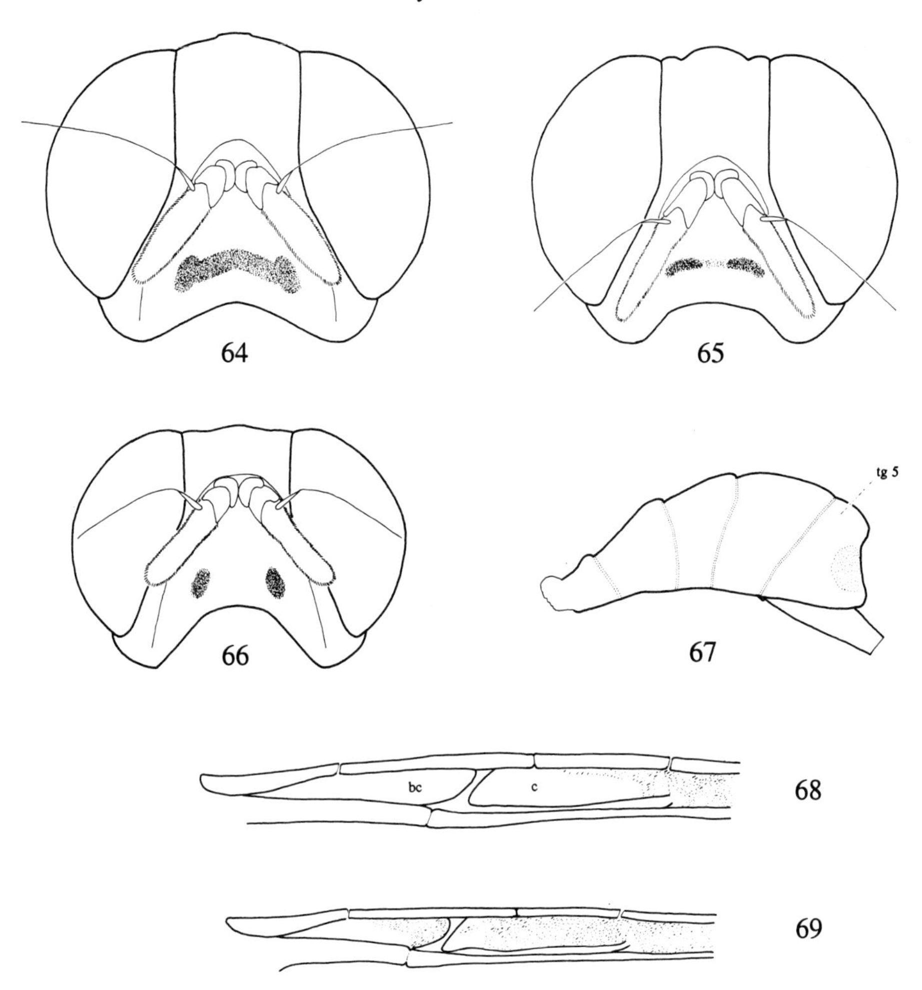

Figs 64-69. Dacini; 64-66, heads, anterior views showing facial markings; 64, *Bactrocera (Zeugodacus) caudata*; 65, *B. (B.) correcta*; 66, *B. (B.) zonata*; 67, *Dacus (Callantra) solomonensis*, lateral view of abdomen; 68-69, wing cells c and sc, showing distribution of microtrichia; 68, *B. (B.) dorsalis*; 69, *B. (B.) tryoni*.

[Some subgenera within this group can only be separated using male characters and a single key (Key 6) to species is provided.]

- Scutellum with only 2 marginal setae (the apical pair) (e.g. figs 177-196). 3

3 Scutum with a medial yellow or orange stripe (vitta) and prescutellar acrostichal setae (e.g. figs 198, 206). [Abdomen with all tergites separate (view from side to see overlapping sclerites). Larvae usually develop in the flowers or fruits of Cucurbitaceae, but the pest spp. will sometimes attack other hosts, particularly tomatoes. Most specimens of *B. (Zeugodacus) cucurbitae* and *B. (Hemigymnodacus) diversa* have only 2 scutellar setae and will run here.]
. . . *B. (Zeugodacus)* group of subgenera (Keys 5 [males], 6 [females]; p. 84)

[Some subgenera within this group can only be separated using male characters and a single key (Key 6) to species is provided.]

- Scutum never with both a medial yellow/orange stripe (vitta) and prescutellar acrostichal setae.
. . . . 4

[Note that this simple combination of characters does not apply to some spp. which have been excluded from the present work.]

4 Scutum with prescutellar acrostichal setae (e.g. figs 177-196). [Abdomen with all tergites separate (view from side to see overlapping sclerites). Tergite 3 of male with a pecten (fig. 9). Larvae develop in a wide variety of fruits, but very rarely in those of Apocynaceae, Asclepiadaceae or Cucurbitaceae.]
. . . . *B. (Bactrocera)* group of subgenera . . 5

- Scutum without prescutellar acrostichal setae (figs 197, 203, 204).
. . . . 6

5 Scutum with anterior supra-alar setae (figs 177-196).
. . . . *B. (Bactrocera)* (Key 2; p. 76)

- Scutum without anterior supra-alar setae (fig. 175). [Wing length 5.3-6.4mm. Aculeus length 1.6mm; apex shape similar to *B. dorsalis* (fig. 70) but preapical sensilla very short. Males attracted to cue lure in some areas only. Australia.]
. . . . *B. (Afrodacus) jarvisi* (p. 166)

6 Abdomen with all tergites separate (view from side to see overlapping sclerites). Larvae of the pest spp. either develop in the fruits of olive or of citrus fruits. [Tergite 3 of male with a pecten.]
. . . . 7

- Abdomen with all tergites fused into a single plate, at most with smooth transverse lines marking the boundaries of each segment (view from side to check that no sclerites overlap the next). Larvae of the pest spp. usually develop in the fruits or flowers of Cucurbitaceae, but they sometimes attack other hosts, particularly tomatoes. Most other (non-pest) spp. either attack the flower buds or fruits of Passifloraceae, or the pods, stems or leaves of Asclepiadaceae or Apocynaceae. [All spp. included in the present work have a pecten on tergite 3.]
. . . . *Dacus* . . 8

7 Scutum with a medial yellow stripe (vitta) and lateral yellow stripes which are usually incurved anteriorly (figs 203, 204). Oviscape round in cross section and usually with a bulbous base. Segment 1 of abdomen roughly parallel sided and about as broad at apex as at base, and with segment 2 forming part of the abdominal pedicel (wasp waist) and distinct from the rounded shape formed by segments 3-5. Area of wing close to apex of vein A_1+CuA_2 without a dense patch of microtrichia. [Wing with a deep costal band, covering the whole of cell r_1 and r_{2+3}; this band is yellow in the basal two thirds (not visible at all in teneral specimens) and then abruptly darkened in the apical third to give the impression of an apical spot. Larvae develop in citrus fruits. Northern India to Japan.]
. . . . *B. (Tetradacus)* (Key 4; p. 83)

\- Scutum without any yellow or orange stripes (vittae) (fig. 197), although specimens with a distinct black area on the scutum may be red-brown laterally, giving the appearance of lateral stripes. Oviscape slightly flattened in cross section, giving distinct lateral margins. Segment 1 of abdomen (basal part of syntergite 1+2) about twice as broad at apex as at base, and with segment 2 being part of the general rounded shape formed by segments 2-5. Area of wing close to apex of vein A_1+CuA_2 with a dense patch of microtrichia. [Wing hyaline, except for a small mark at the apex of vein R_{4+5}, and sometimes a slight darkening of cell sc. Scutum without anterior supra-alar setae. Wing length 4.3-5.2mm. Aculeus length 1.0mm (fig. 78). Larvae develop in olives (*Olea*). Africa and southern Europe to northern India.]
. . . . *B. (Daculus) oleae* (p. 241)

8 Combined length of pedical and 1st flagellomere greater than length of ptilinal suture along one side of face; combined length of scape and pedicel almost as long as frons (fig. 24). Abdominal tergites 1+2 longer than broad, giving a strongly wasp-waisted appearance (figs 219-221) (although all the tergites are fused, transverse shiny lines mark the area of each tergite). [Scutum with anterior supra-alar setae (except in some spp. not included in the present work). Larvae develop in cucurbit fruits. Oriental to Australasia]
. . . . *Dacus (Callantra)* (Key 7; p. 88)

\- Combined length of pedical and 1st flagellomere equal to, or less than, length of ptilinal suture along one side of face; combined length of scape and pedicel at most half as long as frons (similar to fig. 19). Abdominal tergites 1+2 broader than long (figs 222-229).
. . . . 9

9 Scutum with anterior supra-alar setae (figs 222-225).
. . . . *Dacus (Dacus)* (Key 8; p. 89)

\- Scutum without anterior supra-alar setae (figs 226-229).
. . . . *Dacus (Didacus)* (Key 9; p. 91)

[The other major subgenera of *Dacus* are *D. (Metidacus)* and *D. (Leptoxyda)*; *D. (Metidacus)* differs from *D. (Dacus)*, and *D. (Leptoxyda)* differs from *D. (Didacus)*, by the lack of the pecten on tergite 3 of the male.]

Key 2: Key to Species of *Bactrocera (Bactrocera)*

This key separates 18 pest species of *B. (Bactrocera)*; it also separates the *B. dorsalis* species complex from other members of this subgenus.

1 Wing with a pattern made up of three crossbands, each of which extends across the whole of the wing, from costal band to hind margin (fig. 195). [Wing length 5.5-8.1mm. Aculeus length 2.0mm; apex broadly rounded (fig. 73). Males attracted to methyl eugenol. Larvae usually develop in the fruits of *Artocarpus* spp. South east Asia to New Guinea and New Caledonia.]
. . . . *B. (B.) umbrosa* (p. 236)

\- Wing usually without any crossbands; at most with one complete crossband (figs 177, 181, 184), or with a short crossband covering crossvein r-m (fig. 180).
. . . . 2

2 Crossveins r-m and dm-cu covered by a single crossband (figs 177, 181, 184). [Males attracted to cue lure.]
. . . . 3

\- Crossveins r-m and dm-cu not covered by a single crossband, at most with a short crossband covering crossvein r-m (fig. 180).
. . . . 5

3 Costal band distinct from wing base to apex, and broad enough to reach R_{4+5} (fig. 181). Crossband slightly 'stepped' to follow r-m, part of M and dm-cu. Scutellum uniformly orange-yellow. [Wing length 6.2-7.3mm. Aculeus length 1.8mm (fig. 76). Males attracted to cue lure. South Pacific.]
. . . . *B. (B.) distincta* (p. 184)

\- Costal band beyond vein R_1 very faint (figs 177, 184). Crossband evenly curved. Scutellum often with a triangular black mark, which may reach the apex of the scutellum.
. . . . 4

4 Postpronotal lobe predominantly black, sometimes with a small yellow area in posterior half (fig. 184). [Wing length 4.6-5.9mm. Aculeus length 1.3mm; apex shape similar to *B. dorsalis* (fig. 70). Males attracted to cue lure. New Guinea, Australia and the South Pacific.]
. . . . *B. (B.) frauenfeldi* (p. 203)

\- Postpronotal lobe with anterior third largely black, posterior two-thirds largely yellow (fig. 177). [Wing length 4.7-5.7mm. Aculeus length 0.9mm (insufficient material was available to determine if the aculeus length is of value in the separation from *B. frauenfeldi*); apex shape similar to *B. dorsalis* (fig. 70) but preapical sensilla slightly closer to apex. Males attracted to cue lure. Malaysia and Indonesia (west of Irian Jaya).]
. . . . *B. (B.) albistrigata* (p. 173)

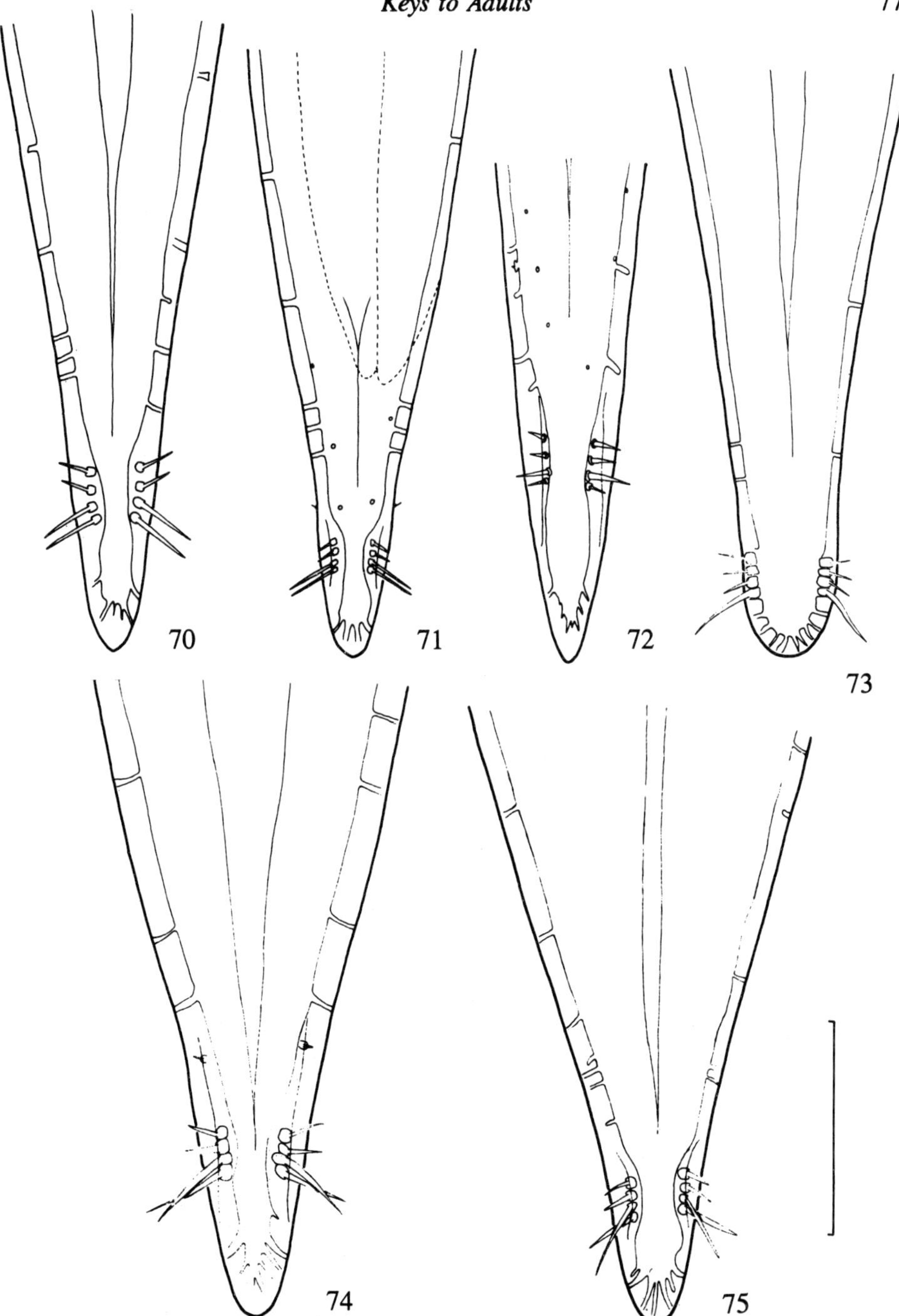

Figs 70-75. *Bactrocera* aculei, dorsal views (optical sections) of apices; species with a pointed apex in the *B. (Bactrocera)* group of subgenera; 70, *B. (B.) dorsalis* (a syntype); 71, *B. (B.) musae* (dashed lines show ventral lobes); 72, *B. (B.) tryoni*; 73, *B. (B.) umbrosa*; 74, *B. (Notodacus) xanthodes*; 75, *B. (B.) zonata*. Scale=0.1mm.

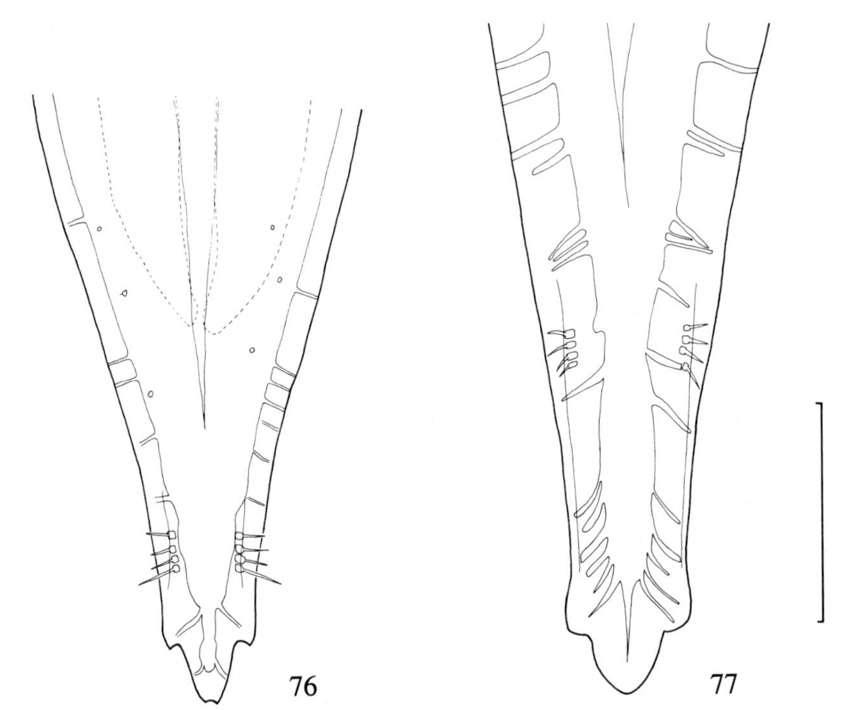

Figs 76-77. *Bactrocera* aculei, dorsal views (optical sections) of apices; species with preapical 'steps' in the *B. (Bactrocera)* group of subgenera; 76, *B. (B.) distincta*; 77, *B. (B.) latifrons*. Scale=0.1mm.

5 Crossvein r-m covered by a short distinct crossband (a spur from the costal band) (fig. 180); scutellum without a triangular black marking. [Wing length 4.7-5.7mm. Aculeus length 1.0mm; apex shape similar to *B. dorsalis* (fig. 70) but preapical sensilla a little shorter. Males attracted to cue lure. Larvae usually develop in citrus fruits. South Pacific.]
. . . . *B. (B.) curvipennis* (p. 182)

\- Crossvein r-m not covered by any distinct marking; if with an indistinct marking (*B. melanota* and *B. psidii* may have a faint infuscation along crossveins r-m and dm-cu), then scutellum black (fig. 187) or with a triangular black mark (fig. 191).
. . . . 6

6 Scutum without lateral yellow or orange stripes (vittae) (figs 185, 187, 190).
. . . . 7

\- Scutum with lateral yellow or orange stripes (vittae) (e.g. fig. 193).
. . . . 9

7 Face with a dark spot in each antennal furrow (similar to figs. 19, 66). [Scutellum black, except at sides (fig. 185). Wing length 4.8-6.0mm. Aculeus length 1.3mm; apex shape similar to *B. dorsalis* (fig. 70). Males attracted to cue lure. South Pacific.]
. . . . *B. (B.) kirki* (p. 206)

- Face entirely orange-yellow, without any black spots in the antennal furrows. 8

8 Scutellum black (fig. 187). Males attracted to methyl eugenol. [Wing length 5.8-7.2mm. Aculeus length 1.6-1.8mm; apex shape similar to *B. dorsalis* (fig. 70). Larvae usually develop in citrus fruits. South Pacific.] . . . .*B. (B.) melanota* (p. 211)

- Scutellum yellow, except at extreme base (fig. 190). Males attracted to cue lure. [Wing length 4.2-5.4mm. Aculeus length 1.4mm; apex shape similar to *B. dorsalis* (fig. 70). South Pacific.] . . . .*B. (B.) passiflorae* (p. 221)

[Some unusually dark specimens of *B. facialis* have very narrow and barely discernible lateral orange stripes (vittae), and those specimens will erroneously run to *B. passiflorae*.]

9 Face entirely yellow, without either a line across the lower facial margin or spots in the antennal furrows. [Scutum predominantly black; lateral stripes (vittae) dark orange; postpronotal lobes and notopleura yellow (fig. 183). Wing length 4.6-6.1mm. Aculeus length 1.6mm; apex shape similar to *B. dorsalis* (fig. 70). Males attracted to cue lure. South Pacific.] . . . .*B. (B.) facialis* (p. 201)

- Face marked, either with a spot in each antennal furrow (figs 19, 66), or with transverse dark markings adjacent to the antennal furrows which usually join to form a line across the lower facial margin (fig. 65). 10

10 Face with transverse dark markings adjacent to the antennal furrows which usually join to form a line across the lower facial margin (fig. 65). [Wing without a complete costal band (fig. 179); area of cell br immediately above cell bm without any microtrichia. Wing length 4.3-6.0mm. Aculeus length 1.1mm; apex shape similar to *B. dorsalis* (fig. 70). Males attracted to methyl eugenol. Sri Lanka to Thailand.] . . . . *B. (B.) correcta* (p. 180)

- Face with a spot in each antennal furrow (figs 19, 66). 11

11 Wing without a distinct costal band; cell sc often yellow, and apex of vein R_{4+5} often with a brown spot (fig. 196); if with an indistinct costal band (*B. psidii* may have faint infuscation along the costal edge of the wing), then scutellum with a triangular black mark (fig. 191). 12

- Wing with a distinct costal band at least from the end of vein Sc to just beyond the end of vein R_{4+5} (e.g. figs 182, 193); scutellum entirely pale coloured, except sometimes for a narrow black line across the base. 14

12 Dorsal surface of scutellum with a large black triangular mark, lateral and apical areas yellow (fig. 191). Area of cell br immediately above cell bm with microtrichia. Wing entirely hyaline, at most with cell sc slightly yellow, or with faint infuscation along the costal edge and along crossveins r-m and dm-cu. [Wing length 5.6-5.8mm. Aculeus length 1.9mm; apex shape similar to *B. dorsalis* (fig. 70). Males attracted to cue lure. South Pacific.]
. . . . *B. (B.) psidii* (p. 225)

- Scutellum entirely pale coloured, except sometimes for a narrow black line across the base (fig. 194, 196). Area of cell br immediately above cell bm without any microtrichia. Cell sc clearly yellow and apex of vein R_{4+5} covered by a brown spot (figs 194, 196), except in teneral specimens which may lack any wing markings.
. . . . 13

13 Thorax and abdomen pale orange-brown to red-brown (fig. 196). [Wing length 5.2-6.1mm. Aculeus length 1.0-1.2mm (fig. 75). Males attracted to methyl eugenol. Sri Lanka to Vietnam; adventive in Mauritius.]
. . . . *B. (B.) zonata* (p. 239)

- Thorax and abdomen black (fig. 194); even teneral specimens of this species, which may lack wing markings, are dark orange-brown and very much darker than *B. zonata*. [Wing length 5.2-6.2mm. Aculeus length 1.6mm; apex shape similar to *B. zonata* (fig. 75) but not so sharply narrowed before apex, although more sharply narrowed than *B. dorsalis* (fig. 70). Males attracted to methyl eugenol. Myanmar to Vietnam.]
. . . . *B. (B.) tuberculata* (p. 233)

14 Costal band extending from wing base to near wing apex, so that cells bc and c are coloured (figs 178, 189, 193); the whole of cell c and at least the anteroapical area of cell bc with a dense covering of microtrichia (fig. 69). [Males attracted to cue lure.]
. . . . 15

- Costal band only extending from the end of vein Sc to near wing apex, so that cells bc and c are hyaline (figs 182, 186, 188, 192); microtrichia in those cells usually restricted to the anteroapical area of cell c (fig. 68); cell bc usually completely bare, except *B. musae*, which sometimes has a few scattered microtrichia in cell bc).
. . . . 17

15 Postpronotal lobe brown, much darker in colour than the yellow lateral stripes (vittae) on the scutum (fig. 189). [Wing length 5.2-5.8mm. Aculeus length 1.3mm; apex shape similar to *B. tryoni* (fig. 72). Males attracted to cue lure. Australia and New Guinea.]
. . . .*B. (B.) neohumeralis* (p. 218)

- Postpronotal lobe yellow, the same colour as the lateral stripes (vittae) on the scutum (figs 178, 193).
. . . . 16

16 Scutum and abdomen predominantly red-brown, except for postpronotal lobe, notopleura and lateral stripes (vittae) which are yellow (fig. 178). Abdomen predominantly red-brown, except for a pale yellowish area across the posterior part of syntergite 1+2. [Wing length 5.2-6.0mm. Aculeus length 1.2mm; apex shape similar to *B. dorsalis* (fig. 70). Males attracted to cue lure. Australia.]
. . . . *.B. (B.) aquilonis* (p. 177)

\- Scutum and abdomen predominantly black, except for postpronotal lobe, notopleura and lateral stripes (vittae) which are yellow (fig. 193). Abdomen varying from predominantly red-brown with a black T-shaped mark on tergites 2-5, to predominantly black. [Wing length 4.8-6.3mm. Aculeus length 1.3mm (fig. 72). Males attracted to cue lure. Australia, New Guinea and South Pacific.]
. . . . *B. (B.) tryoni* (p. 229)

17 Costal band distinctly expanded near the apex of cell r_{2+3} to form a spot which extends below vein R_{4+5} (fig. 186); abdomen predominantly red-brown, usually without any black markings, but sometimes with an indistinct T-shaped mark on tergites 3-5. Aculeus apex with preapical 'steps' (fig. 77). [Wing length 4.5-6.1mm. Aculeus length 1.7mm. Males not attracted to either cue lure or methyl eugenol. Larvae usually develop in the fruits of Solanaceae. Sri Lanka to Taiwan; adventive in Hawaii.]
. . . . *.B. (B.) latifrons* (p. 208)

\- Costal band usually not expanded near apex to form a spot (figs. 182, 188, 192); if slightly expanded (some individuals of *B. occipitalis*) then abdomen with a distinct T-shaped black mark on tergites 3-5 (fig. 182). Aculeus apex evenly tapered to a point (figs 70, 71).
. . . . 18

18 Abdominal tergite 4 broadly black laterally, leaving a central yellow area; without a T-shaped mark (fig. 192). [Wing length 5.7-6.8mm. Aculeus length 1.8mm; apex shape similar to *B. dorsalis* (fig. 70). Males attracted to cue lure. Australia to Sulawesi (Indonesia).]
. . . . *B. (B.) trivialis* (p. 227)

\- Abdominal tergite 4 at most narrowly black laterally (figs 182, 188); if tergites 3-5 marked with black, then with a T-shaped mark (fig. 182).
. . . . 19

19 Abdominal tergites 3-5 without a distinct black T-shaped mark (fig. 188.d-e), although some individuals may have a poorly defined narrow dark medial stripe and narrow dark markings across the base of tergite 3, forming a poorly defined T-shaped mark (fig. 188.a-c). Larvae usually develop in bananas (*Musa*). [Scutum usually black, but sometimes red-brown. Wing length 4.9-6.7mm. Aculeus length 1.7mm (fig. 71). Males attracted to methyl eugenol. Australia and New Guinea.]
. . . . *.B. (B.) musae* (p. 213)

\- Abdominal tergites 3-5 with a distinct black T-shaped mark (fig. 182). Larvae rarely develop in bananas. [Wing length 5.0-6.8mm. Aculeus length 1.3-2.1mm (fig. 70). Males of pest and potential pest spp. attracted to methyl eugenol.]
. . . . *B. (B.) dorsalis* complex (Key 3; p. 82)

Key 3: Simplified Key to Species in the *B. (B.) dorsalis* Complex

Full details of the separation of species belonging to this complex are beyond the scope of the present work and will be provided elsewhere (R.A.I. Drew & D.L. Hancock, in prep.). The following key uses geographic as well as morphological characters to permit preliminary identification of members of the *B. dorsalis* species complex which are known to attack cultivated hosts. Because of the use of geographic characters it should be used cautiously, particularly with regard to quarantine interceptions of uncertain origin. In all cases where identification is critical, specimens should be referred to a specialist for positive determination. A pictorial key to 5 of these species (not *B. caryeae* or sp. D) was presented by Drew (1991), and Ooi (1991) tabulated and illustrated differences between spp. A and B. Aculeus length data were also collected by I.M. White & R.A.I. Drew (unpublished data, 1990), and measurements that fell outside of the typical range for each species are given here in brackets. Those studies showed that the ratio of aculeus length to cell dm length was a more reliable character than aculeus length for the separation of these species, particularly for identifying the two species found sympatrically in Malaysia and Indonesia (spp. A and B).

1 Costal band broad, reaching well below vein R_{2+3} for most of its length. [Black markings on abdomen broad (relative to those in fig. 182). Fore femur without a black spot. Tibiae usually dark. Aculeus less than 1.8mm long. Philippines.]
. . . . *B. (B.) occipitalis* (p. 192)

\- Costal band narrow, usually confluent with vein R_{2+3}, at most just overlapping R_{2+3}, except at apex of R_{2+3} (fig. 182).
. . . . 2

2 From southern India or Sri Lanka. Scutum usually with narrow lateral yellow stripes (vittae) (narrower than in fig. 182; teneral specimens often have the areas lateral to these stripes pale and this character can only be interpreted using mature specimens). [All femora with dark markings.]
. . . . 3

\- Not from southern India or Sri Lanka. Scutum with broad lateral yellow stripes (vittae) (fig. 182). [Black markings on abdomen narrow (fig. 182).] NOTE - female specimens are essential for the identification of these spp.
. . . . 4

3 From southern India. Abdominal tergites 3-5 with a broad medial dark stripe and broad lateral dark stripes (broader than in fig. 182). [aculeus length 1.55-1.80mm; aculeus length 0.65-0.80 times as long as wing cell dm measured along vein CuA_1.]
. . . . *B. (B.) caryeae* (p. 186)

\- From Sri Lanka. Abdominal tergites 3-5 with a narrow medial dark stripe (similar to fig. 182) and no lateral dark stripes. [aculeus length 1.6mm; aculeus length 0.75 times as long as wing cell dm measured along vein CuA_1.]
. . . . sp. near *B. (B.) dorsalis* (D) (p. 198)

4 Aculeus at least 0.77 times as long as wing cell dm measured along vein CuA_1; aculeus usually more than 1.7mm long. [Fore femur without a black spot. Tibiae usually dark. Costal band not overlapping vein R_{2+3}; not sharply broadened at end of vein R_{2+3}, and usually only filling anterior half of cell r_{2+3}, between the end of vein R_{2+3} and R_{4+5}.]
. . . . 5

\- Aculeus less than 0.76 times as long as wing cell dm measured along vein CuA_1; aculeus usually less than 1.7mm long.
. . . . 6

5 From Indonesia, Malaysia or southern Thailand. [Aculeus length (1.52-)1.75-2.12mm (Indonesian specimens tend to have a shorter aculeus than Malaysian specimens); aculeus 0.77-0.93 times as long as wing cell dm measured along vein CuA_1.]
. . . . sp. near *B. (B.) dorsalis* (B) (p. 194)

\- From the Philippines. [Aculeus length (1.68-)1.80-2.08mm; aculeus 0.83-0.93 times as long as wing cell dm measured along vein CuA_1.]
. . . . sp. near *B. (B.) dorsalis* (C) (p. 197)

6 From an area between northern India and Taiwan, Guam or Hawaii. Tibiae pale. Fore femur (both sexes) without a black spot. Costal band not overlapping vein R_{2+3}; not sharply broadened at end of vein R_{2+3}, and usually only filling anterior half of cell r_{2+3}, between the end of vein R_{2+3} and R_{4+5}. [aculeus length (1.28-)1.40-1.60(-1.80mm) (specimens from Taiwan tend to have a very short aculeus and those from India and Hawaii are at the long end of the range); aculeus 0.57-0.74 times as long as wing cell dm measured along vein CuA_1.]
. . . . *B. (B.) dorsalis* (p. 187)

\- From Indonesia, Malaysia, southern Thailand or northern South America. Tibiae usually dark. Fore femur of female usually with a black spot. Costal band usually just overlapping vein R_{2+3}, then sharply broadened at end of vein R_{2+3}, and usually filling anterior three-quarters of cell r_{2+3}, between the end of vein R_{2+3} and R_{4+5}. [aculeus length (1.30-)1.40-1.60mm (Indonesian specimens tend to have a shorter aculeus than Malaysian specimens); aculeus 0.65-0.75 times as long as wing cell dm measured along vein CuA_1.]
. . . . sp. near *B. (B.) dorsalis* (A) (p. 192)

Key 4: Key to Species of *Bactrocera (Tetradacus)*

1 Scutum with anterior supra-alar setae (fig. 204). Ovipositor short, aculeus less than 2.5mm long (fig. 80). [Wing length 9.4-9.6mm. aculeus length 2.1mm. Larvae develop in citrus fruits. China, Japan and Taiwan.]
. . . . *B. (T.) tsuneonis* (p. 260)

[The aberrant condition of the anterior supra-alar setae on one or both sides of the scutum being doubled to give a pair of closely adjacent setae is very common in this species.]

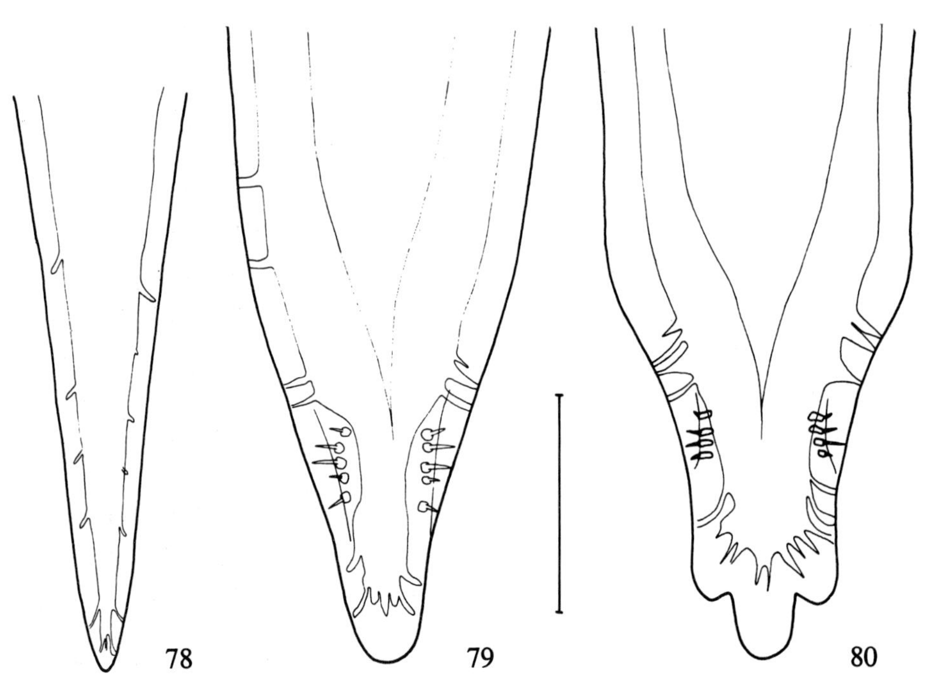

Figs 78-80. *Bactrocera* aculei, dorsal views (optical sections) of apices; species in subgenera *B. (Daculus)* and *B. (Tetradacus)*; 78, *B. (D.) oleae*; 79, *B. (T.) minax*; 80, *B. (T.) tsuneonis*. Scale=0.1mm.

- Scutum without anterior supra-alar setae (fig. 237). Ovipositor long, aculeus more than 3.5mm long (fig. 79). [Wing length 8.9-11.0mm. Aculeus length 3.7-5.0mm. Larvae develop in citrus fruits. China and northern Indian subcontinent.] . . . . *B. (T.) minax* (p. 256)

Key 5: Key to Subgenera in the *Bactrocera (Zeugodacus)* Group (Males Only)

The next key (Key 6) separates species as well as subgenera and should be used for females. This key to subgenera is only included as an additional check when identifying males, and to present details of how the subgenera are defined.

1 Tergite 3 of male with a pecten (fig 9). [Scutum with anterior supra-alar setae.] 2

- Tergite 3 of male without a pecten. 3

2 Scutum with prescutellar acrostichal setae. [Scutellum usually with 4 marginal setae, but most individuals of *B. (Z.) cucurbitae* have only 2 marginal setae (the apical pair).] . . . . *B. (Zeugodacus)* [Key 6; couplet 1]

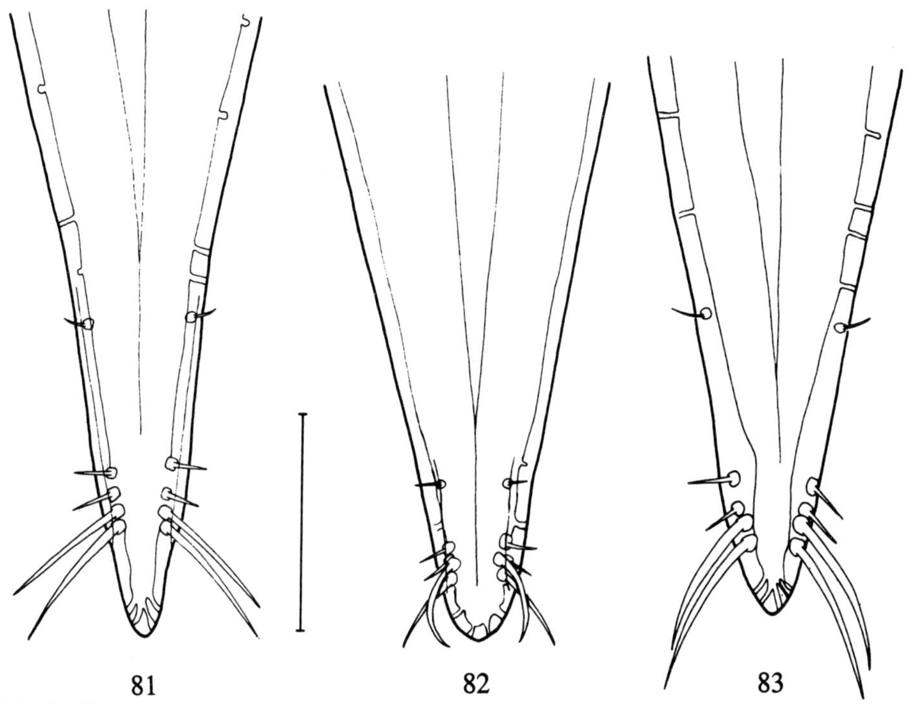

Figs 81-83. *Bactrocera* aculei, dorsal views (optical sections) of apices; species with a pointed apex in the *B. (Zeugodacus)* group of subgenera; 81, *B. (Z.) cucurbitae*; 82, *B. (Hemigymnodacus) diversa*; 83, *B. (Z.) tau*. Scale=0.1mm.

- Scutum without prescutellar acrostichal setae. [Scutellum always with 4 marginal setae.]
  . . . . *B. (Paradacus)* [Key 6; couplet 6]

3 Scutum without anterior supra-alar and without prescutellar acrostichal setae (fig. 176). [Scutellum with 4 marginal setae. Further details in Key 6; couplet 5.]
  . . . . *B. (Austrodacus) cucumis* (p. 170)

- Scutum with anterior supra-alar and prescutellar acrostichal setae.
  4

4 Face with a black spot in each antennal furrow (similar to figs 19, 66). Scutellum with 4 marginal setae. [Further details in Key 6; couplet 4.]
  . . . . *B. (Paratridacus) atrisetosa* (p. 254)

- Face of male without any markings (female has a black line across the mouth opening). Scutellum usually with 2 marginal setae (the apical pair), but sometimes with 4. [Further details in Key 6; couplet 7.]
  . . . . *B. (Hemigymnodacus) diversa* (p. 244)

Key 6: Key to Species of *Bactrocera (Zeugodacus)* Group of Subgenera (Both Sexes)

1 Wing with crossvein dm-cu covered by an infuscate area which is separate from other parts of the wing pattern (fig. 206). [Crossvein r-m usually covered by an infuscate area. Scutellum usually with 2 marginal setae (the apical pair), rarely 4. Scutum with both lateral and medial yellow stripes (vittae). Wing length 4.2-7.1mm. Tergite 3 of male with a pecten (fig. 9). Aculeus length 1.7mm (fig. 81). Males attracted to cue lure. Larvae usually develop in cucurbit fruits but often attack other plants. Tropical Asia to New Guinea; adventive in East Africa, Hawaii, Mauritius and Réunion.]
. . . . *B. (Z.) cucurbitae* (p. 263)

- Wing usually without an infuscate area covering crossvein dm-cu; if dm-cu crossvein covered by an infuscate area, it is linked to other parts of the wing pattern (fig. 200).
. . . . 2

2 Face with a brown or black spot in each antennal furrow (similar to figs 19, 66), without a line across the mouth opening.
. . . . 3

- Face without a brown or black spot in each antennal furrow; face either entirely yellow, or with a brown to black line across mouth opening (figs 64, 65).
. . . . 7

3 Scutum with prescutellar acrostichal setae. [Scutum with anterior supra-alar setae.]
. . . . 4

- Scutum without prescutellar acrostichal setae.
. . . . 5

[Aberrant individuals of *B. depressa* (fig. 201) with one or both prescutellar acrostical setae are common; check other characters carefully.]

4 Costal band expanded near the apex of cell r_{2+3} to form a spot which extends below vein R_{4+5} (fig. 207). Abdomen with a medial black stripe and a transverse black line across tergite 3, together forming a T-shaped mark; tergite 3 of male with a pecten (fig. 9). Aculeus pointed (fig. 83). [Wing length 5.3-8.3mm. Aculeus length 1.5-2.2mm. Males attracted to cue lure. Larvae usually develop in cucurbit fruits. Tropical Asia.]
. . . . *B. (Z.) tau* (p. 271)

- Costal band not expanded near apex to form a spot (fig. 202). Abdominal tergites 3-5 without a black T-shaped mark; tergite 3 of male without a pecten. Aculeus apex truncate, and with a small central point (similar to fig. 85). [Wing length 5.7-7.2mm. Aculeus length 1.8mm. Larvae usually develop in cucurbit fruits. New Guinea.]
. . . . *B. (Paratridacus) atrisetosa* (p. 254)

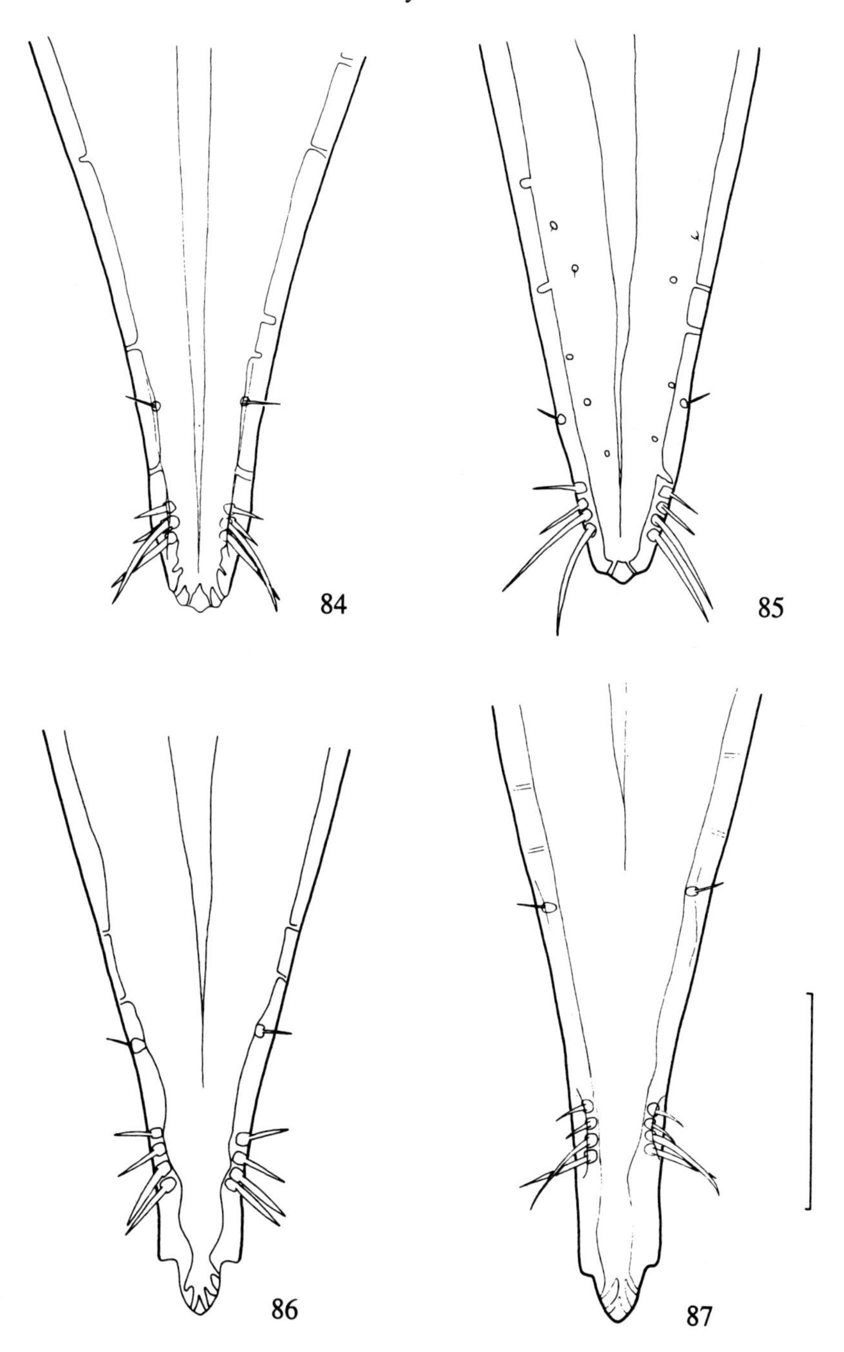

Figs 84-87. *Bactrocera* aculei, dorsal views (optical sections) of apices; species with a complex apex shape in the *B. (Zeugodacus)* group of subgenera; 84, *B. (Z.) caudata*; 85, *B. (Austrodacus) cucumis*; 86, *B. (Paradacus) decipiens*; 87, *B. (P.) depressa*. Scale=0.1mm.

5 Scutum without anterior supra-alar setae (fig. 176). Tergite 3 of male without a pecten. [Wing length 4.7-6.1mm. Aculeus length 1.7mm (fig. 85). Larvae usually develop in cucurbit fruits. Australia.]
. . . . *B. (Austrodacus) cucumis* (p. 170)

- Scutum with anterior supra-alar setae. Tergite 3 of male with a pecten (fig. 9).
. . . . 6

6 Wing with a complex pattern that covers both the r-m and dm-cu crossveins (fig. 200; faint and difficult to see in teneral specimens). [Wing length 8.3-8.7mm. Aculeus length 2.6mm (fig. 86). Larvae develop in cucurbit fruits. New Britain.]
. . . . *B. (Paradacus) decipiens* (p. 250)

- Wing without a pattern that covers crossveins r-m and dm-cu; costal band beyond end of vein R_{2+3}, expanded into a spot (fig. 201). [Wing length 8.9-9.2mm. Aculeus length 3.0mm (fig. 87). Larvae develop in cucurbit fruits. Japan and Taiwan.]
. . . . *B. (P.) depressa* (p. 252)

7 Face of male entirely yellow; face of female with a brown or black line across mouth opening (similar to fig. 65). Abdominal tergite 3 of male without a pecten. Scutellum usually with 2 marginal setae, rarely 4 in males. Aculeus rounded at apex (fig. 82). [Wing length 4.8-5.9mm. Aculeus length 1.2mm. Males attracted to methyl eugenol. Larvae develop in cucurbit flowers and fruits. Sri Lanka to Thailand.]
. . . . *B. (Hemigymnodacus) diversa* (p. 244)

- Face of both sexes with a black line across mouth opening (fig. 64). Abdominal tergite 3 of male with a pecten (fig. 9). Scutellum with 4 marginal setae. Aculeus apex truncate, and with a small central point (fig. 84). [Wing length 4.4-6.5mm. Aculeus length 1.1mm. Males attracted to cue lure. Larvae probably normally develop in cucurbit flowers and fruits. Tropical Asia.]
. . . . *B. (Z.) caudata* (p. 262)

Key 7: Key to Species of *D. (Callantra)*

1 Abdominal tergite 5 with a large hump when viewed in profile (fig. 67). [Wing length 7.5-8.8mm. Aculeus length 3.3mm; apex shape similar to *D. smieroides* (fig. 94) but anterior (small) pair of sensilla further from apex. Males attracted to cue lure. Larvae develop in cucurbit fruits. Solomon Islands area.]
. . . . *D. (C.) solomonensis* (p. 319)

- Abdominal tergite 5 without a hump.
. . . . 2

2 Abdominal tergite 1 (basal part of syntergite 1+2) about 2.5 times as long as broad (fig. 220). [Wing length 8.6-12.0mm. Aculeus length 2.1-2.3mm (fig. 94). Males attracted to cue lure. Larvae develop in luffa. Indonesia area.]
. . . . *D. (C.) smieroides* (p. 317)

- Abdominal tergite 1 not much longer than broad (fig. 219). [Wing length 8.3-9.2mm. Aculeus length 2.5mm; apex shape similar to *D. smieroides* (fig. 94). Males attracted to cue lure. Larvae develop in cucurbit fruits. Australia and New Guinea.]
. . . . *D. (C.) axanus* (p. 314)

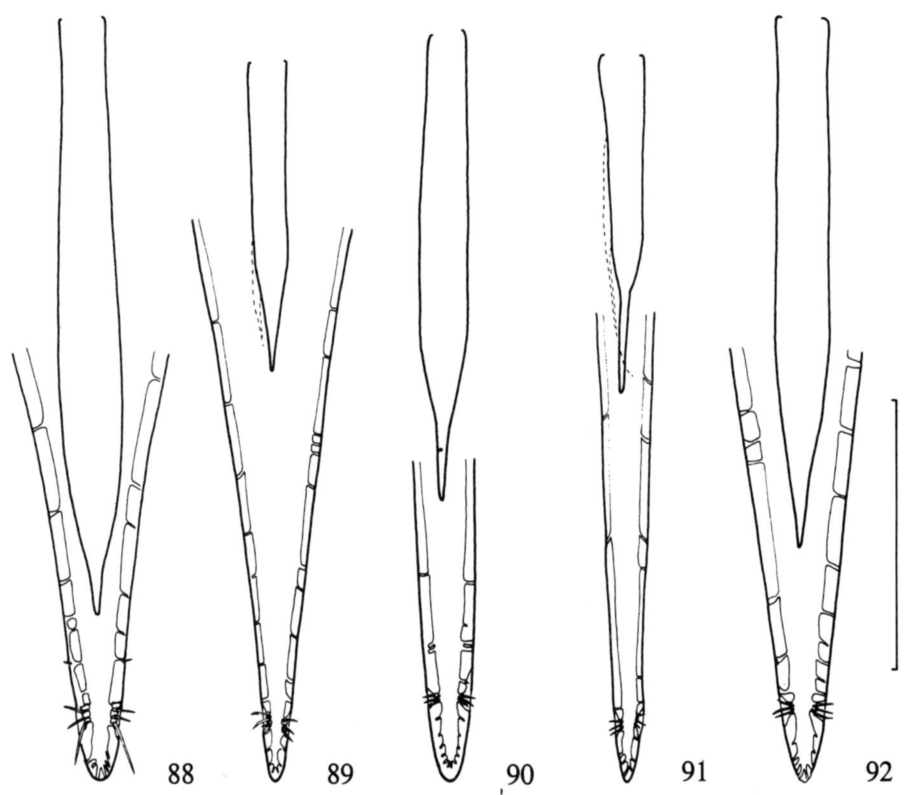

Figs 88-92. *Dacus* aculei, shapes in dorsal/ventral view, with details (optical sections) of apices, and dashed lines showing ventral lobes of aculeus; 88, *D. (D.) bivittatus*; 89, *D. (Didacus) ciliatus*; 90, *Dacus (D.) demmerezi*; 91, *D. (Didacus) frontalis*; 92, *Dacus (Didacus) lounsburyii*. Scale for apex details=0.2mm.

Key 8: Key to Species of *D. (Dacus)*

1 Scutum without lateral yellow or orange stripes (vittae) (fig. 223). [Scutum sometimes with a narrow medial yellow stripe (vitta). Wing length 5.8-6.9mm. Aculeus length 2.3-2.5mm (fig. 90). Males attracted to cue lure. Larvae develop in cucurbit fruits. Indian Ocean Islands.]
. . . . *D. (D.) demmerezi* (p. 323)

\- Scutum with lateral yellow or orange stripes (vittae) (figs 222, 224, 225). [Scutum always with a medial yellow or orange stripe (vitta).]
. . . . 2

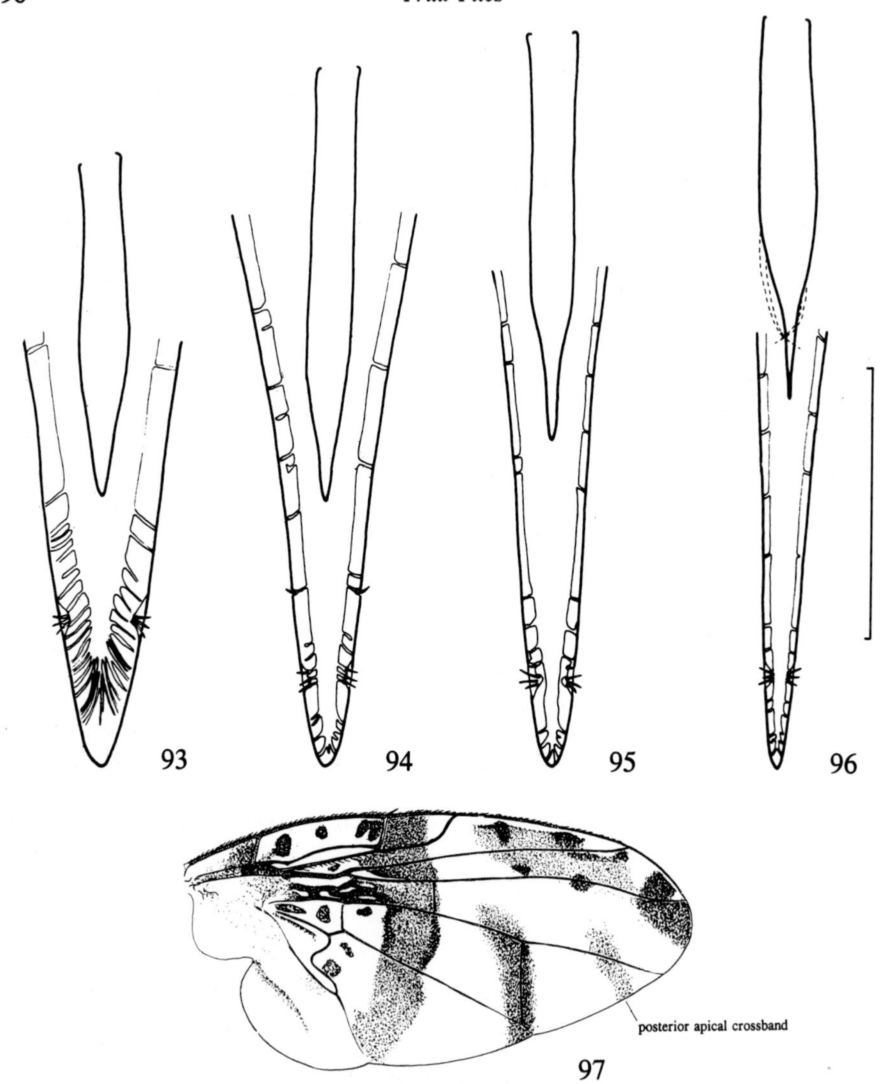

Figs 93-97. *Dacus* and *Ceratitis*; 93-96, *Dacus* aculei, shapes in dorsal/ventral view, with details (optical sections) of apices (scale for apex details=0.2mm), and dashed lines showing ventral lobes of aculeus; 93, *D. (D.) punctatifrons*; 94, *D. (Callantra) smieroides*; 95, *D. (D.) telfaireae*; 96, *D. (Didacus) vertebratus*; 97, *Ceratitis (Pterandrus) rubivora*, wing.

2 Wing with a very broad costal band which extends below vein R_{4+5}, almost reaching vein M (fig. 222); combined depth of cells r_1 and r_{2+3} at r-m crossvein, about equal to length of r-m crossvein. [Wing length 6.4-8.5mm. Aculeus length 2.5-2.9mm (fig. 88). Males attracted to cue lure. Larvae usually develop in cucurbit fruits. Sub-Saharan Africa.]
. . . . *D. (D.) bivittatus* (p. 321)

- Wing without such a broad costal band, at most reaching R_{4+5} (figs 224, 225); combined depth of cells r_1 and r_{2+3} at r-m crossvein only about equal to half the length of the r-m crossvein.
 3

3 General body colour orange-brown (fig. 224). Posterolateral area of thorax with a diagonal yellow stripe below the scutellum which extends across both the katatergite (in front of the haltere base) and the anatergite; stripe only narrowly separated from the scutellum. [Wing length 4.2-7.7mm. Aculeus length 1.6-1.8mm (fig. 93). Males attracted to cue lure. Larvae usually develop in cucurbit fruits. Sub-Saharan Africa and Mauritius.]
 . . . . *D. (D.) punctatifrons* (p. 325)

- General body colour black (fig. 225). Posterolateral area of thorax with a yellow spot in front of the haltere base, which is confined to the katatergite; spot separated from scutellum by at least its own diameter. [Wing length 4.4-6.1mm. Aculeus length 2.0mm (fig. 95). Males attracted to cue lure. Larvae develop in oysternut fruits (*Telfairea*). Zimbabwe to Kenya.]
 . . . . *D. (D.) telfaireae* (p. 327)

Key 9: Key to Species of *Dacus (Didacus)*

1 Scutum with medial and lateral yellow stripes (vittae) (fig. 228). [Wing length 7.1-10.2mm. Aculeus length 2.4-2.6mm (fig. 92). Larvae develop in cucurbit fruits. Southern Africa.]
 . . . . *Dacus (Didacus) lounsburyii* (p. 334)

- Scutum without any yellow or orange stripes (vittae) (figs 226, 227, 229). [Ventral lobes of aculeus extending almost to apex of aculeus and very narrow; often easily visible in slide preparations (figs 89, 91, 96).]
 2

[The following two couplets can only be applied to clean non-teneral specimens.]

2 Posterolateral area of thorax with a yellow spot in front of the haltere base, which is virtually confined to the katatergite; spot separated from scutellum by at least its own diameter. Mid femur entirely yellow or orange-yellow, at most with apical half slightly darker than basal half. Aculeus apex evenly tapered (fig. 89). [Wing length 4.4-6.0mm (fig. 226). Aculeus length 1.5-1.6mm. Larvae usually develop in cucurbit fruits. Sub-Saharan Africa; adventive in the Indian Ocean Islands, Saudi Arabian Peninsula and the Indian subcontinent.]
 . . . . *Dacus (Didacus) ciliatus* (p. 329)

- Posterolateral area of thorax with a diagonal yellow stripe below the scutellum which extends across both the katatergite (in front of the haltere base) and the anatergite; stripe only separated from the scutellum by about one-third its length. Mid femur markedly darker in apical half than basal half. Aculeus tapered sharply well before apex (figs 91, 96).
 3

3 All femora yellow in basal half, orange in apical half. Males attracted to vert lure. Apical section of aculeus tapered (fig. 96). [Wing length 4.8-7.5mm (fig. 229). Aculeus length 1.7-1.8mm. Larvae develop in cucurbit fruits. Sub-Saharan Africa, Madagascar and the Saudi Arabian Peninsula.]
. . . . *Dacus (Didacus) vertebratus* (p. 336)

\- Mid femora yellow in basal half, orange in apical half; fore and hind femora entirely yellow. Males attracted to cue lure. Apical section of aculeus almost parallel sided (fig. 91). [Wing length 4.2-5.8mm (fig. 227). Aculeus length 1.3-1.7mm. Larvae develop in cucurbit fruits. Sub-Saharan Africa, Cape Verde and the Saudi Arabian Peninsula.]
. . . . *Dacus (Didacus) frontalis* (p. 332)

Key to Species of *Carpomya* Associated with *Ziziphus* spp.

The following key separates the two widely distributed species of *Carpomya* that are associated with jujube fruits.

1 Scutum uniformly reddish-yellow; scutellum uniformly pale yellow (fig. 209). Wing with three faint crossbands (subbasal, discal and preapical); without an apical crossband. [Wing length 2.9-3.4mm. Aculeus pointed, length 0.8mm. North-east Africa, Middle East and Italy.]
. . . . *C. incompleta* (p. 281)

\- Scutum, at least in part, and scutellum, bright yellow with black patches (fig. 210). Wing with four distinct crossbands; apical crossband joined to preapical in cells r_1 and r_{2+3}. [Wing length 3.6-4.2mm. Aculeus pointed, length 0.7mm. Italy, southern USSR to Thailand; adventive in Mauritius.]
. . . . *C. vesuviana* (p. 283)

Keys to *Ceratitis* and *Trirhithromyia* Species of Economic Importance

The following two keys separate the 11 species of *Ceratitis* and the one species of *Trirhithromyia* likely to be found in association with fruit crops. Key 2 allows identification of species using specimens of either sex. Key 1 separates the major subgenera of *Ceratitis* using characters that only apply to male specimens (Key 2 is still required for species level identification). Key 1 is included as an additional check when identifying males, and to present details of how the subgenera are defined. Aculeus length values were obtained by examination of specimens; however, values for most *Ceratitis* spp. were based on the examination of only one or two specimens and users of the keys should expect to find a few specimens that do not exactly fit the values given.

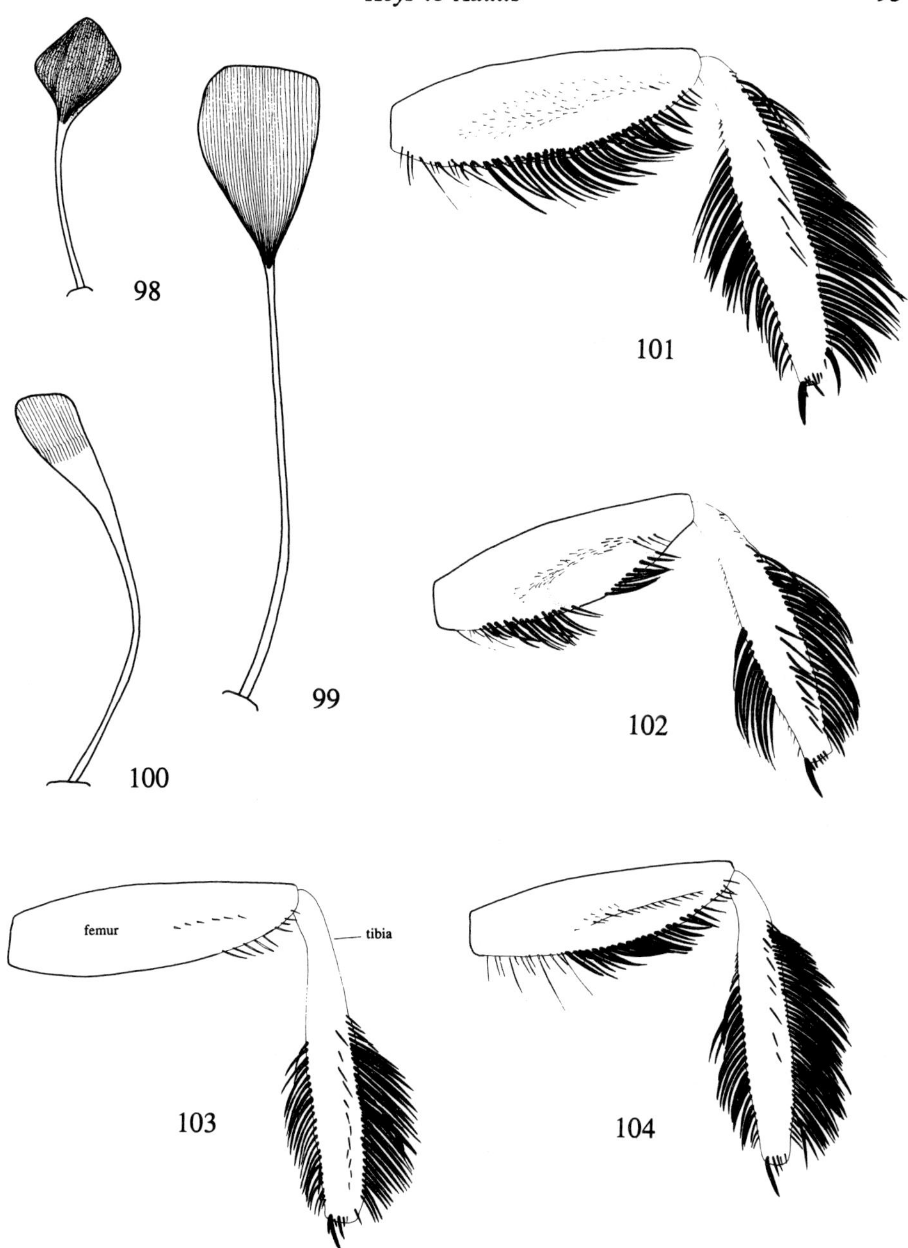

Figs 98-104. *Ceratitis*, heads and legs of males; 98-100, spatulate (capitate) orbital setae; 98, *C. (C.) capitata*; 99, *C. (C.) catoirii*; 100, *C. (C.) malgassa*; 101-104, left mid-legs, femora and tibiae in anterior views; 101, *C. (Pterandrus) anonae*; 102, *C. (P.) colae*; 103, *C. (P.) rosa*; 104, *C. (P.) rubivora*.

Key 1: Key to Subgenera and Some Species of *Ceratitis* (Males Only)

1 Anterior pair of orbital setae modified into spatulate (capitate) appendages (figs 22, 98-100).
. . . . *C. (Ceratitis)* [Key 2; couplet 3]

- Anterior pair of orbital setae not modified in any way (similar to fig. 23).
. . . . 2

2 Mid tibia with rows of stout setae along the anterior and posterior edges giving a feathered appearance (figs 101-104). [Wing bands and general body colour brown (figs 97, 217, 218).]
. . . . *C. (Pterandrus)* (in part) [Key 2; couplet 9]

- Mid tibia without rows of stout setae arranged in such a way as to give a feathered appearance.
. . . . 3

3 Fore femur patterned with black and white on anterior (inner) side. [Further details in Key 2; couplet 8.]
. . . . *C. (P.) pedestris* (p. 305)

- Fore femur not patterned with black and white.
. . . . 4

4 Anepisternum with 2 well developed setae (some spp. not covered by this work have 3). [Wing bands and general body colour brown (fig. 216). Further details in Key 2; couplet 7.]
. . . . *C. (Pardalaspis) punctata* (p. 301)

- Anepisternum with 1 well developed seta.
. . . . 5

5 Wing bands and general body colour yellow (figs 211, 212).
. . . . *C. (Ceratalaspis)* [Key 2; couplet 5]

- Wing bands and general body colour brown (fig. 249). [Further details in Key 2; couplet 2.]
. . . . *Trirhithromyia cyanescens* (p. 393)

Key 2: Key to Species of *Ceratitis* and *Trirhithromyia* (Both Sexes)

1 Scutellum entirely black in apical half (figs 213-215, 249). [Male mid tibia without stout setae arranged in such a way as to give a feathered appearance.]
. . . . 2

- Scutellum with yellow lines or areas meeting margin, such that each apical scutellar seta is based in or adjacent to a yellow stripe (figs 211, 212, 216-218).
. . . . 5

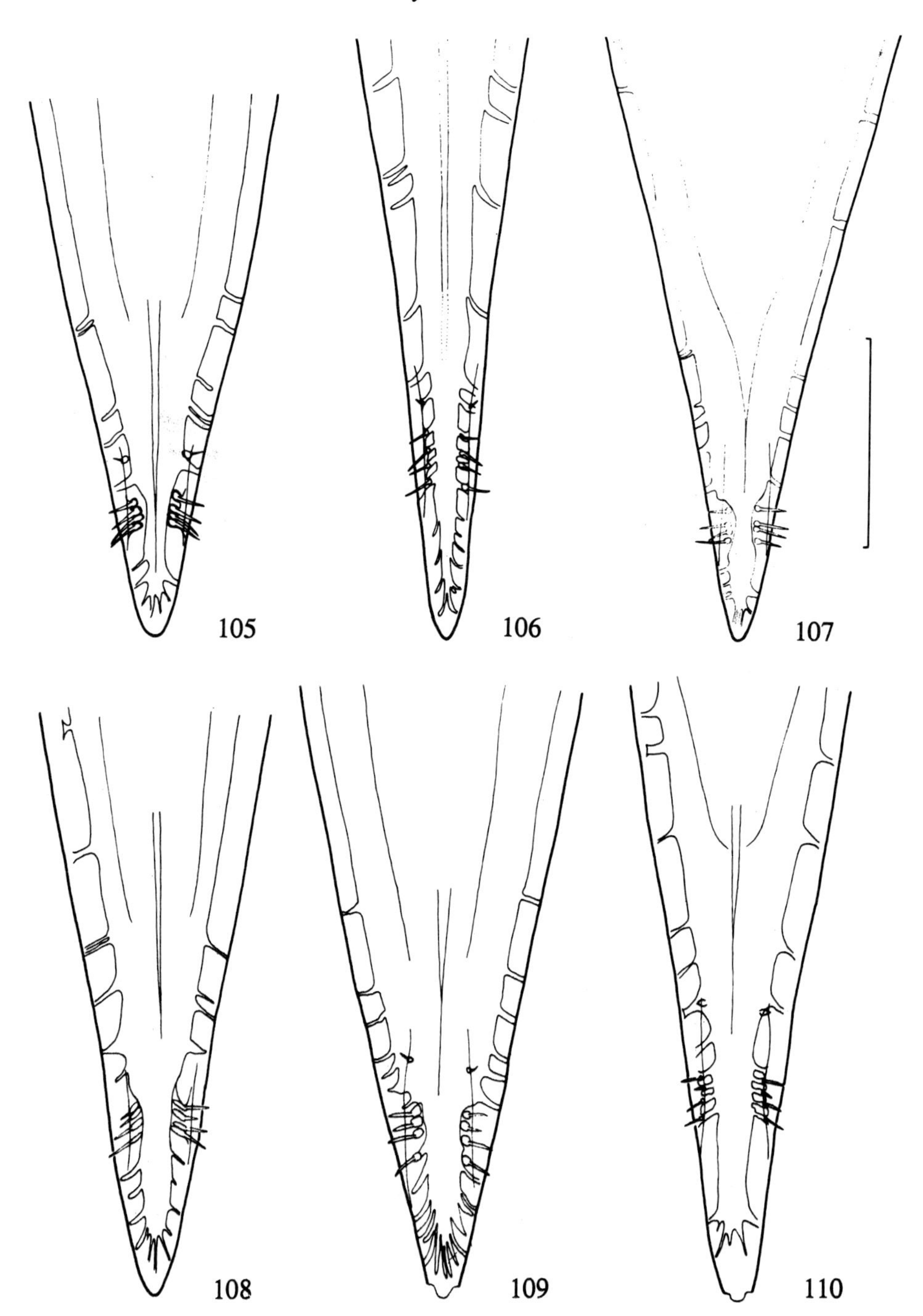

Figs 105-110. ***Ceratitis*** **aculei, dorsal views (optical sections) of apices, grouped by shape;** 105, *C. (Ceratalaspis) cosyra*; 106, *Ceratitis (Pterandrus) pedestris*; 107, *C. (P.) rubivora*; 108, *C. (C.) capitata*; 109, *C. (C.) catoirii*; 110, *C. (C.) malgassa*. Scale=0.1mm.

2 Scutellum with a broad yellow band across base (fig. 249). Male anterior pair of orbital setae not modified in any way. Wing with dark brown crossbands; costal and discal crossbands joined. Aculeus with preapical 'steps' and a pointed apex (fig. 111). [Wing length 4.1-5.8mm. Aculeus length 1.4-1.5mm. Larvae develop in the fruits of Solanaceae. Indian Ocean Islands.]
. . . . *Trirhithromyia cyanescens* (p. 393)

\- Scutellum with a narrow 'wavy' yellow band across base (fig. 213-215). Male anterior pair of orbital setae modified into spatulate (capitate) appendages (figs 22, 98-100). Wing with yellow crossbands; costal band starting beyond the end of vein R_1, and separated from discal crossband by a hyaline area at the end of R_1. Aculeus pointed (fig. 108) or, if with preapical 'steps', then apex rounded (figs 109, 110).
. . . . 3

3 Wing with apex of vein M not covered by a diagonal coloured band (fig. 213). Male anterior pair of orbital setae with a sharp end to the spatulate section (fig. 98), which is black. Aculeus pointed (fig. 108). [Wing length 3.6-5.0mm. Aculeus length 1.0mm. Males strongly attracted to trimedlure; weakly to terpinyl acetate; not to methyl eugenol. Africa; adventive in most tropical and warm temperate areas.]
. . . . *C. (C.) capitata* (p. 291)

\- Wing with apex of vein M covered by a diagonal crossband (figs. 214, 215). Male anterior pair of orbital setae with a blunt end to the spatulate section (figs 99, 100), which is white.
. . . . 4

4 Preapical crossband (i.e. the mark over crossvein dm-cu) connected to the discal crossband by a coloured mark along vein M (fig. 214). Spatulate section of male anterior orbital seta about as long as broad (fig. 99), except in teneral specimens in which it is very elongate, resembling *C. malgassa* (fig. 100). [Wing length 4.4-5.6mm. Aculeus length 1.5mm (fig. 109). Mauritius area.]
. . . . *C. (C.) catoirii* (p. 298)

\- Preapical crossband isolated (fig. 215). Spatulate section of male anterior orbital seta longer than broad (fig. 100). [Wing length 5.2-5.7mm. Aculeus length 1.5mm (fig. 110). Madagascar; adventive, possibly only temporarily, in Puerto Rico.]
. . . . *C. (C.) malgassa* (p. 300)

5 Scutellum with small black areas (fig. 212). [Wing length 2.7-4.1mm. Aculeus length 0.8-0.9mm; apex shape similar to *C. rubivora* (fig. 107). Male mid tibia without stout setae arranged in such a way as to give a feathered appearance. Males strongly attracted to terpinyl acetate; not to trimedlure or methyl eugenol. South Africa to Yemen.]
. . . . *C. (Ceratalaspis) quinaria* (p. 289)

\- Scutellum with large black areas (figs 211, 216-218).
. . . . 6

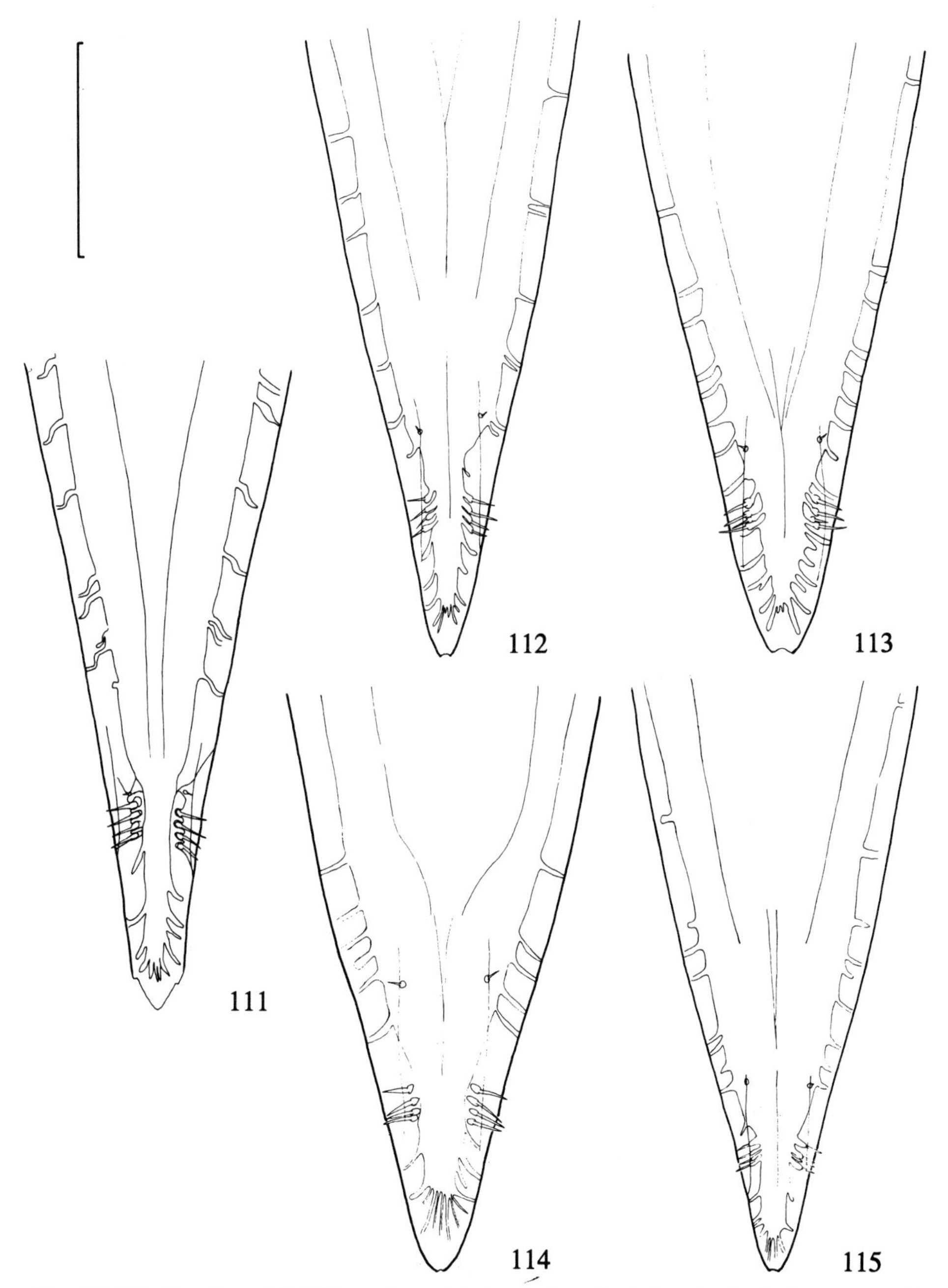

Figs 111-115. *Ceratitis* and *Trirhithromyia*, aculei, dorsal views (optical sections) of apices, grouped by shape; 111, *T. cyanescens*; 112, *C. (Pterandrus) anonae*; 113, *C. (P.) colae*; 114, *C. (Pardalaspis) punctata*; 115, *C. (Pterandrus) rosa*. Scale=0.1mm.

6 Costal band and discal crossband joined (figs 211, 216, 217). Male mid tibia without stout setae arranged in such a way as to give a feathered appearance. 7

- Costal band starting beyond the end of vein R_1, and separated from discal crossband by a hyaline area at the end of R_1 (figs 97, 218). Male mid tibia with anterior and ventral rows of stout setae giving a feathered appearance (figs 101-104). [Arista plumose or long pubescence.] 9

7 Anepisternum with 2 setae. Aculeus with a minute apical notch (fig. 114). [Wing length 5.8-6.7mm (fig. 216). Aculeus length 1.9mm. Males are attracted to methyl eugenol and terpinyl acetate; not to trimedlure. Sub-Saharan Africa.] . . . . *Ceratitis (Pardalaspis) punctata* (p. 301)

- Anepisternum with 1 seta. Aculeus without an apical notch (figs 105, 107). 8

8 Wing bands brown (fig. 217). Scutum predominantly brown to black. Male fore femur patterned with black and white patches on the anterior side. Aculeus very long, 2.2mm (fig. 106). [Postpronotal lobe usually with a large dark spot. Wing length 4.5-5.7mm. Males are attracted to trimedlure and terpinyl acetate; not to methyl eugenol. Southern Africa and Madagascar.] . . . . *C. (Pterandrus) pedestris* (p. 305)

- Wing bands yellow (fig. 211). Scutum predominantly yellow or pale brown, with a pattern of brown to black spots. Fore femur yellow on both sides and in both sexes. Aculeus shorter, 1.3-1.6mm (fig. 105). [Wing length 3.7-5.7mm. Males weakly attracted to terpinyl acetate; not to trimedlure or methyl eugenol. Southern and eastern Africa.] . . . . *C. (Ceratalaspis) cosyra* (p. 287)

9 Apical section of vein M (beyond dm-cu crossvein) crossed by an infuscate area (a posterior apical crossband) (fig. 97) (except in teneral specimens). Aculeus pointed (fig. 107). [Wing length 4.6-5.7mm. Aculeus length 0.9mm. Femora largely yellowish. Male mid femur with stout ventral setae, similar to those on tibia (fig. 104). Male mid tibia feathered along almost entire length. Males attracted to trimedlure and terpinyl acetate; not to methyl eugenol. Larvae develop in the fruits of *Rubus* spp. Sub-Saharan Africa.] . . . . *Ceratitis (P.) rubivora* (p. 311)

- Apical section of vein M not crossed by an infuscate area (fig. 218). Aculeus with an apical notch (figs 112, 113, 115). NOTE - male specimens are essential for positive identification of these spp.; further research is needed to determine if aculeus length, possibly expressed as a ratio to a wing measurement, can be used to reliably separate females. 10

10 Male mid femur without stout ventral setae (fig. 103). Male mid tibia only feathered in distal two-thirds. Male fore femur with a few strong ventral setae, but not as densely arranged as those of mid tibia , and very few anterior setae. [Wing length 4.5-5.8mm. Aculeus length 0.9mm; with a small, shallow apical notch (fig. 115). Males strongly attracted to trimedlure and terpinyl acetate; not to methyl eugenol. Sub-Saharan Africa and Mauritius area.]
. . . . *C. (P.) rosa* (p. 306)

\- Male mid femur with stout ventral setae, similar to those on tibia (figs 101, 102). Male mid tibia feathered along almost entire length. Male fore femur with numerous strong ventral setae which are almost as dense as those of the mid tibia, plus some strong anterodorsal setae.
. . . . 11

11 Male mid tibia broadly feathered and with feathering along most of inner edge (fig. 101). Male mid femur without a gap in the feathering. [Wing length 4.6-6.3mm. Aculeus length 1.2mm; with a shallow apical notch (fig. 112). Male lure response unknown. Sub-Saharan Africa.]
. . . . *C. (P.) anonae* (p. 302)

\- Male mid tibia more narrowly feathered and with feathering restricted to apical half of inner edge (fig. 102). Male mid femur with a gap in the feathering and a silvery-white patch of tomentum on the anterior surface adjacent to the gap. [Wing length 3.8-6.3mm. Aculeus length 1.5mm; with a broad, deep apical notch (fig. 113). Male lure response unknown. Larvae develop in cola fruits (*Cola*). Sub-Saharan Africa.]
. . . . *C. (P.) colae* (p. 304)

Key to *Rhagoletis* Species of Economic Importance

The following key includes the 16 species most likely to be found in association with commercial fruit crops. Unlike the keys to most other genera, host and distribution data is used as the primary method of separating some of the morphologically similar species. *Rhagoletis* spp. have very narrow host plant ranges compared to most other genera of economic importance and host relationship may be of help in identification. The included species of the *R. pomonella* complex are most easily separated using host data; the included species of the *R. cingulata* complex can only be reliably separated using distribution data. Aculeus length values were obtained by examination of specimens, supplemented with values given by Bush (1966).

1 Scutum and abdomen yellow to orange (236, 239, 245). Scutellum entirely cream to yellow white, at most with a very narrow dark band at base, but never extensively black at sides (some individuals, particularly males, of *R. completa* may be marked at the sides; fig. 116). [Wing without an accessory costal crossband. Larvae develop in the husks of walnut (*Juglans*).]
. . . . 2

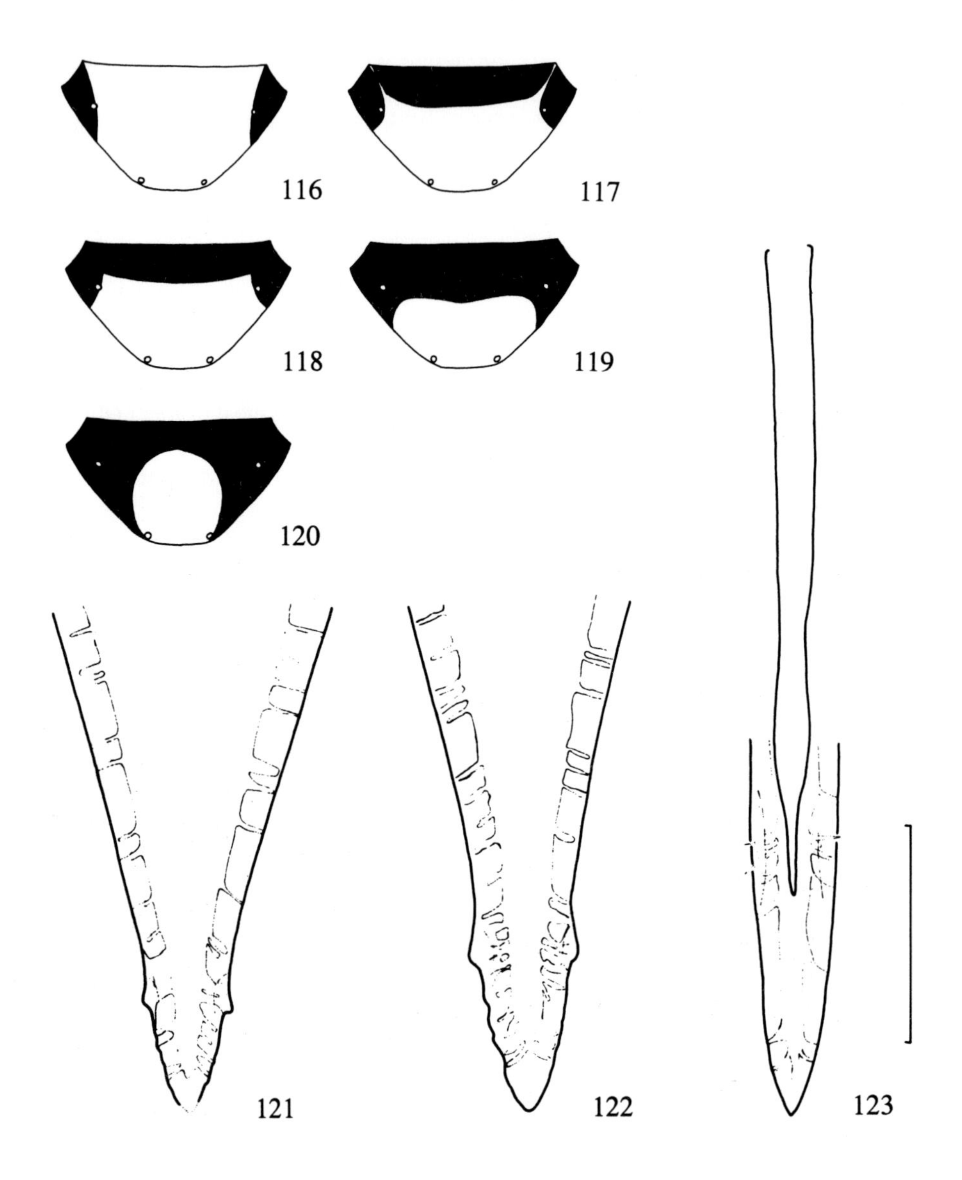

Figs 116-123. *Rhagoletis*; 116-120, scutellum patterns (diagrams of types, not unique to any one sp.), dorsal views; 121-122, aculei, dorsal views (optical sections) of apices, species with preapical 'steps'; 121, *R. conversa*; 122, *R. nova*; 123, *R. striatella*, aculeus, shape in dorsal/ventral view, with detail (optical section) of apex. Scale for aculeus apex details=0.1mm.

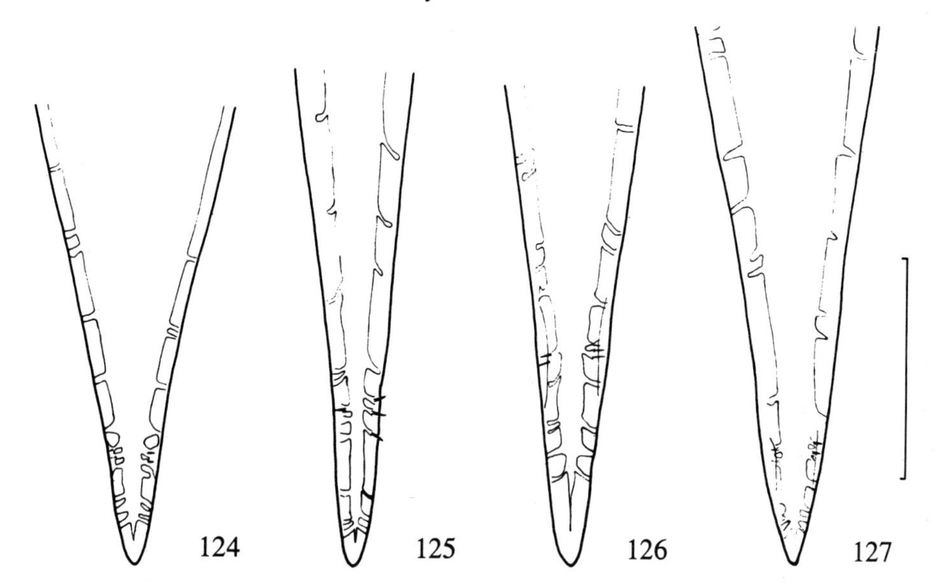

Figs 124-127. *Rhagoletis* aculei, dorsal views (optical sections) of apices; 124, *R. cerasi*; 125, *R. completa*; 126, *R. pomonella*; 127, *R. ribicola*. Scale=0.1mm.

- Scutum and abdomen predominantly black; scutum with 2 or 4 longitudinal bars of tomentum that form grey stripes (figs 234, 235, 237, 238, 240-244, 246, 247). Scutellum marked black at sides and sometimes broadly at base (figs 116-120). 4

2 Wing without a distinct subbasal crossband (fig. 239), but sometimes slightly darkened in the subbasal area, especially at the base of cell c. Preapical and apical crossbands usually separate. Veins R_{4+5} and M usually with small isolated markings along the sides of these veins between the preapical and apical crossbands. [Wing length 2.9-4.1mm. Aculeus length 1.0-1.2mm; apex shape similar to *R. completa* (fig. 125). Mexico and south-western USA.] . . . . *R. juglandis* (p. 365)

- Wing with a distinct subbasal crossband (fig. 236, 245). Preapical and apical crossbands broadly joined between veins C and R_{4+5}. Veins R_{4+5} and M without any isolated markings. 3

3 Wing with discal and preapical crossbands joined across the whole depth of cell dm; preapical crossband forked so that there is a hyaline area running from the hind margin of the wing forwards into cell dm (fig. 245). Mediotergite orange. [Wing length 3.8-5.4mm. Aculeus length 0.8-1.4mm; apex shape similar to *R. completa* (fig. 125). Eastern USA.] . . . . *R. suavis* (p. 382)

- Wing with discal and preapical crossbands usually separate (fig. 236), but sometimes narrowly fused in cell dm or fused posterior to vein CuA_1. Mediotergite either entirely dark brown, or with a pair of vertical dark brown stripes. [Wing length 2.7-4.3mm. Aculeus length 0.8-1.1mm (fig. 125). Western and central USA; adventive in Switzerland.]
  . . . . *R. completa* (p. 359)

4 Scutellum marked at sides only, base yellow or only very narrowly black (fig. 116).
  5

- Scutellum black at base, at least in basal fifth (figs 117-120), but basal and lateral black marks sometimes separated (fig. 117).
  6

5 Wing with a posterior, as well as an anterior, apical crossband (fig. 238). Discal and preapical crossbands joined along veins M and CuA_1, but leaving a hyaline spot in the apical quarter of cell dm. Wing without an accessory costal crossband. Western and northern USA, and southern Canada. [Wing length 3.0-4.2mm. Aculeus length 0.8-0.9mm; apex shape similar to *R. cerasi* (fig. 124) but sensilla further from apex. Larvae develop in cherry fruits (*Prunus*).]
  . . . . *R. fausta* (p. 363)

- Wing without a posterior apical crossband (fig. 234). Discal and preapical crossbands separate. With an accessory costal crossband, except in very small individuals. Northern and central Europe, and east to western Siberia and Kazakhstan. [Wing length 2.5-4.0mm. Aculeus length 0.7-0.8mm (fig. 124). Larvae develop in cherry fruits (*Prunus*).]
  . . . . *R. cerasi* (p. 353)

6 Wing with an accessory costal crossband (figs 237, 240, 241, 247).
  7

- Wing without an accessory costal crossband (figs 235, 242-244, 246).
  10

[The scutum and wing characters used to separate the following four Solanaceae associated spp. are somewhat variable and it is advisable to dissect the aculeus of the female to help confirm identity.]

7 Scutum with all four longitudinal stripes of tomentum united anterior to the transverse suture (fig. 241) (view from above/front; the two medial stripes appear narrowly separated when viewed from above/behind). [Wing without a posterior apical crossband, but sometimes with an isolated infuscate patch on vein M, between the preapical and anterior apical crossbands; anterior apical crossband often reduced to an isolated spot across the apex of vein R_{4+5}. Scutellum (figs 117, 118). Wing length 3.3-4.8mm. Aculeus length 1.1-1.4mm (fig. 122). Larvae develop in the fruit of pepino (*Solanum muricatum*). Chile.]
  . . . . *R. nova* (p. 369)

- Scutum with at least the two medial longitudinal stripes of tomentum well separated (figs 237, 240, 247) (regardless of viewing angle).
  8

8 Scutum with lateral longitudinal stripes of tomentum joined to the medial stripes anterior to the transverse suture (fig. 247). [Wing without a posterior apical crossband or even an isolated mark between the preapical and anterior apical crossbands. Scutellum (figs 117, 118). Wing length 4.0-4.6mm. Aculeus length 1.3mm; apex shape similar to *R. pomonella* (fig. 126) but sensilla closer to apex. Larvae develop in tomato fruit (*Lycopersicon esculentum*). Southern Peru and northern Chile.]
. . . . *R. tomatis* (p. 386)

\- Scutum with all 4 longitudinal stripes of tomentum separate (figs 237, 240).
. . . . 9

9 Wing with a short posterior apical crossband, which is not joined to any other crossband (fig. 240). Lateral stripes on scutum only extending anteriorly as far as transverse suture or a little in front of transverse suture. Aculeus apex with a simple point, similar to *R. pomonella* (fig. 126). [Scutellum (figs 118, 119). Wing length 3.7-4.2mm. Aculeus length 1.2-1.3mm. Larvae develop in tomato fruit (*Lycopersicon esculentum*). Western Peru.]
. . . . *R. lycopersella* (p. 367)

\- Wing without a posterior apical crossband (fig. 237). Lateral stripes on scutum extending anterior to transverse suture. Aculeus with a pair of preapical 'steps' (fig. 121). [Scutellum (fig. 118). Wing length 2.7-4.7mm. Aculeus length 1.1-1.4mm. Larvae develop in fruits of *Solanum* spp. Chile.]
. . . . *R. conversa* (p. 361)

10 Wing with anterior apical crossband separated from vein C leaving a hyaline margin at least across the apices of veins R_{2+3} and R_{4+5} (figs 242, 243, 246).
. . . . 11

\- Wing with anterior apical crossband adjoining vein C at least across the apex of vein R_{4+5} (fig. 235, 244), or with an isolated spot covering the apex of vein R_{4+5} (shown heavily stippled in fig. 235).
. . . . 14

11 Wing with preapical crossband (the band which covers the dm-cu crossvein) running obliquely from a point on the discal crossband near the r-m crossvein, so that it is almost parallel to the apical crossband (fig. 242). [Scutellum (fig. 119). Aculeus tip pointed (fig. 126).]
. . . . *R. pomonella* complex. . 12

\- Wing with preapical crossband crossing the wing transversely (figs 243, 246).
. . . . 13

12 Aculeus usually long; apple race 1.0-1.4mm but individuals from Florida (not from apple) may have an aculeus as short as 0.7mm. Larvae develop in apples (*Malus*) and the fruit of other genera of subfamily Maloideae (Rosaceae). [Wing length 1.5-4.8mm. Eastern half of the USA, south-eastern Canada; it has recently become established in Oregon and southern Washington.]
. . . . *R. pomonella* (p. 375)

- Aculeus usually shorter; length 0.7-0.9mm. Larvae develop in the fruits of *Vaccinium* spp. (Ericaceae). [Wing length 2.3-3.4mm. Eastern USA and south-eastern Canada.]
  . . . . *R. mendax* (p. 371)

13 Subbasal and discal crossbands joined over vein A_1+CuA_2 (fig. 246). Larvae develop in the fruits of cranberry (*Vaccinium macrocarpon*) in western North America and of dogwood (*Cornus*) in most areas of North America. [Scutellum (fig. 120). Wing length 2.4-3.7mm. Aculeus length 0.7-0.8mm; apex shape similar to *R. pomonella* (fig. 126) but sensilla closer to apex.]
  . . . . *R. tabellaria* complex (p. 384)

- Subbasal and discal crossbands not joined (fig. 243). Larvae develop in currants and gooseberries (*Ribes*). [Scutellum (figs 119, 120). Wing length 2.0-3.4mm. Aculeus length 0.7-1.0mm (fig. 127). North-western USA and southern British Columbia.]
  . . . . *Rhagoletis ribicola* (p. 378)

14 Wing with a posterior apical crossband (fig. 244). Basal and lateral black marks on scutellum separated by narrow yellow lines (fig. 117); basal scutellar setae based close to those yellow lines, at the margin of the lateral black marks, and not well within a black area. Aculeus very long, 2.9-3.4mm (fig. 123). Larvae develop in husk tomatoes (*Physalis*). [Wing length 3.7-5.1mm. Aculeus about twice as long as visible part of the oviscape and extending well back into the abdomen; only visible when the abdomen has been cleared in.KOH. Mexico to Canada.]
  . . . . *R. striatella* (p. 381)

- Wing without a posterior apical crossband, although the anterior apical crossband is sometimes divided at its apex, leaving an oblique hyaline stripe across the apex of cell r_{4+5} (fig. 235). Basal and lateral black marks on scutellum usually broadly joined (fig. 119); basal scutellar setae usually based well within black area (some individuals, possibly only when teneral, resemble fig. 117). Aculeus short, 0.7-1.0mm; apex shape similar to *R. cerasi* (fig. 124). Larvae develop in cherries (*Prunus*).
  . . . . *R. cingulata* complex. . 15

[Full details of the separation of species belonging to this complex are beyond the scope of the present work and were provided elsewhere (Bush, 1966). The following couplet is primarily based on geographic characters and is only reliable as long as neither species becomes adventive within the natural range of the other.]

15 Eastern USA and southern Quebec. Wing with anterior apical crossband often reduced to an isolated spot (area shown by heavy stipple in fig. 235). [Wing length 2.6-3.8mm. Aculeus length 0.8-1.0mm.]
  . . . . *R. cingulata* (p. 357)

- North-western USA and southern British Columbia (also adventive in Switzerland). Wing with anterior apical crossband very rarely reduced to an isolated spot (light stipple in fig. 235 shows full extent of crossband). [Wing length 2.6-3.6mm. Aculeus length 0.7-0.9mm; in the specimens dissected for this study the preapical sensilla were slightly further from the aculeus apex in *R. indifferens* than shown in the figure of *R. cerasi* (fig. 124), but further study is needed to confirm if this is a useful character.]
  . . . . *R. indifferens* (p. 358)

Keys to African *Trirhithrum* Species Which Attack Coffee

The two widely distributed African species of *Trirhithrum* which attack coffee fruits (known as coffee 'cherries') are sexually dimorphic and separate keys to male and female flies follow.

Key 1: Key to Males

1 Most of wing infuscate, pale brown, with an ill-defined hyaline patch in the centre of cell r_{4+5} and a contrastingly dark brown mark in the basal third of cell cua_1 (fig. 250). [Wing length 2.9-3.6mm. East and West Africa.]
  . . . . *T. coffeae* (p. 395)
- Anterior half of wing dark brown; distal areas of cells m, dm and cua_1 largely hyaline (fig. 129). [Wing length 3.0-3.2mm. Sub-Saharan Africa and the Comoro Islands.]
  . . . . *T. nigerrimum* (p. 397)

Key 2: Key to Females

1 Cell c with central two-thirds hyaline, at most with a faint marking near the centre of that area (fig. 128). Costal band with deep hyaline indentations in cell r_1. [Wing length 2.8-3.7mm. Aculeus 0.8mm long.]
  . . . . *T. coffeae* (p. 395)
- Cell c mostly dark brown, sometimes with 1 or 2 small marginal hyaline indentations (fig. 130). Costal band without deep hyaline indentations in cell r_1. [Wing length 3.0-3.8mm. Aculeus 0.8mm long.]
  . . . . *T. nigerrimum* (p. 397)

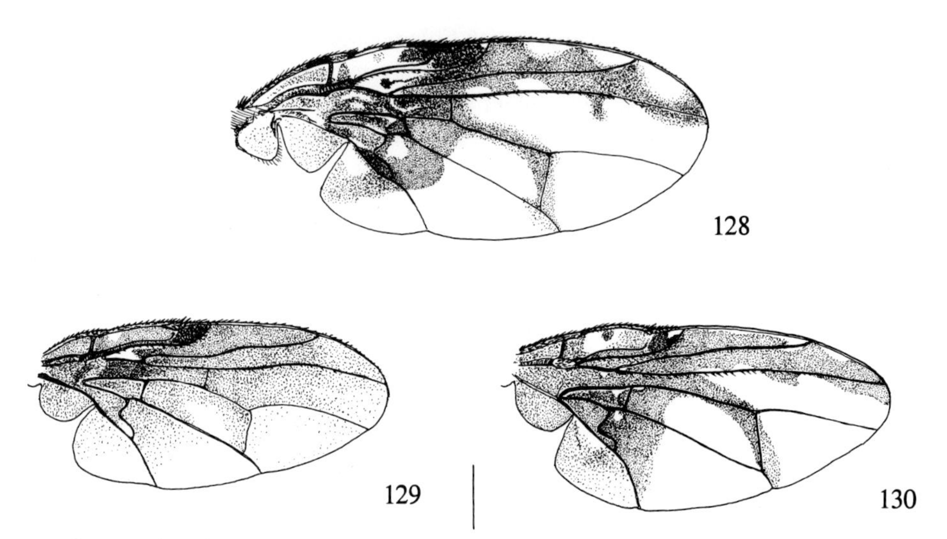

Figs 128-130. *Trirhithrum* wings showing sexual dimorphism; 128, *T. coffeae* female; 129, *T. nigerrimum* male; 130, *T. nigerrimum* female.

Key to Males Caught in Lure Traps

The following key is intended to help with the identification of those *Bactrocera*, *Ceratitis* and *Dacus* spp. which respond to male lures. This key is primarily intended for rapid preliminary identification of fruit flies caught in monitoring traps in areas previously assumed fruit fly free, or from traps placed within large orchard areas.

WARNING: The present work was primarily designed for the identification of fruit flies that have been reared from cultivated fruits. In areas with a diverse fauna of *Bactrocera*, *Ceratitis* or *Dacus* spp., traps baited with male lures are likely to catch many non-pest species which are not covered by this work. The following key should not be used for identifying fruit flies caught in traps placed in areas of natural vegetation, in small orchards surrounded by natural forest or from near the edge of a large orchard. Further characters given in the previous keys, in the following species accounts and as illustrations should be consulted to help verify these identifications. Even then, the possibility that one of the many hundreds of species not covered by this work could erroneously be identified as a pest, cannot be ruled out. It is also very important that no cross contamination of lures has occurred (p. 19). It should also be noted that many flies with no known lure response may in fact respond to a lure; either their response to a well known lure has not yet been noted or they respond to some as yet unidentified lure.

To identify fruit flies bait trapped in or near areas of natural vegetation reference should be made to the following works:

— African Ceratitini: Hancock (1987) listed known lure responses and Hancock (1984; 1987) should be used for identification.

— African Dacini: Hancock (1985b) listed known lure responses and Munro (1984) should be used for identification.
— Australian and South Pacific Dacini: Drew (1974; 1982a; 1989a) and Drew & Hooper (1981) listed known lure responses and Drew (1982a; 1989a) should be used for identification.
— Asian Dacini: No comprehensive lists are available but some lure responses were listed by Hardy (1973) and Tan & Lee (1982); major identification works are Hardy (1973; 1974; 1982b; 1983a), Ito (1983-5) and Zia (1937). Much of the lure response data used in the present work was provided by D.L. Hancock (pers. comm., 1991).

1 Male caught using vert lure.
. . . . *Dacus (Didacus) vertebratus* (p. 336)

[The lure response of most African spp. has not been studied and some others may eventually be shown to respond to vert lure.]

- Male caught using methyl eugenol 2
- Male caught using cue lure 11
- Male caught using trimedlure or terpinyl acetate.
. . . . most *Ceratitis* spp. (key on p. 92)

[Most *Ceratitis* spp. probably respond to one or both of these lures; a few are attracted to methyl eugenol. As the lure responses of many spp. have not yet been established further identification should be carried out using the key on p. 92.]

2 Cell cu*p* broad and the extension short (fig. 42). 1st flagellomere short (fig. 23). Wing pattern includes some coloured crossbands (fig. 216). Scutellum with 4 marginal setae.
. . . . *C. (Pardalaspis) punctata* (p. 301)
- Cell cu*p* very narrow and extension of cell cu*p* very long (fig. 40). 1st flagellomere (3rd segment of antenna) at least 3 times as long as broad (fig. 19). Wing pattern usually confined to a costal band and an anal streak (fig. 36) (except fig. 195). Scutellum usually with 2 marginal setae (except some *B. diversa*).
. . . . 3
3 Scutellum bilobed (fig. 199). Postpronotal lobes each with a single seta.
. . . . *Bactrocera (Notodacus) xanthodes* (p. 247)
- Scutellum not bilobed (e.g. fig. 182). Postpronotal lobes without any setae (sometimes with some small setulae or hairs).
. . . . 4
4 Wing with a pattern made up of three crossbands, each of which extends across the whole of the wing, from costal band to hind margin (fig. 195).
. . . . *B. (B.) umbrosa* (p. 236)
- Wing without any crossbands.
. . . . 5
5 Scutum without lateral yellow or orange stripes (vittae) (fig. 187).
. . . .*B. (B.) melanota* (p. 211)

- Scutum with lateral yellow or orange stripes (vittae) (e.g. fig. 182). 6

6 Face entirely yellow. Scutum with a medial yellow stripe (vitta). Abdominal tergite 3 without a pecten. . . . . *B. (Hemigymnodacus) diversa* (p. 244)

- Face marked with either a spot in each antennal furrow (figs 19, 66) or with a line (or lines) across the lower facial margin (fig. 65). Scutum without a medial yellow stripe (vitta). Abdominal tergite 3 with a pecten. 7

7 Face with dark transverse markings adjacent to the antennal furrows which usually join to form a line across the lower facial margin (fig. 65). . . . . *B. (B.) correcta* (p. 180)

- Face with a spot in each antennal furrow (figs 19, 66). 8

8 Wing without a costal band; cell sc often yellow, and apex of vein R_{4+5} often with a brown spot (fig. 196). 9

- Wing with a costal band at least from the end of vein Sc to just beyond the end of vein R_{4+5} (figs 182, 188). 10

9 Thorax and abdomen pale orange-brown to red-brown (fig. 196). . . . . *B. (B.) zonata* (p. 239)

- Thorax and abdomen dark orange-brown to black (fig. 194). . . . . *B. (B.) tuberculata* (p. 233)

10 Abdominal tergites 3-5 without a distinct black T-shaped mark, although some individuals may have a poorly defined narrow dark medial stripe and narrow dark markings across the base of tergite 3, forming a poorly defined T-shaped mark (fig. 188). . . . . *.B. (B.) musae* (p. 213)

- Abdominal tergites 3-5 with a distinct black T-shaped mark (fig. 182). . . . . *B. (B.) dorsalis* complex (key on p. 82)

[All pest (or potential pest) spp. in this complex are attracted to methyl eugenol and further identification should be carried out using the key on p. 82.]

11 Scutellum with 4 marginal setae (e.g. figs 205, 207). . . . . *B. (Zeugodacus)* group of subgenera (key on p. 84)

[Most spp. in this group which respond to a male lure are attracted to cue lure and further identification should be carried out using the key on p. 84.]

- Scutellum with only 2 marginal setae (the apical pair) (e.g. figs 198, 206). 12

12 Scutum with a medial yellow or orange stripe (vitta) and prescutellar acrostichal setae (figs 198, 206).
. . . . *B. (Zeugodacus)* group of subgenera (key on p. 84)

[Most spp. in this group which respond to a male lure are attracted to cue lure and further identification should be carried out using the key on p. 84.]

\- Scutum never with both a medial yellow/orange stripe (vitta) and prescutellar acrostichal setae.
. . . . 13

13 Scutum without prescutellar acrostichal setae. Abdomen with all tergites fused into a single plate, at most with smooth transverse lines marking the boundaries of each segment (view from side to check that no sclerites overlap the next).
. . . . 14

\- Scutum with prescutellar acrostichal setae. Abdomen with all tergites separate (view from side to see overlapping sclerites).
. . . . 16

14 Combined length of pedical and 1st flagellomere greater than face height (fig. 24). Abdominal tergites 1+2 longer than broad, giving a strongly wasp-waisted appearance (figs 219-221) (although all the tergites are fused, transverse shiny lines mark the area of each tergite).
. . . . *D. (Callantra)* (key on p. 88)

[All spp. in this group which respond to a male lure are attracted to cue lure and further identification should be carried out using the key on p. 88.]

\- Combined length of pedical and 1st flagellomere equal to, or less than, face height (similar to fig. 19). Abdominal tergites 1+2 broader than long (figs 222-229).
. . . . 15

15 Scutum with anterior supra-alar setae (figs 222-225).
. . . . *Dacus (Dacus)* (key on p. 89)

[All spp. in this group which respond to a male lure are attracted to cue lure and further identification should be carried out using the key on p. 89.]

\- Scutum without anterior supra-alar setae (figs 227).
. . . . *D. (Didacus) frontalis* (p. 332)

16 Scutum without anterior supra-alar setae (fig. 175).
. . . . *B. (Afrodacus) jarvisi* (p. 166)

\- Scutum with anterior supra-alar setae (as figs 177-196).
. . . . *B. (Bactrocera)* . . 17

17 Crossveins r-m and dm-cu covered by a single crossband (figs 177, 181, 184).
. . . . 18

\- Crossveins r-m and dm-cu not covered by a single crossband, at most with a short crossband covering crossvein r-m (fig. 180).
. . . . 20

18 Costal band distinct from wing base to apex, and broad enough to reach R_{4+5} (fig. 181). Crossband slightly 'stepped' to follow r-m, part of M and dm-cu. Scutellum uniformly orange-yellow.
. . . . *B. (B.) distincta* (p. 184)

- Costal band beyond vein R_1 very faint (figs 177, 184). Crossband evenly curved. Scutellum often with a triangular black mark, which may reach the apex of the scutellum.
. . . . 19

19 Postpronotal lobe predominantly black, sometimes with a small yellow area in posterior half (fig. 184).
. . . . *B. (B.) frauenfeldi* (p. 203)

- Postpronotal lobe with anterior third black, posterior two-thirds yellow (fig. 177).
. . . . *B. (B.) albistrigata* (p. 173)

20 Crossvein r-m covered by a short crossband (a spur from the costal band) (fig. 180).
. . . . *B. (B.) curvipennis* (p. 182)

- Crossvein r-m not covered by any marking.
. . . . 21

21 Scutum without lateral yellow or orange stripes (vittae) (figs 185, 190).
. . . . 22

- Scutum with lateral yellow or orange stripes (vittae) (e.g. fig. 193).
. . . . 23

22 Face with a dark spot in each antennal furrow (similar to figs. 19, 66).
. . . . *B. (B.) kirki* (p. 206)

- Face without any black markings.
. . . .*B. (B.) passiflorae* (p. 221)

[Some unusually dark specimens of *B. facialis* have very narrow and barely discernible lateral orange stripes (vittae), and these specimens will erroneously run to *B. passiflorae*.]

23 Face entirely yellow, without either a line across the lower facial margin or spots in the antennal furrows.
. . . .*B. (B.) facialis* (p. 201)

- Face with a spot in each antennal furrow (figs 19, 66).
. . . . 24

24 Wing entirely hyaline, at most with cell sc slightly yellow, or with faint infuscation along the costal edge and along crossveins r-m and dm-cu (fig. 191). [Dorsal surface of scutellum with a large black triangular mark, lateral and apical areas yellow.]
. . . .*B. (B.) psidii* (p. 225)

- Wing with a distinct costal band.
. . . . 25

25 Costal band only extending from the end of vein Sc to near wing apex, so that cells bc and c are hyaline (fig. 192).
. . . . *B. (B.) trivialis* (p. 227)

[Note that some non-pest members of the *B. dorsalis* complex are attracted to cue lure and these will also run to here.]

- Costal band extending from wing base to near wing apex, so that cells bc and c are coloured (figs 178, 189, 193).
. . . . 26

26 Postpronotal lobe brown, much darker in colour than the yellow lateral stripes (vittae) on the scutum (fig. 189).
. . . .*B. (B.) neohumeralis* (p. 218)

- Postpronotal lobe yellow, the same colour as the lateral stripes (vittae) on the scutum (figs 178, 193).
. . . . 27

27 Scutum and abdomen predominantly red-brown, except for postpronotal lobe, notopleura and lateral stripes (vittae) which are yellow (fig. 178). Abdomen predominantly red-brown, except for a pale yellowish area across the posterior part of syntergite 1+2.
. . . .*B. (B.) aquilonis* (p. 177)

- Scutum and abdomen predominantly black, except for postpronotal lobe, notopleura and lateral stripes (vittae) which are yellow (fig. 193). Abdomen varying from predominantly red-brown with a black T-shaped mark on tergites 2-5, to predominantly black.
. . . . *B. (B.) tryoni* (p. 229)

Keys to Fruit Pest Tephritidae; 2 - Third Instar Larvae

Many of the larval descriptions given in this work were based on an extensive study of tephritid larvae being carried out at the Queensland Department of Primary Industries (QDPI) (Elson-Harris, 1988; 1991; and other papers in prep.). That study is concentrating on Australian, South Pacific and tropical Asian species, but species from other regions are also being studied when suitable material is available. The present work includes descriptions of 34 species based on that study, plus descriptions, or partial descriptions, of a further 34 species that were derived from published data. The descriptions based on the QDPI study were all derived from scanning electron microscope (SEM) study of long series of specimens, unless otherwise indicated.

Although the keys and descriptions are illustrated by SEM photographs, the key has been written in such a way that most characters can be observed using optical microscope equipment (dissecting and compound microscopes), provided that larvae have been prepared in the manner described earlier in this work (p. 24). It should be stressed however that the available descriptions of many species may be inadequate, and that the larval stages of many pest species remain unknown. Larvae identified using these keys should be carefully checked against the descriptions given in the species accounts, and any identification of a species that has only a "partial description" should be regarded as tentative. The names of species which were not examined during this study are given in square brackets in the following keys.

Key to Genera Associated with Fruit of Economic Importance (Larvae)

The following key separates 12 of the 15 genera of fruit associated tephritids discussed in this work; the missing genera are *Capparimyia*, *Monacrostichus* and *Trirhithromyia*. Host data may assist in the recognition of *Capparimyia savastani* (p. 279), as its larvae only develop in the buds of caper (*Capparis spinosa*), and that host is not known to be attacked by other tephritids. The larvae of *Trirhithromyia* spp. are probably similar to those of *Ceratitis* spp. and any larvae found in Solanaceae fruits running to *Ceratitis* might be *T. cyanescens* (p. 393). *Monacrostichus citricola* (p. 347) is only known to attack citrus fruits and its larva will probably resemble a large *Bactrocera* sp. The major fruit pest genera most likely to be encountered by quarantine entomologists are named in ***BOLD CAPITALS***.

1 Posterior spiracles not on a raised prominence, and with 3 elongate slits, which are usually subparallel (pl. 6.e) (except in some *Rhagoletis* spp. which have the upper and lower slits at about 90°); anterior spiracles with a row of tubules, not retractile (pl. 6.d); mouthhooks usually without preapical teeth (figs 133-138, 140-155, 157-162) (known exceptions are *Dacus* spp. (fig. 159) with well developed preapical teeth, and *Bactrocera cucumis* (fig. 139), *B. cucurbitae* (fig. 156), *Ceratitis cosyra* and *Rhagoletis cerasi* which have vestigial preapical teeth). [A8 truncate and rest of body tapered to anterior end. Length at least 5mm.]
. . . . TEPHRITIDAE 2

\- Posterior spiracles on a raised prominence, and slits often arranged with the upper and lower slits at 180° (Lonchaeidae), or in a circle (*Atherigona* and Neriidae); or with anterior spiracles terminating in many long filamentous processes which can be retracted into the body (*Drosophila*); or mouthhooks with 1 or more well developed preapical teeth (usually only 2nd instar fruit associated Tephritidae, unless in flowers or fruits of Cucurbitaceae; fig. 131); or with only 2 posterior spiracular slits (1st instar Tephritidae; pl. 2.c), or 4 slits (Neriidae).
. . . . Families other than Tephritidae (p. 13), or not 3rd instar

[Unfortunately there is no simple method of separating 2nd and 3rd instars if larvae are removed from an attacked fruit, unless more than one size class is found. In general, 3rd instar larval mouthhooks lack preapical teeth, but some spp. have vestigial teeth, and *Dacus* spp. have well developed teeth. However, *Dacus* spp. are usually only found in cucurbit flowers and fruits, so any larva with well developed preapical teeth in other fruits is probably not a 3rd instar; morphological differences between the instars were more fully described earlier in this work (p. 36). Finding larvae which are emerging from a fruit to pupariate is the only simple practical method of being certain that the larvae are final instar.]

2 Larva large, more than 13.0mm long; A8 lacking prominent tubercles; anterior spiracles with 22-28 tubules; dorsal margin of ventral cornua prominently sclerotised. [In papaya (*Carica papaya*). New World tropics.]
. . . . [*Toxotrypana curvicauda* (p. 390)]

\- Larva usually smaller, less than 13.0mm long; if more than 13.0mm long, then not matching the above combination of characters.
. . . . 3

3 Larva 10.0-12.0mm long; all body segments encircled by spinules; with small scattered depressions laterally and ventrally of posterior spiracles. [In peppers (*Capsicum*). Southern North America.]
. . . . [*Zonosemata electa* (p. 399)]

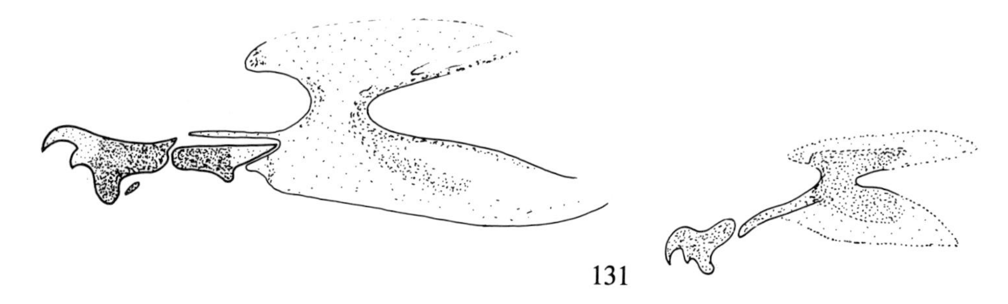

Figs 131-132. Cephalopharyngeal skeletons of early instar *Bactrocera* larvae; 131, 2nd instar; 132, 1st instar. Scale= 0.5mm.

- Larva usually less than 10.0mm long; if more than 10.0mm long, then not matching the above combination of characters. 4

4 With 1 or more preoral teeth near stomal sensory organ (pl. 37.b). 5

- Without preoral teeth. 7

5 Oral ridges unserrated (pl. 36.b). Area between posterior spiracles without a raised cellular pattern. [In the fruits of Berberidaceae, Caprifoliaceae, Cornaceae, Cupressaceae, Elaeagnaceae, Ericaceae, Grossulariaceae, Juglandaceae, Oleaceae, Rosaceae and Solanaceae. North temperate regions and South America.] . . . . ***RHAGOLETIS*** (p. 125)

- Oral ridges with long sharply pointed teeth. Area between posterior spiracles with an obvious raised cellular pattern. 6

6 Anterior spiracles with at least 27 tubules. Spiracular hairs rarely branched. [In fruits of Cucurbitaceae. Southern USSR, Middle East and the northern part of the Indian subcontinent.] . . . . [*Myiopardalis pardalina* (p. 349)]

- Anterior spiracles with less than 26 tubules. Spiracular hairs branched. [Species included in the present work develop in the fruits of jujube (*Ziziphus*). Southern Europe, Middle East and India.] . . . . [*Carpomya* (p. 122)]

7 In coffee fruit (*Coffea*); larvae 6.0mm long; head with 4-5 oral ridges; anterior spiracles with 7 tubules. [Africa.] . . . . [*Trirhithrum* (p. 395)]

[Only *T. nigerrimum* has a described larva but *T. coffeae* will probably also run here.]

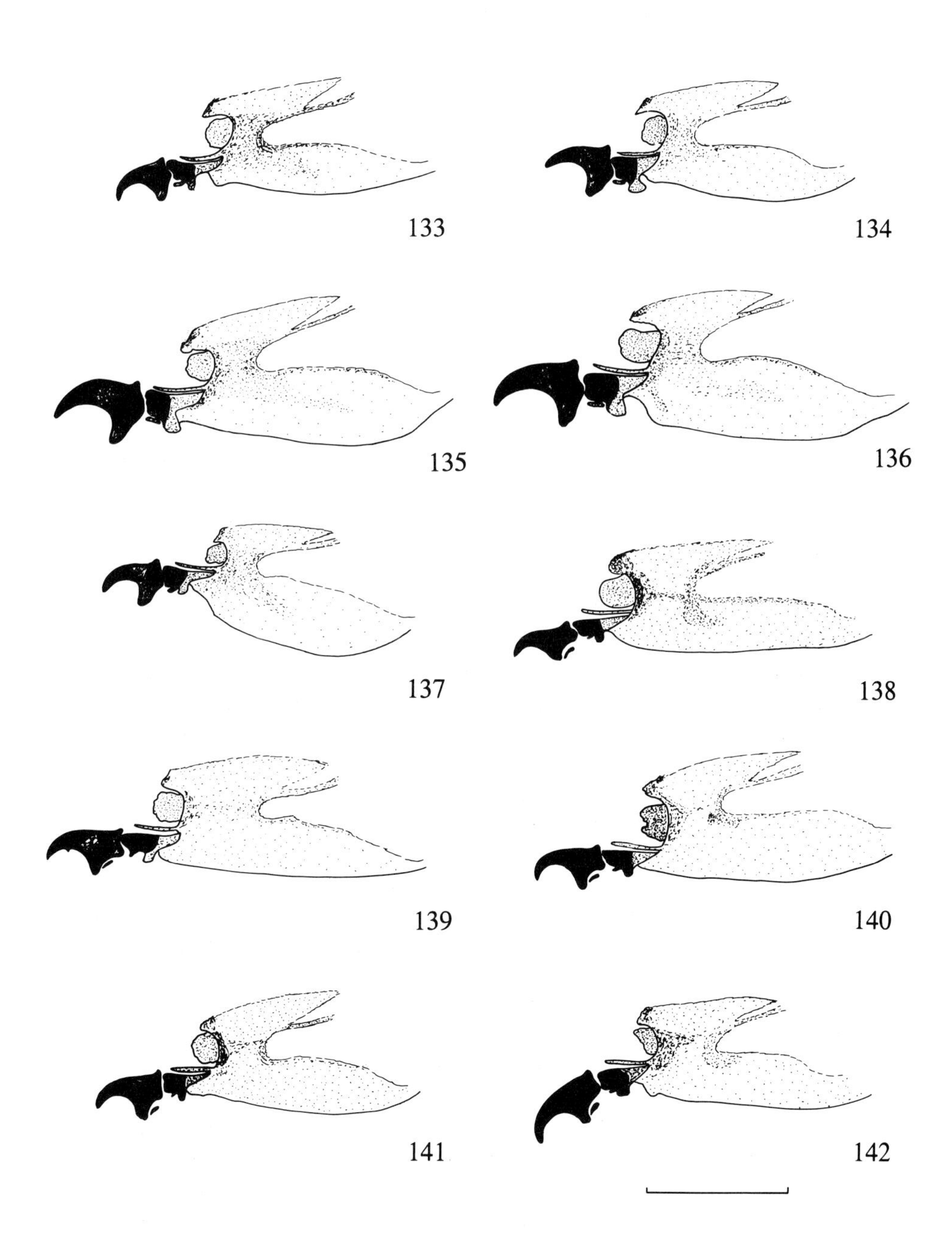

Figs 133-142. Cephalopharyngeal skeletons of 3rd instar larvae; 133, *Anastrepha fraterculus*; 134, *A. obliqua*; 135, *A. serpentina*; 136, *A. striata*; 137, *A. suspensa*; 138, *Bactrocera (Afrodacus) jarvisi*; 139, *B. (Austrodacus) cucumis*; 140, *B. (B.) albistrigata*; 141, *B. (B.) aquilonis*; 142, *B. (B.) dorsalis*. Scale= 1.0mm

- Usually not in coffee fruit; if in coffee, then not matching the above combination of characters.
. . . . 8

8 With accessory plates (pls 14.a, 18.b, 20.c).
. . . . 9

- Without accessory plates.
. . . . 12

9 Accessory plates long, interlocking with outer edge of oral ridges; each anal lobe unequally bilobed (pl. 33.f). [Australia.]
. . . . *Dirioxa* (p. 340)

- Accessory plates usually short; if long (pl. 32.c), then anal lobes not bilobed.
. . . . 10

10 Preapical teeth large (fig. 159; pl. 32.b). [Accessory plates long. In cucurbit and sometimes other fruits. Old World tropics.]
. . . . ***DACUS*** (p. 125)

- Preapical teeth very small (figs 139, 156; pl. 28.d) or absent (figs 133-138, 140-155, 157).
. . . . 11

11 Anterior spiracles concave centrally (pl. 5.c). Anal lobes often bilobed (pl. 4.e). [South America, West Indies and southern USA.]
. . . . ***ANASTREPHA*** (p. 116)

- Anterior spiracles not concave centrally (pl. 23.b). Anal lobes not bilobed (pl. 23.d). [Tropical and warm temperate areas of the Old World; adventive spp. are found in Hawaii and northern South America.]
. . . . ***BACTROCERA*** (p. 118)

12 Oral ridges with long, stout, sharply pointed teeth (pl 31.b). [In currants and gooseberry (*Ribes*). North America.]
. . . . *Epochra canadensis* (p. 344)

- Oral ridges with shorter, bluntly rounded teeth (pl. 31.d). [*Ceratitis capitata* is found in most tropical and warm temperate areas; other spp. are African.]
. . . . ***CERATITIS*** (p. 122)

Key to Some *Anastrepha* Species of Economic Importance (Larvae)

The following key to eight *Anastrepha* spp. provides an alternative to the very detailed key provided by Steck *et al.* (1990), which also included *A. distincta* and *A. leptozona*. Other pests or potential pests not included in the following key are unknown in their larval stages, and they are as follows: *A. antunesi*, *A. macrura*, *A. ocresia*, *A. ornata* and *A. pseudoparallela*. Host data may be of value in identifying *A. distincta*, as it normally only attacks the pods of *Inga* spp., which are seldom hosts of any other species.

1 Anterior spiracles with at least 26 tubules. [Anal lobes bifid. In fruits of cucurbits. Argentina to Panama.]
. . . . [*A. grandis* (p. 136)]

- Anterior spiracles with less than 25 tubules.
. . . . 2

2 Stomal sensory organ and preoral lobes with small stout spinules (pl. 7.c). [Posterior (lower) margins of oral ridges unserrated (pl. 7.a, b). Brazil to southern USA.]
. . . . *A. striata* (p. 155)

- Stomal sensory organ and preoral lobes without small stout spinules.
. . . . 3

3 Anal lobes grooved or bilobed (pls 4.e, 6.f).
. . . . 4

- Anal lobes not grooved or bilobed.
. . . . 7

4 Dorsal and ventral posterior spiracular hair bundles with only 6-9 hairs per bundle (pl. 6.e); oral ridges of 8-12 rows. [Argentina to southern USA.]
. . . . *A. serpentina* (p. 152)

- Dorsal and ventral posterior spiracular hair bundles usually with more than 9 hairs per bundle; if with less than 9 hairs (*A. ludens*) then oral ridges of 11-17 rows.
. . . . 5

5 Dorsal side of A1 and A2 with spinules. [In guava fruit (*Psidium*). Southern Brazil.]
. . . . [*A. bistrigata* (in part; p. 131)]

- Dorsal side of A1 and A2 without spinules.
. . . . 6

6 Oral ridges of 11-17 rows. Posterior spiracular hairs short; dorsal and ventral bundles with 6-13 hairs. [Costa Rica to southern USA.]
. . . . [*A. ludens* (p. 140)]

- Oral ridges of 7-10 rows (pl. 4.b). Posterior spiracular hairs long; dorsal and ventral bundles with 12-16 hairs (pl. 4.d). [Argentina to southern USA.]
. . . . *A. fraterculus* complex (in part; p. 133)

7 Dorsal side of A1 and A2 with spinules. [In guava fruit (*Psidium*). Southern Brazil.]
. . . . [*A. bistrigata* (in part; p. 131)]

- Dorsal side of A1 and A2 without spinules.
. . . . 8

8 Anal lobes with a heavily sculptured surface (pl. 5.e, f). Oral ridges and accessory plates with few serrations or teeth (pl. 5.a). [Argentina to southern USA.]
. . . . *A. obliqua* (p. 144)

- Anal lobes without a distinctly sculptured surface (sometimes with a lightly reticulate surface). Oral ridges with well defined teeth (pls 4.b, 8.b).
. . . . 9

9 Oral ridges with stout, tapered, bluntly rounded, widely spaced teeth (pl. 8.a, b). [West Indies and southern USA.]
. . . . *A. suspensa* (p. 158)
- Oral ridges with a few short, irregularly spaced, teeth (pl. 4.a, b). [Argentina to southern USA.]
. . . . *A. fraterculus* complex (in part; p. 133)

Key to Some *Bactrocera* Species of Economic Importance (Larvae)

The following key separates 22 of the 24 *Bactrocera* spp. that have at least partial descriptions in the species accounts given later in the present work. The species that could not be included here were *B. (B.) zonata* and *B. (Hemigymnodacus) diversa*. Pest and potential pest species whose larvae are unknown are as follows: *B. (B.) caryeae*, *B. (B.) correcta*, *B. (B.) curvipennis*, *B. (B.) distincta*, *B. (B.)* sp. near *B. dorsalis* (D) (see p. 186), *B. (B.) facialis*, *B. (B.) kirki*, *B. (B.) occipitalis*, *B. (B.) psidii*, *B. (B.) trivialis*, *B. (Paradacus) decipiens*, *B. (P.) depressa*, *B. (Paratridacus) atrisetosa*, *B. (Tetradacus) tsuneonis* and *B. (Zeugodacus) caudata*.

1 Large, 13.0-15.0mm long. [Oral ridges of 16-18 rows (pl. 27.a). Anterior spiracles, 17-19 tubules (pl. 27.c). In citrus fruits. China and northern Indian subcontinent.]
. . . . *B. (Tetradacus) minax* (p. 256)
- Smaller, less than 13mm long.
. . . . 2
2 Mouthhooks each with a small preapical tooth (figs 139, 156; pl. 28.d).
. . . . 3
- Mouthhooks without preapical teeth.
. . . . 4

[Note that *B. (Hemigymnodacus) diversa* (in cucurbit flowers and fruits; p. 244) will also run to couplet 3, but there is insufficient information to separate it from other spp.]

3 Mature larva with an obvious pigmented transverse line between the intermediate lobes. [Usually in cucurbit fruits. Tropical Asia to New Guinea; adventive in East Africa, Hawaii, Mauritius and Réunion.]
. . . . *B. (Zeugodacus) cucurbitae* (p. 263)
Mature larva rarely with a pigmented transverse line between the intermediate lobes. [Usually in cucurbit fruits. Australia.]
. . . . *B. (Austrodacus) cucumis* (p. 170)

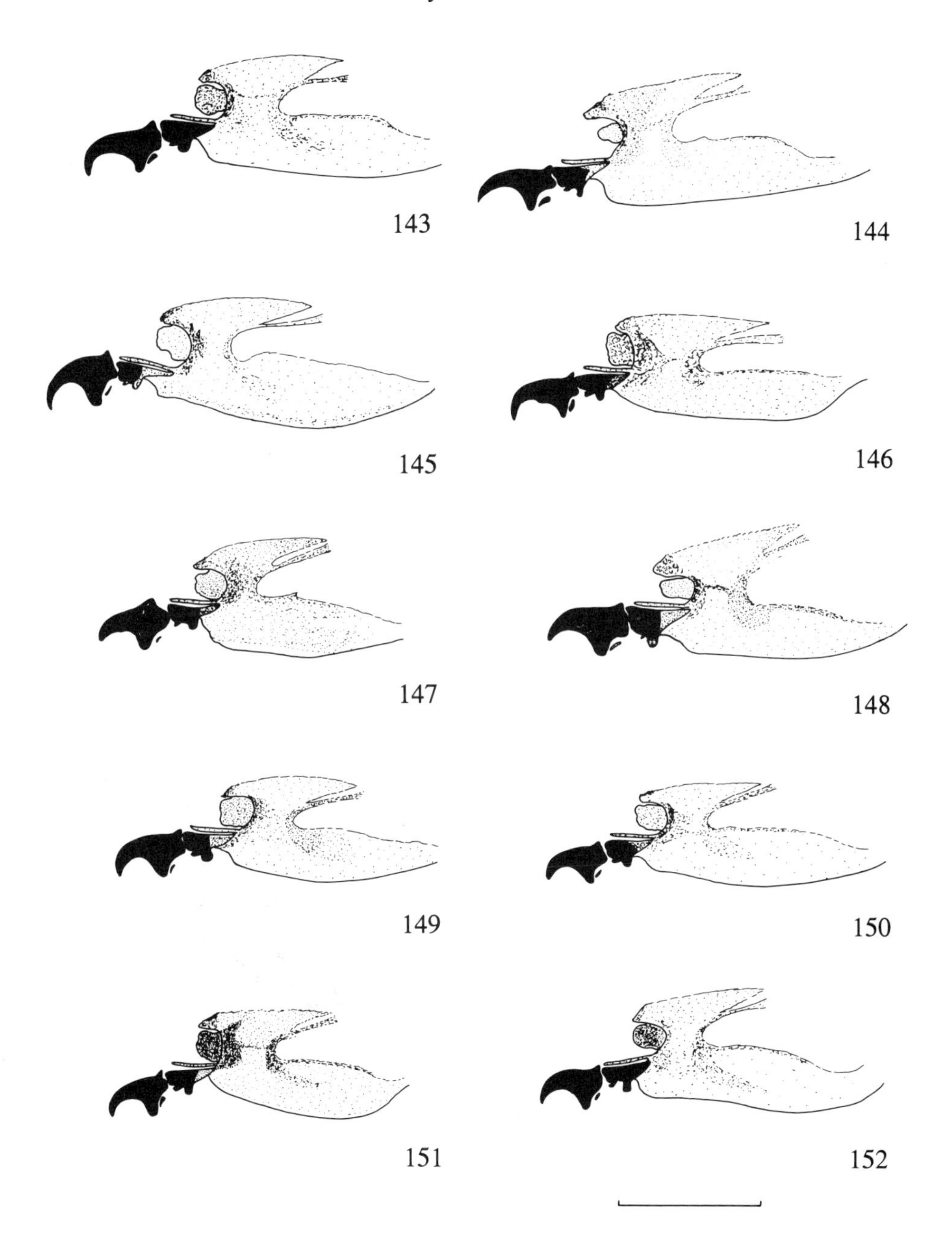

Figs 143-152. Cephalopharyngeal skeletons of 3rd instar larvae; 143, *Bactrocera* sp. near *B. (B.) dorsalis* (A); 144, *Bactrocera* sp. near *B. (B.) dorsalis* (B); 145, *Bactrocera* sp. near *B. (B.) dorsalis* (C); 146, *B. (B.) frauenfeldi*; 147, *B. (B.) latifrons*; 148, *B. (B.) melanota*; 149, *B. (B.) musae*; 150, *B. (B.) neohumeralis*; 151, *B. (B.) passiflorae*; 152, *B. (B.) tryoni*. Scale= 1.0mm.

4 In olive fruit (*Olea*); larva 6.5-7.0mm long; oral ridges of 10-12 rows; dorsal spinules on A1-A5. [Africa, and southern Europe to northern India.]
. . . . [*B. (Daculus) oleae* (p. 241)]

- Usually not in olive fruit; if in olive (records for *B. tryoni* and possibly other polyphagous spp.), then not matching the above combination of characters.
. . . . 5

5 Oral ridges with 17-21 rows of very short teeth (less than one-sixth depth of ridge) (pl. 25.b). Ventral surface of T1 with a heart-shaped pigmented patch. [Usually in the fruits of *Artocarpus* spp. South-east Asia to New Guinea and New Caledonia.]
. . . . *B. (B.) umbrosa* (p. 236)

- Oral ridges with longer teeth (more than one-sixth ridge depth; pls 9.a, 22.b). Ventral surface of T1 without a pigmented patch.
. . . . 6

6 Oral ridges with medium length (¼ to ½ ridge depth), sharply pointed, widely spaced teeth creating a jagged appearance (pl. 9.a). [Australia.]
. . . . *B. (Afrodacus) jarvisi* (p. 166)

- Oral ridges without sharply pointed teeth that have a jagged appearance.
. . . . 7

7 Oral ridges with at least 17 rows of medium length teeth (¼ to ½ depth of ridge; pls 26.b, 29.b). A8 of mature larva with an obvious pigmented transverse line between intermediate lobes; anal lobes surrounded by stout spinules. [T1 encircled by a wide (8-12 row) band of stout spinules (pls 26.c, 29.d).]
. . . . 8

- Oral ridges with less than 16 rows of teeth. A8 of mature larva without pigmented transverse line.
. . . . 9

8 Preoral lobes with serrated edges (pl. 29.c). [Usually in cucurbit fruits. Tropical Asia.]
. . . . *B. (Zeugodacus) tau* (p. 271)

- Preoral lobes without serrated edges. [South Pacific.]
. . . . *B. (Notodacus) xanthodes* (p. 247)

9 Oral ridges with long teeth (more than ½ ridge depth; pls 14.a, 22.b).
. . . . 10

- Oral ridges with short (pl. 13.b) to medium length (pl. 16.b) teeth (less than ½ ridge depth).
. . . . 15

10 Oral ridges with well spaced teeth (pl. 14.a). [Accessory plates usually shell-shaped (occasionally elongate), with strongly serrated edges (pl. 14.a).]
. . . . 11

- Oral ridges with closely spaced teeth (pl. 22.b).
. . . . 13

11 Teeth on oral ridges almost parallel-sided, with bluntly rounded tips (pl. 14.a). [Indonesia, Malaysia, southern Thailand or northern South America.]
. . . . sp. near *B. (B.) dorsalis* (A) (p. 192)

- Teeth on oral ridges obviously tapered, with gently rounded tips (pl. 18.a). 12

12 Preoral lobes with stout, well spaced teeth on some lobes; accessory plates shell-shaped (pl. 18.a). Usually in the fruits of Solanaceae. [Sri Lanka to Taiwan; adventive in Hawaii.]
. . . .*B. (B.) latifrons* (p. 208)

- Preoral lobes without stout teeth; accessory plates elongate (pl. 20.b). Usually in bananas (*Musa*). [Australia and New Guinea.]
. . . .*B. (B.) musae* (p. 213)

13 Larva at least 8.0mm long. Some sensilla on A8 long and tapering (pl. 22.c). [South Pacific (Fiji, Niue and Tonga).]
. . . .*B. (B.) passiflorae* (p. 221)

- Larva less than 8.0mm long. A8 without long tapering sensilla. [May be South Pacific, but not known from Fiji, Niue or Tonga.]
. . . . 14

14 Tubules in anterior spiracles 2 times as long as broad (pl. 11.c). Posterior spiracular slits 3.0 times as long as broad (pl. 11.d). [Malaysia and Indonesia (west of Irian Jaya).]
. . . . *B. (B.) albistrigata* (p. 173)

- Tubules in anterior spiracles almost as long as broad (pl. 17.d). Posterior spiracular slits 3.5-4.0 times as long as broad (pl. 17.e). [New Guinea, Australia and the South Pacific.]
. . . . *B. (B.) frauenfeldi* (p. 203)

15 Anal lobes surrounded by long, stout, sharply pointed spinules (pl. 15.f). Some preoral lobes with obviously serrated edges. [Indonesia, Malaysia and southern Thailand.]
. . . . sp. near *B. (B.) dorsalis* (B) (p. 194)

- Anal lobes surrounded by small fine spinules. Preoral lobes without serrated edges. 16

16 Oral ridges with medium length teeth (¼ to ½ ridge depth; pl. 16.b). 17

- Oral ridges with short teeth (less than ¼ ridge depth; pl. 13.b). 19

17 Teeth on oral ridges obviously tapered, and well spaced (pl. 16.b). Lateral spiracular hair bundles of 4-7 hairs (pl. 16.c). [Philippines.]
. . . . sp. near *B. (B.) dorsalis* (C) (p. 197)

- Teeth on oral ridges not as obviously tapered, and closely spaced (pls 21.b, 23.a). Lateral spiracular hair bundles of 5-9 hairs (pls 21.d, 23.e). 18

18 Posterior spiracular slits 3.0-3.5 times as long as broad (pl. 21.d). [Posterior spiracular hair bundles, with 16-20 hairs in dorsal and ventral bundles, and 6-12 hairs in lateral bundles. Australia and New Guinea.]
. . . .*B. (B.) neohumeralis* (p. 218)

- Posterior spiracular slits 2.5-3.0 times as long as broad (pl. 23.e). [Posterior spiracular hair bundles, with 12-17 hairs in dorsal and ventral bundles, and 5-9 in lateral bundles. Australia, New Guinea and South Pacific.]
  . . . . *B. (B.) tryoni* (p. 229)

19 Posterior spiracular hair bundles with few hairs (pl. 24.d); dorsal and ventral bundles with 11-14 hairs, lateral bundles with 4-7 hairs. [Burma to Vietnam.]
  . . . . *B. (B.) tuberculata* (p. 233)

- Posterior spiracular hairs numerous (pls 12.e, 13.d, 19.d); dorsal and ventral bundles with 16-24 hairs, lateral bundles with 6-12 hairs.
  20

20 Oral ridges with 8-10 rows of teeth (pl. 12.b). Posterior spiracular slits 3.0-3.5 times as long as broad (pl. 12.e). [Australia.]
  . . . .*B. (B.) aquilonis* (p. 177)

- Oral ridges with 11-15 rows of teeth (pls 13.b, 19.b). Posterior spiracular slits 2.5-3.0 times as long as broad (pls 13.e, 19.d). [Not known from Australia.]
  21

21 Posterior spiracular hairs with dorsal and ventral bundles of 18-24 hairs (pl. 19.d). South Pacific (only known from Cook Islands).
  . . . .*B. (B.) melanota* (p. 211)

- Posterior spiracular hairs with dorsal and ventral bundles of 17-20 hairs (pl. 13.e). Asia between India and Taiwan; Pacific (Guam or Hawaii).
  . . . . *B. (B.) dorsalis* (p. 187)

Key to Species of *Carpomya* Associated with *Ziziphus* spp.(Larvae)

The following key separates the two widely distributed species of *Carpomya* that are associated with jujube (*Ziziphus*) fruits.

1 Anterior spiracles with 20-23 short tubules. Oral ridges of 3 rows. [Italy, southern USSR to Thailand; adventive in Mauritius.]
  . . . . [*C. vesuviana* (p. 283)]

- Anterior spiracles with 15-17 tubules. Oral ridges of 4 rows. [North-east Africa, Middle East and Italy.]
  . . . . [*C. incompleta* (p. 281)]

Key to Some *Ceratitis* Species of Economic Importance (Larvae)

The following key to seven species is largely based on old and inadequate descriptions. A major omission from this key is *C. (Pterandrus) rosa*, whose larval stages are undescribed, despite its being a pest of quarantine importance. Other pest or potential pest species whose larvae are unknown are as follows: *C. (Ceratalaspis) quinaria*, *Ceratitis (C.) catoirii* and *C. (C.) malgassa*.

1 Anterior spiracles with at least 21 tubules. [Sub-Saharan Africa.]
. . . . [*C. (Pardalaspis) punctata* (p. 301)]
- Anterior spiracles with less than 20 tubules.
. . . . 2
2 Anterior spiracles with at least 15 tubules (pl. 31.e).
. . . . 3
- Anterior spiracles with less than 14 tubules (pl. 30.c).
. . . . 4
3 Larva 8.0-8.5mm long; width 1.5-1.6mm. [In cola fruits (*Cola*). Sub-Saharan Africa.]
. . . . [*C. (Pterandrus) colae* (p. 304)]
- Larva 7.0mm long; width 1.2mm. [Southern Africa and Madagascar.]
. . . . *C. (P.) pedestris* (p. 305)
4 Anterior spiracles with at least 11 tubules.
. . . . 5

[*C. (P.) rosa* probably fits here; Orian & Moutia (1960) illustrated an anterior spiracle with 11 tubules.]

- Anterior spiracles with less than 11 tubules (pl. 30.c).
. . . . 7
5 Larva more than 7.5mm long. Anterior spiracles with 11-13 tubules. [Sub-Saharan Africa.]
. . . . [*C. (P.) anonae* (p. 302)]
- Larva less than 7.5mm long. Anterior spiracles with 10-12 tubules.
. . . . 6
6 Each mouthhook with a small preapical tooth. [Southern and eastern Africa.]
. . . . [*C. (Ceratalaspis) cosyra* (p. 287)]
- Each mouthhook without a small preapical tooth. [In the fruits of *Rubus* spp. Sub-Saharan Africa.]
. . . . [*Ceratitis (P.) rubivora* (in part; p. 311)]
7 Anterior spiracles with 10-11 tubules. [Larva 7.0mm long; 2.0mm wide. In the fruits of *Rubus* spp. Sub-Saharan Africa.]
. . . . [*C. (P.) rubivora* (in part; p. 311)]
- Anterior spiracles usually with less than 10 tubules (pl. 30.c) (sometimes with 10). [Larva 6.6-9.0mm long; 1.2-1.5mm wide. Africa; adventive in most tropical and warm temperate areas.]
. . . . *C. (C.) capitata* (p. 291)

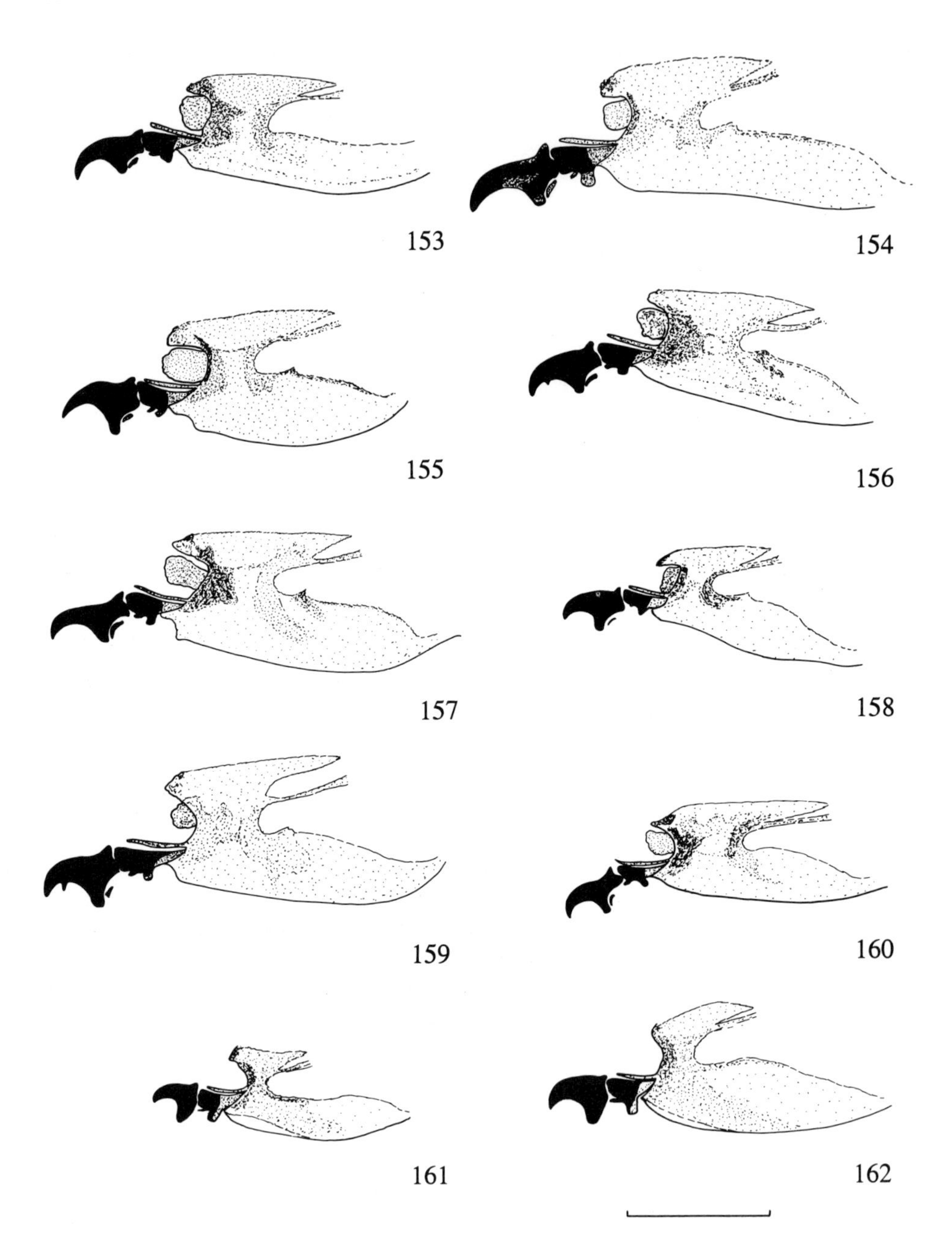

Figs 153-162. Cephalopharyngeal skeletons of 3rd instar larvae; 153, *Bactrocera (B.) tuberculata*; 154, *B. (B.) umbrosa*; 155, *B. (Notodacus) xanthodes*; 156, *B. (Zeugodacus) cucurbitae*; 157, *B. (Z.) tau*; 158, *Ceratitis (C.) capitata*; 159, *Dacus (Callantra) axanus*; 160, *Dirioxa pornia*; 161, *Rhagoletis mendax*; 162, *R. pomonella*. Scale= 1.0mm.

Key to Some *Dacus* Species of Economic Importance (Larvae)

The following key to three species, or species groups, is largely based on old and inadequate descriptions. *D. (D.) bivittatus* is the greatest omission and other pest, or potential pest, species whose larvae are still undescribed are as follows: *D. (Callantra) smieroides*, *D. (C.) solomonensis*, *D. (D.) demmerezi*, *D. (D.) punctatifrons* and *D. (D.) telfaireae*. Host data may be used for tentative identification of *D. telfaireae* as it only attacks oysternut (*Telfairea pedata*), and that is rarely attacked by other fruit flies.

1 Anterior spiracles with at least 23 tubules. Larva 13.0mm long. [In cucurbit fruits. Southern Africa.]
 . . . . [*D. (Didacus) lounsburyii* (p. 334)]

\- Anterior spiracles with less than 22 tubules. Larva less then 13.0mm long.
 2

2 Oral ridges with 17-21 rows of teeth (pl. 32.a). Anterior spiracles with 19-21 tubules (pl. 32.d). [In cucurbit fruits. Australia and New Guinea.]
 . . . . *Dacus (Callantra) axanus* (p. 314)

\- Oral ridges with 11-16 rows of teeth. Anterior spiracles with 14-19 tubules. [Usually in cucurbit fruits. Africa.]
 . . . . [*D. (Didacus) ciliatus* spp. group]

[The available details of *Dacus ciliatus*, *D. frontalis* and *D. vertebratus* were insufficient for separation of these spp.]

Key to Some *Rhagoletis* Species of Economic Importance (Larvae)

Rhagoletis larvae have not been sufficiently well described to allow construction of a key based primarily on morphology, and instead, host relationships provide better primary characters. However, that will cause a few misidentifications as exceptions to the normal pattern of host relationships may occur. This key only includes 13 of the major pest and potential pest species. Pests or potential pests whose larvae are still undescribed are as follows: *R. conversa*, *R. lycopersella*, *R. tabellaria* and *R. tomatis*. These and other *Rhagoletis* spp. may be tentatively identified from their host plants and the host list at the end of this work (p. 433) should be consulted.

1 In fruits of Solanaceae. [New World.]
 2

\- Not in fruits of Solanaceae. [North temperate regions of New and Old World, and Mexico.]
 3

[The larvae of *R. conversa*, *R. lycopersella* and *R. tomatis* will also be found in the fruits of Solanaceae, but they have not been described.]

2 Anterior spiracles with 33-35 tubules. [In husk tomato (*Physalis*). Mexico and Canada.]
. . . . [*R. striatella* (p. 381)]

\- Anterior spiracles with 23 tubules. [In the fruit of pepino (*Solanum muricatum*). Chile.]
. . . . [*R. nova* (p. 369)]

3 In walnut (*Juglans*) husks. [*R. completa* and *R. suavis* have also been known to attack peach (*Prunus persica*).]
. . . . 4

\- Not in walnut husks.
. . . . 6

4 Anterior spiracles with at least 23 tubules. Oral ridges of 9-12 rows. [Upper and lower slits of posterior spiracles at about 90° to each other. Eastern USA.]
. . . . [*R. suavis* (p. 382)]

\- Anterior spiracles with less than 22 tubules. Oral ridges of 7 rows.
. . . . 5

5 Anterior spiracles with at least 16 tubules; tubules in 2 groups. Upper and lower slits of posterior spiracles at about 60° to each other. [Western and central USA; adventive in Switzerland.]
. . . . [*R. completa* (p. 359)]

\- Anterior spiracles with less than 15 tubules; tubules in a single row. Upper and lower slits of posterior spiracles at about 90° to each other. [Mexico and south-western USA.]
. . . . [*R. juglandis* (p. 365)]

6 In cherry fruit (*Prunus*). [*R. indifferens* may also attack some plum (*Prunus*) species.]
. . . . 7

\- Not in cherry fruit. [In Utah *R. pomonella* has adapted to attacking sour cherry (*Prunus cerasus*).]
. . . . 10

7 Anterior spiracles with at least 21 tubules. [Eastern USA and southern Quebec.]
. . . . [*R. cingulata* (p. 357)]

\- Anterior spiracles with less than 20 tubules.
. . . . 8

8 Larva less than 7.0mm long. [Northern and central Europe, and east to western Siberia and Kazakhstan.]
. . . . *R. cerasi* (p. 353)

\- Larva more than 7.0mm long. [North America; *R. indifferens* is also adventive in Switzerland.]
. . . . 9

9 Posterior spiracular slits 4.0 times as long as broad (requires confirmation). [Western and northern USA, and southern Canada.]
. . . . [*R. fausta* (p. 363)]

- Posterior spiracular slits 3.5 times as long as broad (requires confirmation). [North-western USA and southern British Columbia; also adventive in Switzerland.]
  . . . . [*R. indifferens* (p. 358)]

10 In *Cornus* sp. or cranberry (*Vaccinium macrocarpon*). [Larva not described. North America; only attacks *Vaccinium* in western North America.]
  . . . . [*R. tabellaria* (p. 384)]

- In other Ericaceae, particularly blueberry (*Vaccinium*). [Eastern USA and south-eastern Canada.]
  . . . . *R. mendax* (p. 371)

- In apple (*Malus*). [Eastern half of the USA, south-eastern Canada; it has recently become established in Oregon and southern Washington.]
  . . . . *R. pomonella* (p. 375)

- In currant or gooseberry (*Ribes*). [North-western USA and southern British Columbia.]
  . . . . [*Rhagoletis ribicola* (p. 378)]

Species Accounts;
1 - Fruit Pests

About 1500 (38%) of the described species of Tephritidae are probably fruit associated, although this section includes almost all of the pest species. The larvae of almost all Dacinae (e.g. *Bactrocera*, *Ceratitis* and *Dacus* spp.) and most Trypetinae (e.g. *Anastrepha* and *Rhagoletis* spp.) develop in fruit. The few *Bactrocera* and *Dacus* spp. whose larvae develop in flowers, and *Capparimyia* spp. whose larvae develop in flower buds, are also included in this section. They were included here because there is no clear taxonomic division between flower, fruit and bud associated species in those groups, and some cucurbit associated *Bactrocera* spp. may attack either flowers or fruit.

GENUS *Anastrepha* Schiner

(figs 18, 35, 39)

Taxonomic notes

Some species have formerly been placed in the genera *Acrotoxa* Loew, *Lucumaphila* Stone, *Phobema* Aldrich and *Pseudodacus* Hendel, which are synonyms of *Anastrepha*.

Distribution and biology

A genus of about 180 species which attack a wide range of fruits in South and Central America, the West Indies and the extreme south of the USA. No *Anastrepha* spp. have become established in the Old World, although a few individuals have been intercepted in fruit arriving at European ports and some species could become established in the Old World tropics. Most of the pest species attack fruits belonging to several unrelated families of plants. The larvae of most *Anastrepha* spp. of known biology develop in the flesh of the fruit, but a few species are known to be able to feed on the seeds, either in part or exclusively on seeds (Norrbom & Kim, 1988b). It has been suggested that many of the species with an exceptional long aculeus may have seed feeding larvae, the long ovipositor allowing eggs to be placed deep within the fruit (Norrbom & Kim, 1988b). The major pest species are multivoltine and mating can take place away from suitable hosts, for example in the top of the tallest tree in the area (Fletcher, 1989b; 1989c).

Diagnosis of third instar larva
Antennal sensory organ with a broad, stout basal segment and a smaller cone-shaped distal segment; maxillary sensory organ large, flat with 2 distinct groups of sensilla surrounded by small cuticular folds (pl. 4.a); preoral lobes fused forming a stout bridge above mouthhooks; oral ridges arranged in 6-17 short rows with posterior margins serrated or unserrated; accessory plates small, shell-like, often unserrated. Mouthhooks each without a preapical tooth (figs 133-137). Anterior spiracles pale amber, concave centrally, with 9-37 long, stout, finger-like tubules on a large protuberant base. Anal lobes large, protuberant; not grooved, grooved or bilobed; sometimes with a sculptured surface.

Literature
The adults of most species can be identified using a pictorial key by Steyskal (1977). The key by Stone (1942) is a conventional key that is sometimes easier to follow than Steyskal's pictorial key. The larvae of thirteen species were separated in a key by Steck *et al.* (1990) and the six most serious pest species were also keyed by Berg (1979). A list of host records was produced by Norrbom & Kim (1988b), and there have been recent reviews of *Anastrepha* taxonomy by Norrbom & Foote (1989) and pest status by Enkerlin *et al.* (1989). A modern revision of the genus is being carried out in a series of papers (Norrbom & Kim, 1988a; Norrbom, 1991).

Anastrepha antunesi Lima

(figs 54, 163)

Pest status and commercial hosts
A pest of hog-plum (*Spondias mombin*) (Norrbom & Foote, 1989), which has also been recorded from common guava (*Psidium guajava*), ketembilla (*Dovyalis hebecarpa*), marmalade-box (*Genipa americana*), red mombin (*S. purpurea*) and sapodilla (*Manilkara zapota*) (Norrbom & Kim, 1988b).

Adult identification
A species with a typical *Anastrepha* type wing pattern (fig. 163) which is easily separated from other species by the small number of large tooth-like serrations on the aculeus apex (fig. 54).

Distribution
Central America: Costa Rica, Panama.
South America: Brazil, Peru, Venezuela.
West Indies: Trinidad.

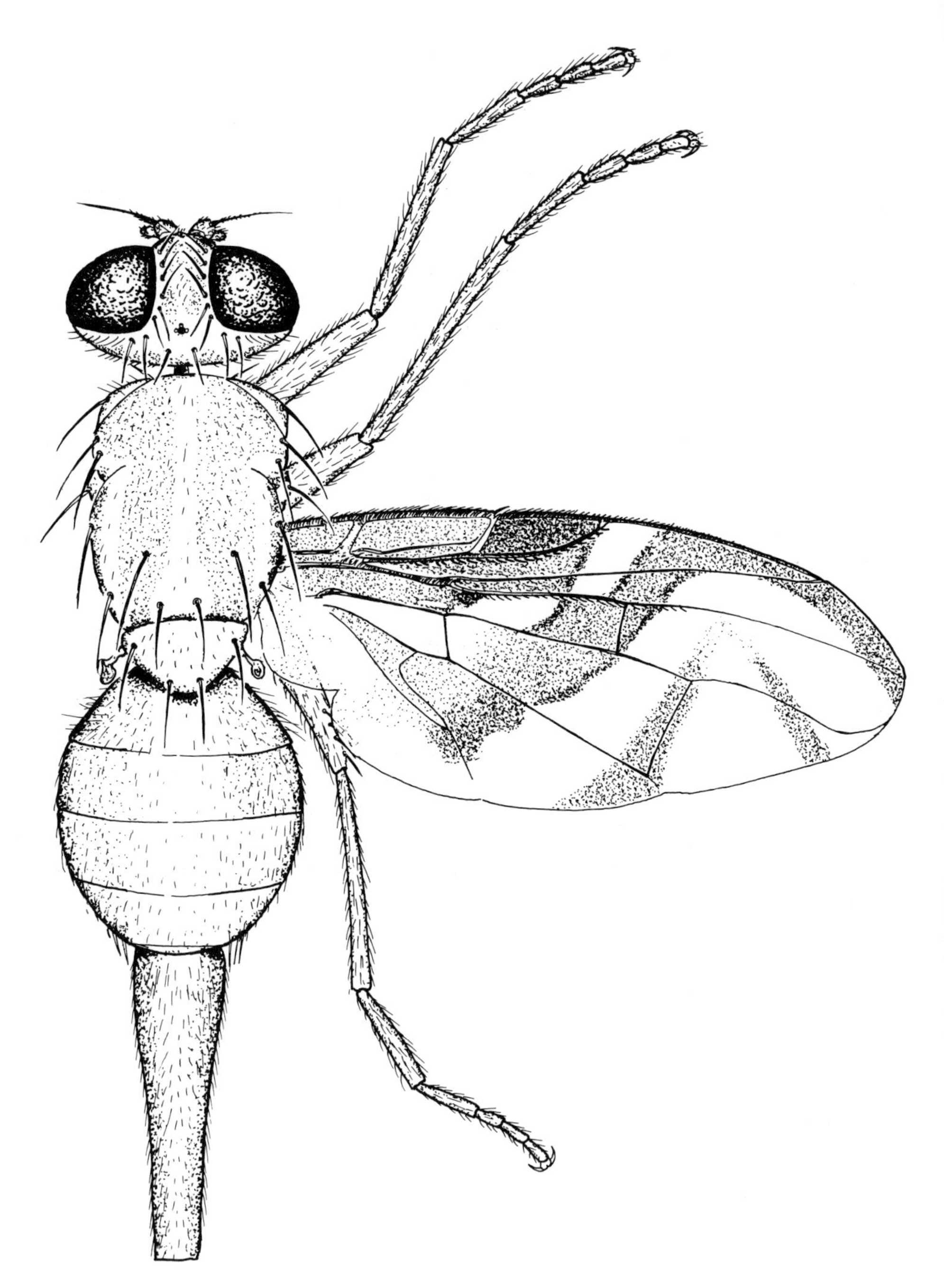

Fig. 163. *Anastrepha antunesi*, adult female.

Anastrepha bistrigata Bezzi

(fig. 50)

Pest status and commercial hosts
A pest of common guava (*Psidium guajava*) (Norrbom & Foote, 1989) which has also been recorded from Brazilian guava (*P. guineense*) (Norrbom & Kim, 1988b).

Adult identification
A species with a typical *Anastrepha* type wing pattern (similar to fig. 173) which is separated from other species by the combination of a long (over 3.0mm) aculeus with a non-serrate apex (fig. 50) and the U-shaped pattern on the scutum.

Description (not original) of third instar larva
Larvae medium-sized, length 6.0-9.6mm; width 1.1-2.2mm. *Head*: Stomal sensory organ small, rounded; oral ridges in 7-10 rows; mouthhooks moderately to heavily sclerotised, each with a long slender strongly curved apical tooth. *Thoracic and abdominal segments*: Spinules forming small, encircling, discontinuous rows on most segments; T1 with 3-5 rows dorsally, 6-8 rows laterally and 2-4 rows ventrally; T2 with 4-5 rows dorsally and laterally, and 3-6 rows ventrally; T3 with 3-4 rows encircling segment; A1 with 1-3 rows dorsally and laterally, and 6-9 rows ventrally; A2-A8 with 0-3 rows dorsally and laterally and 8-13 rows ventrally; A8 with 7 pairs of tubercles bearing sensilla. *Anterior spiracles*: Irregular row of 13-20 tubercles. *Posterior spiracles*: Spiracular slits 3.3-4.4 times as long as broad, with heavily sclerotised rimae; spiracular hairs short (less than length of a spiracular slit), most hairs branched in apical third; dorsal and ventral hair bundles of 14-21 hairs, lateral hair bundles of 6-14 hairs. *Anal area*: Lobes very large, protuberant, usually bifid (occasionally entire) and surrounded by a few small discontinuous rows of spinules. *Source*: Based on Steck & Malavasi (1988).

Distribution
South America: Southern Brazil; possibly Peru.

Other references
BEHAVIOUR, Silva *et al.* (1985); LARVA: Steck *et al.* (1990); PHEROMONES, Teles & Polloni (1989).

Anastrepha distincta Greene

(fig. 55)

Inga Fruit Fly

Taxonomic notes

This species has also been known as *Anastrepha silvai* Lima.

Pests status and commercial hosts

A pest of *Inga* spp. (Fabaceae) (Norrbom & Foote, 1989), including caspirol (*I. laurina*), guavo real (*I. spectabilis*), ice cream bean (*I. edulis*) and paterna (*I. paterna*); it has also been recorded from mango (*Mangifera indica*) and star-apple (*Chrysophyllum cainito*) (Norrbom & Kim, 1988b).

Wild hosts

Recorded from a wide range of *Inga* spp., plus *Eugenia nesiotica* Standley (Norrbom & Kim, 1988b).

Adult identification

One of many difficult to recognise species that have a complete *Anastrepha* type wing pattern (similar to fig. 166), and a relatively long aculeus (2.1-3.4mm) with a finely serrate apex (fig. 58); see key for details.

Distribution

Central America: Costa Rica, Guatemala, Mexico, Panama.
North America: USA (Texas).
South America: Brazil, Colombia, Guyana, Peru, Venezuela.

Other references

BRAZIL, Nascimento & Zucchi (1981); LARVA: Steck *et al.* (1990); MEXICO, Malo *et al.* (1987); PERU, Herrera & Vinas, 1977.

Anastrepha fraterculus (Wiedemann) Species Complex

South American Fruit Fly

(figs 18, 47, 56, 133; pl. 4)

Taxonomic notes

This is a species complex that has not yet been studied in sufficient detail to permit a clear separation of the included species. Recent work by Steck (1991), indicated that in Venezuela, Andean and lowland populations are distinct species, and populations from southern and north-eastern Brazil also have marked genetic differences.

Members of this species complex have also been known as *Acrotoxa fraterculus* (Wiedemann), *Anastrepha braziliensis* Greene, *A. costarukmanii* Capoor, *A. peruviana* Townsend, *A. pseudofraterculus* Capoor, *A. scholae* Capoor, *A. soluta* Bezzi, *Anthomyia frutalis* Weyenburgh, *Dacus fraterculus* Wiedemann, *Tephritis mellea* Walker, *Trypeta fraterculus* (Wiedemann) and *T. unicolor* Loew. The recently described *Anastrepha sororcula* Zucchi is also a member of this species complex (p. 164).

Pest status and commercial hosts

This species complex includes pests of citrus (*Citrus* spp.), common guava (*Psidium guajava*), *Eugenia*, *Prunus* and *Syzygium* spp. (Norrbom & Foote, 1989). Members of this complex have also been recorded from the following hosts (Norrbom & Kim, 1988b): Andean walnut (*Juglans neotropica*), Andes berry (*Rubus glaucus*), apple (*Malus domestica*), apricot (*Prunus armeniaca*), arabica coffee (*Coffea arabica*), avocado (*Persea americana*), Brazil cherry (*Eugenia brasiliensis*), Brazilian guava (*Psidium guineense*), carambola (*Averrhoa carambola*), cashew (*Anacardium occidentale*), cherimoya (*Annona cherimola*), cocoa (*Theobroma cacao*), common fig (*Ficus carica*), English walnut (*J. regia*), feijoa (*Feijoa sellowiana*), grapefruit (*Citrus x paradisi*), hog-plum (*Spondias mombin*), ice cream bean (*Inga edulis*), Japanese persimmon (*Diospyros kaki*), Jew plum (*Spondias cytherea*), ketembilla (*Dovyalis hebecarpa*), liberica coffee (*Coffea liberica*), loquat (*Eriobotrya japonica*), lucmo (*Pouteria obovata*), Malay-apple (*Syzygium malaccense*), mango (*Mangifera indica*), naranjilla (*Solanum quitoense*), peach (*Prunus persica*), pear (*Pyrus communis*), pomegranate (*Punica granatum*), pummelo (*Citrus maxima*), quince (*Cydonia oblonga*), red mombin (*Spondias purpurea*), rose-apple (*Syzygium jambos*), round kumquat (*Fortunella japonica*), sapodilla (*Manilkara zapota*), sour orange (*Citrus aurantium*), soursop (*Annona muricata*), strawberry (*Fragaria vesca*), strawberry guava (*Psidium littorale*), sugar-apple (*A. squamosa*), Surinam cherry (*Eugenia uniflora*), sweet lime (*Citrus limetta*), sweet orange (*C. sinensis*), tangerine (*C. reticulata*), tropical almond (*Terminalia catappa*) and wine grape (*Vitis vinifera*). The following additional hosts require confirmation: citron (*C. medica*) and plum (*Prunus domestica*) (USDA, *Pests not known to occur in the United States or of limited distribution*, No.18, undated).

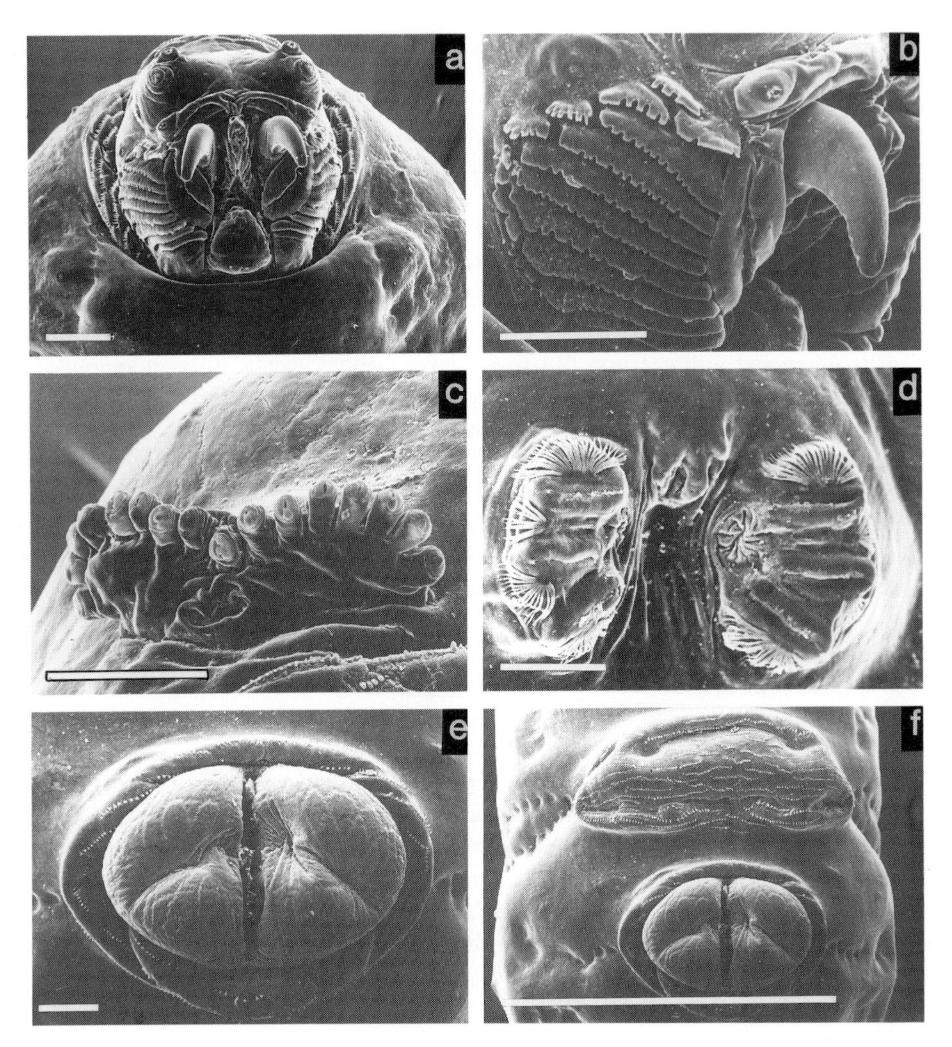

Plate 4. *Anastrepha fraterculus*, SEMs of 3rd instar larva; a, head, ventral view (scale=0.1mm); b, oral ridges and accessory plates, lateral view (scale=0.1mm); c, anterior spiracle (scale=0.1mm); d, posterior spiracles (scale=0.1mm); e, anal lobes (scale=0.1mm); f, A8, creeping welt and anal lobes (scale=1.0mm).

Wild hosts

Members of this species complex have been recorded from numerous species of Annonaceae, Euphorbiaceae, Myrtaceae, Olacaceae, Rosaceae, Sapotaceae and Staphyleaceae (Norrbom & Kim, 1988b).

Adult identification

This complex belongs to a wider group of difficult to recognise species that have a complete *Anastrepha* type wing pattern (similar to fig. 168) and a relatively short aculeus (1.4-1.7mm) with a finely serrate apex (fig. 56); see key for further details.

[Stone (1942) said the aculeus may be up to 1.95mm long but we could not confirm that from specimens available to us].

Description of third instar larva

Larvae medium-sized, length 8.0-9.5mm; width 1.4-1.8mm. *Head*: Stomal sensory organ rounded, protuberant with 2-3 peg-like sensilla; oral ridges (pl. 4.a, b) of 7-10 short rows of ridges with irregular serrations along posterior margins; accessory plates small, anterior plates with serrated edges; mouthhooks heavily sclerotised, each with a strongly curved slender apical tooth (fig. 133). *Thoracic and abdominal segments*: T1 with an anterior, broad, encircling band of 4-11 discontinuous rows of small, sharply pointed spinules; T2 and T3 with 3-7 rows of smaller spinules encircling each segment. Dorsal spinules occasionally on A1-A3 but absent from A4-A8. Creeping welts (pl. 4.f) with 7-12 rows of small spinules. A8 with dorsal tubercles and sensilla well developed; intermediate areas obvious, with large sensilla; ventral sensilla present. *Anterior spiracles* (pl. 4.c): 14-18 tubules. *Posterior spiracles* (pl. 4.d): Spiracular slits with heavily sclerotised, dark brown rimae; about 3 times as long as wide; with large spiracular hair bundles; dorsal and ventral bundles of 12-16 long hairs, many branched in apical third; lateral bundles of 6-9 hairs similarly branched. *Anal area* (pl. 4.e, f): Lobes large, protuberant; not grooved, grooved or bilobed; surrounded by 2-4 discontinuous rows of small, sharp spinules. *Source*: Based on 5 specimens from Ecuador (ex *Annona cherimola*).

Distribution

Central America: Belize, Costa Rica, Guatemala, Honduras, Mexico, Nicaragua, Panama.
North America: USA (southern Texas).
South America: Argentina, Bolivia, Brazil, Colombia, Ecuador (including the Galapagos Islands), Guyana, Peru, Suriname, Uruguay, Venezuela; there are records from Chile but it is not currently established there (A.L. Norrbom, pers. comm., 1991).
West Indies: Trinidad and Tobago.
A distribution map was produced by CIE (1958a).

Other references

BEHAVIOUR, Barros *et al.* (1983), Katsoyannos (1989b), Morgante *et al.* (1983) and Souza *et al.* (1983); BIBLIOGRAPHY, Zwölfer (1985); BIOLOGY, Fletcher (1989b); DATASHEET, Weems (1980); ECOLOGY, Espinosa (1980); HOST SELECTION, Malavasi *et al.* (1983); LARVA, Berg (1979), Steck *et al.* (1990) and Weems (1980); MEXICO, Aluja *et al.* (1987b) and Malo *et al.* (1987); PARASITOIDS, Narayanan & Chawla (1962) and Loiacono (1981); PERU, Herrera & Vinas (1977); PHEROMONES, Averill & Prokopy (1989a) and Nation (1981; 1989a); TRAPPING, Katsoyannos (1989b).

Anastrepha grandis (Macquart)

South American Cucurbit Fruit Fly

(figs 53, 164)

Taxonomic notes

This species has also been known as *Acrotoxa grandis* (Macquart), *Anastrepha latifascia* Hering, *A. schineri* Hendel, *Tephritis grandis* Macquart and *Trypeta grandis* (Macquart).

Pest status and commercial hosts

A pest of cucurbits (Norrbom & Foote, 1989), which has been recorded from cucumber (*Cucumis sativus*), pumpkin (*Cucurbita maxima*, *C. moschata* and *C. pepo*), watermelon (*Citrullus lanatus*) and white-flowered gourd (*Lagenaria siceraria*); it has also been known to attack common guava (*Psidium guajava*) (Norrbom & Kim, 1988b) but this was almost certainly an aberrant host association (Norrbom, 1991).

Adult identification

Easily recognised by its very large size (wing length 10mm), long aculeus (over 5mm; fig. 53) and rather diffuse wing markings which are not of the typical *Anastrepha* type (fig. 164).

Description (not original) of third instar larva

Larvae medium-sized, length 6.6-7.0mm; width 1.6-2.7mm. *Head*: Stomal sensory organ small; oral ridges of 8-13 rows; mouthhooks heavily sclerotised, each with a large strongly curved apical tooth. *Thoracic and abdominal segments*: Spinules forming small, encircling discontinuous rows on most segments; T1 with 6-8 big rows dorsally, increasing to 7-10 rows ventrally; T2 and T3 with 5-7 and 4-6 rows respectively, encircling segments; A1 with 2-5 rows dorsally increasing to 7-11 rows ventrally; A2-A5 with scattered spinules dorsally, increasing to 16-24 rows ventrally; A6-A8 without spinules dorsally, but 16-24 rows ventrally; A8 with tubercles not well developed but sensilla very obvious. *Anterior spiracles*: Irregular row of 28-37 tubules. *Posterior spiracles*: Spiracular slits 3.0-5.3 times as long as broad, each with a heavily sclerotised rima; spiracular hairs short (less than half the length of a spiracular slit), sometimes branched in apical third; dorsal and ventral hair bundles of 11-22 hairs; lateral hair bundles of 6-13 hairs. *Anal area*: Lobes bifid, prominently protruding ventrally, with stout spinules in 4-8 anterior rows and an irregular patch of 4-8 posterior rows. *Source*: Based on Steck & Wharton (1988).

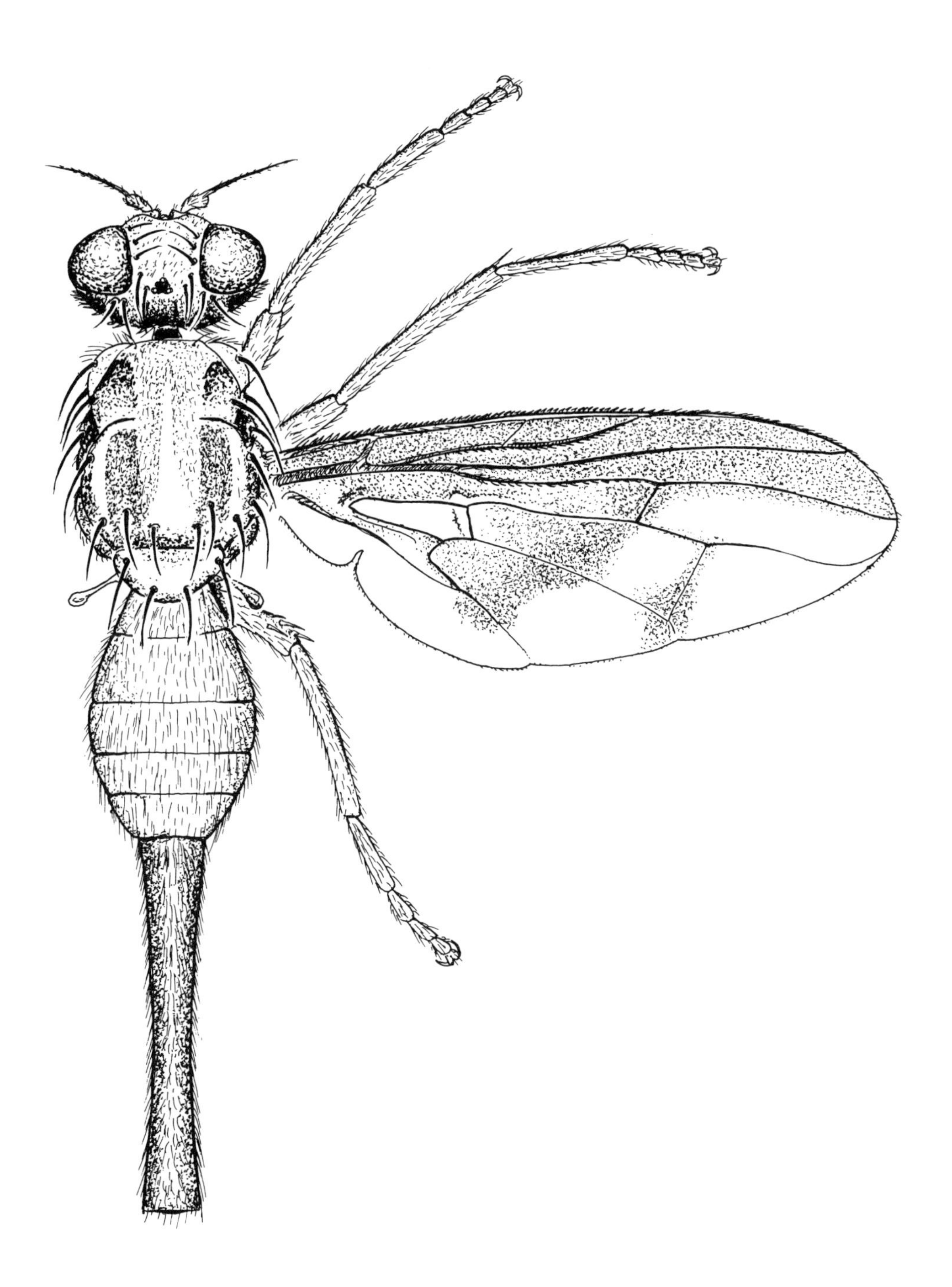

Fig. 164. *Anastrepha grandis*, adult female.

Distribution

Central America: Possibly Panama.
South America: Argentina, Bolivia, Brazil (except north), Colombia, Ecuador, Paraguay, Peru, Venezuela.

Other references

BAITS, Malavasi *et al.* (1990); DATASHEET, Whittle & Norrbom (1987); LARVA, Steck *et al.* (1990) and Whittle & Norrbom (1987); SEPARATION FROM SIMILAR SPP., Norrbom (1991).

Anastrepha leptozona Hendel

(figs 57, 165)

Commercial hosts

Most known hosts (wild and commercial) are species of Sapotaceae and it is recorded from abiu (*Pouteria caimito*), egg-fruit tree (*P. campechiana*) and star-apple (*Chrysophyllum cainito*) (Norrbom & Kim, 1988b). This is a potential pest of cultivated Sapotaceae spp.

Wild hosts

Recorded from a wild species of Sapotaceae (*Micropholis mexicana* Gilly) and Rosaceae (*Crataegus* sp.) (Norrbom & Kim, 1988b).

Adult identification

One of many difficult to recognise species that have a complete *Anastrepha* type wing pattern, but with the apical curve of vein M particularly strong (fig. 165); aculeus long (2.3-3.1mm) with a finely serrate apex (fig. 57); see key for details.

Distribution

Central America: Guatemala, Mexico and Panama.
South America: Bolivia, Brazil, Guyana and Venezuela.

Other reference

LARVA, Steck *et al.* (1990).

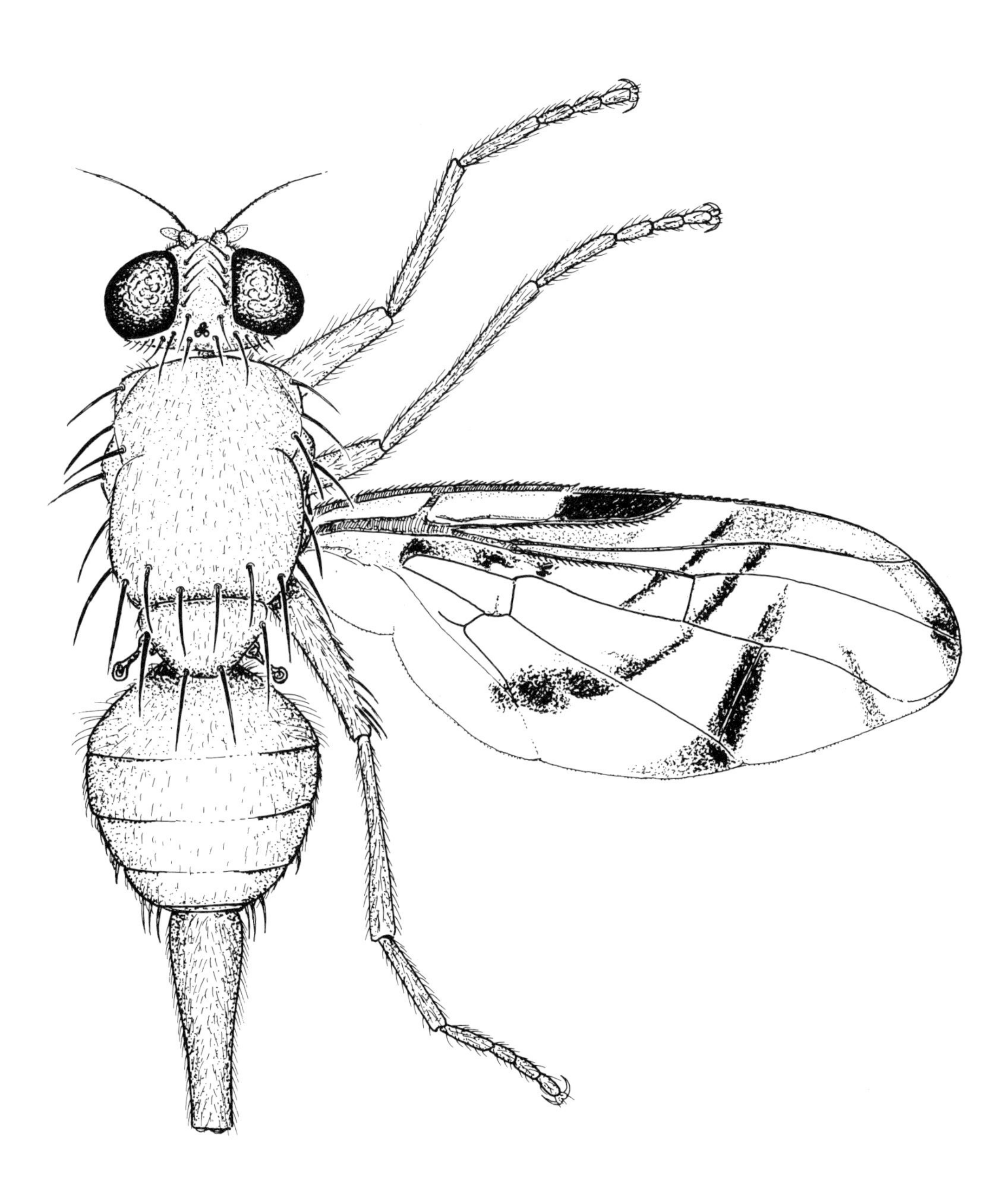

Fig. 165. *Anastrepha leptozona*, adult female.

Anastrepha ludens (Loew)

Mexican Fruit Fly

(figs 48, 58, 166)

Taxonomic notes

This species has also been known as *Acrotoxa ludens* Loew and *Trypeta ludens* (Loew).

Pest status and commercial hosts

A pest of citrus (*Citrus* spp.), mango (*Mangifera indica*) and peach (*Prunus persica*) (Norrbom & Foote, 1989), although it does not appear to attack citrus or mango in Costa Rica (Jiron *et al.*, 1988). *A. ludens* has also been recorded from the following hosts by Norrbom & Kim (1988b): apple (*Malus domestica*), arabica coffee (*Coffea arabica*), avocado (*Persea americana*), cashew (*Anacardium occidentale*), cherimoya (*Annona cherimola*), citron (*Citrus medica*), common guava (*Psidium guajava*), custard-apple (*A. reticulata*), feijoa (*Feijoa sellowiana*), grapefruit (*C. x paradisi*), Japanese persimmon (*Diospyros kaki*), lime (*C. aurantiifolia*), mammy-apple (*Mammea americana*), matasano (*Casirmiroa tetrameria*), papaya (*Carica papaya*), paradise apple (*Malus pumila*), pear (*Pyrus communis*), pomegranate (*Punica granatum*), pummelo (*Citrus maxima*), purple granadilla (*Passiflora edulis*), quince (*Cydonia oblonga*), red mombin (*Spondias purpurea*), rose-apple (*Syzygium jambos*), sapote (*Pouteria sapota*), sour orange (*Citrus aurantium*), strawberry guava (*Psidium littorale*), sugar-apple (*A. squamosa*), sweet lime (*C. limetta*), sweet orange (*C. sinensis*), tangelo (*C. x tangelo*), tangerine (*C. reticulata*) and white sapote (*Casimiroa edulis*). The following additional hosts require confirmation: Brazilian guava (*Psidium guineense*), plum (*Prunus domestica*), soursop (*Annona muricata*) and Texas persimmon (*Diospyros texana*) (USDA, *Pests not known to occur in the United States or of limited distribution*, No.19, undated).

Wild hosts

Recorded from some *Inga* spp., and *Sargentia greggii* Coult. (Rutaceae) (Norrbom & Kim, 1988b).

Adult identification

One of many difficult to recognise species that have a complete *Anastrepha* type wing pattern (fig. 166), and a relatively long aculeus (3.3-4.7mm) with a finely serrate apex (fig. 58); see key for details.

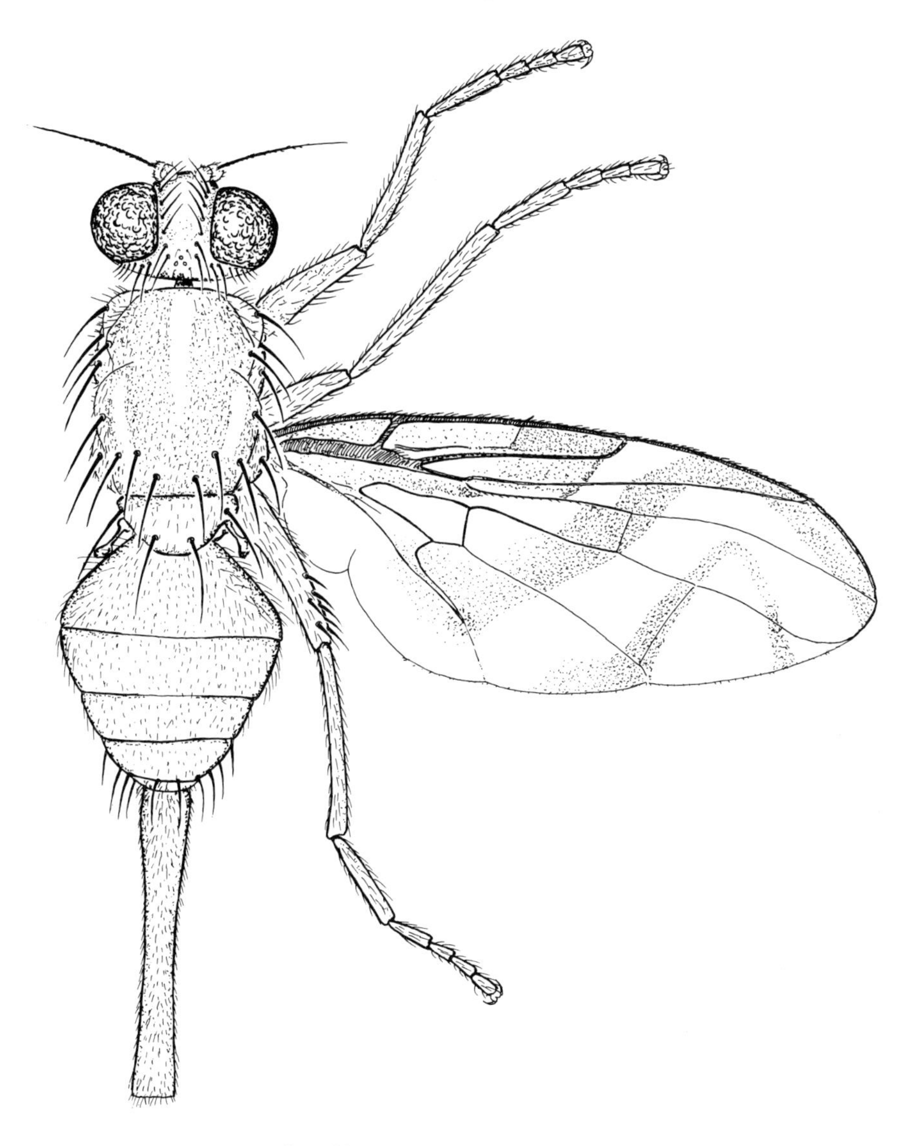

Fig. 166. *Anastrepha ludens*, adult female.

Description (not original) of third instar larva

Larvae medium-large sized, length 5.8-11.1mm; width 1.2-2.5mm. *Head*: Stomal sensory organ with 5 small sensilla; oral ridges of 11-17 rows with margins entire or slightly undulant; accessory plates small, interlocking with outer edges of oral ridges. *Thoracic and abdominal segments*: T1-T3 with spinulose areas on anterior margins; middorsally, with 4-6, 3-5 and 1-2 rows of spinules respectively, and 4-6 rows each midventrally. A1-A8 with creeping welts; A1 of 7-9 rows; A2-A8 of 9-17 rows. A8 with intermediate lobes moderately developed; tubercles and sensilla small but obvious. *Anterior spiracles*: Irregular row of 12-21 tubules. *Posterior spiracles*: Spiracular slits about 3.5 times as long as broad, with moderately sclerotised rimae; spiracular hairs short (about one-third to one-fifth the length of a spiracular slit), often branched in apical third; dorsal and ventral hair bundles of 6-13 hairs; lateral hair bundles of 4-7 hairs. *Anal area*: Lobes large, protuberant, usually distinctly bifid; surrounded by 3-4 discontinuous rows of small spinules. *Source*: Based on Carroll & Wharton (1989).

Distribution

Central America: Belize, Costa Rica, El Salvador, Guatemala, Honduras, Mexico, Nicaragua.
North America: USA (Texas).
A distribution map was produced by CIE (1958b).

Other references

BEHAVIOUR, Katsoyannos (1989b), Robacker & Hart (1985b), Robacker & Moreno (1988), Robacker & Wolfenbarger (1988), Sivinski & Burk (1989) and P.H. Smith (1989); BIOCONTROL, Wharton (1989a); BIOLOGY, Fletcher (1989b); CUTICULAR HYDROCARBONS, Carlson & Yocom (1986); DATASHEET, Weems (1963); ECOLOGY, Berrigan *et al.* (1988), Celedonio-Hurtado *et al.* (1988), Fletcher (1989c) and Meats (1989a); HOST RESISTANCE, Greany (1989); IMPACT, E.J. Harris (1989); LARVA, Berg (1979), Heppner (1984) and Steck *et al.* (1990); MEXICO, Aluja *et al.* (1987a; 1987b), Malo *et al.* (1987), Hernandez & Tejada (1980) and Williamson & Hart (1989); PARASITOIDS, Aluja *et al.* (1990), Narayanan & Chawla (1962), Hernandez & Tejada (1979) and Wharton (1989a); PHEROMONES, Dickens *et al.* (1988), Nation (1981; 1989a), Robacker & Hart (1985a; 1986), Robacker & Moreno (1988), Robacker & Wolfenbarger (1988) and Robacker *et al.* (1985; 1986); POST-HARVEST DISINFESTATION, Heather (1989), Leyva-Vazquez (1988) and Rigney (1989); REARING, Fay (1989) and Tsitsipis (1989); STERILE INSECT TECHNIQUE, Calkins (1989); TRAPPING, Katsoyannos (1989b) and Rhode & Sanchez (1982).

Anastrepha macrura Hendel

(figs 45, 167)

Taxonomic notes

This species has also been known as *Pseudodacus macrurus* (Hendel).

Commercial hosts

Recorded from star-apple (*Chrysophyllum cainito*) (Norrbom & Kim, 1988b); it is a potential pest of cultivated Sapotaceae.

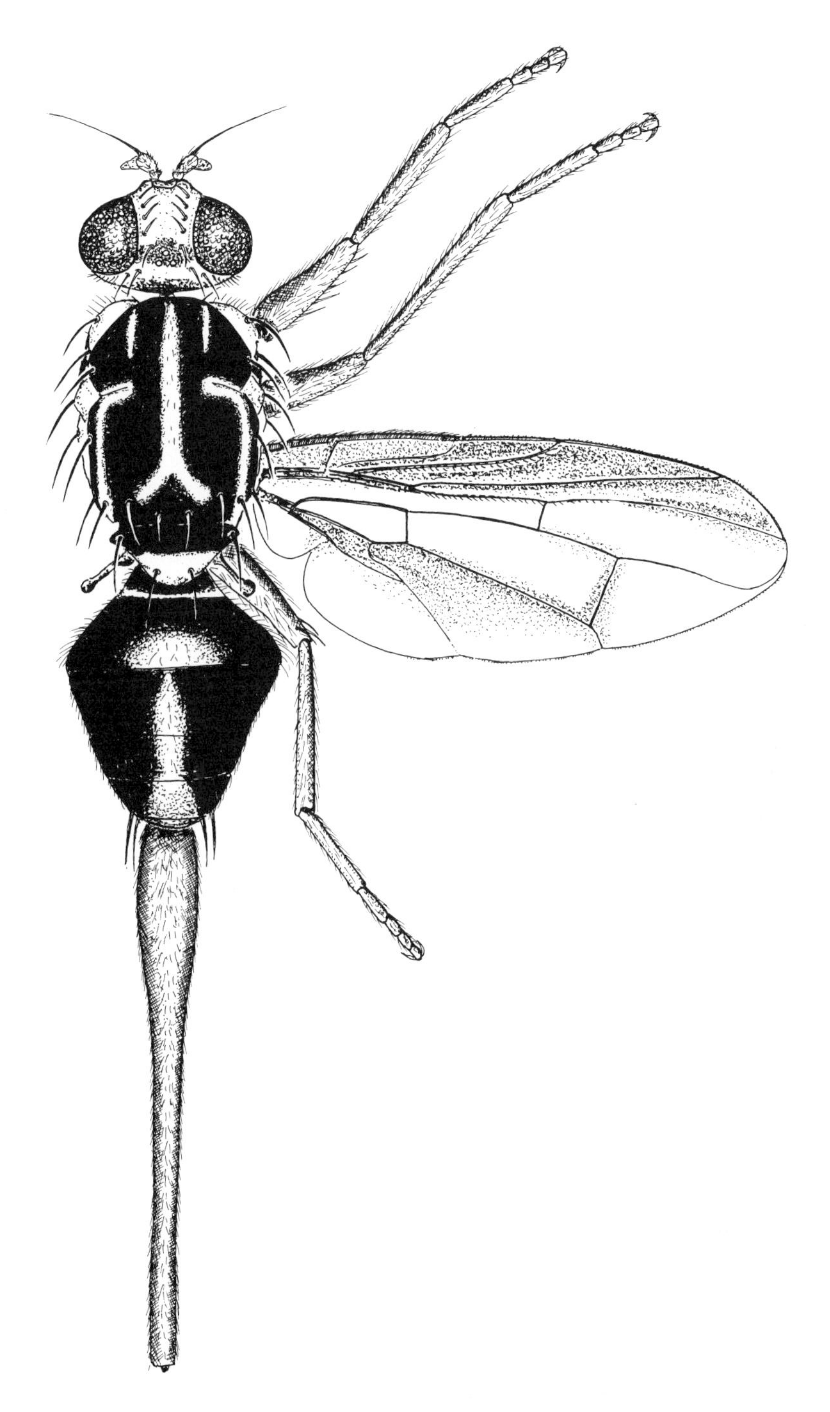

Fig. 167. *Anastrepha macrura*, adult female.

Wild hosts

Recorded from another species of Sapotaceae (*Pouteria lactescens* Vell.) (Norrbom & Kim, 1988b).

Adult identification

Easily recognised by the lack of a typical *Anastrepha* type wing pattern (fig. 167) and the remarkably long aculeus (over 5.5mm).

Distribution

South America: Argentina, Brazil, Paraguay, Venezuela.

Anastrepha obliqua (Macquart)

West Indian Fruit Fly
Antillean Fruit Fly

(figs 49, 59, 134, 168; pl. 5)

Taxonomic notes

This species has also been known as *Acrotoxa obliqua* (Macquart), *Anastrepha fraterculus* var. *ligata* Lima, *A. fraterculus* var. *mombinpraeoptans* Seín, *A. mombinpraeoptans* Seín, *A. trinidadensis* Greene, *Tephritis obliqua* Macquart and *Trypeta obliqua* (Macquart). In addition, most records of *A. acidusa* (Walker) were based on misidentifications of this species.

Pest status and commercial hosts

A pest of *Spondias* spp. (Norrbom & Foote, 1989), with a preference for that and other Anacardiaceae (Whervin, 1974). This is the major pest of mango (*Mangifera indica*) in most of the New World tropics, but mango is never known to be attacked in Trinidad or St Lucia (CPPC, unpublished data, 1991). Although recorded from *Citrus* spp., they do not appear to be important hosts (Enkerlin *et al.*, 1989); citrus is sometimes attacked in Dominica but never in Cuba, Guyana or Trinidad and Tobago (CPPC, unpublished data, 1991). It was also recorded from the following hosts by Norrbom & Kim (1988b): almond (*Prunus dulcis*), arabica coffee (*Coffea arabica*), black sapote (*Diospyros digyna*), carambola (*Averrhoa carambola*), cashew (*Anacardium occidentale*), common guava (*Psidium guajava*), giant granadilla (*Passiflora quadrangularis*), grapefruit (*Citrus x paradisi*), green sapote (*Pouteria viridis*), hog-plum (*Spondias mombin*), huesito (*Malpighia glabra*), Japanese plum (*Prunus salicina*), Jew plum (*Spondias cytherea*), ketembilla (*Dovyalis hebecarpa*), loquat (*Eriobotrya japonica*), Malay-apple (*Syzygium malaccense*), pear (*Pyrus communis*), ramón (*Brosimum alicastrum*), red mombin (*Spondias purpurea*), rose-apple (*Syzygium jambos*), sapodilla (*Manilkara zapota*), sapote (*Pouteria sapota*), sour orange (*C. aurantium*), strawberry guava (*Psidium littorale*), sweet lime (*C. limetta*) and sweet orange (*C. sinensis*).

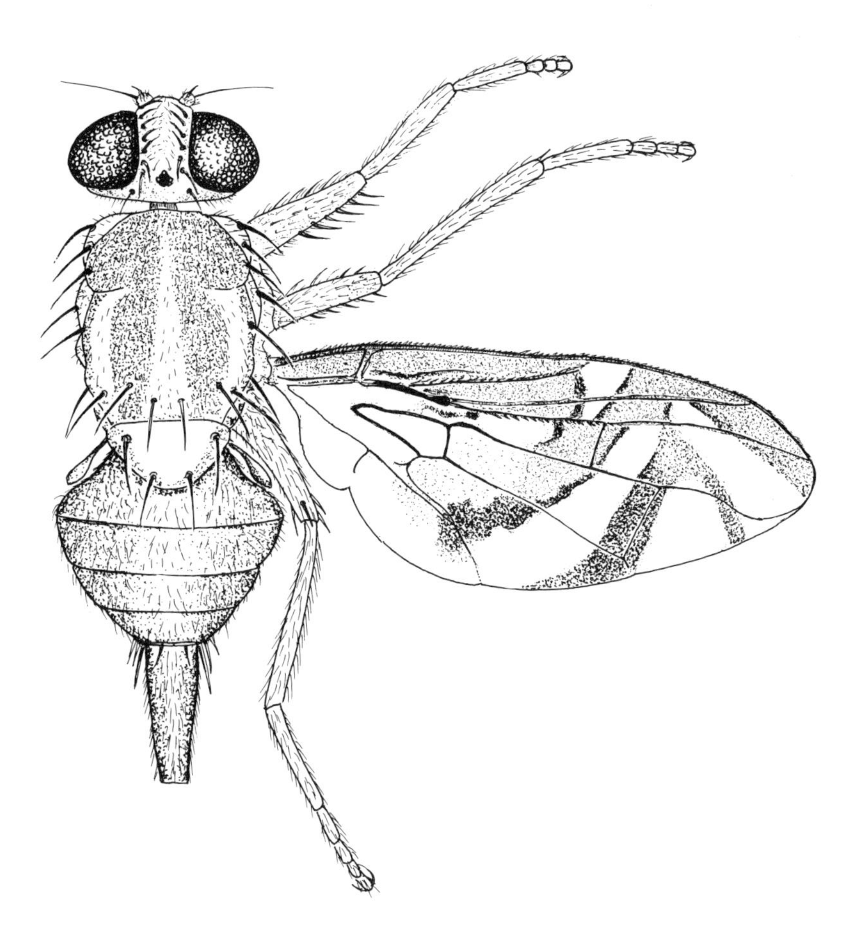

Fig. 168. *Anastrepha obliqua*, adult female.

Wild hosts

Wild growing hog-plum provides an important reservoir host in many areas (C.Y.L. Schotman, pers. comm., 1991). *A. obliqua* is also recorded from many unrelated hosts belonging to the families Anacardiaceae, Annonaceae, Bignoniaceae, Euphorbiaceae, Fabaceae, Myrtaceae and Rosaceae (Norrbom & Kim, 1988b).

Adult identification

One of several difficult to recognise species that have a complete *Anastrepha* type wing pattern (fig. 168), and a relatively short aculeus (1.3-1.6mm) with a finely serrate apex (fig. 59); see key for further details.

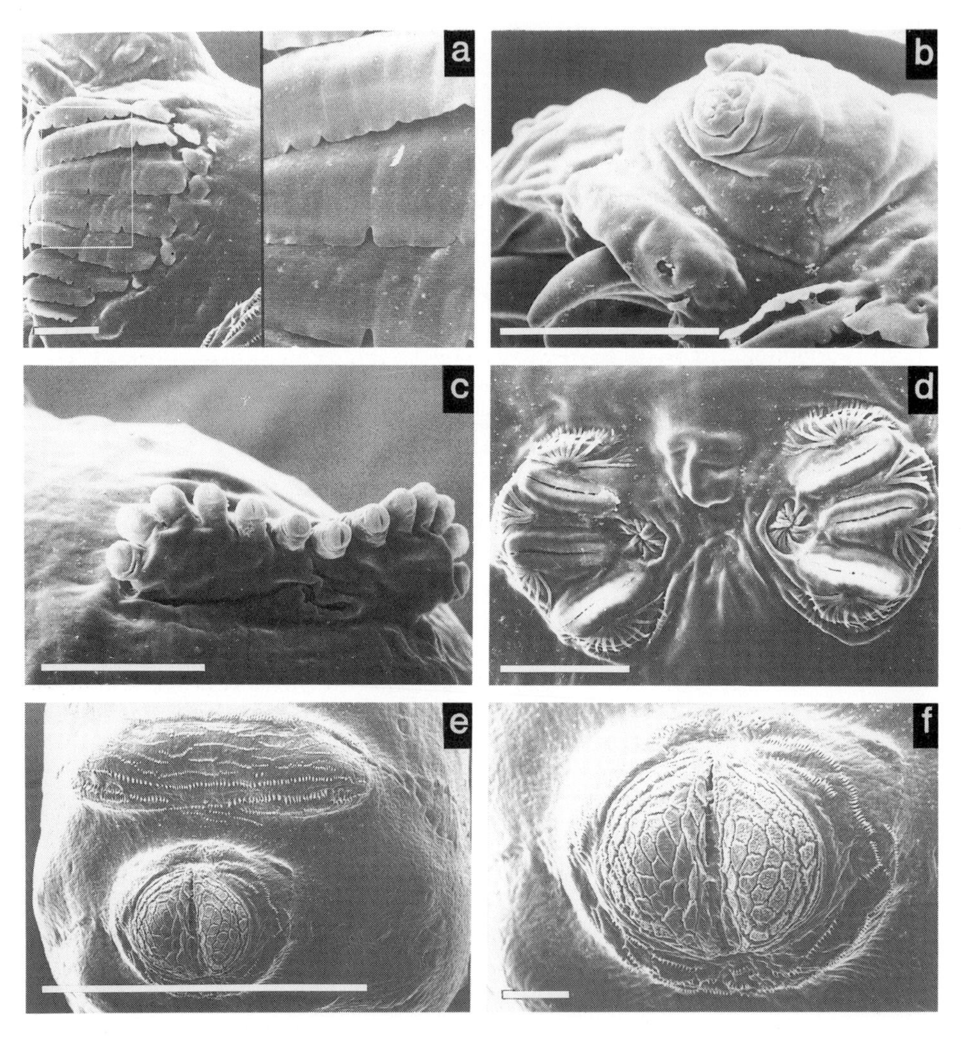

Plate 5. *Anastrepha obliqua*, SEMs of 3rd instar larva; a, oral ridges, lateral view (scale=50μm); b, antennal, maxillary and stomal sensory organs, lateral view (scale=0.1mm); c, anterior spiracle scale=0.1mm); d, posterior spiracles (scale=0.1mm); e, A8, creeping welt and anal lobes (scale=1.0mm); f, anal lobes (scale=0.1mm).

Description of third instar larva

Larvae medium-sized, length 7.5-9.0mm, width 1.4-1.8mm. *Head*: Stomal sensory organ (pl. 5.b) rounded, only slightly protuberant, with 2-3 very small sensilla; oral ridges (pl. 5.a) of 7-10 short rows of ridges with irregular serrations along posterior margins; accessory plates small with very few serrations on posterior margins; mouthhooks moderately to heavily sclerotised, each with a large slender curved apical tooth (fig. 134). *Thoracic and abdominal segments*: A broad, anterior band of 5-10 discontinuous rows of small, sharply pointed spinules encircling T1, spinule patches widely spaced dorsally; T2 and T3 with 2-5 rows of stouter, slightly smaller, spinules forming rows dorsally and ventrally, but reduced or absent laterally. Dorsal spinules absent from A1-A8. Creeping welts (pl. 5.e) on A1-A8 with 7-11 rows of stout spinules with slightly larger spinules in some posterior rows. A8 with large dorsal tubercles and stout sensilla; intermediate areas well developed with obvious sensilla;ventral sensilla small but well defined. *Anterior spiracles* (pl. 5.c): 12-16 tubules. *Posterior spiracles* (pl. 5.d): Spiracular slits about 3 times as long as broad with heavily sclerotised, dark brown rimae. Spiracular hair bundles well defined with dorsal and ventral bundles of 10-16 stout hairs branched in apical third, lateral bundles of 3-6 hairs similarly branched. *Anal area* (pl. 5.e, f): Lobes very large, protuberant, heavily sculptured, not grooved; surrounded by 2-5 discontinuous rows of small, sharp spinules. *Source*: Based on specimens from Suriname (ex *Spondias mombin*).

Distribution

Central America: Belize, Costa Rica, Guatemala, Honduras, Mexico, Nicaragua, Panama.

North America: Has been recorded in USA (Florida, Texas) but not currently established (A.L. Norrbom, pers. comm., 1991).

South America: Argentina, Brazil, Colombia, Ecuador, Guyana, Peru, Suriname, Venezuela.

West Indies and nearby islands: Bahamas, Bermuda, Cuba, Dominica, Dominican Republic, Guadeloupe, Haiti, Jamaica, Martinique, Montserrat, Nevis, Puerto Rico, St Christopher, St Lucia, Trinidad and Tobago, Virgin Islands.

A distribution map was produced by CIE (1988b).

Other references

BEHAVIOUR, Silva *et al.* (1985) and P.H. Smith (1989); COSTA RICA, Jiron *et al.* (1988); ECOLOGY, Celedonio-Hurtado *et al.* (1988); LARVA, Berg (1979) and Steck *et al.* (1990); MEXICO, Aluja *et al.* (1987a; 1987b); PARASITOIDS, Aluja *et al.* (1990), Narayanan & Chawla (1962) and Wharton *et al.* (1981); PHEROMONES, Teles & Polloni (1989); POST-HARVEST DISINFESTATION, Sharp *et al.* (1988a); PUERTO RICO, Segarra (1988); REARING, Tsitsipis (1989), Message & Zucoloto (1980) and Braga & Zucoloto (1981); TRAPPING, Hedstrom & Jimenez (1988).

Anastrepha ocresia (Walker)

(figs 60, 169)

Taxonomic notes

This species has also been known as *Acrotoxa ocresia* (Walker), *Anastrepha tricincta* (Loew), *Trypeta ocresia* Walker and *T. tricincta* Loew. It is sometimes misspelt *A. ochresia.*

Commercial hosts

A potential pest which has been recorded from common guava (*Psidium guajava*) and sapodilla (*Manilkara zapota*) (Norrbom & Kim, 1988b). A record from grapefruit (*Citrus x paradisi*) (Weems, 1968) requires confirmation.

Adult identification

One of many difficult to recognise species that have a complete *Anastrepha* type wing pattern (fig. 169), and a relatively long aculeus (2.7-3.3mm) with a finely serrate apex (fig. 60); see key for details.

Distribution

North America: USA (Florida).
West Indies: Cuba, Dominican Republic, Jamaica, Puerto Rico.

Other reference

DATASHEET, Weems (1968).

Anastrepha ornata Aldrich

(figs 51, 170)

Commercial hosts

A potential pest species which has been recorded from common guava (*Psidium guajava*) and pear (*Pyrus communis*) (Norrbom & Kim, 1988b).

Adult identification

Easily recognised by its characteristic wing pattern (fig. 170), which is not typical of the genus, combined with a non-serrate aculeus apex (4mm long; fig. 51).

Distribution

South America: Ecuador.

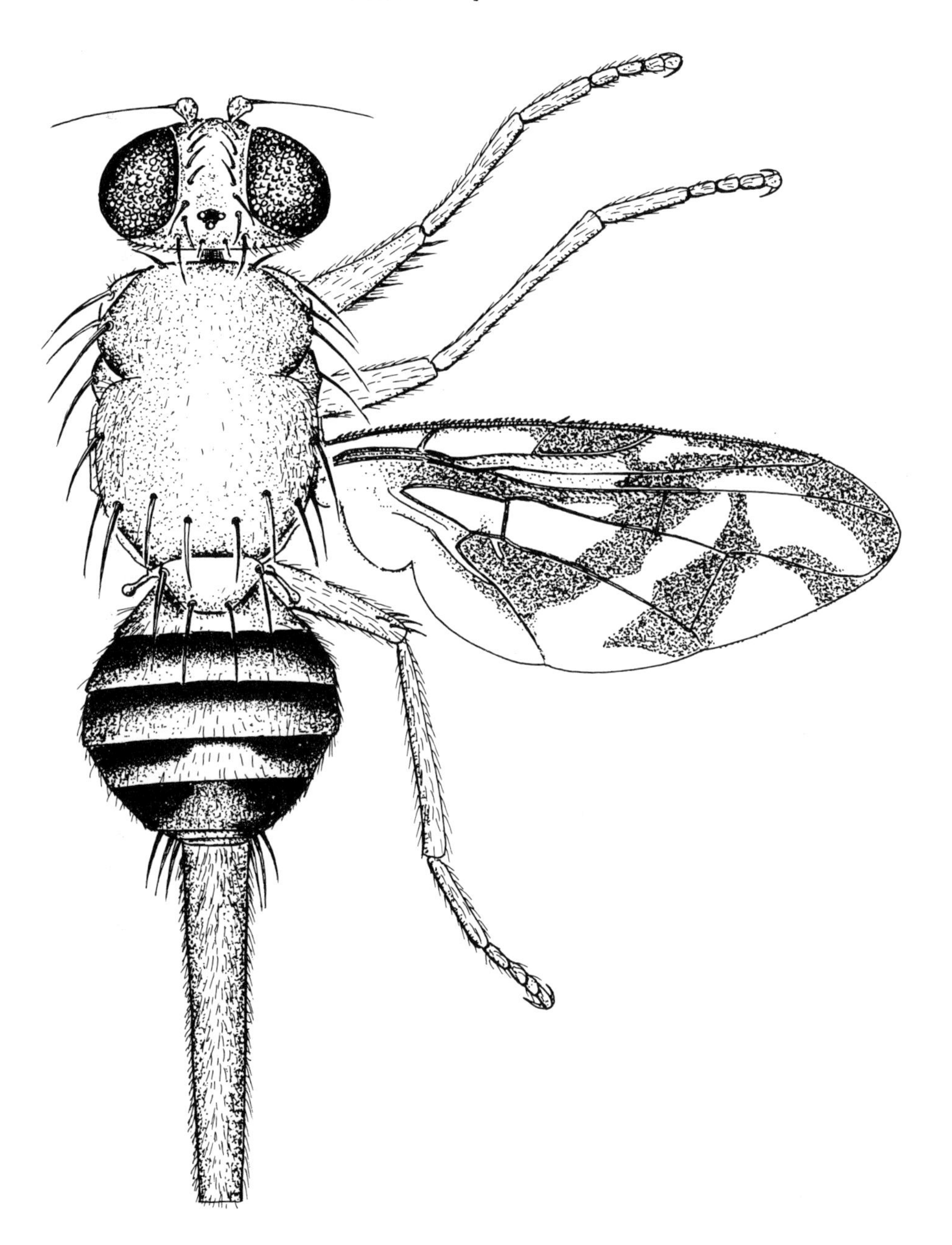

Fig. 169. *Anastrepha ocresia*, adult female.

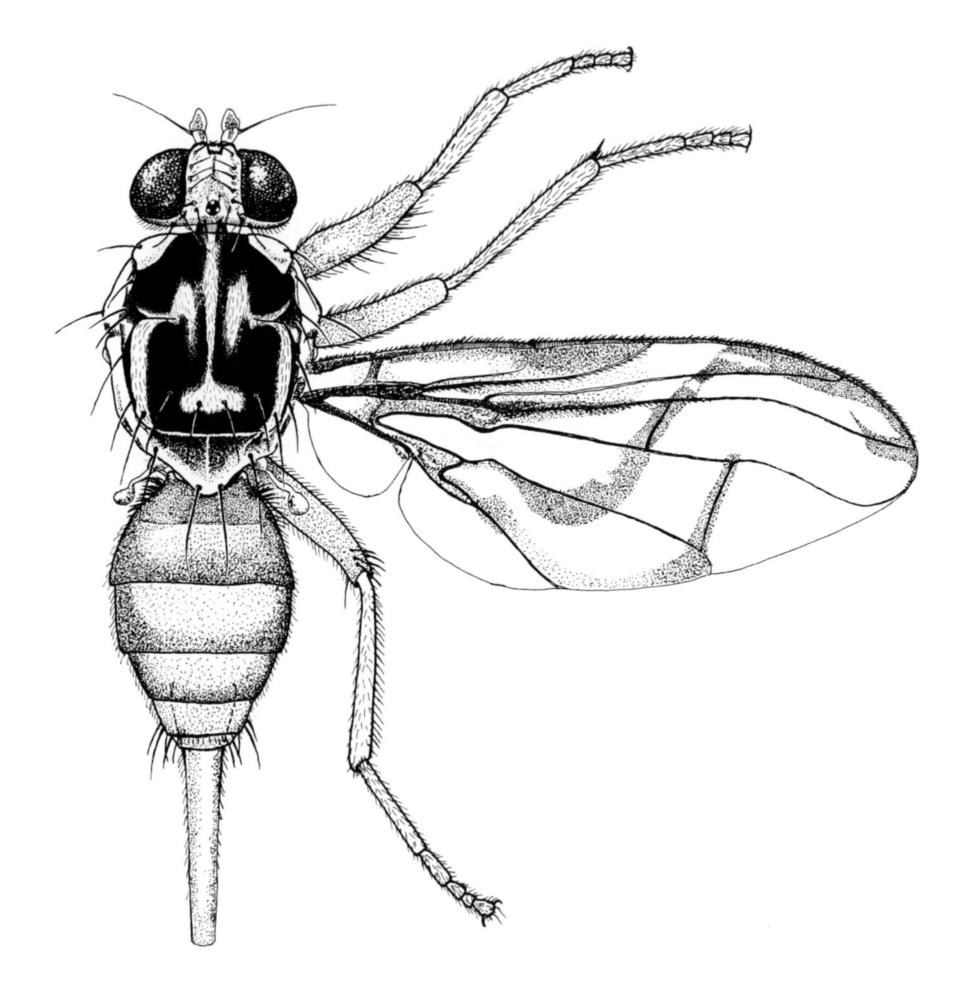

Fig. 170. *Anastrepha ornata*, adult female.

Anastrepha pseudoparallela (Loew)

(figs 63, 171)

Taxonomic notes
This species has also been known as *Acrotoxa pseudoparallela* Loew and *Trypeta pseudoparallela* (Loew).

Commercial hosts
A potential passion fruit pest which has been recorded from several *Passiflora* spp., namely blue passion fruit (*P. caerulea*), giant granadilla (*P. quadrangularis*), maracuja grande (*P. alata*), purple granadilla (*P. edulis*) and sweet granadilla (*P. ligularis*); there is also a record from mango (*Mangifera indica*) (Norrbom & Kim, 1988b).

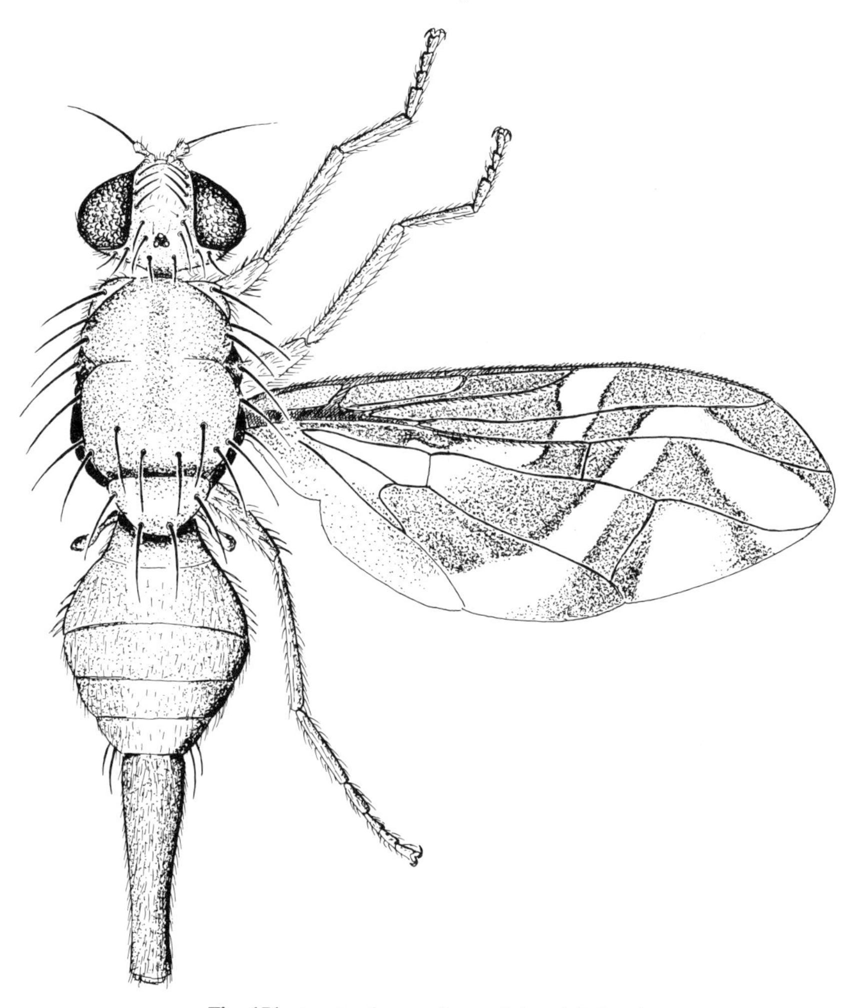

Fig. 171. *Anastrepha pseudoparallela*, adult female.

Adult identification

One of many difficult to recognise species that have a complete *Anastrepha* type wing pattern (fig. 171), and a relatively long aculeus (2.6-3.2mm) with a finely serrate apex (fig. 63); see key for details.

Distribution

South America: Argentina, Brazil, Peru.

Other references

BEHAVIOUR, Silva *et al.* (1985); PHEROMONES, Teles & Polloni (1989).

Anastrepha serpentina **(Wiedemann)**

Sapote Fruit Fly
Serpentine Fruit Fly

(figs 46, 62, 135, 172; pl. 6)

Taxonomic notes
This species has also been known as *Acrotoxa serpentina* (Wiedemann), *Dacus serpentinus* Wiedemann, *Leptoxys serpentina* (Wiedemann), *Trypeta serpentina* (Wiedemann) and *Urophora vittithorax* Macquart.

Pest status and commercial hosts
A pest of citrus (*Citrus* spp.), sapodilla (*Manilkara zapota*), star-apple (*Chrysophyllum cainito*) and *Pouteria* spp. (Norrbom & Foote, 1989), which was also recorded from the following hosts by Norrbom & Kim (1988b): abiu (*P. caimito*), apple (*Malus domestica*), avocado (*Persea americana*), black sapote (*Diospyros digyna*), common guava (*Psidium guajava*), egg-fruit tree (*Pouteria campechiana*), grapefruit (*Citrus x paradisi*), green sapote (*P. viridis*), hog-plum (*Spondias mombin*), ketembilla (*Dovyalis hebecarpa*), lucmo (*P. obovata*), mammy-apple (*Mammea americana*), mango (*Mangifera indica*), nance (*Byrsonima crassifolia*), Panama orange (*x Citrofortunella x mitis*), peach (*Prunus persica*), pear (*Pyrus communis*), pond-apple (*Annona glabra*), pummelo (*Citrus maxima*), quince (*Cydonia oblonga*), red mombin (*Spondias purpurea*), sapote (*Pouteria sapota*), sour orange (*Citrus aurantium*), sweet orange (*C. sinensis*) and tangerine (*C. reticulata*).

Wild hosts
Recorded from several genera of Sapotaceae, plus single species of Apocynaceae, Euphorbiaceae and Moraceae (Norrbom & Kim, 1988b).

Adult identification
Easily recognised by its characteristic wing and abdominal markings (fig. 172), which are not typical of the genus, combined with a finely serrate aculeus apex (2.8-3.7mm long; fig. 62).

Description of third instar larva
Larvae medium-sized, length 7.5-9.0mm, width 1.0-1.5mm. *Head*: Stomal sensory organ (pl. 6.c) large, rounded, protuberant with 3 large sensilla (2 long and tapering, 1 short and peg-like); smaller sensilla around edge of depression. Oral ridges (pl. 6.a, b) of 8-12 rows of small ridges with irregular serrations along posterior margins; accessory plates large, anterior ones with small serrations along margins; mouthhooks moderately sclerotised, each with a large curved apical tooth (fig. 135). *Thoracic and abdominal segments*: T1 (pl. 6.a) with a broad band of 4-9 discontinuous rows of

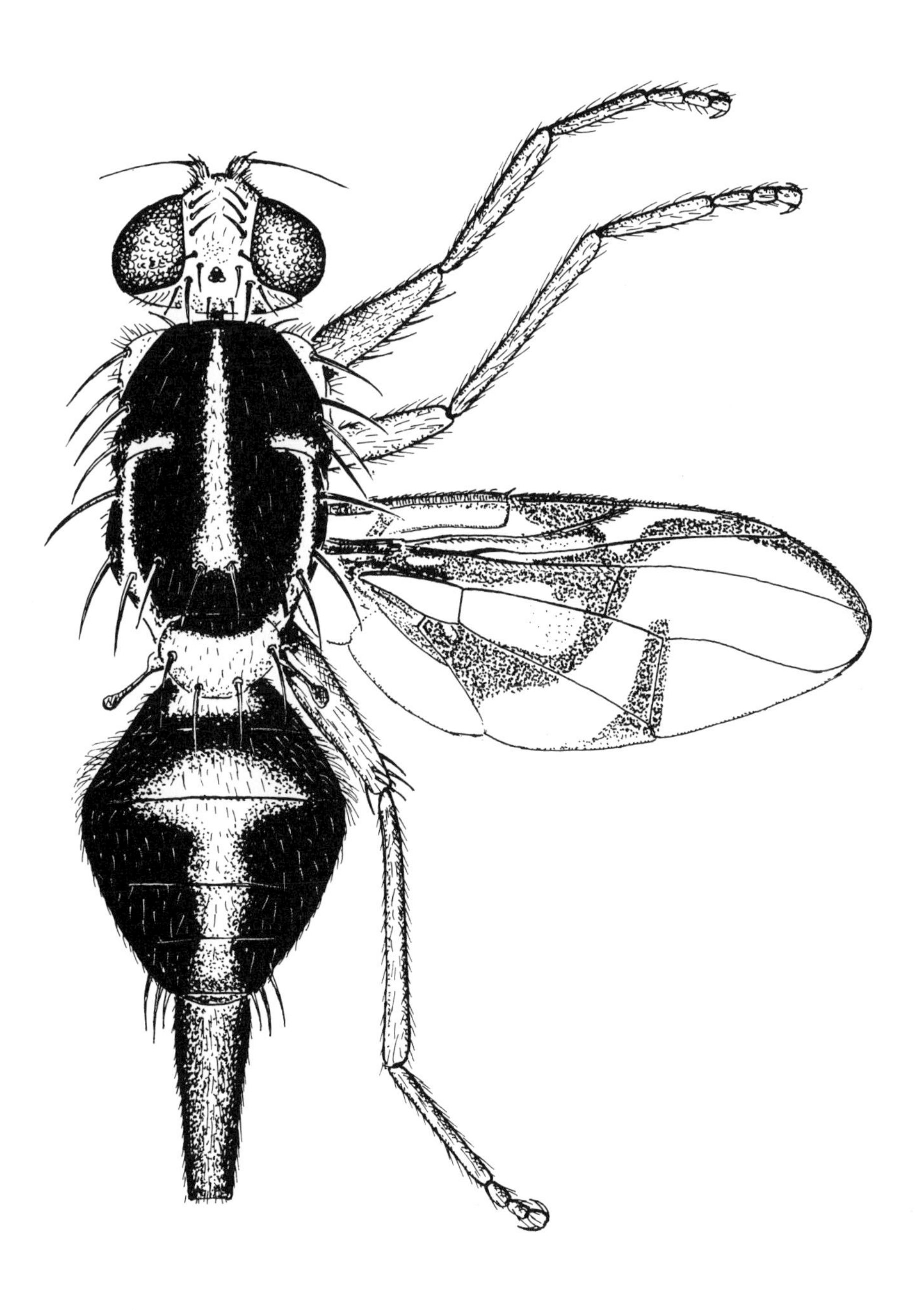

Fig. 172. *Anastrepha serpentina*, adult female.

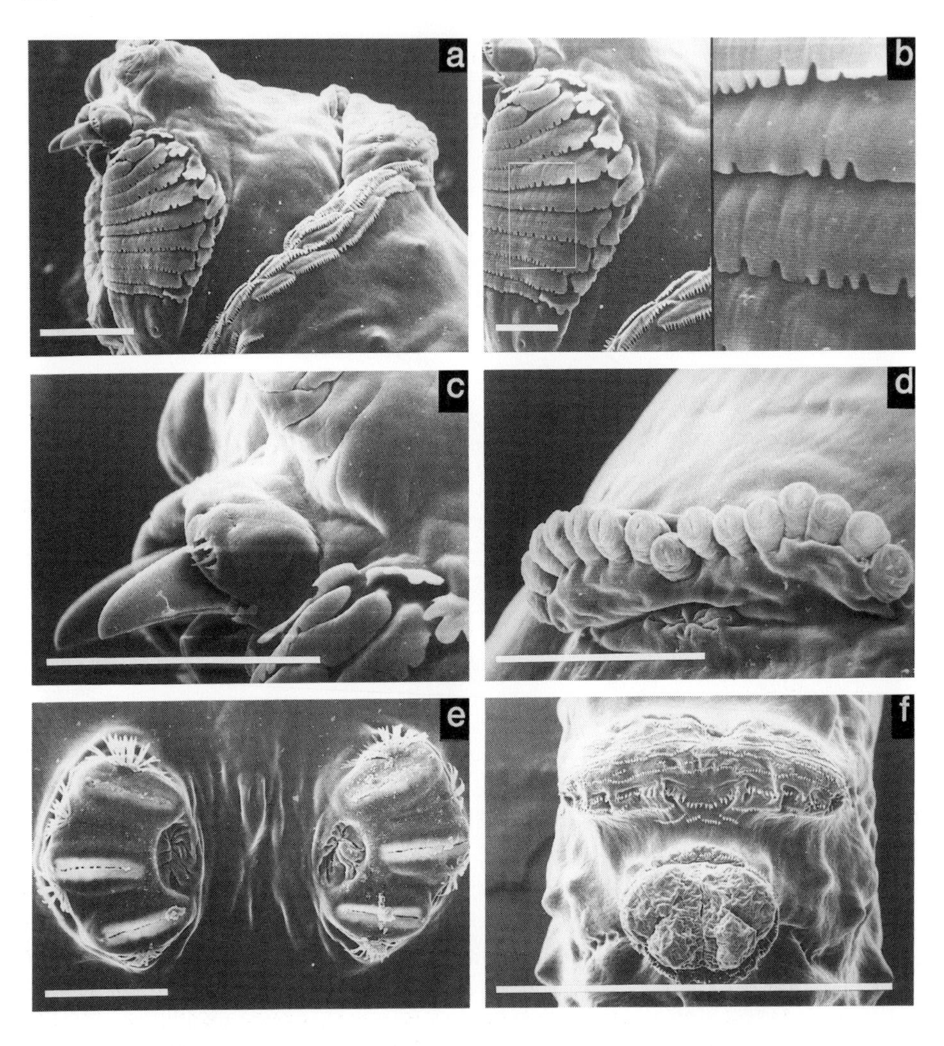

Plate 6. *Anastrepha serpentina*, SEMs of 3rd instar larva; a, head and T1, lateral view (scale=0.1mm); b, oral ridges, lateral view (scale=50μm); c, stomal sensory organ, lateral view (scale=0.1mm); d, anterior spiracle (scale=0.1mm); e, posterior spiracles (scale=0.1mm); f, A8, creeping welt and anal lobes (scale=1.0mm).

small, sharply pointed spinules surrounding anterior margin; T2 with 2-5 discontinuous rows of slightly smaller spinules dorsally and ventrally, but none midlaterally; T3 similar to T2 but no spinules laterally. Dorsal spinules absent from A1-A8. Creeping welts (pl. 6.f) on A1-A8 large, with 7-9 rows of small, stout spinules. A8 with area around spiracles protuberant, with obvious intermediate areas. Dorsal and intermediate tubercles and sensilla very obvious, ventral sensilla smaller. *Anterior spiracles* (pl. 6.d): 13-18 tubules. *Posterior spiracles* (pl. 6.e): Spiracular slits about 2.5-3.0 times as long as broad with heavily sclerotised, dark brown rimae. Spiracular hairs relatively short (less than length of a spiracular slit), broad, mostly branched in apical third;

dorsal and ventral bundles of 6-9 hairs, lateral bundles of 4-6 hairs. *Anal area* (pl. 6.f): Lobes very large, protuberant, obviously grooved or bilobed; surrounded by 2-4 discontinuous rows of small, sharp spinules. *Source*: Based on specimens from Suriname (ex *Chrysophyllum cainito*).

Distribution

Central America: Costa Rica, Guatemala, Mexico, Panama.
North America: USA (Texas).
South America: Argentina, Brazil, Colombia, Ecuador, Guyana, Peru, Suriname, Venezuela.
West Indies: Dominica, Trinidad.

Other references

ECOLOGY, Celedonio-Hurtado *et al.* (1988); LARVA, Berg (1979) and Steck *et al.* (1990); PARASITOIDS, Aluja *et al.* (1990), Narayanan & Chawla (1962) and Wharton *et al.* (1981).

Anastrepha striata Schiner

Guava Fruit Fly

(figs 52, 136, 173; pl. 7)

Pest status and commercial hosts

A pest of common guava (*Psidium guajava*) (Norrbom & Foote, 1989), which has a preference for Myrtaceae (Jiron *et al.*, 1988). *A. striata* has also been recorded from the following hosts by Norrbom & Kim (1988b) and Jiron *et al.* (1988): avocado (*Persea americana*), black sapote (*Diospyros digyna*), Brazilian guava (*Psidium guineense*), cassava (*Manihot esculenta*), green sapote (*Pouteria viridis*), hog-plum (*Spondias mombin*), Malay-apple (*Syzygium malaccense*), mango (*Mangifera indica*), peach (*Prunus persica*), red mombin (*Spondias purpurea*), rose-apple (*Syzygium jambos*), soursop (*Annona muricata*), star-apple (*Chrysophyllum cainito*), strawberry guava (*Psidium littorale*), sweet orange (*Citrus sinensis*), tropical almond (*Terminalia catappa*) and wild guava (*P. friedrichsthalianum*).

Wild hosts

Recorded from several species of Myrtaceae, and other families, including Sapotaceae and Solanaceae (Norrbom & Kim, 1988b).

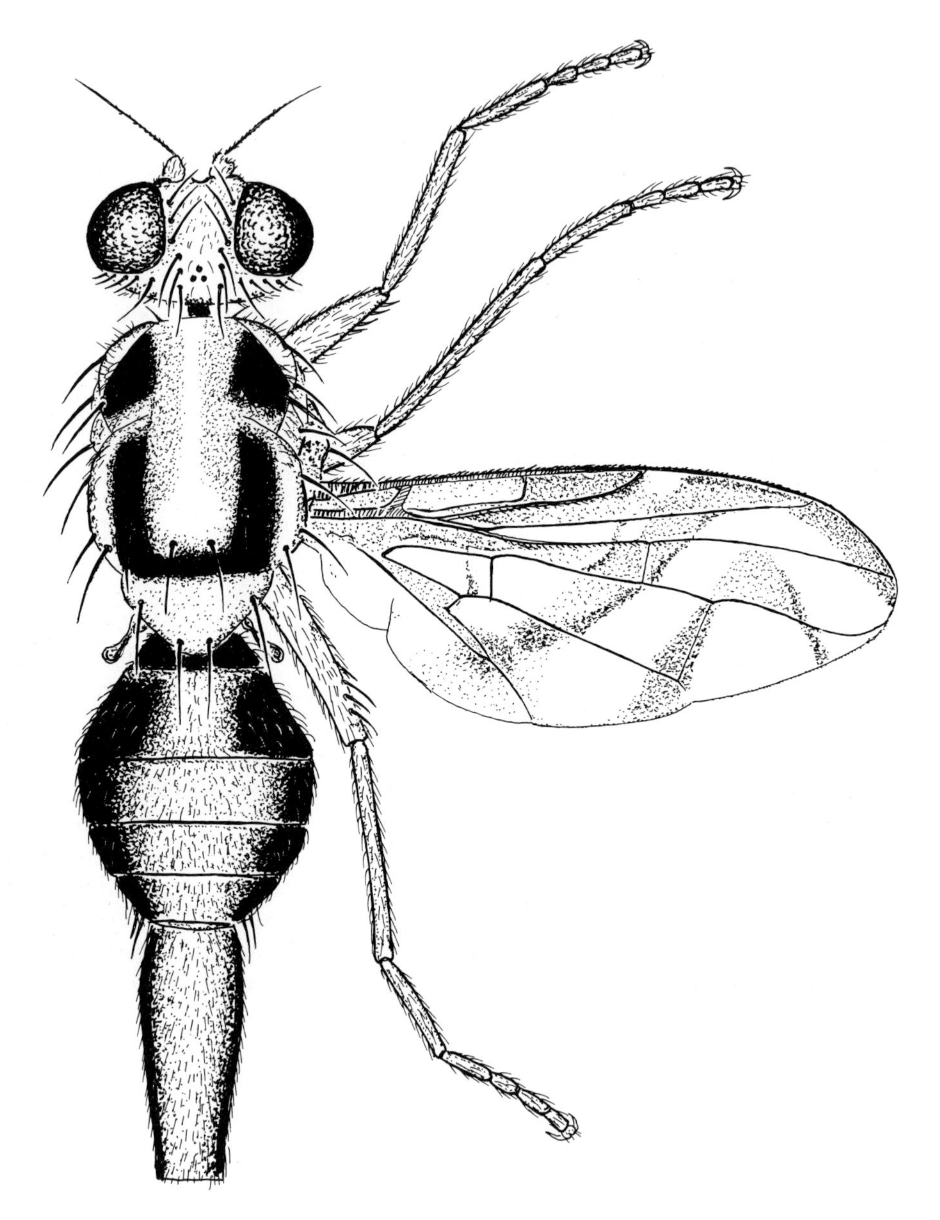

Fig. 173. *Anastrepha striata*, adult female.

Adult identification

Separated from the other species with a complete *Anastrepha* type wing pattern (fig. 173), by the short (under 3.0mm) aculeus with a non-serrate apex (fig. 52), and U-shaped pattern on the scutum.

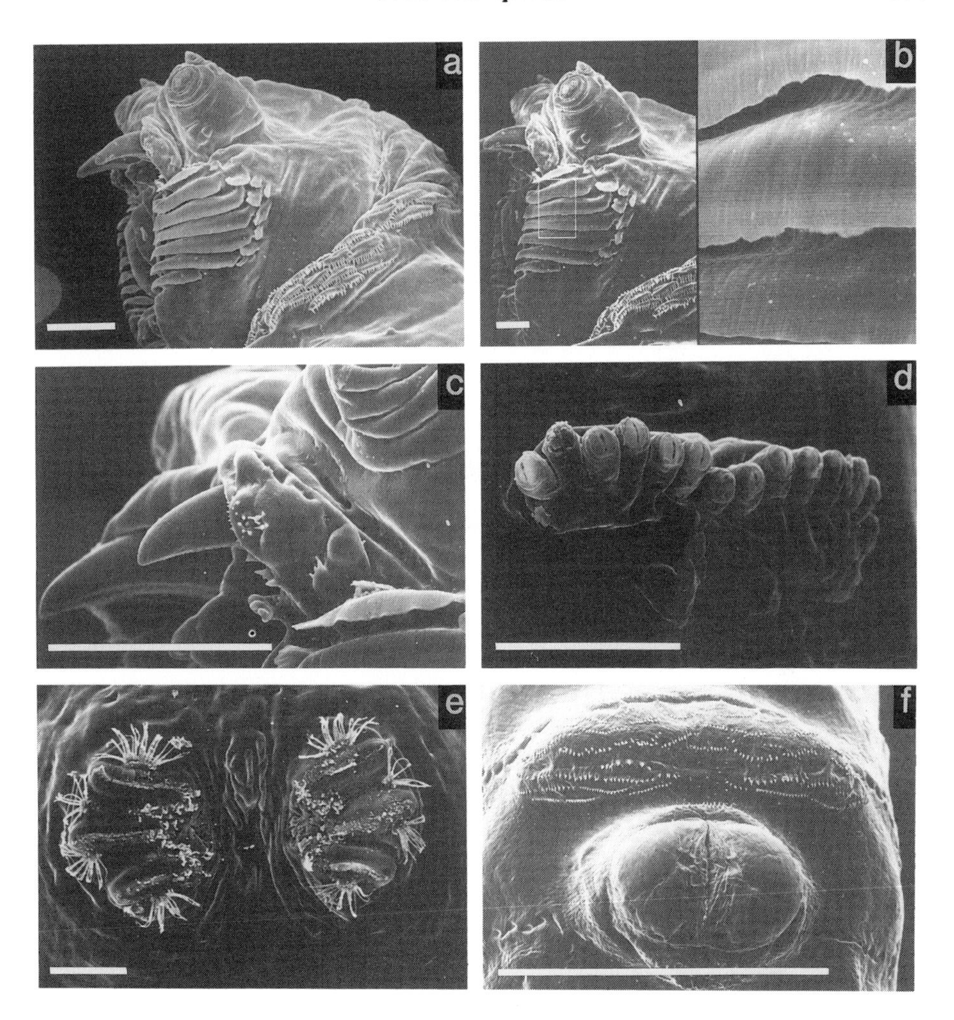

Plate 7. *Anastrepha striata*, SEMs of 3rd instar larva; a, head and T1, lateral view (scale=0.1mm); b, oral ridges, lateral view (scale=50μm); c, stomal sensory organ, lateral view (scale=0.1mm); d, anterior spiracle (scale=0.1mm); e, posterior spiracles (scale=0.1mm); f, A8, creeping welt and anal lobes (scale=1.0mm).

Description of third instar larva

Larvae medium-sized, length 7.0-9.0mm; width 1.2-1.5mm. *Head*: Stomal sensory organ (pl. 7.c) large, rounded, with 2-3 peg-like sensilla, with small, sharply pointed spinules scattered over surface of stomal sensory organ and adjacent preoral lobe; oral ridges (pl. 7.a, b) of 6-9 ridges with unserrated posterior margins; accessory plates well defined with unserrated margins; mouthhooks moderately sclerotised, each with a large curved apical tooth (fig. 136). *Thoracic and abdominal segments*: T1, anterior margin with a broad, encircling band of 6-9 discontinuous rows of stout, sharply pointed spinules; T2 and T3 with 3-5 rows of smaller spinules encircling anterior margins of each segment. Dorsal spinules absent from A1-A8. Creeping welts on A1-A8 with 6-10 rows of small spinules. A8, area around spiracles slightly protuberant with well defined intermediate areas. Dorsal and intermediate tubercles and sensilla well developed, ventral sensilla smaller. *Anterior spiracles* (pl. 7.d): 14-18 tubules. *Posterior spiracles* (pl. 7.e): Spiracular slits large, about 5 times as long as broad, with heavily sclerotised dark brown rimae. Spiracular hair bundles dense, with long, slender, branched hairs almost as long as spiracular slits; dorsal and ventral hair bundles of 14-20 hairs, lateral bundles of 6-10 hairs. *Anal area* (pl. 7.f): Lobes large, slightly grooved or bilobed, surrounded by 2-4 discontinuous rows of small, sharply pointed spinules concentrating into a sharper, stouter patch just below anal opening. *Source*: Based on specimens from Suriname (ex *Psidium guajava*).

Distribution

Central America: Costa Rica, Guatemala, Honduras, Mexico, Panama.
North America: USA (Texas).
South America: Bolivia, Brazil, Colombia, Ecuador, Guyana, Peru, Suriname, Venezuela.
West Indies: Trinidad.

Other references

BEHAVIOUR, Saravia & Freidberg (1989); DATASHEET, Weems (1982); ECOLOGY, Espinosa (1980); LARVA, Berg (1979) and Steck *et al.* (1990); PARASITOIDS, Aluja *et al.* (1990), Narayanan & Chawla (1962) and Wharton *et al.* (1981); TRAPPING, Hedstrom & Jimenez (1988).

Anastrepha suspensa (Loew)

Caribbean Fruit Fly
Greater Antillean Fruit Fly

(figs 61, 137, 174; pl. 8)

Taxonomic notes

This species has also been known as *Acrotoxa suspensa* Loew, *Anastrepha longimacula* Greene, *A. unipuncta* Seín and *Trypeta suspensa* (Loew).

Pest status and commercial hosts

A pest of common guava (*Psidium guajava*), *Eugenia* spp., *Syzygium* spp., *Annona* spp. and tropical almond (*Terminalia catappa*) (Norrbom & Foote, 1989), and Whervin (1974) noted a preference for species of Myrtaceae. It was also recorded from the following hosts by Norrbom & Kim (1988b): akee (*Blighia sapida*), allspice (*Pimenta dioica*), apple (*Malus domestica*), avocado (*Persea americana*), balsam-apple (*Momordica balsamina*), bell pepper (*Capsicum annuum*), bitter gourd (*M. charantia*), Brazil cherry (*E. brasiliensis*), Brazilian guava (*Psidium guineense*), calabur (*Muntingia

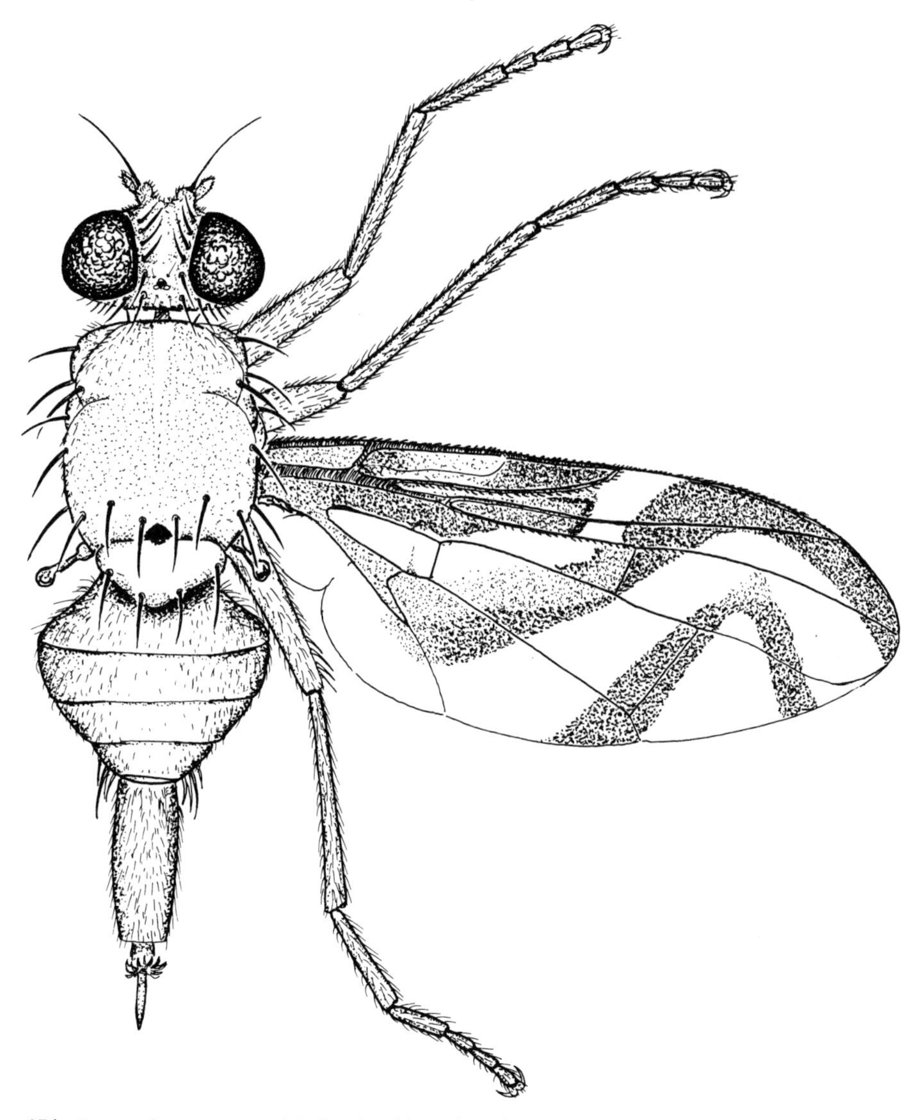

Fig. 174. *Anastrepha suspensa*, adult female with aculeus tip and large 'teeth' in basal area of eversible ovipositor sheath showing.

calabura), carambola (*Averrhoa carambola*), cocoplum (*Chrysobalanus icaco*), common fig (*Ficus carica*), common persimmon (*Diospyros virginiana*), custard-apple (*Annona reticulata*), date palm (*Phoenix dactylifera*), egg-fruit tree (*Pouteria campechiana*), governor's plum (*Flacourtia indica*), grapefruit (*Citrus x paradisi*), hog-plum (*Spondias mombin*), huesito (*Malpighia glabra*), imbe (*Garcinia livingstonei*), jaboticaba (*Myrciaria cauliflora*), Japanese persimmon (*Diospyros kaki*), Java plum (*Syzygium cumini*), Jew plum (*Spondias cytherea*), kei apple (*Dovyalis caffra*), ketembilla (*D. hebecarpa*), lemandarin (*C. x limonia*), lime (*C. aurantiifolia*), limeberry (*Triphasia trifolia*), loquat (*Eriobotrya japonica*), Malay-apple (*Syzygium

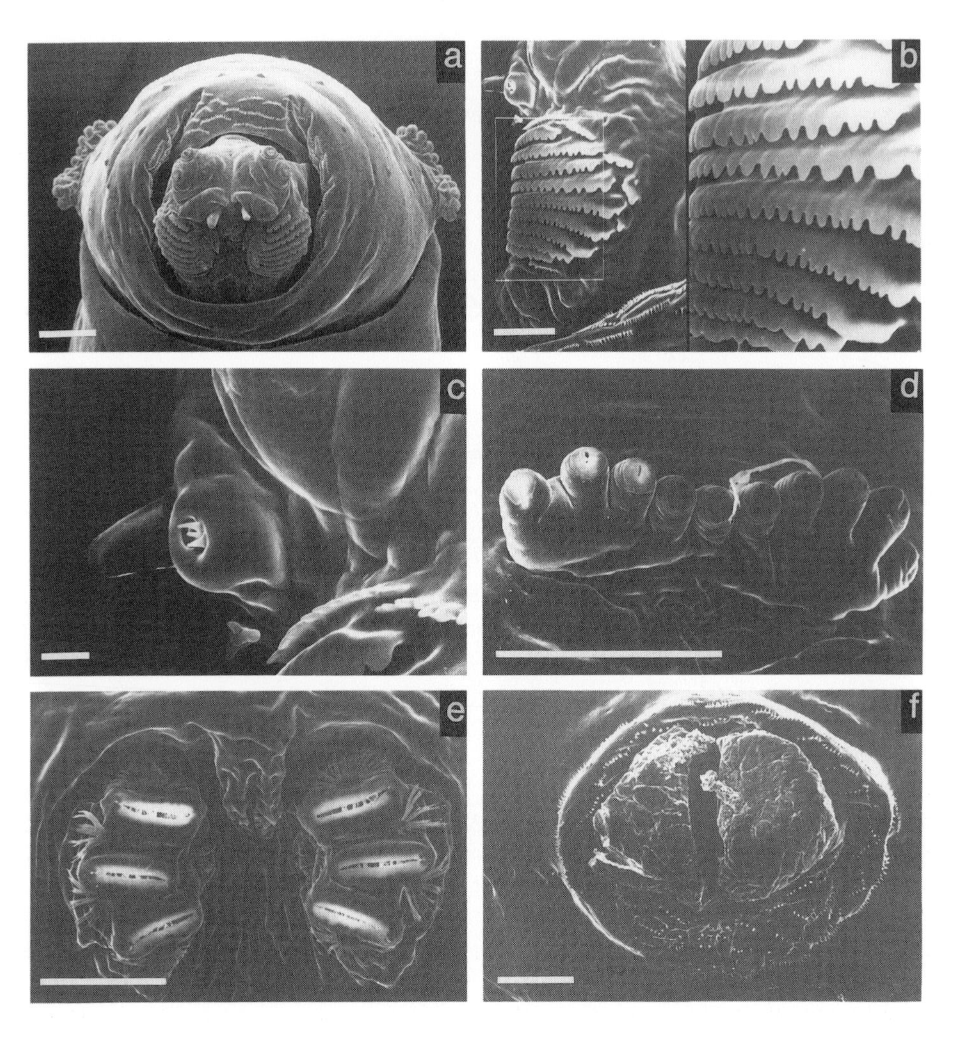

Plate 8. Anastrepha suspensa, SEMs of 3rd instar larva; a, head and T1, anterior view (scale=0.1mm); b, oral ridges, lateral view (scale=50μm); c, stomal sensory organ, lateral view (scale=10μm); d, anterior spiracle (scale=0.1mm); e, posterior spiracles (scale=0.1mm); f, anal lobes (scale=0.1mm).

malaccense), mango (*Mangifera indica*), meiwa kumquat (*Fortunella x crassifolia*), miraculous berry (*Synsepalum dulcificum*), Natal plum (*Carissa macrocarpa*), orange jessamine (*Murraya paniculata*), oval kumquat (*F. margarita*), Panama orange (*x Citrofortunella x mitis*), papaya (*Carica papaya*), peach (*Prunus persica*), pear (*Pyrus communis*), pomegranate (*Punica granatum*), pond-apple (*Annona glabra*), pummelo (*Citrus maxima*), Rio Grande cherry (*Eugenia aggregata*), rose-apple (*Syzygium jambos*), sand pear (*Pyrus pyrifolia*), sapodilla (*Manilkara zapota*), satinleaf (*Chrysophyllum oliviforme*), seagrape (*Coccoloba uvifera*), sour orange (*Citrus aurantium*), star-apple (*Chrysophyllum cainito*), strawberry guava (*Psidium littorale*), sugar-apple (*Annona squamosa*), Surinam cherry (*E. uniflora*), sweet lime (*Citrus limetta*), sweet orange (*C. sinensis*), tangelo (*C. x tangelo*), tangerine (*C. reticulata*), tomato (*Lycopersicon esculentum*), velvet apple (*Diospyros blancoi*), wampi (*Clausena lansium*), water apple (*S. samarangense*), white sapote (*Casimiroa edulis*), wild cinnamon (*Canella winteriana*) and wild guava (*P. friedrichsthalianum*). Norrbom & Kim (1988b) also recorded "a probable incidental infestation" of lychee (*Litchi chinensis*).

Citrus (*Citrus* spp.) is not known to be attacked in the West Indies (Enkerlin *et al.*, 1989; Whervin, 1974), except in Puerto Rico (Norrbom & Kim, 1988b), and in Florida only very ripe citrus is attacked (A.L. Norrbom, pers. comm., 1991).

Wild hosts

Recorded from several unrelated hosts belonging to the families Araliaceae, Clusiaceae, Combretaceae, Euphorbiaceae, Moraceae, Myrtaceae, Rosaceae, Rutaceae and Sapotaceae (Norrbom & Kim, 1988b).

Adult identification

A species that has a complete *Anastrepha* type wing pattern (fig. 174), and a relatively short aculeus (1.4-1.6mm) with a finely serrate apex (fig. 61), plus a black spot that overlays the base of the scutellum and the hind margin of the scutum (except in some specimens from Jamaica).

Description of third instar larva

Larvae medium-sized, length 7.5-9.0mm; width 1.0-1.5mm. *Head*: Stomal sensory organ (pl. 8.c) large, rounded, with 3 short stout sensilla; oral ridges (pl. 8.a, b) with 8-11 rows of short, widely spaced, bluntly rounded teeth; accessory plates of 3-4 small, shell-like plates, some with unserrated margins; mouthhooks moderately sclerotised, each with a large slender curved apical tooth (fig. 137). *Thoracic and abdominal segments*: Anterior margin of T1 with a broad, encircling band of 5-10 discontinuous rows of stout, sharply pointed spinules; T2 with 3-5 discontinuous rows of smaller spinules almost encircling segment but absent middorsally; T3 similar to T2. Dorsal spinules absent from A1-A8. Creeping welts on A1-A8 large, with 8-10 rows of small stout spinules. A8, area around spiracles protuberant with large intermediate areas. Bases of dorsal and intermediate tubercles large, well developed, with large sensilla; ventral sensilla smaller. *Anterior spiracles* (pl. 8.a, d): 9-15 tubules. *Posterior spiracles* (pl. 8.e): Spiracular slits about 3 times as long as broad with heavily sclerotised, dark brown rimae. Spiracular hairs short (less than length of spiracular slit), some branched in apical third; dorsal and ventral bundles of 9-16 hairs, lateral bundles of 4-7 hairs. *Anal area* (pl. 8.f): Lobes large, protuberant, not grooved; surrounded by 2-5 discontinuous rows of small, sharply pointed spinules forming small groups and a concentration of spinules just below anal opening. *Source*: Based on specimens from Florida, USA (lab. colony).

Distribution
North America: USA (southern Florida).
West Indies and nearby islands: Bahamas, Cuba, Dominican Republic, Haiti, Jamaica, Puerto Rico.

Other references
ACOUSTIC DETECTION, Calkins & Webb (1988) and Sharp *et al.* (1988b); ACOUSTIC SIGNALS, Burk & Webb (1983), Sivinski (1987; 1988), Sivinski & Webb (1985a; 1986) and Webb *et al.* (1983a; 1984); BAITS, Sharp (1987) and Sharp & Chambers (1983); BEHAVIOUR, Burk (1983; 1984), Dodson (1982), Katsoyannos (1989b), Sivinski (1984; 1989), Sivinski & Burk (1989) and P.H. Smith (1989); BIBLIOGRAPHY, Zwölfer (1985); BIOCONTROL, Glenn & Baranowski (1987) and Wharton (1989a; 1989b); BIOLOGY, Fletcher (1989b); CUTICULAR HYDROCARBONS, Carlson & Yocom (1986); DATASHEET, Weems (1965b); ECOLOGY, Meats (1989a); HOST RESISTANCE, Greany (1989); IMPACT, E.J. Harris (1989); LARVA, Berg (1979), Heppner (1984), Lawrence (1979) and Steck *et al.* (1990); PARASITOIDS, Ashley & Chambers (1979), Glenn & Baranowski (1987), Lawrence (1986; 1988), Narayanan & Chawla (1962) and Wharton (1989a; 1989b); PHEROMONES, Averill & Prokopy (1989a), Nation (1989a; 1989b), Robacker & Hart (1987) and Webb *et al.* (1983a); POST-HARVEST DISINFESTATION, Moshonas & Shaw (1984), Sharp (1986), Sharp & Chew (1987), Sharp *et al.* (1988a), Spalding *et al.* (1988), Rigney (1989) and Windeguth *et al.* (1976); REARING, Carroll (1986), Fay (1989), Kamasaki *et al.* (1970), Singh *et al.* (1988) and Tsitsipis (1989); STERILE INSECT TECHNIQUE, Calkins (1989); TRAPPING, Burditt (1982), Calkins *et al.* (1984), Katsoyannos (1989b) and Witherell (1982).

Other *Anastrepha* Species of Economic Interest

Three *Anastrepha* spp. attack cassava (*Manihot esculenta*), namely *A. manihoti* Lima (Brazil to Costa Rica), *A. montei* Lima (Argentina to Mexico) and *A. pickeli* Lima (Argentina to Costa Rica). *A. manihoti* has been reared from the stem, and there is a doubtful record from fruit. *A. montei* and *A. pickeli* attack the seed capsules (Norrbom & Kim, 1988b). Identification: Steyskal (1977) and Stone (1942).

A. acris Stone, from Mexico, Panama, Tobago and Venezuela, has been recorded from manchineel (*Hippomane mancinella*) (Norrbom & Kim, 1988b). Identification: Steyskal (1977) and Stone (1942).

A. bahiensis Lima, from Brazil, Colombia, Panama and Trinidad, is recorded from Andean walnut (*Juglans neotropica*), arabica coffee (*Coffea arabica*) and English walnut (*J. regia*) (Norrbom & Kim, 1988b). Records from common guava (*Psidium guajava*) and hog-plum (*Spondias mombin*) require confirmation (Anon, 1975). Identification: Steyskal (1977) and Stone (1942).

A. bezzii Lima, from Panama and Venezuela, is recorded from *Sterculia apetala*; a record from tropical almond (*Terminalia catappa*) requires confirmation (Norrbom

& Kim, 1988b). Identification: Steyskal (1977) and Stone (1942) (as *A. balloui* Stone), and Norrbom (1991).

A. consobrina (Loew), from Argentina, Brazil and Guyana, is recorded from giant granadilla (*Passiflora quadrangularis*) (Norrbom & Kim, 1988b). Identification: Steyskal (1977) and Stone (1942).

A. daciformis Bezzi, from Argentina, Bolivia, southern Brazil, Paraguay and Peru, is recorded from guava (*Psidium* sp.) and peach (*Prunus persica*) (Norrbom & Kim, 1988b). Identification: Steyskal (1977).

A. ethalea (Walker), from Brazil and Trinidad, is recorded from giant granadilla (*Passiflora quadrangularis*) and yellow granadilla (*P. laurifolia*) (Norrbom & Kim, 1988b). Identification: Steyskal (1977) and Stone (1942).

A. limae Stone, from and Venezuela to southern USA, is recorded from giant granadilla (*P. quadrangularis*) (Norrbom & Kim, 1988b). Identification: Steyskal (1977) and Stone (1942).

A. margarita Caraballo, from Venezuela, is recorded from star-apple (*Chrysophyllum cainito*) (Norrbom & Kim, 1988b). Identification: Caraballo (1985).

A. mburucuyae Blanchard, from Argentina, is recorded from blue passion fruit (*P. caerulea*) (Norrbom & Kim, 1988b). Identification: Steyskal (1977).

A. minensis Lima, from Brazil and Peru, is recorded from common guava (*Psidium guajava*), loquat (*Eriobotrya japonica*) and peach (*Prunus persica*) (Norrbom & Kim, 1988b). Identification: Steyskal (1977).

A. nigrifascia Stone, from Florida (USA), is recorded from sapodilla (*Manilkara zapota*) (Norrbom & Kim, 1988b). Identification: Steyskal (1977) and Stone (1942).

A. pallens Coquillett, from Central America and Texas, is recorded from gum bumelia (*Bumelia languinosa*) (Wasbauer, 1972). Identification: Steyskal (1977).

A. pallidipennis Greene, from Colombia and Panama, is recorded from giant granadilla (*Passiflora quadrangularis*) (Norrbom & Kim, 1988b). Identification: Steyskal (1977) and Stone (1942).

A. panamensis Greene, from Costa Rica and Panama, is recorded from star-apple (*C. cainito*) (Norrbom & Kim, 1988b). Identification: Steyskal (1977) and Stone (1942).

A. parishi Stone, from Costa Rica, Guyana and Venezuela, is recorded from common guava (*Psidium guajava*) (Norrbom & Kim, 1988b). Identification: Steyskal (1977).

A. passiflorae Greene, from Panama, is recorded from a species of granadilla (*Passiflora vitifolia*) (Norrbom & Kim, 1988b). Identification: Steyskal (1977) and Stone (1942).

A. perdita Stone, from Brazil and Peru, is recorded from hog-plum (*Spondias mombin*) (Norrbom & Kim, 1988b). Identification: Steyskal (1977) and Stone (1942).

A. punctata Hendel, from Argentina, Brazil (central and southern) and Paraguay, is recorded from common guava (*Psidium guajava*) (Norrbom & Kim, 1988b). Identification: Steyskal (1977) and Stone (1942).

A. rheediae Stone, from Brazil, Columbia, Panama, Trinidad and Venezuela, is recorded from bakupari (*Rheedia braziliensis*) and madrono (*R. acuminata*) (Norrbom & Kim, 1988b). Identification: Steyskal (1977) and Stone (1942).

A. sagittata (Stone), from Mexico, Panama, and Texas (USA), is recorded from the egg-fruit tree (*Pouteria campechiana*) (Norrbom & Kim, 1988b). Identification: Steyskal (1977).

A. schultzi Blanchard, from Argentina, is recorded from the common guava (*Psidium guajava*) (Norrbom & Kim, 1988b). Identification: Steyskal (1977).

A. sororcula Zucchi, from Brazil, is a pest of common guava (*Psidium guajava*) and Surinam cherry (*Eugenia uniflora*) (Norrbom & Foote, 1989), and it has also been recorded from arabica coffee (*Coffea arabica*) (Norrbom & Kim, 1988b). It is very similar to *A. fraterculus* and its separation from that species was described by Zucchi (1979).

A. steyskali Korytkowski, from Peru, is recorded from star-apple (*Chrysophyllum cainito*) (Norrbom & Kim, 1988b). Identification: Steyskal (1977).

A. turicai Blanchard, from Argentina, is recorded from peach (*Prunus persica*) (Norrbom & Kim, 1988b). Identification: Steyskal (1977).

A. zenildae Zucchi, from Brazil, is recorded from common guava (*Psidium guajava*) (Zucchi, 1979). Identification: Zucchi (1979).

GENUS *Bactrocera* Macquart

(figs 19, 36, 40)

Taxonomic notes

Prior to the work of Drew (1989a) most authors placed all *Bactrocera* spp. within the genus *Dacus*, although a few authors treated the present subgenera as distinct genera (recent examples being Cogan & Munro, 1980; Munro, 1984; Ito, 1983-5). The reasons for separating *Bactrocera* from *Dacus* were explained by Drew (1989a; 1989c) and that separation is further supported by the larval data presented in this work.

Distribution and biology

A genus of about 440 species which attack a wide range of fruits in the tropical and warm temperate regions of the Old World. Most species are found in tropical Asia, the South Pacific and Australia, with very few species in Africa and only *B. oleae* in southern Europe. A few species have become established outside of their native range, namely *B. cucurbitae* in eastern Africa and Hawaii, *B. latifrons* in Hawaii, and members of the *B. dorsalis* species complex in both Hawaii and Suriname. Most of the pest species are multivoltine and attack fruits belonging to several unrelated families of plants, their larvae developing in the fruit flesh. The exceptions are a few species of subgenus *Zeugodacus*, which like many *Dacus* spp., are primarily associated with the fruit and flowers of Cucurbitaceae. Mating usually occurs on or near the host fruits, but it can occur at some distance, and pupariation usually takes place in the soil beneath the host tree. The biology of *Bactrocera* and *Dacus* spp. was reviewed by Fletcher (1987; 1989b; 1989c). The genus is divided into 23 subgenera, all of which are represented in the area from Asia to Australia, and only five of which occur in Africa.

Diagnosis of third instar larva

Antennal sensory organ with a stout basal segment and cone-shaped distal segment; maxillary sensory organ broad and flat, with 2 distinct groups of sensilla surrounded by small, discontinuous cuticular folds. Stomal sensory organ small, rounded, with small sensilla, surrounded by several preoral lobes; oral ridges with 6-26 rows of teeth; accessory plates along outer edges of oral ridges. Mouthhooks usually each without a preapical tooth (known exceptions are some spp. of the *B. (Zeugodacus)* group of subgenera which have a vestigial tooth). Anterior margin of T1 with several rows of small, sharply pointed spinules; anterior spiracles with a single row of 8-25 uniform, stout, rounded tubules, each with a distinct apical slit; posterior spiracular slits about 2.5-4.5 times as long as broad; spiracular hairs in 4 bundles either side of spiracular slits; anal lobes protuberant, surrounded by 3-8 discontinuous rows of small, sharply pointed spinules with a small concentration of spinules below anal opening.

Attractants

Males of most species are attracted to either cue lure or methyl eugenol, but never both. Members of subgenus *Zeugodacus* are only known to respond to cue lure, a feature they share in common with most *Dacus* spp. Known responses of Australian and Pacific species to male lures were listed by Drew (1974; 1982a; 1989a) and Drew & Hooper (1981); no comprehensive lists exist for species in other regions (see p. 107).

Literature

The adults of most species can be identified using the following key works: Africa (Munro, 1984); Australia (Drew, 1982a; 1989a); Indonesia (Hardy, 1982b; 1983a); Japan (Ito, 1983-5; Shiraki, 1968); Papua New Guinea area (Drew, 1989a); Philippines (Hardy, 1974); South Pacific Region (Drew, 1989a); and Thailand (Hardy, 1973). The larvae of 16 Australian species, including all of the important pest species from that area, were keyed by Exley (1955). The larvae of four Malaysian pest species can be separated using a key by Rohani (1987).

Subgenus *AFRODACUS* BEZZI

A subgenus of 18 species belonging to the *B. (Bactrocera)* group of subgenera. Most *B. (Afrodacus)* spp. are found in Asia, the South Pacific and Australia, but the subgenus also includes a few African species.

Bactrocera (Afrodacus) jarvisi (Tryon)

(figs 138, 175; pl. 9)

Taxonomic notes

This species has also been known as *Afrodacus jarvisi* (Tryon), *Chaetodacus jarvisi* Tryon, *Dacus australis* Hendel and *D. jarvisi* (Tryon).

Pest status and commercial hosts

A major pest in Queensland, Australia, where it attacks common guava (*Psidium guajava*), Japanese persimmon (*Diospyros kaki*) and mango (*Mangifera indica*) (Drew, 1982a); guava is also heavily attacked in the Northern Territory, Australia (Smith *et al.*, 1988). In addition, *B. jarvisi* may attack apricot (*Prunus armeniaca*), bilimbi (*Averrhoa bilimbi*), cavendish banana (*Musa* sp.), dwarf banana (*M. acuminata*), grapefruit (*Citrus x paradisi*), Jew plum (*Spondias cytherea*), Malay-apple (*Syzygium malaccense*), papaya (*Carica papaya*), peach (*Prunus persica*), pear (*Pyrus communis*), pomegranate (*Punica granatum*), quince (*Cydonia oblonga*), soursop (*Annona muricata*), strawberry guava (*Psidium littorale*), sweet orange (*Citrus sinensis*) and tropical almond (*Terminalia catappa*) (Drew, 1989a; Smith *et al.*, 1988).

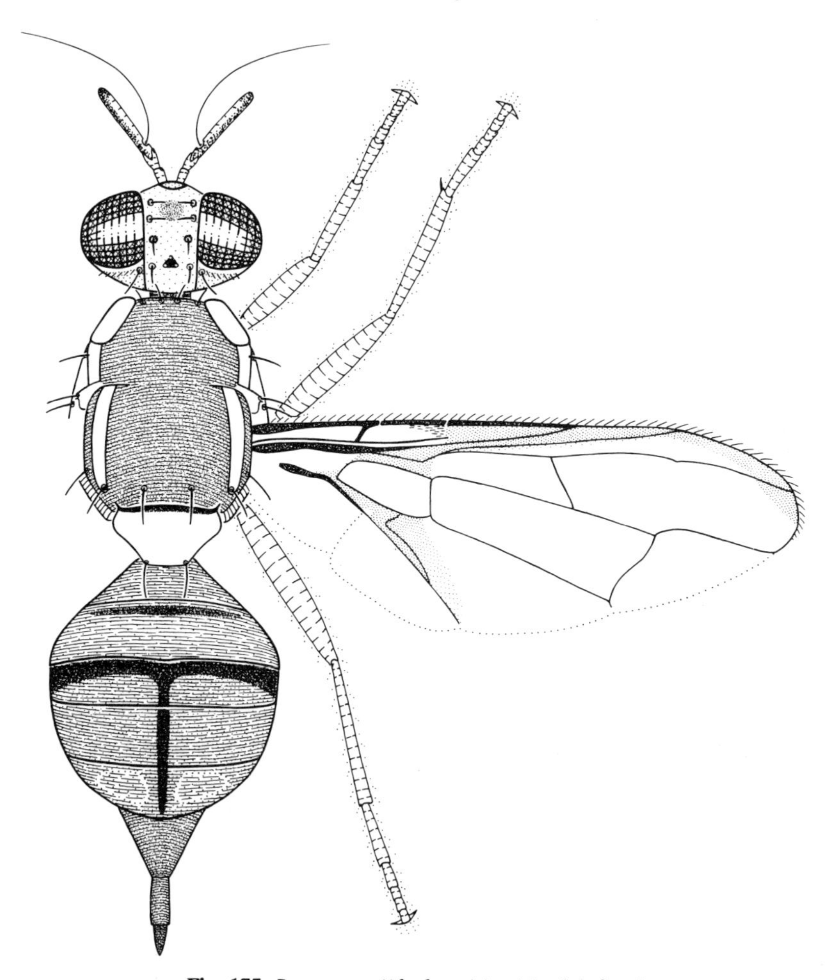

Fig. 175. *Bactrocera (Afrodacus) jarvisi*, adult female.

Wild hosts

This is a highly polyphagous species which has been recorded from wild fruits belonging to the families Anacardiaceae, Barringtoniaceae, Chrysobalanaceae, Combretaceae, Cucurbitaceae, Lecythidaceae, Meliaceae, Myrtaceae, Oleaceae and Sapotaceae (Drew, 1989a; Smith *et al.*, 1988). Fitt (1986) showed that individuals reared from a wild host (*Planchonia careya* (F. Muell.) Knuth, Lecythidaceae) had a strong preference for that plant over cultivated hosts offered in a laboratory test, but they readily attacked cultivated fruits (apple, pear, tomato) when the wild fruit was not offered. Correspondingly, *B. jarvisi* usually only attacks cultivated fruit outside of the fruiting season of *P. careya* (Smith *et al.*, 1988).

Adult identification

A yellow to orange-brown species with lateral yellow stripes on the scutum, facial spots, prescutellar acrostichal setae, 2 scutellar setae, and a typical dacine wing pattern (fig. 175); male with a pecten. Without anterior supra-alar setae.

Description of third instar larva

Larvae medium sized, length 8.0-10.0mm; width 1.2-1.5mm. *Head*: Stomal sensory organs (pl. 9.b) each with 4-5 long sensilla surrounded by 5 large unserrated preoral lobes; oral ridges (pl. 9.a) with 11-16 rows of medium to long, sharply pointed teeth; accessory plates (pl. 9.a) small and numerous, with serrated edges; mouthhooks heavily sclerotised, apically black, basally brown to black, without preapical teeth (fig. 138). *Thoracic and abdominal segments*: Anterior portions of each thoracic segment with encircling, discontinuous rows of small spinules. T1 with 7-10 rows of small, sharply pointed spinules, all posteriorly directed; T2 with 4-6 rows dorsally, 1-2 rows laterally, 3-4 rows ventrally; T3 with 3-4 rows dorsally, 1-2 rows laterally, 1-3 rows ventrally. A1-A8 with creeping welts of slightly raised fusiform areas with small, stout spinules arranged in transverse rows, with 2-3 rows anteriorly directed, the remaining 3-7 rows posteriorly inclined. A8 with obvious intermediate areas and 7 small paired sensilla arranged 2 dorsal, 1 lateral, 3 intermediate and 1 ventral. Ecdysial scar obvious. *Anterior spiracles* (pl. 9.c): 11-14 tubules. *Posterior spiracles* (pl. 9.d): Placed just above midline; 3 pairs of spiracular slits arranged in a radiating pattern from midline, each about 3 times as long as broad and surrounded by a sclerotised rima; dorsal and ventral spiracular hair bundles of 17-24 hairs, lateral bundles of 8-12 hairs. *Anal area* (pl. 9.e): Lobes surrounded by 3-5 discontinuous rows of small, stout spinules, concentrating into stouter spinules just below anal opening. *Source*: Based on specimens from Queensland, Australia.

Distribution

Australia: Across the north from Broome in Western Australia to the Sydney area of New South Wales.
Distribution mapped by Drew (1982a).

Attractant

Males are weakly attracted to cue lure in north-west Western Australia, but show no lure response elsewhere.

Other references

BACTERIAL SYMBIONTS, Fitt & O'Brien (1985); BEHAVIOUR, Fitt (1984) and P.H. Smith (1989); BIOLOGY, Fletcher (1989b); ECOLOGY, Fitt (1989); LARVA, Exley (1955); PHEROMONES, Fitt (1981); POST-HARVEST DISINFESTATION, Heather (1985).

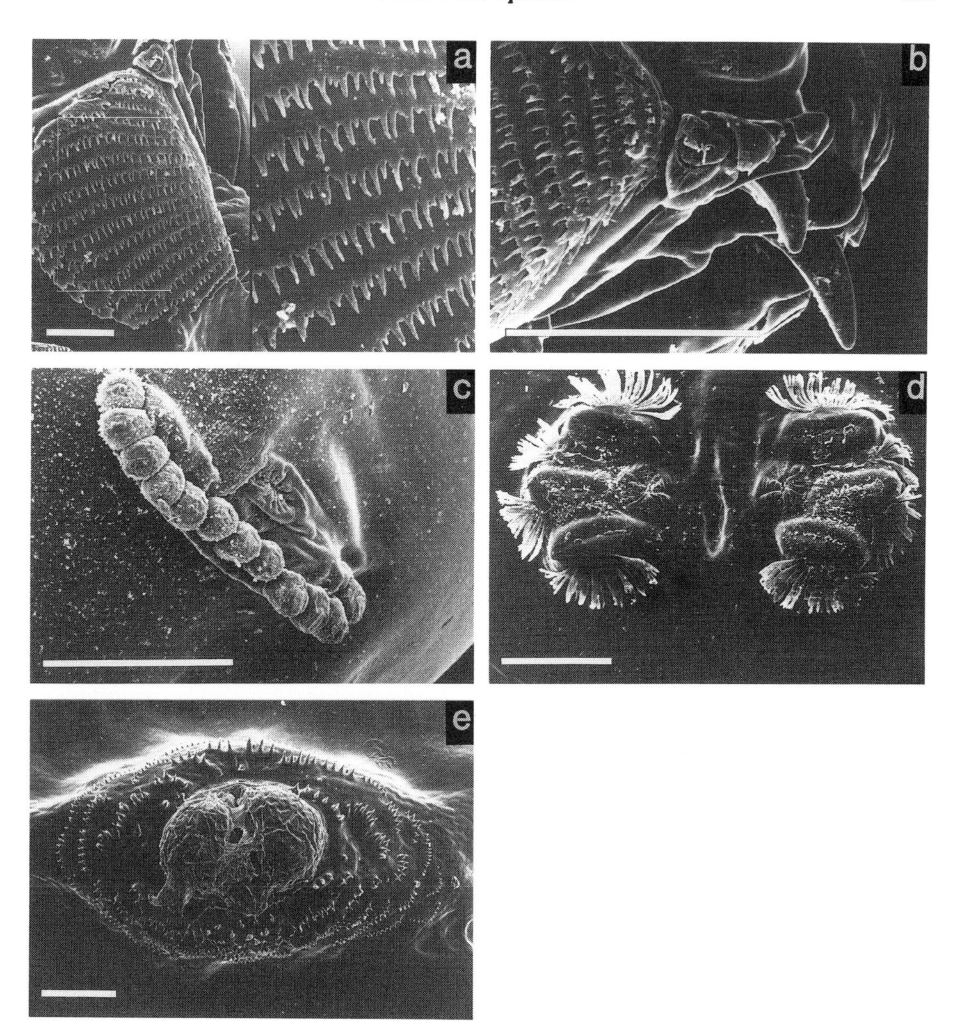

Plate 9. *Bactrocera (Afrodacus) jarvisi*, SEMs of 3rd instar larva; a, oral ridges, lateral view (scale=50μm); b, stomal sensory organ, lateral view (scale=0.1mm); c, anterior spiracle (scale=0.1mm); d, posterior spiracles (scale=0.1mm); e, anal lobes (scale=0.1mm).

Subgenus *AUSTRODACUS* PERKINS

Bactrocera (Austrodacus) cucumis (French)

Cucumber Fruit Fly

(figs 85, 139, 176; pl. 10)

Taxonomic notes

Subgenus *Bactrocera (Austrodacus)* belongs to the *B. (Zeugodacus)* group of subgenera. This is the only species of subgenus *B. (Austrodacus)* and it has also been known as *A. cucumis* (French), *Dacus cucumis* French and *D. tryoni* var. *cucumis* French.

Pest status and commercial hosts

A serious pest of cucurbits, tomato (*Lycopersicon esculentum*) and papaya (*Carica papaya*). The cucurbits attacked are bitter gourd (*Momordica charantia*), cucumber (*Cucumis sativus*), luffa (*Luffa aegyptiaca*), rockmelon (*C. melo*), snakegourd (*Trichosanthes cucumerina*) and marrow, pumpkin and squash (*Cucurbita pepo*) (Drew, 1982a; 1989a). It is also recorded from purple granadilla (*Passiflora edulis*) (Smith *et al.*, 1988).

Wild hosts

Known wild hosts are *Diplocyclos palmatus* (L.) C. Jeffrey (Cucurbitaceae) and *Glochidion harveyanum* Domin (Euphorbiaceae) (Drew, 1989a). Under laboratory conditions, *B. cucumis* will not readily accept fruits outside of its normal host range when deprived of preferred hosts (Fitt, 1986).

Adult identification

A yellow to orange-brown species with both medial and lateral yellow stripes on the scutum, facial black spots, 4 scutellar setae, and a typical dacine wing pattern (fig. 176). Without anterior supra-alar setae and prescutellar acrostichal setae; male without a pecten.

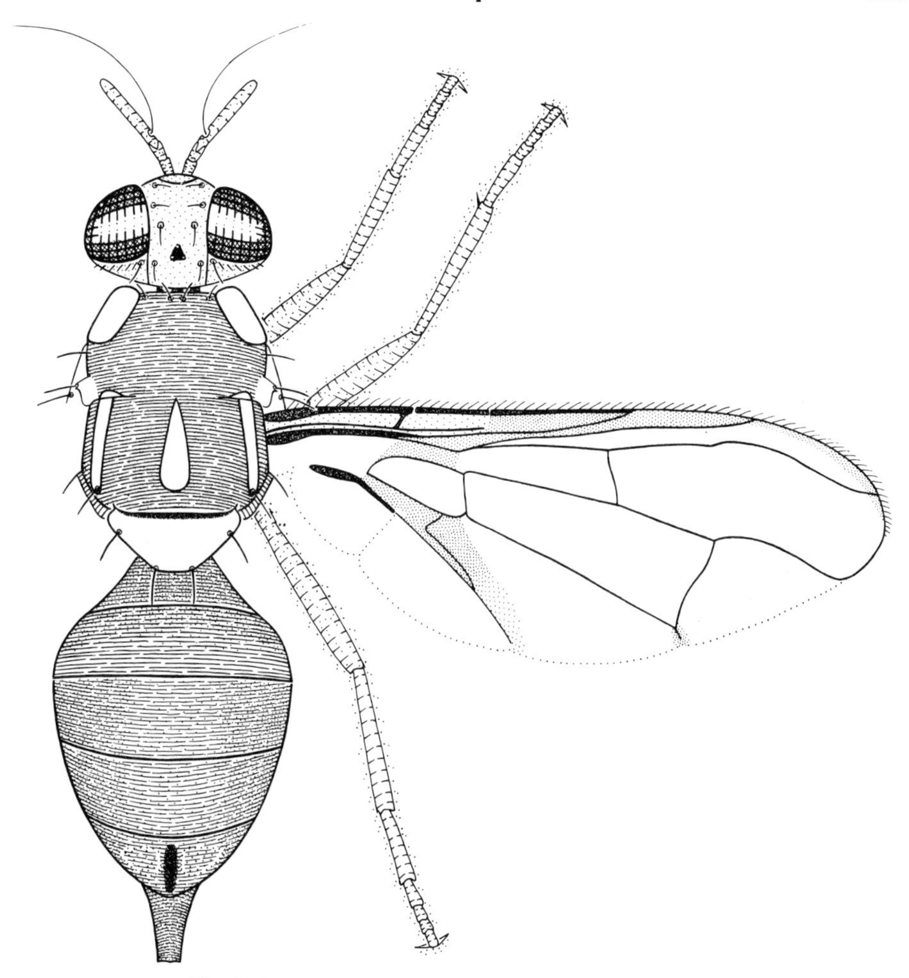

Fig. 176. *Bactrocera (Austrodacus) cucumis*, adult female.

Description of third instar larva

Larvae large, length 9.0-12.0mm; width 1.5-2.0mm. *Head*: Each stomal sensory organ (pl. 10.a) large, surrounded by 6-8 large, serrated preoral lobes resembling short oral ridges; oral ridges (pl. 10.b) with 14-20 rows of large, deeply serrated, bluntly rounded teeth; accessory plates elongate, large, serrated, interlocking with outer edges of oral ridges; mouthhooks large, heavily sclerotised, each with a small preapical tooth (fig. 139). *Thoracic and abdominal segments*: A broad band of discontinuous rows of very stout, sharp spinules encircling anterior margin of each thoracic segment. T1 with 7-10 rows of stout spinules dorsally and laterally, becoming finer ventrally; T2 with stout, broad spinules, 5-8 rows dorsally, 3-4 rows laterally and 5-8 rows ventrally; T3

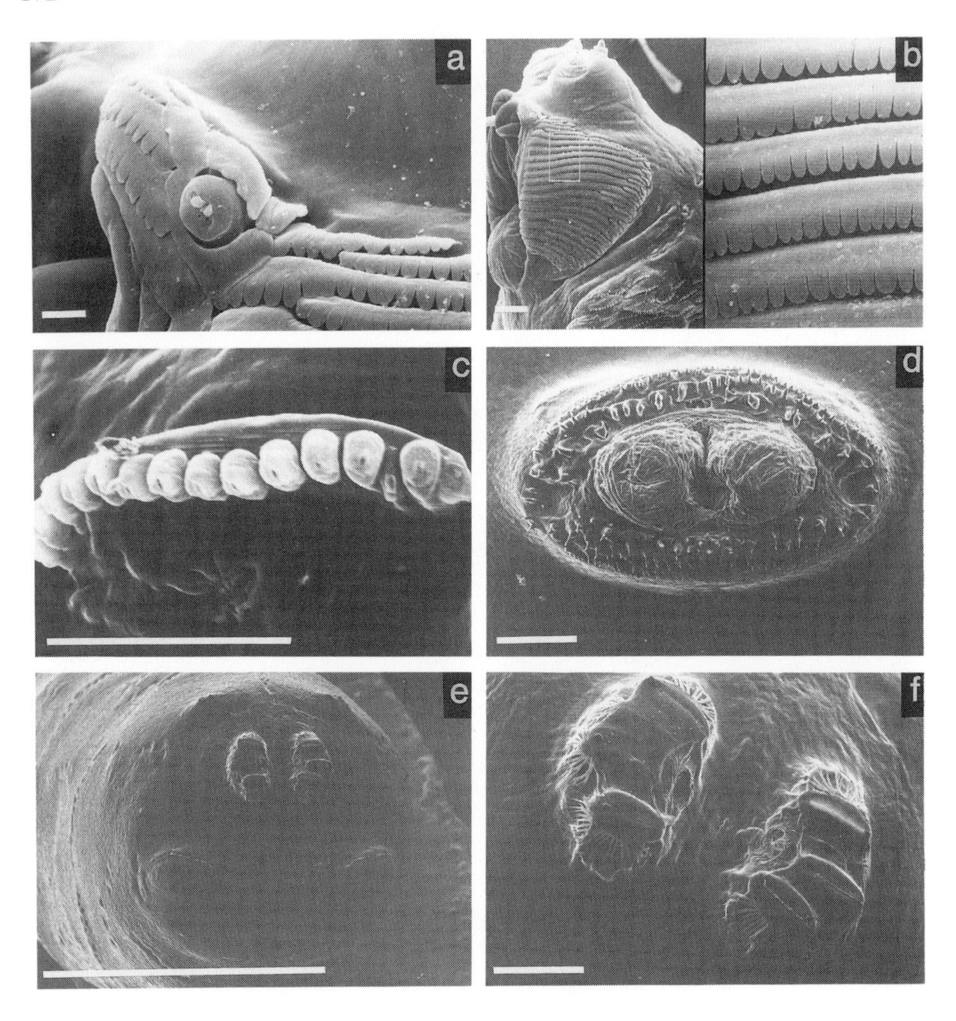

Plate 10. *Bactrocera (Austrodacus) cucumis*, SEMs of 3rd instar larva; a, stomal sensory organ and preoral lobes, lateral view (scale=10μm); b, head and oral ridges, lateral view (scale=50μm); c, anterior spiracle (scale=0.1mm); d, anal lobes (scale=0.1mm); e, A8, posterior view (scale=1.0mm); f, posterior spiracles (scale=0.1mm).

spinules similar to those on T2, with 4-6 rows dorsally, 0-4 rows laterally and 5-7 rows ventrally. Creeping welts large, obvious, with broad, stout spinules in discontinuous transverse rows. A8 with large, protuberant intermediate areas (pl. 10.e) and obvious sensilla. *Anterior spiracles* (pl. 10.c): 13-19 tubules. *Posterior spiracles* (pl. 10.e, f): Each spiracular slit large, thick walled, about 3 times as long as broad; dorsal and ventral spiracular hair bundles of 9-15 hairs each, lateral bundles of 4-16 hairs; each spiracular hair with a broad trunk and branched in apical third to a quarter. *Anal area* (pl. 10.d): Lobes large, surrounded by 3-5 discontinuous rows of large, stout spinules concentrating in a cluster of spinules below anal opening. *Source*: Based on specimens from Queensland, Australia.

Distribution
Australia: The entire east coast of Queensland, and the extreme north-east of New South Wales. Also recorded from the Northern Territory (possibly not established) and from Prince of Wales Island in the Torres Straits.
Distribution mapped by Drew (1982a).

Attractant
Males are not known to be attracted to either cue lure or methyl eugenol.

Other references
BEHAVIOUR, P.H. Smith (1989); BIOLOGY, Fletcher (1989b) and Vuttanatungum & Hooper (1974); LARVA, Exley (1955); PHEROMONES, Kitching *et al.* (1989); POST-HARVEST DISINFESTATION, Heather (1985); REARING, Tsitsipis (1989).

Subgenus *BACTROCERA* MACQUART

Taxonomic notes
Some species have formerly been placed in the subgenera *Apodacus* Perkins, *Chaetodacus* Bezzi, *Dasyneura* Saunders, *Marquesadacus* Malloch and *Strumeta* Walker, which are synonyms of *Bactrocera*. Species belonging to the genera *Aglaodacus* and *Mauritidacus*, described by Munro (1984), should also be included in subgenus *B. (Bactrocera)*.

Distribution
A subgenus of about 260 species which are found from tropical Asia to Australia and the South Pacific Region, with two species spread to other regions, namely, *B. latifrons* in Hawaii, and members of the *B. dorsalis* species complex in both Hawaii and Suriname.

Bactrocera (Bactrocera) albistrigata (de Meijere)

(figs 140, 177; pl. 11)

Taxonomic notes
This species has also been known as *Dacus albistrigatus* de Meijere. All records of *B. frauenfeldi* (p. 203) from north and west of Irian Jaya were probably based on misidentifications of *B. albistrigata* (see Hardy, 1983a). Other members of the *B. frauenfeldi* species group (see Drew, 1989a), including *B. trilineola* Drew (p. 277) from Vanuatu, only differ in the degree of darkening of the postpronotal lobes, scutellum and face, and further investigation is required to confirm that they are all distinct species.

Commercial hosts

A potential pest of cultivated *Syzygium* spp. and possibly other Myrtaceae, and of tropical almond (*Terminalia catappa*). In Malaysia it attacks water apple (*S. samarangense*) and watery rose-apple (*S. aqueum*) (Tan & Lee, 1982; Yunus & Ho, 1980) and in Java it is recorded from Malay-apple (*S. malaccense*) (Hardy, 1983a).

Wild hosts

Wild plants of tropical almond are the major wild host in Malaysia (R.A.I. Drew, unpublished data, 1990).

Adult identification

A predominantly black species; with facial spots, scutum with lateral yellow stripes, postpronotal lobe yellow in posterior half, with anterior supra-alar setae and prescutellar acrostichal setae, 2 scutellar setae, and a characteristic wing pattern (fig. 177); scutellum often with a triangular black mark (similar to fig. 184); male with a pecten.

Description of third instar larva

Larvae medium-sized, length 6.0-7.5mm; width 1.0-1.2mm. *Head*: Stomal sensory organ (pl. 11.a) with 2-3 short, squat, branched sensilla on a rounded base, surrounded by 5-6 unserrated, preoral lobes; oral ridges (pl. 11.b) with 7-12 rows of moderately long, bluntly rounded teeth; 4-7 large, serrated accessory plates; mouthhooks heavily sclerotised, with bluntly rounded apical teeth but lacking preapical teeth (fig. 140). *Thoracic and abdominal segments*: An encircling, anterior band of small, sharply pointed spinules on each thoracic segment. T1 with 6-14 rows gradually increasing dorsally to ventrally; T2 with 4-6 rows surrounding segment; T3 with 0-3 rows dorsally, 2-4 rows ventrally, none laterally. Creeping welts (pl. 11.f) with 9-13 rows of small, sharply pointed spinules. A8 well rounded with obvious intermediate areas and well defined sensilla. *Anterior spiracles* (pl. 11.c): 9-13 tubules. *Posterior spiracles* (pl. 11.d): Each spiracular slit with a thick sclerotised rima, and about 3 times as long as broad with large spiracular hair bundles along outer edge; dorsal and ventral bundles of 12-18 hairs, some branched; lateral bundles of 4-9 hairs, some branched. *Anal area* (pl. 11.e): Lobes well developed; surrounded by 4-7 rows of small, sharply pointed spinules, becoming stouter and more concentrated below anal opening. *Source*: Based on specimens from Malaysia (ex *Terminalia catappa*).

Distribution

Oriental Asia: Indonesia (Java, Lombok, Sulawesi, Sumatra), Malaysia (Peninsular), Thailand (southern).

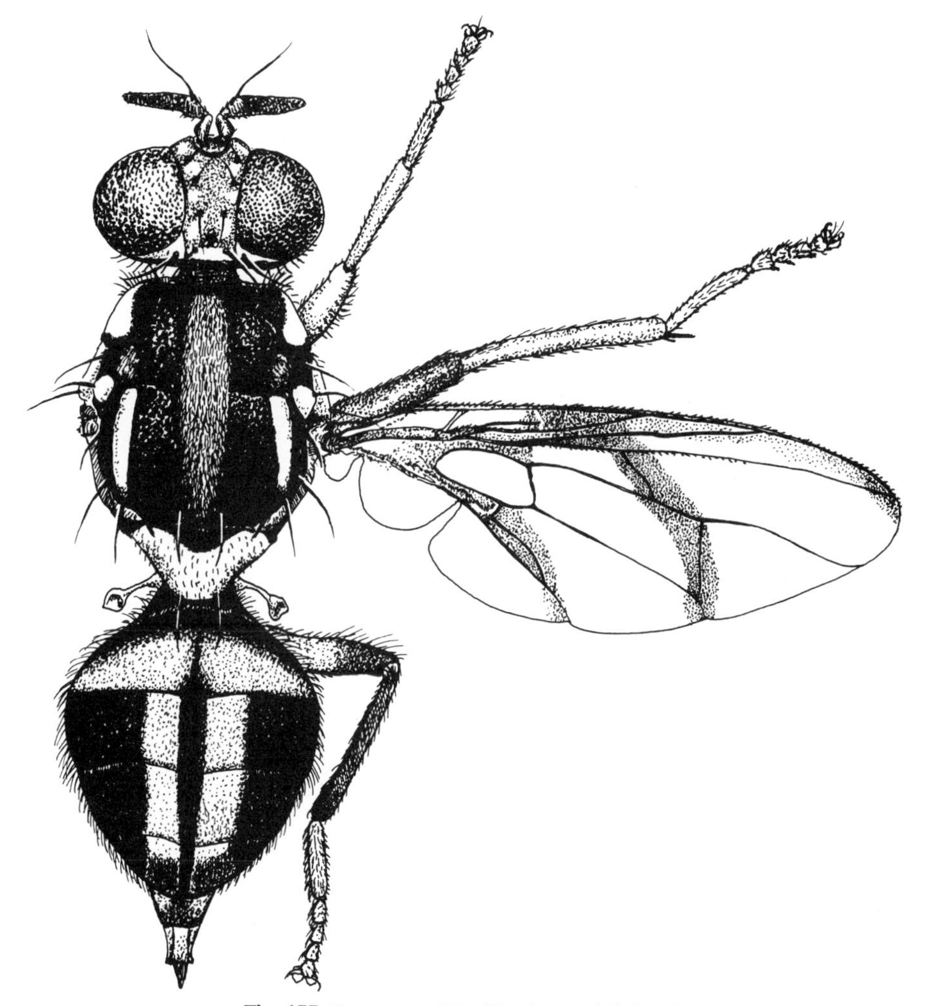

Fig. 177. *Bactrocera (B.) albistrigata*, adult female.

Attractant

Males attracted to cue lure.

Other references

ECOLOGY, Tan & Lee (1982; as *Dacus frauenfeldi*); KRAKATAU, Yukawa (1984).

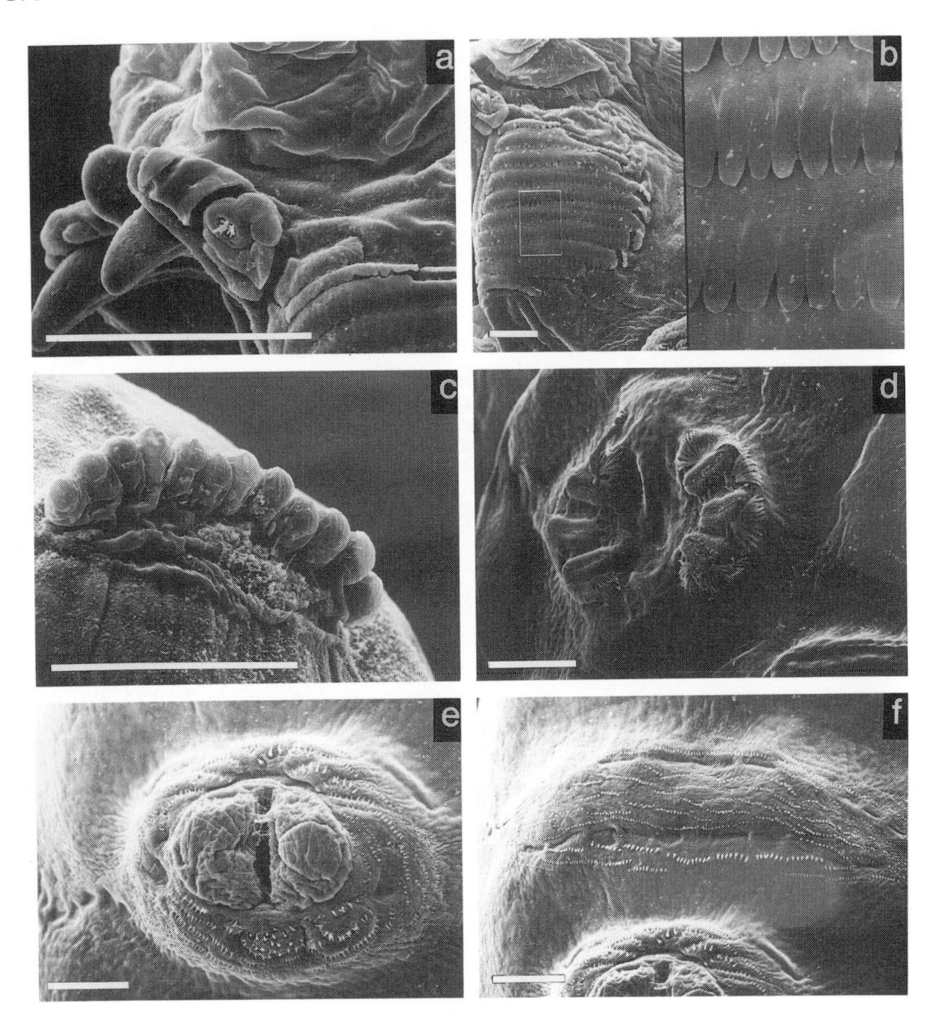

Plate 11. *Bactrocera (B.) albistrigata*, SEMs of 3rd instar larva; a, stomal sensory organ and preoral lobes, lateral view (scale=0.1mm); b, oral ridges, lateral view (scale=50μm); c, anterior spiracle (scale=0.1mm); d, posterior spiracles (scale=0.1mm); e, anal lobes (scale=0.1mm); f, creeping welt (scale=0.1mm).

Bactrocera (Bactrocera) aquilonis (May)

(figs 141, 178; pl. 12)

Taxonomic notes

This species has also been known as *Dacus aquilonis* (May) and *Strumeta aquilonis* May.

Commercial hosts

This potential pest species has expanded its hosts in the Northern Territory of Australia since 1985, for example, such widely planted fruit trees as carambola (*Averrhoa carambola*) and mango (*Mangifera indica*) have only been infested since that date (Smith *et al.*, 1988). Other important hosts in that area are grapefruit (*Citrus x paradisi*), huesito (*Malpighia glabra*) and soursop (*Annona muricata*) (Smith *et al.*, 1988). *B. aquilonis* is also recorded from akee (*Blighia sapida*), apple (*Malus domestica*), avocado (*Persea americana*), bell pepper (*Capsicum annuum*), biribá (*Rollinia pulchrinervis*), cashew (*Anacardium occidentale*), common guava (*Psidium guajava*), custard apple (*Annona reticulata*), dwarf banana (*Musa acuminata*), Indian jujube (*Ziziphus mauritiana*), Indian plum (*Flacourtia jangomas*), Indian prune (*Flacourtia rukam*), Jew plum (*Spondias cytherea*), lemon (*Citrus limon*), loquat (*Eriobotrya japonica*), Malay-apple (*Syzygium malaccense*), meiwa kumquat (*Fortunella x crassifolia*), Otaheite gooseberry tree (*Phyllanthus acidus*), peach (*Prunus persica*), pummelo (*C. maxima*), rose-apple (*S. jambos*), sapodilla (*Manilkara zapota*), star-apple (*Chrysophyllum cainito*), strawberry guava (*Psidium littorale*), sugar-apple (*A. squamosa*), Surinam cherry (*Eugenia uniflora*), tangerine (*Citrus reticulata*), tomato (*Lycopersicon esculentum*), tropical almond (*Terminalia catappa*), watery rose-apple (*S. aqueum*) and wild sweetsop (*Rollinia mucosa*) (Drew, 1982a; 1989a; Smith *et al.*, 1988).

Wild hosts

A polyphagous species recorded from species of Annonaceae, Apocynaceae, Arecaceae, Chrysobalanaceae, Combretaceae, Ebenaceae, Elaeocarpaceae, Euphorbiaceae, Lauraceae, Meliaceae, Myrtaceae, Rosaceae, Rubiaceae and Rutaceae (Drew, 1989a).

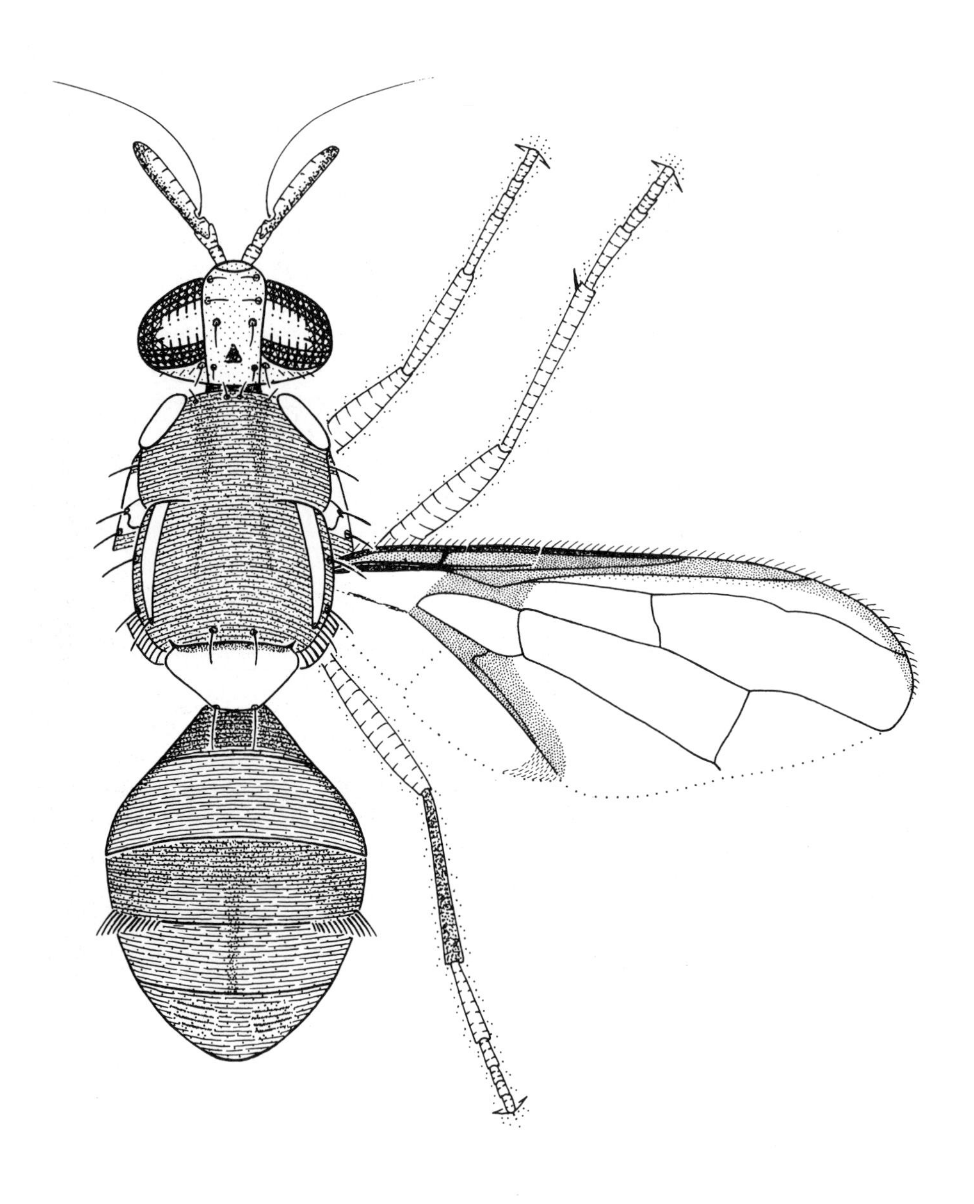

Fig. 178. *Bactrocera (B.) aquilonis*, adult male.

Adult identification

An orange-red species; with facial spots, scutum with lateral yellow stripes, postpronotal lobe yellow, anterior supra-alar setae, prescutellar acrostichal setae, 2 scutellar setae, wing cell c completely covered in microtrichia, and the broad costal band extending from the wing base to the wing apex (fig. 178); male with a pecten.

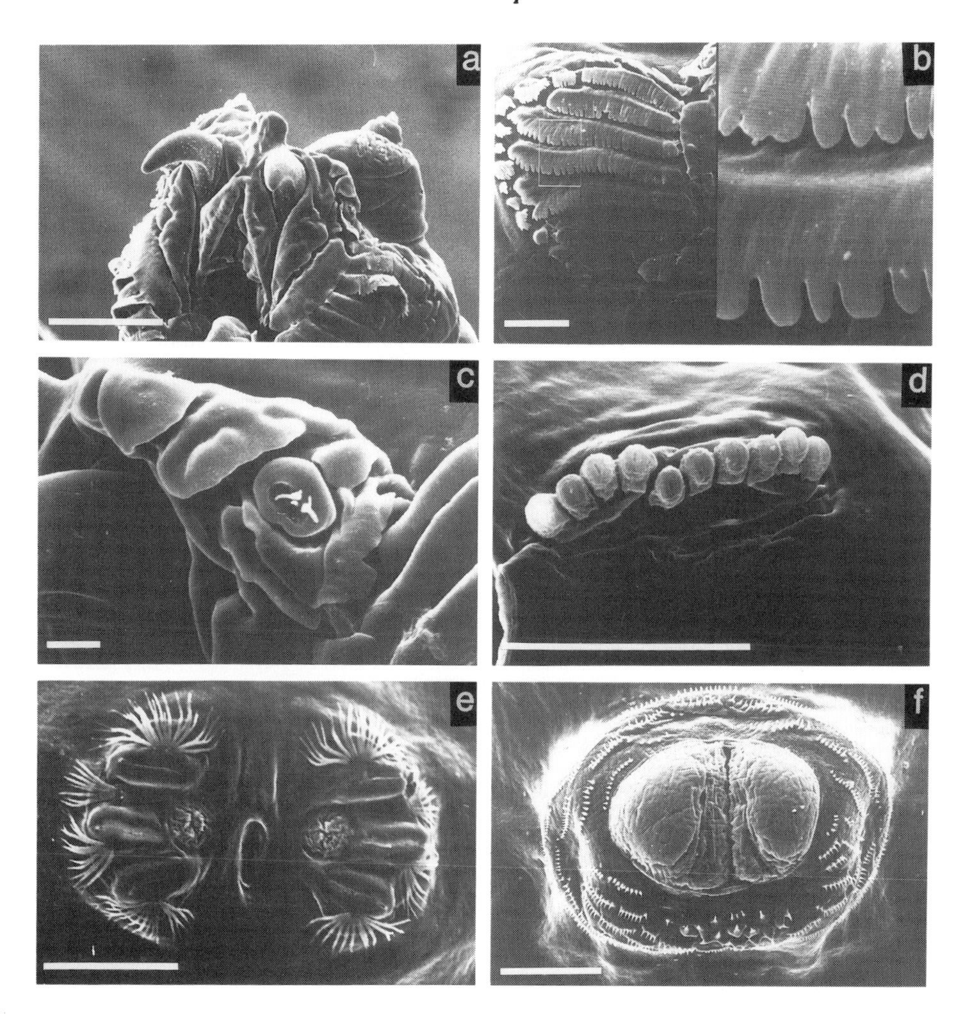

Plate 12. *Bactrocera (B.) aquilonis*, SEMs of 3rd instar larva; a, head, ventral view (scale=0.1mm); b, oral ridges, lateral view (scale=50μm); c, stomal sensory organ and preoral lobes, lateral view (scale=10μm); d, anterior spiracle (scale=0.1mm); e, posterior spiracles (scale=0.1mm); f, anal lobes (scale=0.1mm).

Description of third instar larva

Larvae are medium-sized, length 7.5-10.0mm, width 1.0-2.0mm. *Head*: Stomal sensory organ (pl. 12.c) well defined, with 2-4 branched sensilla, and surrounded by 5-6 large, unserrated preoral lobes; oral ridges (pl. 12.b) with 8-10 rows of short, well defined, bluntly rounded teeth; 8-12 small, serrated accessory plates. Anterior margin of oral cavity with a large rounded projection between heavily sclerotised, large, black mouthhooks which are without preapical teeth (fig. 141). *Thoracic and abdominal segments*: Each thoracic segment with an anterior band of discontinuous rows of small, sharply pointed spinules. T1 with 7-10 rows dorsally, increasing to 11-16 rows laterally and ventrally; T2 and T3 with 3-8 discontinuous rows of small spinules encircling segments. Creeping welts well defined with a combination of 2-4 anteriorly directed and 5-9 posteriorly directed rows of small, sharp spinules. A8 well rounded with obvious sensilla and intermediate areas. *Anterior spiracles* (pl. 12.d): 9-12 tubules. *Posterior spiracles* (pl. 12.e): Each spiracular slit about 3.0-3.5 times as long as broad with a sclerotised rima. Spiracular hairs, long, fine, branched in apical half; dorsal and ventral bundles of 12-20 hairs, lateral bundles of 6-10 hairs. *Anal area* (pl. 12.f): Lobes very large, protuberant, surrounded by 5-7 discontinuous rows of large, sharply pointed spinules, on stout bases, becoming larger below anal opening. *Source*: Based on specimens from Northern Territory, Australia (ex *Terminalia ferdinandiana* Exell).

Distribution

Australia: Northern areas of Western Australia and the Northern Territory.

Attractant

Males attracted to cue lure.

Other references

BEHAVIOUR, P.H. Smith (1989); PHEROMONES, Fitt (1981); SEPARATION FROM *B. TRYONI*, Drew & Lambert, 1986; STYLOPS PARASITOID, Drew & Allwood (1985).

Bactrocera (Bactrocera) correcta (Bezzi)

Guava Fruit Fly

(figs 65, 179)

Taxonomic notes

This species has also been known as *Chaetodacus correctus* Bezzi and *Dacus correctus* (Bezzi).

Commercial hosts

In India this potential pest often occurs with serious pest species such as *B. zonata* and *B. dorsalis* (Kapoor, 1989). In Thailand it frequently attacks common jujube (*Ziziphus jujube*) and tropical almond (*Terminalia catappa*) (D.L. Hancock, pers. comm., 1991). It is also recorded from common guava (*Psidium guajava*), mango (*Mangifera indica*), peach (*Prunus persica*), rose-apple (*Syzygium jambos*) and sapodilla (*Manilkara zapota*) (Clausen *et al.*, 1965; Fletcher, 1919; Satoh *et al.*, 1985; Shah & Vora, 1975). Other recorded hosts, requiring confirmation, are castor-oil plant (*Ricinus communis*), Indian bael (*Aegle marmelos*), karanda (*Carissa carandas*), orange (*Citrus* sp.), robusta coffee (*Coffea canephora*), sandalwood (*Santalum album*) and Surinam cherry (*Eugenia uniflora*) (Kapoor, 1989; Kapoor & Agarwal, 1983; Weems, 1987).

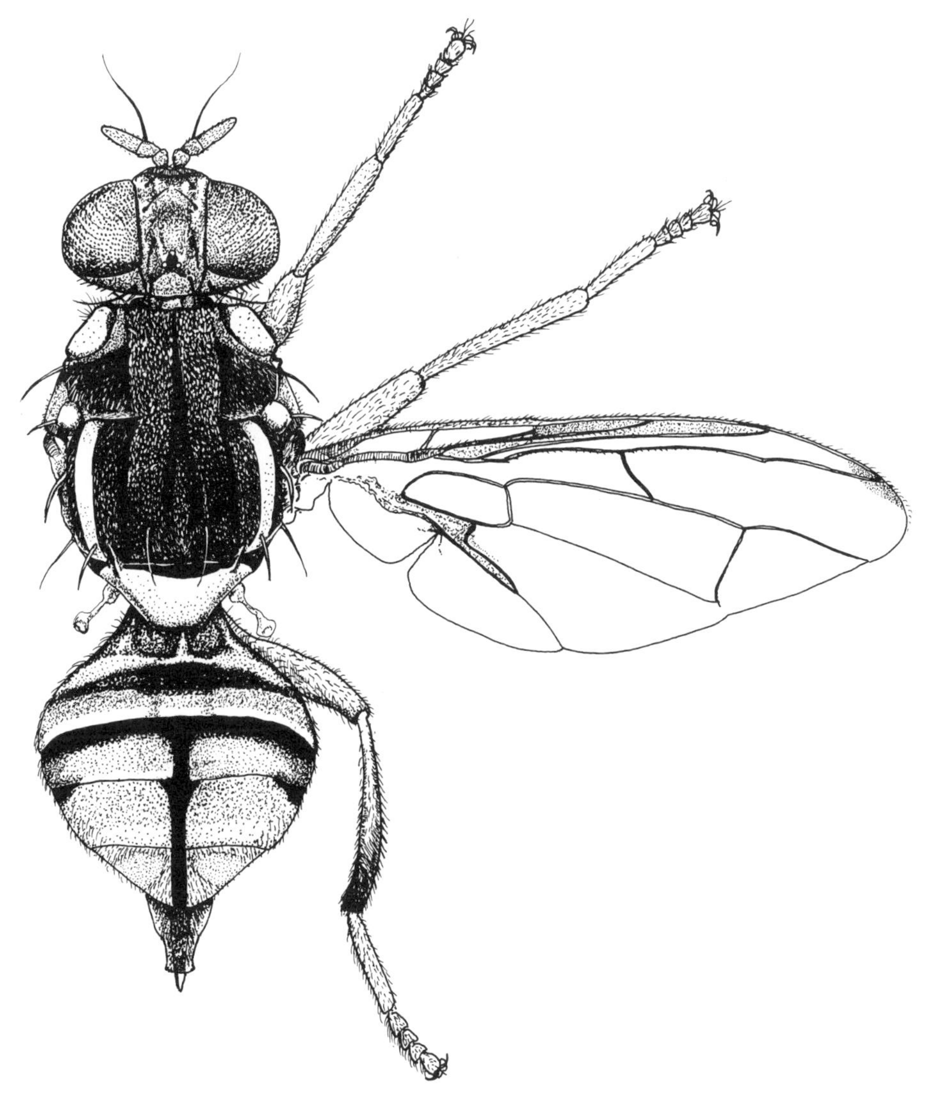

Fig. 179. *Bactrocera (B.) correcta*, adult female.

Adult identification

Scutum predominantly black with lateral yellow stripes; face with lines above the mouth opening which usually join in the centre (fig. 65); with anterior supra-alar setae, prescutellar acrostichal setae, 2 scutellar setae, and wing with a reduced pattern (fig. 179); male with a pecten.

Distribution
North America: A few individuals were trapped in California during 1986, but that adventive population does not appear to have become established (Weems, 1987).
Oriental Asia: India, Nepal, Pakistan, Sri Lanka, Thailand.

Attractant
Males attracted to methyl eugenol; in India Shah & Patel (1976) found that males were also attracted to tulsi plant (*Ocimum sanctum*, Lamiaceae) which yields aromatic oils, 40% of which are methyl eugenol.

Other references
DATASHEET, Weems (1987); MALE TERMINALIA, Singh & Premlata (1985); PARASITOIDS, Narayanan & Chawla (1962).

Bactrocera (Bactrocera) curvipennis (Froggatt)

(fig. 180)

Taxonomic notes
This species has also been known as *Dacus curvipennis* Froggatt and *Strumeta curvipennis* (Froggatt).

Commercial hosts
This is a potential pest of citrus (*Citrus* spp.) which was recorded from mandarin orange (*C. reticulata*) by Drew (1989a). There is also a record from banana (*Musa x paradisiaca*) (Froggatt, 1909) but that was derived from an error caused by confused collection data (Perkins, 1939).

Adult identification
Scutum predominantly black with lateral yellow stripes; with facial spots, anterior supra-alar setae, prescutellar acrostichal setae, 2 scutellar setae, and a characteristic wing pattern (fig. 180); male with a pecten.

Distribution
South Pacific: New Caledonia, Vanuatu.

Attractant
Males attracted to cue lure.

Other reference
PARASITOIDS, Cochereau (1970).

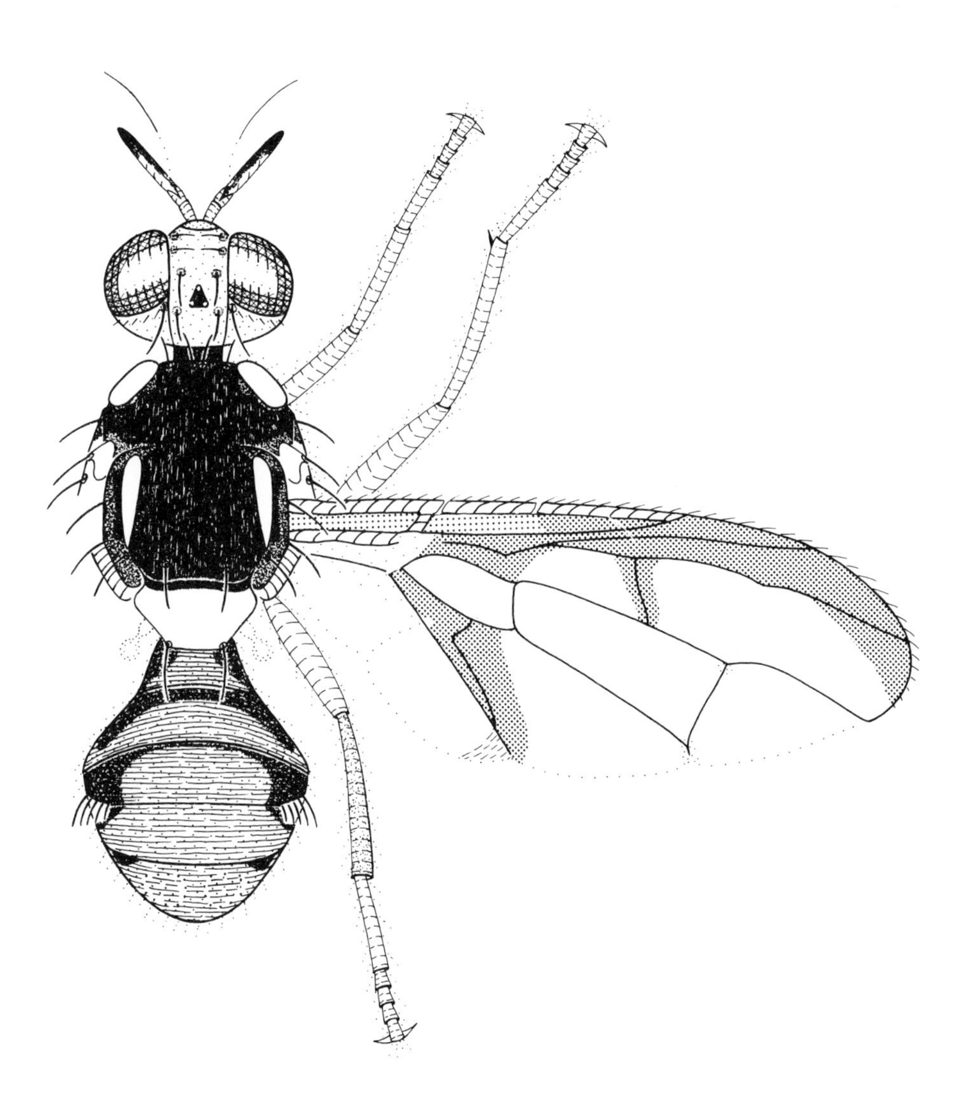

Fig. 180. *Bactrocera (B.) curvipennis*, adult male.

Bactrocera (Bactrocera) distincta (Malloch)

(figs 76, 181)

Taxonomic notes

This species has also been known as *Dacus distinctus* Malloch and *Strumeta distincta* (Malloch).

Commercial hosts

A potential pest which has been recorded from breadfruit (*Artocarpus altilis*), sea hibiscus (*Hibiscus tiliaceus*) and star-apple (*Chrysophyllum cainito*) (Drew, 1989a; Litsinger *et al.*, 1991); the sea hibiscus record has been questioned (A.J. Allwood, pers. comm., 1991) and that may have been an aberrant host. There is also an unconfirmed record from Pacific lychee (*Pometia pinnata*) (SPC, unpublished data, 1988).

Wild hosts

Recorded from *Burckella richii* (A. Gray) H.J. Lam (Sapotaceae) (Drew, 1989a) but it probably has a wider range of hosts.

Adult identification

Scutum predominantly black with lateral yellow stripes; with facial spots, anterior supra-alar setae, prescutellar acrostichal setae, 2 scutellar setae, and a characteristic wing pattern (fig. 181); male with a pecten.

Distribution

South Pacific: American and Western Samoa, Fiji, Tonga.

Attractant

Males attracted to cue lure.

Other reference

PARASITOIDS, Arambourg & Onillon (1971).

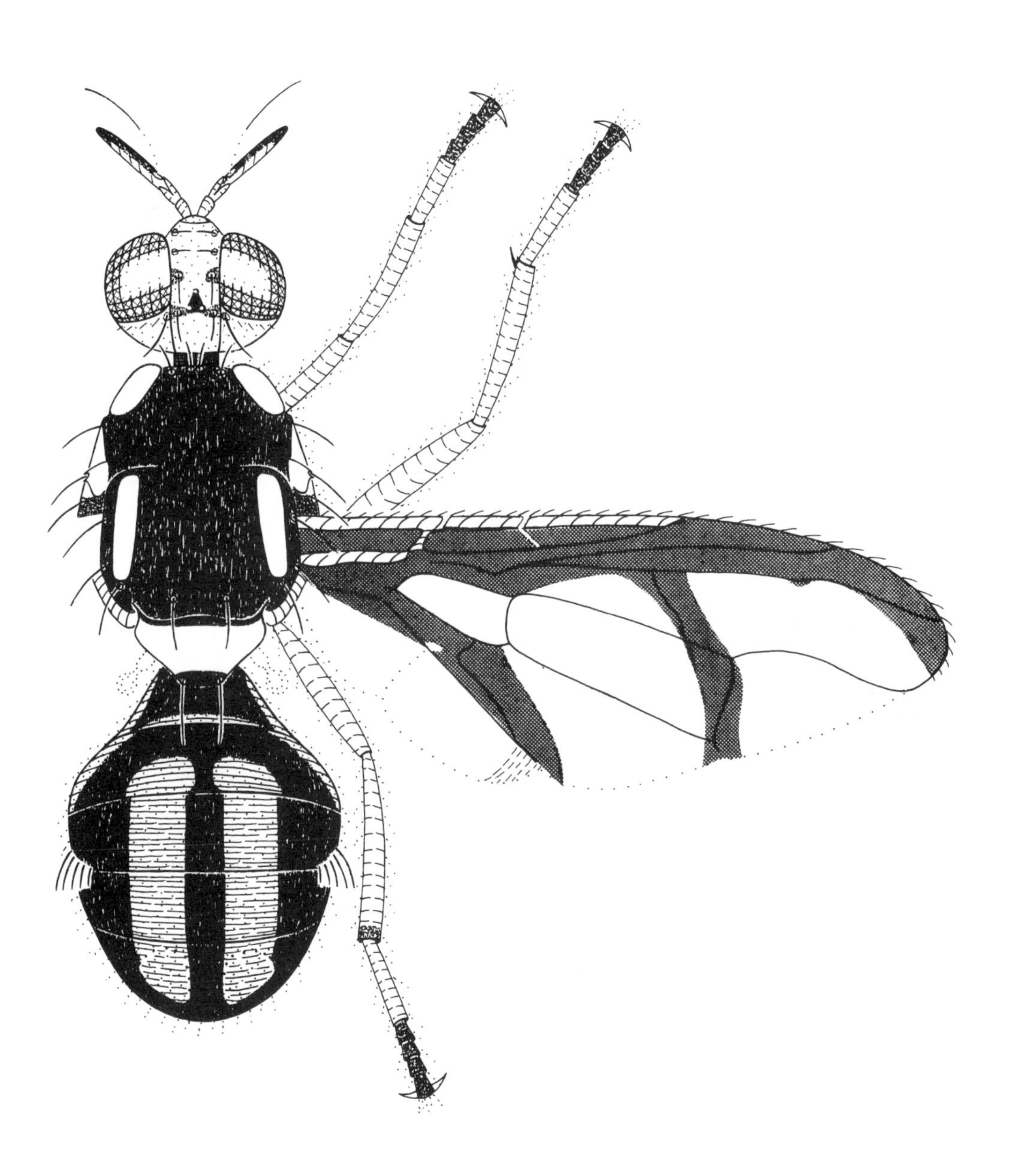

Fig. 181. *Bactrocera (B.) distincta*, adult male.

Bactrocera (Bactrocera) dorsalis (Hendel) Species Complex

Oriental Fruit Fly Species Complex

Taxonomic notes

The fact that the well known *B. dorsalis* is part of a species complex has been known for a long time and the first serious attempt to resolve the problem was made by Hardy (1969), although even Bezzi (1916) appears to have perceived the problem. However, it is only with the application of modern techniques such as gel electrophoresis, pheromone analysis and the study of acoustic signals, that it has been possible to recognize species limits amongst the many varied populations that have been observed in tropical Asia and Australia. Initially, Drew & Hardy (1981) recognized *B. opiliae* (Drew & Hardy) (p. 276) from northern Australia, and more recently Drew (1989a) keyed the eight Australian and Papuan species. Work by R.A.I. Drew and colleagues (Drew & D.L. Hancock, in prep.; M.M. Elson-Harris & Drew, in prep.) has led to the recognition of over 40 species in this complex, including seven Asian species which have been known to attack cultivated hosts. Most of that work has yet to be published, but data so far available include pheromone analysis of five of the species associated with commercial hosts (Perkins *et al.*, 1990), and three symposium papers reporting the early stages of the research (Drew, 1991; Elson-Harris, 1991; Ooi, 1991). Other results of that work have been made available to us and the seven pest (or at least potential pest) species are differentiated in the present work, although full morphological details of the separations are not included. Four of those species will be formally described by R.A.I. Drew & D.L. Hancock (in prep.) and they are labelled as species "A-D" in the present work.

Adult identification

Scutum predominantly black with lateral yellow stripes; with facial spots, anterior supra-alar setae, prescutellar acrostichal setae, 2 scutellar setae, a black T-shaped mark on the abdomen, and a typical dacine wing pattern (fig. 182); male with a pecten.

Attractant

Males of the pest species within this complex are attracted to methyl eugenol; some non-pest species in this complex are attracted to cue lure, and *B. arecae* (Hardy & Adachi) (p. 274), from the Malay Peninsula, is not attracted to either of those lures. Some non-pest species may be found in traps, for example, *B. cacuminata* (Hering) (p. 275) will be trapped in large numbers in methyl eugenol traps set in eastern Australia.

B. (B.) dorsalis Complex: *B. (B.) caryeae* Kapoor

Taxonomic notes

This species has also been known as *Dacus caryeae* Kapoor. It was first recognized by Bezzi (1916) who misidentified it and called it *Chaetodacus ferrugineus incisus* Walker, and many subsequent records of *B. incisa* probably also refer to *B. caryeae* (R.A.I. Drew & D.L. Hancock, in prep.). The true *B. incisa* (Walker) (p. 275) is a Burmese species which is not part of the *B. dorsalis* species complex. Kapoor (1971) described *B. caryeae* from material collected by T.B. Fletcher in 1914, and the same collector supplied some of the material examined by Bezzi (1916).

Commercial hosts

The only confirmed hosts of *B. caryeae* are citrus (*Citrus* sp.), common guava (*Psidium guajava*) and mango (*Mangifera indica*) (R.A.I. Drew & D.L. Hancock, in prep.). Clausen *et al.* (1965) also recorded the following hosts (as *B. dorsalis* and *B. incisa*) which almost certainly refer to this species: robusta coffee (*Coffea canephora*), tangerine (*Citrus reticulata*) and *Ficus* sp. Another record that probably relates to this species is a report of "*B. dorsalis*" in banana (*Musa x paradisiaca*) but the author regarded that as an unusual host (Rao, 1956). Records requiring confirmation are as follows: Chempedak (*Artocarpus integer*) (Bezzi, 1916), plus jackfruit (*A. heterophyllus*) and plum (presumably *Prunus domestica*) (labelled as "on" in NHM collection, and possibly not reared).

Wild hosts

The only recorded wild hosts are bug tree (*Solanum auriculatum* Ait.), patana oak (*Careya arborea* Roxb.; Lecythidaceae) and *S. verbascifolium* L. (Bezzi, 1916; Clausen *et al.*, 1965; Kapoor, 1971).

Distribution

Oriental Asia: Southern India.

B. (B.) dorsalis Complex: *B. (B.) dorsalis* (Hendel)

Oriental Fruit Fly

(figs 19, 68, 70, 142, 182; pl. 13)

Taxonomic notes

This species has also been known as *Chaetodacus dorsalis* (Hendel), *C. ferrugineus dorsalis* (Hendel), *C. ferrugineus okinawanus* Shiraki, *Dacus dorsalis* Hendel and *Strumeta dorsalis* (Hendel). It has also been known as *D. ferrugineus* Fabricius but that name is unavailable under the rules of nomenclature. The names *C. ferrugineus* var. *versicolor* Bezzi and *D. ferrugineus* var. *mangiferae* Cotes have been listed as synonyms of *B. dorsalis*, but they actually belong to species outside of the *B. dorsalis* species complex (R.A.I. Drew & D.L. Hancock, in prep.).

Pest status and commercial hosts

B. dorsalis is a serious pest of a wide range of fruit crops in Taiwan, southern Japan, China and in the northern areas of the Indian subcontinent, and it has also been

Pest status and commercial hosts

B. dorsalis is a serious pest of a wide range of fruit crops in Taiwan, southern Japan, China and in the northern areas of the Indian subcontinent, and it has also been established in the Hawaiian Islands since about 1945 (Pemberton, 1946). Due to the confusion between *B. dorsalis* and related species in Malaysia, the Philippines, Indonesia, southern India and Sri Lanka, there are very little published host data which definitely refer to *B. dorsalis*, as opposed to misidentifications of related species within the *B. dorsalis* species complex.

In China it is a pest of apple (*Malus domestica*), common guava (*Psidium guajava*), mango (*Mangifera indica*), peach (*Prunus persica*) and pear (*Pyrus communis*) (X.-j. Wang, unpublished data, 1988). Other recorded hosts in the eastern part of its native range (China, Japan and Taiwan) are as follows: banana (*Musa x paradisiaca*), carambola (*Averrhoa carambola*), chilli pepper (*Capsicum annuum*), orange (*Citrus* sp.), papaya (*Carica papaya*), plum (*Prunus domestica*), sugar-apple (*Annona squamosa*), tomato (*Lycopersicon esculentum*), wax apple (*Syzygium samarangense*) and wampi (*Clausena lansium*) (Clausen *et al.*, 1965; Koyama, 1989a). Cheng & Lee (1991) gave a list of 89 hosts from Taiwan, but that list included some doubtful hosts, e.g. durian (*Durio zibethinus*), which is not known to be attacked by any fruit flies elsewhere [those records, all of which require confirmation, were received too late for full inclusion in this work].

Host data in the western and southern parts of the range of *B. dorsalis* are confused by misidentifications of other members of the complex. Syed (1970) listed the following hosts from Pakistan: apricot (*Prunus armeniaca*), citrus (*Citrus* spp), common guava, date-plum (*Diospyros lotus*), loquat (*Eriobotrya japonica*), mango, paradise apple (*Malus pumila*), peach, pear and watermelon (*Citrullus lanatus*). There is also an old record, requiring confirmation, from tabasco pepper (*Capsicum frutescens*) (Bezzi, 1916).

The Hawaiian population is said to have been recorded from over 173 species of host plant (Drew, 1982a). However, Carter (1950) said "life-history studies showed that the fruit-fly can infest 128 plants", which indicates that many of those host records may have been based on experimental rearing rather than field observation; no published list of those plants could be traced. Recent studies by Harris & Lee (1986) and Nishida *et al.* (1985) listed the following hosts: coffee (*Coffea* sp.), fig (*Ficus* sp.), loquat, mango, orange, orange jessamine (*Murraya paniculata*), peach, plum, pummelo (*Citrus maxima*), sapote (*Pouteria sapote*), soursop (*A. muricata*), strawberry guava (*Psidium littorale*), Surinam cherry (*Eugenia uniflora*), tangerine (*C. reticulata*), tropical almond (*Terminalia catappa*) and Valencia orange (*C. sinensis*). After 1950, populations of *B. dorsalis* declined in the Hawaiian Islands (E.J. Harris, 1989) and many of the initially attacked hosts were not recorded in recent lists. Maehler (1951), Nakagawa *et al.* (1968) and Pemberton (1946) listed the following additional hosts: akee (*Blighia sapida*), banana, banana passion fruit (*Passiflora mollisima*), Cape gooseberry (*Physalis peruviana*), feijoa (*Feijoa sellowiana*), Hinds' walnut (*Juglans hindsii*), hog-plum (*Spondias mombin*), huesito (*Malpighia glabra*), Malay-apple (*Syzygium malaccense*), paradise tree (*Quassia simarouba*), red mombin (*Spondias

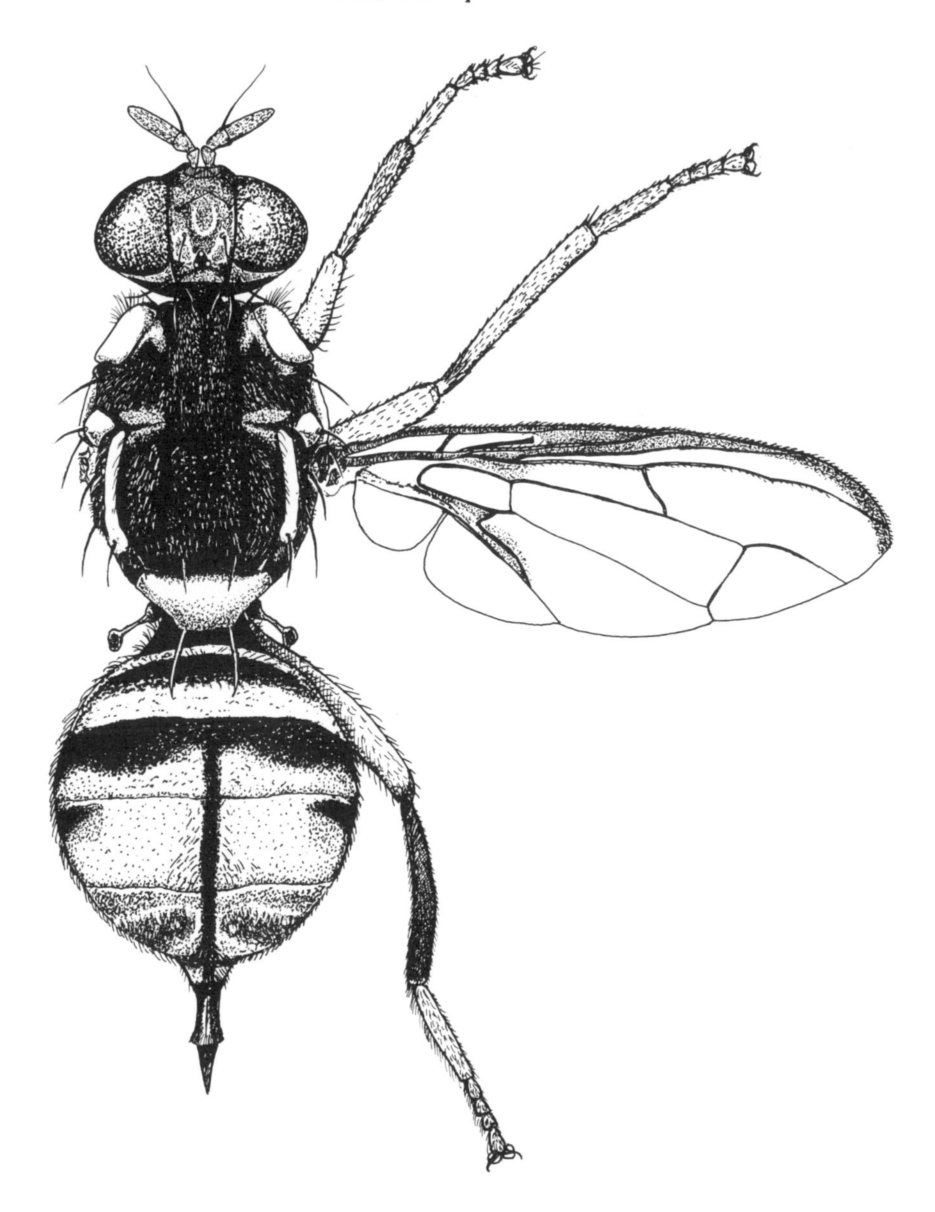

Fig. 182. *Bactrocera (B.) dorsalis*, adult female from the type locality (Taiwan).

purpurea), rose-apple (*Syzygium jambos*), watery rose-apple (*S. aqueum*) and *Solanum seaforthianum*. Haramoto & Bess (1970) indicated that avocado (*Persea americana*) and papaya were also attacked during the first few years following the arrival of *B. dorsalis* in the mid 1940s. The varieties of avocado and banana now planted in Hawaii are not subject to attack (Armstrong, 1983; Armstrong *et al.*, 1983).

There are some unusual horticultural hosts recorded in Hawaii, for example one

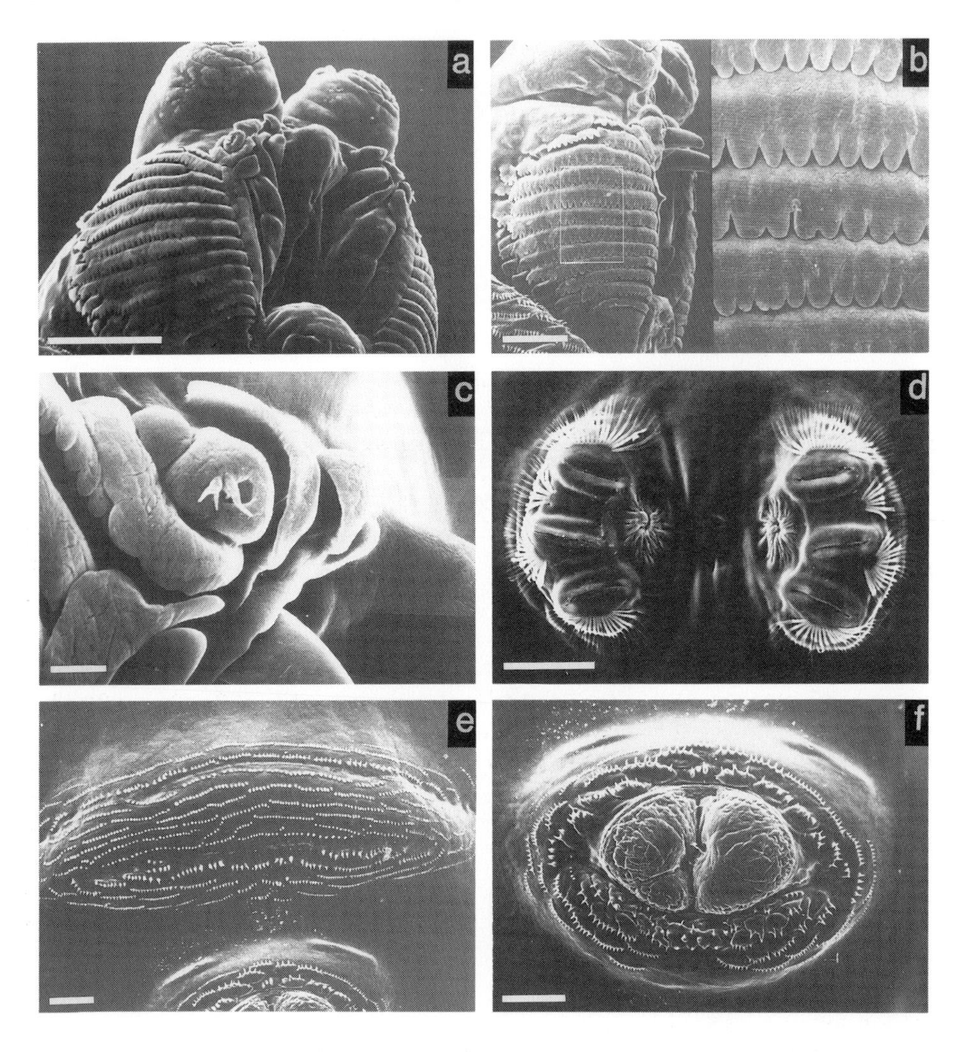

Plate 13. *Bactrocera (B.) dorsalis*, SEMs of 3rd instar larva; a, head, ventral view (scale=0.1mm); b, oral ridges, lateral view (scale=50μm); c, stomal sensory organ and preoral lobes, lateral view (scale=10μm); d, posterior spiracles (scale=0.1mm); e, A8, creeping welt (scale=0.1mm); f, anal lobes (scale=0.1mm).

individual was reared from a sample of 27,000 flowers of Miss Joaquim orchid (*Vanda teres x V. hookerana*) (Flitters, 1951). Similarly, it has been reared from buds of *Sesbania grandiflora* (Nakagawa & Yamada, 1965).

B. dorsalis became established in Guam at about the same time as in Hawaii, and its major host in Guam was custard-apple (*A. reticulata*) (Maehler, 1948). In Rota, Mariana Islands, it was recorded from calabur (*Muntingia calabura*) and limeberry (*Triphasia trifolia*) (Nakagawa *et al.*, 1968).

A member of the *B. dorsalis* species complex, which was probably true *B. dorsalis*, became established in California, USA, during 1974, but was subsequently eradicated. The initially reported hosts were lemon (*C. limon*) and peach (Anon., 1975).

Wild hosts

In Pakistan, Syed (1970) recorded *B. dorsalis* from a wild species of Euphorbiaceae, two species of Rhamnaceae and three of Rosaceae. In Hawaii wild plants of strawberry guava are a major reservoir host (E.J. Harris, 1989) and there are many other less important wild hosts recorded in the Islands.

Description of third instar larva

Larvae medium-sized, length 7.5-10.0mm; width 1.5-2.0mm. *Head*: Stomal sensory organ (pl. 13.c) with 3-4 sensilla, surrounded by 5 large, unserrated preoral lobes; oral ridges (pl. 13.a, b) with 11-14 rows of blunt edged, short teeth; accessory plates 12-15, shell-shaped with small rounded teeth; mouthhooks moderately sclerotised, without preapical teeth (fig. 142). *Thoracic and abdominal segments*: Anterior portion of each thoracic segment with an encircling band of several discontinuous rows of small spinules. T1 with 9-11 rows of large, sharply pointed spinules; T2 spinules small, stout, sharply pointed with 5-6 rows dorsally, 3-4 rows laterally, 5-7 rows ventrally; T3 spinules similar to those on T2, 2-4 rows dorsally, 1-3 rows laterally, 3-5 rows ventrally. Creeping welts (pl. 13.c) with small, stout spinules, with one posterior row of spinules larger and stouter than remainder. A8 rounded with prominent intermediate areas and obvious sensilla. *Anterior spiracles*: 8-12 tubules. *Posterior spiracles* (pl. 13.d): Spiracular slits thick walled, about 2.5-3.0 times as long as broad. Spiracular hairs just longer than a spiracular slit; dorsal and ventral bundles with 17-20 broad, flat hairs, branched apically; lateral bundles with 8-12 similarly shaped hairs. *Anal area* (pl. 13.f): Lobes protuberant, surrounded by 3-5 discontinuous rows of spinules. The inner rows of spinules stout, slightly curved, sharply pointed becoming larger just below anal opening, outer rows with smaller spinules. *Source*: Based on specimens from Hawaii (lab. colony).

Distribution

North America: Outbreaks in California eradicated (Anon., 1987).
Pacific Ocean: Guam since 1947 and Hawaiian Islands since about 1945. An outbreak on Rota, Mariana Islands, was eradicated (Nakagawa *et al.*, 1968).
Oriental Asia: Bhutan, southern China, northern India, Myanmar, northern Thailand.

Other references

BAITS, Mitchell *et al.* (1985) and Wong *et al.* (1989); BEHAVIOUR, Arakaki *et al.* (1984); BIOCONTROL, Huffaker & Caltagirone (1986); COMPETITION WITH *CERATITIS CAPITATA*, Fitt (1989), Nishida *et al.* (1985) and Wong *et al.* (1983); ECOLOGY, E.J. Harris (1989), Harris & Lee (1986), Nishida (1980) and Vargas *et al.* (1983c); HOST RESISTANCE, Armstrong (1983), Armstrong & Vargas (1982) and Armstrong *et al.* (1979; 1983); LARVA, Elson-Harris (1991) and Hardy (1949); PARASITOIDS, Wong *et al.* (1984b) and Wong & Ramadan (1987); PHEROMONES, Perkins *et al.* (1990); POST-HARVEST DISINFESTATION, Armstrong & Couey

(1989), Armstrong *et al.* (1988), Burditt & Balock (1985), Couey *et al.* (1985), Couey & Hayes (1986), Hansen *et al.* (1988), Hayes *et al.* (1984), Heather (1989), Jang (1986), Rigney (1989) and Saul *et al.* (1987); STERILE INSECT TECHNIQUE, Habu *et al.* (1984); TRANSMISSION OF PLANT PATHOGENS, Ito *et al.* (1979) and Nishijima *et al.* (1987); TRAPPING, Cunningham & Suda (1985).

B. (B.) dorsalis Complex: *B. (B.) occipitalis* (Bezzi)

Taxonomic notes

This species has also been known as *Chaetodacus ferrugineus* var. *occipitalis* Bezzi, *C. occipitalis* Bezzi, *Dacus dorsalis* var. *occipitalis* (Bezzi), *D. occipitalis* (Bezzi) and *Strumeta pedestris* var. *occipitalis* (Bezzi). This species is only known from the Philippines and some Philippine records of *B. dorsalis* were probably based on misidentifications of this species. *B. occipitalis* males are attracted to methyl eugenol; records of *B. occipitalis* from Malaysia were based on misidentifications of a superficially similar cue lure attracted species which is only known to attack the fruits of wild species of Melastomataceae (R.A.I. Drew & D.L. Hancock, in prep.).

Commercial hosts

Available pest status data for the *B. dorsalis* complex in the Philippines do not distinguish between *B. occipitalis* and a sp. near *B. dorsalis* (C). The only confirmed host for this species is mango (*Mangifera indica*) (R.A.I. Drew & D.L. Hancock, unpublished data, 1990).

Distribution

Oriental Asia: Philippines.

Other reference

PHEROMONES, Perkins *et al.* (1990).

B. (B.) dorsalis complex: *B. (B.)* sp. near *B. dorsalis* (A)

(fig. 143; pl. 14)

Carambola Fly

Taxonomic notes

The name *Bactrocera conformis* Doleschall is unavailable under the rules of nomenclature, although it may belong to this species. Most records of *B. dorsalis* from Malaysia and Indonesia were probably based on misidentifications of this species which will be formally described by R.A.I. Drew & D.L. Hancock (in prep.).

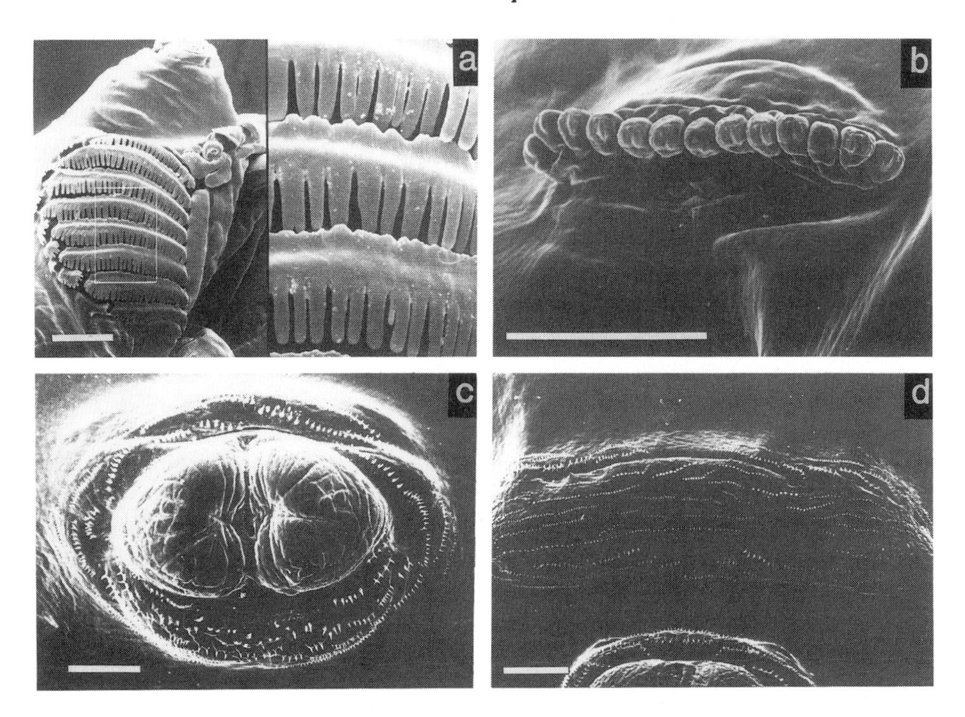

Plate 14. *Bactrocera* sp. near *B. (B.) dorsalis* (A), SEMs of 3rd instar larva; a, oral ridges, lateral view (scale=50μm); b, anterior spiracle (scale=0.1mm); c, anal lobes (scale=0.1mm); d, A8, creeping welt (scale=0.1mm).

Pest status and commercial hosts

This is a serious pest of carambola (*Averrhoa carambola*), which also heavily attacks rose-apple (*Syzygium jambos*) and watery rose-apple (*S. aqueum*). Other hosts in South-east Asia (R.A.I. Drew, unpublished data, 1990) are as follows: banana (*Musax paradisiaca*), bilimbi (*A. bilimbi*), breadfruit (*Artocarpus altilis*), chilli pepper (*Capsicum annuum*), common guava (*Psidium guajava*), jackfruit (*Artocarpus heterophyllus*), Malay-apple (*S. malaccense*), mango (*Mangifera indica*), rambai (*Baccaurea motleyana*), sapodilla (*Manilkara zapota*), sugar palm (*Arenga pinnata*), tomato (*Lycopersicon esculentum*), tropical almond (*Terminalia catappa*), *Artocarpus elasticus* and *Solanum ferox*.

The adventive population in Suriname is also a pest of carambola, and other heavily attacked hosts are common guava, huesito (*Malpighia glabra*), mango and water apple (*Syzygium samarangense*). Other known hosts in Suriname are as follows: cashew (*Anacardium occidentale*), common jujube (*Ziziphus jujube*), grapefruit (*Citrus x paradisi*) [probably only damaged fruit], Malay-apple, sapodilla, star-apple (*Chrysophyllum cainito*) [damaged fruit only], Surinam cherry (*Eugenia uniflora*), sweet orange (*Citrus sinensis*) [probably only damaged fruit] and tangerine (*C. reticulata*) [probably only damaged fruit] (D.L. Hancock & FAO, unpublished data, 1989).

Wild hosts

In South-east Asia the most important wild host is luna nut (*Lepisanthes fruticosa* (Roxb.) Leenh.; Sapindaceae) and it is also recorded from wild species of Clusiaceae, Euphorbiaceae, Meliaceae, Moraceae, Myrtaceae, Olacaceae, Rhizophoraceae, Rutaceae and Symplocaceae (R.A.I. Drew, unpublished data, 1990).

Description of third instar larva

Larvae medium-sized, length 7.5-9.5mm; width 1.5-2.0mm. *Head*: Stomal sensory organ large, with 5 preoral lobes (at most 1 lobe with small serrations); oral ridges (pl. 14.a) with 8-10 rows of large, deeply serrated, blunt-edged teeth; 8-11 small accessory plates with strongly serrated edges; mouthhooks sclerotised, without preapical teeth (fig. 143). *Thoracic and abdominal segments*: Encircling bands of discontinuous rows of small spinules on anterior portion of each thoracic segment. T1 with 11-17 rows of large, sharply pointed spinules, forming small groups dorsally which gradually become discontinuous rows laterally and ventrally; T2 and T3 with 5-7 rows of smaller, stouter spinules. Creeping welts (pl. 14.d) with small, stout spinules similar to those on T2, with 1 posterior row of slightly larger spinules. A8 with well defined intermediate areas and obvious sensilla. *Anterior spiracles* (pl. 14.b): 9-15 prominent tubules. *Posterior spiracles*: Spiracular slits thick walled; about 2.5-3.0 times as long as broad. Spiracular hair bundles of 10-15 hairs in dorsal and ventral groups and 4-7 in lateral bundles; each hair with a broad trunk, branched in apical third and about the same length as a spiracular slit. *Anal area* (pl. 14.c): Lobes large, protuberant, with 3-7 surrounding rows of small, stout, slightly curved spinules forming a small concentration just below anal opening. *Source*: Based on specimens from Malaysia (ex *Averrhoa carambola*).

Distribution

South America: French Guiana (detected 1989); Suriname since about 1975.
Oriental Asia: Indonesia (Lombok, Sumbawa and probably Kalimantan), Malaysia (Peninsular and Sabah), southern Thailand.

Other references

LARVA, Elson-Harris (1991); PHEROMONES, Perkins *et al.* (1990); SURINAME, Enkerlin *et al.* (1989).

B. (B.) dorsalis complex: *B. (B.)* sp. near *B. dorsalis* (B)

(fig. 144; pl. 15)

Taxonomic notes

The name *Bactrocera conformis* Doleschall is unavailable under the rules of nomenclature, although it may belong to this species. Many records of *B. pedestris* (Bezzi) from Malaysia and Indonesia were probably based on misidentifications of this species; the true *B. pedestris* is a very rare species which is only known from one locality in the Philippines. Many Malaysian, Indonesian and southern Thailand records of *B. dorsalis* are also likely to have been based on misidentifications of this species which will be formally described by R.A.I. Drew & D.L. Hancock (in prep.).

Pest status and commercial hosts

A pest of banana (*Musa x paradisiaca*), mango (*Mangifera indica*) and papaya (*Carica papaya*), which also heavily attacks Surinam cherry (*Eugenia uniflora*) and sugar palm (*Arenga pinnata*). Other recorded hosts are as follows: bitter gourd (*Momordica charantia*), carambola (*Averrhoa carambola*), chilli pepper (*Capsicum annuum*), common guava (*Psidium guajava*), Indian jujube (*Ziziphus mauritiana*), jackfruit (*Artocarpus heterophyllus*), Jew plum (*Spondias cytherea*), langsat (*Lansium domesticum*), Malay-apple (*Syzygium malaccense*), monkey-jack (*A. rigidus*), pond-apple (*Annona glabra*), purple granadilla (*Passiflora edulis*), rambai (*Baccaurea motleyana*), rambutan (*Nephelium lappaceum*), rose-apple (*S. jambos*), sapodilla (*Manilkara zapota*), soursop (*A. muricata*), sweet orange (*Citrus sinensis*), terongan (*Solanum torvum*), velvet apple (*Diospyros blancoi*), watery rose-apple (*Syzygium aqueum*) and *Solanum ferox* (R.A.I. Drew, unpublished data, 1990).

Wild hosts

An important wild host in South-east Asia is seashore mangosteen (*Garcinia hombroniana* Pierre) (Yong, 1990b). It is also recorded from other wild species of Clusiaceae, Moraceae, Rutaceae and Sapindaceae (R.A.I. Drew, unpublished data, 1990).

Description of third instar larva

Larvae medium-sized, length 7.0-9.0mm; width 1.5-1.8mm. *Head*: Stomal sensory organ (pl. 15.a) large, rounded and protuberant; with 3-4 sensilla (1 short and peg-like, 2-3 long and tapered), surrounded by 5-6 large preoral lobes, some with obviously serrated edges; oral ridges (pl. 15.b) with 10-15 rows of short, gently tapered, bluntly rounded teeth; 15-19 small irregular-shaped accessory plates with small serrations along posterior margins; mouthhooks moderately sclerotised, without preapical teeth (fig. 144). *Thoracic and abdominal segments*: Broad bands of long, sharply pointed spinules in discontinuous rows surrounding anterior margins of thoracic segments. T1 with 7-10 rows dorsally and laterally, increasing to 11-15 rows ventrally; T2 with 4-8 rows dorsally and laterally, and 7-9 rows ventrally; T3 with 3-5 rows dorsally and 3-4 rows laterally (sometimes absent in midline), and 3-6 rows ventrally. Dorsal spinules absent from A1-A8. Creeping welts (pl. 15.e) with long, stout, sharply pointed spinules, with several rows of spinules posteriorly directed. A8 with large intermediate areas and obvious dorsal and lateral sensilla. *Anterior spiracles* (pl. 15.c): 11-15 tubules. *Posterior spiracles* (pl. 15.d): Spiracular slits about 3 times as long as broad, with heavily sclerotised rimae. Spiracular hairs very distinct with broad trunks, branched in apical half; dorsal and ventral bundles of 15-25 hairs, lateral bundles of 6-10 hairs. *Anal area* (pl. 15.f): Lobes large, protuberant, surrounded by 3-6 rows of very long, sharply pointed spinules forming discontinuous rows dorsally and laterally, and becoming a small concentration of individual spinules below anal opening. *Source*: Based on specimens from Malaysia (ex *Psidium guajava*).

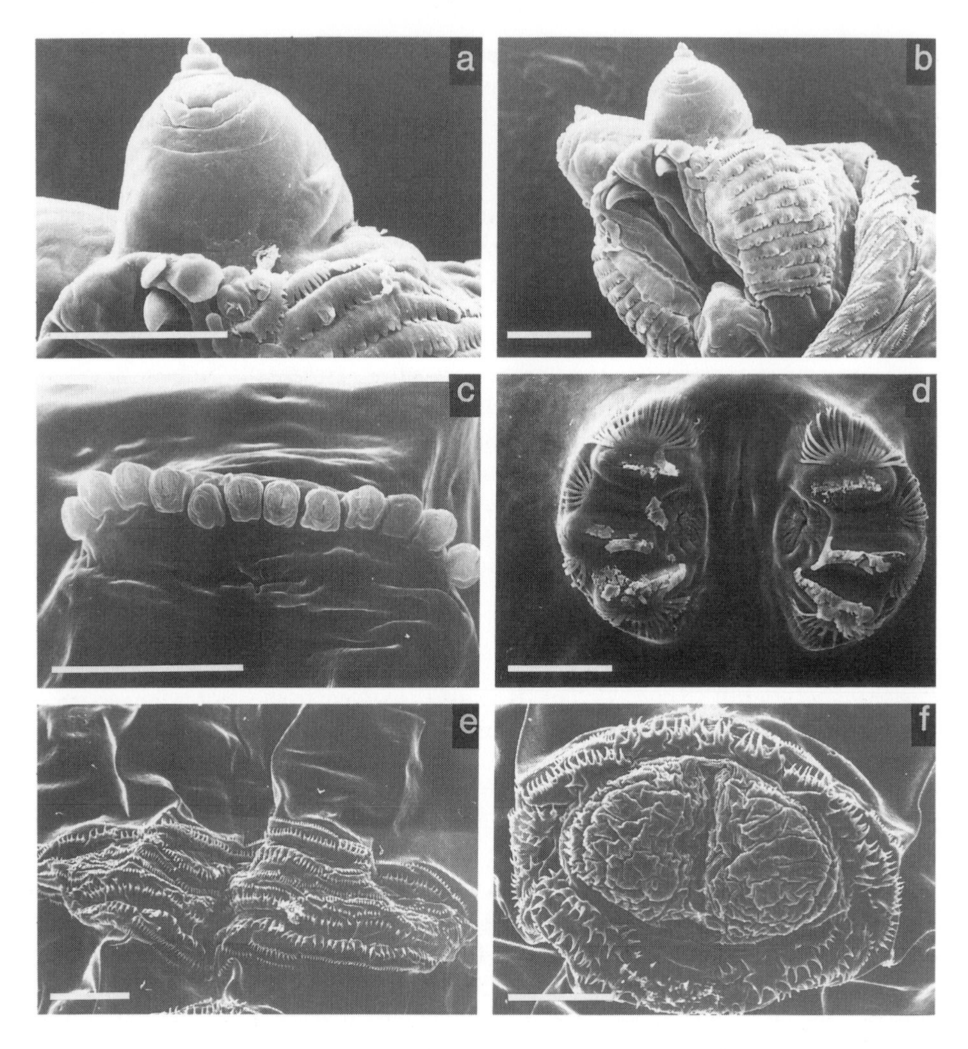

Plate 15. *Bactrocera* sp. near *B. (B.) dorsalis* (B), SEMs of 3rd instar larva; a, antennal, maxillary and stomal sensory organs, lateral view; b, head, ventrolateral view; c, anterior spiracle; d, posterior spiracles; e, A8, creeping welt; f, anal lobes. (scales=0.1mm).

Distribution

Oriental Asia: Malaysia (Peninsular, Sabah), Indonesia (Lombok, Sumbawa and probably Kalimantan), southern Thailand.

Other reference

LARVA, Elson-Harris (1991); PHEROMONES, Perkins *et al.* (1990).

B. (B.) *dorsalis* complex: B. (B.) sp. near B. *dorsalis* (C)

(fig. 145; pl. 16)

Taxonomic notes

Most Philippine records of *B. pedestris* and many Philippine records of *B. dorsalis* were probably based on misidentifications of this species which will be formally described by R.A.I. Drew & D.L. Hancock (in prep.).

Commercial hosts

Available pest status data for the *B. dorsalis* complex in the Philippines do not distinguish between *B. occipitalis* and a sp. near *B. dorsalis* (C). The only confirmed hosts for this species are breadfruit (*Artocarpus altilis*), Malay-apple (*Syzygium malaccensis*) and mango (*Mangifera indica*) (R.A.I. Drew & D.L. Hancock, unpublished data, 1990). Rejesus *et al.* (1991) gave an extensive list of host data from the Philippines which probably related to this species (as *B. dorsalis*). However, their data were based on laboratory host tests and included fruits that females would oviposit in, but from which no adults emerged.

Wild hosts

The only recorded wild host is *Pouteria duklitan* (Blanco) Baehni (R.A.I. Drew & D.L. Hancock, unpublished data, 1990) but it probably has a wider range of hosts.

Description of third instar larva

Larvae are medium-sized, length 7.0-8.5mm; width 1.2-1.5mm. *Head*: Stomal sensory organs, each with 3 sensilla (1 short and peg-like, 2 long and tapering), surrounded by 5-6 large, unserrated preoral lobes; oral ridges (pl. 16.b) with 10-12 rows of moderately long, serrated, tapered, rounded teeth; accessory plates of 9-14 small, irregular shaped plates, several with shallow serrations; mouthhooks well developed, sclerotised, without preapical teeth (fig. 145). *Thoracic and abdominal segments*: Each thoracic segment with a broad anterior band of small spinules in discontinuous rows. T1 with 9-13 rows of stout, sharply pointed teeth; T2 with 5-8 rows dorsally, increasing to 7-9 rows ventrally; T3 with 2-4 rows dorsally, 0-4 rows laterally and 4-6 rows ventrally. Creeping welts (pl. 16.d) with small, stout spinules; 1-3 rows anteriorly directed, the remainder posteriorly directed; some abdominal segments, especially A7, with 1 posterior row of larger spinules. A8 with very well defined intermediate areas. *Anterior spiracles*: 9-13 tubules. *Posterior spiracles* (pl. 16.c): Each spiracular slit about 2.5-3.0 times as long as broad, with a heavily sclerotised rima; dorsal and ventral hair bundles of 12-20 hairs, lateral bundles of 4-7 hairs; hairs about as long as spiracular slits, branched in apical third to a half. *Anal area* (pl. 16.d): Lobes large, protuberant, with 2-5 rows of small, stout spinules forming discontinuous rows around anal lobes. Below anal opening, spinules forming a small concentration of slightly stouter spinules. *Source*: Based on specimens from the Philippines (lab. colony).

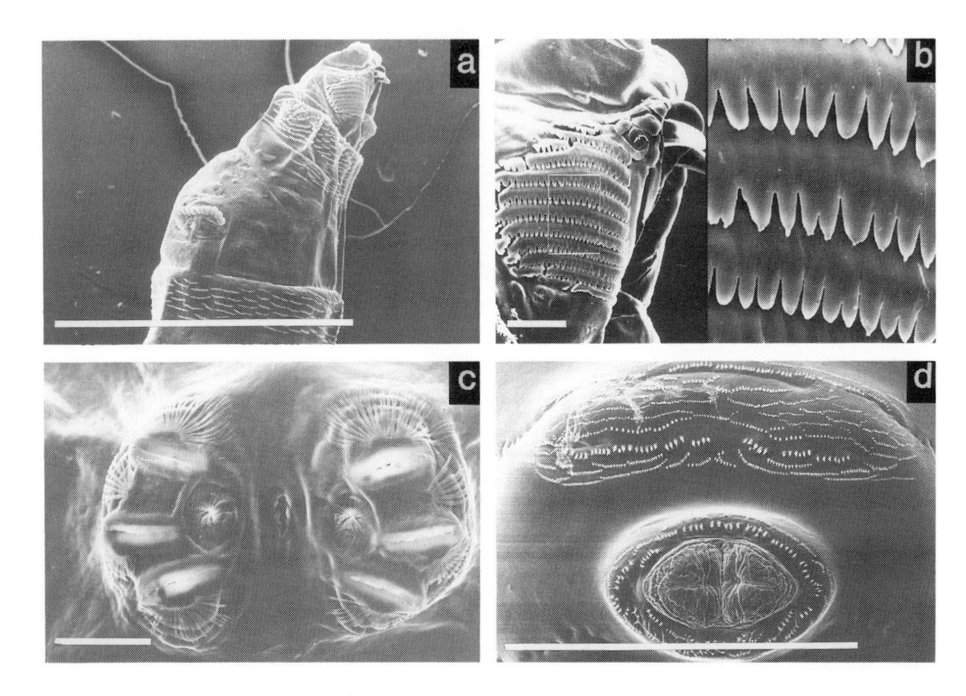

Plate 16. *Bactrocera* sp. near *B. (B.) dorsalis* (C), SEMs of 3rd instar larva; a, head, T1 and T2, lateral view (scale=1.0mm); b, oral ridges, lateral view (scale=50μm); c, posterior spiracles (scale=0.1mm); d, A8, creeping welt and anal lobes (scale=1.0mm).

Distribution

Oriental Asia: Philippines.

Other references

LARVA, Elson-Harris (1991); PHEROMONES, Perkins *et al.* (1990).

B. (B.) dorsalis complex: *B. (B.)* sp. near *B. dorsalis* (D)

Taxonomic notes

Most Sri Lankan records of *B. dorsalis* were probably based on misidentifications of this species which will be formally described by R.A.I. Drew & D.L. Hancock (in prep.).

Commercial hosts

The only confirmed host is mango (*Mangifera indica*) (R.A.I. Drew & D.L. Hancock, unpublished data, 1990). There is a report of *B. dorsalis* heavily attacking grapefruit (*Citrus x paradisi*) in Sri Lanka (USDA, *Pests not known to occur in the United States or of limited distribution*, No.20, undated), and that may refer to this species. There is also an old record of *B. dorsalis* from eggplant (*Solanum melongena*) (Fletcher, 1920), which was probably based on a misidentification of this species.

Distribution
Oriental Asia: Sri Lanka.

B. (B.) dorsalis Complex: Data on Unspecified Species

Taxonomic notes
Due to the past confusion between member species of the *B. dorsalis* species complex, much published data cannot be attributed to individual species. Recorded hosts which have not been listed under at least one member of the complex, and further references, are listed here.

Additional commercial hosts
Additional host records from Malaysia and Indonesia (Perkins, 1938; Tan & Lee, 1982; Yunus & Ho, 1980), which probably refer to spp. near *B. dorsalis*, (A) or (B), are as follows: avocado (*Persea americana*), bachang (*Mangifera foetida*), biribá (*Rollinia pulchrinervis*), cherimoya (*Annona cherimola*), citron (*Citrus medica*), eggplant (*Solanum melongena*), grapefruit (*C. x paradisi*), kwini (*Mangifera odorata* Griffith), sour orange (*C. aurantium*), soursop (*A. muricata*), tabasco pepper (*Capsicum frutescens*) and water apple (*Syzygium samarangense*). Yunus & Ho (1980) also recorded angled luffa (*Luffa acutangula*) and Levant cotton (*Gossypium herbaceum* L.), which are doubtful hosts, and their record from betel nut (*Areca catechu*) was probably based on a misidentification of *B. arecae* (p. 274).

Additional host records from the Philippines (Clausen *et al.*, 1965; Koyama, 1989a), which probably refer to either *B. occipitalis* or to sp. near *B. dorsalis* (C), are as follows: avocado (*Persea americana*), carambola, common guava (*Psidium guajava*), jackfruit (*Artocarpus heterophyllus*), papaya (*Carica papaya*), red mombin (*Spondias purpurea*), water apple and ylang-ylang (*Cananga odorata*).

Many of the additional hosts recorded by Clausen *et al.* (1965), Kapoor (1970) and USDA (*Pests not known to occur in the United States or of limited distribution*, No.20, undated) cannot be placed to a particular member of the *B. dorsalis* species complex; they are as follows: allspice (*Pimenta dioica*), arabica coffee (*Coffea arabica*), bay rum tree (*Pimenta racemosa* (Miller) J.W. Moore), coconut (*Cocos nucifera* L.; Arecaceae), cocoplum (*Chrysobalanus icaco*), common fig (*Ficus carica*), date palm (*Phoenix dactylifera*), egg-fruit tree (*Pouteria campechiana*), egg tree (*Garcinia xanthochymus*), governor's plum (*Flacourtia indica*), gympie nut (*Macadamia ternifolia* F. Muell.), Indian laurel fig (*Ficus microcarpa* L. f.), Indian laurel (*Calophyllum inophyllum*), Indian bael (*Aegle marmelos*), ivy gourd (*Coccinia grandis*), Japanese plum (*Prunus salicina*), Japanese persimmon (*Diospyros kaki*), ketembilla (*Dovyalis hebecarpa*), liberica coffee (*Coffeae liberica*), lychee (*Litchi chinensis*), mammy-apple (*Mammea americana*), Manila tamarind (*Pithecellobium dulce* (Roxb.) Benth.; Fabaceae), mission pricklypear (*Opuntia megacantha* Salm-Dyck), myrobalan plum (*Prunus cerasifera*), Natal plum (*Carissa macrocarpa*), olive (*Olea europaea*), pepino (*Solanum muricatum*), pineapple (*Ananas comosus*), pomegranate (*Punica granatum*), quince (*Cydonia oblonga*), robusta coffee (*Coffea canephora*), round kumquat

(*Fortunella japonica*), sandalwood (*Santalum album*), satinleaf (*Chrysophyllum oliviforme*), Sea Island cotton (*Gossypium barbadense* L.), Spanish cherry (*Mimusops elengi*), strawberry (*Fragaria vesca*), sweet lime (*Citrus limetta*), tangor (*C. x nobilis*), teruah (*Momordica cochinchinensis*), white mulberry (*Morus alba*), white sapote (*Casimiroa edulis*), wine grape (*Vitis vinifera*) and yellow granadilla (*Passiflora laurifolia*). The pineapple record was probably based on a chance oviposition in damaged fruit, as it has been shown that pineapple is not normally attacked (Armstrong & Vargas, 1982).

References to sp. near *B. dorsalis* (A) or (B)
DATASHEET, Vijaysegaran (1989); ECOLOGY, Hong & Serit (1988), Tan (1985) and Tan & Lee (1982); LARVA, Rohani (1987); IMPACT ON STARFRUIT, Vijaysegaran (1984); PARASITOIDS, Vijaysegaran (1984); PHEROMONES, Nishida *et al.* (1988); PUPARIATION, Ibrahim & Mohamad (1978); TRAPPING, Ibrahim *et al.* (1979) and Tan (1984).

References to unspecified members of the complex
ANTENNAL SENSILLA, Dickens *et al.* (1988); BEHAVIOUR, Katsoyannos (1989b) and P.H. Smith (1989); BIBLIOGRAPHY, Zwölfer (1985); BIOLOGICAL CONTROL, Wharton (1989a; 1989b); BIOLOGY, Fletcher (1989b); CUTICULAR HYDROCARBONS, Carlson & Yocom (1986); DATASHEET, Weems (1964a); DEVELOPMENT RATE, Fletcher (1989a); ECOLOGY, Carey (1989), Debouzie (1989), Fletcher (1989c) and Meats (1989a); ERADICATION FROM CALIFORNIA, Anon. (1987); DISTRIBUTION MAP OF SOME OF THE PEST SPECIES IN THE COMPLEX, CIE (1986b) and Drew (1982a); EGGS, Williamson (1989); HOSTS, Koyama (1989a); HOST RESISTANCE, Greany (1989); INCIDENCE ON CASHEW, Gowda & Ramaiah (1979); LARVA, Berg (1979), Heppner (1988), Khan & Khan (1987) and Kandybina (1977); MALE ANNIHILATION, Cunningham (1989c) and Shiga (1989); MALE TERMINALIA, Singh & Premlata (1985); MECHANICAL FORCE IN OVIPOSITION, Bose & Mehrotra (1986); PARASITOID IDENTIFICATION, Wharton & Gilstrap (1983); PARASITOIDS, Ben-Salah *et al.* (1989), Fry (1990), Narayanan & Chawla (1962) and Wharton (1989a; 1989b); PHEROMONES, Fitt (1981) and Nation (1981); PREDATION, Wong & Wong (1988); REARING, Fay (1989), Tsitsipis (1989) and Vargas (1989); STERILE INSECT TECHNIQUE, Calkins (1989), Gilmore (1989), Shiga (1989), Thakur & Kumar (1986; 1988) and Vargas (1989); TRAPPING, Economopoulos (1989) and Katsoyannos (1989b).

Bactrocera (Bactrocera) facialis (Coquillett)

(fig. 183)

Taxonomic notes

This species has also been known as *Chaetodacus facialis* (Coquillett), *Dacus facialis* Coquillett, *D. tongensis* Froggatt and *Strumeta facialis* (Coquillett).

Commercial hosts

This species has the potential to become a serious pest if introduced into any major fruit and vegetable producing countries (Drew, 1982a). It attacks avocado (*Persea americana*), bell pepper (*Capsicum annuum*), breadfruit (*Artocarpus altilis*), cashew (*Anacardium occidentale*), common guava (*Psidium guajava*), giant granadilla (*Passiflora quadrangularis*), grapefruit (*Citrus x paradisi*), lemon (*C. limon*), mandarin (*C. reticulata*), mango (*Mangifera indica*), Pacific lychee (*Pometia pinnata*), peach (*Prunus persica*), pummelo (*C. maxima*), rose-apple (*Syzygium jambos*), sea hibiscus (*Hibiscus tiliaceus*), Surinam cherry (*Eugenia uniflora*), sweet orange (*C. sinensis*), tabasco pepper (*Capsicum frutescens*), Tahiti chestnut (*Inocarpus fagifer*), tomato (*Lycopersicon esculentum*) and tropical almond (*Terminalia catappa*) (Drew, 1989a; Litsinger *et al.*, 1991); the sea hibiscus record has been questioned (A.J. Allwood, pers. comm., 1991) and that may have been an aberrant host. There are also records, which require confirmation, from Malay-apple (*S. malaccense*) and pineapple (*Ananas comosus*) (SPC, unpublished data, 1988).

Wild hosts

Recorded from Apocynaceae, Myrtaceae and Rubiaceae by Litsinger *et al.* (1991).

Adult identification

Scutum predominantly black, usually with lateral yellow stripes; with anterior supra-alar setae, prescutellar acrostichal setae, 2 scutellar setae, and a typical dacine wing pattern (fig. 183); male with a pecten; without facial spots.

Distribution

South Pacific: Tonga (Tongatapu and Ha'apai Group).
Distribution mapped by Drew (1982a).

Attractant

Males attracted to cue lure.

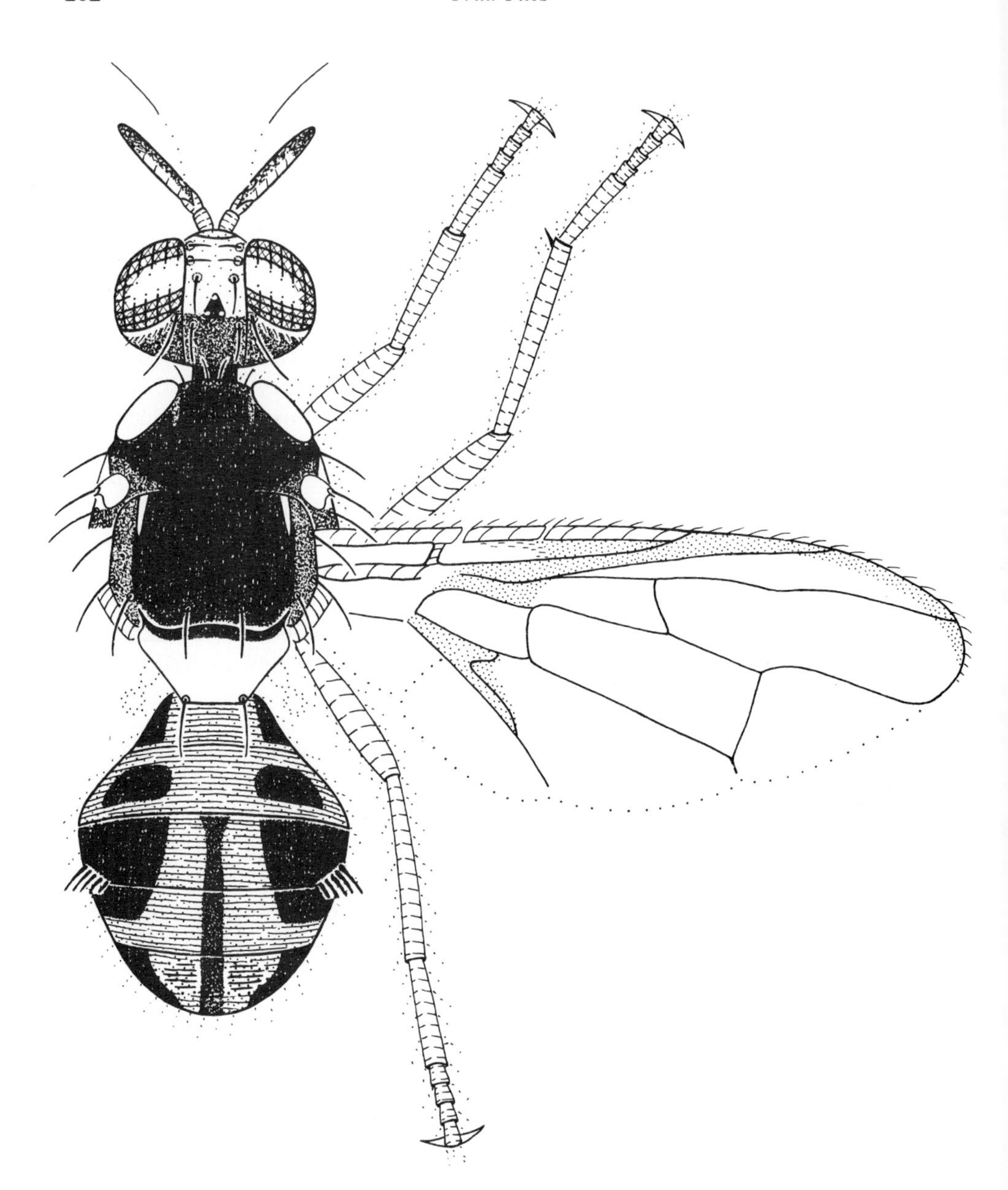

Fig. 183. *Bactrocera (B.) facialis*, adult male.

Bactrocera (Bactrocera) frauenfeldi (Schiner)

(figs 146, 184; pl. 17)

Taxonomic notes

This species has also been known as *Dacus frauenfeldi* Schiner and *Strumeta frauenfeldi* (Schiner). All records of *B. frauenfeldi* from north and west of Irian Jaya were probably based on misidentifications of *B. albistrigata* (p. 173) (see Hardy, 1983a). Records from Vanuatu were based on misidentifications of *B. trilineola* Drew (p. 277), a related species which has a completely black face.

Commercial hosts

This species has the potential to become a serious pest when commercial fruit growing develops within its range (see Drew, 1982a). It attacks breadfruit (*Artocarpus altilis*), common guava (*Psidium guajava*), Malay-apple (*Syzygium malaccense*), mango (*Mangifera indica*), sauh (*Manilkara kauki*) and tropical almond (*Terminalia catappa*) (Drew, 1982a; 1989a). There are also records requiring confirmation, from banana (*Musa x paradisiaca*), papaya (*Carica papaya*) and rose-apple (*S. jambos*) (Malloch, 1939b; SPC, unpublished data, 1988).

Wild hosts

Wild plants of tropical almond are the major wild host in Australia (R.A.I. Drew, unpublished data, 1990).

Adult identification

A predominantly black species; scutum with lateral yellow stripes and postpronotal lobe black; with facial spots, anterior supra-alar setae, prescutellar acrostichal setae, 2 scutellar setae, and a characteristic wing pattern (fig. 184); scutellum with a triangular black mark, which may reach the apex of the scutellum; male with a pecten.

Description of third instar larva

Larvae medium-sized, length 5.0-7.5mm; width 1.0-1.5mm. *Head*: Stomal sensory organ large, protuberant, with short, stout, branched sensilla and surrounded by 5 unserrated preoral lobes; oral ridges (pl. 17.a, b) with 9-12 moderately long, deeply serrated rows of long, parallel-sided, bluntly rounded teeth; accessory plates (pl. 17.b), a mixture of long and short interlocking plates, some with small serrations; mouthhooks moderately sclerotised, with large curved apical teeth (fig. 146). *Thoracic and abdominal segments*: Anterior margin of each segment (pl. 17.c) with an encircling band of discontinuous rows of small, sharply pointed spinules. T1 with 6-9 rows of spinules dorsally, expanding to 8-13 rows laterally and ventrally; T2 with 3-6 rows dorsally and laterally, expanding to 5-8 rows ventrally; T3 with 2-5 rows encircling segment. A1-A8 without spinules dorsally. Creeping welts with small spinules in 6-8 discontinuous rows, with 1 posterior row of larger, stouter spinules. A8 with

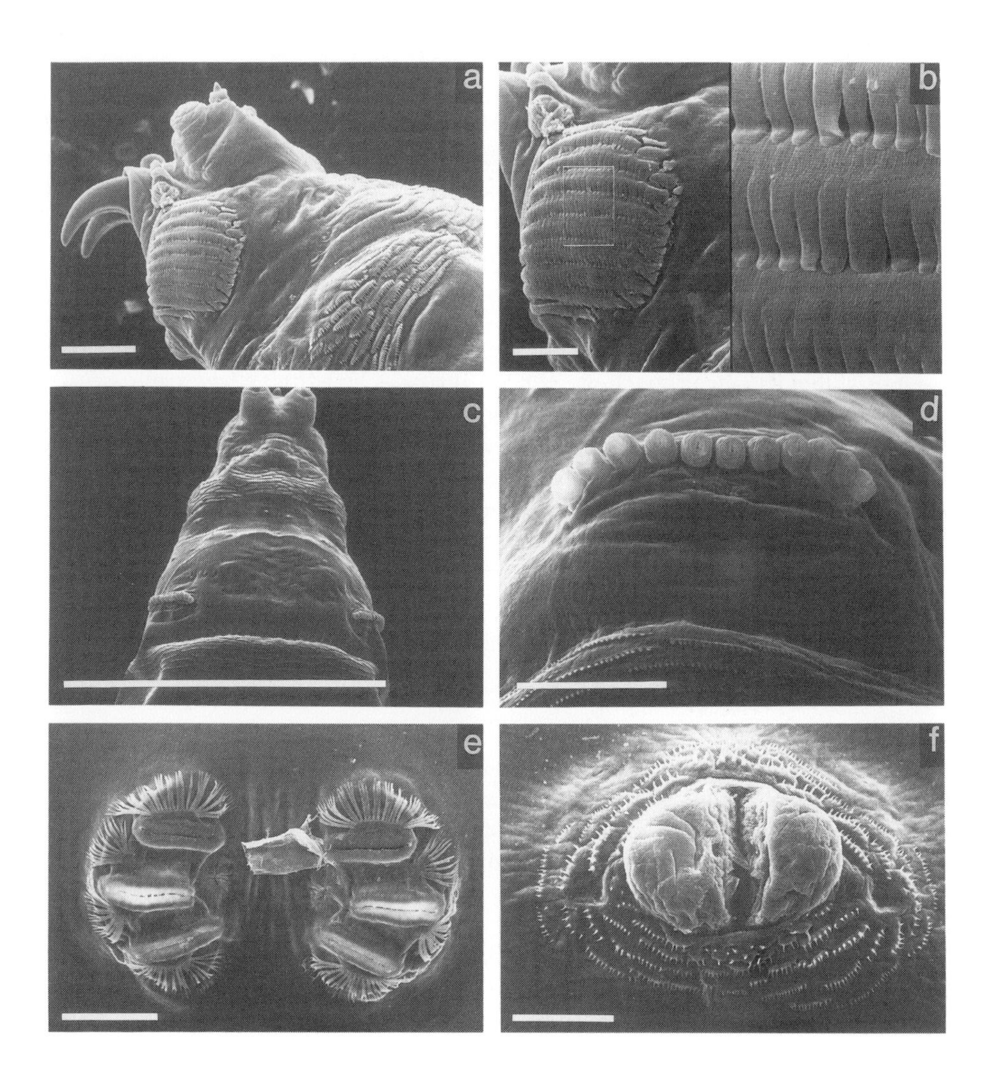

Plate 17. *Bactrocera (B.) frauenfeldi*, SEMs of 3rd instar larva; a, head and T1, lateral view (scale=0.1mm); b, oral ridges, lateral view (scale=50μm); c, T1 and T2, dorsal view (scale=1.0mm); d, anterior spiracle (scale=0.1mm); e, posterior spiracles (scale=0.1mm); f, anal lobes (scale=0.1mm).

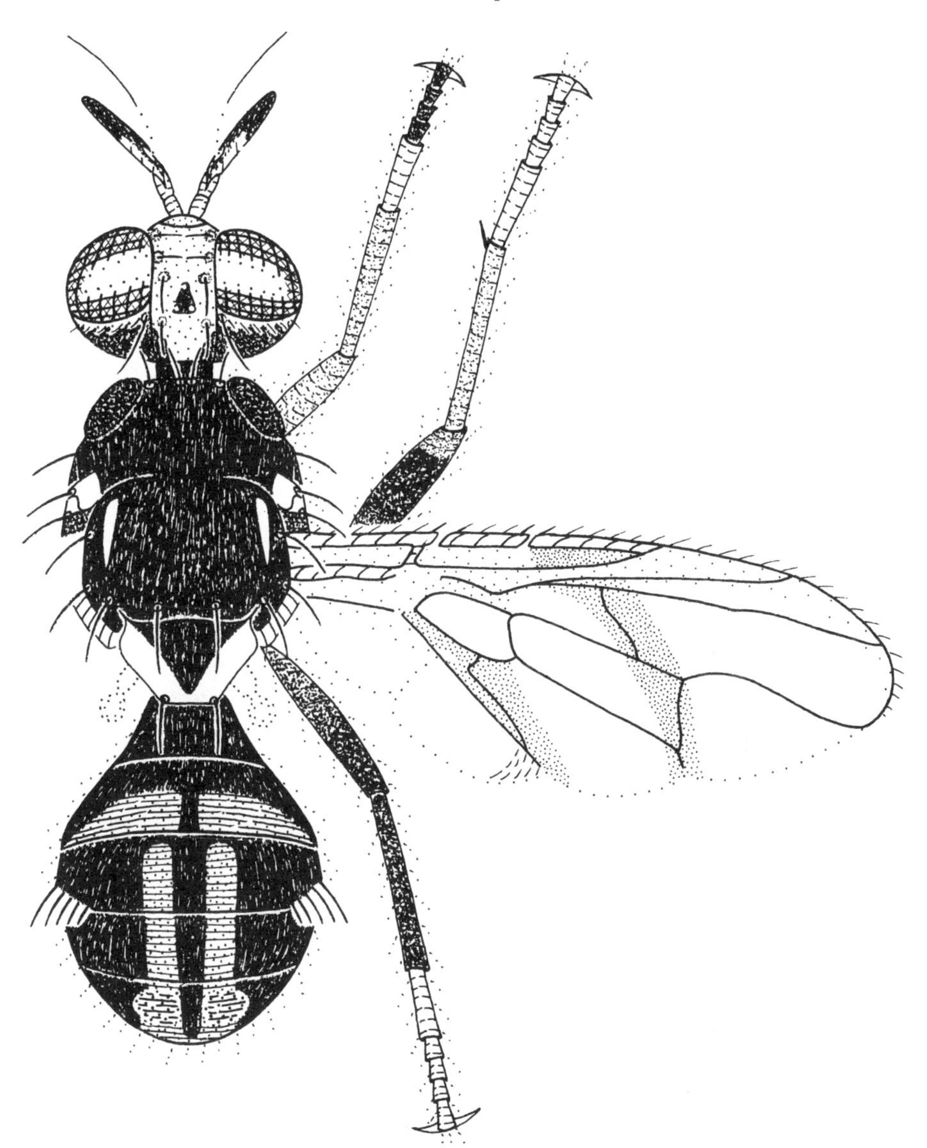

Fig. 184. *Bactrocera (B.) frauenfeldi*, adult male.

intermediate areas only moderately developed; sensilla in dorsal and lateral areas small but obvious. *Anterior spiracles* (pl. 17.d): 9-12 tubules. *Posterior spiracles* (pl. 17.e): Each spiracular slit about 3.5-4.0 times as long as broad, each with a sclerotised rima. Spiracular hairs, with 12-16 broad, flat, branched hairs in dorsal and ventral bundles; 6-9 in each lateral bundle. *Anal area* (pl. 17.f): Lobes large, protuberant and surrounded by 5-8 discontinuous rows of spinules; spinules concentrated below anal opening; those just anterior to lobes stouter and more strongly curved. *Source*: Based on specimens from Queensland, Australia (ex *Terminalia catappa*).

Distribution
Australia: Queensland (Northern part of Cape York Peninsula).
New Guinea area: Bismarck Archipelago, Bougainville Island, Papua New Guinea, Solomon Islands, Stuart Island.
South Pacific: Belau, Kiribati, Marshall Islands, Micronesia, Northern Marianas.
Distribution mapped by Drew (1982a).

Attractant
Males attracted to cue lure.

Other reference
LARVA, Hardy & Adachi (1956) [description may have been based on a misidentification of another species].

Bactrocera (Bactrocera) kirki (Froggatt)

(fig. 185)

Taxonomic notes
This species has also been known as *Dacus kirki* Froggatt and *Strumeta kirki* (Froggatt).

Commercial hosts
A potential pest which has been recorded from bell pepper (*Capsicum annuum*), carambola (*Averrhoa carambola*), common guava (*Psidium guajava*), lime (*Citrus aurantiifolia*), Malay-apple (*Syzygium malaccense*), mandarin (*C. reticulata*), mango (*Mangifera indica*), peach (*Prunus persica*), pineapple (*Ananas comosus*), purple granadilla (*Passiflora edulis*), rose-apple (*S. jambos*), sea hibiscus (*Hibiscus tiliaceus*), Surinam cherry (*Eugenia uniflora*), sweet orange (*C. sinensis*), tabasco pepper (*Capsicum frutescens*), Tahiti chestnut (*Inocarpus fagifer*) and tropical almond (*Terminalia catappa*) (Drew, 1989a; Litsinger *et al.*, 1991); the sea hibiscus record has been questioned (A.J. Allwood, pers. comm., 1991) and that may have been an aberrant host.

Wild hosts
Recorded from some *Syzygium* spp. and from fao (*Ochrosia oppositifolia* (Lam.) K. Schum.; Apocynaceae) (Litsinger *et al.*, 1991).

Adult identification
A predominantly black species; with large facial spots, anterior supra-alar setae, prescutellar acrostichal setae, 2 scutellar setae, and a reduced wing pattern (fig. 185); male with a pecten; scutum without yellow stripes.

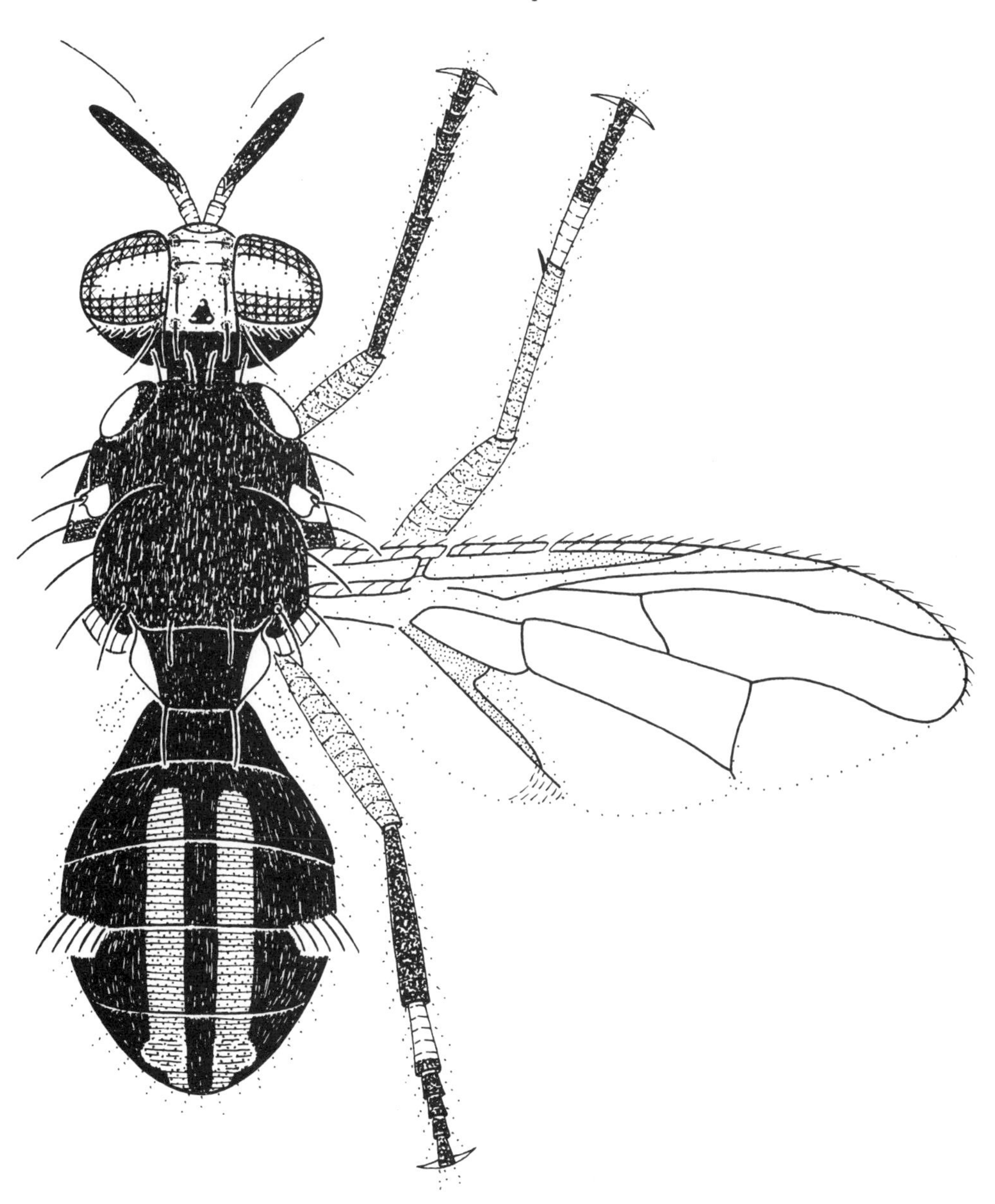

Fig. 185. *Bactrocera (B.) kirki*, adult male.

Distribution
South Pacific: Austral Islands, Niue, American and Western Samoa, Tahiti, Tonga. Distribution mapped by Drew (1982a).

Attractant
Males attracted to cue lure.

Bactrocera (Bactrocera) latifrons (Hendel)

Solanum Fruit Fly

(figs 77, 147, 186; pl. 18)

Taxonomic notes
This species has also been known as *Chaetodacus latifrons* Hendel and *Dacus latifrons* (Hendel).

Pest status and commercial hosts
A pest of solanaceous crops, including chilli (*Capsicum annuum*) in Malaysia and *Solanum incanum* in Taiwan (Hardy, 1973). In South-east Asia it has also been reared from black nightshade (*S. nigrum*) and terongan (*S. torvum*) (R.A.I. Drew, unpublished data, 1990). Additional hosts recorded in Hawaii are eggplant (*S. melongena*), currant tomato (*Lycopersicon pimpinellifolium*) and tomato (*L. esculentum*) (E.J. Harris, 1989; Vargas & Nishida, 1985b). In India and Pakistan it has been reared from eggplant and *S. sisymbriifolium* (Clausen *et al.*, 1965; Syed, 1970). The only reliable rearing records from non-solanaceous plants are from Hardy (1973) who recorded it from rambai (*Baccaurea motleyana*), and R.A.I. Drew (unpublished data, 1990) who reared it from apple (*Malus domestica*), but those were probably aberrant host associations. There are some doubtful records from non-solanaceous plants, namely banana (*Musa x paradisiaca*), carambola (*Averrhoa carambola*), coffee (*Coffea* sp.), common guava (*Psidium guajava*), cucumber (*Cucumis sativus*), lemon (*Citrus limon*), lychee (*Litchi chinensis*), mango (*Mangifera indica*), snakegourd (*Trichosanthes cucumerina*) and sweet orange (*C. sinensis*) (Kapoor & Agarwal, 1983; Satoh *et al.*, 1985; Syed, 1970; Yunus & Ho, 1980).

Wild hosts
Recorded from *Solanum indicum* L., *S. sarmentosum* Nees, *S. verbascifolium* and *S. virginianum* L. (as *S. surattense*) (Hardy, 1973; Syed, 1970; Vargas & Nishida, 1985b).

Adult identification
Scutum predominantly black with lateral yellow stripes; with facial spots, anterior supra-alar setae, prescutellar acrostichal setae, 2 scutellar setae, a predominantly orange

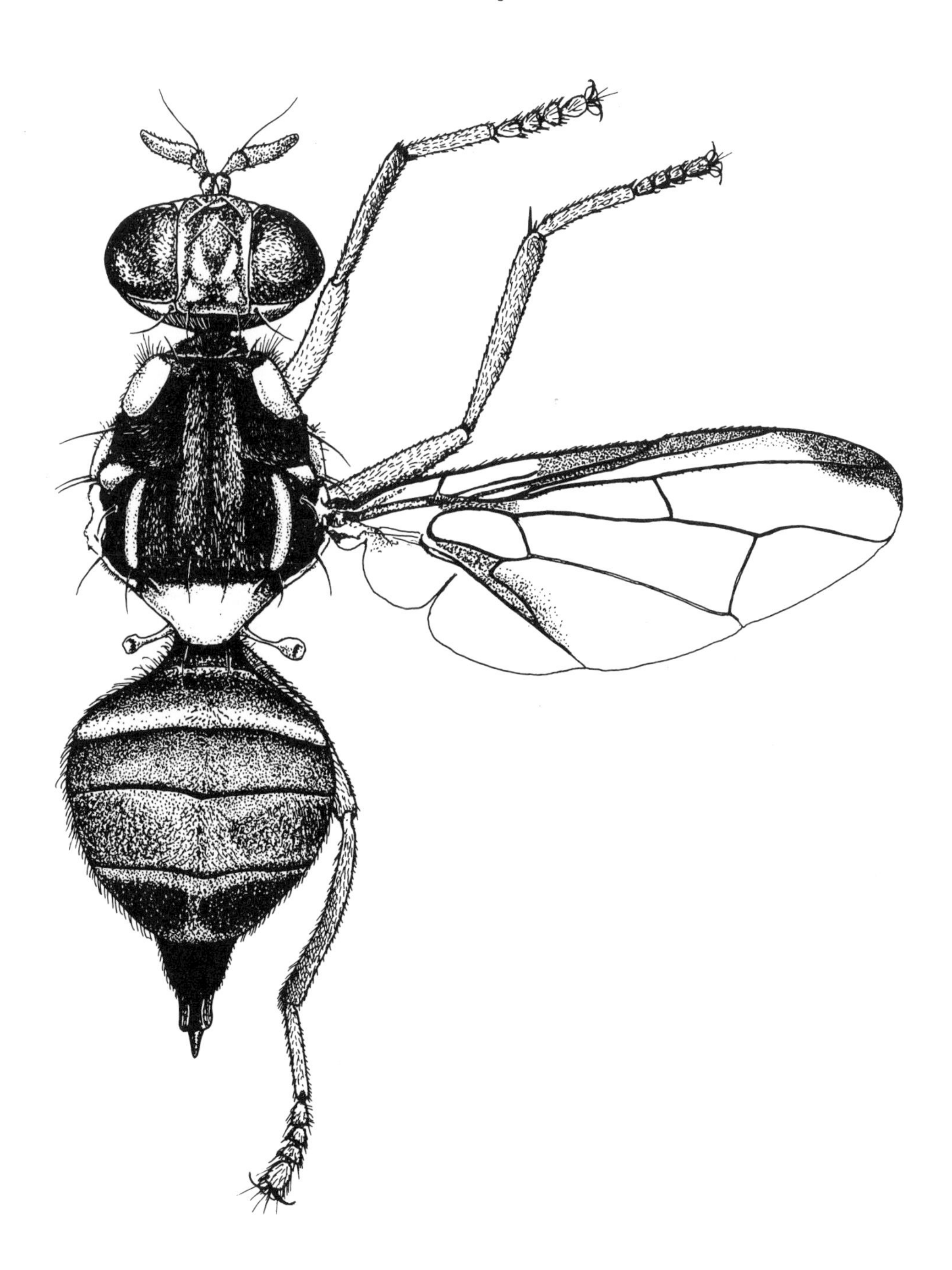

Fig. 186. *Bactrocera (B.) latifrons*, adult female.

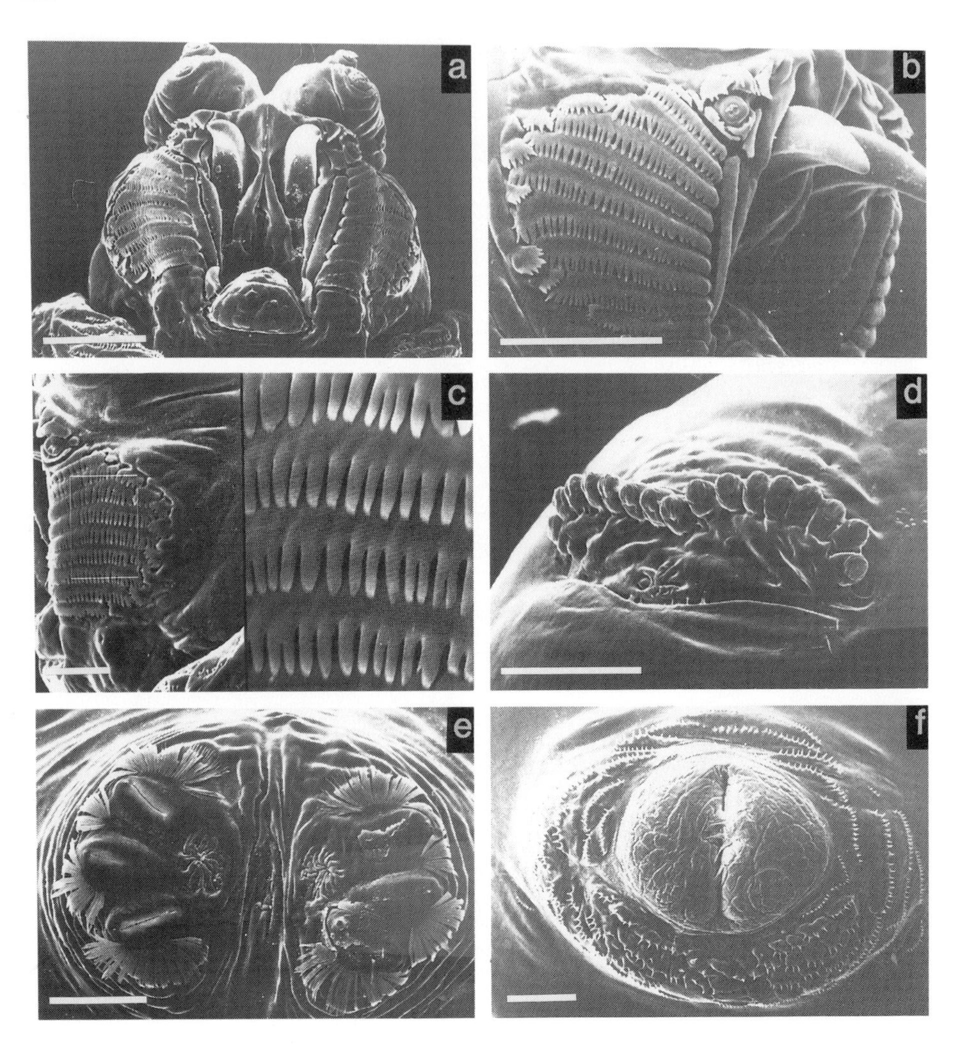

Plate 18. *Bactrocera (B.) latifrons*, SEMs of 3rd instar larva; a, head, ventral view (scale=0.1mm); b, stomal sensory organs, preoral lobes and oral ridges, lateral views (scale=0.1mm); c, oral ridges, lateral view (scale=50μm); d, anterior spiracle (scale=0.1mm); e, posterior spiracles (scale=0.1mm); f, anal lobes (scale=0.1mm).

abdomen, and with the costal band expanded into an apical spot (fig. 186); male with a pecten; female with a distinctive aculeus tip (fig. 77).

Description of third instar larva

Larvae medium-sized, length 7.0-8.5mm; width 1.2-1.5mm. *Head*: Stomal sensory organ (pl. 18.a-c) with 3-4 small, peg-like sensilla on a protuberant base surrounded by 5-6 large preoral lobes, some with small serrations; oral ridges (pl. 18.b) with 9-14 rows of moderately long, tapering, bluntly rounded teeth; accessory plates with 6-10 small, serrated plates along the outer edge; mouthhooks moderately sclerotised, with slender, curved apical teeth (fig. 147). *Thoracic and abdominal segments*: A broad, encircling, anterior band of discontinuous rows of small spinules surrounding each thoracic segment. T1 with 6-10 rows of small, sharply pointed spinules; T2 and T3 with 3-7 rows of small spinules decreasing laterally. Creeping welts with stout spinules, 1 posterior row slightly larger. A8 with intermediate areas large and sensilla well developed. *Anterior spiracles* (pl. 18.d): 13-18 tubules. *Posterior spiracles* (pl. 18.e): Each spiracular slit about 3 times as long as broad, with a thick rima. Spiracular hairs broad, flat; dorsal and ventral bundles of 16-22 hairs branched in apical third to a quarter; lateral bundles of 6-11 hairs. *Anal area* (pl. 18.f): Lobes large, protuberant, surrounded by 3-6 rows of small, sharply pointed spinules, becoming more concentrated and stouter below anal opening. *Source*: Based on specimens from Hawaii, USA (lab. colony).

Distribution

Pacific Ocean: Adventive population discovered in Hawaiian Islands (Oahu) in 1983. *Oriental Asia*: China, India, Laos, Malaysia (Peninsular), Pakistan, Sri Lanka, Taiwan, Thailand.

Attractant

Not attracted to cue lure or methyl eugenol.

Other references

ECOLOGY, Carey (1989) and Vargas & Nishida (1985b); HAWAII, Vargas & Nishida (1985a), E.J. Harris (1989); PAKISTAN, Syed (1970); PARASITOIDS, Hardy 1973, Narayanan & Chawla (1962) and Syed 1970; REARING, Vargas & Mitchell (1987).

Bactrocera (Bactrocera) melanota **(Coquillett)**

(figs 148, 187; pl. 19)

Taxonomic notes

This species has also been known as *Chaetodacus melanotus* (Coquillett), *Dacus melanotus* Coquillett, *D. rarotongae* Froggatt and *Strumeta melanota* (Coquillett).

Pest status and commercial hosts

A serious pest of citrus (*Citrus* sp.) (Drew, 1982a), although the only specific published record is from sour orange (*C. aurantium*) (Bezzi, 1928), and that requires confirmation; it is also recorded from common guava (*Psidium guajava*) and mango (*Mangifera indica*) (Drew, 1982a).

Adult identification

A predominantly black species; with anterior supra-alar setae, prescutellar acrostichal setae, and 2 scutellar setae; wing with costal band often indistinct, and often with faint infuscation along crossveins r-m and dm-cu (fig. 187); male with a pecten; scutum without yellow stripes and face without spots.

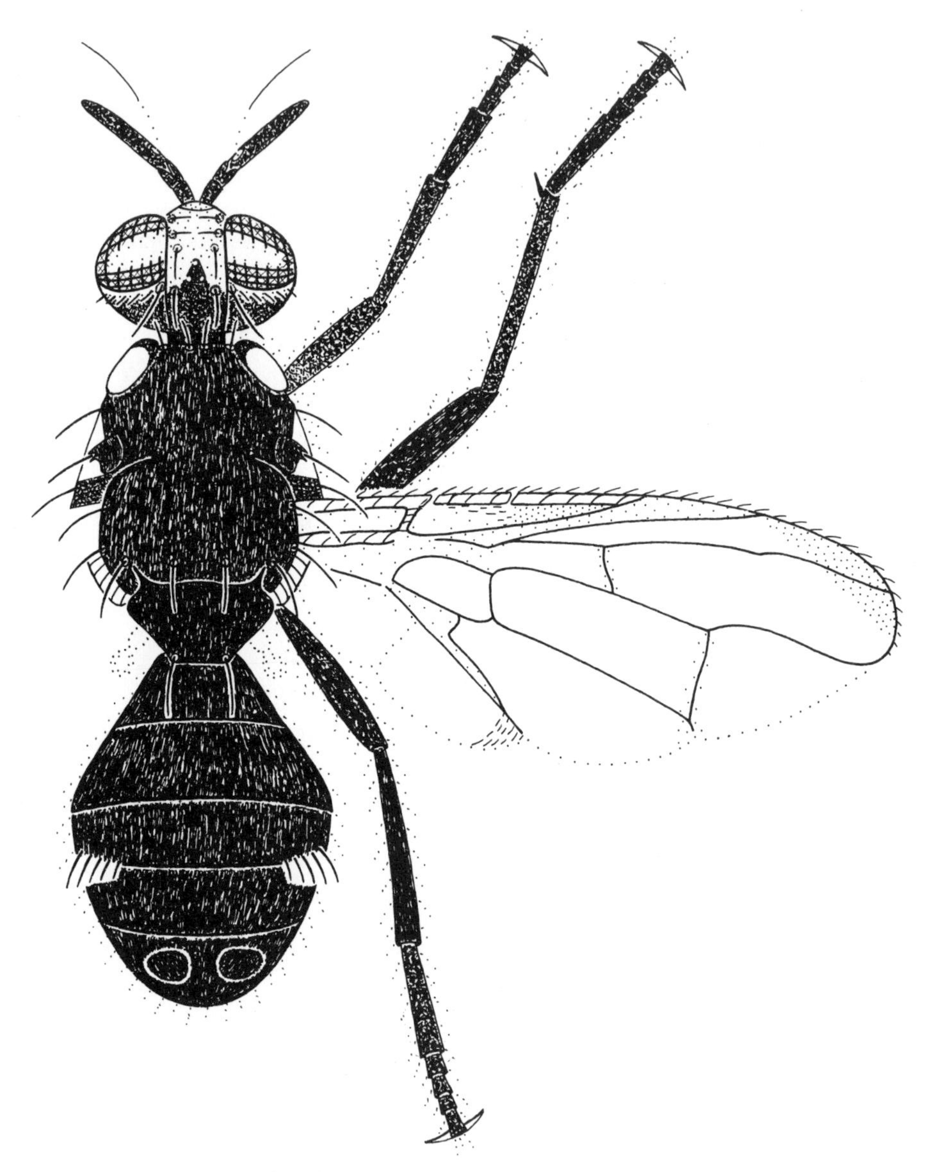

Fig. 187. *Bactrocera (B.) melanota*, adult male.

Description of third instar larva

Larvae medium-sized, length 7.5-9.5mm; width 1.3-1.8mm. *Head*: Stomal sensory organ with a mixture of long tapering and short peg-like sensilla; organ surrounded by 5-6 large, unserrated preoral lobes; oral ridges (pl. 19.a, b) with 11-15 rows of relatively short, stout, gently tapering teeth; accessory plates elongate with stout teeth on posterior margin; mouthhooks heavily sclerotised and strongly curved, each without a preapical tooth (fig. 148). *Thoracic and abdominal segments*: Anterior portion of each thoracic segment with an encircling band of several discontinuous rows of spinules. T1 with 7-10 rows of small, sharply pointed spinules; T2 spinules (pl. 19.c) small, stout, sharply pointed, with 4-7 rows surrounding segment; T3 spinules similar to those on T2 but only 3-4 rows dorsally, reduced to no spinules mid laterally and only 2-3 rows ventrally. Creeping welts (pl. 19.e) with small spinules similar to those on T2 in transverse rows, with 2-3 rows anteriorly directed, the remainder posteriorly directed. A8 with intermediate areas well defined and sensilla obvious. *Anterior spiracles* (pl. 19.c): 9-14 tubules. *Posterior spiracles* (pl. 19.a): Each spiracular slit thick walled, about 2.5 times as long as broad. Spiracular hairs long, branched in apical half; 18-24 hairs in dorsal and ventral bundles and 8-12 hairs in lateral bundles. *Anal area* (pl. 19.e, f): Lobes small, slightly retracted, with small, stout spinules in 4-6 discontinuous rows surrounding anal lobes. Larger, curved spinules forming a small concentration below anal opening. *Source*: Based on specimens from the Cook Islands.

Distribution
South Pacific: Cook Islands.
Distribution mapped by Drew (1982a).

Attractant
Males attracted to cue lure; methyl eugenol records are incorrect, see Drew (1989a).

Bactrocera (Bactrocera) musae (Tryon)

Banana Fruit Fly

(figs 71, 149, 188; pl. 20)

Taxonomic notes
This species has also been known as *Chaetodacus musae* Tryon, *C. tryoni* var. *musae* Tryon, *Dacus musae* (Tryon) and *Strumeta musae* (Tryon). Some records of *D. ornatissimus* Froggatt, a synonym of *B. psidii*, were based on misidentifications of this species.

Pest status and commercial hosts
A major pest of banana (*Musa* x *paradisiaca*) and dwarf banana (*M. acuminata*) in northern Queensland and Papua New Guinea (Drew, 1982a). *B. musae* has also been known to attack common guava (*Psidium guajava*) and papaya (*Carica papaya*) (Drew, 1989a).

Wild hosts
Recorded from *Musa banksii* F. Muell., and species of Capparidaceae and Myrtaceae (Drew, 1989a).

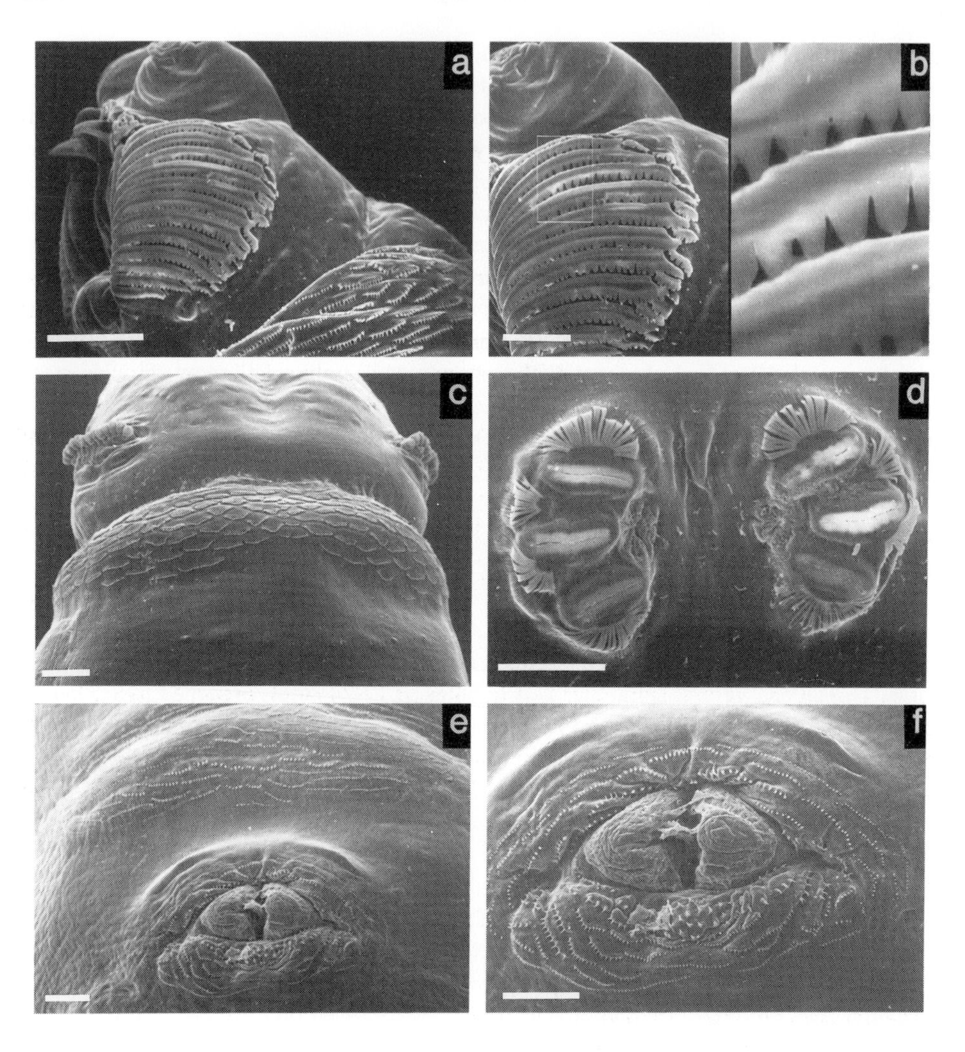

Plate 19. *Bactrocera (B.) melanota*, SEMs of 3rd instar larva; a, head, lateral view (scale=0.1mm); b, oral ridges, lateral view (scale=50μm); c, T1 and T2, dorsal view (scale=0.1mm); d, posterior spiracles (scale=0.1mm); e, A8, creeping welt and anal lobes (scale=0.1mm); f, anal lobes (scale=0.1mm).

Adult identification

An orange to black species; scutum with lateral yellow stripes; with facial spots, anterior supra-alar setae, prescutellar acrostichal setae, 2 scutellar setae, and a typical dacine wing pattern (fig. 188); male with a pecten.

Description of third instar larva

Larvae medium-sized, length 7.5-9.0mm, width 1.2-1.5mm. *Head*: Stomal sensory organ, with 2-3 long, tapering sensilla, surrounded by 4 large smooth-edged preoral lobes; oral ridges (pl. 20.a, b) with 10-15 rows of large, long, bluntly rounded teeth; accessory plates (pl. 20.c) with 10-15 elongate, deeply serrated plates, some of the more posterior plates interlocking with oral ridges; mouthhooks large, black, heavily sclerotised, without preapical teeth (fig. 149). *Thoracic and abdominal segments*: Anterior margin of T1 with an encircling broad band of 13-17 discontinuous rows of large, sharply pointed spinules. T2 spinules large, in discontinuous rows around anterior portion of segment, 7-9 rows dorsally, 5-7 rows laterally and 7-9 rows ventrally; T3 spinules slightly smaller, 4-7 rows dorsally, 1-3 rows laterally and 2-4 rows ventrally. Creeping welts with rows of small spinules, 2-3 rows anteriorly and 7-9 rows posteriorly directed. A8 with obvious intermediate areas and well defined sensilla. *Anterior spiracles* (pl. 20.d): 13-17 tubules. *Posterior spiracles* (pl. 20.e): Spiracles large; spiracular slits about 3 times as long as broad, each with a very thick sclerotised rima. Dorsal and ventral spiracular hair bundles of 17-24 stout, branched hairs; lateral bundles of 7-9 stout, branched hairs. *Anal area* (pl. 20.f): Lobes large, protuberant, surrounded by 3-4 rows of stout spinules in discontinuous rows. Below anal opening, spinules concentrating into a small cluster of slightly stouter projections. *Source*: Based on specimens from Queensland, Australia.

Distribution

Australia: Eastern coastal Queensland, north of Townsville, and some isolated localities in the west of the Cape York Peninsula.
New Guinea area: Bismarck Archipelago, Papua New Guinea, Solomon Islands.
Oriental Asia: Possibly Indonesia (Sulawesi).
Distribution in Australia mapped by Drew (1982a).

Attractant

Males attracted to methyl eugenol.

Other references

BEHAVIOUR, P.H. Smith (1989); BIOLOGY, Fletcher (1989b); LARVA, Exley (1955); PAPUA NEW GUINEA, Smith (1977); POST-HARVEST DISINFESTATION, Heather (1989); STYLOPS PARASITOID, Drew & Allwood (1985).

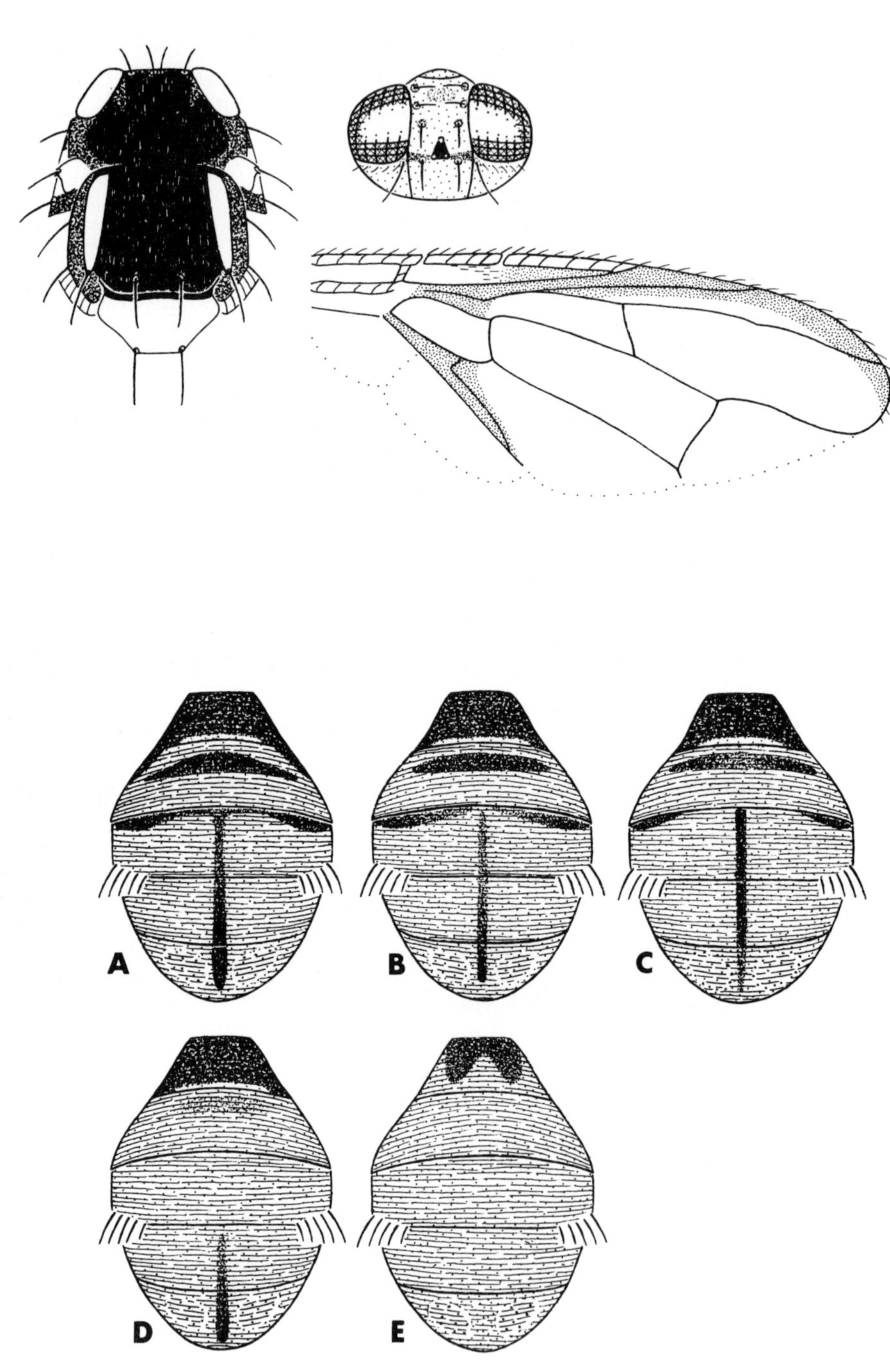

Fig. 188. *Bactrocera (B.) musae*, adult male; insets A-E show variation in abdominal patterning.

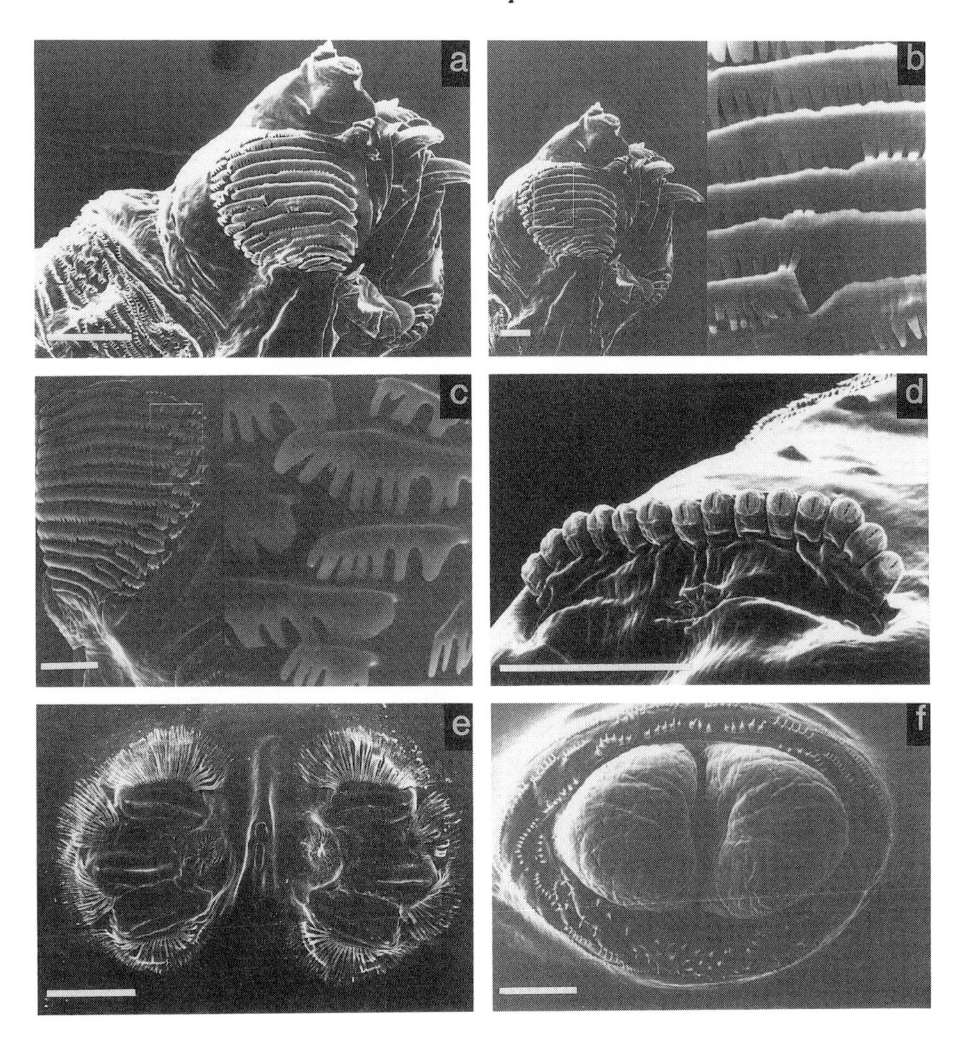

Plate 20. *Bactrocera (B.) musae*, SEMs of 3rd instar larva; a, head and T1, lateral view (scale=0.1mm); b, oral ridges, lateral view (scale=50μm); c, oral ridges and accessory plates, lateral view (scale=50μm); d, anterior spiracle (scale=0.1mm); e, posterior spiracles (scale=0.1mm); f, anal lobes (scale=0.1mm).

Bactrocera (Bactrocera) neohumeralis (Hardy)

(figs 150, 189; pl. 21)

Taxonomic notes

This species has also been known as *Chaetodacus humeralis* Perkins, *C. tryoni* var. *sarcocephali* Tryon, *Dacus neohumeralis* Hardy and *D. tryoni* var. *neohumeralis* Hardy. Care should be taken not to confuse the name *C. humeralis* Perkins (a homonym) with the African *D. humeralis* Bezzi (p. 339).

Pest status and commercial hosts

A major pest in Queensland where it attacks a similar list of crops to *B. tryoni* (Drew, 1982a), to which it is very closely related. Unlike most other *Bactrocera* spp., including *B. tryoni*, *B. neohumeralis* mates in the middle of the day rather than at dusk, and that is the only known isolating mechanism between those two species (Smith, 1979; 1989). It is recorded from the following commercial hosts: apple (*Malus domestica*), apricot (*Prunus armeniaca*), arabica coffee (*Coffea arabica*), black mulberry (*Morus nigra*), common guava (*Psidium guajava*), date palm (*Phoenix dactylifera*), feijoa (*Feijoa sellowiana*), fox grape (*Vitis labrusca*), grapefruit (*Citrus x paradisi*), kangaroo apple (*Solanum laciniatum*), lemon (*C. limon*), loquat (*Eriobotrya japonica*), Jew plum (*Spondias cytherea*), mandarin (*C. reticulata*), mango (*Mangifera indica*), Mauritius raspberry (*Rubus rosifolius*), peach (*Prunus persica*), pear (*Pyrus communis*), plum (*Prunus domestica*), pummelo (*C. maxima*), round kumquat (*Fortunella japonica*), sacred garlic-pear (*Crateva religiosa*), strawberry (*Fragaria* sp.), strawberry guava (*Psidium littorale*), sugar-apple (*Annona squamosa*), Surinam cherry (*Eugenia uniflora*), sweet orange (*Citrus sinensis*), tomato (*Lycopersicon esculentum*), tree tomato (*Cyphomandra betacea*), tropical almond (*Terminalia catappa*), watery rose-apple (*Syzygium aqueum*), weeping fig (*Ficus benjamini*), white sapote (*Casimiroa edulis*), ylang-ylang (*Cananga odorata*), *Passiflora suberosa*, and *Solanum seaforthianum* (Drew, 1989a; M.M. Elson-Harris, unpublished data, 1991).

Wild hosts

Recorded from a wide variety of hosts belonging to the families Annonaceae, Capparidaceae, Celastraceae, Combretaceae, Euphorbiaceae, Lauraceae, Oleaceae, Passifloraceae, Naucleaceae, Rutaceae, Sapindaceae, Sapotaceae and Solanaceae.

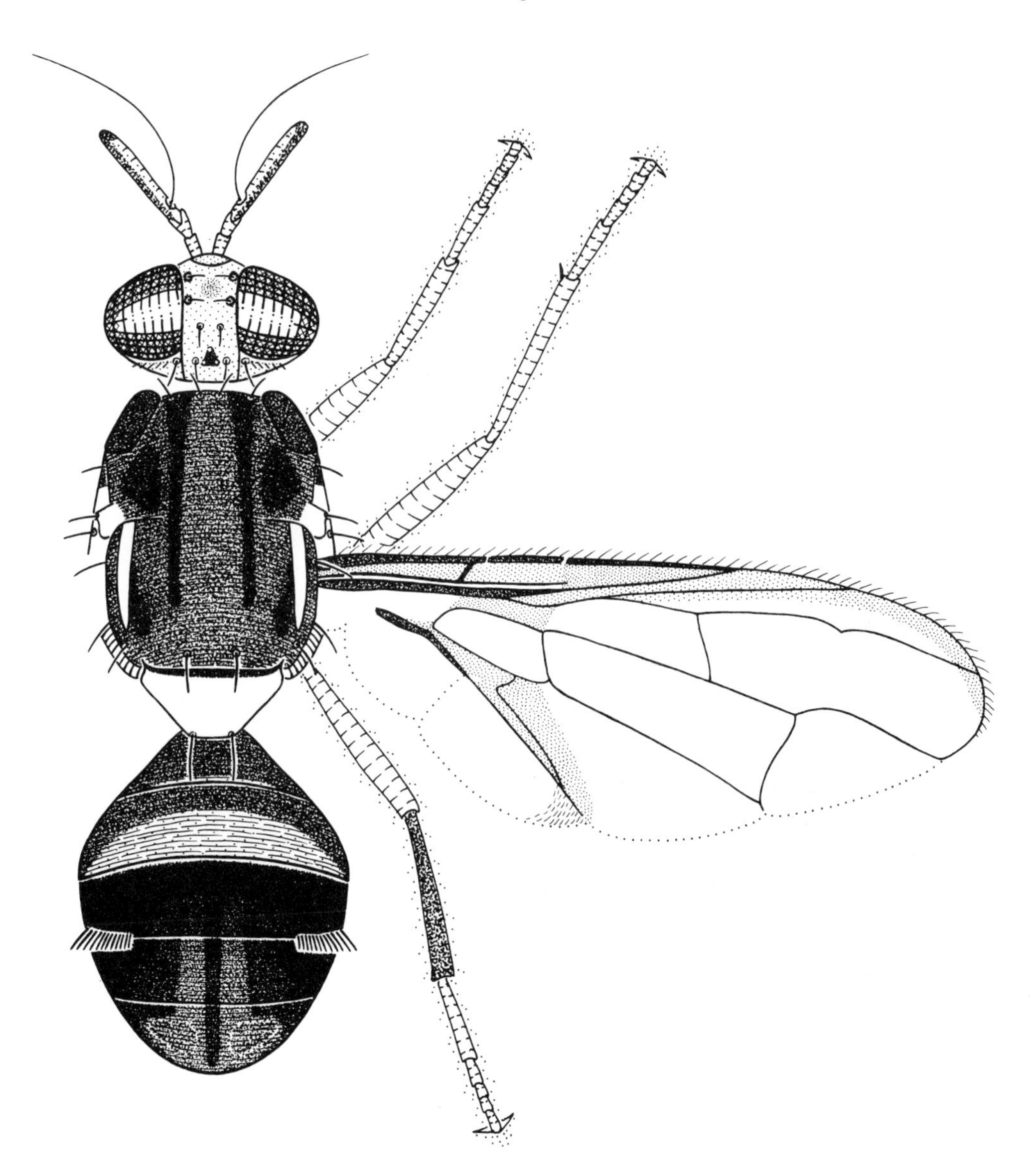

Fig. 189. *Bactrocera (B.) neohumeralis*, adult male.

Adult identification

An orange to black species; scutum with lateral yellow stripes; postpronotal lobe not yellow; with facial spots, anterior supra-alar setae, prescutellar acrostichal setae, 2 scutellar setae, wing cell c completely covered in microtrichia, and the costal band extending from the wing base to the wing apex (fig. 189); male with a pecten.

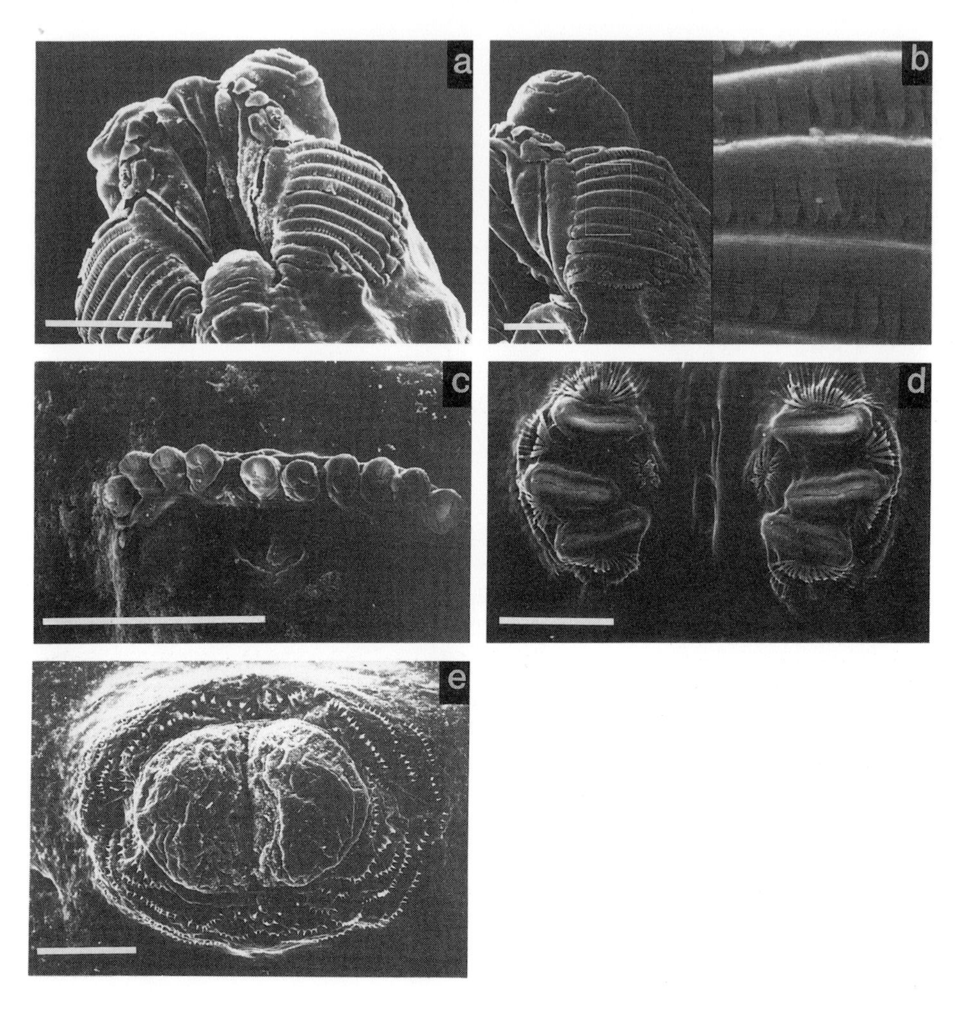

Plate 21. *Bactrocera (B.) neohumeralis*, SEMs of 3rd instar larva; a, head, ventral view (scale=0.1mm); b, oral ridges, lateral view (scale=50μm); c, anterior spiracle (scale=0.1mm); d, posterior spiracles (scale=0.1mm); e, anal lobes (scale=0.1mm).

Description of third instar larva

Larvae medium-sized, length 7.0-9.5mm; width 1.2-1.5mm. *Head*: Stomal sensory organ with 2-3 sensilla (often 2 long, tapered and branched), surrounded by 5-6 unserrated preoral lobes; oral ridges (pl. 21.a, b) with 8-12 rows of short, tapering, bluntly rounded teeth; accessory plates of 6-8 large, serrated plates; mouthhooks large, with heavily sclerotised apical teeth (fig. 150). *Thoracic and abdominal segments*: Anterior portion of each thoracic segment with an encircling band of discontinuous rows of small, sharp spinules. T1 with 7-12 rows of small, sharply pointed spinules grouped dorsally and ventrally but changing to discontinuous rows ventrally; T2, 4-6 rows dorsally, 3-4 rows laterally and 5-7 rows ventrally; T3, 1-3 rows dorsally, 0-3 rows laterally and 3-5 rows ventrally. Large creeping welts of small sharply pointed spinules, with 1-4 rows anteriorly directed, the remainder posteriorly directed. *Anterior spiracles* (pl. 21.c): 9-12 tubules. *Posterior spiracles* (pl. 21.d): Each spiracular slit about 2.5-3.0 times as long as broad. Dorsal and ventral spiracular hair bundles of 16-20 hairs; lateral bundles of 6-12 hairs; several spiracular hairs branched in apical third to half. *Anal area* (pl. 21.e): Lobes large, protuberant, surrounded by several rows of small, sharply pointed spinules concentrating into a small group below anal opening. *Source*: Based on specimens from Queensland, Australia.

Distribution

Australia: Eastern coastal areas of Queensland, plus some isolated areas on the west of the Cape York Peninsula; north eastern New South Wales.
New Guinea area: Papua New Guinea.

Attractant

Males attracted to cue lure.

Other references

BACTERIAL SYMBIONTS, Fitt & O'Brien (1985); BEHAVIOUR, Katsoyannos (1989b); BIOLOGY, Fletcher (1989b); ECOLOGY, Fitt (1989); LARVA, Exley (1955); PHEROMONES, Bellas & Fletcher (1979) and Koyama (1989b); SEPARATION FROM *B. TRYONI* USING ENZYME POLYMORPHISM, McKechnie (1975); STYLOPS PARASITOID, Drew & Allwood (1985); TRAPPING, Hill (1986a; 1986b), Hill & Hooper (1984) and Katsoyannos (1989b).

Bactrocera (Bactrocera) passiflorae (Froggatt)

Fijian Fruit Fly

(figs 151, 190; pl. 22)

Taxonomic notes

This species has also been known as *Chaetodacus passiflorae* (Froggatt), *Dacus passiflorae* Froggatt and *Strumeta passiflorae* (Froggatt).

Commercial hosts

A potential pest which has been recorded from avocado (*Persea americana*), breadfruit (*Artocarpus altilis*), cashew (*Anacardium occidentale*), cocoa (*Theobroma cacao*), common guava (*Psidium guajava*), eggplant (*Solanum melongena*), giant granadilla (*Passiflora quadrangularis*), lime (*Citrus aurantiifolia*), purple granadilla (*P. edulis*), mandarin (*C. reticulata*), mango (*Mangifera indica*) and papaya (*Carica papaya*) (Clausen *et al.*, 1965; Drew, 1989a; Litsinger *et al.*, 1991); the eggplant record has been questioned (A.J. Allwood, pers. comm., 1991) and that may have been an aberrant host. There are also records, requiring confirmation, from coffee (*Coffea* sp.),

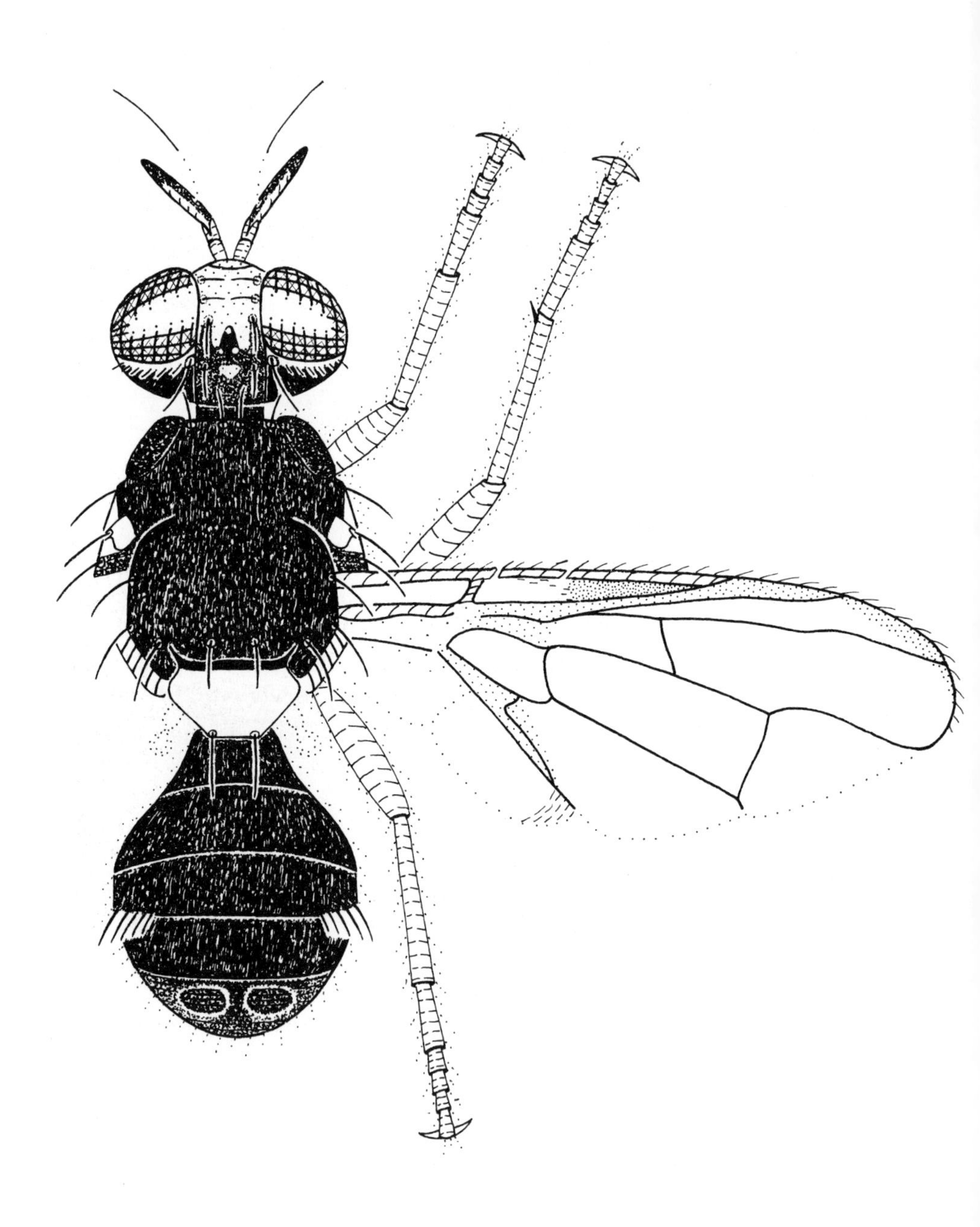

Fig. 190. *Bactrocera (B.) passiflorae*, adult male.

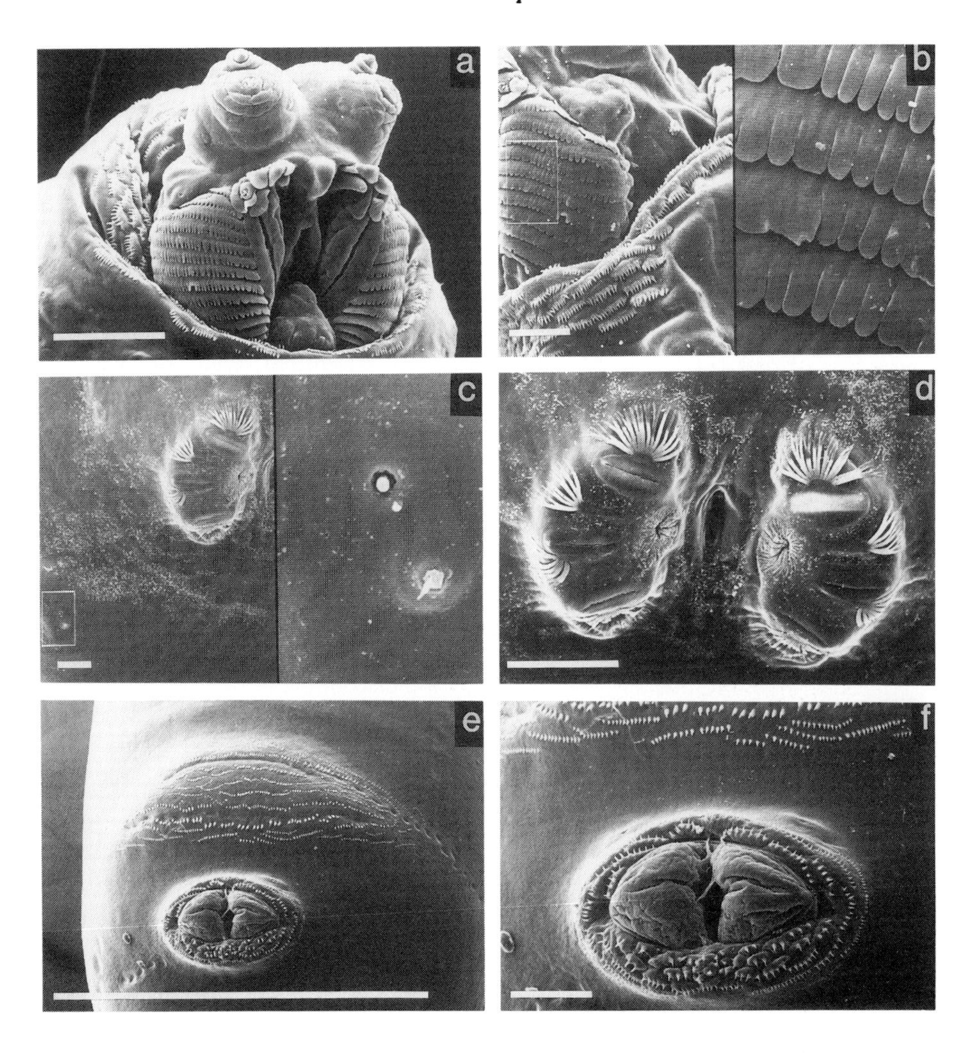

Plate 22. *Bactrocera (B.) passiflorae*, SEMs of 3rd instar larva; a, head, ventral view (scale=0.1mm); b, oral ridges, lateral view (scale=50μm); c, A8, posterior view showing sensilla (scale=50μm); d, posterior spiracles (scale=0.1mm); e, A8, creeping welt and anal lobes (scale=1.0mm); f, anal lobes (scale=0.1mm).

Malay-apple (*Syzygium malaccense*), Pacific lychee (*Pometia pinnata*) and strawberry guava (*Psidium littorale*) (Bezzi, 1928; Le Pelley, 1968; Simmonds, 1936; SPC, unpublished data, 1988).

Wild hosts

Recorded from fao (*Ochrosia oppositifolia*; Apocynaceae) (Litsinger *et al.*, 1991) but it probably has a wider range of wild hosts.

Adult identification

A predominantly dark orange-brown to black species; with anterior supra-alar setae, prescutellar acrostichal setae, 2 scutellar setae, and a typical dacine wing pattern (fig. 190); male with a pecten; scutum without yellow stripes and face without spots.

Description of third instar larva

Larvae medium-sized, length 8.0-9.5mm; width 1.2-1.5mm. *Head*: Stomal sensory organs (pl. 22.a) each with 4-6 sensilla (some long, some short) surrounded by 5-6 large unserrated preoral lobes; oral ridges (pl. 22.a, b) with 9-13 rows of stout, almost parallel-sided, bluntly rounded teeth; accessory plates small, serrated, arranged along outer edges of oral ridges; mouthhooks moderately sclerotised, each with a curved, slender apical tooth (fig. 151). *Thoracic and abdominal segments*: Anterior margin of T1 with an encircling band of sharply pointed spinules, with 5-8 rows dorsally, 4-6 rows laterally and 4-8 rows ventrally; T2 with 3-5 rows of slightly smaller spinules encircling segment; T3 without spinules. Creeping welts of 10-13 rows of small sharply pointed spinules with 1 posterior row of slightly stouter spinules (pl. 22.e). A8 with small intermediate areas and small, well defined tubercles and sensilla, some of which are long and tapering. *Anterior spiracles*: 9-13 tubules. *Posterior spiracles* (pl. 22.d): Each spiracular slit about 3 times as long as broad with a sclerotised rima. Dorsal and ventral spiracular hair bundles of 9-16 hairs, lateral bundles of 4-8 hairs. *Anal area* (pl. 22.f): Lobes large, protuberant, surrounded by 3-6 discontinuous rows of small, stout, sharply pointed spinules concentrating into stouter spinules just below anal opening. *Source*: Based on specimens collected in Fiji.

Distribution

South Pacific: Fiji, Niue Island and Tonga. In 1990 a few specimens were trapped in New Zealand, but they probably emerged from illegally imported fruit and did not become established (R.A.I. Drew, pers. comm., 1990).
Distribution mapped by Drew (1982a).

Attractant

Males attracted to cue lure.

Other references

BEHAVIOUR, P.H. Smith (1989); BIOCONTROL, Wharton (1989a; 1989b); PARASITOIDS, Narayanan & Chawla (1962), Simmonds (1936), Wharton (1989a; 1989b) and Wharton & Gilstrap (1983).

Bactrocera (Bactrocera) psidii (Froggatt)

(fig. 191)

Taxonomic notes
This species has also been known as *Dacus psidii* (Froggatt), *D. ornatissimus* Froggatt, *D. virgatus* Coquillett, *Strumeta psidii* (Froggatt) and *Tephritis psidii* Froggatt. The original description of *D. ornatissimus* was based on a mixed series of *B. musae* and *B. psidii*.

Commercial hosts
Recorded from citrus (*Citrus* sp.), common guava (*Psidium guajava*), giant granadilla (*Passiflora quadrangularis*), mango (*Mangifera indica*) and strawberry guava (*Psidium littorale*) (Drew, 1989a). Drew (1982a) noted that *B. psidii* has the potential to become a serious pest of those crops if they are developed in New Caledonia. Records from Tahiti chestnut (*Inocarpus fagifer*) and *Nephelium* sp. (includes rambutan, pulasan, and formerly lychee) (Malloch, 1938) were not from New Caledonia and were probably based on misidentifications of another species.

Adult identification
A predominantly dark orange-brown to black species; scutum with short lateral yellow stripes; with small facial spots, anterior supra-alar setae, prescutellar acrostichal setae and 2 scutellar setae; wing pattern very reduced, at most faintly infuscate along the costal edge and along crossveins r-m and dm-cu (fig. 191); male with a pecten.

Distribution
South Pacific: New Caledonia.
Distribution mapped by Drew (1982a).

Attractant
Males attracted to cue lure.

Other reference
PARASITOIDS, Cochereau (1970).

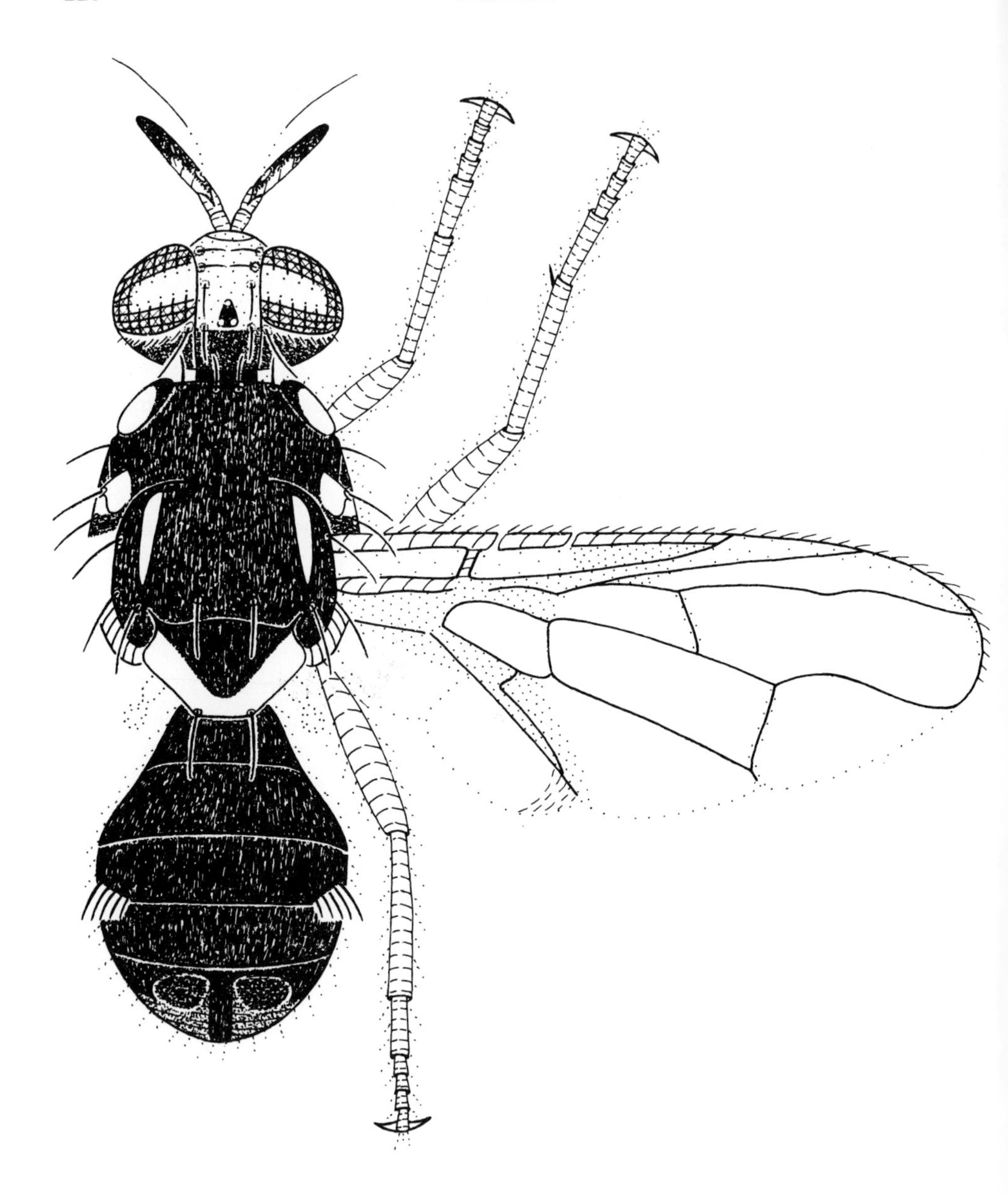

Fig. 191. *Bactrocera (B.) psidii*, adult male.

Bactrocera (Bactrocera) trivialis (Drew)

(fig. 192)

Taxonomic notes
This species has also been known as *Dacus trivialis* Drew.

Commercial hosts
Recorded from common guava (*Psidium guajava*), grapefruit (*Citrus x paradisi*), peach (*Prunus persica*) and tabasco pepper (*Capsicum frutescens*) (Drew, 1989a). Drew (1982a) noted that *B. trivialis* has the potential to become a serious pest if peach and guava crops are developed in Papua New Guinea.

Adult identification
Scutum with lateral yellow stripes; with facial spots, anterior supra-alar setae, prescutellar acrostichal setae, 2 scutellar setae, abdomen orange centrally and black laterally, and wing with a broad costal band (fig. 192); male with a pecten.

Distribution
Australia: Torres Strait islands.
New Guinea area: Indonesia (Irian Jaya), Papua New Guinea.
Oriental Asia: Indonesia (Sulawesi).

Attractant
Males attracted to cue lure.

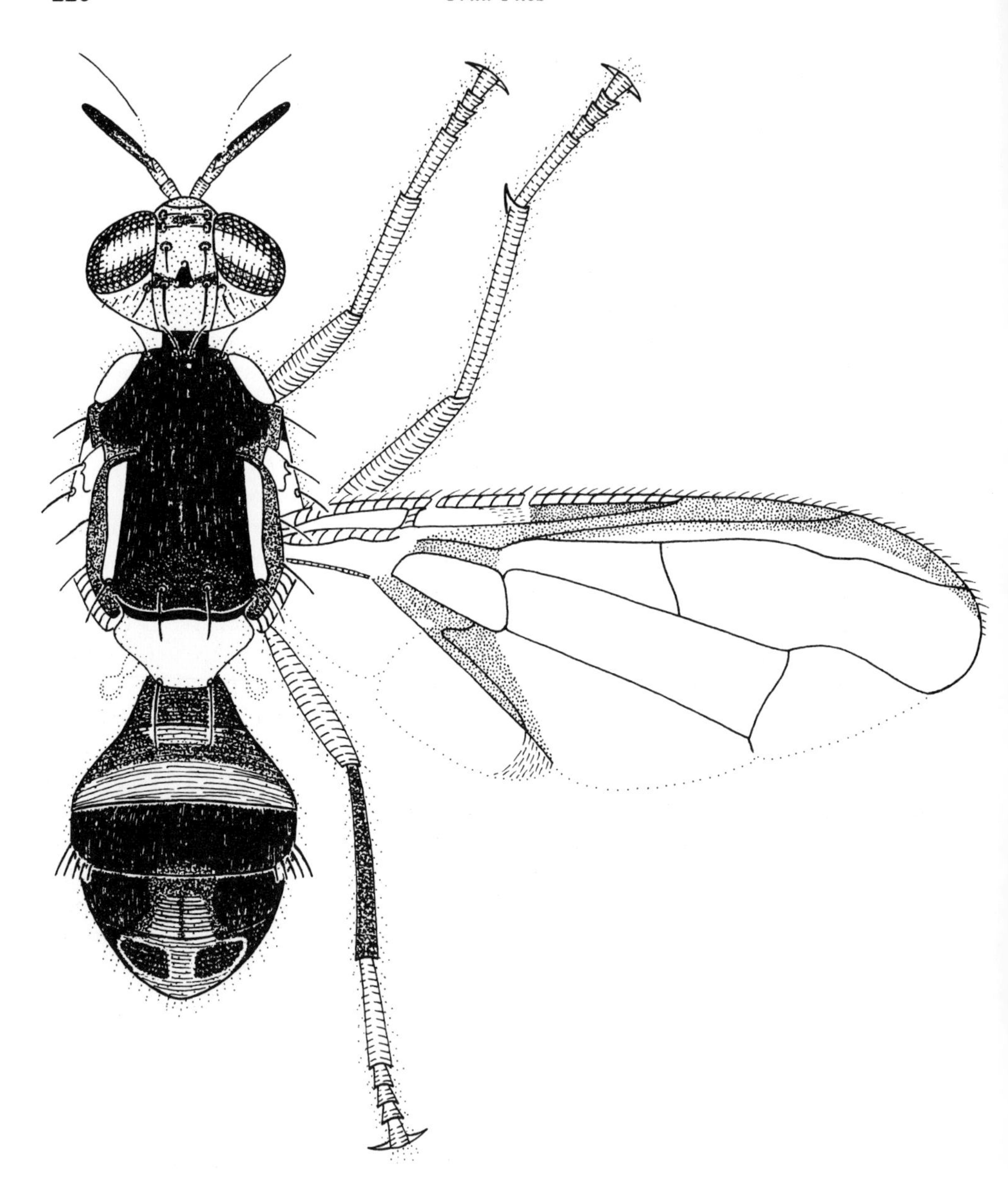

Fig. 192. *Bactrocera (B.) trivialis*, adult male.

Bactrocera (Bactrocera) tryoni **(Froggatt)**

Queensland Fruit Fly
Qfly

(figs 69, 72, 152, 193; pl. 23)

Taxonomic notes

This species has also been known as *Chaetodacus tryoni* (Froggatt), *Dacus ferrugineus tryoni* (Froggatt), *D. tryoni* (Froggatt), *Strumeta tryoni* (Froggatt) and *Tephritis tryoni* Froggatt.

Pest status and commercial hosts

This is the most serious insect pest of fruit and vegetable crops in Australia, and it infests all commercial fruit crops, other than pineapple and strawberry (Drew, 1982a). It is recorded from the following commercial hosts: apple (*Malus domestica*), apricot (*Prunus armeniaca*), arabica coffee (*Coffea arabica*), Australian desert lime (*Eremocitrus glauca*), avocado (*Persea americana*), black mulberry (*Morus nigra*), blackberry (*Rubus fruticosa*), California berry (*R. ursinus*), Cape gooseberry (*Physalis peruviana*), carambola (*Averrhoa carambola*), cashew (*Anacardium occidentale*), citron (*Citrus medica*), cluster fig (*Ficus racemosa*), common fig (*F. carica*), common guava (*Psidium guajava*), custard apple (*Annona reticulata*), date palm (*Phoenix dactylifera*), dwarf banana (*Musa acuminata*), English walnut (*Juglans regia*), fox grape (*Vitis labrusca*), giant granadilla (*Passiflora quadrangularis*), grapefruit (*C. x paradisi*), Indian fig prickly pear (*Opuntia ficus-indica*), Indian jujube (*Ziziphus mauritiana*), Indian plum (*Flacourtia jangomas*), Jew plum (*Spondias cytherea*), kangaroo apple (*Solanum laciniatum*), kei apple (*Dovyalis caffra*), lemon (*C. limon*), loquat (*Eriobotrya japonica*), mandarin (*C. reticulata*), mango (*Mangifera indica*), myrobalan plum (*Prunus cerasifera*), olive (*Olea europaea*), papaya (*Carica papaya*), peach and nectarine (*P. persica*), pear (*Pyrus communis*), persimmon (*Diospyros kaki*), plum (*Prunus domestica*), pomegranate (*Punica granatum*), pummelo (*Citrus maxima*), purple granadilla (*Passiflora edulis*), quince (*Cydonia oblonga*), rose-apple (*Syzygium jambos*), round kumquat (*Fortunella japonica*), Spanish cherry (*Mimusops elengi*), strawberry guava (*Psidium littorale*), sugar-apple (*Annona squamosa*), Surinam cherry (*Eugenia uniflora*), sweet cherry (*Prunus avium*), sweet orange (*Citrus sinensis*), tabasco pepper (*Capsicum frutescens*), tomato (*Lycopersicon esculentum*), tropical almond (*Terminalia catappa*), watery rose-apple (*S. aqueum*), weeping fig (*Ficus benjamini*), white mulberry (*Morus alba*), white sapote (*Casimiroa edulis*), wine grape (*Vitis vinifera*), ylang-ylang (*Cananga odorata*) and *Solanum seaforthianum* (Drew, 1989a; M.M. Elson-Harris, unpublished data, 1991). There are also old records, requiring confirmation, from banana (*Musa x paradisiaca*) and sour orange (*Citrus aurantium*) (Froggatt, 1909).

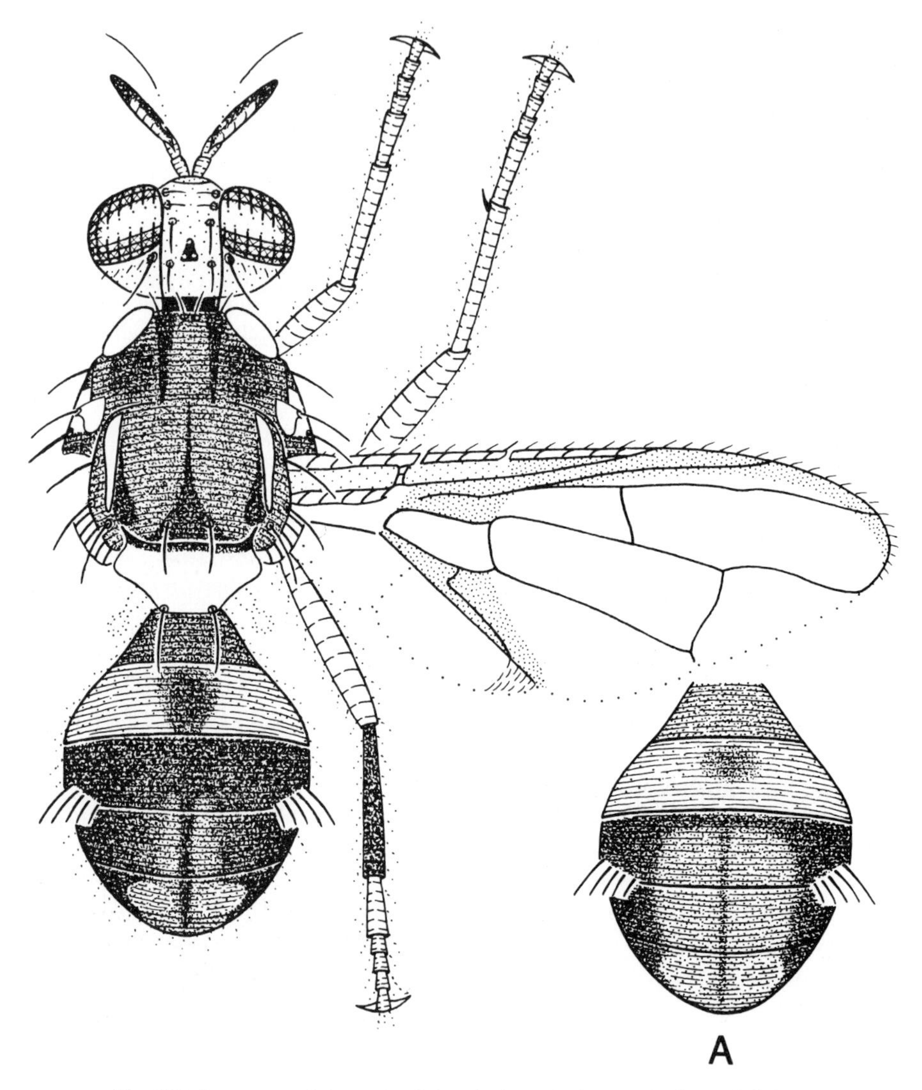

Fig. 193. *Bactrocera (B.) tryoni*, adult male; inset A shows pale form of abdomen.

Wild hosts

Recorded from 60 wild hosts by Drew (1989a), belonging to the following families: Anacardiaceae, Annonaceae, Apocynaceae, Capparidaceae, Celastraceae, Combretaceae, Cunoniaceae, Davidsoniaceae, Ebenaceae, Euphorbiaceae, Lauraceae, Meliaceae, Moraceae, Myrtaceae, Naucleaceae, Oleaceae, Passifloraceae, Rhamnaceae, Rutaceae, Sapindaceae, Sapotaceae, Siphonodontaceae, Smilacaceae, Solanaceae and Vitaceae. This very wide host range enables *B. tryoni* to build up large populations in forest areas, which then act as reservoirs for invading crops (see Fitt, 1986).

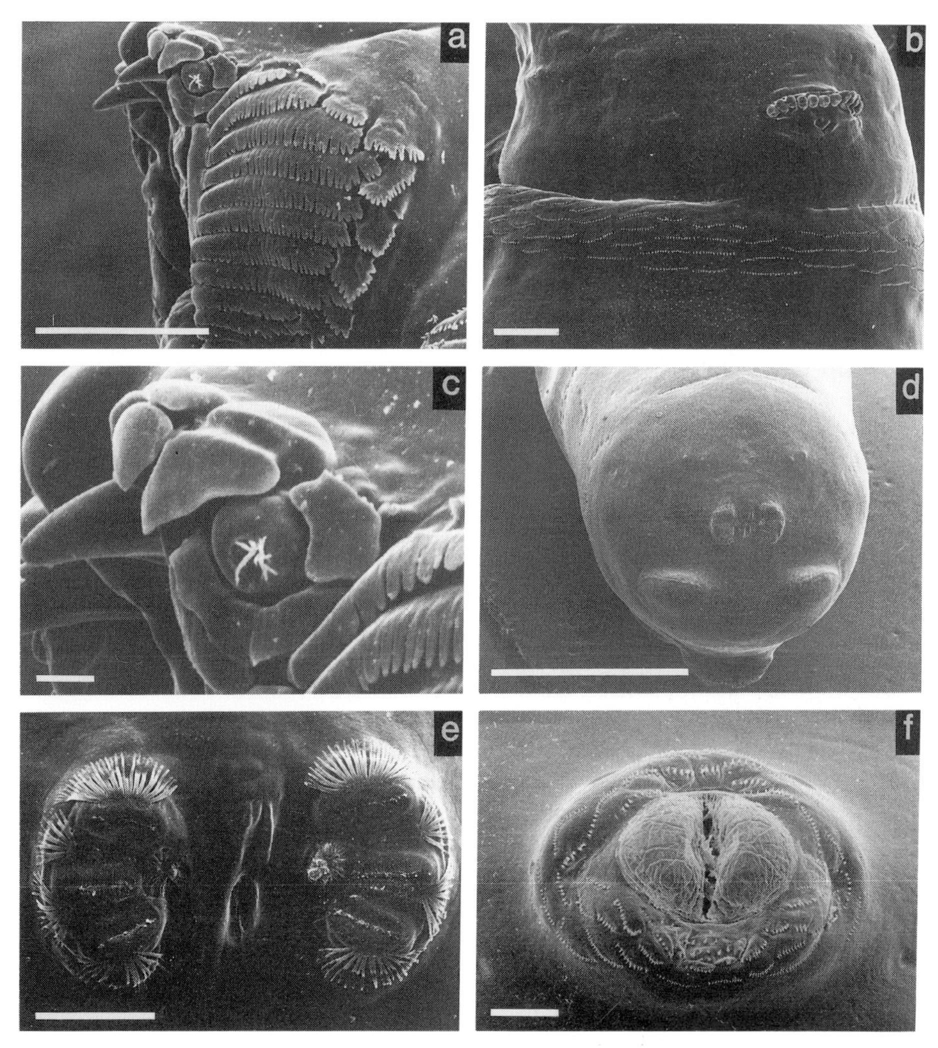

Plate 23. *Bactrocera (B.) tryoni*, SEMs of 3rd instar larva; a, oral ridges, lateral view (scale=0.1mm); b, T1 and T2, lateral view (scale=0.1mm); c, stomal sensory organ and preoral lobes, lateral view (scale=10μm); d, A8, posterior view (scale=1.0mm); e, posterior spiracles (scale=0.1mm); f, anal lobes (scale=0.1mm).

Adult identification

A predominantly orange to black species; scutum with lateral yellow stripes and postpronotal lobe yellow; with facial spots, anterior supra-alar setae, prescutellar acrostichal setae, 2 scutellar setae, wing cell c completely covered in microtrichia, and the costal band extending from the wing base to the wing apex (fig. 193); male with a pecten.

Description of third instar larva

Larvae medium-sized, length 8.0-11.0mm; width 1.2-1.5mm. *Head*: Stomal sensory organs large, rounded, each with 3 sensilla and surrounded by 6 large unserrated preoral lobes (pl. 23.c); oral ridges (pl. 23.a) with 9-12 rows of deeply serrated, bluntly rounded teeth; 8-12 small, serrated accessory plates; mouthhooks large, heavily sclerotised, without preapical teeth (fig. 152). *Thoracic and abdominal segments*: A band of small posteriorly directed spinules encircling anterior portion of each thoracic segment. T1 with 9-13 discontinuous rows; T2 (pl. 23.b) with 4-7 rows dorsally and laterally, and 4-8 rows ventrally; T3 with 3-6 rows dorsally and laterally, and 3-5 rows ventrally. Creeping welts with 2-3 anteriorly directed and 3-8 posteriorly directed rows of spinules. A8 with well defined intermediate areas and large sensilla (pl. 23.d). *Anterior spiracles* (pl. 23.b): 9-12 tubules. *Posterior spiracles* (pl. 23.e): Placed just above midline; each spiracular slit about 3 times as long as broad. Dorsal and ventral spiracular hair bundles of 12-17, broad, stout, often branched hairs; lateral bundles of 5-9 similar hairs. *Anal area* (pl. 23.f): Lobes well defined, surrounded by 3-5 discontinuous rows of spinules, becoming longer and stouter below anal opening. *Source*: Based on specimens from Queensland, Australia.

Rapid diagnostic technique Dadour *et al.* (in press) have described two rapid methods for separation of larvae and puparia of *B. tryoni* from those of *Ceratitis capitata*, using morphology and electrophoresis; these methods were used to monitor numbers of those flies in the recent Perth eradication campaign (Yeates, 1990).

Distribution

Australia: Throughout the eastern half of Queensland, eastern New South Wales, and the extreme east of Victoria. In 1989 *B. tryoni* became established in the Perth area of Western Australia and there have been outbreaks in South Australia. Following an eradication campaign using baits, male lures and the sterile insect technique (Yeates, 1990), the Perth outbreak appears to have been eradicated; no wild flies have been collected since November 1990 (D.K. Yeates, pers. comm., Sept. 1991).
New Guinea area: A few males have been trapped in Papua New Guinea but it is unlikely to be established there (Drew, 1989a).
South Pacific: Adventive in French Polynesia (Austral and Society Islands) and New Caledonia. Twice adventive in Easter Island, but eradicated (Bateman, 1982).
Distribution mapped by Drew (1982a) and IIE (1991a).

Attractant

Males attracted to cue lure.

Other references

ANTENNAL SENSILLA, Giannakakis & Fletcher (1985); BACTERIAL SYMBIONTS, Fitt & O'Brien (1985); BEHAVIOUR, Drew (1987a), Fitt (1984), Katsoyannos (1989b), Smith (1979; 1989) and Tychsen & Bateman (1977); BIBLIOGRAPHY, Zwölfer (1985); BIOCONTROL, Wharton (1989a; 1989b); BIOLOGY, Fletcher (1989b); CONTROL, Hargreaves *et al.* (1986); DATASHEETS,

Gellatley & Turpin (1987) and Weems (1965a); DEVELOPMENT RATE, Fletcher (1989a); ECOLOGY, Debouzie (1989), Fitt (1989), Fletcher (1989c) and Meats (1989a; 1989b); HOST RESISTANCE, Greany (1989); LARVA, Anderson (1963), Elson-Harris (1988), Exley (1955) and Kandybina (1977); MALE ANNIHILATION, Cunningham (1989c); PARASITOIDS, Narayanan & Chawla (1962), Wharton (1989a; 1989b) and Wharton & Gilstrap (1983); PASSION FRUIT, Hargreaves *et al.* (1986); PHEROMONES, Bellas & Fletcher (1979), Fletcher & Giannakakis (1973), Giannakakis *et al.* (1978), Koyama (1989b) and Nation (1981); POST-HARVEST DISINFESTATION, Heather (1985; 1989), Hill *et al.* (1988) and Rigney (1989); REARING, Fay (1989), Hooper (1978a) and Tsitsipis (1989); SENSILLA ON ACULEUS, Eisemann & Rice (1989); SEPARATION FROM *B. AQUILONIS*, Drew & Lambert (1986); SEPARATION FROM *B. NEOHUMERALIS* USING ENZYME POLYMORPHISM, McKechnie (1975); STERILE INSECT TECHNIQUE, Calkins (1989); SYMBIONTS, Drew & Lloyd (1989); STYLOPS PARASITOIDS, Drew & Allwood (1985); TRAPPING, Economopoulos (1989), Hill (1986a; 1986b), Hill & Hooper (1984), Katsoyannos (1989b) and O'Loughlin *et al.* (1983).

Bactrocera (Bactrocera) tuberculata (Bezzi)

(figs 153, 194; pl. 24)

Taxonomic notes

This species has also been known as *Chaetodacus tuberculatus* Bezzi and *Dacus tuberculatus* (Bezzi).

Pest status and commercial hosts

A pest of peach (*Prunus persica*), which has also been recorded from mango (*Mangifera indica*), in Thailand (D.L. Hancock, pers. comm., 1991).

Adult identification

A predominantly black species; scutum with lateral yellow stripes; with facial spots, anterior supra-alar setae, prescutellar acrostichal setae, 2 scutellar setae, legs yellow, and a reduced wing pattern (fig. 194); male with a pecten.

Description of third instar larva

Larvae medium-sized, length 7.5-9.0mm; width 1.2-1.7mm. *Head*: Stomal sensory organs (pl. 24.b) rounded, with 3 long, tapered sensilla, surrounded by 4-5 unserrated preoral lobes; oral ridges (pl. 24.a, b) with 11-15 rows of short, bluntly rounded teeth; accessory plates small, serrated, shell-shaped; mouthhooks moderately sclerotised, with large, curved apical teeth (fig. 153). *Thoracic and abdominal segments*: Anterior portion of each thoracic segment with a surrounding band of several discontinuous rows of small spinules. T1 with 5-7 rows of small, sharply pointed spinules; T2 with 4-7

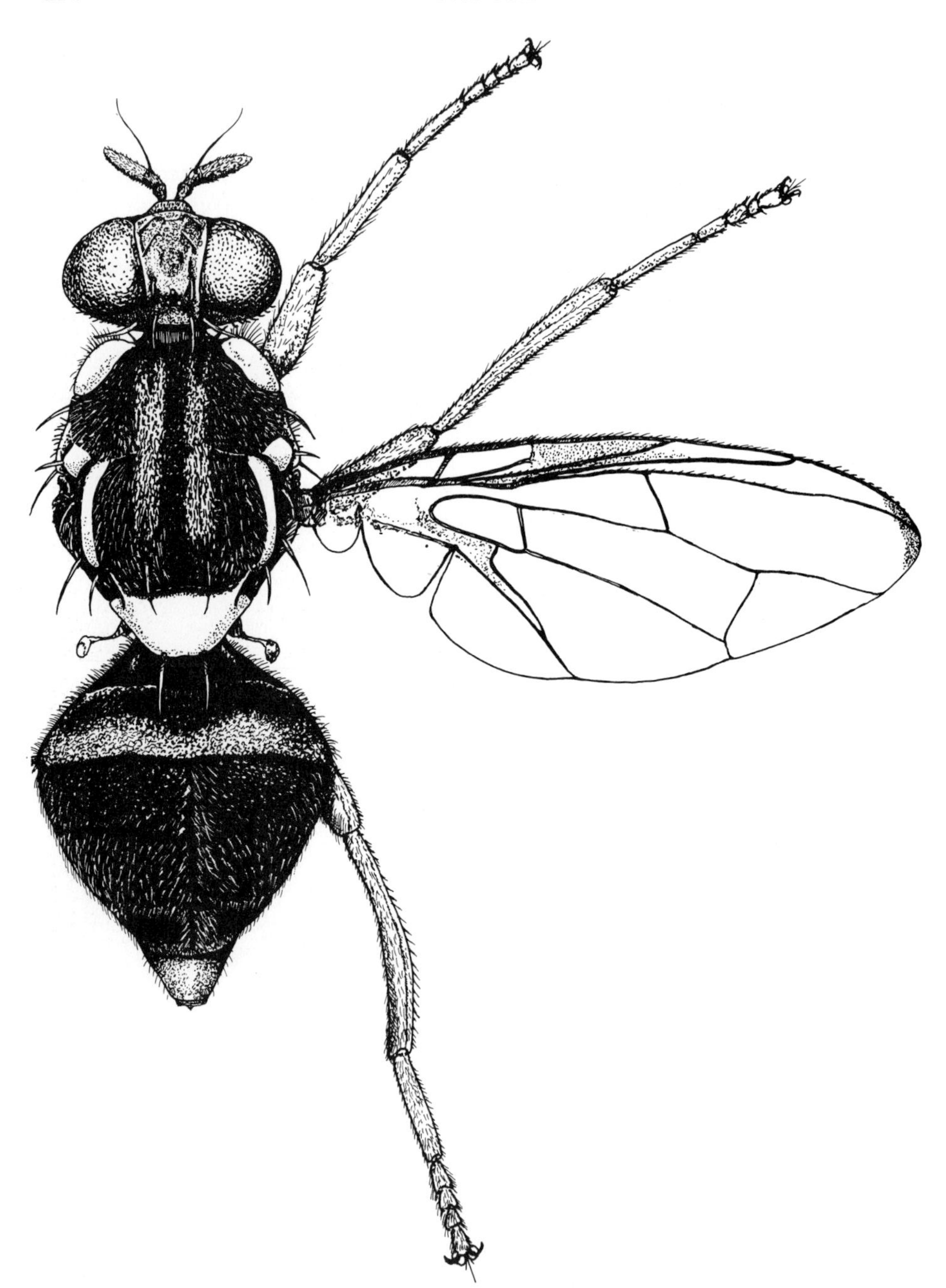

Fig. 194. *Bactrocera (B.) tuberculata*, adult female.

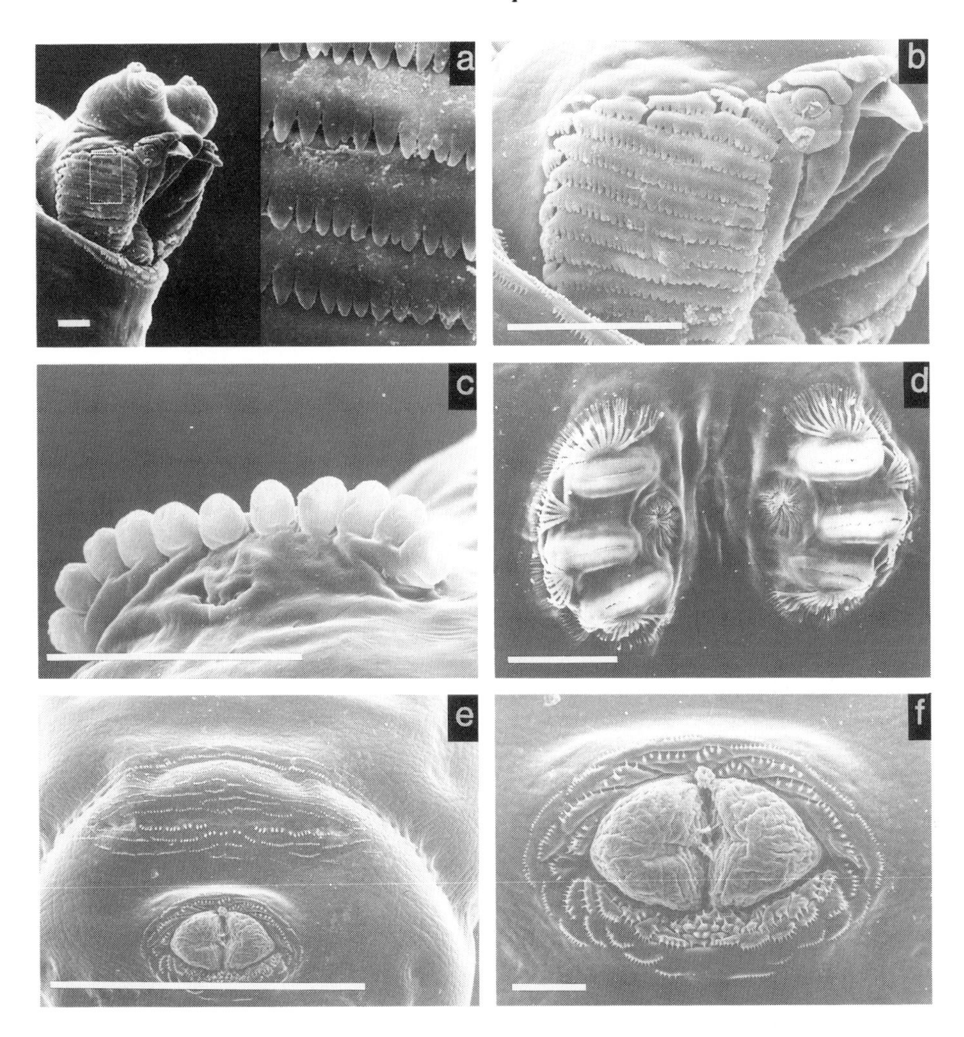

Plate 24. *Bactrocera (B.) tuberculata*, SEMs of 3rd instar larva; a, head and oral ridges, lateral view (scale=50μm); b, stomal sensory organ, preoral lobes and oral ridges, lateral view (scale=0.1mm); c, anterior spiracle (scale=0.1mm); d, posterior spiracles (scale=0.1mm); e, A8, creeping welt and anal lobes (scale=1.0mm); f, anal lobes (scale=0.1mm).

rows surrounding segment; T3 with 3-4 rows. Creeping welts (pl. 24.e) large with 2-3 rows of small, anteriorly directed, and 8-10 posteriorly directed, rows of spinules. A8 with obvious intermediate lobes and sensilla. *Anterior spiracles* (pl. 24.c): 10-14 tubules. *Posterior spiracles* (pl. 24.d): Placed just above the midline; each spiracular slit about 3 times as long as broad, with a heavily sclerotised rima. Dorsal and ventral spiracular hair bundles of 11-14 hairs; lateral bundles with 8-12 hairs; most spiracular hairs branched in apical third. *Anal area* (pl. 24.e, f): Lobes large, protuberant, surrounded by 2-5 discontinuous rows of small spinules becoming stouter and more concentrated below anal opening. *Source*: Based on specimens from Thailand (lab. colony).

Distribution
Oriental Asia: Myanmar, Thailand, Vietnam.

Attractant
Males attracted to methyl eugenol.

Bactrocera (Bactrocera) umbrosa (Fabricius)

(figs 73, 154, 195; pl. 25)

Taxonomic notes
This species has also been known as *Bactrocera fasciatipennis* Doleschall, *Dacus conformis* Walker, *D. diffusus* Walker, *D. fascipennis* Wiedemann, *D. frenchi* Froggatt, *D. umbrosus* Fabricius, *Strumeta frenchi* (Froggatt) and *S. umbrosa* (Fabricius).

Pest status and commercial hosts
A serious pest of bread fruit (*Artocarpus altilis*) and jackfruit (*A. heterophyllus*) (Yukawa, 1984). In Malaysia it has also been recorded from chempedak (*A. integer*) (Yunus & Ho, 1980). There are also some records from hosts other than *Artocarpus* spp.; Hardy (1973) recorded it from bitter gourd (*Momordica charantia*) in Kalimantan, and there are old records, requiring confirmation, from giant granadilla (*Passiflora quadrangularis*), pummelo (*Citrus maxima*) and sour orange (*C. aurantium*) (Froggatt, 1909; Perkins, 1938; Yunus & Ho, 1980).

Adult identification
Scutum predominantly black with lateral yellow stripes; with facial spots, anterior supra-alar setae, prescutellar acrostichal setae, 2 scutellar setae, and a characteristic wing pattern (fig. 195); abdomen very variable, sometimes broadly black laterally; male with a pecten.

Description of third instar larva
Larvae large, length 8.0-11.0mm; width 1.8-2.2mm. *Head*: Stomal sensory organ (pl. 25.c) large, with several small, often branched sensilla; surrounded by 5 preoral lobes with the anterior lobe very large, forming a single lobe above the mouthhook; oral ridges (pl. 25.a, b) with 16-21 rows of very short, broad, blunt teeth; accessory plates small, shell shaped, with very few serrations; mouthhooks heavily sclerotised, with strongly curved apical teeth (fig. 154). *Thoracic and abdominal segments*: Mature larvae with a small heart-shaped patch of pigment below cuticle surface on ventral side of T1. A band of discontinuous rows of small spinules surrounding the anterior portion of each thoracic segment. Spinules on T1 (pl. 25.a) very long, stout and sharply

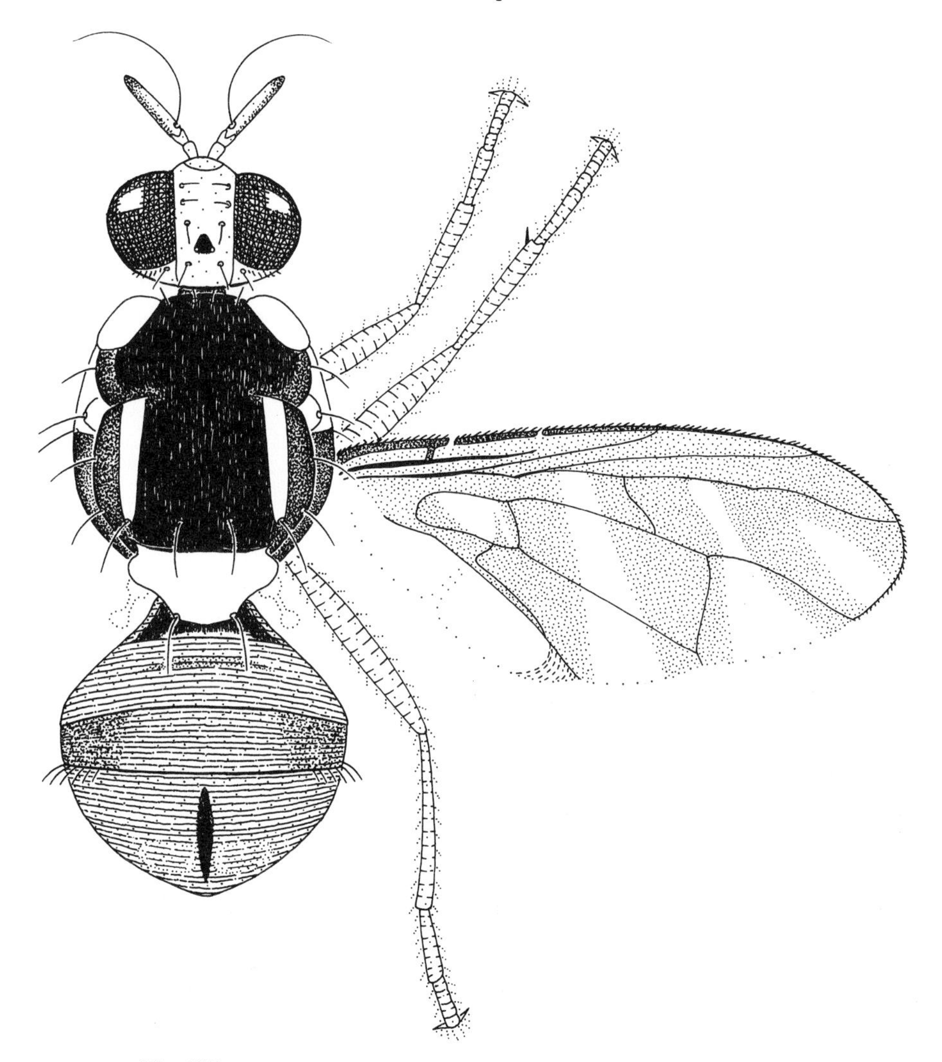

Fig. 195. *Bactrocera (B.) umbrosa*, adult male (abdomen pattern variable).

pointed, forming 8-10 rows of small plates dorsally and laterally, becoming slightly smaller and forming 5-9 rows ventrally; T2 spinules much shorter and in 5-7 rows dorsally, 2-4 rows laterally and 6-8 rows ventrally; T3 spinules similar to those on T2, with only 2-4 rows of spinules around segment, often none mid-laterally. Creeping welts (pl. 25.f) large, with small, stout spinules. A8 with intermediate areas very large, sensilla obvious. *Anterior spiracles* (pl. 25.d): 17-21 tubules. *Posterior spiracles* (pl. 25.e): Large, thick walled spiracular slits set in a radiating pattern just above midline; each about 3.5 times as long as broad. Spiracular hairs long, fine, branched in apical two-thirds. *Anal area* (pl. 25.f): Lobes very large, protuberant, with 4-6 rows of small, stout spinules surrounding the outer edge of anal pad; slightly larger, curved spinules forming a small concentration just below anal opening. *Source*: Based on specimens from Malaysia (ex *Artocarpus integer*).

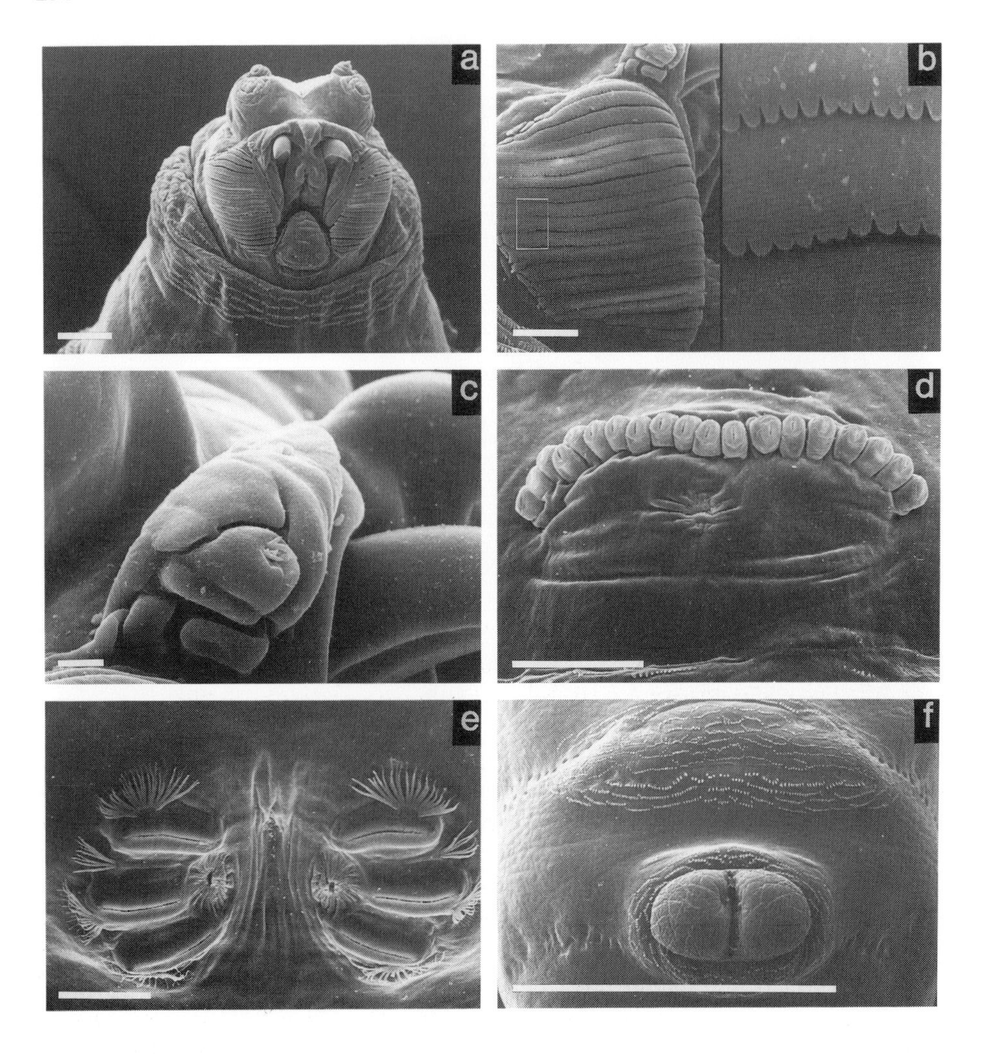

Plate 25. *Bactrocera (B.) umbrosa*, SEMs of 3rd instar larva; a, head and T1, ventral view (scale=0.1mm); b, oral ridges, lateral view (scale=50μm); c, stomal sensory organ and preoral lobes (scale=10μm); d, anterior spiracle (scale=0.1mm); e, posterior spiracles (scale=0.1mm); f, A8, creeping welt and anal lobes (scale=1.0mm).

Distribution

New Guinea area: Bismarck Archipelago, Bougainville Island, Papua New Guinea, Solomon Islands.
Oriental Asia: Indonesia, Malaysia, Philippines, Palau Islands, Thailand (southern).
South Pacific: New Caledonia and Vanuatu.

Attractant
Males attracted to methyl eugenol.

Other references
ECOLOGY, Hong & Serit (1988), Tan (1985) and Tan & Lee (1982); GEL ELECTROPHORESIS, Yong (1988); LARVA, Rohani (1987); PARASITOIDS, Cochereau (1970) and Narayanan & Chawla (1962); TRAPPING, Tan (1984).

Bactrocera (Bactrocera) zonata (Saunders)

Peach Fruit Fly

(figs 66, 75, 196)

Taxonomic notes
This species has also been known as *Bactrocera maculigera* Doleschall, *Dacus zonatus* (Saunders), *Dasyneura zonata* Saunders and *Rivellia persicae* Bigot. The name *Dacus ferrugineus* var. *mangiferae* Cotes, which has been listed as a synonym of *B. dorsalis*, probably belongs to this species (R.A.I. Drew & D.L. Hancock, in prep.).

Pest status and commercial hosts
A pest of peach (*Prunus persica*) and sugar-apple (*Annona squamosa*) in India (Butani, 1976; Grewal & Malhi, 1987), and common guava (*Psidium guajava*) and mango (*Mangifera indica*) in Pakistan (Syed *et al.*, 1970a). In Pakistan it has also been reared from the following hosts: apple (*Malus domestica*), bitter gourd (*Momordica charantia*), date palm (*Phoenix dactylifera*), okra (*Abelmoschus esculentus*), papaya (*Carica papaya*), paradise apple (*Malus pumila*), peach (*Prunus persica*), phalsa (*Grewia asiatica*), pomegranate (*Punica granatum*), quince (*Cydonia oblonga*) and sweet orange (*Citrus sinensis*) (Syed *et al.*, 1970a). The adventive population in Mauritius attacks mango and tropical almond (*Terminalia catappa*), and possibly melon (*Cucumis melo*) and watermelon (*Citrullus lanatus*) (Orian & Moutia, 1960). Kapoor & Agarwal (1983) listed the following hosts, but many of their records were derived from Bezzi (1916) (see Kapoor, 1970), and some were based on casual observation rather than rearing: common fig (*Ficus carica* - not reared), common jujube (*Ziziphus jujube*), eggplant (*Solanum melongena*), Indian bael (*Aegle marmelos*), *Luffa* sp., sapodilla (*Manilkara zapota* - not reared), tomato (*Lycopersicon esculentum*) and white-flowered gourd (*Lagenaria siceraria*). Syed *et al.* (1970a) examined egg plant, *Luffa* and tomato fruits, but failed to find any *B. zonata*; they suggested that records from those plants may have been based on misidentifications of *Dacus ciliatus*.

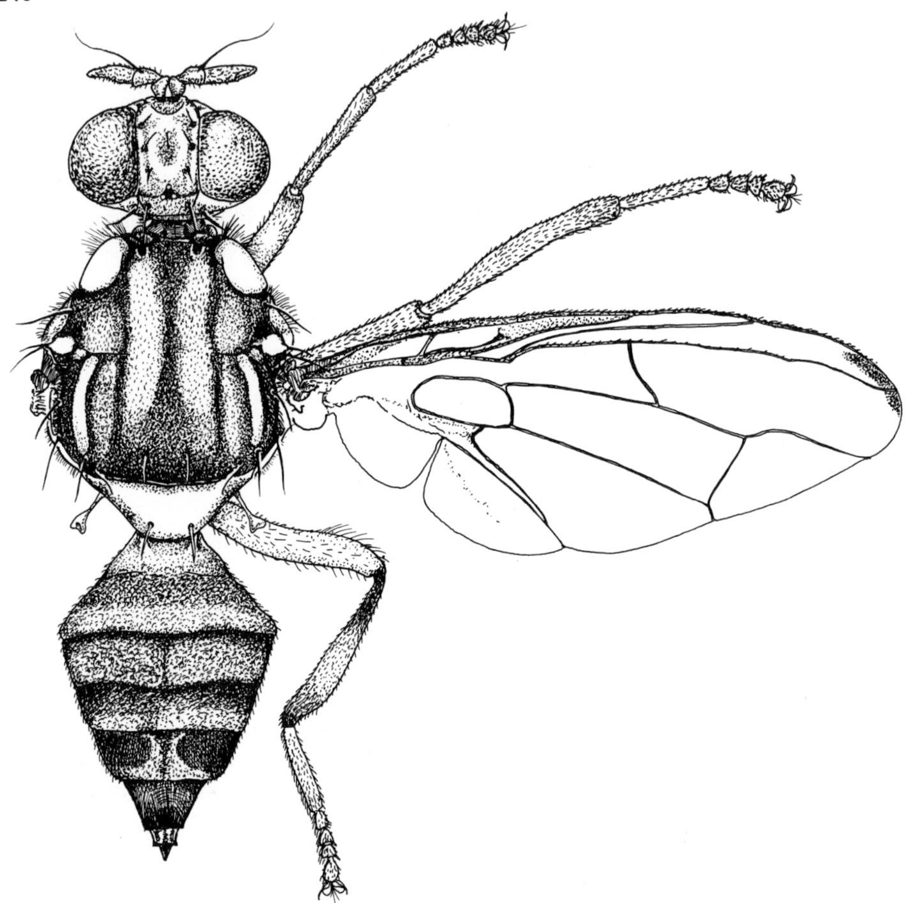

Fig. 196. *Bactrocera (B.) zonata*, adult female.

Wild hosts

Recorded from some Euphorbiaceae, Lecythidaceae and Rhamnaceae (Kapoor & Agarwal, 1983; Syed *et al.*, 1970a).

Adult identification

A predominantly pale orange-brown to red-brown species; scutum with lateral yellow stripes; with facial spots, anterior supra-alar setae, prescutellar acrostichal setae, 2 scutellar setae, and a reduced wing pattern (fig. 196); male with a pecten.

Partial description of third instar larva

Larvae large, length 10.0-11.0mm. *Head*: Stomal sensory organ small, rounded; oral ridges with 10-11 deep, clearly defined rows; mouthhooks moderately sclerotised, each with a long thin apical tooth. *Thoracic and abdominal segments*: T1 with 6-9 rows of small spinules encircling anterior portion of segment; T2 with fewer rows encircling anterior portion of segment; T3 with a few spinules dorsally but forming rows laterally and ventrally; A1-A8 with rows of spinules ventrally forming creeping welts, with 1 anterior and 1-2 posterior rows of slightly larger spinules. *Anterior spiracles*: 13-15 short tubules. *Posterior spiracle*: Spiracular slits 3.0-3.5 times as long as broad, each with a moderately sclerotised rima; spiracular hairs slightly longer than half the length of a spiracular slit, frequently branched; dorsal and ventral bundles of 3-17 hairs, lateral bundles of 6-8 hairs. *Anal area*: Lobes well developed and surrounded by discontinuous rows of small spinules. *Source*: Based on Kandybina (1977).

Distribution

Indian Ocean: Adventive population in Mauritius; records from Réunion were based on misidentifications of *B. (B.) montyana* (Munro) (D.L. Hancock, pers. comm., 1991).
North America: Has been trapped in the wild in California (Carey & Dowell, 1989), but eradicated (Spaugy, 1988).
Tropical Asia: India, Indonesia (Sumatra), Laos, Sri Lanka, Thailand, Vietnam.
A distribution map was produced by CIE (1961b).

Attractant

Males attracted to methyl eugenol.

Other references

BEHAVIOUR, Qureshi *et al.* (1974b), P.H. Smith (1989) and Syed *et al.* (1970a); BIOLOGY, Fletcher (1989b) and Narayanan (1953); ECOLOGY, Fletcher (1989c); INDIA, Kapoor (1989); LARVA, Khan & Khan (1987); MALE TERMINALIA, Singh & Premlata (1985); PARASITOIDS, Agarwal & Kapoor (1989), Narayanan & Chawla (1962) and Syed *et al.* (1970b); POST-HARVEST DISINFESTATION, Rigney (1989); REARING, Qureshi *et al.* (1974b) and Tsitsipis (1989).

Subgenus *DACULUS* SPEISER

Bactrocera (Daculus) oleae (Gmelin)

Olive Fly
Olive Fruit Fly

(figs 78, 197)

Taxonomic notes

This species has also been known as *Daculus oleae* (Gmelin), *Dacus oleae* (Gmelin) and *Musca oleae* Gmelin. It is the only species which definitely belongs to subgenus *B. (Daculus)*, although Drew (1989a) included it in *B. (Polistomimetes)* together with species which are here placed in *B. (Tetradacus)*.

Pest status and commercial hosts

A serious pest of olive (*Olea europaea*) which typically infests one-third of a crop. However, most olives are only used for oil production and an infestation level of 30% at harvest time does not reduce oil value unless they are stored for a prolonged period. When stored for four weeks before processing, Neuenschwander & Michelakis (1978) found that batches in which all fruits had emergence holes produced oil that was up to twelve times as acidic as normal. Parlati *et al.* (1986) also reported increases in peroxide number, and in areas with a 60% infestation they found a small increase in the proportion of saturated fatty acids. Both groups of workers also reported differences in damage caused between olive varieties. The losses caused in Mediterranean countries were listed by Fimiani (1989). Cultivated olive is also attacked by *B. oleae* in South Africa but it is not a serious pest in that area (Annecke & Moran, 1982; Hancock, 1989).

Wild hosts

Recorded from wild olive (*Olea europaea* spp. *africana* (Miller) P. Green) in Africa (Munro, 1984).

Adult identification

Scutum predominantly black (dark red-brown in some areas of the Middle East), but with black area sometimes not extending to the lateral margins, leaving the lateral areas red-brown (especially in Africa); with facial spots, 2 scutellar setae, abdomen orange medially and black laterally, and wing with a reduced pattern (fig. 197). Scutum without yellow stripes, anterior supra-alar setae and prescutellar acrostichal setae; male without a pecten.

Partial description of third instar larva

Larvae medium sized, length 6.5-7.0mm; width 1.2-1.7mm. *Head*: Oral ridges in 10-12 shallow, short rows; mouthhooks heavily sclerotised, each with a short, slender, curved apical tooth. *Thoracic and abdominal segments*: Anterior portion of T1-T3, A1 and A2 with 3-5 rows of small spinules encircling each segment; A3-A5 with a few spinules dorsally and a heavier concentration ventrally; A6-A8 with spinules ventrally, none dorsally and laterally; spinules in creeping welts smaller in central rows. *Anterior spiracles*: 8-12 short tubules. *Posterior spiracles*: Spiracular slits 3.5-4.0 times as long as broad, with a thick rima; spiracular hairs about half the length of a spiracular slit, frequently branched, dorsal and ventral bundles of 7 hairs, lateral bundles of 2-4 hairs. *Anal area*: Lobes small, slightly protuberant, surrounded by several discontinuous rows of small spinules. *Source*: Based on Kandybina (1977) and Phillips (1946).

Distribution

Africa: Throughout the olive-zone of the Mediterranean; Algeria, Canary Islands, Egypt (coast and Nile Valley), Ethiopia (Eritrea), Kenya, Libya, Morocco, South Africa, Sudan, Tunisia.

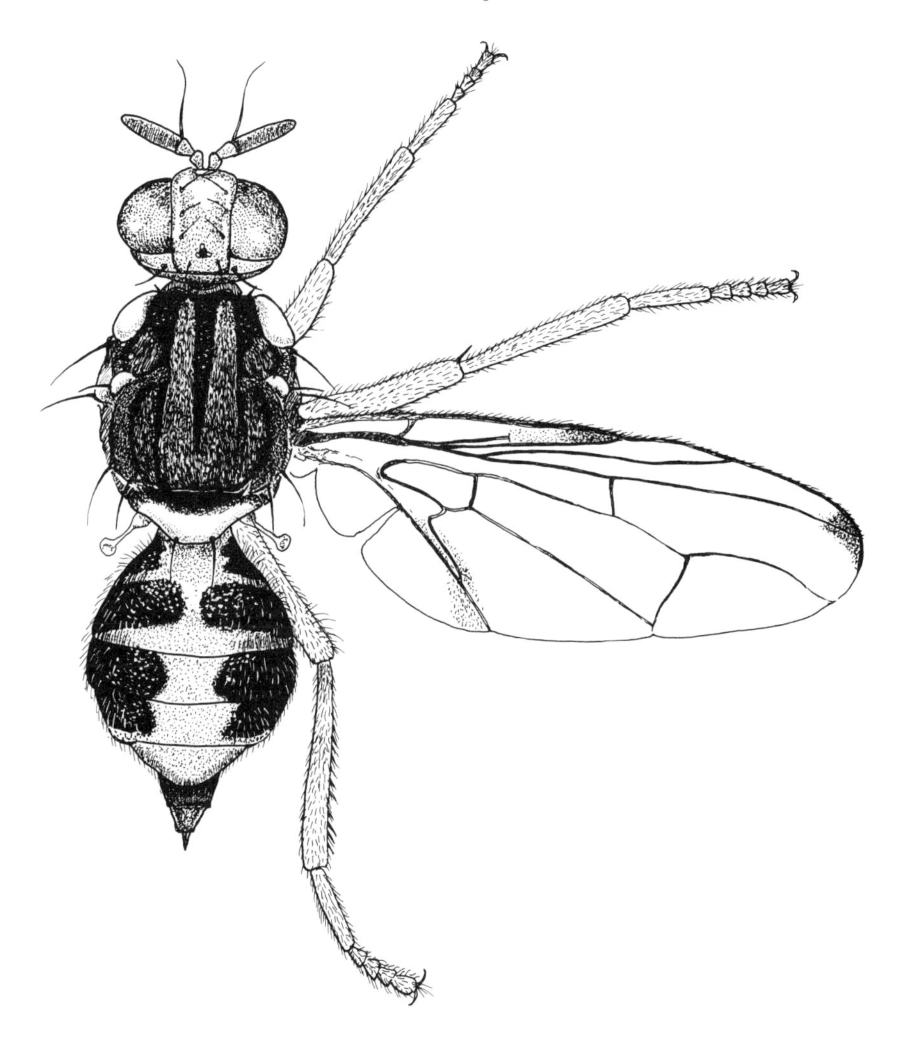

Fig. 197. *Bactrocera (Daculus) oleae*, adult female.

Europe: Throughout the olive-zone of the Mediterranean; southern coastal areas of France, Portugal, Spain, Yugoslavia; all Mediterranean Islands; throughout Greece; extreme south of Switzerland (Neuenschwander, 1984); USSR (Georgia).
Middle East: Israel, Lebanon, western Syria, Turkey.
Oriental Asia: Northern India, north-western Pakistan.
A distribution map was produced by CIE (1957).

Attractant

No synthetic male lure known (no record traced of either methyl eugenol or cue lure having been tested).

Other references

BACTERIAL SYMBIONTS, Manousis & Ellar (1988); BEHAVIOUR, Katsoyannos (1989b) and P.H. Smith (1989); BIBLIOGRAPHY, Delrio (1979) and Zwölfer (1985); BIOCONTROL, Kapatos *et al.* (1977a), Liaropoulos *et al.* (1977) and Wharton (1989a; 1989b); BIOLOGY, Fletcher (1989b); DATASHEET, Weems (1966a); DEVELOPMENT RATE, Fletcher (1989a); ECOLOGY, Debouzie (1989), Kapatos & Fletcher (1984), Kapatos *et al.* (1977b; 1977c), McFadden *et al.* (1977), Meats (1989a), Mustafa & Zaghal (1987) and Pappas *et al.* (1977); EGG MORPHOLOGY, Margaritis (1985); FUNCTIONAL ANATOMY OF FEMALE TERMINALIA, Solinas & Nuzzaci (1984); GENETICS, Zouros & Loukas (1989); INTEGRATED PEST MANAGEMENT, Kapatos (1989); LARVA, Berg (1979) and Sabatino (1974); OVIPOSITION DETERRENT, Averill & Prokopy (1989a); PARASITOIDS, Ben-Salah *et al.* (1989), Bigler & Delucchi (1981), Bigler *et al.* (1986), Delrio & Prota (1977), Economopoulos (1989), Fry (1990), Hancock (1989), Louskas *et al.* (1980), Monaco (1978), Mustafa & Zaghal (1987), Narayanan & Chawla (1962), Neuenschwander (1982), Neuenschwander *et al.* (1983), Ranaldi & Santoni (1987), Wharton (1989a; 1989b) and Wharton & Gilstrap (1983); PHEROMONES, Bueno & Mata (1985), Gariboldi *et al.* (1982), Haniotakis (1977; 1987), Haniotakis *et al.* (1986; 1989), Haniotakis & Vassilio-Waite (1987) and Mazomenos (1989); POST-HARVEST DISINFESTATION, Rigney (1989); REARING, Tsitsipis (1977; 1989), Tzanakakis (1989) and Tzanakakis & Economopoulos (1967); STERILE INSECT TECHNIQUE, Calkins (1989); TRAPPING, Cirio & Vita (1980), Economopoulos (1989), Economopoulos *et al.* (1986), Haniotakis & Vassilio-Waite (1987), Jones *et al.* (1983), Katsoyannos (1989b) and Zervas (1982); VARIETIES OF OLIVE, Neuenschwander *et al.* (1985); VIRAL CONTROL, Manousis & Moore (1987).

Subgenus *HEMIGYMNODACUS* HARDY

Bactrocera (Hemigymnodacus) diversa (Coquillett)

(figs 82, 198)

Taxonomic notes

Subgenus *B. (Hemigymnodacus)* is part of the *B. (Zeugodacus)* group of subgenera. The only species, *B. diversa*, has also been known as *Asiadacus diversus* (Coquillett) and *Dacus diversus* Coquillett.

Commercial hosts

A potential pest of Cucurbitaceae which was recorded from the flowers of the following species by Syed (1970): angled luff (*Luffa acutangula*), luffa (*Luffa aegyptiaca*), pumpkin (*Cucurbita maxima* and *C. pepo*) and white-flowered gourd (*Lagenaria siceraria*). It has also been reared from the fruit of pumpkin (*C. maxima*), but in that case the flower was still attached to the young fruit and the larvae had probably bored into the fruit from the flower (Syed, 1970). In cage experiments, Syed (1970) found that *B. diversa* would readily attack, and develop in, cucurbit fruits, and Batra (1964) found that cut flowers of white-flowered gourd were never attacked, but ovaries and young fruits were.

Other records of this species attacking fruit may refer to hosts only attacked under exceptional conditions or be based on casual observations, or misidentifications. Kapoor & Agarwal (1983) listed the following additional hosts: banana (*Musa* x *paradisiaca*), common guava (*Psidium guajava*), Java plum (*Syzygium cumini*), mango (*Mangifera indica*), nutmeg (*Myristica* sp.) and sour orange (*Citrus aurantium*). It is possible that none of those records were confirmed by rearing; Kapoor & Agarwal (1983) also listed radish (*Raphanus* sp.) as a host, but Brassicaceae lack fleshy fruits. The banana record may derive from a detailed account of *B. diversa* in "plantain" (presumably *Musa* sp.) by Fletcher (1919), which suggests that it may be correct, assuming that Fletcher did not misidentify this species. *B. diversa* has been known to form winter swarms, congregating on the undersides of guava leaves (Batra, 1964).

Wild hosts

Recorded from *Tabernaemontana dichotoma* Roxb. (as *Cerbera manghas*) (Apocynaceae) and *Solanum verbascifolium* (Kapoor & Agarwal, 1983), but these should be regarded as doubtful hosts.

Adult identification

Scutum predominantly black with lateral yellow stripes; with anterior supra-alar setae, prescutellar acrostichal setae, 2 or 4 scutellar setae, and a typical dacine wing pattern (fig. 198); face of male entirely yellow, without facial spots; face of female with a black transverse line above the mouth opening; male without a pecten.

Partial description of third instar larva

Head: Mouthhooks sclerotised, small, broad; each with a small preapical tooth. *Thoracic and abdominal segments*: Posterior spiracular plate small; spiracular slits each with a narrow rima and openings traversed by 10 cross-bars. *Source*: Based on Menon *et al.* (1968).

Distribution

Oriental Asia: China, southern India, Sri Lanka, Thailand.

Attractant

Males attracted to methyl eugenol.

Other references

BEHAVIOUR, P.H. Smith (1989) and Syed (1970); ECOLOGY, Fletcher (1989c); MALE TERMINALIA, Singh & Premlata (1985).

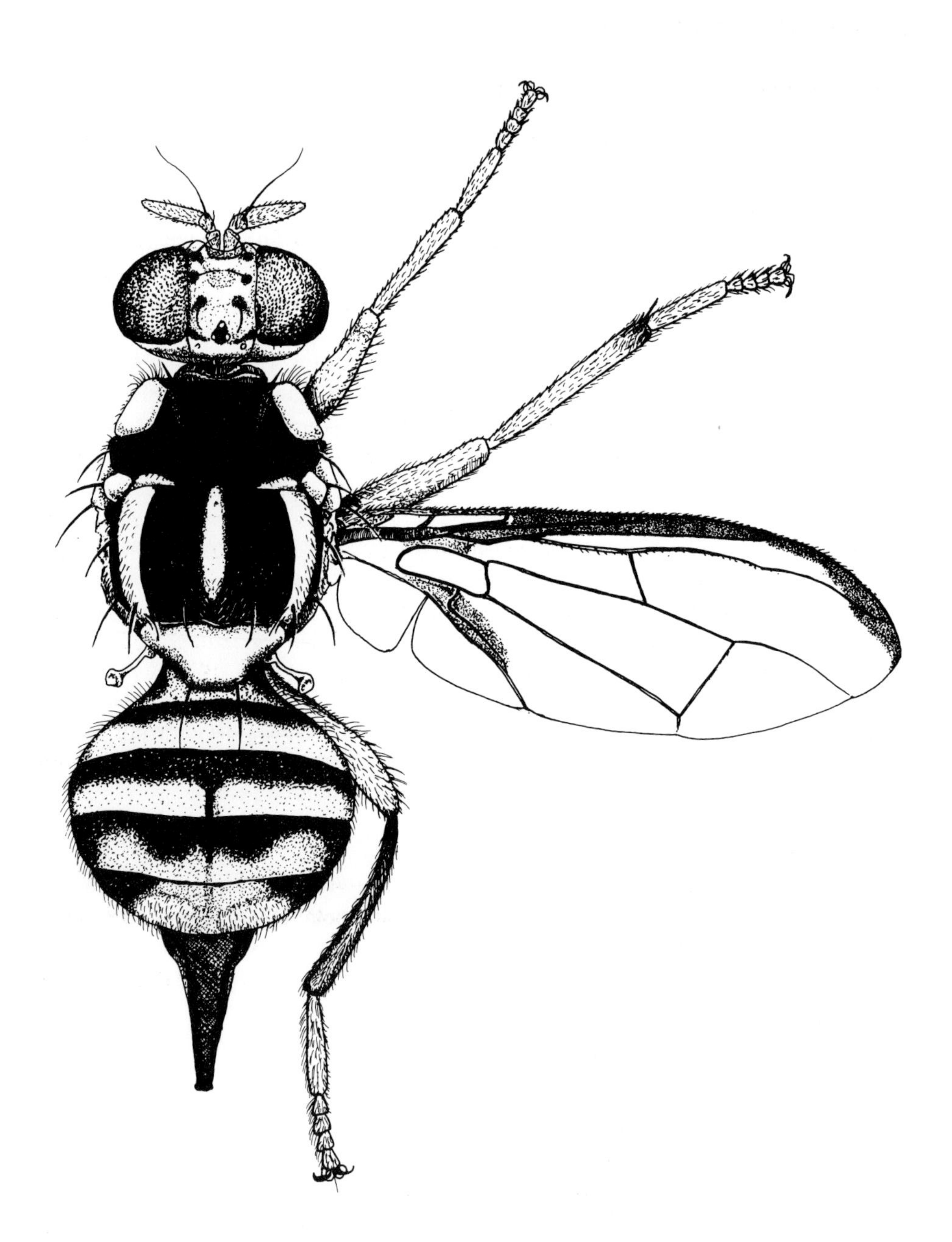

Fig. 198. *Bactrocera (Hemigymnodacus) diversa*, adult female.

Subgenus *NOTODACUS* PERKINS

Bactrocera (Notodacus) xanthodes (Broun)

(figs 74, 155, 199; pl. 26)

Taxonomic notes
Subgenus *B. (Notodacus)* is a member of the *B. (Bactrocera)* group of subgenera. *B. xanthodes* is the only species, and it has also been known as *Chaetodacus xanthodes* (Broun), *Dacus xanthodes* (Broun), *Notodacus xanthodes* (Broun) and *Tephritis xanthodes* Broun.

Commercial hosts
A potential pest which has been recorded from bell pepper (*Capsicum annuum*), breadfruit (*Artocarpus altilis*), common guava (*Psidium guajava*), giant granadilla (*Passiflora quadrangularis*), mandarin (*Citrus reticulata*), papaya (*Carica papaya*), pineapple (*Ananas comosus*), sea hibiscus (*Hibiscus tiliaceus*), tomato (*Lycopersicon esculentum*) and watermelon (*Citrullus lanatus*) (Drew, 1989a; Litsinger *et al.*, 1991); the sea hibiscus record has been questioned (A.J. Allwood, pers. comm., 1991) and that may have been an aberrant host. There are also records, requiring confirmation, from mammy-apple (*Mammea americana*), mango (*Mangifera indica*) and Pacific lychee (*Pometia pinnata*) (Froggatt, 1909; SPC, unpublished data, 1988).

Wild hosts
Recorded from Apocynaceae, Euphorbiaceae, Lecythidaceae (or Barringtoniaceae) and Sapotaceae (Drew, 1989a; Litsinger *et al.*, 1991).

Adult identification
A predominantly orange species; scutum with lateral yellow stripes which are continuous from the side of the scutellum to the postpronotal lobes; scutum sometimes also with a medial yellow stripe; with facial spots, a seta on each postpronotal lobe, anterior supra-alar setae, prescutellar acrostichal setae, a bilobed scutellum with 2 setae, and a typical dacine wing pattern (fig. 199); male with a pecten.

Description of third instar larva
Larvae medium-sized, length 8.0-10.0mm; width 1.5-2.0mm. *Head*: Stomal sensory organ with short, tapered sensilla surrounded by 6-7 small to large preoral lobes; oral ridges (pl. 26.a, b) with 18-24 rows of moderately long, evenly spaced, tapering teeth; accessory plates, with deeply serrated, evenly spaced teeth, interlocking with oral ridges; mouthhooks moderately sclerotised, with slender curved apical teeth but lacking preapical teeth (fig. 155). *Thoracic and abdominal segments*: Anterior portion of each segment with an encircling band of discontinuous rows of small, sharply pointed spinules. T1 (pl. 26.a) with a wide band of 8-12 rows of very long spinules arranged

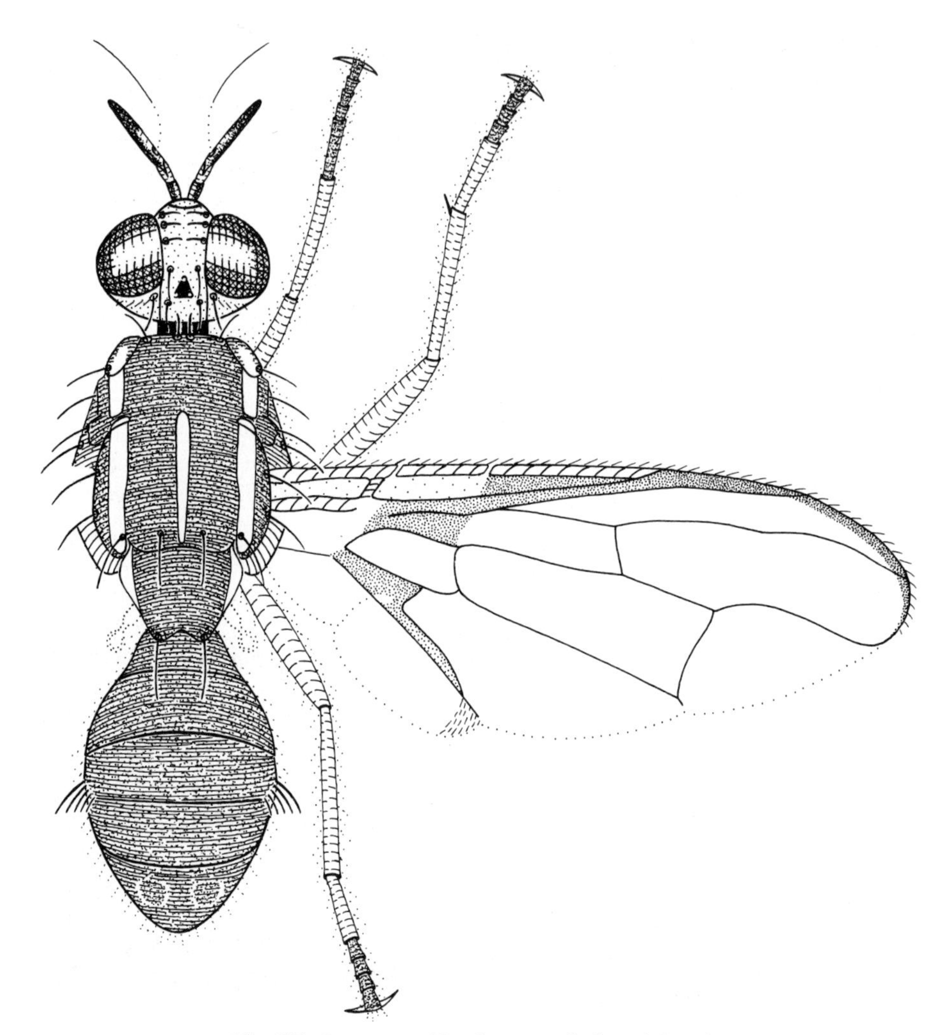

Fig. 199. *Bactrocera (Notodacus) xanthodes*, adult male.

in small groups forming plate-like structures, which are very obvious dorsally and laterally, and less so ventrally; T2 (pl. 26.c) spinules smaller but still sharply pointed; 5-7 rows dorsally, 4-8 rows laterally and 7-10 rows ventrally. T3 spinules similar to those on T2; 1-4 rows dorsally, 2-4 rows laterally and 3-4 rows ventrally. Creeping welts with stout, sharply pointed spinules; 1-3 rows anteriorly directed, the remainder posteriorly directed. A7 (pl. 26.e) with 1 posterior row of slightly larger spinules on broad, stout bases similar to those around anal lobes. A8 with large, well defined intermediate areas, almost linked by a long, slightly curved pigmented transverse line (mature larvae only). *Anterior spiracles* (pl. 26.c): 11-16 tubules. *Posterior spiracles* (pl. 26.d): Spiracular slits set in usual radiating pattern. Spiracular hairs relatively short, broad and branched in apical third to a half; dorsal and ventral bundles of 10-16 hairs, lateral bundles of 4-7 hairs. Spiracular slits very thick walled and about 2.5 times

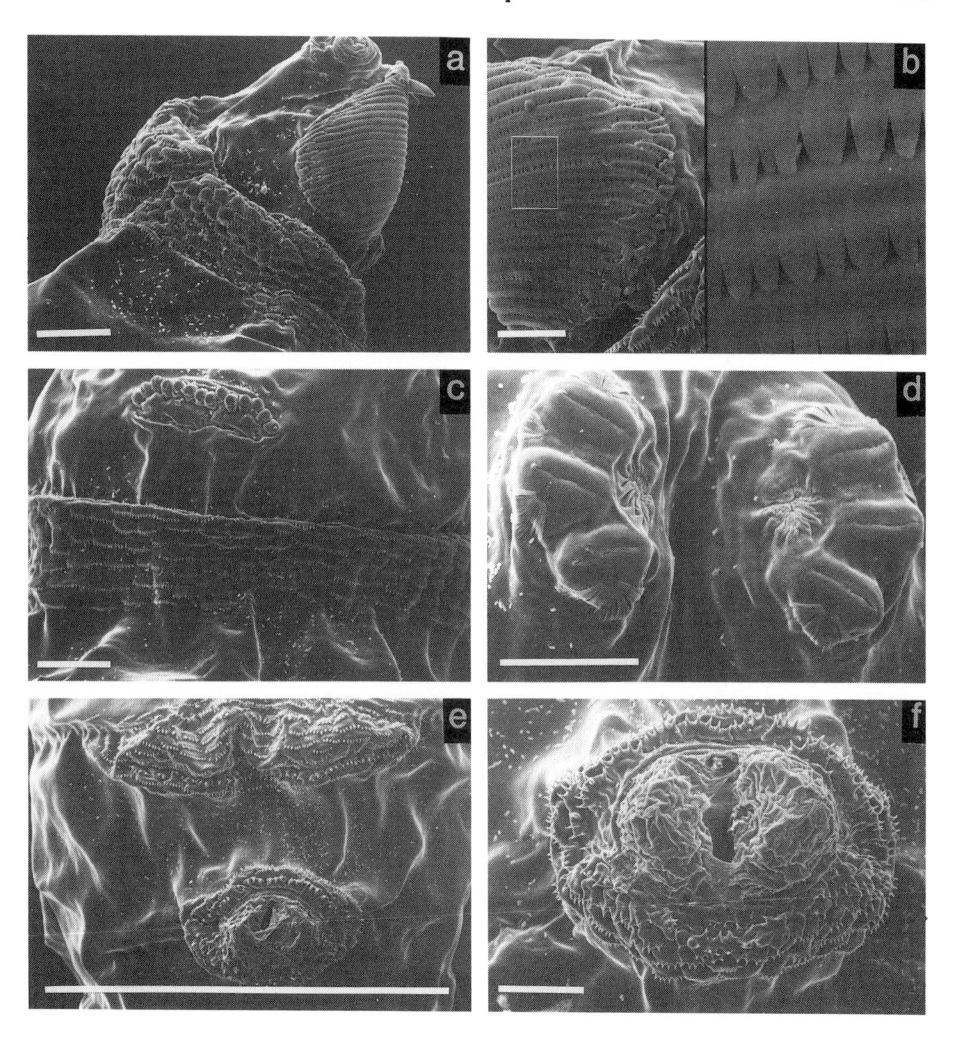

Plate 26. *Bactrocera (Notodacus) xanthodes*, SEMs of 3rd instar larva; a, head and T1, lateral view (scale=0.1mm); b, oral ridges, lateral view (scale=50μm); c, T1 and T2, lateral view (scale=0.1mm); d, posterior spiracles (scale=0.1mm); e, A8, creeping welt and anal lobes (scale=1.0mm); f, anal lobes (scale=0.1mm).

as long as broad. *Anal area* (pl. 26.e, f): Lobes only moderately protuberant and surrounded by 3-5 rows of large, stout spines, with an obvious concentration below the anal opening. *Source*: Based on specimens from the Cook Islands (ex *Carica papaya*).

Distribution

South Pacific: Cook Islands, Fiji, Tonga, Vanuatu, Western Samoa.
Distribution mapped by Drew (1982a).

Attractant
Males attracted to methyl eugenol.

Other references
PARASITOIDS, Narayanan & Chawla (1962); POST-HARVEST DISINFESTATION, Stechmann *et al.* (1988).

Subgenus *PARADACUS* PERKINS

A subgenus of 12 species found in tropical Asia, Indonesia, Japan and Papua New Guinea, which belongs to the *B. (Zeugodacus)* group of subgenera.

Bactrocera (Paradacus) decipiens (Drew)

(figs 86, 200)

Taxonomic notes
This species has also been known as *Dacus decipiens* Drew.

Pest status and commercial hosts
A serious pest of pumpkin (*Cucurbita pepo*), which has the potential to become a pest of other cucurbits (Drew, 1982a).

Adult identification
Scutum orange with black stripes, and lateral and medial yellow stripes; abdomen orange; with facial spots, anterior supra-alar setae, 4 scutellar setae, and a characteristic wing pattern (fig. 200); male with a pecten. Without prescutellar acrostichal setae.

Distribution
New Guinea area: New Britain.
Distribution mapped by Drew (1982a).

Attractant
Males not known to be attracted to any synthetic lures.

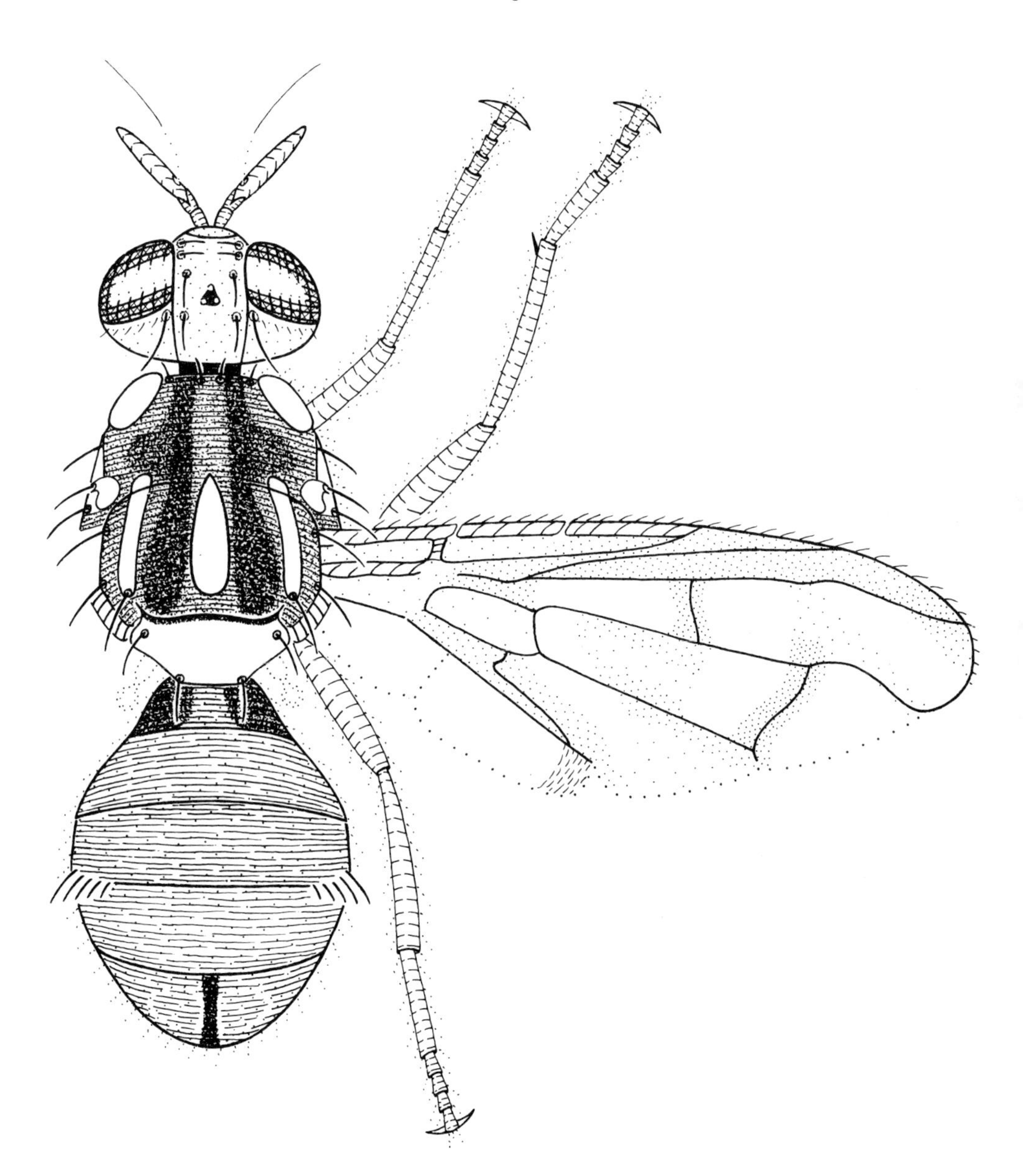

Fig. 200. *Bactrocera (Paradacus) decipiens*, adult male.

Bactrocera (Paradacus) depressa (Shiraki)

(figs 87, 201)

Taxonomic notes
This species has also been known as *Dacus depressus* (Shiraki), *Paradacus depressus* (Shiraki) and *Zeugodacus depressus* Shiraki.

Pest status and commercial hosts
A pest of pumpkin (*Cucurbita moschata*) in Japan (Shiraki, 1968). There are also records, which require confirmation, from African horned cucumber (*Cucumis metuliferus*), cucumber (*C. sativus*), watermelon (*Citrullus lanatus*) and white-flowered gourd (*Lagenaria siceraria*); there is also a doubtful record from tomato (*Lycopersicon esculentum*) (Kandybina, 1977).

Wild hosts
In Amani-Oshima Island (Ryukyu Islands) *B. depressa* has been found in mountain areas where there are no cucurbits, suggesting that it has other hosts (Shiraki, 1968).

Adult identification
Scutum predominantly black with lateral and medial orange stripes; with facial spots, anterior supra-alar setae, 4 scutellar setae, and wing with costal band expanded into an apical spot (fig. 201); male with a pecten. Without prescutellar acrostichal setae (aberrant individuals with one or both prescutellar acrostical setae are common).

Distribution
Temperate Asia: Japan (all major islands).
Tropical Asia: Japan (Ryukyu Islands), Taiwan.

Attractant
Males not known to be attracted to any synthetic lures.

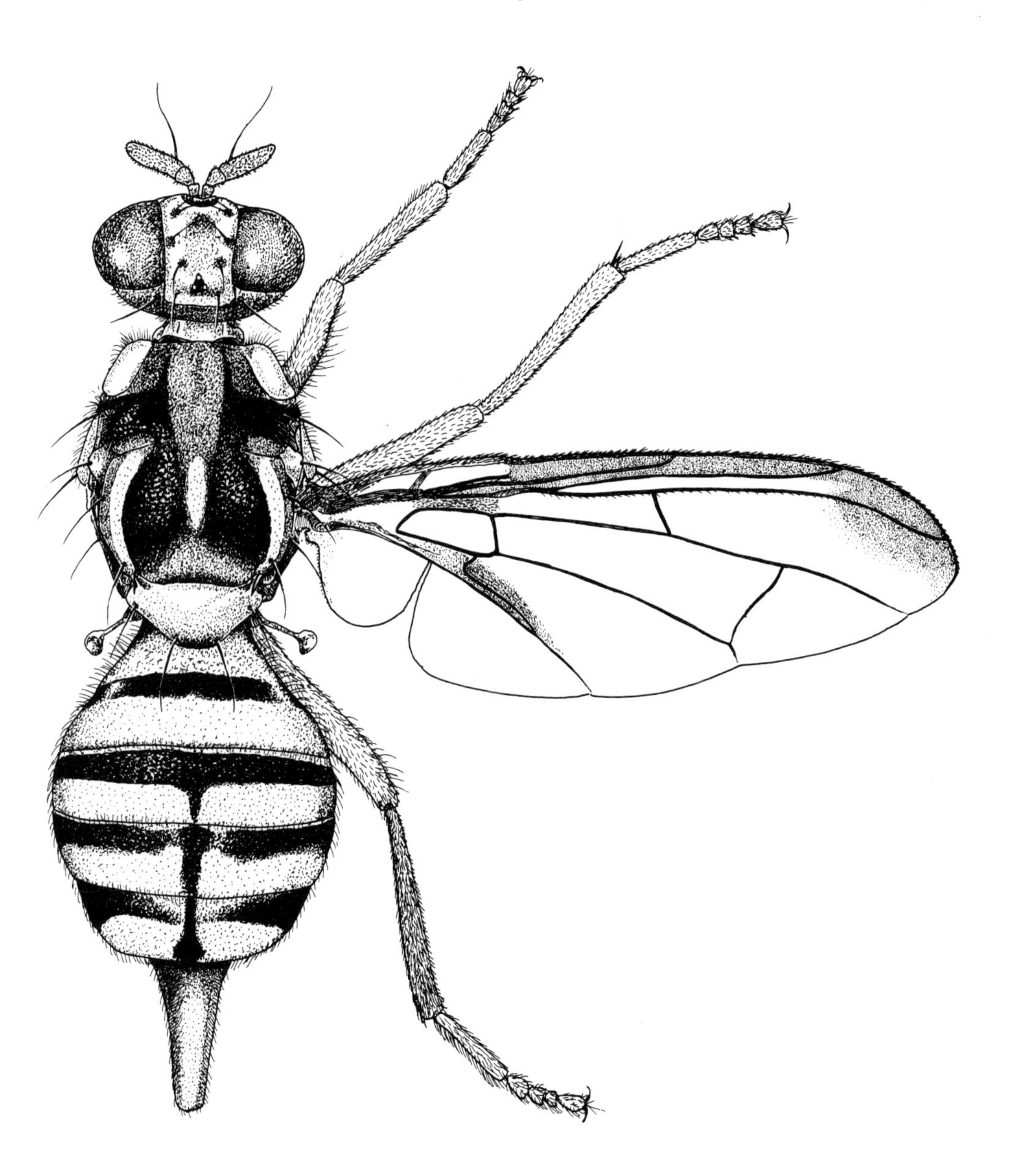

Fig. 201. *Bactrocera (Paradacus) depressa*, adult female.

Subgenus *PARATRIDACUS* SHIRAKI

A subgenus of eight species found in tropical Asia, Australia, Indonesia, Japan and Papua New Guinea, which belongs to the *B. (Zeugodacus)* group of subgenera.

Bactrocera (Paratridacus) atrisetosa (Perkins)

(fig. 202)

Taxonomic notes
This species has also been known as *Dacus atrisetosus* (Perkins), *D. papuaensis* Malloch, *Melanodacus atrisetosus* (Perkins), *M. rubidus* May and *Zeugodacus atrisetosus* Perkins.

Commercial hosts
A potential pest which has been recorded from cucumber (*Cucumis sativus*), pumpkin (*Cucurbita pepo*), tomato (*Lycopersicon esculentum*) and watermelon (*Citrullus lanatus*) (Drew, 1989a).

Wild hosts
Recorded from *Aglaia sapindina* Harms (Meliaceae) (Drew, 1989a) but it probably has a wider range of hosts.

Adult identification
A predominantly orange to orange-brown species; scutum with lateral and medial yellow stripes; with facial spots, anterior supra-alar setae, prescutellar acrostichal setae, 4 scutellar setae; a typical dacine wing pattern, but costal band broad (fig. 202); male without a pecten.

Distribution
New Guinea area: Papua New Guinea (between 1200m and 1650m altitude).

Attractant
Males not attracted to cue lure or methyl eugenol.

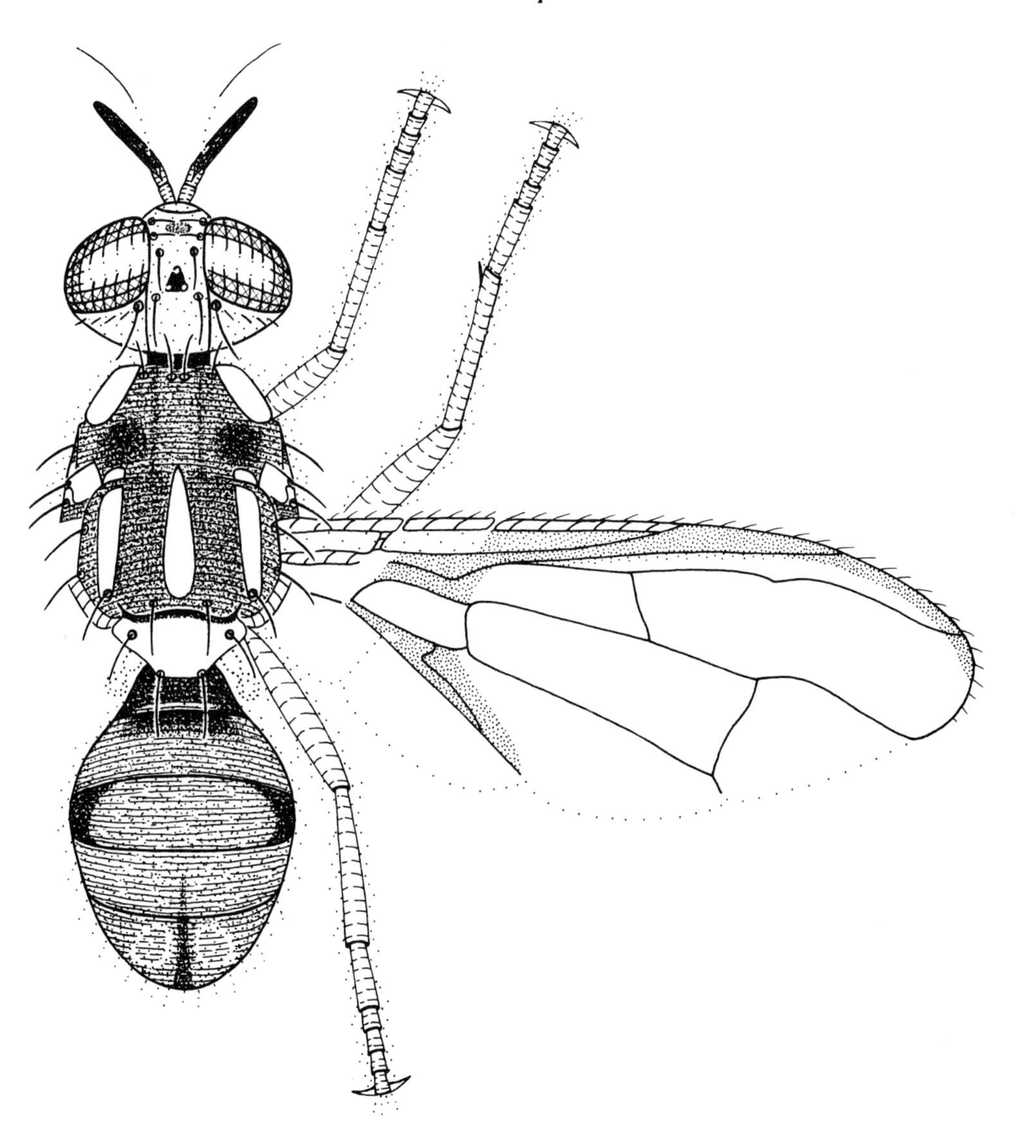

Fig. 202. *Bactrocera (Paratridacus) atrisetosa*, adult male.

Subgenus *TETRADACUS* MIYAKE

This subgenus includes the two Oriental citrus fruit flies plus three other Indonesian and Papuan species. Most authors have placed *B. minax* in subgenus *Polistomimetes* and many have regarded *Daculus*, which includes the olive fly (*B. oleae*), as being a synonym of *Polistomimetes* (e.g. Drew, 1989a) because no clear separation of these subgenera had been recognised. We have placed the olive fly and the citrus flies in separate subgenera, and adopted the view expressed by Chinese workers (e.g. Chen & Zia, 1955) that both citrus flies belong to *Tetradacus*.

Bactrocera (Tetradacus) minax (Enderlein)

Chinese Citrus Fly

(figs 79, 203; pl. 27)

Taxonomic notes

This species has also been known as *Callantra minax* (Enderlein), *Dacus citri* (Chen), *D. minax* (Enderlein), *Mellesis citri* Chen, *Polistomimetes minax* Enderlein and *Tetradacus citri* (Chen). The new synonymy of *M. citri* with *B. minax* will be formally published elsewhere (White & X.-j. Wang, in press).

Pest status and commercial hosts

Recorded as causing a high incidence of damage to tangerine (*Citrus reticulata*) in northern India (Nath, 1973). In China (as *B. citri*) it has been recorded from citron (*C. medica*), lemon (*C. limon*), meiwa kumquat (*Fortunella x crassifolia*), pummelo (*C. maxima*), sour orange (*C. aurantium*), sweet orange (*C. sinensis*) and *C. tangerina* (Kandybina, 1977; Yang, 1988; X.-j. Wang, unpublished data, 1988).

Adult identification

A predominantly orange to brown species; scutum with lateral and medial yellow stripes; with facial spots, 2 scutellar setae, and wing with a broad costal band that is darkened apically (fig. 203); male with a pecten. Without anterior supra-alar and prescutellar acrostichal setae.

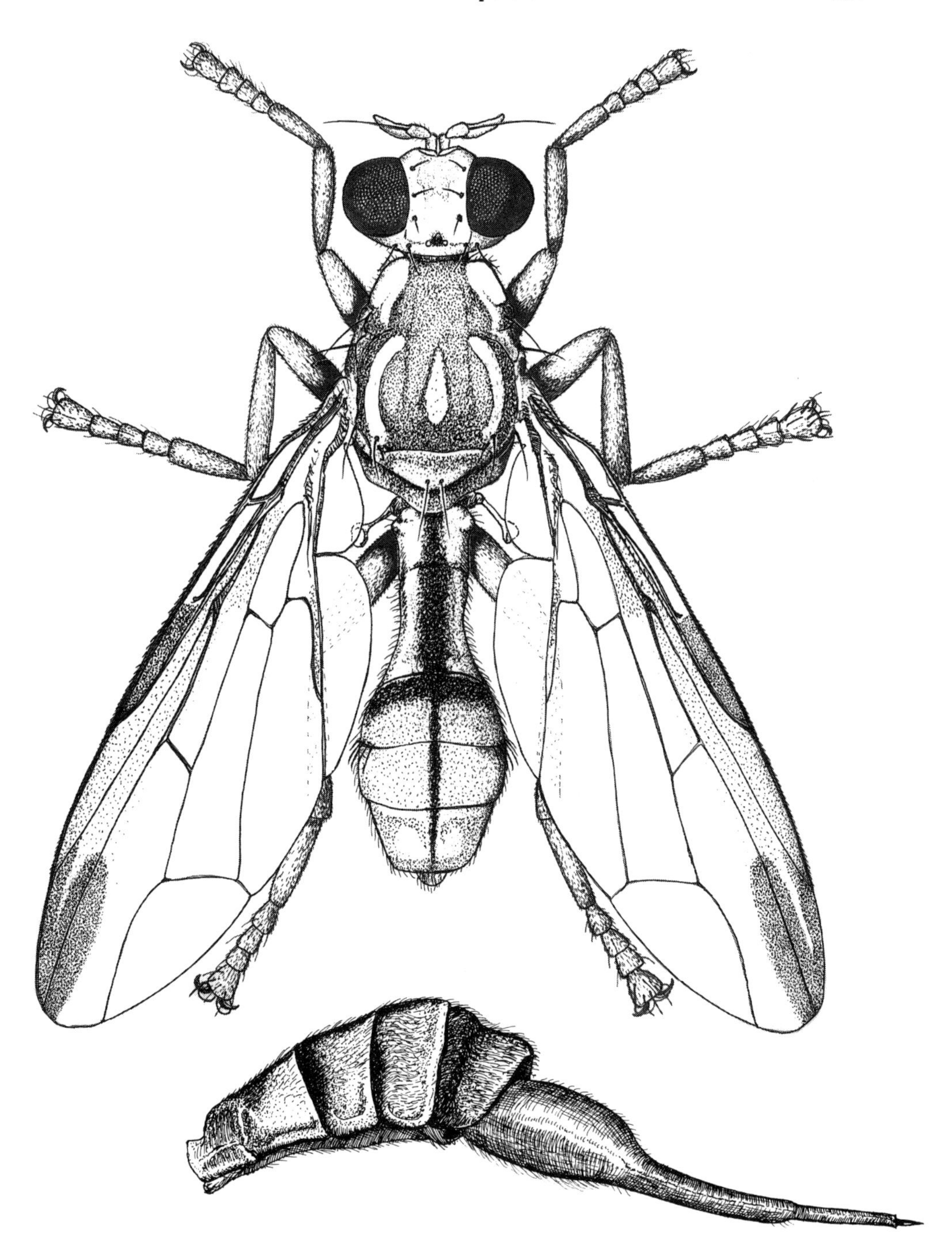

Fig. 203. *Bactrocera (Tetradacus) minax*, adult male with inset of female abdomen in lateral view.

Description of third instar larva

Larvae large, length 13.0-15.0mm; width 2.5-3.0mm. *Head*: Stomal sensory organ (pl. 27.b) with small sensilla surrounded by elongate preoral lobes, appearing similar to small oral ridges; oral ridges (pl. 27.a) in 16-18 long, stout, unserrated rows; accessory plates numerous, small, shell-shaped, interlocking with oral ridges along outer margin; mouthhooks heavily sclerotised, each with a strongly curved apical tooth. *Thoracic and abdominal segments*: Anterior margin of each thoracic segment with a broad encircling band of spinules. T1 with 6-9 rows of stout, sharply pointed spinules dorsally, 3-5 rows laterally and 5-8 rows ventrally; T2 with 4-6 rows of smaller spinules dorsally, and 3-5 rows laterally and ventrally; T3 with 7-8 rows dorsally and 1-3 rows laterally and ventrally. Dorsal spinules absent from A1-A8. Creeping welts (pl. 27.e) large, with 13-29 transverse rows of small, stout spinules. A8 with very large intermediate areas, and well defined dorsal and lateral tubercles and sensilla. *Anterior spiracles* (pl. 27.c): 17-19 tubules. *Posterior spiracles* (pl. 27.d): Each spiracular slit about 2.5-3.0 times as long as broad, with a heavily sclerotised rima. Spiracular hairs long, branched in apical half to a third, with dorsal and ventral bundles of 9-16 hairs and lateral bundles of 4-9 hairs. *Anal area* (pl. 27.f): Lobes large, protuberant, with 3-6 discontinuous rows of small stout spinules surrounding lobes. Below anal opening, a very well defined patch of slightly larger spinules. *Source*: Based on 3 specimens from Bhutan (ex *Citrus* sp.).

Distribution

Oriental Asia: Bhutan, China (south), India (West Bengal and Sikkim).
A distribution map was produced by IIE (1991c).

Attractant

Males not known to be attracted to any synthetic lures.

Other references

BIOLOGY AND CONTROL, Sun *et al.* (1958); LARVA, Kandybina (1977; as *D. citri*).

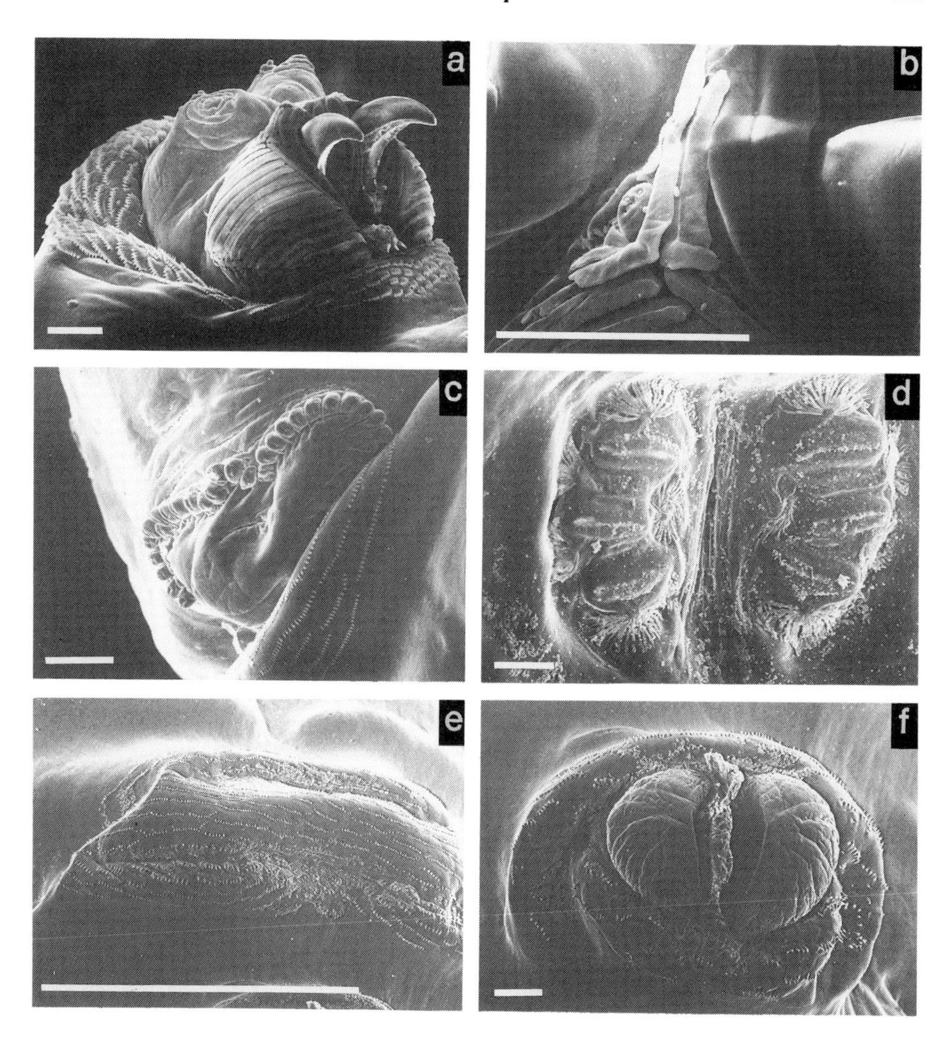

Plate 27. *Bactrocera (Tetradacus) minax*, SEMs of 3rd instar larva; a, head and T1, ventrolateral view (scale=0.1mm); b, stomal sensory organ and preoral lobes, lateral view (scale=0.1mm); c, anterior spiracle (scale=0.1mm); d, posterior spiracles (scale=0.1mm); e, A8, creeping welt (scale=1.0mm); f, anal lobes (scale=0.1mm).

Bactrocera (Tetradacus) tsuneonis (Miyake)

Japanese Orange Fly

(figs 80, 204)

Taxonomic notes

This species has also been known as *Dacus cheni* Chao, *D. tsuneonis* Miyake and *Tetradacus tsuneonis* (Miyake). Some previous workers have placed *B. citri* as a synonym of this species (e.g. R.H. Foote, 1984), but it is actually a synonym of *B. minax*. The new synonymy of *D. cheni* with *B. tsuneonis* will be formally published elsewhere (White & X.-j. Wang, in press).

Pest status and commercial hosts

A serious pest of citrus (*Citrus* spp.) which has been recorded from meiwa kumquat (*Fortunella x crassifolia*), oval kumquat (*Fortunella margarita*), round kumquat (*F. japonica*), sour orange (*C. aurantium*), sweet orange (*C. sinensis*), tangerine (*C. reticulata*) and *C. tangerina* (Chao, 1987; Ito, 1983-5; Yang, 1988; X.-j. Wang, unpublished data, 1988).

Adult identification

A predominantly orange to orange-brown species; scutum with lateral and medial yellow stripes; with facial spots, anterior supra-alar setae, 2 scutellar setae, and wing with a pale yellow costal band that is darkened apically (fig. 204); male with a pecten. Without prescutellar acrostichal setae.

Distribution

Oriental Asia: China (Guangxi, Hunan, Jiangsu, Sichuan), Japan (Ryukyu Islands); probably Taiwan (recorded by Ito, 1983-5 but not listed by Cheng & Lee, 1991).
Temperate Asia: Japan (Kyushu Island).
A distribution map was produced by IIE (1991b).

Attractant

Males not known to be attracted to any synthetic lures.

Other references

BEHAVIOUR, P.H. Smith (1989); DATASHEET, Weems (1967); GENERAL ACCOUNT, Miyake (1919) [a substantial work which is still of value]; HOST RESISTANCE, Greany (1989).

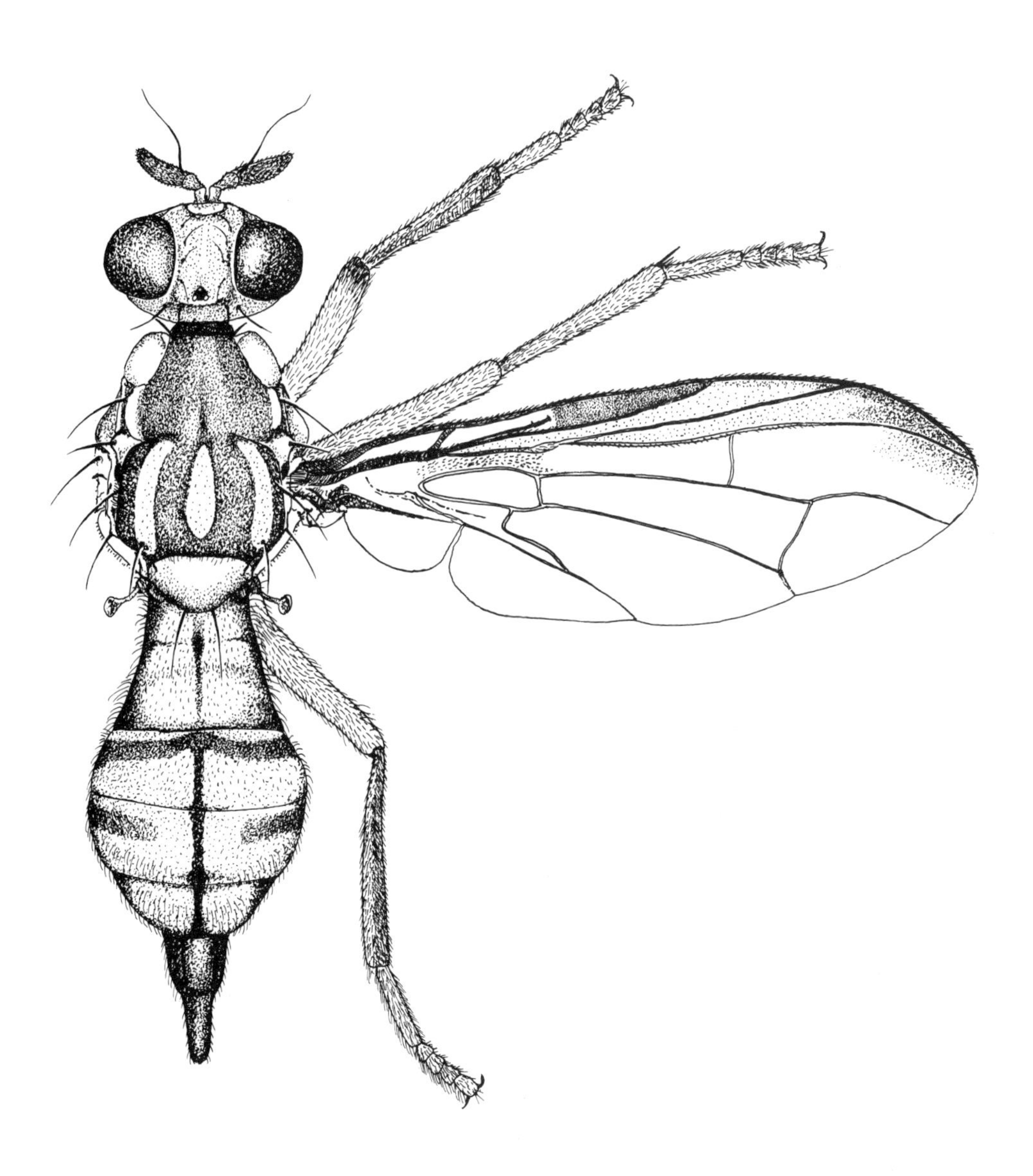

Fig. 204. *Bactrocera (Tetradacus) tsuneonis*, adult female.

Subgenus *ZEUGODACUS* HENDEL

A subgenus of about 70 species found in tropical Asia, Australia, Indonesia and Papua New Guinea. One species, *B. cucurbitae*, has been spread to Africa, Guam, Hawaii, the Indian Ocean Islands and Iran. In contrast to most other members of the genus *Bactrocera*, most species of the *B. (Zeugodacus)* group of subgenera have a strong preference for attacking Cucurbitaceae, often attacking the flowers rather than the fruit. This pattern of host relationships is more typical of *Dacus* than of *Bactrocera* spp.

Bactrocera (Zeugodacus) caudata (Fabricius)

(figs 64, 84, 205)

Taxonomic notes

This species has also been known as *Bactrocera maculipennis* Doleschall, *Chaetodacus caudatus* (Fabricius), *Dacus caudatus* Fabricius and *Zeugodacus caudatus* (Fabricius).

Commercial hosts

A potential pest of Cucurbitaceae which Hardy (1973) recorded as being reared from the male flowers of pumpkin (*Cucurbita* sp.). There are also unconfirmed records from several other species of Cucurbitaceae, namely angled luffa (*Luffa acutangula*), bitter gourd (*Momordica charantia*), cucumber (*Cucumis sativus*), luffa (*Luffa aegyptiaca*), pumpkin (*Cucurbita pepo*), snakegourd (*Trichosanthes cucumerina*), wax gourd (*Benincasa hispida*) and white-flowered gourd (*Lagenaria siceraria*) (Perkins, 1938; Yunus & Ho, 1980). Recorded non-cucurbit hosts are chilli pepper (*Capsicum annuum*), common guava (*Psidium guajava*), peach (*Prunus persica*), pummelo (*Citrus maxima*), sapodilla (*Manilkara zapota*), sour orange (*C. aurantium*), tomato (*Lycopersicon esculentum*) and water apple (*Syzygium samarangense*) (Fletcher, 1920; Jepson, 1935; Kapoor & Agarwal, 1983; Perkins, 1938; Tan & Lee, 1982; Yunus & Ho, 1980); those records should be regarded as doubtful until the normal pattern of host relations of this species have been confirmed. Kapoor & Agarwal (1983) also listed grain sorghum (*Sorghum bicolor*) and that record must have been derived from Bezzi (1916) who noted that it was found "on" that plant. That record is obviously implausible, but there is no way of knowing how many plausible records were also based on casual observations of resting adults.

Adult identification

Scutum predominantly black; with lateral and medial yellow stripes; face with a black line across mouth opening (fig. 64), or with the black spots in the antennal furrows extended laterally and almost forming a line across the mouth opening; with anterior supra-alar setae, prescutellar acrostichal setae, 4 scutellar setae, and wing with costal band expanded into an apical spot (fig. 205); male with a pecten.

Distribution
Oriental Asia: Brunei, India, Indonesia (Java, Sumatra), Malaysia, Myanmar, Taiwan, Thailand, Vietnam. Kapoor (1989) suggested that Indian records may have been based on misidentifications of *B. cucurbitae*; however, we have examined Indian specimens (NHM, collection data).

Attractant
Males attracted to cue lure.

Other references
ECOLOGY, Tan & Lee (1982); MALE TERMINALIA, Singh & Premlata (1985).

Bactrocera (Zeugodacus) cucurbitae **(Coquillett)**

Melon Fly
Melon Fruit Fly

(figs 81, 156, 206; pl. 28)

Taxonomic notes
This species has also been known as *Chaetodacus cucurbitae* (Coquillett), *Dacus cucurbitae* Coquillett, *Strumeta cucurbitae* (Coquillett) and *Zeugodacus cucurbitae* (Coquillett).

Pest status and commercial hosts
This is a very serious pest of cucurbit crops. According to Weems (1964b) it has been recorded from over 125 plants, including members of families other than Cucurbitaceae; however, many of those records may have been based on casual observation of adults resting on plants or caught in traps set in non-host trees. In common with some other species of subgenus *B. (Zeugodacus)* it can attack flowers as well as fruit, and additionally, will sometimes attack stem and root tissue. In Hawaii pumpkin and squash fields (varieties of *Cucurbita pepo*) have been known to be heavily attacked before fruit had even set, with eggs being laid into unopened male and female flowers, and larvae even developing successfully in the taproots, stems and leaf stalks (Back & Pemberton, 1914).

Syed (1971) recorded *B. cucurbitae* from the following species of Cucurbitaceae (he also included habitat and attack rate details): angled luffa (*Luffa acutangula*), balsam-apple (*Momordica balsamina*), bitter gourd (*M. charantia*), colocynth (*Citrullus colocynthis*), cucumber (*Cucumis sativus* - fruit and stem), luffa (*Luffa aegyptiaca*), melon (*C. melo*), pumpkin (*Cucurbita maxima* and *C. pepo* - fruit and stem), watermelon (*Citrullus lanatus*), wax gourd (*Benincasa hispida*) and white-flowered gourd (*Lagenaria siceraria*). Drew (1989a) recorded it from cucumber, marrow (a variety of *Cucurbita pepo*) and melon in the South Pacific region. Tan & Lee (1982) listed the following hosts for Malaysia: bitter gourd, cucumber and watermelon, and

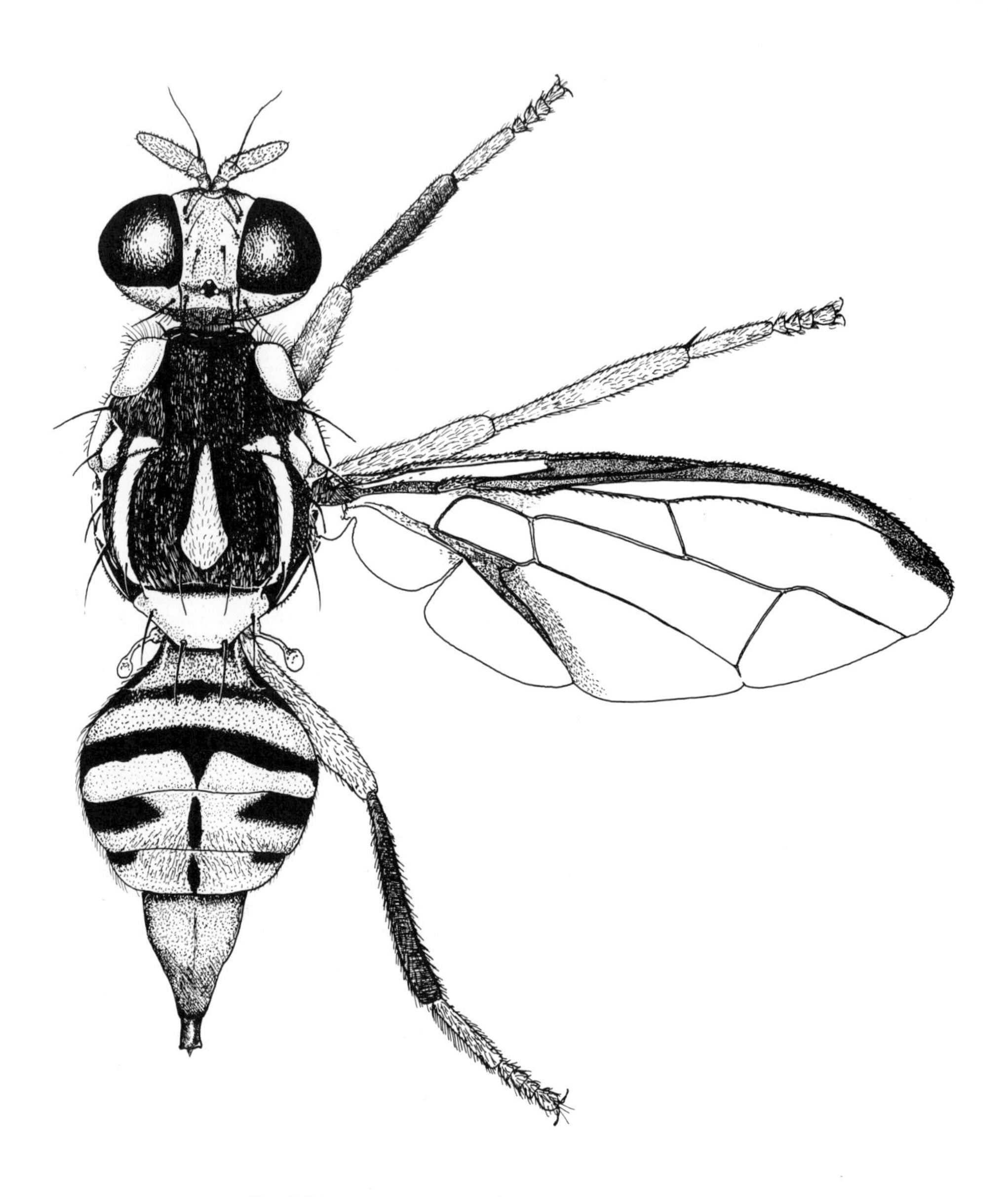

Fig. 205. *Bactrocera (Zeugodacus) caudata*, adult female.

in Thailand it has also been reared from teruah (*Momordica cochinchinensis*) (Clausen *et al.*, 1965). In Taiwan *B. cucurbitae* has a preference for melon and colocynth (Lee, 1972).

There are adventive populations in Mauritius, Réunion, eastern Africa, Iran and Hawaii. In Mauritius it has been recorded from cucumber, melon, pumpkin (*C. maxima*), snakegourd (*Trichosanthes cucumerina*), squash (a variety of *C. pepo*), watermelon and white-flowered gourd (Orian & Moutia, 1960). In Réunion it is recorded from angled luffa, bitter gourd, cucumber, melon and squash (*C. pepo*) (Étienne, 1972). Harris *et al.* (1986b) listed cucumber, squash (*Cucurbita* sp.) and watermelon as hosts in Kauai, Hawaiian Islands; they also reported an old record from chayote (*Sechium edule*).

The following additional hosts require confirmation: African horned cucumber (*Cucumis metuliferus*), pointed gourd (*Trichosanthes dioica*) and pumpkin (*Cucurbita moschata*) (Kandybina, 1977; Kapoor & Agarwal, 1983; Syed, 1971). *B. cucurbitae* has been known to develop in stem galls induced by another insect, namely those of *Lasioptera toombii* (Grover) (Diptera, Cecidomyiidae) on ivy gourd (*Coccinia grandis*) (Bhatia & Mahto, 1968). It may attack cucurbit flowers, at least when caged, and adults have been known to congregate in swarms on the underside of guava leaves during the winter (Batra, 1964).

B. cucurbitae has been reared from a few non-cucurbit hosts, for example, Syed (1971) reared a few *B. cucurbitae* from quince (*Cydonia oblonga*) and tomato (*Lycopersicon esculentum*). From Malaysia, Tan & Lee (1982) listed pummelo (*Citrus maxima*) and yard-long bean (*Vigna unguiculata*). R.A.I. Drew (unpublished data, 1990) gave another bean record, namely garden bean (*Phaseolus vulgaris*) and in Hawaii it has been reared from the buds of *Sesbania grandiflora* (Nakagawa & Yamada, 1965). In South-east Asia and Borneo it has been reared from jackfruit (*Artocarpus heterophyllus*) and water apple (*Syzygium samarangense*) (Clausen *et al.*, 1965). Other Hawaiian records are avocado (*Persea americana*), common fig (*Ficus carica*), granadilla (*Passiflora* sp.), Hinds' walnut (*Juglans hindsii*), mango (*Mangifera indica*), papaya (*Carica papaya*), peach (*Prunus persica*), sweet orange (*Citrus sinensis*) and tree tomato (*Cyphomandra crassicaulis*), and in the Marianas it was recorded from limeberry (*Triphasia trifolia*) (Harris *et al.*, 1986b; Maehler, 1951; Nakagawa *et al.*, 1968). The following additional non-cucurbit host records require confirmation: apple (*Malus domestica*), apricot (*Prunus armeniaca*), Argus pheasant tree (*Dracontomelon dao*), bell pepper (*Capsicum annuum*), carambola (*Averrhoa carambola*), Chilean strawberry (*Fragaria chiloensis*), common guava (*Psidium guajava*), custard apple (*Annona reticulata*), date palm (*Phoenix dactylifera*), eggplant (*Solanum melongena*), giant granadilla (*Passiflora quadrangularis*), grapefruit (*Citrus x paradisi*), hyacinth bean (*Lablab purpureus*), Indian laurel (*Calophyllum inophyllum*), lemon (*Citrus limon*), Lima bean (*Phaseolus lunatus*), loquat (*Eriobotrya japonica*), mung bean (*Vigna radiata*), okra (*Abelmoschus esculentus*), pear (*Pyrus communis*), pigeon pea (*Cajanus cajan*), purple granadilla (*Passiflora edulis*), satinleaf (*Chrysophyllum oliviforme*), sour orange (*Citrus aurantium*), soursop (*Annona muricata*), star-apple (*Chrysophyllum cainito*), strawberry guava (*Psidium littorale*), strychnine tree (*Strychnos nux-vomica*), sugar-apple (*A. squamosa*), tabasco pepper

(*Capsicum frutescens*), wampi (*Clausena lansium*), watery rose-apple (*Syzygium aqueum*), white sapote (*Casimiroa edulis*), yellow granadilla (*Passiflora laurifolia*) and wild waterlemon (*P. foetida*) (McBride & Tanada, 1949; Kandybina, 1977; Kapoor & Agarwal, 1983; Syed, 1971; Yunus & Ho, 1980; USDA, *Pests not known to occur in the United States or of limited distribution*, No.33, undated). Lee (1972) noted that in Taiwan *B. cucurbitae* adults fed on honeydew on citrus (*Citrus* spp.), guava and mango trees; that habit may have caused some workers to record erroneous hosts. Syed (1971) also cited some old Hawaiian records, namely banana (*Musa x paradisiaca*), longan (*Dimocarpus longan*) and tangerine (*Citrus reticulata*), which require confirmation, plus onion (*Allium cepa*), which is a doubtful host.

There are also some unexpected hosts of *B. cucurbitae*, which it probably only attacks under unusual circumstances. In Malaysia it can attack the stems of tomato grown in hydroponics (Carey & Dowell, 1989) and it has been reported as causing severe damage to kai choy (a variety of *Brassica juncea*) in Hawaii (Hardy, 1948). Syed (1971) listed records from broccoli and kohlrabi (vars of *B. oleracea*), but the original source of those data is uncertain.

Recent work has shown that, at least up to harvesting stage, banana and avocado varieties grown in Hawaii are not attacked (Armstrong, 1983; Armstrong *et al.*, 1983). Similarly, pineapple (*Ananas comosus*) has been shown to be unsuitable as a host (Armstrong *et al.*, 1979; Armstrong & Vargas, 1982).

Wild hosts

Recorded from several wild species of Cucurbitaceae. The major host in Kauai, Hawaii, is balsam apple and Harris *et al.* (1986b) also reported an old record for *Sicyos* sp. In Okinawa, Ryukyu Islands, Japan, the wild host is *Diplocyclos palmatus* (Okinawa Prefecture, 1987). In Pakistan wild plants of balsam apple, colocynth and *Cucumis trigonus* Roxb. are important wild hosts (Syed, 1971).

Adult identification

A predominantly orange-brown species; scutum with lateral and medial yellow stripes; with facial spots, anterior supra-alar setae, prescutellar acrostichal setae, 2 (or rarely 4) scutellar setae, and a characteristic wing pattern (fig. 206); male with a pecten.

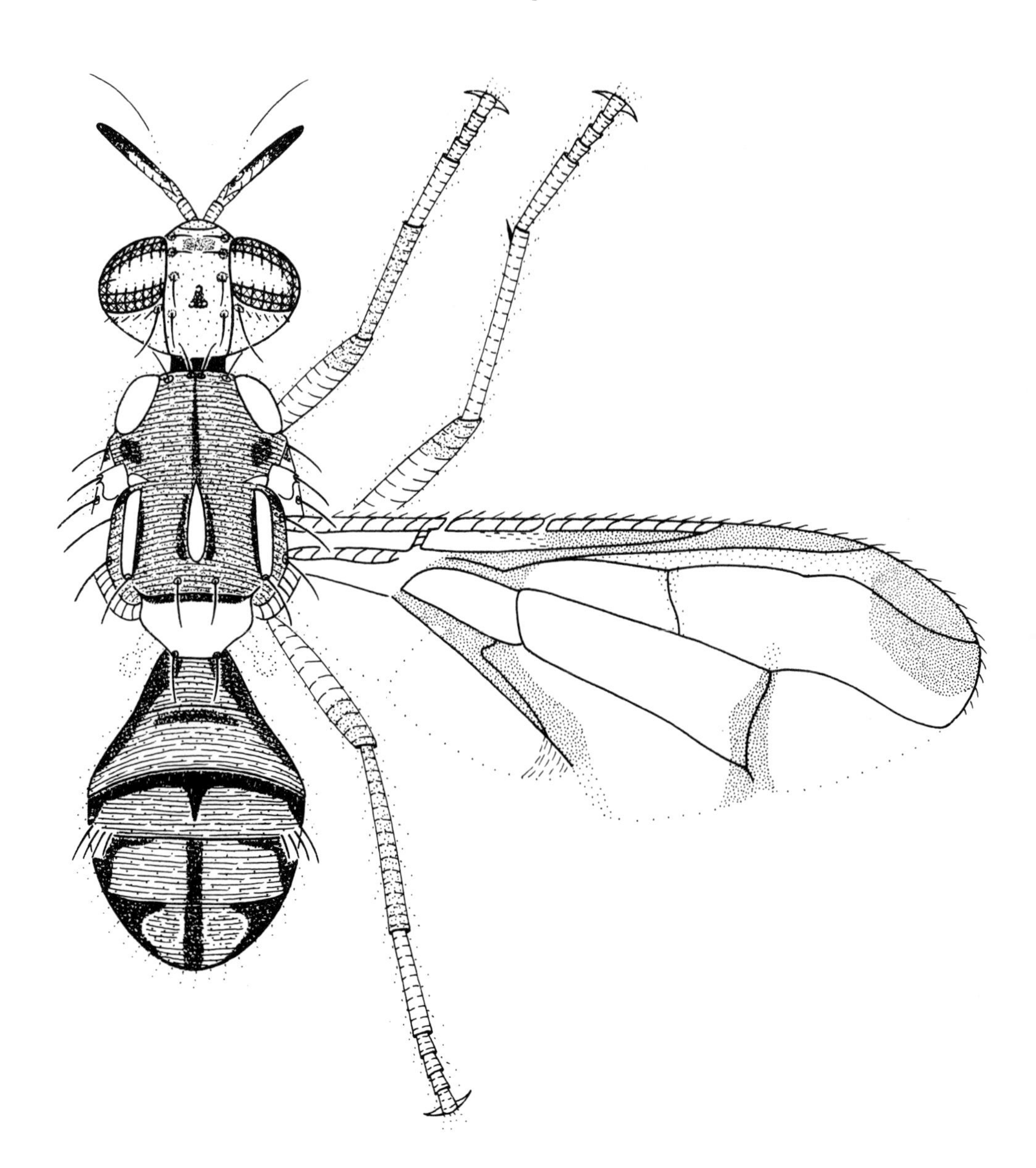

Fig. 206. *Bactrocera (Zeugodacus) cucurbitae*, adult male.

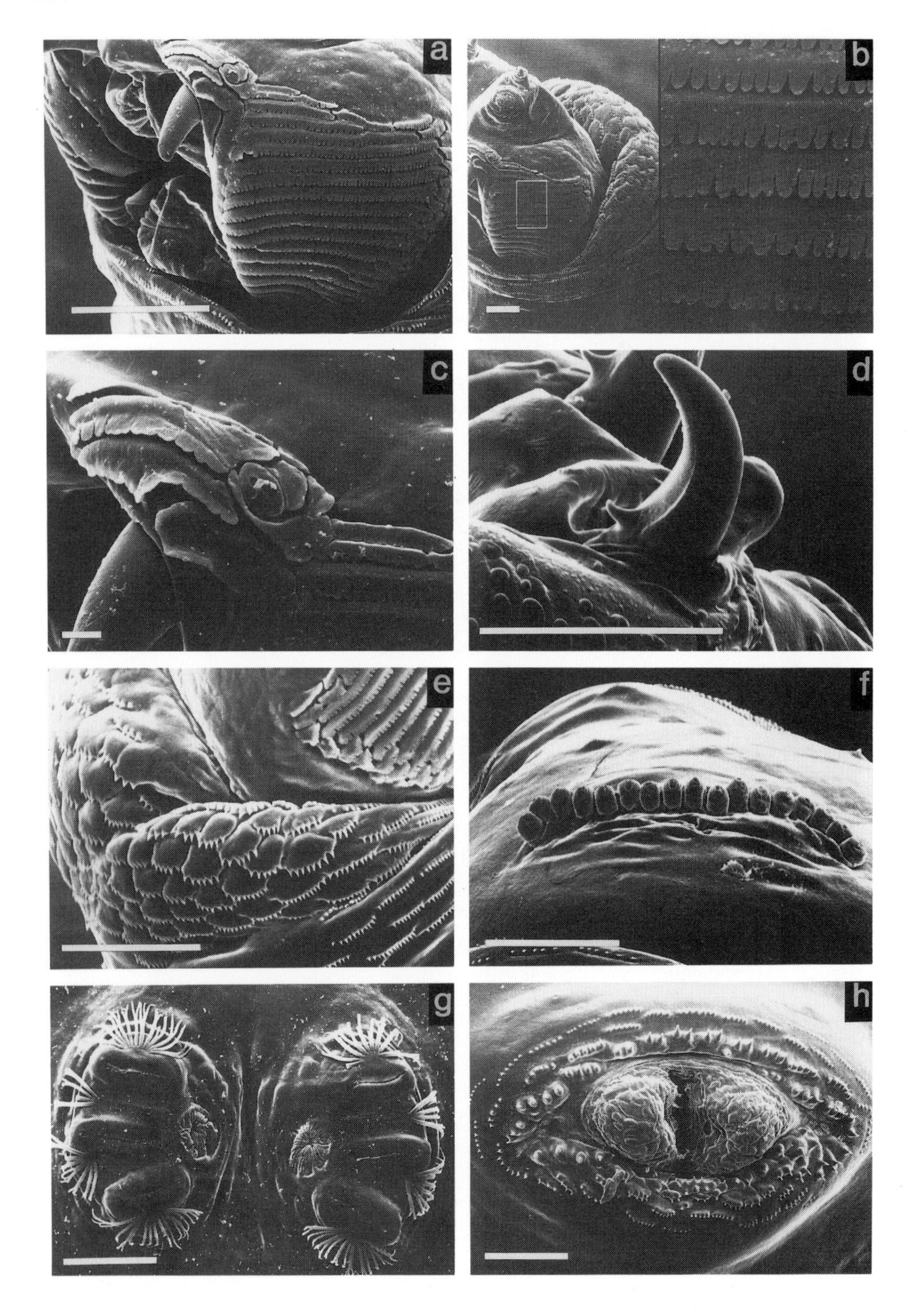
a
b
c
d
e
f
g
h

Description of third instar larva

Larvae large, length 9.0-11.0mm; width 1.0-2.0mm. *Head*: Stomal sensory organ (pl. 28.c) small, completely surrounded by 6-7 large preoral lobes, some bearing serrated edges similar to oral ridges; oral ridges (pl. 28.a, b) with 17-23 rows of moderately long, uniform, bluntly rounded teeth; accessory plates numerous, with serrated edges and interlocking with oral ridges; mouthhooks large, heavily sclerotised, each with a small but well defined preapical tooth (fig. 156; pl. 28.d). *Thoracic and abdominal segments*: Anterior portion of T1 (pl. 28.e) with an encircling, broad band of spinules which dorsally and laterally form small plates 7-10 rows deep, becoming discontinuous rows ventrally; T2 with smaller, stouter spinules, forming 5-7 discontinuous rows around anterior portion of segment; T3 similar to T2, but reduced to 4-6 rows. Creeping welts obvious, with 9-13 rows of small spinules. A8 with large well rounded intermediate areas, almost linked by a large, slightly curved, pigmented transverse line (mature larvae only). Tubercles and sensilla well defined. *Anterior spiracles* (pl. 28.f): 16-20 tubules. *Posterior spiracles* (pl. 28.g): Spiracular slits large, with heavily sclerotised rimae; about 3 times as long as broad. Spiracular hairs long, fine and often branched in apical half; dorsal and ventral bundles of 6-12 spiracular hairs; lateral bundles of 4-6 hairs. *Anal area* (pl. 28.h): Lobes large with a lightly sculptured surface, surrounded by 3-7 rows of spinules. Around outer edges spinules small, in discontinuous rows; closer to anal lobes, spinules becoming stouter, and forming small groups below anal opening. *Source*: Based on specimens from Hawaii, USA (lab. colony).

Distribution

Africa: Adventive populations in Egypt (Lower Nile Valley), Kenya and Tanzania.
Indian Ocean: Adventive populations in Mauritius and Réunion.
New Guinea area: Bougainville Islands, Indonesia (Irian Jaya), Papua New Guinea, New Britain, New Ireland. In the Solomon Islands it became established in the Shortland Islands group, where it has been subjected to an eradication campaign using a combination of bait spraying and male annihilation with cue lure traps (Eta, 1986).
North America: Has been trapped in the wild in California (Carey & Dowell, 1989), but eradicated (Spaugy, 1988).
Oriental Asia: Bangladesh, Brunei, Cambodia, China, Hong Kong, India, Indonesia (Java, Kalimantan, Sulawesi, Sumatra, Timor), Japan (Ryukyu Islands), Laos, Malaysia, Myanmar, Nepal, Pakistan, Philippines, Sri Lanka, Taiwan, Thailand, Vietnam. In the Ryukyu Islands it was eradicated from Kume Island in 1978 and from the Miyako Islands in 1987, using the sterile insect technique (Iwahashi, 1977; Okinawa Prefecture, 1987).

Plate 28. *Bactrocera (Zeugodacus) cucurbitae*, SEMs of 3rd instar larva; a, oral ridges and stomal sensory organ, lateral view (scale=0.1mm); b, oral ridges, lateral view (scale=50μm); c, stomal sensory organ and preoral lobes (scale=10μm); d, mouthhook showing preapical tooth (scale=0.1mm); e, T2, lateral view showing spinules (scale=0.1mm); f, anterior spiracle (scale=0.1mm); g, posterior spiracles (scale=0.1mm); h, anal lobes (scale=0.1mm).

Pacific: Adventive populations in Guam and the Hawaiian Islands (since 1890s); eradicated from the Mariana Islands using the sterile insect technique, but re-established on Rota in 1981 (Cunningham, 1989c).
Temperate Asia: Recently established in Iran (Fischer-Colbrie & Busch-Petersen, 1989).
A distribution map was produced by CIE (1978) and Drew (1982a) mapped its distribution in the Asia-Pacific regions.

Attractant
Males attracted to cue lure.

Other references
ANTENNAL SENSILLA, Dickens *et al.* (1988); BEHAVIOUR, Iwahashi & Majima (1986), Kanmiya (1988), Kanmiya *et al.* (1987), Katsoyannos (1989b), Kuba & Koyama (1985), Kuba *et al.* (1984), Sivinski & Burk (1989) and P.H. Smith (1989); BIBLIOGRAPHY, Zwölfer (1985); BIOCONTROL, Roy (1977) and Wharton (1989a; 1989b); BIOLOGY, Fletcher (1989b) and Tsubaki & Sokei (1988); COMPETITION, Prokopy & Koyama (1982) and Qureshi *et al.* (1987); CONTROL, Agarwal *et al.* (1987) and Ravindranath & Pillai (1986); CUTICULAR HYDROCARBONS, Carlson & Yocom (1986); DATASHEET, Weems (1964b); DEVELOPMENT RATE, Fletcher (1989a); ECOLOGY, Carey (1989), Fang & Chang (1984), Fletcher (1989c), Meats (1989a), Nishida (1980), Su (1986) and Tan & Lee (1982); EGG, Williamson (1989); HOST RESISTANCE, Greany (1989); INDIA, Kapoor (1989); INSECTICIDE RESISTANCE, Keiser (1989); HAWAII, E.J. Harris (1989) and Harris *et al.* (1986b); IMPACT, Fang & Chang (1984); LARVA, Berg (1979), Hardy (1949), Heppner (1989), Khan & Khan (1987), Kandybina (1977), Menon *et al.* (1968), Rohani (1987) and Zaka-ur-Rab (1978a; 1978b); MALE ANNIHILATION, Cunningham (1989c); MALE TERMINALIA, Singh & Premlata (1985); MELONS, Ullah (1987); PAKISTAN, Qureshi *et al.* (1974a); PARASITOIDS, Ben-Salah *et al.* (1989), Étienne (1973b), Narayanan & Chawla (1962), Roy (1977), Syed (1971), Wharton (1989a; 1989b) and Wharton & Gilstrap (1983); PHEROMONES, Koyama (1989b), Kuba & Sokei (1988) and Nation (1981); PLANT PATHOGENS, Ito *et al.* (1979); POST-HARVEST DISINFESTATION, Armstrong & Couey (1989), Burditt & Balock (1985), Couey *et al.* (1985), Heather (1989), Jang (1986), Rigney (1989) and Sungawa *et al.* (1988); REARING, Fay (1989), Paripurna & Srivastava (1988), Tsitsipis (1989) and Vargas (1989); STERILE INSECT TECHNIQUE, Calkins (1989), Gilmore (1989), Kamikado *et al.* (1987), Shiga (1989) and Vargas (1989); TRAPPING, Economopoulos (1989), Katsoyannos (1989b) and Ramsamy *et al.* (1987).

Bactrocera (Zeugodacus) tau (Walker)

(figs 83, 157, 207; pl. 29)

Taxonomic notes

This species has also been known as *Chaetodacus tau* (Walker), *Dacus caudatus* var. *nubilus* Hendel, *D. hageni* de Meijere, *D. nubilus* Hendel, *D. tau* (Walker), *Dasyneura tau* Walker, *Zeugodacus bezzianus* Hering, *Z. nubilus* (Hendel) and *Z. nubilus* ssp. *heinrichi* Hering. According to Hardy (1973) *B. nubilus* (Hendel) is a distinct species, distinguished from *B. tau* by having the apex of the aculeus "trilobed". Although Hardy (1973) examined specimens from Taiwan, he did not indicate that he examined the type, which was also from Taiwan. We have examined the aculeus of another specimen from the original Taiwanese series (a paralectotype) and found it to be a typical specimen of *B. tau* (NHM collection; the lectotype, in NHMV, has lost its aculeus). It is clear from Hardy (1973) that there is another species similar to *B. tau* in Taiwan, but further study is needed to determine its identity, and most records of *B. nubilus* from Taiwan and elsewhere were probably based on misidentifications of *B. tau*.

Commercial hosts

This species appears to show a preference for attacking the fruits of Cucurbitaceae, but it has also been reared from the fruits of several other plant families and it is a potential pest species. It is recorded from bachang (*Mangifera foetida*), cucumber (*Cucumis sativus*) and angled luffa (*Luffa acutangula*) in Malaysia (Rohani, 1987; Tan & Lee, 1982). Other South-east Asian hosts are bitter gourd (*Momordica charantia*), calabur (*Muntingia calabura*), Malay-apple (*Syzygium malaccense*) and pumpkin (*Cucurbita maxima*) (R.A.I. Drew, unpublished data, 1990). There are also unconfirmed records from mulberry (*Morus* sp.) and watermelon (*Citrullus lanatus*) (Butani, 1978; Yunus & Ho, 1980). Hardy (1973) listed several other hosts, but did not indicate the origin of those records which should therefore be regarded as unconfirmed; they are common guava (*Psidium guajava*), plus *Artocarpus*, *Averrhoa*, *Dracontomelon*, *Luffa*, *Manilkara* and *Trichosanthes* spp.

Batra (1968) discussed the bionomics of a species called "*D. hageni* (=*D. caudatus*)" which probably refers to *B. tau*, rather than *B. caudata*. Syed (1970) reared "*D. hageni*" from luffa (*Luffa aegyptiaca*) and pumpkin (*Cucurbita pepo*). The following unconfirmed records for "*D. hageni*" probably refer to *B. tau*: colocynth (*Citrullus colocynthis*), mango (*Mangifera indica*), pummelo (*Citrus maxima*), sapodilla (*Manilkara zapota*; as *Achras sapota*), snakegourd (*T. cucumerina*), tomato (*Lycopersicon esculentum*), water apple (*S. samarangense*), wax gourd (*Benincasa hispida*) and white-flowered gourd (*Lagenaria siceraria*) (Batra, 1968; Fletcher, 1920; Syed, 1970).

Shiraki (1933) recorded "*Z. nubilus*" from luffa, melon (*Cucumis melo*), pumpkin (*Cucurbita moschata*) and white-flowered gourd (*Lagenaria siceraria*) and, as those records probably derived from Taiwan, it is possible that some of them refer to the so called "*nubilus*" of Hardy (1973) rather than *B. tau*.

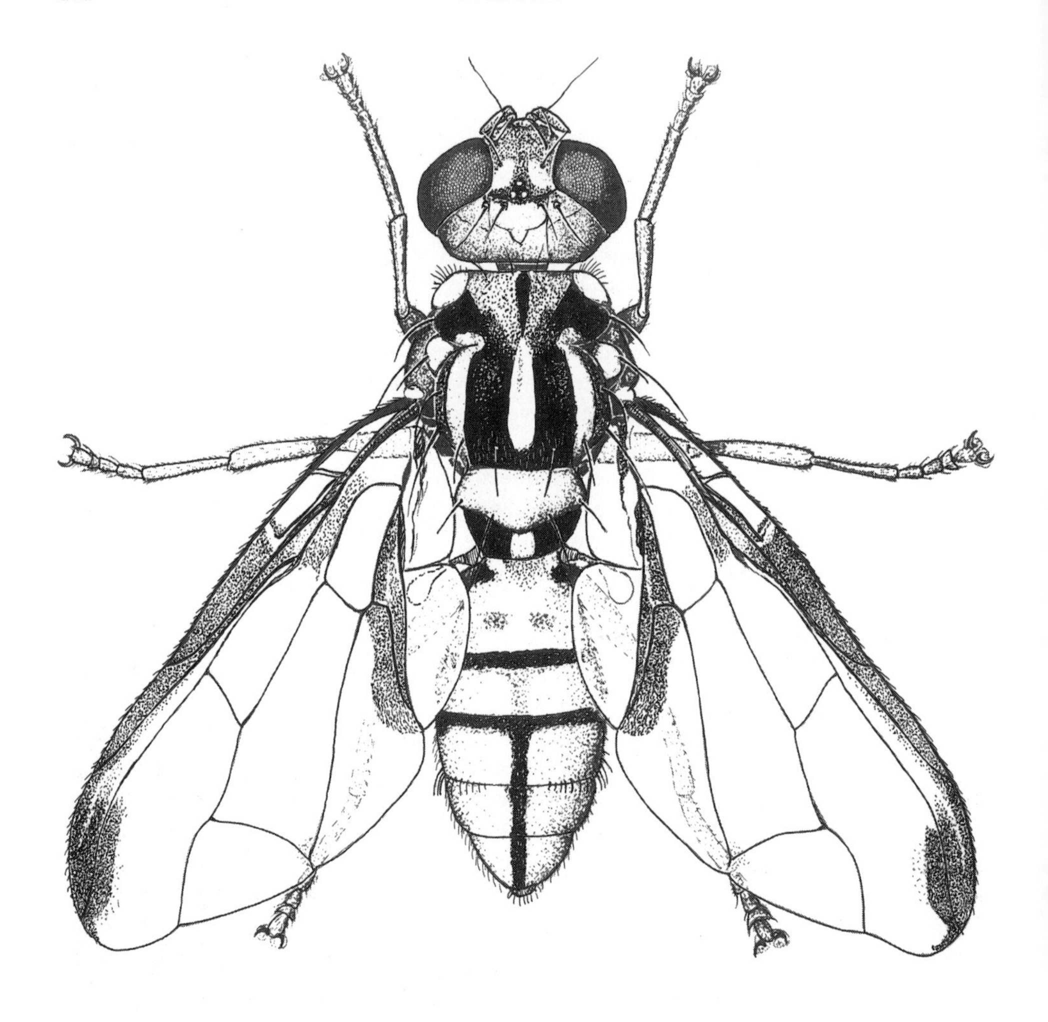

Fig. 207. *Bactrocera (Zeugodacus) tau*, adult male.

Adult identification

Scutum orange brown and marked with black, and with lateral and medial yellow stripes; with facial spots, anterior supra-alar setae, prescutellar acrostichal setae, 4 scutellar setae, and wing with a costal band expanded into an apical spot (fig. 207); male with a pecten.

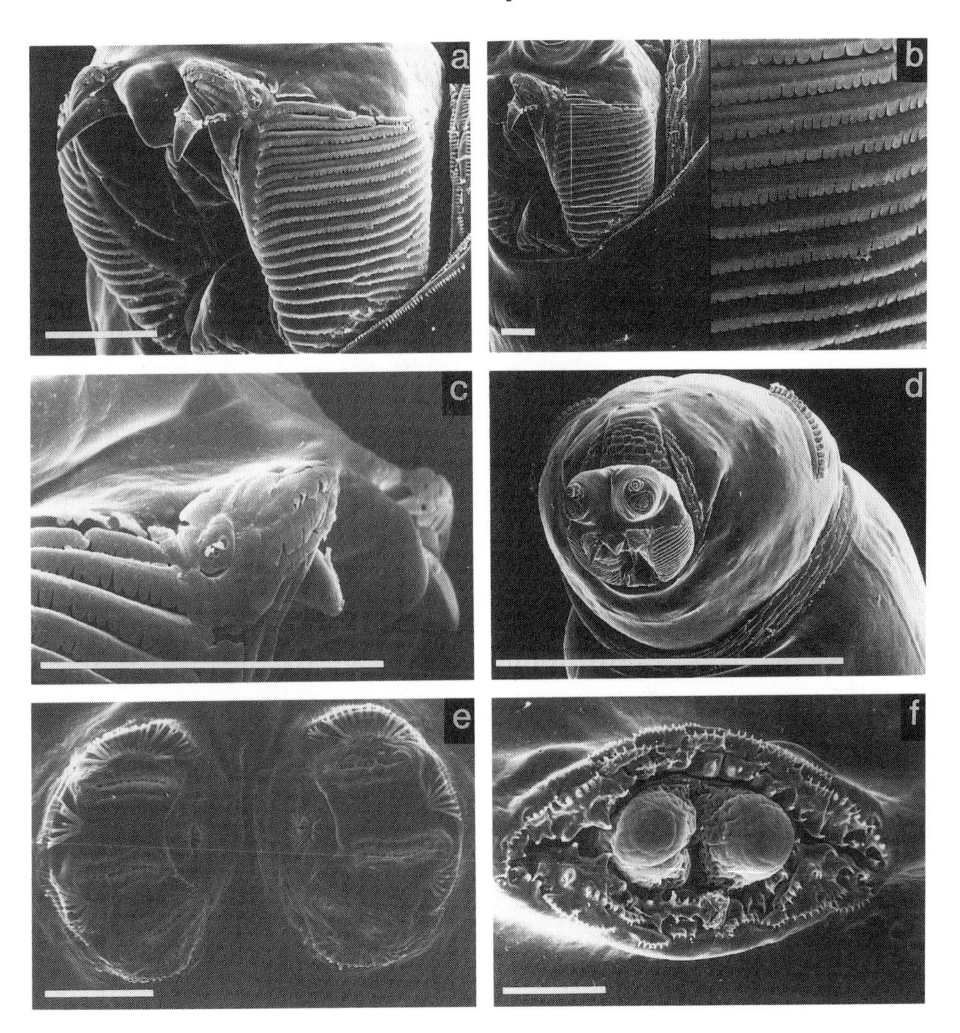

Plate 29. *Bactrocera (Zeugodacus) tau*, SEMs of 3rd instar larva; a, head, ventro-lateral view (scale=0.1mm); b, oral ridges, lateral view (scale=50μm); c, stomal sensory organ, lateral view and preoral lobes (scale=0.1mm); d, head and T1, ventro-lateral view showing left anterior spiracle (scale=1.0mm); e, posterior spiracles (scale=0.1mm); f, anal lobes (scale=0.1mm).

Description of third instar larva

Larvae medium-sized, length 7.5-9.0mm, width 1.0-1.5mm. *Head*: Stomal sensory organ (pl. 29.c) rounded, with small sensilla; surrounded by 6-9 preoral lobes, those closest to mouth opening appearing similar to small oral ridges; oral ridges (pl. 29.a, b) with 17-23 long rows of moderately long, bluntly rounded teeth; accessory plates forming numerous, serrated, long and short interlocking rows; mouthhooks large, heavily sclerotised, each with a strong apical tooth (fig. 157). *Thoracic and abdominal segments*: Anterior margin of each thoracic segment with an encircling, broad band of spinules forming discontinuous rows. T1 spinules stout, sharply pointed and arranged dorsally and ventrally in small groups or plates, becoming discontinuous rows ventrally; T2 with short stout spinules, arranged in 6-9 discontinuous rows; T3 spinules similar to T2, arranged in 5-7 rows. A1-A8 without spinules dorsally, but with spinules forming creeping welts ventrally. Each creeping welt with small stout spinules arranged in 9-13 rows, with 2-5 rows anteriorly directed, the remainder posteriorly directed. A8 with intermediate areas large and protuberant (in mature larvae, almost linked by a long slightly curved pigmented transverse line), with obvious sensilla; dorsal and lateral areas also large and well defined. *Anterior spiracles* (pl. 29.d): 14-16 tubules. *Posterior spiracles* (pl. 29.e): Spiracular slits large, about 3.0-3.5 times as long as broad, arranged in a slightly radiating pattern and bordered by a strongly sclerotised rima. Spiracular hairs long, almost as long as spiracular slits, each with a broad trunk and branched in apical third to a half; hairs arranged in 4 large bundles of 14-18 in dorsal and ventral bundles, and 5-9 in each lateral bundle. *Anal area* (pl. 29.f): Lobes large, protuberant, surrounded by 3-6 discontinuous rows of small, sharply pointed spinules. Spinules closest to anal lobes stout, long, curved and sharply pointed. *Source*: Based on specimens from Thailand.

Distribution

Oriental Asia: Bhutan, Cambodia, China, India, Indonesia (Java, Sulawesi, Sumatra), Laos, Malaysia, Philippines, Sri Lanka, Taiwan, Thailand, Vietnam.

Attractant

Males attracted to cue lure.

Other references

BEHAVIOUR, P.H. Smith (1989) and Syed (1970); DIET, Bala (1987); ECOLOGY, Tan & Lee (1982); LARVA, Rohani (1987).

Other *Bactrocera* Species of Economic Interest

B. (Afrodacus) biguttula (Bezzi), from southern Africa, attacks wild olives (*Chionanthus* and *Olea* spp., including *O. europaea* ssp. *africana*) (Munro, 1924; 1984) but it has not been recorded from cultivated olives (which are believed to be derived from *O. europaea* ssp. *africana*). Identification: Munro (1984).

B. (B.) arecae (Hardy & Adachi), from the Malay Peninsula, is a member of the *B. dorsalis* complex which probably only attacks the betel nut palm (*Areca catechu*); the larvae feed in the nuts (Hardy, 1973). There is a doubtful record from ylang-ylang (*Cananga odorata*) (Hardy, 1969). Male attractant: Not attracted to cue lure or methyl eugenol. Identification: Hardy (1973).

B. (B.) breviaculeus (Hardy), from eastern Australia, has been recorded from common guava (*Psidium guajava*) (Hardy, 1969), but that record is doubtful (Drew, 1989a). Male attractant: Cue lure. Identification: Drew (1982a; 1989a).

B. (B.) bryoniae (Tryon), from eastern Australia, Indonesia and Papua New

Guinea, is associated with a wild species of Cucurbitaceae (*Diplocyclos palmatus*). It has also been known to attack a few other plants, including wild waterlemon (*Passiflora foetida*) and *P. suberosa* (Drew, 1989a; M.M. Elson-Harris, unpublished data, 1991). Records from banana (*Musa* sp.) and tabasco pepper (*Capsicum frutescens*) were probably based on misidentifications of other spp. (Drew, 1989a). Male attractant: Cue lure. Identification: Drew (1982a; 1989a).

B. (B.) cacuminata (Hering), from eastern Australia, is a member of the *B. dorsalis* complex which is associated with wild tobacco (*Solanum mauritianum* Scop.). It has also been recorded from tabasco pepper (*Capsicum frutescens*) and tomato (*Lycopersicon esculentum*) by May (1953), but those records were regarded as doubtful by Drew (1989a). Male attractant: Methyl eugenol. Identification: Drew (1982a; 1989a).

B. (B.) cilifer (Hendel), from South-east Asia, has been recorded from sour orange (*Citrus aurantium*), tangerine (*C. reticulata*) and teruah (*Momordica cochinchinensis*) (Shiraki, 1933; Yang, 1988), but those records require confirmation. Male attractant: Cue lure. Identification: Shiraki (1933).

B. (B.) endiandrae (Perkins & May), from Australia and Papua New Guinea, is a member of the *B. dorsalis* complex which has been recorded from Indian laurel (*Calophyllum inophyllum*) (M.M. Elson-Harris, unpublished data, 1991) and ylang-ylang (*Cananga odorata*) (Drew, 1989a). Male attractant: Methyl eugenol. Identification: Drew (1982a; 1989a).

B. (B.) froggatti (Bezzi), from the Solomon Islands area, has been recorded from mango (*Mangifera indica*) (SPC, unpublished data, 1988), but that record requires confirmation. R.A.I. Drew (pers. comm., 1990) noted that any records of this species from commercial hosts were probably derived from misidentifications of other species. Male attractant: Methyl eugenol. Identification: Drew (1989a).

B. (B.) halfordiae (Tryon), from eastern Australia, has been recorded from cluster fig (*Ficus racemosa*), feijoa (*Feijoa sellowiana*), grapefruit (*Citrus x paradisi*), loquat (*Eriobotrya japonica*), mandarin (*C. reticulata*), round kumquat (*Fortunella japonica*), Surinam cherry (*Eugenia uniflora*) and sweet orange (*C. sinensis*), but those records were probably based on misidentifications of *B. tryoni* (Drew, 1989a). Identification: Drew (1982a; 1989a).

B. (B.) incisa (Walker), from Myanmar, has been recorded from jackfruit (*Artocarpus heterophyllus*), common guava (*Psidium guajava*) and mango (*Mangifera indica*) in India (Fletcher, 1920), but those records were probably based on misidentifications of *B. caryeae*. The true identity of *B. incisa* was discussed by Hardy (1973).

B. (B.) kraussi (Hardy), from north-eastern Australia, has been recorded from lucky nut (*Thevetia peruviana*) and ylang-ylang (*Cananga odorata*) (May, 1953), but those records require confirmation. Male attractant: Cue lure. Identification: Drew (1989a).

B. (B.) limbifera (Bezzi), from the Philippines and Indonesia, has been reared from the fruit of Argus pheasant tree (*Dracontomelon dao*) (Clausen *et al.*, 1965). Male attractant: Cue lure. Identification: Hardy (1974; 1982b).

B. (B.) mayi (Hardy), from eastern Australia, has been recorded from apricot

(*Prunus armeniaca*) (May, 1953; as *D. bilineata* Perkins & May), but that was probably based on a misidentification of *B. tryoni*. Male attractant: Methyl eugenol. Identification: Drew (1982a; 1989a).

B. (B.) mcgregori (Bezzi), from Singapore and the Philippines, has been reared from buko (*Gnetum gnemon*) (Hardy, 1974). Identification: Hardy (1973; 1974).

B. (B.) melas (Perkins & May), from eastern Australia, is very similar to *B. tryoni*, but darker in colour, and to *B. neohumeralis*, but with pale coloured postpronotal lobes; it is possible that many records of *B. melas* refer to aberrant individuals of those species and all records should be regarded cautiously. *B. melas* is recorded from many of the same commercial fruits as *B. tryoni*, namely apple (*Malus domestica*), common fig (*Ficus carica*), common guava (*Psidium guajava*), date palm (*Phoenix dactylifera*), grapefruit (*Citrus x paradisi*), loquat (*Eriobotrya japonica*), peach (*Prunus persica*), pear (*Pyrus communis*), plum (*Prunus domestica*), round kumquat (*Fortunella japonica*) and sweet orange (*C. sinensis*) (Drew, 1982a; 1989a). Male attractant: Cue lure. Identification: Drew (1982a; 1989a).

B. (B.) moluccensis (Perkins), from Java and the Papua New Guinea area, is recorded from Tahiti chestnut (*Inocarpus fagiferus*) (Drew, 1989a; Hardy, 1983a). Male attractant: Cue lure. Identification: Drew (1989a).

B. (B.) mutabilis (May), from eastern Australia, has been recorded from round kumquat (*Fortunella japonica*) (May, 1953), but that requires confirmation. Identification: Drew (1982a; 1989a).

B. (B.) nigrotibialis (Perkins), from South-east Asia, has been recorded from robusta coffee (*Coffea canephora*) (Clausen *et al.*, 1965). Male attractant: Cue lure. Identification: Hardy (1973).

B. (B.) nigrovittata Drew, from Papua New Guinea, has been reared from terongan (*Solanum torvum*) (Drew, 1989a). Identification: Drew (1989a).

B. (B.) obliqua (Malloch), from the New Britain area, is recorded from Malay-apple (*Syzygium malaccense*) (Drew, 1989a). Identification: Drew (1989a).

B. (B.) ochrosiae (Malloch), from Guam and Northern Marianas, has been recorded from huesito (*Malpighia glabra*), rose-apple (*Syzygium jambos*) and tropical almond (*Terminalia catappa*) (Hardy & Adachi, 1956), but it is not clear if those were rearing records. Identification: Hardy & Adachi (1956).

B. (B.) opiliae (Drew & Hardy), from northern Australia, is a member of the *B. dorsalis* complex (Drew & Hardy, 1981) which is normally only associated with a wild host (*Opilia amentacea* Roxb.; Opiliaceae). It has also been recorded from a few other hosts including mango (*Mangifera indica*); however, there have been no records from mango since 1969 (Smith *et al.*, 1988). Male attractant: Methyl eugenol. Identification: Drew (1989a); Drew & Hardy (1981).

B. (B.) pallida (Perkins & May), from eastern Australia, is recorded from sea hibiscus (*Hibiscus tiliaceus*) (Smith *et al.*, 1988). Male attractant: Methyl eugenol. Identification: Drew (1989a).

B. (B.) parvula (Hendel), from Taiwan, is recorded as "on flowers" of white-flowered gourd (*Lagenaria siceraria*), in India, by Philip (1950); that probably derived from a misidentification of another species. Identification: Shiraki (1933) supplemented by Munro (1935c).

B. (B.) simulata (Malloch), from the Papua New Guinea area and Vanuatu, has been recorded from chilli (*Capsicum annuum*) (R.A.I. Drew, pers. comm., 1990) and there is an unconfirmed record from Pacific lychee (*Pometia pinnata*) (SPC, unpublished data, 1988). Male attractant: Cue lure. Identification: Drew (1989a).

B. (B.) trilineola Drew (formerly known as *Dacus triseriatus* Drew), from Vanuatu, has been recorded from the following hosts, all of which require confirmation: avocado (*Persea americana*), banana (*Musa x paradisiaca*), breadfruit (*Artocarpus altilis*), common guava (*Psidium guajava*), Malay-apple (*Syzygium malaccense*), mango (*Mangifera indica*), papaya (*Carica papaya*) and pummelo (*Citrus maxima*) (SPC, unpublished data, 1988). Male attractant: Cue lure. Identification: Drew (1989a).

B. (B.) versicolor (Bezzi), from southern India, has been erroneously interpreted as teneral *B. dorsalis*. Mature specimens are similar to *B. zonata*, differing in the presence of a black spot at the apex of the scutellum and a narrow, but complete, costal band. It is recorded from sapodilla (*Manilkara zapota*) and common guava (*Psidium guajava*) (Bezzi, 1916), but the guava record requires confirmation. A record from mango (*Mangifera indica*) (Bezzi, 1916) was based on a misidentification, probably of *B. caryeae*. Male attractant: Methyl eugenol.

B. (B.) sp. near *B. (B.) zonata* (Saunders): A few specimens collected on custard-apple (*Annona reticulata*) leaves from India (NHM, collection data) differ from normal *B. zonata* in the shape of their aculeus tip, which has a pair of preapical 'steps'. The true identity of those specimens will be established in a future study of Asian Dacini (R.A.I. Drew, D.L. Hancock & I.M. White, in prep.). Grewal & Kapoor (1989) recognized karyotype differences between some populations of *B. zonata* and it is possible that they were referring to the same unidentified species.

B. (B.) sp. from Bhutan: The true identity of a pest of peach (*Prunus persica*) from Bhutan will be established in a future study of Asian Dacini (R.A.I. Drew, D.L. Hancock & I.M. White, in prep.). The available specimens lack a costal band (similar to the *B. correcta*, *B. tuberculata*, *B. versicolor* and *B. zonata* group), have dark tibiae, and microtrichia at the base of cell br (a feature absent in the *B. zonata* group).

B. (Gymnodacus) calophylli (Perkins & May), from Belau, eastern Australia and Malaysia, has been reared from Indian laurel (*Calophyllum inophyllum*) (Drew, 1989a). Identification: Drew (1989a).

B. (Hemisurstylus) melanoscutata Drew, from New Britain, has been reared from egg tree (*Garcinia xanthochymus*) (Drew, 1989a). Identification: Drew (1989a).

B. (Javadacus) trilineata (Hardy), from southern India, has been recorded from ivy gourd (*Coccinia grandis*) (Kapoor, 1970), but that requires confirmation. Male attractant: Cue lure. Identification: Hardy (1955).

B. (Melanodacus) nigra (Tryon), from eastern Australia, has been recorded from a wild olive (*Olea paniculata* R. Br.), but it is a polyphagous species which also attacks members of other plant families and is unlikely to be a specific threat to commercial olive production. Identification: Drew (1989a).

B. (Paratridacus) expandens (Walker), from tropical Asia and north-eastern Australia, is recorded from egg tree (*G. xanthochymus*) and mundur (*G. dulcis*) (Drew,

1989a), and there is an old record, requiring confirmation, from pumpkin (*Cucurbita moschata*) (Shiraki, 1933; as *P. yayeyamanus* (Matsumura)). Identification: Drew (1989a); Hardy (1973).

B. (Zeugodacus) duplicata (Bezzi), from India, was recorded from peach (*Prunus persica*) by Kapoor & Agarwal (1983); however, that record appears to be derived from Bezzi (1916) who only recorded it as "on" peach, suggesting that it was not reared. Identification: Bezzi (1916).

B. (Z.) munda (Bezzi), from Taiwan and the Philippines, has been recorded from pumpkin (*Cucurbita maxima*) (Shiraki, 1933; as *Dacus tibialis* Shiraki). Identification: Hardy (1974).

B. (Z.) scutellaris (Bezzi), from south-east Asia, has been reared from the flowers of pumpkin (*C. maxima*) and white-flowered gourd (*Lagenaria siceraria*) (Syed, 1970); it was sometimes reared from small fruits to which flowers were still attached. Male attractant: Cue lure. Identification: Hardy (1973).

B. (Z.) scutellata (Hendel), from China, Taiwan and the Ryukyu Islands of Japan, has been reared from the male flower buds of pumpkin (*Cucurbita* sp.) and from unidentified flower and stem galls on *Melothria liukiuensis* Nakai (Cucurbitaceae) (Sugimoto *et al.*, 1988). It has also been reared from *Trichosanthes cucumeroides* (Tanaka, 1936). There is a doubtful record from pear (*Pyrus communis*) (Yang, 1988). Male attractant: Cue lure. Identification: Shiraki (1968).

B. (Z.) trimaculata (Hardy & Adachi), from the Philippines, has been reared from bitter gourd (*Momordica charantia*) and teruah (*M. cochinchinensis*) (Hardy & Adachi, 1954); as the authors did not state whether the flowers or the fruit were attacked, it was presumably reared from fruit. Identification: Hardy (1974).

GENUS *Capparimyia* Bezzi

A genus of three species, two of which are African (Hancock, 1987), and one which is found in the Mediterranean area and Pakistan.

Capparimyia savastani (Martelli)

(figs 20, 41, 208)

Taxonomic notes
This species has also been known as *Ceratitis savastani* Martelli and sometimes misspelt as *Capparimyia savastanii* or *C. savastanoi.*

Commercial hosts
Larvae develop in the edible buds of caper (*Capparis spinosa*) (Freidberg & Kugler, 1989) and it is a potential pest of caper.

Adult identification
A predominantly yellow species with shiny black markings on the scutum and scutellum, very small ocellar setae and a characteristic wing pattern (fig. 208).

Distribution
Africa: Tunisia.
Europe: France, Italy, Malta.
Middle East: Egypt, Israel.
Oriental Asia: Pakistan.

Other references
BEHAVIOUR, Longo & Siscaro (1989); PARASITOIDS, Ben-Salah *et al.* (1989) and Soria & Yana (1962).

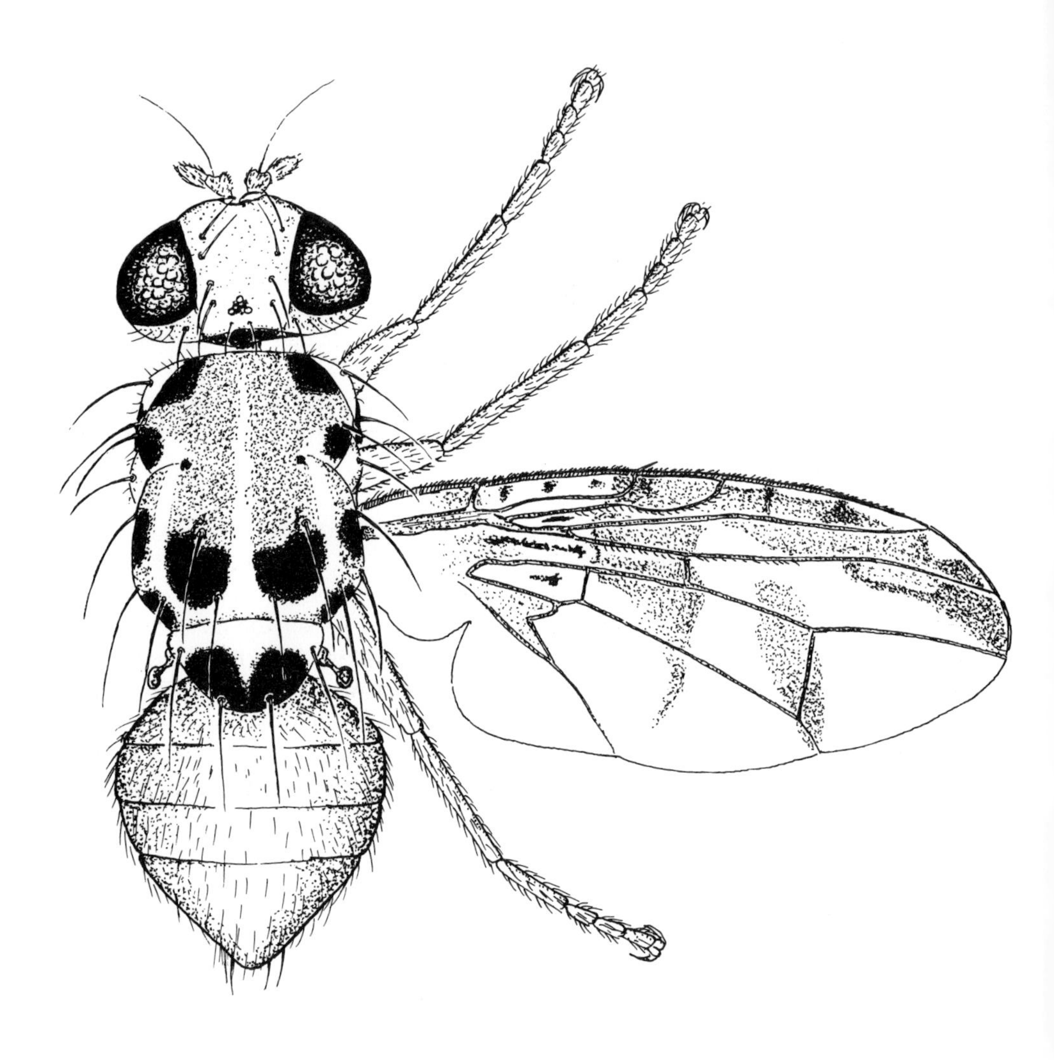

Fig. 208. *Capparimyia savastani*, adult male.

GENUS *Carpomya* Costa

(fig. 21)

Taxonomic notes
The name of this genus has often been misspelt as *Carpomyia*.

Distribution and host relationships
There are five described species, known from central and southern Europe, north Africa, the Middle East and from the cooler areas of the Indian subcontinent; one species is also adventive in Mauritius. One species is associated with the fruits of *Rosa* spp. (Rosaceae) and three species attack the fruits of *Ziziphus* spp. (Rhamnaceae), including common and Indian jujube. A key to the larvae of the three most widespread species was presented by Kandybina (1965; 1977).

Diagnosis of third instar larva
Antennal sensory organ with a ring-shaped basal segment and a rounded, flattened apical segment; maxillary sensory organ rounded, slightly protuberant, with small sensilla. Stomal sensory organ large with 3-4 blunt, sclerotised, brown coloured preoral teeth at its base; oral ridges with 3-9 rows of long, sharply pointed teeth; accessory plates absent. Mouthhooks each very strongly sclerotised with an arched apical tooth but lacking a preapical tooth. Small spinules arranged in discontinuous rows surround most thoracic and abdominal segments; anterior spiracles with 15-23 short, closely-spaced tubules in grape-like clusters; posterior spiracular slits 4-5 times as long as broad, with a raised cellular pattern between spiracles.

Carpomya incompleta (Becker)

(fig. 209)

Taxonomic notes
This species has also been known as *Trypeta incompleta* Becker.

Pest status and commercial hosts
A pest of common jujube (*Ziziphus jujube*) (Awadallah *et al.*, 1975), which is also recorded from lotus (of ancient man) (*Z. lotus*) (Freidberg & Kugler, 1989).

Wild hosts
Recorded from Christ's thorn (*Ziziphus spina-christi* (L.) Willd.) and *Z. sativus* Gaertn. (Hendel, 1927).

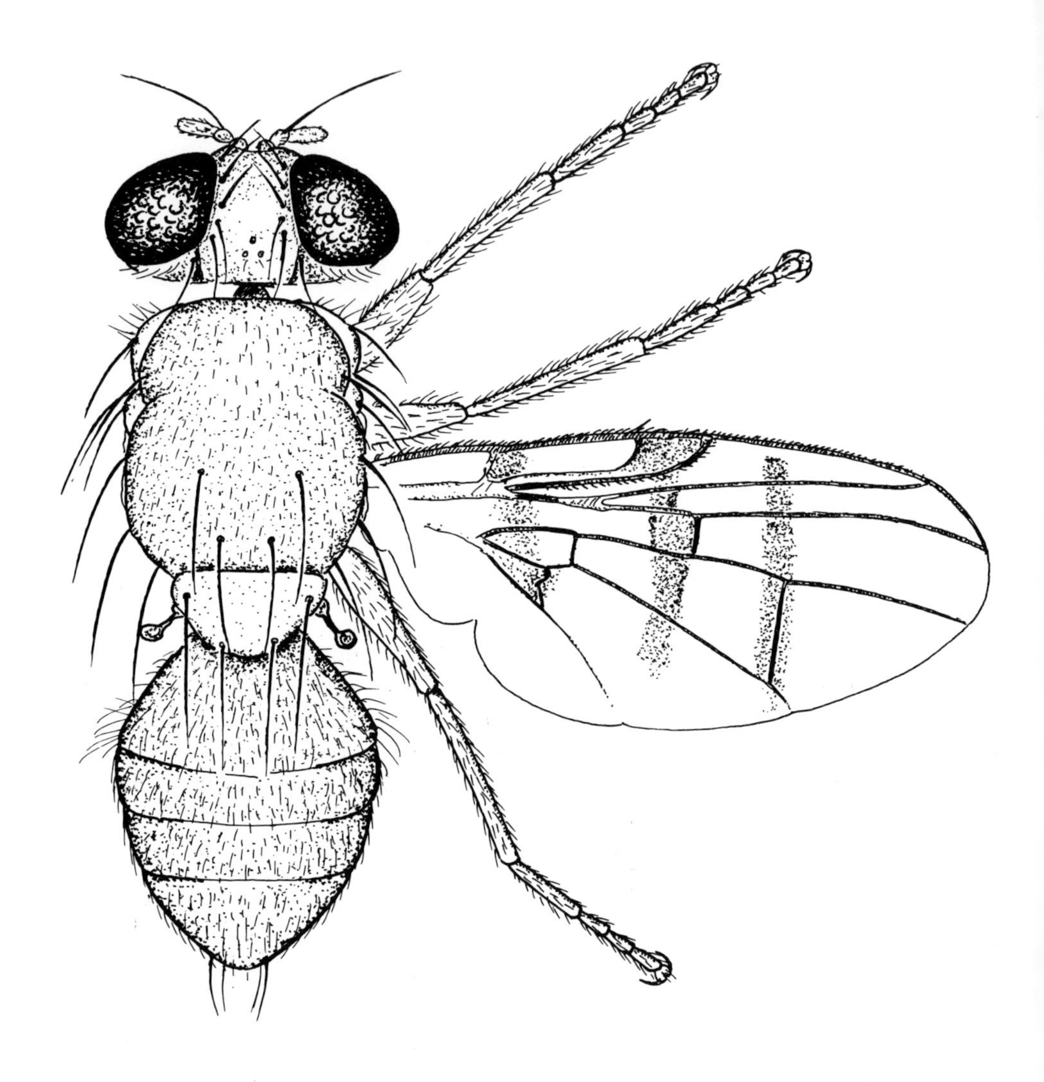

Fig. 209. *Carpomya incompleta*, adult male.

Adult identification

A small yellow to reddish yellow species with infuscate wing patterning (fig. 209).

Partial description of third instar larva

Larvae medium-sized, length 6.8-7.2mm; width 1.5-1.7mm. *Head*: Stomal sensory organ with 3-4 preoral teeth; oral ridges with 4 rows of sharply pointed teeth; mouthhooks each with a large rounded apical tooth. *Thoracic and abdominal segments*: Anterior spiracles with 15-18 tubules; posterior spiracular slits 4 times as long as broad, with small, spinous, generally single spiracular hairs. *Source*: Based on Kandybina (1965; 1977).

Distribution
Africa: Egypt, Ethiopia, Sudan.
Europe: Italy.
Middle East: Iraq, Israel.

Other references
ECOLOGY, Ali *et al.* (1975); PARASITOIDS, Fry (1990) and Kandybina (1977).

Carpomya vesuviana Costa

Ber Fruit Fly

(figs 21, 210)

Taxonomic notes
This species has also been known as *Orellia bucchichi* Frauenfeld.

Pest status and commercial hosts
A pest of Indian jujube (*Ziziphus mauritiana*) (Lakra & Singh, 1984) which has also been recorded from common jujube (*Z. jujube*) and *Z. nummularia* (Lakra & Singh, 1984). There is also a record of this species from common guava (*Psidium guajava*) (Clausen *et al.*, 1965), but that is very doubtful, and probably derived from data being confused with samples of common jujube collected from the same locality.

Wild hosts
Recorded from *Ziziphus rotundifolius* Lam. and *Z. sativus* (Hendel, 1927; Lakra & Singh, 1984).

Distribution
Europe: Italy, USSR (Caucasus area).
Indian Ocean Islands: Mauritius (S. Facknath, pers. comm., 1991), presumably adventive.
Oriental Asia: India, Pakistan, Thailand.
Temperate Asia: Afghanistan, USSR (Tadzikistan, Turkmenistan, Uzbekistan).

Other references
BEHAVIOUR, Lakra & Singh (1983); BIOLOGY, Lakra & Singh (1989); ECOLOGY, Berdyeva (1978) and Lakra & Singh (1985); HOST RESISTANCE, Singh (1984) and Singh & Vashishta (1985); PARASITOIDS, Narayanan & Chawla (1962) and Singh (1989).

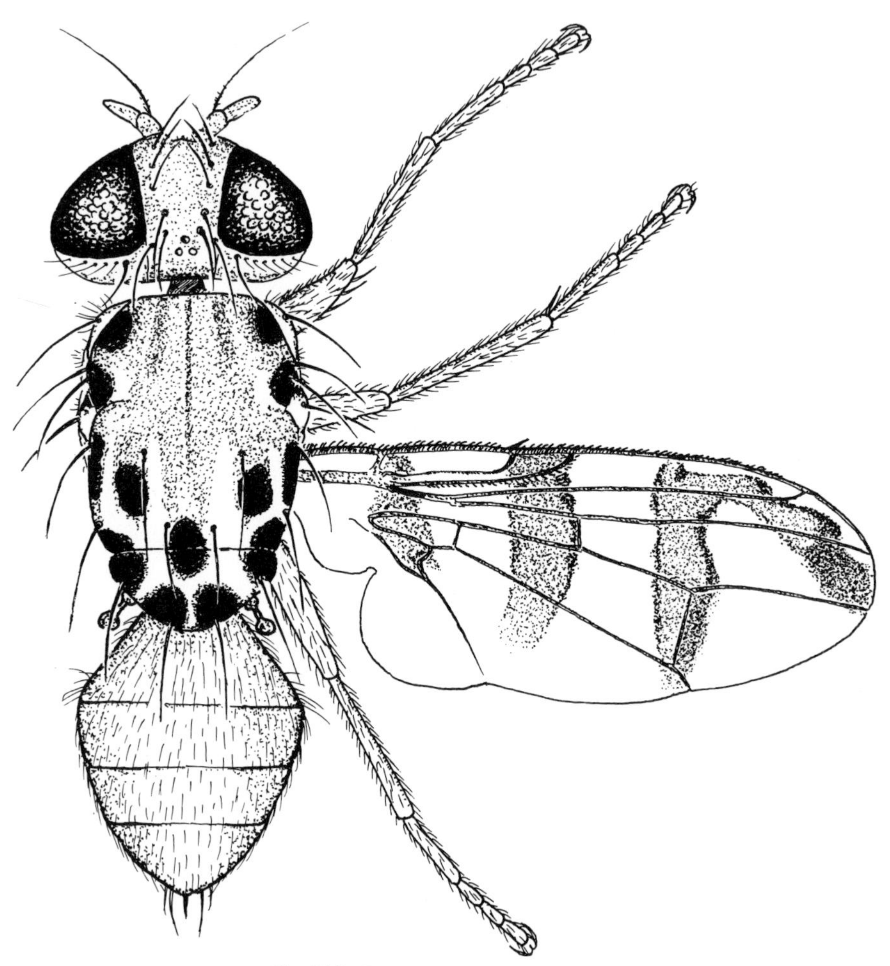

Fig. 210. *Carpomya vesuviana*, adult male.

Adult identification

A small yellow species which has a distinctive pattern of black marks on its scutum and scutellum, and a distinctive wing pattern (fig. 210).

Partial description of third instar larva

Larvae medium-sized, length 8.0mm; width 1.9-2.0mm. *Head*: Stomal sensory organ with 4 preoral teeth; oral ridges with 3 rows of sharply-pointed, closely spaced teeth; mouthhooks each with a large, arched apical tooth. *Thoracic and abdominal segments*: T1 with spinules ventrally; T2, T3 and A1 spinules encircle segment; A3-A7 with ventral creeping welts; A8 with several pairs of large tubercles; anterior spiracles with 21-23 tubules; posterior spiracular slits large, 4-5 times as long as broad, each with a finely sclerotised rima. *Source*: Based on Kandybina (1965; 1977).

Other *Carpomya* Species of Economic Interest

C. schineri (Loew), from Europe and central Asia, attacks the hips (fruit) of some roses (*Rosa* spp.), including some cultivated species; namely apple rose (*R. villosa*), damask rose (*R. damascena*), dog rose (*R. canina*), French rose (*R. gallica*) and Japanese rose (*R. rugosa*) (Freidberg & Kugler, 1989; Hendel, 1927; Kandybina, 1977; Sengalevich, 1970). In Bulgaria it has been regarded as a pest of damask rose (Tanev, 1967) and it has been considered as a potential biocontrol agent of sweet brier (*R. rubiginosa* L.) growing as an adventive weed in New Zealand (Eichhorn, 1967). Identification: Freidberg & Kugler (1989).

C. zizyphae Agarwal & Kapoor is only known from a single female reared from a *Ziziphus* sp. in the Punjab area of India. Agarwal & Kapoor (1985) reared several specimens of *C. vesuviana* from the same fruit collection (same host, locality, and collecting date). Under those circumstances, the collection of a single individual of a different species seems unlikely. It is therefore likely that the single specimen described as a new species called *C. zizyphae* is simply an aberrant individual of *C. vesuviana*.

GENUS *Ceratitis* MacLeay

(figs 22, 23, 37, 42)

Taxonomic notes

Cogan & Munro (1980), and some earlier authors, regarded *Pardalaspis* and *Pterandrus* as genera separate from *Ceratitis*, rather than subgenera within *Ceratitis*. However, these generic divisions were not generally followed by economic entomologists, and the subgeneric divisions of *Ceratitis* used here derive from Hancock (1984).

Distribution and biology

A genus of about 65 species found in tropical and southern Africa. One species, *C. capitata*, has been spread to almost all tropical and warm temperate areas of the world, and *C. malgassa* has been at least temporarily established in the West Indies. The most serious pest species are multivoltine and attack a wide range of unrelated fruits, but the few other species of known biology appear to be associated with single plant families. Mating takes place on the host plant and pupariation occurs in the soil (Fletcher, 1989b).

Diagnosis of third instar larva

Antennal sensory organ with a large basal segment and a smaller cone-shaped distal segment (pl. 31.a); maxillary sensory organ broad, flat, with well defined sensilla in 2 distinct groups, surrounded by small cuticular folds (pl. 31.a). Stomal sensory organ with small peg-like sensilla, surrounded by 4-5 unserrated preoral lobes (pl. 30.a); oral ridges with 7-13 rows of short, bluntly rounded teeth; accessory plates absent (pl. 30.b). Anterior portion of each thoracic segment with an encircling band of small spinules in discontinuous rows; anterior spiracles (pl. 30.c) with 8-25 uniform, large, stout, rounded, well spaced tubules, each with a distinct apical slit; posterior spiracular hair bundles with 4-13 hairs; anal lobes protuberant, surrounded by 3-6 discontinuous rows of small stout spinules.

Attractants

Males of tested species belonging to subgenera *C. (Ceratitis)* and *C. (Pterandrus)* are attracted to trimedlure and terpinyl acetate, but not methyl eugenol. Conversely, males of tested *C. (Pardalaspis)* spp. respond to methyl eugenol and terpinyl acetate. Males of tested *C. (Ceratalaspis)* spp. are only attracted to terpinyl acetate. No species of Ceratitini is known to be attracted to cue lure and the responses to baits of 16 *Ceratitis* spp. were tabulated by Hancock (1987).

Literature

Despite the importance of this group, there has been no comprehensive key work since Bezzi (1924a). The most useful modern works are Hancock (1984; 1987), and Munro (1935a) may be consulted for *C. (Ceratalaspis)* and *C. (Pardalaspis)* spp.

Subgenus *CERATALASPIS* HANCOCK

A subgenus of about 30 species all of which were included in *C. (Pardalaspis)* prior to the work of Hancock (1984).

Ceratitis (Ceratalaspis) cosyra (Walker)

Mango Fruit Fly
Marula Fruit Fly
Marula Fly

(figs 105, 211)

Taxonomic notes
This species has also been known as *Pardalaspis cosyra* (Walker), *P. parinarii* Hering and *Trypeta cosyra* Walker.

Pest status and commercial hosts
In Kenya and Zambia this is the main pest of mango (*Mangifera indica*) (Javaid, 1986; Malio, 1979). In Kenya it also attacks common guava (*Psidium guajava*) and sour orange (*Citrus aurantium*) (Malio, 1979), and in Zimbabwe it has been recorded from avocado (*Persea americana*), maroola plum (*Sclerocarya birrea*) and wild custard-apple (*Annona senegalensis*) (Hancock, 1987). In South Africa it also attacks early peaches (*Prunus persica*) (Annecke & Moran, 1982).

Wild hosts
Wild maroola plum trees are an important host in South Africa, but this is a highly polyphagous species that has been recorded from wild species of Ebenaceae, Euphorbiaceae and Fabaceae in Zimbabwe (Hancock, 1987). There are also old records from Apocynaceae, Canellaceae and Chrysobalanaceae (Hering, 1935; Le Pelley, 1959; Munro, 1929a).

Adult identification
Separated from most other species by its characteristic pattern of yellow wing bands, and the three black areas in the apical half of the scutellum (fig. 211). The male orbital setae are not expanded at the apex and the tibiae are not feathered. Similar species may be separated using a key by Munro (1935a).

Partial description of third instar larva
Larvae medium-sized, length 6.5-7.0mm; width 1.5mm. *Head*: Oral ridges in 10-12 rows; mouthhooks with a small preapical tooth. *Thoracic and abdominal segments*: Anterior portion of each thoracic segment with an encircling band of spinules; A8 with 2 pairs of small tubercles and sensilla, intermediate areas small; anterior spiracles with 11-12 short tubercles; posterior spiracles with spiracular slits 3.0-3.5 times as long

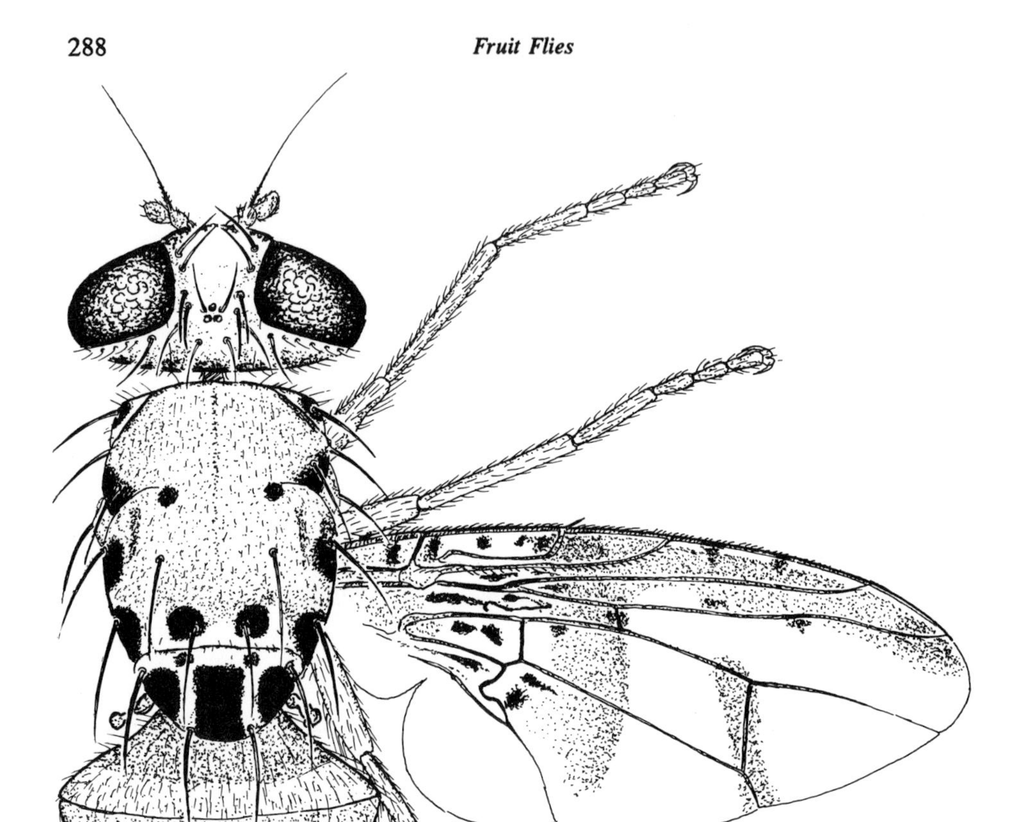

Fig. 211. *Ceratitis (Ceratalaspis) cosyra*, adult male.

as broad; spiracular hairs short, about half the length of a spiracular slit, most branched with 7-13 hairs per bundle; anal lobes well defined, surrounded by small spinules. *Source*: Based on Kandybina (1977).

Distribution

Africa: Kenya, Malawi, Mozambique, South Africa, Sudan, Tanzania, Zaire, Zambia, Zimbabwe.

Attractant

Males are attracted to terpinyl acetate, but not trimedlure or methyl eugenol.

Other reference

PARASITOIDS, Steck *et al.* (1986).

Ceratitis (Ceratalaspis) quinaria (Bezzi)

Five Spotted Fruit Fly
Rhodesian Fruit Fly
Zimbabwean Fruit Fly

(fig. 212)

Taxonomic notes
This species has also been known as *Pardalaspis quinaria* Bezzi.

Commercial hosts
A potential pest which has been recorded from apricot (*Prunus armeniaca*), citrus (*Citrus* spp.), common guava (*Psidium guajava*) and peach (*Prunus persica*) (Annecke & Moran, 1982; Hancock, 1987; Venkatraman & Khidir, 1967). There is also an old record from fig (*Ficus* sp.) (listed by Hancock, 1989), but that requires confirmation.

Wild hosts
Recorded from Christ's thorn (*Ziziphus spina-christi*) in the Sudan (Venkatraman & Khidir, 1967) but it probably has a wider range of wild hosts.

Adult identification
Recognised by its small black areas on the scutellum, and its characteristic pattern of yellow wing bands (fig. 212). The male orbital setae are not expanded at the apex and the tibiae are not feathered.

Distribution
Africa: Botswana, Malawi, Namibia, South Africa, Sudan, Zimbabwe.
Middle East: Yemen People's Democratic Republic.
A distribution map was produced by CIE (1963b).

Attractant
Males are attracted to terpinyl acetate, but not trimedlure or methyl eugenol.

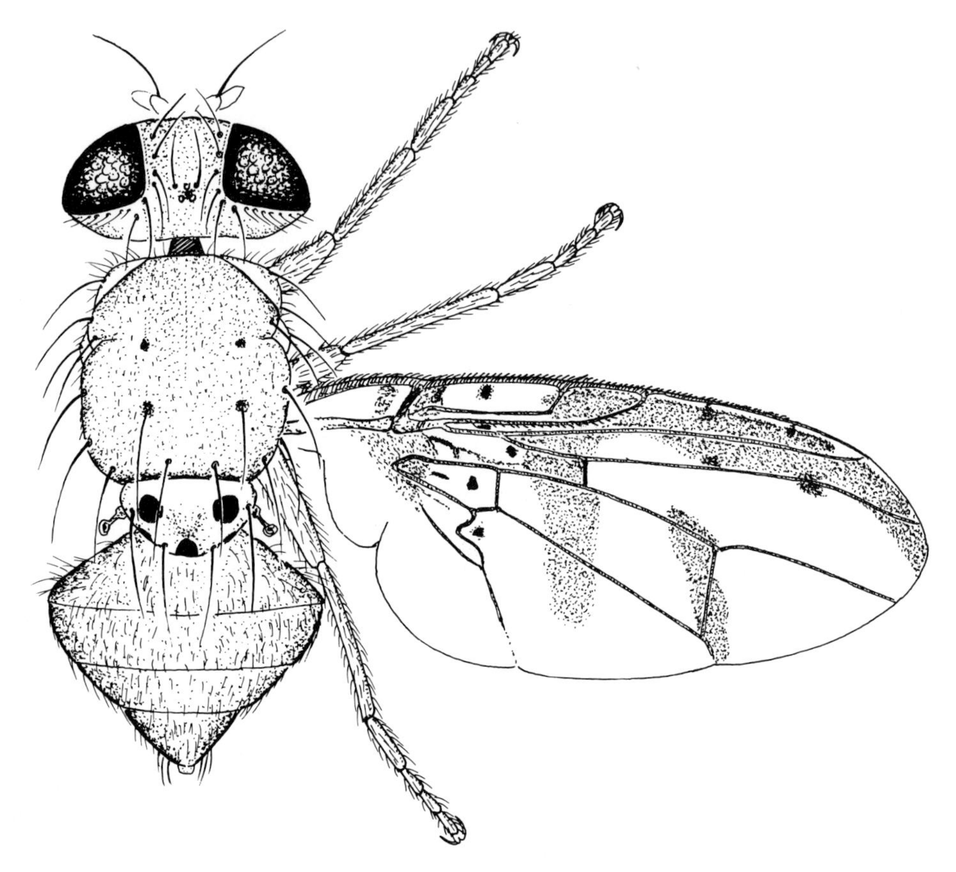

Fig. 212. *Ceratitis (Ceratalaspis) quinaria*, adult male.

Subgenus *CERATITIS* MACLEAY

(fig. 22)

Taxonomic notes

Some species have formerly been placed in the genera *Petalophora* Macquart and *Pinacochaeta* Munro, which are synonyms of *Ceratitis*; *Halterophora* Rondani was an unnecessary replacement name for *Ceratitis*.

Distribution

A subgenus of eight species found in tropical and southern Africa, but *C. capitata* and *C. malgassa* have been spread to other areas.

Ceratitis (Ceratitis) capitata (Wiedemann)

Mediterranean Fruit Fly
Medfly

(figs 22, 23, 98, 108, 158, 213; pl. 30)

Taxonomic notes
This species has also been known as *Ceratitis citriperda* MacLeay, *C. hispanica* De Brême, *Pardalaspis asparagi* Bezzi and *Tephritis capitata* Wiedemann.

Pest status and commercial hosts
This is the most widespread and probably the most serious pest species in the entire family, and the oldest populations are probably those found in tropical Africa. It is recorded from the following hosts in southern Africa: apple (*Malus domestica*), apricot (*Prunus armeniaca*), blue passion fruit (*Passiflora caerulea*), coffee (*Coffea* sp.), common guava (*Psidium guajava*), fig (*Ficus* sp.), granadilla (*Passiflora* sp.), imbe (*Garcinia livingstonei*), lychee (*Litchi chinensis*), mango (*Mangifera indica*), mulberry (*Morus* sp.), navel and Valencia oranges (*Citrus sinensis*), peach (*Prunus persica*), pear (*Pyrus communis*), plum (*Prunus domestica*), quince (*Cydonia oblonga*), wine grape (*Vitis vinifera*) and youngberry (believed to be *Rubus flagellaris* x *R. loganobaccus*) (Annecke & Moran, 1982; Clausen *et al.*, 1965; Hancock, 1987; 1989). Le Pelley (1959) recorded a similar range of hosts from East Africa, as follows: arabica coffee (*Coffea arabica*), common guava, Egyptian carissa (*Carissa edulis*), fig, kei apple (*Dovyalis caffra*), mango, peach, quince, sapodilla (*Manilkara zapota*) and sour orange (*Citrus aurantium*). The garden ornamental, kaffir plum (*Harpephyllum caffrum*), is also attacked in southern Africa, and it may act as a reservoir host for pest populations (Willers, 1979). *Ceratitis capitata* has been reared from a "red asparagus berry" (Bezzi, 1924c), but there was no confirmation that it was garden asparagus (*Asparagus officinalis*), although some workers may have assumed that to be the case.

The oldest adventive populations are found around the Mediterranean area and Freidberg & Kugler (1989) recorded the following hosts from Israel and neighbouring areas: apple, avocado (*Persea americana*), common fig (*Ficus carica*), common guava, date palm (*Phoenix dactylifera*), English walnut (*Juglans regia*), feijoa (*Feijoa sellowiana*), grapefruit (*Citrus x paradisi*), huesito (*Malpighia glabra*), Java plum (*Syzygium cumini*), kei apple, loquat (*Eriobotrya japonica*), lucky nut (*Thevetia peruviana*), mango, papaya (*Carica papaya*), peach, pear, plum, quince, sapodilla, sour orange, strawberry guava (*Psidium littorale*), Surinam cherry (*Eugenia uniflora*), sweet orange (*Citrus sinensis*), tangerine (*C. reticulata*), water apple (*Syzygium samarangense*), white sapote (*Casimiroa edulis*) and *Solanum incanum*. Fimiani (1989) recorded the following additional hosts from the Mediterranean area: apricot, bell pepper (*Capsicum annuum*), lemon (*Citrus limon*), medlar (*Mespilus germanica*), persimmon (*Diospyros* sp.), pomegranate (*Punica granatum*) and prickly pear (*Opuntia vulgaris*). Larvae have also been found in cherry (*Prunus* sp.) imported from Cyprus into the United Kingdom (IIE, unpublished data, 1988).

Another very old adventive population is found in the Canary Islands and Madeira, and Fimiani (1989) listed the following hosts: apricot, coffee (*Coffea* sp.), common fig, feijoa, grapefruit, medlar, orange (*Citrus* sp.), peach, pear, persimmon, strawberry guava and Surinam cherry.

Adventive populations are found in all other continental regions, except tropical Asia and North America (eradicated, although re-established at the time of writing). Malavasi *et al.* (1980) recorded the following hosts from Brazil: apple, arabica coffee, avocado, jaboticaba (*Myrciaria cauliflora*), Japanese persimmon (*Diospyros kaki*), lemandarin (*Citrus x limonia*), loquat, mango, peach, pear, plum, pummelo (*C. maxima*), rose-apple (*Syzygium jambos*), sapodilla, sour orange, Surinam cherry, sweet orange and tangerine. Malavasi & Morgante (1980) tabulated the rates of attack on several major hosts and it is clear that *Ceratitis capitata*, although highly polyphagous in the area studied, did not attack some hosts favoured by *Anastrepha* spp., for example some *Inga*, *Passiflora* and *Spondias* spp.

The population in Central America is of especial quarantine importance to North America and extensive host fruit surveys have been carried out in Guatemala. Kolbe & Eskafi (1990) noted that arabica coffee, was the preferred host, with loquat and oranges also being very important. Eskafi & Cunningham (1987) also recorded *C. capitata* from the following hosts: avocado, calabur (*Muntingia calabura*), cashew (*Anacardium occidentale*), common fig, common guava, cuachilote (*Parmentiera aculeata*), grapefruit, green sapote (*Pouteria viridis*), Japanese persimmon, lime (*Citrus aurantiifolia*), mango, papaya, paradise apple (*Malus pumila*), peach, pear, plum, pummelo, red mombin (*Spondias purpurea*), rose-apple, sour orange, star-apple (*Chrysophyllum cainito*), strawberry guava, sweet orange, sweet lime (*Citrus limetta*), tangelo (*C. x tangelo*), tangerine, tropical almond (*Terminalia catappa*) and white sapote (*Casimiroa edulis*). Those authors found *Anastrepha* spp., but not *Ceratitis capitata* in the fruits of the following plants: Panama orange (*x Citrofortunella mitis*), pomegranate (*Punica granatum*), purple granadilla (*Passiflora edulis*), sapodilla and Surinam cherry. Eskafi & Cunningham (1987) also listed plants with fleshy fruits in which neither *Ceratitis capitata* nor *Anastrepha* spp. was found, and they included akee (*Blighia sapida*), breadfruit (*Artocarpus altilis*), carambola (*Averrhoa carambola*), cherimoya (*Annona cherimola*), citron (*Citrus medica*), cocoplum (*Chrysobalanus icaco*), date palm, huesito, Java plum, orange jessamine (*Murraya paniculata*), quince, raspberry (*Rubus idaeus*), soursop (*Annona muricata*), strawberry (*Fragaria vesca*), sugar-apple (*Annona squamosa*), sweet cherry (*Prunus avium*), tomato (*Lycopersicon esculentum*), white mulberry (*Morus alba*) and wine grape (*Vitis vinifera*).

The Hawaiian population is another well studied adventive population which has been established since 1910. E.J. Harris (1989) listed 39 hosts that were attacked in 1910, and following a decline in the population, only 14 of those were found to be attacked in the same area of Honolulu in 1985 (see also Vargas *et al.*, 1983b; Harris & Lee, 1986). The hosts of economic importance recorded in the study area in 1985 were as follows: avocado, breadfruit, common fig, Indian laurel (*Calophyllum inophyllum*), liberica coffee (*Coffea liberica*), loquat, lychee, mango, papaya, sour orange, star-apple, strawberry guava and tropical almond. The additional hosts recorded in 1910 were apricot, arabica coffee, Brazil cherry (*Eugenia brasiliensis*), carambola, citron, common guava, custard apple (*Annona reticulata*), grapefruit, Java plum, Jew plum (*Spondias cytherea*), lemon, Malay-apple (*Syzygium malaccense*), mangosteen (*Garcinia mangostana*), peach, pear, pomegranate, rose-apple, round kumquat (*Fortunella japonica*), satinleaf (*Chrysophyllum oliviforme*), soursop, Surinam cherry, tangor (*Citrus x nobilis*) and white sapote. Many of the 1910 hosts in the Honolulu area are still attacked elsewhere in the Hawaiian Islands, together with the following additional hosts: akee, Cape gooseberry (*Physalis peruviana*), huesito, loquat, Japanese persimmon, Jerusalem cherry (*Solanum pseudocapsicum*), orange jessamine, plum (presumably *Prunus domestica*), sandalwood (*Santalum album*), sapote (*Pouteria sapote*), sweet orange and tree tomato (*Cyphomandra crassicaulis*) (Nakagawa *et al.*, 1968; Vargas *et al.*, 1983b; Vargas & Nishida, 1989).

The Réunion population was studied by Étienne (1972) who noted that the major hosts were arabica coffee, black nightshade (*Solanum nigrum*) and Japanese persimmon; other hosts were common guava, mango, orange jessamine, sour orange, Surinam cherry, tabasco pepper (*Capsicum frutescens*), tangerine, tropical almond and wine grape.

Many additional hosts have been recorded, including the following, all of which require confirmation: allspice (*Pimenta dioica*), Barbados gooseberry (*Pereskia aculeata*), belladonna (*Atropa bella-donna*), black sapote (*Diospyros digyna*), cherimoya, cocoa (*Theobroma cacao*), cocoplum, cucumber (*Cucumis sativus*), dwarf papaya (*Carica quercifolia*), egg-fruit tree (*Pouteria campechiana*), eggplant (*Solanum melongena*), elephant-ear prickly pear (*Opuntia tuna*), garden bean (*Phaseolus vulgaris* - doubtful), giant granadilla (*Passiflora quadrangularis*), Indian fig-prickly pear (*O. ficus-indica*), Indian jujube (*Ziziphus mauritiana*), hog-plum (*Spondias mombin*), ketembilla (*Dovyalis hebecarpa*), kiwi fruit (*Actinidia chinensis*), longan (*Dimocarpus longan*), mammy-apple (*Mammea americana*), myrobalan nut (*Terminalia chebula*), Natal plum (*Carissa macrocarpa*), olive (*Olea europaea*), Panama orange (*x Citrofortunella x mitis*), purple granadilla (*Passiflora edulis*), sapote, sour cherry (*Prunus cerasus*), Spanish cherry (*Mimusops elengi*), strawberry tree (*Arbutus unedo*), sugar-apple, tomato, white mulberry, wild waterlemon (*Passiflora foetida*) and *Cucurbita* sp. (Doss, 1989; Enkerlin *et al.*, 1989; Hancock, 1981; 1989; Hendel, 1927; Kandybina, 1977; Karpati, 1983; Munro, 1925; 1929b; 1935b; Russell, 1936; Séguy, 1934; Silvestri, 1913; USDA, *Pests not known to occur in the United States or of limited distribution*, No.26, undated). Karpati (1983) also listed cotton (*Gossypium* sp.) and that record was probably derived from studies of the Hawaiian population which showed a low incidence of attack on a wild cotton (*Gossypium tomentosum* Nutt.) in 1910 (E.J. Harris, 1989). A recent datasheet (Doss, 1989) also listed pineapple (*Ananas comosus*) as a host, without any indication of the source of that data; however, it has been shown to be unsuitable as a host (Armstrong *et al.*, 1979).

Recent work has shown that, at least up to harvesting stage, banana (*Musa x paradisiaca*) and avocado varieties grown in Hawaii are not attacked (Armstrong, 1983; Armstrong *et al.*, 1983). Similarly, some varieties of lemon have been shown to be unsuitable as hosts (Spitler *et al.*, 1984). Some recorded hosts are only attacked when already damaged, namely avocado, banana (*Musa* sp.), lemon, papaya and pomegranate (USDA, *Pests not known to occur in the United States or of limited distribution*, No.26, undated).

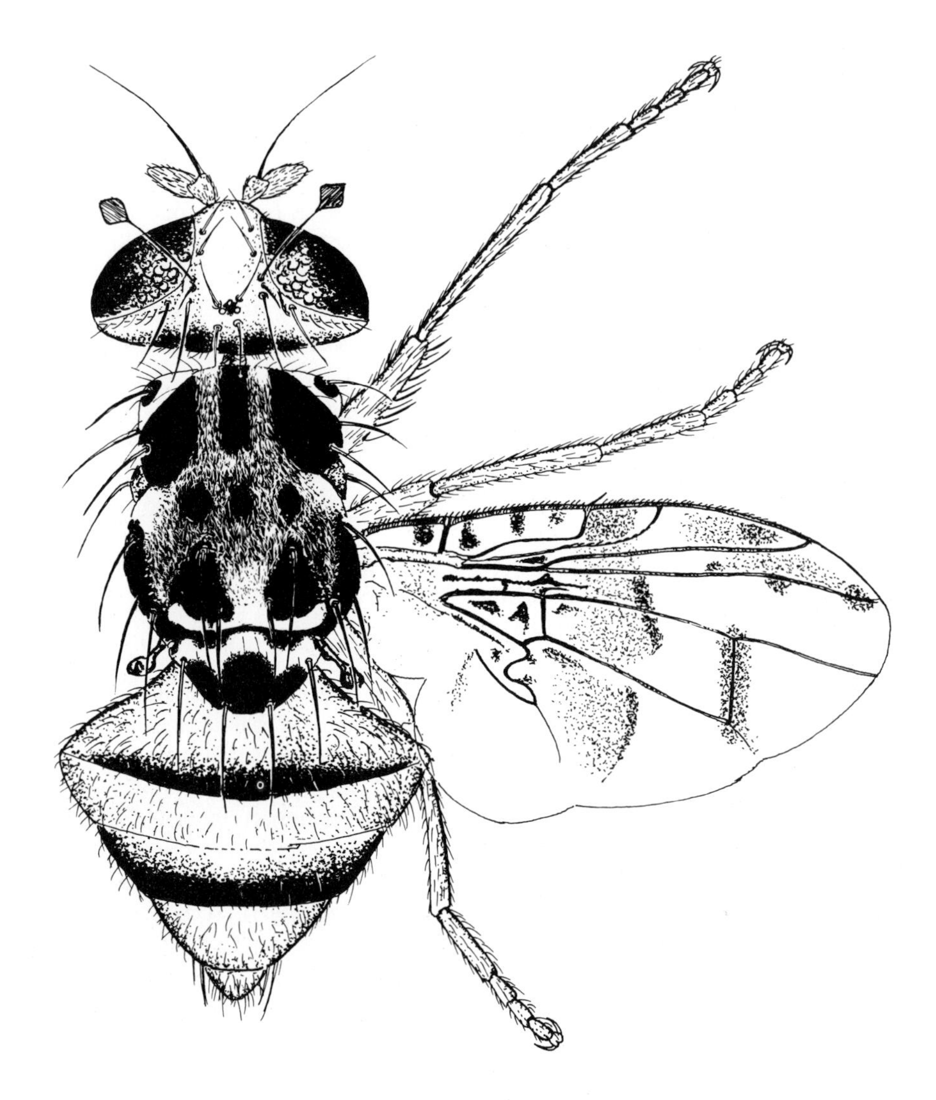

Fig. 213. *Ceratitis (C.) capitata*, adult male.

Wild hosts

A highly polyphagous species which has been recorded from wild hosts belonging to a large number of families, including the following: Anacardiaceae, Chrysobalanaceae, Cucurbitaceae, Ebenaceae, Loganiaceae, Malpighiaceae, Meliaceae, Oleaceae, Podocarpaceae, Rosaceae, Rubiaceae, Rutaceae, Sapotaceae and Solanaceae.

Adult identification

The males are easily separated from all other members of the family by the black pointed expansion at the apex of the anterior pair of orbital setae (figs 22, 98). The females can be separated from most other species by the characteristic yellow wing pattern and the apical half of the scutellum being entirely black (fig. 213).

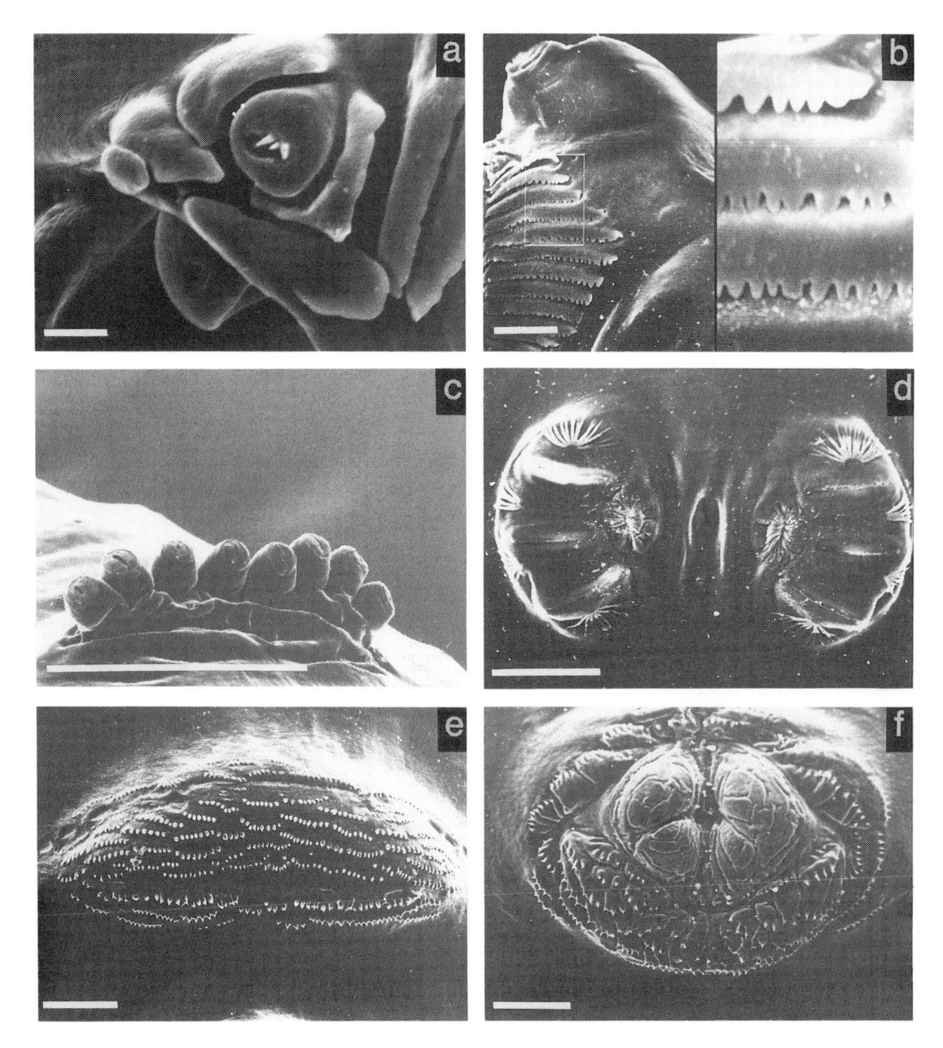

Plate 30. ***Ceratitis (C.) capitata*, SEMs of 3rd instar larva; a, stomal sensory organ and preoral lobes, lateral** view (scale=10μm); b, oral ridges, lateral view (scale=50μm); c, anterior spiracle (scale=0.1mm); d, posterior spiracles (scale=0.1mm); e, A8, creeping welt (scale=0.1mm); f, anal lobes (scale=0.1mm).

Description of third instar larva

Larvae medium-sized, length 6.5-9.0mm; width 1.2-1.5mm. *Head*: Stomal sensory organ (pl. 30.a) with 3 small peg-like sensilla; oral ridges (pl. 30.b) with 9-13 rows of short, bluntly rounded teeth; mouthhooks black, heavily sclerotised, without preapical teeth (fig. 158). *Thoracic and abdominal segments*: T1 spinules small, sharply pointed, forming 3-4 rows dorsally, 8-10 rows laterally and 8-11 rows ventrally; T2 with stout spinules forming small plate-like concentrations of 3-5 rows dorsally, 0-3 rows laterally and 4-7 rows ventrally; T3 with spinules similar to those on T2, 1-3 rows dorsally, 0-3 rows laterally and 4-7 rows ventrally. A few small rows of spinules on A1 dorsally but absent from A2-A8. Large creeping welts (pl. 30.e) of 9-13 rows of small, stout spinules, 1-3 rows anteriorly directed, the remainder posteriorly directed, some posterior rows with slightly stouter spinules. A8 with large intermediate lobes and well defined sensilla. *Anterior spiracles* (pl. 30.c): 8-10 tubules. *Posterior spiracles* (pl. 30.d): Each spiracular slit with a heavily sclerotised rima and about 2.5-3 times as long as broad. Dorsal and ventral spiracular bundles of 6-9 hairs, branched in apical half, lateral bundles with 4-6 hairs. *Anal area* (pl. 30.f): Lobes surrounded by 3-6 discontinuous rows of small, stout spinules becoming stouter and more concentrated below anal opening. *Source*: Based on specimens from Western Australia and Hawaii, USA (lab. colonies).

Rapid diagnostic technique Dadour *et al.* (in press) have described two rapid methods for separation of larvae and puparia of *Ceratitis capitata* from those of *Bactrocera tryoni*, using morphology and electrophoresis.

Distribution

Africa: Algeria, Angola, Benin, Burkina Faso, Burundi, Cameroun, Congo, Egypt, Ethiopia, Gabon, Ghana, Guinea, Ivory Coast, Kenya, Liberia, Libya, Malawi, Mali, Morocco, Mozambique, Niger, Nigeria, Senegal, South Africa, Sudan, Tanzania, Togo, Tunisia, Uganda, Zaire, Zimbabwe. Karpati (1983) listed some other African countries but did not give the source of his data.

Atlantic Islands (both Afrotropical and Palaearctic): Adventive in Azores, Canary Islands, Cape Verde Islands, Madeira, St Helena, São Tomé.

Australia: Adventive in Western Australia; temporary outbreaks in the Adelaide area in most summers (D.K. Yeates, pers. comm., 1991).

Central America: Adventive in Costa Rica, El Salvador, Guatemala, Nicaragua, Panama; eradicated from Mexico.

Europe: Adventive in Albania, Austria, Cyprus, France, Greece (including Crete), Italy, Malta, Portugal, Spain, USSR (southern Ukraine), Yugoslavia. Karpati (1983) also listed Belgium, Germany, Hungary and Switzerland, but those records were probably based on non-established introductions or quarantine interceptions. Fischer-Colbrie & Busch-Petersen (1989) described temporary outbreaks in several European countries.

Indian Ocean: Adventive in Madagascar, Mauritius, Réunion, Seychelles.

Middle East: Adventive in Israel, Jordan, Lebanon, Saudi Arabia, Syria, Turkey.

North America: Currently (1991) established in California; temporary establishments in Florida and Texas have been eradicated (Cunningham, 1989b; Lorraine & Chambers, 1989).

Oriental Asia: Recorded from India, but that record was probably derived from a quarantine interception (Kapoor, 1989).

Pacific Ocean: Adventive in Hawaiian Islands, Mariana Islands.

South America: Adventive in Argentina, Bolivia, Brazil, Chile, Colombia, Ecuador, Paraguay, Peru, Uruguay, Venezuela.

West Indies and nearby islands: Adventive in Jamaica; recently eradicated from Bermuda (Hilburn & Dow, 1990).

A distribution map was produced by CIE (1988a) and Gonzalez (1978) mapped the history of introduction and eradication in the New World.

Attractant

Males are attracted to trimedlure and terpinyl acetate, but not methyl eugenol.

Other references

BACTERIAL SYMBIONTS, Girolami (1986); BEHAVIOUR, Arita & Kaneshiro (1989), Burk & Calkins (1983), Cooley *et al.* (1986), Katsoyannos (1989b), Papaj *et al.* (1989a; 1989b), Prokopy *et al.* (1986; 1989b), Sivinski & Burk (1989), P.H. Smith (1989) and Webb *et al.* (1983b); BIBLIOGRAPHY, Zwölfer (1985); BIOCONTROL, Étienne (1973b), Gingrich (1987) and Wharton (1989a; 1989b); BIOLOGY, Fletcher (1989b); BRAZIL, Malavasi & Morgante (1980), Malavasi *et al.* (1980) and Martins & Alves (1988); CONTROL, Mitchell & Saul (1990); CUTICULAR HYDROCARBONS, Carlson & Yocom (1986); DATASHEETS, Doss (1989) and Weems (1981); ECOLOGY, Avilla & Albajes (1984), Carey (1989), Debouzie (1989), Filho *et al.* (1979), Fitt (1989), Fletcher (1989c), Hashem *et al.* (1987), Krainacker *et al.* (1987), Meats (1989a), Podoler & Mendel (1977), and Wong *et al.* (1984a); EGG STRUCTURE, Margaritis (1985) and Williamson (1989); DEVELOPMENT RATE, Fletcher (1989a); GENETICS, IAEA (1990), Lifschitz & Cladera (1989), Milani *et al.* (1989), Robinson (1989) Roessler (1989a) and Wood & Harris (1989); GUATEMALA, Eskafi & Cunningham (1987); HAWAII, Harris & Lee (1986), Nishida (1980), Nishida *et al.* (1985), Vargas *et al.* (1983b) and Wong *et al.* (1983); HOST RESISTANCE, Greany (1989); HOST STIMULI, Freeman & Carey (1990) and Warthen & McInnes (1989); IMPACT, Fimiani (1989) and Hancock (1989); INSECTICIDE RESISTANCE, Keiser (1989); LARVA, Berg (1979), Hardy (1949), Heppner (1985), Orian & Moutia (1960), Sabatino (1974) and K.G.V. Smith (1989); NICARAGUA, Daxl (1978); PARASITOIDS, Araoz & Nasca (1985), Avilla & Albajes (1984), Ben-Hamouds & Ben-Salah (1984), Ben-Salah *et al.* (1989), Cals-Usciati (1972), Cals-Usciati *et al.* (1985), Doss (1989), Étienne (1973b), Fry (1990), Greathead (1972), Jirón & Mexzon (1989), Narayanan & Chawla (1962), Podoler & Mazor (1981a; 1981b), Podoler & Mendel (1977; 1979), Steck *et al.* (1986), Wharton (1983; 1989a; 1989b), Wharton & Gilstrap (1983), Wharton *et al.* (1981), Wong & Ramadan (1987) and Wong *et al.* (1984b); PERU, Herrera & Vinas (1977); PHEROMONES, Arita & Kaneshiro (1986), Averill & Prokopy (1989a), Delrio & Ortu (1988), Dickens *et al.* (1988), Howse & Foda (1986), Jang *et al.* (1989), Jones (1989), Levinson *et al.* (1987; 1989), McDonald (1987), Mayo *et al.* (1987), Nation (1981), Papaj *et al.* (1990), Prokopy *et al.* (1978) and Scalera *et al.* (1987); POST-HARVEST DISINFESTATION, Armstrong & Couey (1989), Armstrong *et al.* (1984), Couey *et al.*, (1985), Heather (1989), Hill *et al.* (1988), Jang (1986), Rigney (1989) and Saul *et al.* (1985; 1987); PREDATION, Wong & Wong (1988); REARING, Barnes (1979), Fay (1989), Schwarz *et al.* (1985), Tsitsipis (1989), Vargas (1989) and Vargas *et al.* (1983a); REVIEW, Amador (1988) and Yount (1981); TRANSMISSION OF FRUIT ROT, Ito *et al.* (1979); STERILE INSECT TECHNIQUE, Baker *et al.* (1988), Calkins (1989), Gilmore (1989), Harris *et al.* (1986a), IAEA (1990), Plant

(1986), Schwarz *et al.* (1989a), Vargas (1989) and Wong *et al.* (1986); TRAPPING & BAITS, Cirio & Vita (1980), Cunningham (1989a), Delrio & Ortu (1988), Economopoulos (1989), Hill (1986a; 1986b; 1987), Katsoyannos (1987; 1989b), Katsoyannos *et al.* (1986), Leonhardt *et al.* (1984; 1987), McGovern & Cunningham (1988), McGovern *et al.* (1987) Mazor *et al.* (1987) and Rice *et al.* (1984); UGANDA, Greathead (1972).

Ceratitis (Ceratitis) catoirii Guérin-Méneville

Mascarene Fruit Fly

(figs 99, 109, 214)

Taxonomic notes

This species name is sometimes misspelt as *Ceratitis catoirei* or *C. catoiri*.

Commercial hosts

A potential pest of a wide range of fruits which has been recorded from avocado (*Persea americana*), bell and chilli pepper (*Capsicum annuum*), carambola (*Averrhoa carambola*), common guava (*Psidium guajava*), common jujube (*Ziziphus jujube*), custard-apple (*Annona reticulata*), loquat (*Eriobotrya japonica*), mango (*Mangifera indica*), peach (*Prunus persica*), pomegranate (*Punica granatum*), strawberry guava (*Psidium littorale*), Surinam cherry (*Eugenia uniflora*), tangerine (*Citrus reticulata*), tomato (*Lycopersicon esculentum*), tropical almond (*Terminalia catappa*) and watery rose-apple (*Syzygium aqueum*) (Étienne, 1972; Orian, 1962; Orian & Moutia, 1960). Orian & Moutia (1960) also noted that *Ceratitis catoirii* can attack over-ripe papaya (*Carica papaya*), and they listed doubtful records from governor's plum (*Flacourtia indica)*, Java plum (*S. cumini*), Jew plum (*Spondias dulcis*), pummelo (*Citrus maxima*) and sapodilla (*Manilkara zapota*; as "sapodilla, *Achras sapota*").

Adult identification

The males are easily recognised by the broad white expansion at the apex of the anterior pair of orbital setae (fig. 99). The females can be separated from most other species by the characteristic yellow wing pattern and the apical half of the scutellum being entirely black (fig. 214).

Distribution

Indian Ocean: Mauritius, Réunion, Seychelles.
A distribution map was produced by CIE (1966).

Attractant

Not known.

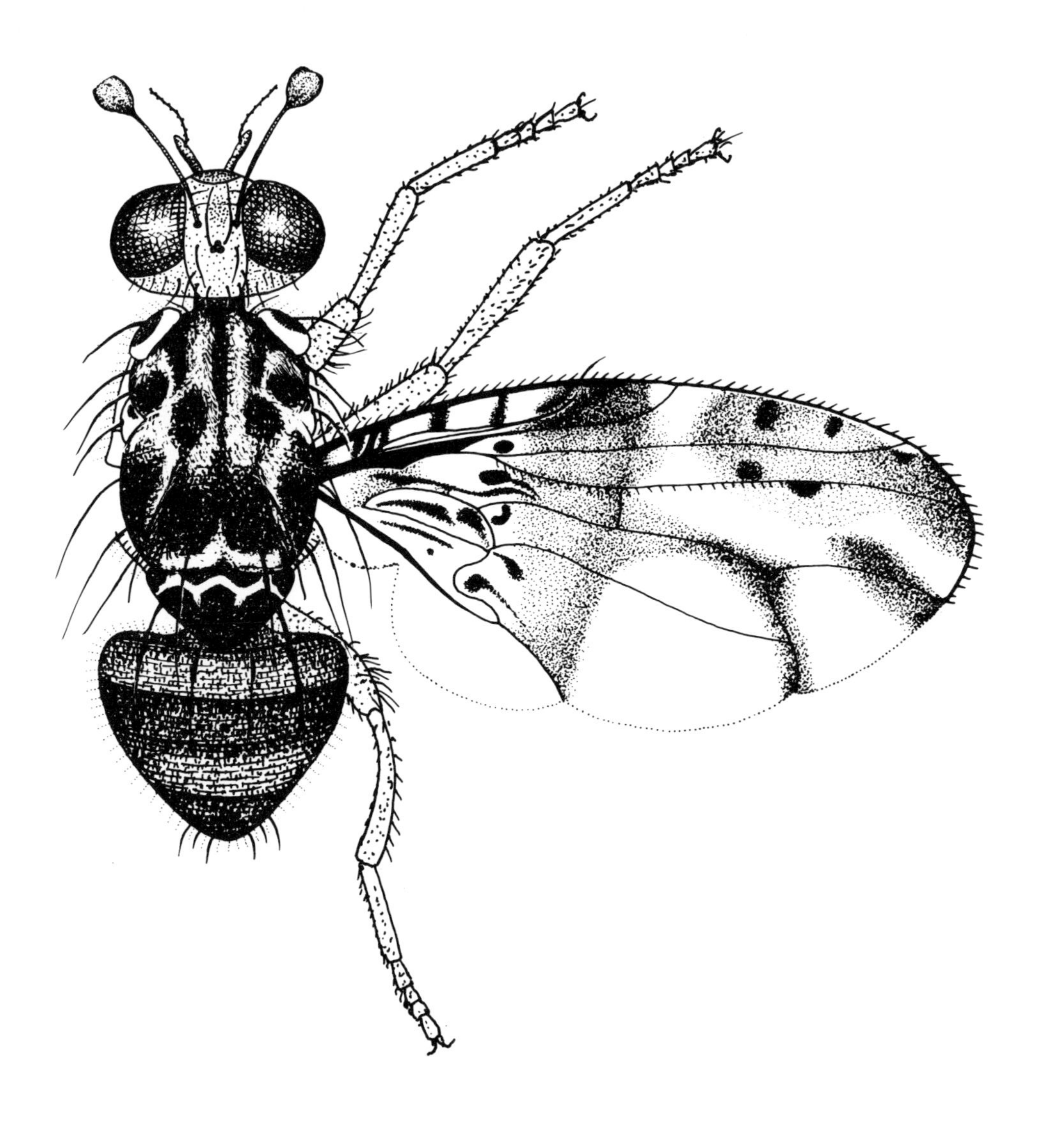

Fig. 214. *Ceratitis (C.) catoirii*, adult male.

Other reference

PARASITOIDS, Orian & Moutia (1960).

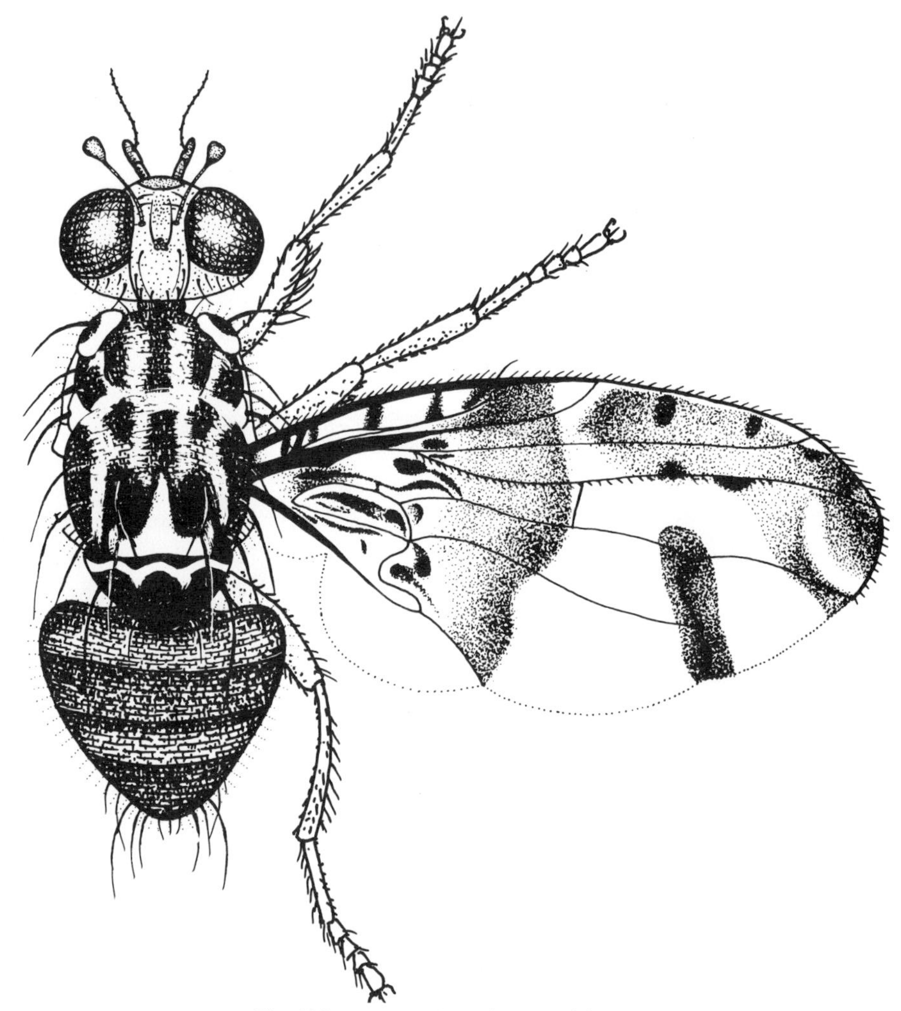

Fig. 215. *Ceratitis (C.) malgassa*, adult male.

Ceratitis (Ceratitis) malgassa Munro

Madagascan Fruit Fly

(figs 100, 110, 215)

Taxonomic note
Sometimes misspelt as *C. malagassa*.

Pest status and commercial hosts
In Madagascar this species is the main restriction on the expansion of the citrus industry (Dubois, 1965a; 1965b). It has been reared from common guava (*Psidium guajava*), mandarin orange (*Citrus reticulata*), maroola plum (*Sclerocarya birrea*) and orange (*Citrus* sp.) (Paulian, 1953). It has also been reared from nutmeg (*Myristica fragrans*) in Puerto Rico (Steyskal, 1982).

Adult identification
The males are easily recognised by the narrow white expansion at the apex of the anterior pair of orbital setae (fig. 100). The females can be separated from most other species by the characteristic yellow wing pattern and the apical half of the scutellum being entirely black (fig. 215).

Distribution
Indian Ocean: Madagascar.

West Indies: Puerto Rico; apparently established during 1968 (Steyskal, 1982), but not recorded there since.

Attractant
Not known.

Subgenus *PARDALASPIS* BEZZI

A subgenus of six species all of which are found in Africa.

Ceratitis (Pardalaspis) punctata (Wiedemann)

(figs 114, 216)

Taxonomic notes
This species has also been known as *Pardalaspis punctata* (Wiedemann), *Tephritis punctata* Wiedemann and *T. senegalensis* Macquart.

Commercial hosts
This used to be a pest of cocoa (*Theobroma cacao*) in Uganda and to a lesser extent in West Africa (Entwistle, 1972); he also recorded *C. punctata* from common guava (*Psidium guajava*), granadilla (*Passiflora* sp.), mango (*Mangifera indica*) and melon (*Cucumis melo*), but those records require confirmation as they were probably based on old information. The likely sources of Entwistle's data also listed arabica coffee (*Coffea arabica*) and white star-apple (*Chrysophyllum albidum*) (Le Pelley, 1959; Munro, 1925; Silvestri, 1913), which also require confirmation as hosts.

Wild hosts

Recorded from some wild species of Apocynaceae and Passifloraceae (Bezzi, 1912; Munro, 1925; Silvestri, 1913).

Adult identification

Separated from most other species by its characteristic pattern of yellow wing bands, and the three black areas in the apical half of the scutellum (fig. 216). The male orbital setae are not expanded at the apex and the tibiae are not feathered.

Partial description of third instar larva

Thoracic and abdominal segments: Anterior portion of each thoracic segment with a small, encircling band of spinules; A1 with a few spinules dorsally; A2 with some dorsal and lateral spinules; A3 with some lateral spinules. Anterior spiracles with 25 tubules. Posterior spiracles with spiracular slits 3 times as long as broad. Anal lobes well defined, surrounded by small spinules. *Source*: Based on Silvestri (1913).

Distribution

Africa: Cameroun, Côte d'Ivoire, Gambia, Ghana, Kenya, Nigeria, Sierra Leone, South Africa, Uganda, Zaire, Zambia, Zimbabwe.

Attractant

Males are attracted to methyl eugenol and terpinyl acetate; the record from trimedlure is an error (Hancock, 1987).

Subgenus *PTERANDRUS* BEZZI

A subgenus of 17 species all of which are found in Africa.

Ceratitis (Pterandrus) anonae Graham

(figs 101, 112)

Taxonomic notes

This species has also been known as *Ceratitis pennipes* Bezzi and *Pterandrus anonae* (Graham).

Commercial hosts

A potential pest of fruit crops which has been recorded from mango (*Mangifera indica*), robusta coffee (*Coffea canephora*) and tropical almond (*Terminalia catappa*) (Clausen *et al.*, 1965; Steck *et al.*, 1986). There are also old records, requiring confirmation, from the following hosts: avocado (*Persea americana*), common guava (*Psidium guajava*), soursop (*Annona muricata*) and strawberry guava (*P. littorale*) (Le

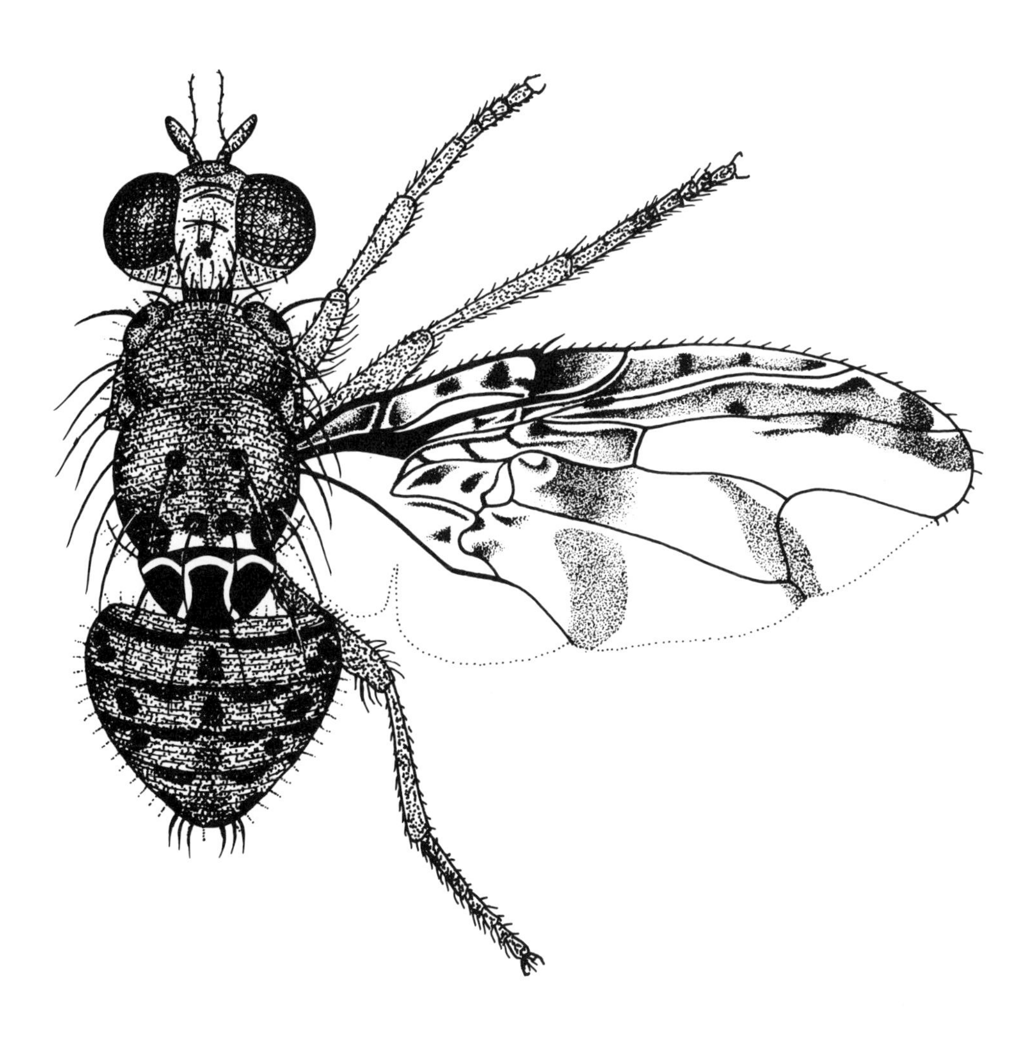

Fig. 216. *Ceratitis (Pardalaspis) punctata*, adult male.

Pelley, 1959; Munro, 1925; Silvestri, 1913). There is also a record from cocoa (*Theobroma cacao*), but Entwistle (1972) discussed its origin and it is likely to have been based on a misidentification.

Wild hosts
Recorded from *Myrianthus arboreus* Pal. (Cecropiaceae) (Steck *et al.*, 1986) but it probably has a wider range of hosts.

Adult identification
Recognised by its characteristic pattern of brown wing bands (similar to fig. 218), the three black areas in the apical half of the scutellum, and by the male having broad feathering on the mid tibia (fig. 101) and feathering on the mid femur.

Partial description of third instar larva
Larvae medium-sized, length 8.0mm; width 1.5mm. *Thoracic and abdominal segments*: Dorsal spinules on T1 and T2. Anterior spiracles with 11-13 tubules. Posterior spiracles with spiracular slits 3 times as long as broad. *Source*: Based on Silvestri (1913).

Distribution
Africa: Cameroun, Congo, Côte d'Ivoire, Ghana, Nigeria, Tanzania, Togo, Uganda, Zaire.

Attractant
Not known.

Other reference
PARASITOIDS, Steck *et al.* (1986).

Ceratitis (Pterandrus) colae Silvestri

(figs 102, 113)

Taxonomic notes
This species has also been known as *Pterandrus colae* (Silvestri).

Pest status and commercial hosts
A pest of abata cola (*Cola acuminata*); Owusu-Manu & Bonku (1987) gave an account of its population size and damage caused in Ghana.

Adult identification
Recognised by its characteristic pattern of brown wing bands (similar to fig. 218), the three black areas in the apical half of the scutellum, and by the male having narrow feathering on the mid tibia (fig. 102) and feathering on the mid femur.

Partial description of third instar larva
Larvae medium-sized, length 8.0-8.5mm; width 1.5-1.6mm. *Thoracic and abdominal segments*: Dorsal spinules on T1 and T2. Anterior spiracles with 15 tubules. Posterior spiracles with spiracular slits 3 times as long as broad. *Source*: Based on Silvestri (1913).

Distribution
Africa: Cameroun, Côte d'Ivoire, Ghana, Nigeria, Sierra Leone, Zaire.

Attractant
Not known.

Other references
BIOLOGY, Daramola & Ivbijaro (1975); PARASITOIDS, Steck *et al.* (1986).

***Ceratitis (Pterandrus) pedestris* (Bezzi)**

Strychnos Fruit Fly

(figs 106, 217; pl. 31)

Taxonomic notes
This species has also been known as *Pardalaspis pedestris* Bezzi.

Pest status and commercial hosts
Sometimes a minor pest of tomato (*Lycopersicon esculentum*) (Hancock, 1984; 1989).

Wild hosts
Normally associated with *Strychnos* spp. (Hancock, 1984).

Adult identification
Recognised by its characteristic pattern of brown wing bands (fig. 217), the three black areas in the apical half of the scutellum, and by the male having black and white patches on the inner side of the fore femur.

Description of third instar larva
Larva medium-sized, length 7.0mm; width 1.2mm. *Head*: Stomal sensory organ (pl. 31.b) with small, peg-like sensilla, surrounded by 5 unserrated preoral lobes; oral ridges (pl. 31.c, d) with 7 rows of short, bluntly rounded teeth; mouthhooks heavily sclerotised, without preapical teeth. *Thoracic and abdominal segments*: Spinules stout, sharply pointed [number of bands on T1 not counted as segment slightly retracted]; T2 and T3 with 3-5 rows of similar spinules. A1 with 1-2 rows of spinules dorsally but absent from A2-A8. Creeping welts with 9-13 rows of small, stout spinules, a mixture of anteriorly and posteriorly directed rows. *Anterior spiracles*: 15 tubules. *Posterior*

spiracles: Each spiracular slit with a heavily sclerotised rima and about 3 times as long as broad. Dorsal and ventral spiracular hair bundles of 9-13 hairs branched in apical half; lateral bundles of 4-6 hairs. *Anal area*: Similar to *C. capitata* (pl. 30.f). *Source*: Based on one specimen from South Africa (ex *Strychnos pungens* Soler.).

Distribution

Africa: Angola, South Africa, Zambia, Zimbabwe.
Indian Ocean: Madagascar.

Attractant

Males are attracted to trimedlure and terpinyl acetate, but not methyl eugenol.

Ceratitis (Pterandrus) rosa Karsch

Natal Fruit Fly
Natal Fly

(figs 103, 115, 218)

Taxonomic notes

This species has also been known as *Pterandrus flavotibialis* Hering and *P. rosa* (Karsch).

Pest status and commercial hosts

This is a major pest species which Hancock (1987) recorded from the following hosts in Zimbabwe: apple (*Malus domestica*), common guava (*Psidium guajava*), mango (*Mangifera indica*), papaya (*Carica papaya*), peach (*Prunus persica*), pear (*Pyrus communis*), quince (*Cydonia oblonga*), tomato (*Lycopersicon esculentum*) and wine grape (*Vitis vinifera*). In South Africa it damages apricot (*Prunus armeniaca*), avocado (*Persea americana*), common fig (*Ficus carica*), common guava, lychee (*Litchi chinensis*), mango, navel and Valencia oranges (*Citrus sinensis*), papaya, peach, pear, plum (*Prunus domestica*), quince and wine grape (Annecke & Moran, 1982). In Kenya *Ceratitis rosa* is almost as common a pest of arabica coffee (*Coffea arabica*) as *Ceratitis capitata*, and its status as a coffee pest in Uganda was reviewed by Greathead (1972). It is also recorded from Natal plum (*Carissa macrocarpa*) (Clausen *et al.*, 1965).

In the early 1950s *Ceratitis rosa* became established in Mauritius, and by the late 1950s it had become a serious pest there, where it was recorded from the following hosts by Orian & Moutia (1960): avocado, common guava, common jujube (*Ziziphus jujuba*), loquat (*Eriobotrya japonica*), mango, papaya, peach, sapodilla (*Manilkara zapota*), tropical almond (*Terminalia catappa*) and watery rose-apple (*Syzygium aqueum*).

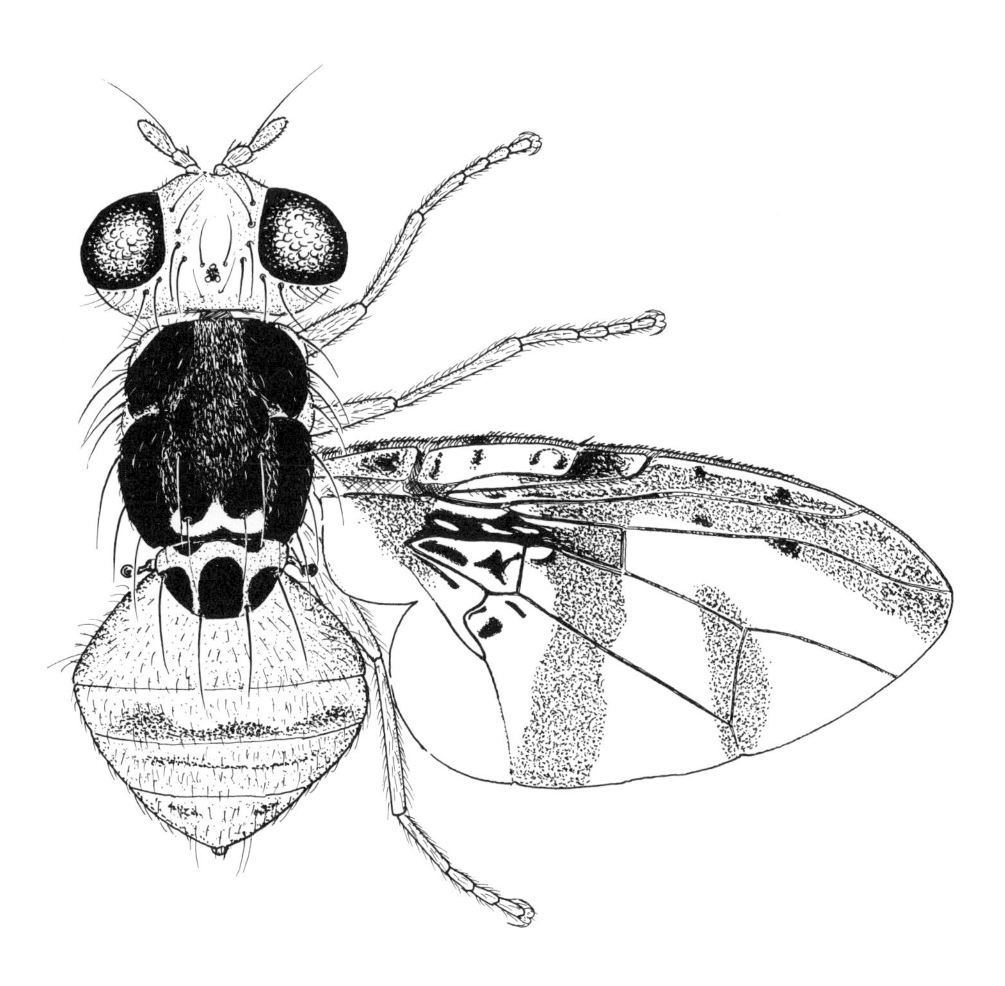

Fig. 217. *Ceratitis (Pterandrus) pedestris*, adult male.

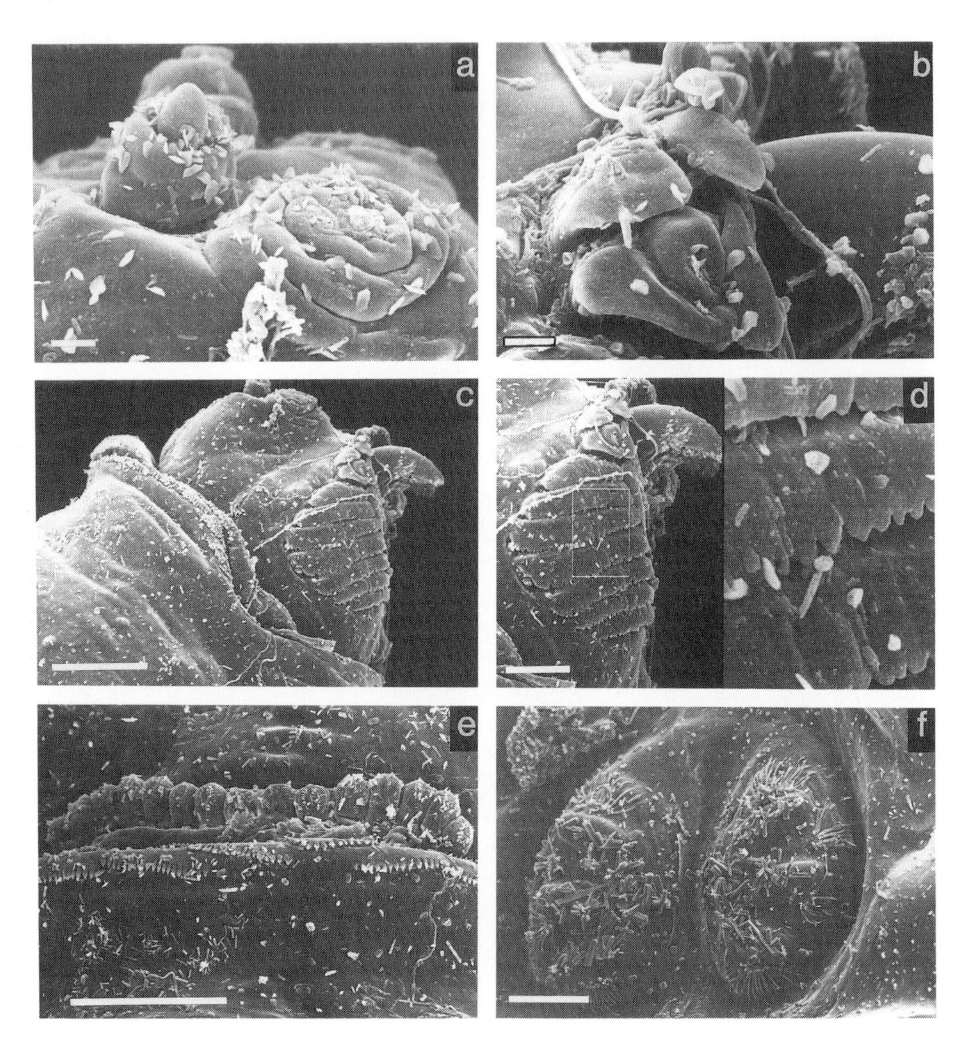

Plate 31. *Ceratitis (Pterandrus) pedestris*, SEMs of 3rd instar larva; a, antennal and maxillary sensory organs and preoral lobes, lateral view (scale=10μm); b, stomal sensory organ, lateral view (scale=10μm); c, head and T1, lateral view (scale=0.1mm); d, oral ridges, lateral view (scale=50μm); e, anterior spiracle (scale=0.1mm); f, posterior spiracles (scale=0.1mm).

In the late 1950s *C. rosa* became established in Réunion where it was studied by Étienne (1972); major hosts were apricot (*Prunus armeniaca*), arabica coffee, common guava, custard-apple (*Annona reticulata*), loquat, mango, peach and rose-apple (*S. jambos*); other hosts were apple, carambola (*Averrhoa carambola*), cocoa (*Theobroma cacao*), common fig, common jujube, Java plum (*S. cumini*), lychee, Malay-apple (*S. malaccense*), mangosteen (*Garcinia mangostana*), papaya, pear, plum, quince, sour orange (*Citrus aurantium*), strawberry guava (*Psidium littorale*), Surinam cherry (*Eugenia uniflora*), tabasco pepper (*Capsicum frutescens*), tangerine (*Citrus reticulata*) and tropical almond. Étienne (1972) also listed *Musa nana* and *Persea gratissima*; although it is possible that unusual banana and avocado species are grown in Réunion, it is also possible that these were errors for *M. x paradisiaca* and *P. americana*.

There are also old records, requiring confirmation, from the following hosts: blackberry (*Rubus fruticosa*), common persimmon (*Diospyros virginiana*), elephant-ear prickly pear (*Opuntia tuna*), grapefruit (*C. x paradisi*), imbe (*G. livingstonei*), Indian fig prickly pear (*O. ficus-indica*), kei apple (*Dovyalis caffra*), soursop (*Annona muricata*) and white star-apple (*Chrysophyllum albidum*) (Le Pelley, 1959; Munro, 1925; 1929b; 1935b; Orian & Moutia, 1960; Weems, 1966b).

Wild hosts

In South Africa, bug tree (*Solanum auriculatum*), which was introduced from South America, provides a host for the first spring generation before cultivated fruits become available (Ripley & Hepburn, 1930). The adventive population in Réunion is also recorded from bug tree (Étienne, 1972) but it does not attack that plant in Mauritius (Orian & Moutia, 1960). *Ceratitis rosa* has also been recorded from wild species of Cecropiaceae, Euphorbiaceae, Flacourtiaceae, Loganiaceae, Myrtaceae, Podocarpaceae, Rubiaceae, Rutaceae and Sapotaceae (Hancock, 1987; Le Pelley, 1959; Munro, 1925; 1935b; Weems, 1966b).

Adult identification

Recognised by its characteristic pattern of brown wing bands (fig. 218), the three black areas in the apical half of the scutellum, and by the male having feathering on the mid tibia (fig. 103), but no feathering on the mid femur.

Distribution

Africa: Angola, Ethiopia, Kenya, Malawi, Mali, Mozambique, Nigeria, Rwanda, South Africa, Swaziland, Tanzania, Uganda, Zaire, Zambia, Zimbabwe.
Indian Ocean: Adventive in Mauritius, Réunion.
A distribution map was produced by CIE (1986a).

Attractant

Males are attracted to trimedlure and terpinyl acetate, but not methyl eugenol.

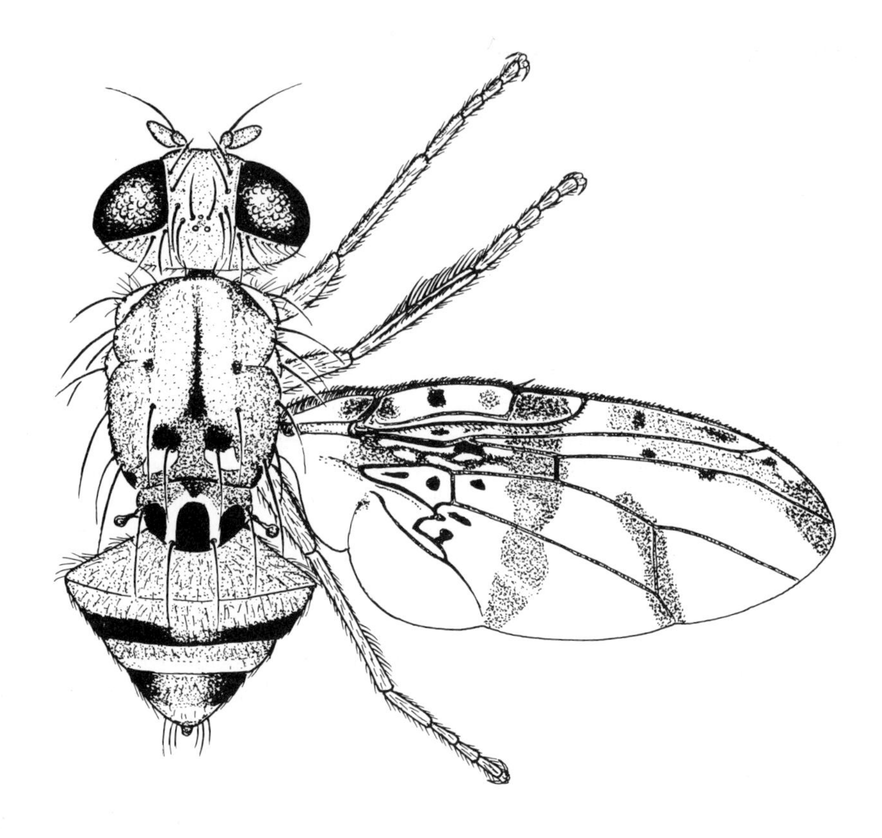

Fig. 218. *Ceratitis (Pterandrus) rosa*, adult male.

Other references

BEHAVIOUR, P.H. Smith (1989); BIBLIOGRAPHY, Zwölfer (1985); BIOCONTROL, Wharton (1989a); CUTICULAR HYDROCARBONS, Carlson & Yocom (1986); DATASHEET, Weems (1966b); IMPACT, Hancock (1989); LARVA, Orian & Moutia (1960) [inadequate illustrations only]; PARASITOIDS, Anon. (1979), Étienne (1973b), Greathead (1972), Roy (1977) and Wharton (1989a); PHEROMONES, Nation (1981); REARING, Barnes (1976; 1979), Monty (1973) and Tsitsipis (1989).

***Ceratitis (Pterandrus) rubivora* (Coquillett)**

Blackberry Fruit Fly

(figs 97, 104, 107)

Taxonomic notes
This species has also been known as *Pterandrus rubivorus* Coquillett.

Pest status and commercial hosts
In Zimbabwe this species has been recorded as a pest of youngberry (believed to be *Rubus flagellaris* x *R. loganobaccus*) (Jack, 1943) and it is also recorded from several other *Rubus* spp., including blackberry (*R. fruticosa*), loganberry (*R. loganobaccus*) and raspberry (*R. idaeus*) (Hancock, 1987; 1989). Le Pelley (1959) also listed star-apple (*Chrysophyllum cainito*) as a host, but he noted a pers. comm. from H.K. Munro, that records that were not from *Rubus* spp., may have been based on misidentifications.

Adult identification
Recognised by its characteristic pattern of brown wing bands (fig. 97), the three black areas in the apical half of the scutellum, and by the male having feathering on the mid tibia (fig. 104) and feathering on the mid femur.

Partial description of third instar larva
Larvae medium-sized, length 7.0mm; width 2.0mm. *Thoracic and abdominal segments*: Dorsal spinules on T1 and T2. Anterior spiracles with 10-11 lobes. Posterior spiracles with spiracular slits 3 times as long as broad. *Source*: Based on Silvestri (1913).

Distribution
Africa: Cameroun, Kenya, Malawi, South Africa, Uganda, Zimbabwe.

Attractant
Males are attracted to trimedlure and terpinyl acetate, but not methyl eugenol.

Other *Ceratitis* Species of Economic Interest

C. (Ceratalaspis) aliena (Bezzi), from Ethiopia and southern Africa, has been reared from black nightshade (*Solanum nigrum*) (Munro, 1935b). Identification: Hancock (1987).

Ceratitis (Ceratalaspis) discussa (Munro), from southern Africa, has been recorded from some *Citrus* and wild *Annona* spp. (Hancock, 1987; 1989; Munro, 1935a), and it could become a pest of cultivated *Annona* fruits (Hancock, 1989). However, the differences between *Ceratitis discussa* and *C. cosyra* described by Munro (1935a) and

Hancock (1987) appear to be trivial, and it is doubtful if this is a species distinct from *C. cosyra*.

Ceratitis (Ceratalaspis) giffardi Bezzi, from East and West Africa, is recorded from maroola plum (*Sclerocarya birrea*) (Paulian, 1953); it is also recorded from cherimoya (*A. cherimola*) (Le Pelley, 1959), but that requires confirmation. Identification: Munro (1935a).

Ceratitis (Ceratalaspis) morstatti Bezzi, from West Africa, has been reared from the nuts of "kola", presumably abata cola (*Cola acuminata*), in which it was destroying the pulp (Bezzi, 1912). Identification: Bezzi (1924a).

Ceratitis (Ceratalaspis) silvestrii Bezzi, from West Africa, has been recorded from shea butter (*Vitellaria paradoxa*) (Silvestri, 1913), but that requires confirmation. Identification: Bezzi (1924a).

Ceratitis (Ceratalaspis) turneri Munro, from East Africa, has been reared from *Solanum nodiflorum* (Munro, 1937). Identification: Munro (1937).

Ceratitis (Pterandrus) flexuosa (Walker), from West Africa and Uganda, has been recorded from mango (*Mangifera indica*) (Le Pelley, 1959), but that requires confirmation. There is also a record from rose-apple (*Syzygium jambos*), which requires confirmation (Hargreaves, 1924). Identification: Bezzi (1924a).

C. (P.) penicillata Bigot, from Côte l'Ivoire and Zaire, has been recorded from "kola", presumably abata cola (*Cola acuminata*) (Munro, 1938b; as *P. fumitactus* Munro). Identification: Bezzi (1924a).

GENUS *Dacus* Fabricius

Taxonomic notes

The recent work of Munro (1984) elevated this genus to family level (Dacidae) and divided it into fifty genera, many of which contained only one species and were defined by trivial characters. Drew (1989a) proposed that most Dacini should be divided between two genera, namely *Bactrocera* and *Dacus*, and that system is used here. The genus *Dacus* is divided into seven subgenera, including *D. (Callantra)*, which all authors prior to Drew (1989a) regarded as a distinct genus.

Anyone wishing to identify African *Dacus* spp. using the work of Munro (1984) should note that we regard species belonging to the following genera and subgenera as being members of subgenus *Dacus (Leptoxyda)* Macquart: *Aoptodacus*, *Athlodacus*, *Guyodacus*, *Janseidacus*, *Nebrodacus*, *Oligodacus*, *Pionodacus*, *Saccodacus*, *Timiodacus* and *Xylenodacus*, all described by Munro (1984), plus *Lophodacus* Collart and *Psilodacus* Collart. Species belonging to the genera *Andriadacus*, *Anomoiodacus* and *Coccinodacus*, all described by Munro (1984), are regarded as being members of subgenus *D. (Metidacus)* Munro. Munro (1984) genera which should be regarded as synonyms of the subgenera *D. (Dacus)* and *D. (Didacus)* are listed in the sections covering those subgenera.

Distribution and biology

A genus of about 240 species most of which show a strong preference for attacking the pods of Asclepiadaceae and Apocynaceae, or the fruits and flowers of Cucurbitaceae. Most species are found in Africa, with only a few species in other areas of the Old World. This contrasts with *Bactrocera* spp., most of which are found in an area from tropical Asia to Australia and the South Pacific, and relatively few of them attack Cucurbitaceae, and none are known to attack Asclepiadaceae. The biology of *Dacus* spp. is probably similar to that of *Bactrocera* spp., but it has not been as well studied; see Fletcher (1987; 1989b; 1989c) for reviews of dacine biology. *Bactrocera* spp. normally mate on or near their host fruit. However, *D. frontalis* mates on rest plants (Steffens, 1983) and the host plant is visited only by the females at oviposition time. Known rest plants included maize (*Zea mays* L.), pigeon pea (*Cajanus cajan*) and sunflower (*Helianthus annuus*).

Diagnosis of third instar larva

Antennal sensory organ with a broad, stout basal segment and a smaller, rounded, cone-shaped distal segment (pl. 32.a). Maxillary sensory organ, broad, flat, with 2 distinct groups of sensilla surrounded by several discontinuous cuticular folds (pl. 32.a). Stomal sensory organ small, rounded, completely surrounded by preoral lobes; those closest to mouth opening resembling small oral ridges (pl. 32.b); oral ridges with 12-21 very long rows of long, deeply serrated, parallel-sided, bluntly rounded teeth (pl. 32.a); numerous long, deeply serrated accessory plates (pl. 32.c). Mouthhooks each with a strong preapical tooth (fig. 159). Anterior spiracles with a long row of 14-25 short, stout tubules. Posterior spiracles 2-4 times as long as broad; with short, stout, branched spiracular hairs. Anal lobes very large, surrounded by 5-7 discontinuous rows of short, stout, sharply pointed spinules.

Attractants

Male lures have not been widely used in Africa and so the response is only known for a few species which were listed by Hancock (1985b). Unlike *Bactrocera* spp., very few *Dacus* spp. are known to be attracted to methyl eugenol, which in Africa attracts *Ceratitis (Pardalaspis)* spp. However, the males of many species are attracted to cue lure and *D. vertebratus* is attracted to vert lure.

Literature

The adults of most species can be identified using the following key works: Africa (Munro, 1984); Australia (Drew, 1982a; 1989a); Indonesia (Hardy, 1982b; 1983a); Japan (Ito, 1983-5; Shiraki, 1968); New Guinea area (Drew, 1989a); Philippines (Hardy, 1974); and Thailand (Hardy, 1973). The larvae of three important pest species of *D. (Didacus)* were distinguished by Malan & Giliomee (1969).

Subgenus *CALLANTRA* WALKER

(fig. 24)

Taxonomic notes

Some species have formerly been placed in the genus *Mellesis* Bezzi, a synonym of *Callantra*, and species of *Paracallantra* Hendel should also be regarded as belonging to *D. (Callantra)*.

Distribution and host relationships

A subgenus of about 40 species found in Oriental Asia, Australia, Indonesia and the New Guinea area. Species of known biology attack the pods of Asclepiadaceae and the fruits of Cucurbitaceae.

Dacus (Callantra) axanus (Hering)

(figs 159, 219; pl. 32)

Taxonomic notes

This species has also been known as *Callantra auricoma* May and *C. axana* Hering. References to *C. smieroides* Walker by Drew (1972a; 1972b; 1973) were based on misidentifications of *Dacus axanus* rather than the true *D. (C.) smieroides* (Walker) (Drew, 1989a).

Commercial hosts

A potential pest of cucurbits which has been recorded from luffa (*Luffa aegyptiaca*) and snakegourd (*Trichosanthes cucumerina*) (Drew, 1989a; Smith *et al.*, 1988).

Adult identification

A predominantly orange to orange-brown species; with long antennae (similar to fig. 24), facial spots, anterior supra-alar setae, 2 scutellar setae, abdominal tergites covered with golden pubescence, and wing with a very broad costal band from base to apex (fig. 219); male with a pecten. Scutum without yellow stripes and prescutellar acrostichal setae.

Description of third instar larva

Larvae medium to large, length 9.0-10.5mm; width 1.5-2.0mm. *Head*: Stomal sensory organ (pl. 32.a) small, rounded, with small sensilla and completely surrounded by several preoral lobes; those lobes closest to mouth bearing small serrations similar to oral ridges; oral ridges (pl. 32.a, c) with 17-21 very long rows of long, deeply serrated, parallel-sided, bluntly rounded teeth; numerous long, deeply serrated accessory plates interlocking with outer edges of oral ridges (pl. 32.c); mouthhooks heavily sclerotised, each with a large preapical tooth (pl. 32.b; fig. 159). *Thoracic and abdominal segments*: Anterior margins of T1-T3 completely encircled by a broad band of small, stout, sharply pointed spinules in 8-13 discontinuous rows. A1-A8 without spinules dorsally, but with spinules forming large creeping welts ventrally; with 9-14 rows of small spinules, some rows anteriorly directed. A8 with intermediate areas large, with obvious tubercles and sensilla. *Anterior spiracles* (pl. 32.d): 19-21 tubules. *Posterior spiracles* (pl. 32.e): Spiracular slits stout, broad, about 2.5 times as long as broad, with very thick rimae. Spiracular hairs short (less than length of a spiracular slit), thin, branched apically; dorsal and ventral spiracular hairs in bundles of 12-16 hairs, lateral bundles of 5-9 hairs. *Anal area* (pl. 32.f): Lobes very large, protuberant and surrounded by 5-7 discontinuous rows of short, stout, sharply pointed spinules. *Source*: Based on specimens from Queensland, Australia.

Distribution

Australia: North-eastern and north-western areas.
New Guinea area: Indonesia (Moluccas, Lesser Sunda Islands), Kei, Lihir, New Britain, New Ireland, Papua New Guinea.

Attractant

Males attracted to cue lure.

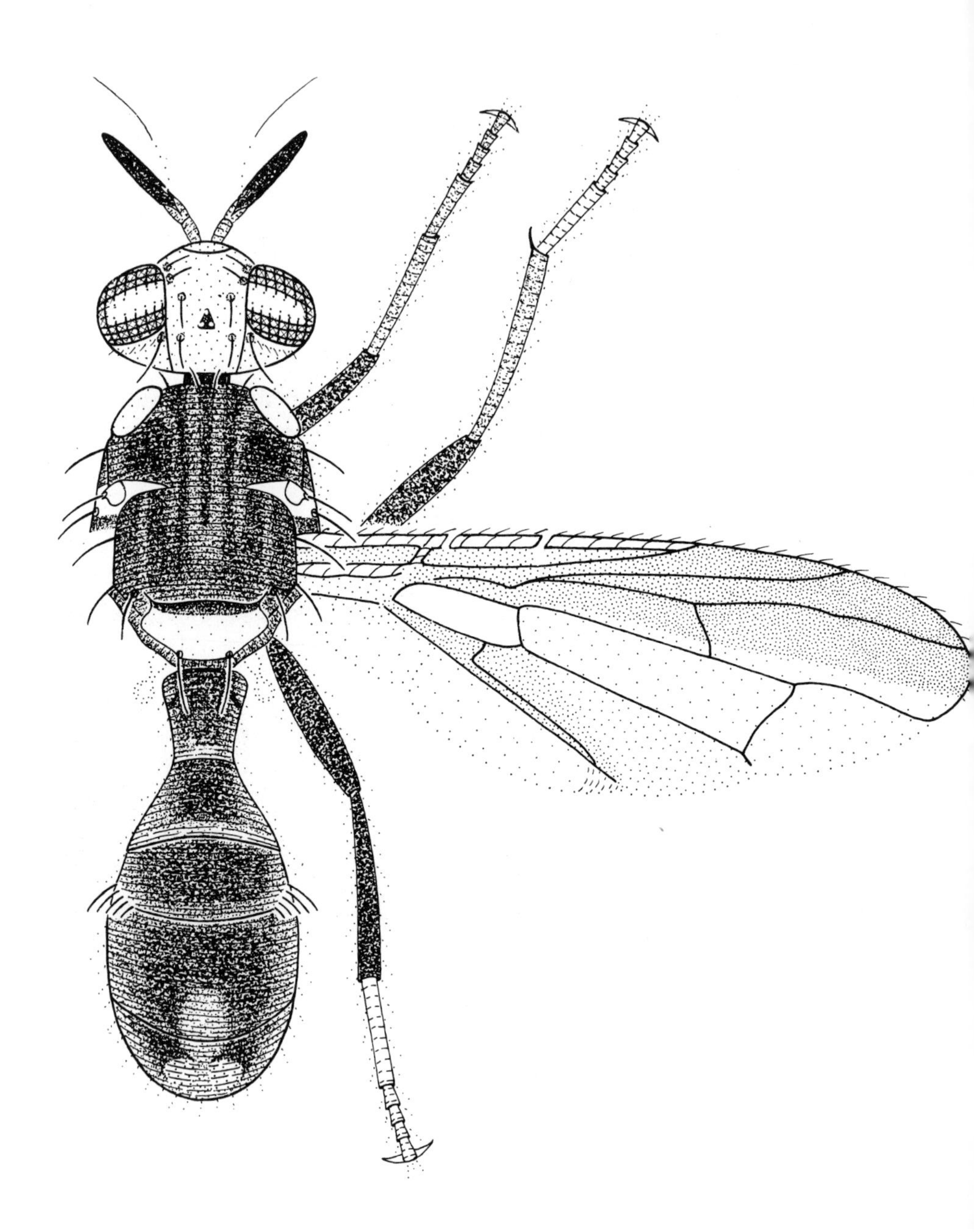

Fig. 219. *Dacus (Callantra) axanus*, adult male.

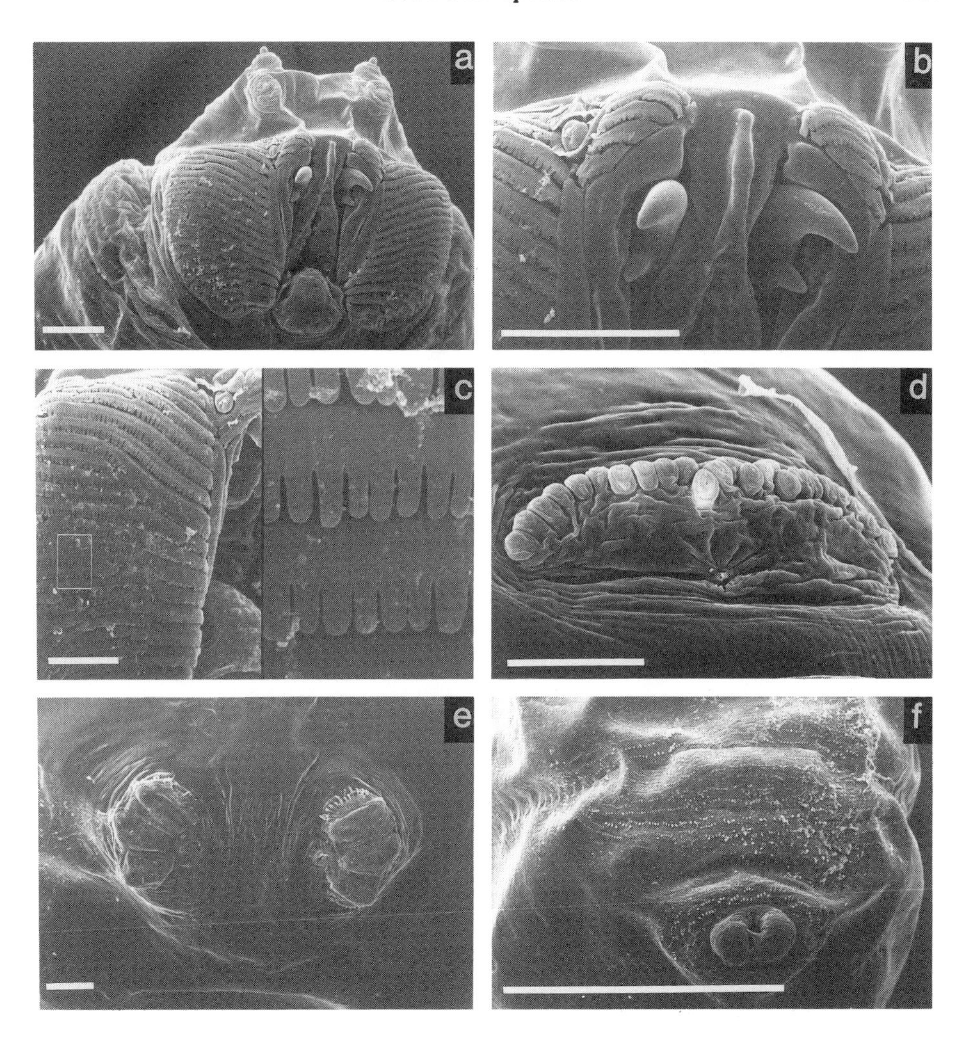

Plate 32. *Dacus (Callantra) axanus*, SEMs of 3rd instar larva; a, head, ventral view (scale=0.1mm); b, mouthhooks, showing preapical teeth, and preoral lobes, anterior view (scale=0.1mm); c, oral ridges, lateral view (scale=50μm); d, anterior spiracle (scale=0.1mm); e, posterior spiracles (scale=0.1mm); f, A8, creeping welt and anal lobes (scale=1.0mm).

Dacus (Callantra) smieroides (Walker)

(figs 24, 94, 220)

Taxonomic notes

This species has also been known as *Callantra smieroides* Walker. References to *C. smieroides* by Drew (1972a; 1972b; 1973) were based on misidentifications of *Dacus (C.) axanus* (Drew, 1989a).

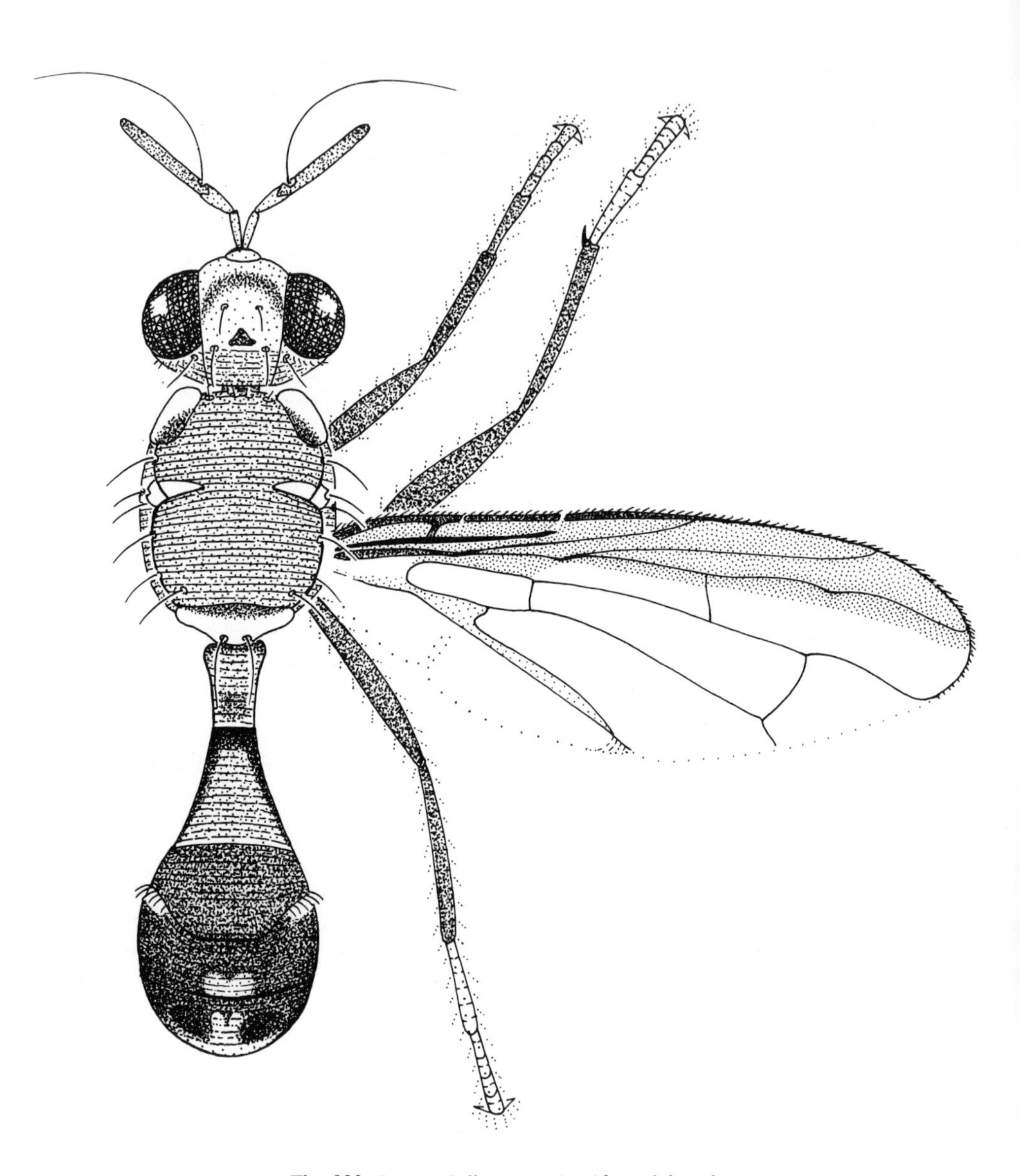

Fig. 220. *Dacus (Callantra) smieroides*, adult male.

Commercial hosts

A potential pest of cucurbits which has been recorded from angled luffa (*Luffa acutangula*) (Drew, 1989a).

Adult identification

A predominantly orange to dark orange-brown species; with long antennae (fig. 24), facial spots, anterior supra-alar setae, 2 scutellar setae, abdominal tergites with golden pubescence, and wing with a very broad costal band from base to apex (fig. 220); male with a pecten. Scutum without yellow stripes and prescutellar acrostichal setae.

Distribution

Oriental Asia: Brunei, Indonesia (Kalimantan, Makassar, Sulawesi), Malaysia (Sabah, Sarawak).

Attractant

Males attracted to cue lure.

Dacus (Callantra) solomonensis Malloch

(figs 67, 221)

Taxonomic notes

This species has also been known as *Callantra solomonensis* (Malloch).

Pest status and commercial hosts

A pest of cucumber (*Cucumis sativus*) and pumpkin (*Cucurbita pepo*) (Drew, 1989a; R.A.I. Drew, pers. comm., 1990).

Adult identification

A predominantly dark red-brown to orange-brown species; with long antennae (similar to fig. 24), facial spots, anterior supra-alar setae, 2 scutellar setae, abdominal tergite 5 hump-shaped when viewed in profile, and wing with a very broad costal band from base to apex (fig. 221); male with a pecten. Scutum without yellow stripes and prescutellar acrostichal setae.

Distribution

New Guinea area: Bougainville Island, Solomon Islands.

Attractant

Males attracted to cue lure.

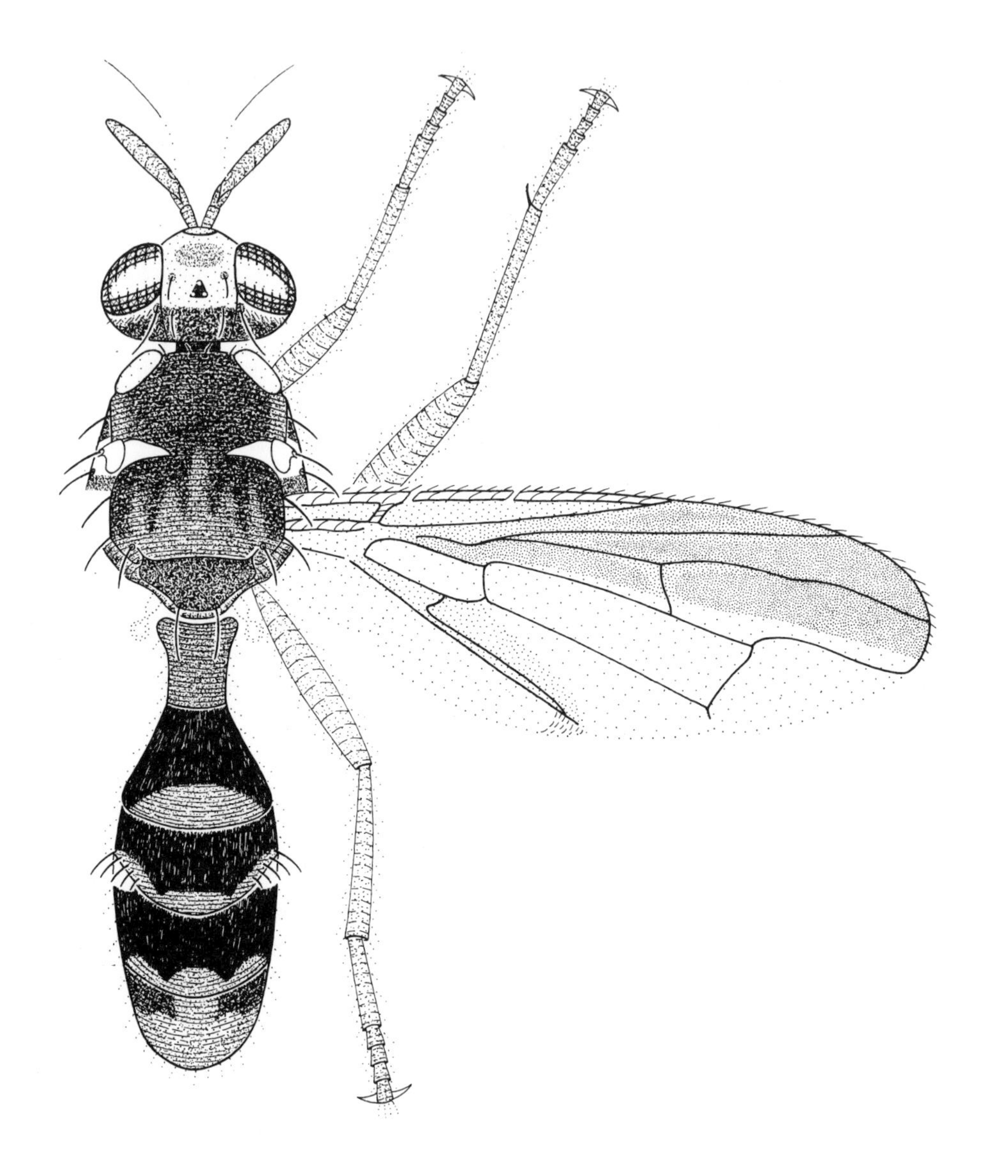

Fig. 221. *Dacus (Callantra) solomonensis*, adult male.

Subgenus *DACUS* FABRICIUS

Taxonomic notes

Some species were formerly included in the subgenera *Neodacus* Perkins and *Tridacus* Bezzi, which are synonyms of *Dacus*. Species belonging to the genera *Ancylodacus*, *Desmodacus*, *Dorylodacus* and *Rhamphodacus*, described by Munro (1984), should also be included in subgenus *Dacus (Dacus)*.

Distribution and host relationships

A subgenus of about 60 species, most of which are found in Africa, but with a few representatives in the area from tropical Asia to Australia. Most species are either exclusively associated with the pods of Asclepiadaceae, or have a strong preference for the fruits of Cucurbitaceae, and a few African species attack the fruits of Passifloraceae.

Dacus (Dacus) bivittatus **(Bigot)**

Pumpkin Fly
Greater Pumpkin Fly
Two Spotted Pumpkin Fly

(figs 88, 222)

Taxonomic notes

This species has also been known as *Dacus bipartitus* Graham, *D. cucumarius* Sack, *D. pectoralis* Walker, *D. rubiginosus* Hendel, *Leptoxys bivittatus* Bigot and *Tridacus pectoralis* (Walker).

Pest status and commercial hosts

A pest of cucurbit crops, particularly in moist forest areas (Hancock, 1989), which has been recorded from the following hosts in Nigeria (Matanmi, 1975): canterloupe (*Cucumis melo*), cucumber (*C. sativus*), squash (*Cucurbita maxima*), watermelon (*Citrullus lanatus*) and white egusi (*Cucumeropsis mannii*); cucumber was only attacked when over-ripe. Other recorded hosts are African horned cucumber (*Cucumis metuliferus*), bitter gourd (*Momordica charantia*), chayote (*Sechium edule*), luffa (*Luffa aegyptiaca*), oysternut (*Telfairea pedata*), pumpkin (*Cucurbita pepo*) and white-flowered gourd (*Lagenaria siceraria*); (Hancock, 1989; Munro, 1984; Steck *et al.*, 1986). There are some other records from non-cucurbit hosts, namely coffee (*Coffea* sp.), giant granadilla (*Passiflora quadrangularis*), papaya (*Carica papaya*) and tomato (*Lycopersicon esculentum*) (Annecke & Moran, 1982; Munro, 1984; Schmidt, 1967), and these should be regarded as doubtful. The coffee record (Schmidt, 1967) refers to trapped adults and larvae, but the larvae were not reared to confirm that they were this species. Matanmi (1975) examined papaya and tomato in an area where cucurbits were heavily infested and found no attack, although Munro (1926) reported heavy attack on tomato.

Wild hosts

Many species of Cucurbitaceae, e.g. wild *Momordica* and *Peponium* spp. (Matanmi, 1975; Munro, 1984).

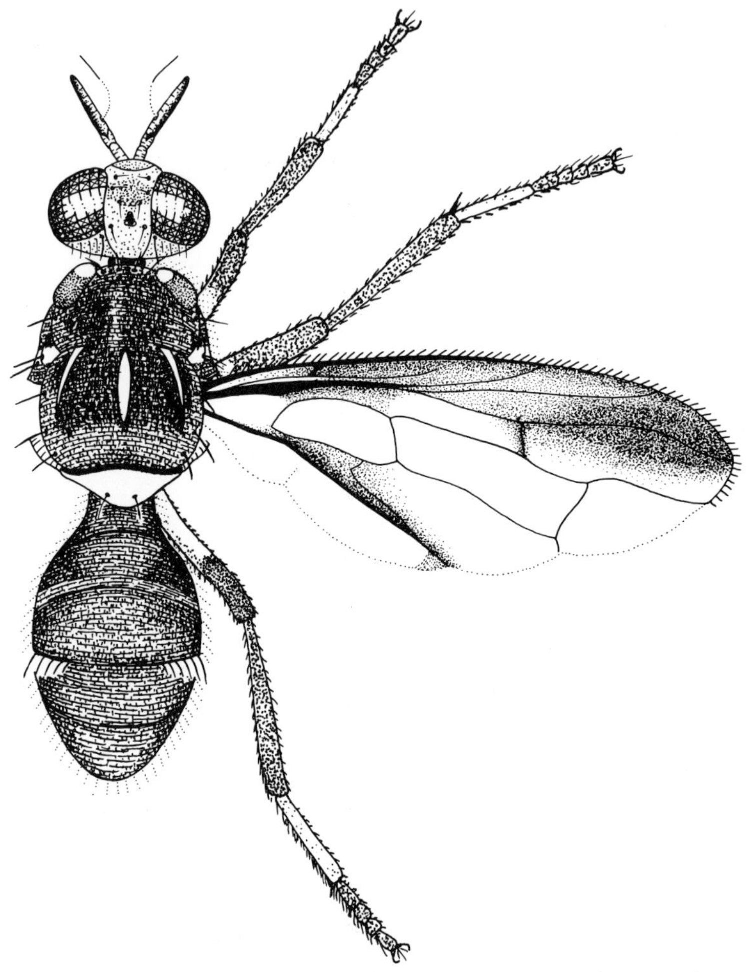

Fig. 222. *Dacus (D.) bivittatus*, adult male.

Adult identification

A predominantly dark orange to red-brown species; scutum with lateral and medial yellow stripes; with facial spots, anterior supra-alar setae, 2 scutellar setae, and wing with a very broad costal band from base to apex (fig. 222); male with a pecten. Without prescutellar acrostichal setae.

Distribution

Africa: Angola, Cameroun, Kenya, Malawi, Mozambique, Nigeria, Sierra Leone, South Africa, Tanzania, Uganda, Zaire, Zimbabwe; possibly Senegal; probably all other countries south of the Sahal.

Attractant
Males attracted to cue lure.

Other references
ECOLOGY, Fletcher (1989c); PARASITOIDS, Narayanan & Chawla (1962) and Steck *et al.* (1986).

Dacus (Dacus) demmerezi (Bezzi)

(figs 90, 223)

Taxonomic notes
This species has also been known as *Tridacus demmerezi* Bezzi and it is sometimes misspelt as *Dacus d'emmerezi* or *D. emmerezi*.

Pest status and commercial hosts
A pest of cucurbits which has been recorded from bitter gourd (*Momordica charantia*), chayote (*Sechium edule*), cucumber (*Cucumis sativus*), luffa (*Luffa acutangula*), melon (*C. melo*), pumpkin (*Cucurbita maxima* and *C. pepo*), snakegourd (*Trichosanthes cucumerina*), watermelon (*Citrullus lanatus*) and white-flowered gourd (*Lagenaria siceraria*) (Étienne, 1972; Munro, 1984; Orian & Moutia, 1960).

Wild host
Recorded from *Melothria* sp. (Cucurbitaceae) (Munro, 1984) but it is probably associated with a wide range of wild cucurbits.

Adult identification
A predominantly orange-brown to dark orange-brown species; scutum sometimes with a narrow medial yellow stripe; with facial spots, anterior supra-alar setae, 2 scutellar setae, and wing with crossvein r-m covered by an infuscate mark (fig. 223); male with a pecten. Without prescutellar acrostichal setae.

Distribution
Indian Ocean: Madagascar, Mauritius, Réunion.

Attractant
Males attracted to cue lure.

Other reference
BIOCONTROL, Anon. (1979); PARASITOIDS, Anon. (1979), Étienne (1973b) and Narayanan & Chawla (1962); TRAPPING, Ramsamy *et al.* (1987).

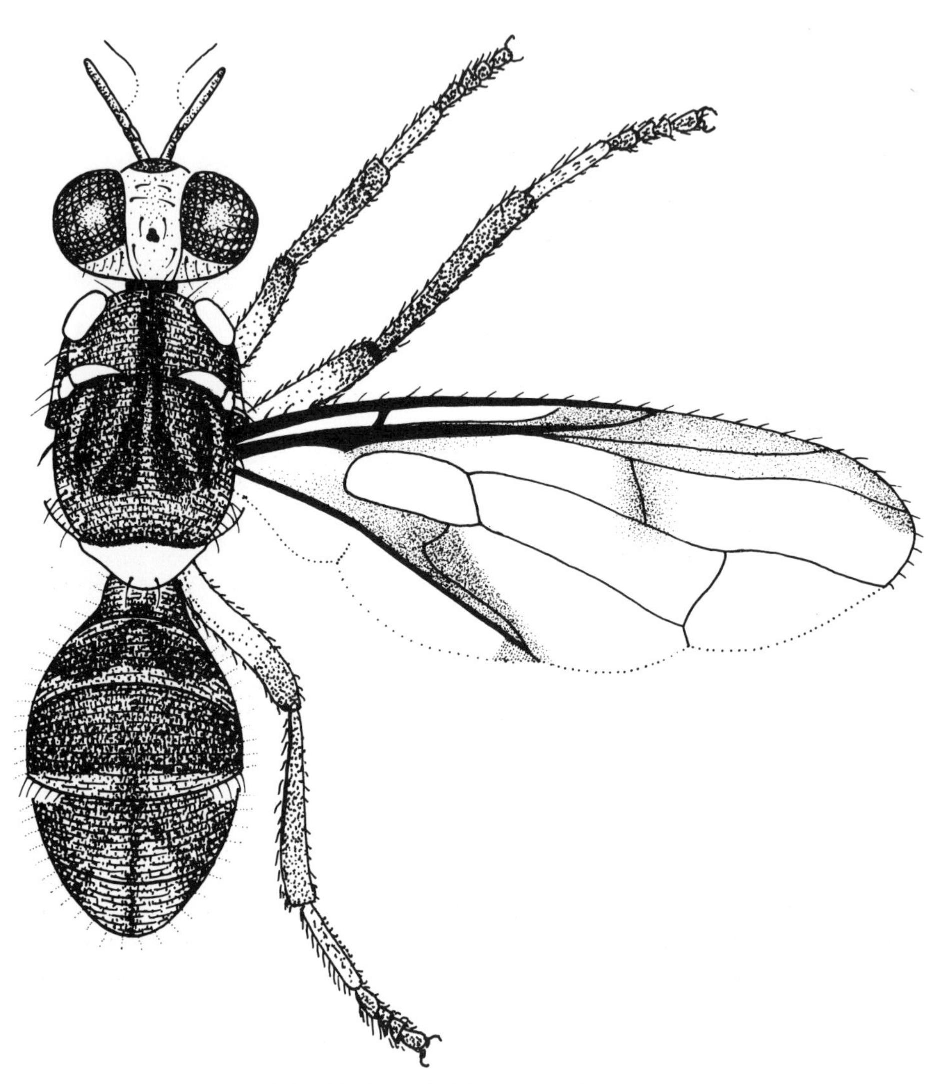

Fig. 223. *Dacus (D.) demmerezi*, adult male.

Dacus (Dacus) punctatifrons Karsch

(figs 93, 224)

Taxonomic notes

This species has also been known as *Dacus furcatus* Hendel. Munro (1984) also regarded *D. zimmermanni* Hendel as a synonym, but syntypes (NHM collection) examined by us are clearly *D. telfaireae*.

Commercial hosts

This is a rare species which sometimes attacks cultivated cucurbit crops (Hancock, 1989; Munro, 1964a; 1984), although only pumpkin (*Cucurbita pepo*) was listed by Munro (1984). It is also recorded from bitter gourd (*Momordica charantia*) and cucumber (*Cucumis sativus*) (NHM, collection data) and there is an old record for chayote (*Sechium edule*) (Le Pelley, 1959). It has been reared from wild waterlemon (*Passiflora foetida*) (Passifloraceae) (NHM, collection data), but that may have been an aberrant host association. In addition, it has been recorded from *Gloriosa* sp. (Liliaceae) (Munro, 1984), but that record may have been based on casual observation rather than rearing.

Wild hosts

Munro (1984) listed some *Bryanopsis*, *Melothria*, *Momordica* and *Peponium* spp. (all Cucurbitaceae).

Adult identification

A predominantly orange-brown species; scutum with a medial black stripe or predominantly black; scutum with lateral and medial yellow stripes; with facial spots, anterior supra-alar setae, both anatergite and katatergite largely covered by a single yellow stripe, 2 scutellar setae, and wing with crossvein r-m sometimes covered by an infuscate mark (fig. 224); male with a pecten. Without prescutellar acrostichal setae.

Distribution

Africa: Angola, Cameroun, Ghana, Nigeria, Kenya, Sierra Leone, South Africa, Tanzania, Uganda, Zambia, Zimbabwe; probably other countries south of the Sahal.
Indian Ocean: Adventive in Mauritius (Roy, 1977).
Middle East: Yemen People's Democratic Republic (probably adventive).

Attractant

Males attracted to cue lure.

Other reference

PARASITOIDS, Narayanan & Chawla (1962).

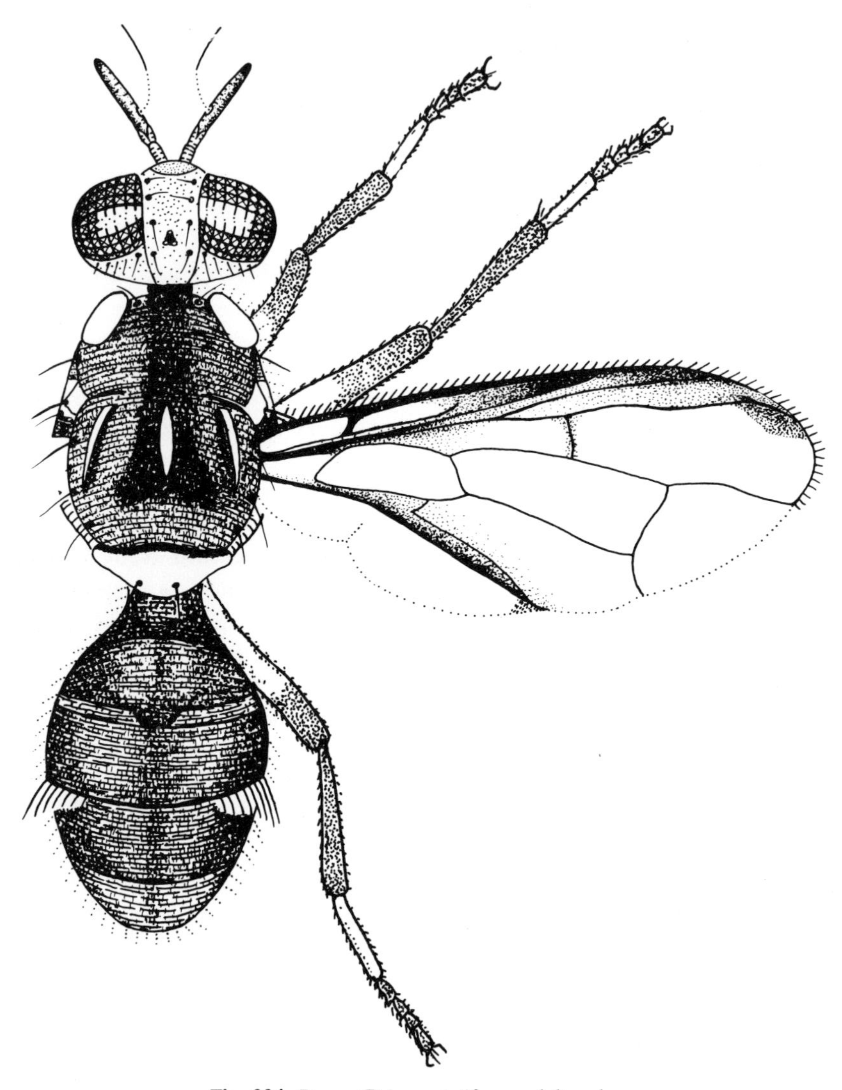

Fig. 224. *Dacus (D.) punctatifrons*, adult male.

Dacus (Dacus) telfaireae (Bezzi)

Oyster Nut Fly

(figs 95, 225)

Taxonomic notes
This species has also been known as *Tridacus telfaireae* Bezzi. Some records of *Dacus zimmermanni* Hendel also refer to this species.

Pest status and commercial hosts
A pest of oysternut (*Telfairea pedata*) (Hancock, 1989). There is also a record from a gourd (*Lagenaria* sp.) (Munro, 1984), but that was probably based on a misidentification (D.L. Hancock, pers. comm., 1990).

Adult identification
A predominantly black species; scutum sometimes with very narrow lateral and medial yellow stripes; face with diffuse poorly defined spots (none at all in teneral specimens); with anterior supra-alar setae, 2 scutellar setae, a yellow mark across the centre of the katatergite, anatergite black, and wing with crossvein r-m covered by a narrow infuscate mark (fig. 225); male with a pecten. Without prescutellar acrostichal setae.

Distribution
Africa: Kenya, Malawi, Tanzania, Zimbabwe; a record from Zaire (Munro, 1984) was probably based on a misidentification (D.L. Hancock, pers. comm., 1990).

Attractant
Males attracted to cue lure.

Subgenus *DIDACUS* COLLART

Taxonomic notes
Species belonging to the genera *Abebaiodacus*, *Acanodacus*, *Ambitidacus*, *Baucidacus*, *Blaxodacus*, *Dixoodacus*, *Ectopodacus*, *Fusodacus*, *Karphodacus*, *Lactodacus*, *Mictodacus* and *Myrmecodacus*, described by Munro (1984), should also be included in subgenus *Dacus (Didacus)*.

Distribution and host relationships
A subgenus of about 70 species found in Africa, with *Dacus ciliatus* found in Africa, the Indian subcontinent and the Indian Ocean Islands. Most species are associated with the fruits of Cucurbitaceae, or with the pods of Asclepiadaceae and Apocynaceae.

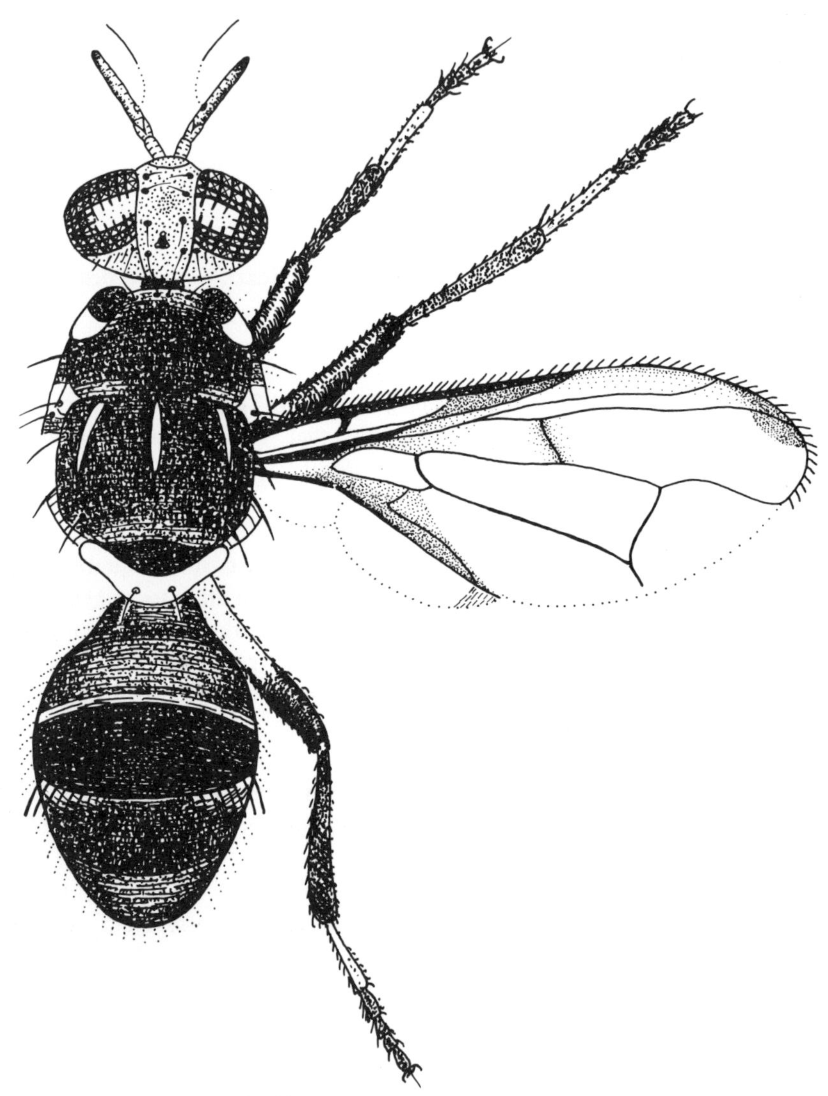

Fig. 225. *Dacus (D.) telfaireae*, adult male.

Dacus (Didacus) ciliatus Loew

Ethiopian Fruit Fly
Lesser Pumpkin Fly
Cucurbit Fly

(figs 89, 226)

Taxonomic notes
This species has also been known as *Dacus appoxanthus* var. *decolor* Bezzi, *D. brevistylus* Bezzi, *D. insistens* Curran, *D. sigmoides* Coquillett, *Didacus ciliatus* (Loew), *Leptoxyda ciliata* (Loew) and *Tridacus mallyi* Munro.

Pest status and commercial hosts
A pest of cucurbit crops (Hancock, 1989) which was recorded from the following hosts in Nigeria by Matanmi (1975): cantaloupe (*Cucumis melo*), cucumber (*C. sativus*), squash (*Cucurbita maxima*) and watermelon (*Citrullus lanatus*); cucumber was only attacked when over-ripe. It has also been recorded from African horned cucumber (*Cucumis metuliferus*), colocynth (*Citrullus colocynthis*) and balsam apple (*Momordica balsamina*) (Munro, 1984; Shaheen *et al.*, 1973).

There are adventive populations of this African species in Mauritius, Réunion and the Indian subcontinent. In Mauritius it has been recorded from cucumber, melon (*Cucumis melo*), pumpkin (*Cucurbita maxima*), snakegourd (*Trichosanthes cucumerina*), squash (*C. pepo*), watermelon and white-flowered gourd (*Lagenaria siceraria*) (Orian & Moutia, 1960). In Réunion it has been recorded from angled luffa (*Luffa acutangula*), bitter gourd (*Momordica charantia*), chayote (*Sechium edule*), cucumber, melon, squash (*C. pepo*) and white-flowered gourd (Étienne, 1972). In Pakistan it attacks bitter gourd, cucumber, luffa (*L. aegyptiaca*), melon and watermelon (Qureshi *et al.*, 1974a). Kapoor (1970) listed the following hosts for India: bitter gourd, cucumber, angled luffa, luffa, melon, pumpkin (*C. maxima*), watermelon and white-flowered gourd.

Munro (1984) also listed some non-cucurbit hosts, namely, beans (*Phaseolus* spp.), cotton (*Gossypium* sp.), okra (*Abelmoschus esculentus*) and tomato (*Lycopersicon esculentum*), but those are unusual hosts (Hancock, 1981), and Kapoor (1970) listed a pepper (*Capsicum* sp.). However, Matanmi (1975) examined tomato in an area where cucurbits were heavily infested and found no attack and Orian & Moutia (1960) noted that attack was very rare. Papers by Clausen *et al.* (1965), Kandybina (1977) and Munro (1926; 1932) gave additional host records, but many of those were likely to have been based on misidentifications of *Dacus frontalis* and *D. vertebratus*. *D. ciliatus* has also been known to develop in the stem gall of another insect, namely *Lasioptera toombii* (Grover) (Diptera, Cecidomyiidae) on ivy gourd (*Coccinia grandis*) (Bhatia & Mahto, 1968). Hancock (1989) noted that *D. ciliatus* has a preference for fruits measuring less than 50mm diameter. A report that in Egypt this species (as *D. brevistylus*) was originally only associated with Sodom apple (*Calotropis procera*) (Kapoor, 1989) was based on a misinterpretation of data presented by Zoheiry (1950).

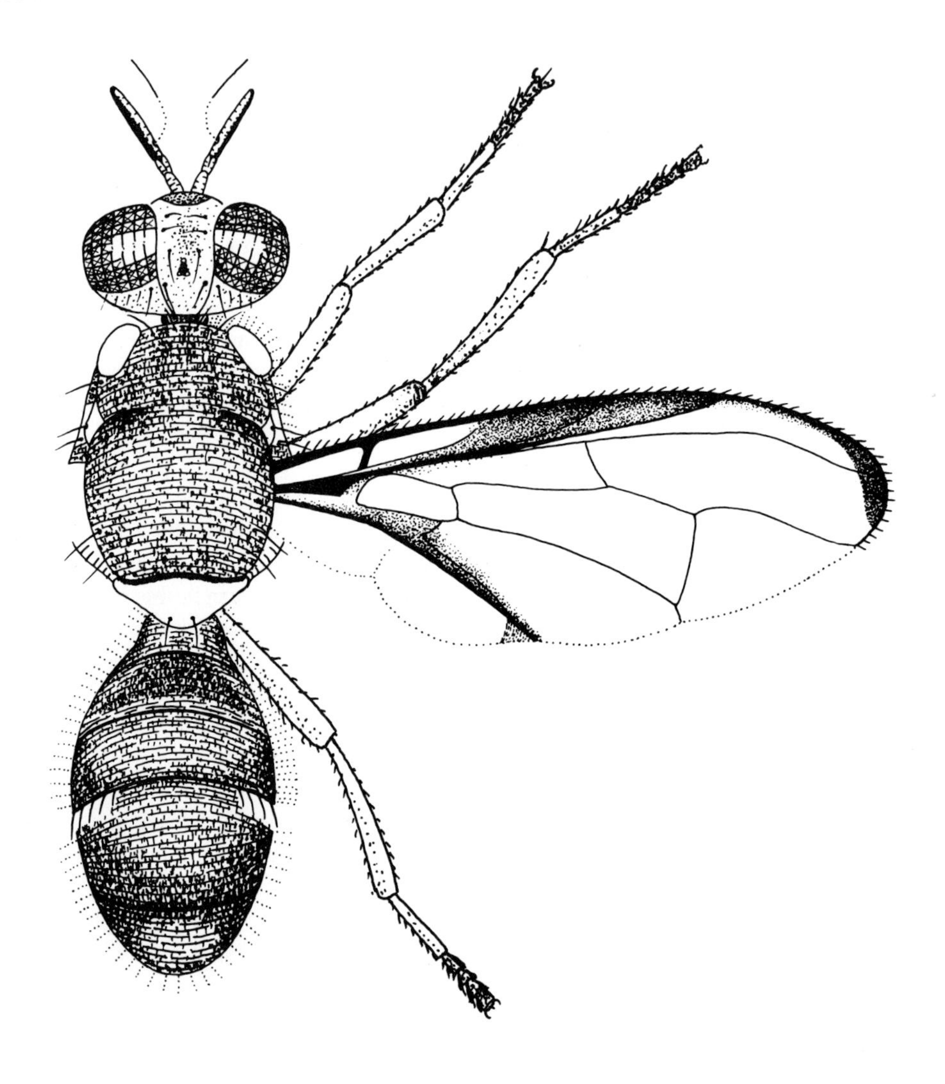

Fig. 226. *Dacus (Didacus) ciliatus*, adult male.

Wild hosts

Many species of Cucurbitaceae, e.g. *Momordica* spp. (Matanmi, 1975; Munro, 1984). In Egypt, wild growing colocynth is believed to be the main reservoir host from which crops become infested (Shaheen *et al.*, 1973).

Adult identification

A predominantly orange species; with facial spots, 2 scutellar setae, a yellow spot covering most of the katatergite, anatergite orange, mid femur yellow or orange-yellow and wing with a costal band that is expanded apically to form an apical spot (fig. 226);

male with a pecten. Scutum without yellow stripes, anterior supra-alar setae and prescutellar acrostichal setae.

Partial description of third instar larva

Larvae medium to large, length 9.0-10.5mm; width 1.5-2.0mm. *Head*: Oral ridges in 12-13 rows, some branched; mouthhooks with a stout preapical tooth. *Thoracic and abdominal segments*: Rows of spinules encircling the anterior portions of T1, T2, T3 and A1; A8 with 2 pairs of small tubercles and sensilla, intermediate lobes small. Anterior spiracles with 14-16 stout tubules. Posterior spiracles with spiracular slits 3.5-4.0 times as long as broad; spiracular hairs short (less than half the length of a spiracular slit), some branched, with 4-19 hairs per bundle. Anal lobes small, surrounded by small spinules. *Source*: Based on Malan & Giliomee (1969) and Kandybina (1977).

Distribution

Africa: Angola, Botswana, Cameroun, Chad, Dahomey, Egypt, Ethiopia, Ghana, Guinea, Kenya, Lesotho, Malawi, Mozambique, Nigeria, Senegal, Sierra Leone, Somalia, South Africa, South West Africa, Sudan, Tanzania, Uganda, Zaire, Zambia, Zimbabwe.

Atlantic Islands: St Helena (possibly a quarantine record only).

Indian Ocean: Madagascar, Mauritius, Réunion.

Middle East: Saudi Arabia, Yemen People's Democratic Republic, Yemen Arab Republic; probably in Iran (see note about *D. persicus*; p. 338).

Oriental Asia: Bangladesh, India, Pakistan, Sri Lanka.

The distribution was mapped by CIE (1974), but that included a Cape Verde Island record which was based on a misidentification of *D. frontalis*.

Attractant

Not attracted to cue lure, methyl eugenol or vert lure.

Other references

BEHAVIOUR, P.H. Smith (1989); BIOLOGY, Fletcher (1989b), Hancock (1989) and Narayanan (1953); COMPETITION, Qureshi *et al.* (1987); EGYPT, Nahal *et al.* (1971); NIGERIA, Matanmi (1975); ECOLOGY, Fletcher (1989c); LARVA, Azab *et al.* (1971) and Menon *et al.* (1968); PARASITOIDS, Narayanan & Chawla (1962) and Steck *et al.* (1986).

Dacus (Didacus) frontalis Becker

(figs 91, 227)

Taxonomic notes

This species has also been known as *Dacus ciliatus* var. *duplex* Munro, *D. ciliatus* form *frontalis* Becker, *D. scopatus* Munro and *Didacus frontalis* (Becker).

Pest status and commercial hosts

A pest of cucurbits (Hancock, 1989). In the Cape Verde Islands the major hosts are cucumber (*Cucumis sativus*), pumpkin (*Cucurbita pepo*), sweet melon (*Cucumis melo*) and watermelon (*Citrullus lanatus*), and colocynth (*Citrullus colocynthis)* is an important alternative host (Steffens, 1983). In South Africa it has been recorded from a gourd (*Coccinia* sp.) and a squash (*Cucurbita* sp.) (Munro, 1984).

Wild hosts

Recorded from wild growing colocynth and some *Cucumis* spp. (Munro, 1984; Steffens, 1983).

Adult identification

A predominantly orange species; with facial spots, 2 scutellar setae, a yellow stripe covering most of the anatergite and katatergite, only mid femur darkened in apical half, and wing with a costal band that is expanded apically to form an apical spot (fig. 227); male with a pecten. Scutum without yellow stripes, anterior supra-alar setae and prescutellar acrostichal setae.

Partial description of third instar larva

Larvae medium-sized, length 8.0-9.5mm; width 1.5-2.0mm. *Head*: Oral ridges in 11-12 rows, some branched; mouthhooks with a stout preapical tooth. *Thoracic and abdominal segments*: Small bands of spinules encircling anterior portions of T1, T2, T3 and A1; A8 with intermediate lobes small, sensilla absent. Anterior spiracles with 14-15 stout tubules. Posterior spiracles with spiracular slits 4 times as long as broad; spiracular hairs short, (less than half the length of a spiracular slit), some branched with 4-17 hairs per bundle. Anal lobes prominent, surrounded by small spinules. *Source*: Based on Malan & Giliomee (1969).

Distribution

Africa: Egypt, Kenya, Lesotho, South Africa, South West Africa, Sudan, Tanzania, Zimbabwe.
Atlantic Ocean Islands: Cape Verde Islands.
Middle East: Saudi Arabia, Yemen Arab Republic.

Attractant

Males attracted to cue lure.

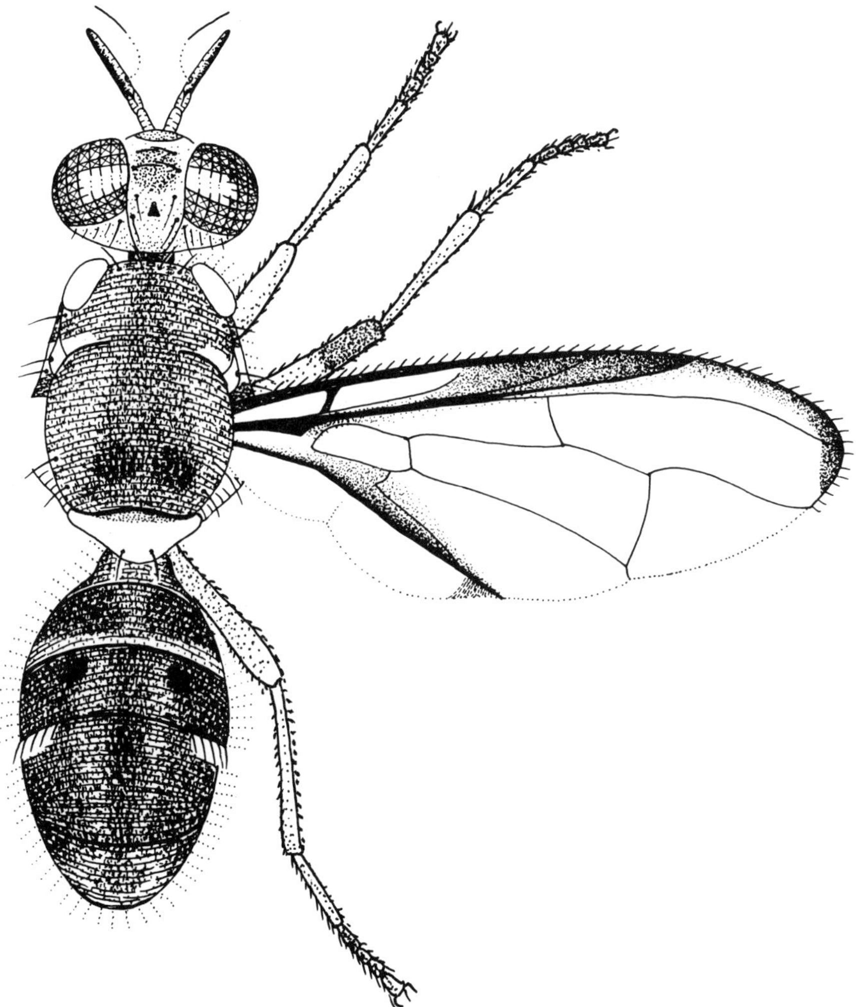

Fig. 227. *Dacus (Didacus) frontalis*, adult male.

Other references

CAPE VERDE ISLANDS, Harten (1987) and Steffens (1983); ECOLOGY, Fletcher (1989c); PARASITOIDS, Fry (1990); YEMEN, Angood (1977).

Dacus (Didacus) lounsburyii Coquillett

(figs 92, 228)

Taxonomic notes
This species has also been known as *Callantra lounsburyii* (Coquillett), *Mictodacus lounsburyii* (Coquillett) and *Tridacus lounsburyii* (Coquillett); it is usually misspelt as *Dacus lounsburyi*.

Commercial hosts
A rare species which sometimes infests cultivated cucurbits according to Hancock (1989) and Munro (1964a; 1984), but specific hosts were not listed by those authors. There are old host records of this potential cucurbit pest from melon (*Cucumis melo*), pumpkin (*Cucurbita pepo*) and watermelon (*Citrullus lanatus*) (Munro, 1925), but they require confirmation.

Adult identification
A predominantly orange-brown species; scutum with a broad medial black stripe and both lateral and medial yellow stripes; with facial spots, 2 scutellar setae, and wing with a costal band that is expanded apically to form a large apical spot (fig. 228); often with one or two small irregular yellow spots in the centre of the katatergite; anatergite orange-brown; male with a pecten. Without anterior supra-alar setae and prescutellar acrostichal setae.

Partial description of third instar larva
Larvae large, length 13.0mm; width 2.8mm. *Head*: Oral ridges in 17-19 rows, some branched; mouthhooks heavily sclerotised, with a large preapical tooth. *Thoracic and abdominal segments*: Small bands of spinules on dorsal surfaces of T1-T3, fewer spinules on A1, sometimes present on A2. Anterior spiracles with 23-25 tubules. *Source*: Based on Silvestri (1913).

Distribution
Africa: Angola, South Africa, Zimbabwe.

Attractant
Males not known to be attracted to any synthetic lure.

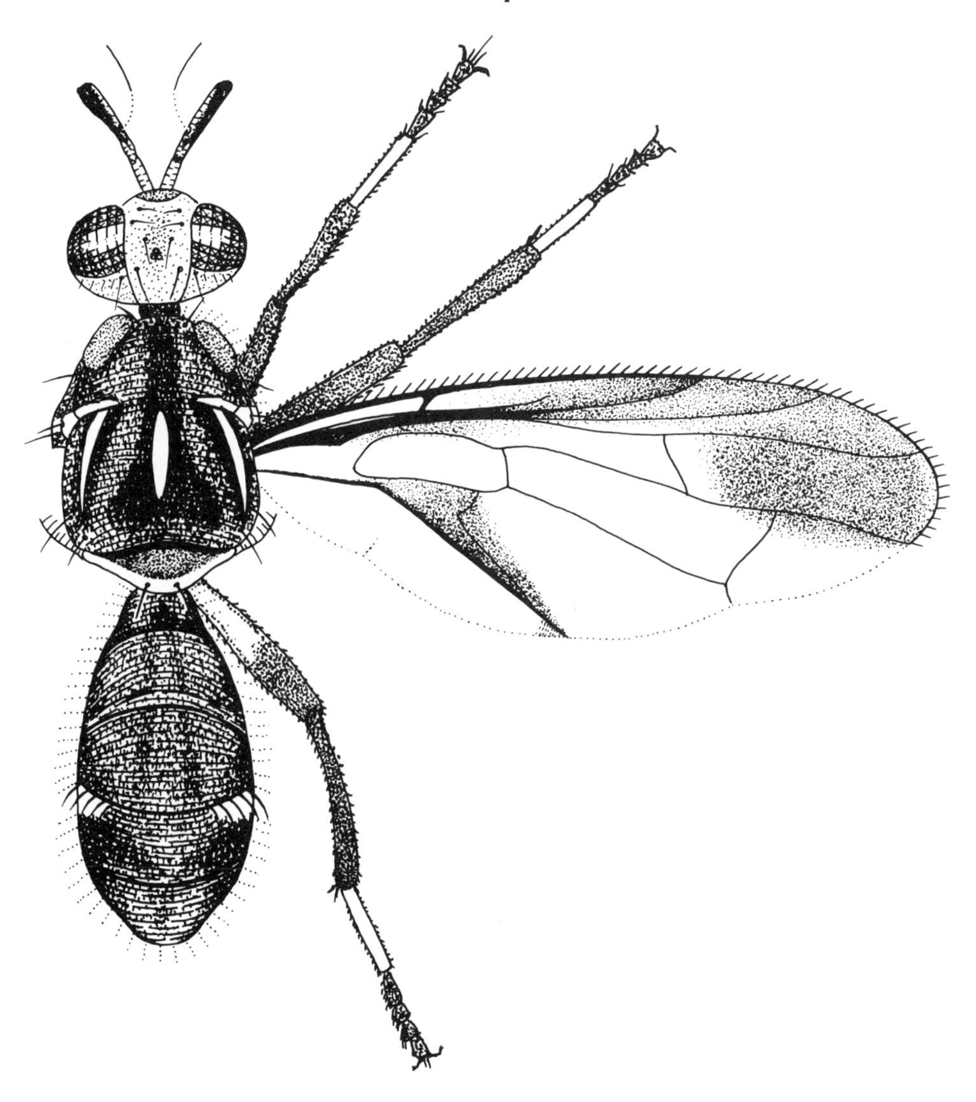

Fig. 228. *Dacus (Didacus) lounsburyii*, adult male.

Dacus (Didacus) vertebratus Bezzi

Jointed Pumpkin Fly
Melon Fly

(figs 96, 229)

Taxonomic notes

This species has also been known as *Dacus marginalis* Bezzi, *D. mimeticus* Collart, *D. vertebratus* var. *marginalis* Bezzi, *D. triseriatus* Curran and *Didacus vertebratus* (Bezzi).

Pest status and commercial hosts

A pest of cucurbit crops with a strong preference for watermelon (*Citrullus lanatus*) (Hancock, 1989). It has been recorded from the following hosts in Nigeria (Matanmi, 1975): cantaloupe (*Cucumis melo*), cucumber (*C. sativus*), squash (*Cucurbita maxima*), watermelon and white egusi (*Cucumeropsis mannii*); cucumber was only attacked when over-ripe. Munro (1984) noted that *D. vertebratus* may also make abortive attacks on granadilla (*Passiflora* spp.).

Wild hosts

Many species of Cucurbitaceae, e.g. *Momordica* spp. (Matanmi, 1975; Munro, 1984).

Adult identification

A predominantly orange species; with facial spots, 2 scutellar setae, a yellow stripe covering most of the anatergite and katatergite, all femora darkened in apical half, and wing with a costal band that is expanded apically to form an apical spot (fig. 229); male with a pecten. Scutum without yellow stripes, anterior supra-alar setae and prescutellar acrostichal setae.

Partial description of third instar larva

Larvae large, length 11.0-12.5mm; width 2.0mm. *Head*: Oral ridges in 13-16 rows, some branched; mouthhooks with stout preapical teeth. *Thoracic and abdominal segments*: Small bands of spinules encircling anterior portion of T1-T3; A8 with intermediate lobes small, sensilla absent. Anterior spiracles with 17-19 stout tubules. Posterior spiracles with spiracular slits 4 times as long as broad; spiracular hairs short (less than half the length of a spiracular slit), some branched with 4-18 hairs per bundle. Anal lobes prominent, surrounded by small spinules. *Source*: Based on Malan & Giliomee (1969).

Distribution

Africa: Angola, Botswana, Ethiopia, Gambia, Ghana, Kenya, Liberia, Malawi, Nigeria, Senegal, South Africa, South West Africa, Tanzania, Zambia, Zimbabwe.
Indian Ocean: Madagascar.
Middle East: Saudi Arabia, Yemen Arab Republic.

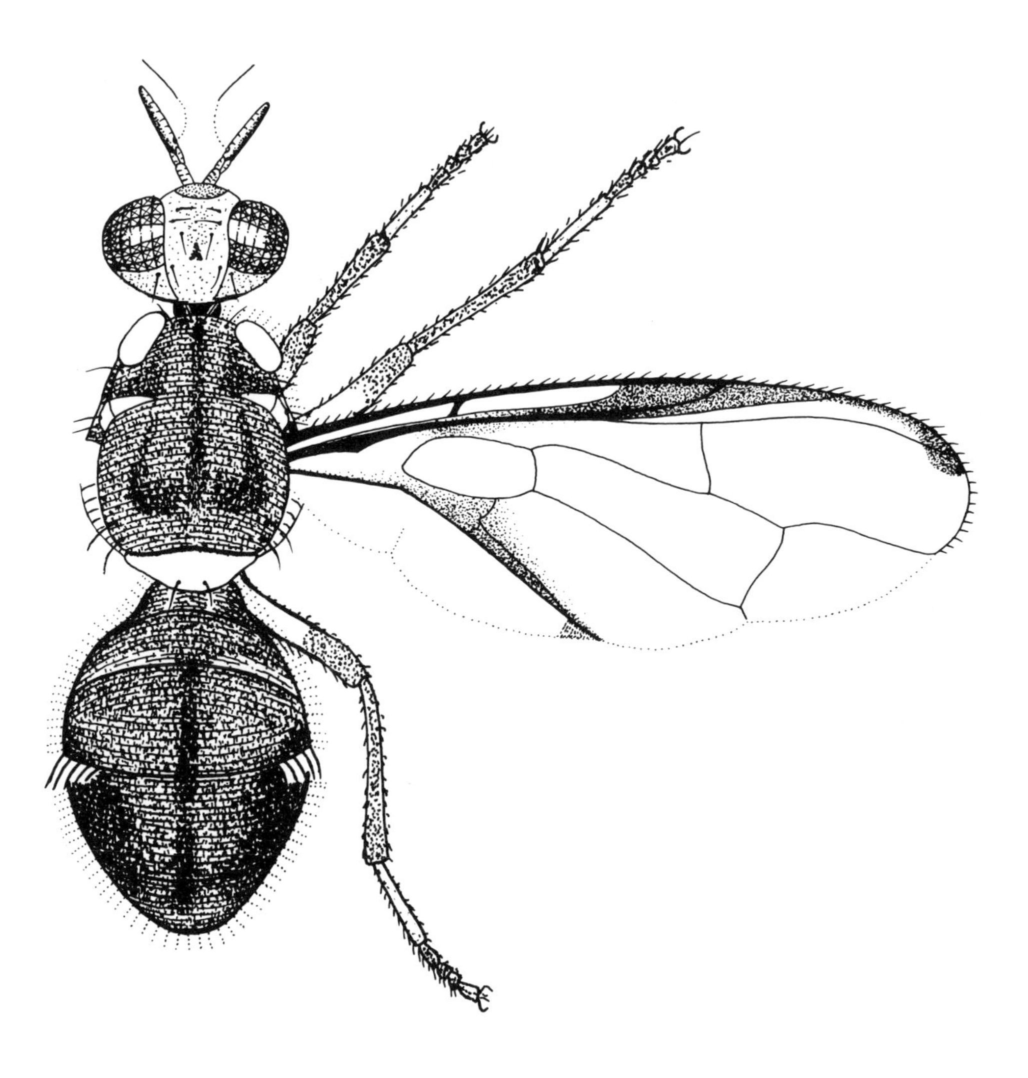

Fig. 229. *Dacus (Didacus) vertebratus*, adult male.

Attractant

Males attracted to vert lure.

Other references

ECOLOGY, Fletcher (1989c); NIGERIA, Matanmi (1975).

Other *Dacus* Species of Economic Interest

About half of the African species of *Dacus* that have known hosts are exclusively associated with the plant family Asclepiadaceae (data from Munro, 1984). Some of these species attack asclepiads of minor economic importance, but they do not cause serious damage to the utilized parts of those plants. The seed-hairs of Sodom apple (*Calotropis procera*) are used as floss ('French cotton') and that may be damaged by *Dacus (Leptoxyda) longistylus* Wiedemann (Fletcher, 1920); the pods are also attacked by *D. (Didacus) aspilus* Bezzi, *Dacus (Didacus) fuscatus* Wiedemann and *Dacus (L.) obesus* Munro. In India, Sodom apple is grown on a commercial scale and the biology of *D. (L.) persicus* Hendel (misidentified as *D. longistylus*) in India was described by Parihar (1984). Blood flower (*Asclepias curasavica*) is an ornamental and a pantropical weed, whose stems are attacked by *D. (L.) annulatus* (Becker). Other species which attack asclepiads of some economic interest are *D. (Didacus) arcuatus* Munro and *Dacus (Didacus) eminus* Munro, which attack the pods of *Pergularia daemia* and *Gymnema sylvestre*, respectively. Some Asian species are also associated with asclepiads, for example *Dacus (Callantra) sphaeroidalis* (Bezzi), from Thailand and Vietnam, in the fruits of Chinese violet (*Telosma cordata*) (Hardy, 1973). Kapoor (1989) said that shortage of Sodom apple could induce *D. longistylus* to attack cucurbitaceous plants, but that suggestion derived from Zoheiry (1950) who almost certainly misidentified another species. There are records of *D. persicus* attacking cucumber (*Cucumis sativus*) and melon (*C. melo*) in Iran (Ayatollahi, 1971), but a photograph indicates that those records were almost certainly based on misidentifications of *D. ciliatus*. Male attractant: *D. eminus* to cue lure. Identification: Munro (1984).

D. (Callantra) eumenoides (Bezzi), from Myanmar and Thailand, is recorded from snakegourd (*Trichosanthes cucumerina*) (Hardy, 1973; NHM, collection data); there is also an old record, requiring confirmation, from cucumber (*Cucumis sativus*) (Bezzi, 1916; Fletcher, 1920). Identification: Hardy (1973).

D. (Callantra) petioliforma (May), from north-eastern Australia, has been recorded from snakegourd (*T. cucumerina*) (Drew, 1989a). Identification: Drew (1989a).

D. (Didacus) armatus Fabricius, from Zaire, is recorded from cucumber (*Cucumis sativus*) and melon (*C. melo*) (Silvestri, 1913), both Cucurbitaceae, but Munro (1984) only associated *D. armatus* with Passifloraceae. Identification: Munro (1984).

D. (D.) momordicae (Bezzi) is found in an area extending from Cameroun to Kenya and Zaire. It is normally associated with wild cucurbits such as *Momordica* and *Melothria* spp., but there is a single record from pumpkin (*Cucurbita maxima*) in Uganda (Le Pelley, 1959). Identification: Munro (1984).

D. (D.) yangambinus Munro, from Zaire, has been reared from bitter gourd (*Momordica charantia*) (Munro, 1984). Identification: Munro (1984).

D. (D.) aequalis Coquillett, from Queensland, Australia, has been recorded from sour orange (*Citrus aurantium*) (Froggatt, 1909), but that record was regarded as doubtful by Drew (1989a). Male attractant: Cue lure. Identification: Drew (1982a; 1989a).

Dacus (D.) pallidilatus Munro, from Zimbabwe, is listed amongst minor pests of cucurbits by Hancock (1989), but no detailed host records are available. Male attractant: cue lure. Identification: Munro (1984).

It is possible that any *Dacus* spp. associated with wild species of Cucurbitaceae may occasionally attack a cultivated cucurbit. Munro (1984) included in his host list a section for species associated with "Cucurbits generally, cultivated and wild". Species listed in that section, but not described above, were *Dacus (Didacus) africanus* Adams and *Dacus (Didacus) vansomereni* Munro. *Dacus africanus* has been reared from an unidentified gourd (Munro, 1984). *D. vansomereni* is recorded from *Adenia* sp. (Passifloraceae) (Munro, 1938a) and an unidentified cucurbit (NHM, collection data); the record from *Adenium* sp. (Apocynaceae), by Le Pelley (1959), was probably an error for *Adenia*. Other species recorded from wild cucurbits are *D. (D.) humeralis* Bezzi, *D. (Leptoxyda) hamatus* Bezzi (larvae in stamens of male flower), *D. (L.) hyalobasis* Bezzi, *D. (L.) inflatus* Munro, *D. (L.) inornatus* Bezzi (larvae in stamens of male flower), *D. (L.) maynei* Bezzi, *D. (L.) retextus* Munro and *D. (L.) rufoscutellatus* (Hering) (Le Pelley, 1959; Munro, 1984). There is also an unconfirmed record of *D. (D.) bistrigatus* Loew from an unidentified species of Cucurbitaceae (Munro, 1984). Matanmi (1972) also recorded *D. (D.) disjunctus* (Bezzi) from bitter gourd (*Momordica charantia*), but the only confirmed host records for that species are from wild species of Passifloraceae and Munro (1984) regarded the gourd record as doubtful. Male attractants: *D. africanus* and *D. humeralis* to cue lure. Identification: Munro (1984).

GENUS *Dirioxa* Hendel

Taxonomic notes

A genus of four species found in Australia and Vietnam. Hardy (1951) presented a key to the three species found in Australia.

Diagnosis of third instar larva

Antennal sensory organ large, with a stout basal segment and long cone-shaped distal segment. Maxillary sensory organ broad, flat, with small well defined sensilla surrounded by small cuticular folds (pl. 33.b). Stomal sensory organ with small sensilla surrounded by numerous long serrated preoral lobes (pl. 33.b); oral ridges with 13-18 very long serrated rows of bluntly rounded teeth (pl. 33.a); accessory plates long, deeply serrated, interlocking with outer edges of oral ridges. Mouthhooks each without a preapical tooth (fig. 160). T2-T3 without dorsal and lateral spinules. Anterior spiracle fan-shaped with 9-13 tubules. Posterior spiracular slits about 2.5 times as long as broad. Anal lobes large, unequally bilobed.

Dirioxa pornia (Walker)

Island Fruit Fly
Boatman Fly
South Sea Fly

(figs 25, 160, 230; pl. 33)

Taxonomic notes

This species has also been known as *Trypeta musae* Froggatt and *T. pornia* Walker. It was first described by Froggatt (1899), supposedly from a shipment of banana (*Musa x paradisiaca*) from Vanuatu, but Gurney (1910) suggested that the cargo may have been from Queensland.

Commercial hosts

This species is frequently reared from cultivated fruits, but it is not a pest as it usually only attacks damaged fruit. Despite its being recorded from a wide variety of fruits there is very little modern information on this species. The most detailed account of *D. pornia* was by Gurney (1912), who noted that it rarely attacked undamaged fruit, preferring fruit that had been thorn-pricked, fallen or attacked by another fruit fly or codling moth. He recorded it from lemon (*Citrus limon*), mandarin orange (*C. reticulata*), orange (*Citrus* sp.) and peach (*Prunus persica*), and noted that it may also attack apple (*Malus domestica*), pear (*Pyrus communis*) and quince (*Cydonia oblonga*). Allman (1941) reared it from a passion fruit (*Passiflora* sp.), but noted that it did not normally attack such woody fruits. It has also been reared from citrange (*x Citroncirus*

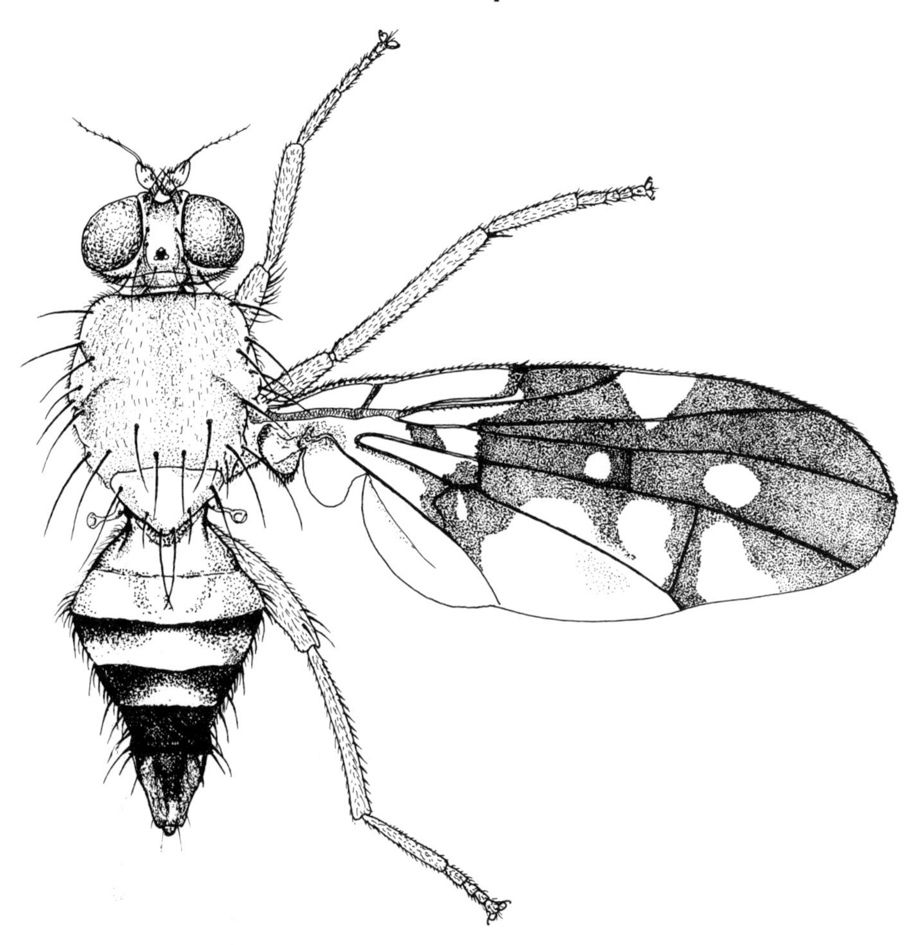

Fig. 230. *Dirioxa pornia*, adult female.

webberi), feijoa (*Feijoa sellowiana*) and star-apple (*Chrysophyllum cainito*), andcollected on (not reared) mango (*Mangifera indica*), mulberry (*Morus* sp.), persimmon (*Diospyros* sp.), plum (*Prunus domestica*), pummelo (*Citrus maxima*), sand pear (*Pyrus pyrifolia*) and trifoliate orange (*Poncirus trifoliata*) (ANIC, collection data; QDPI, unpublished data, 1990).

Wild hosts

Recorded from *Chrysophyllum pruniferum* F. Muell., *Planchonella australis* and *Sideroxylon* sp. (all Sapotaceae); there are also records from wild species of Fabaceae and Lauraceae (Allman, 1941; ANIC, collection data; ANMS, collection data).

Adult identification

Dirioxa spp. are the only tephritids with six setae on the scutellar margin, that are likely to be found in fruit crops; the wing pattern is characteristic (fig. 230).

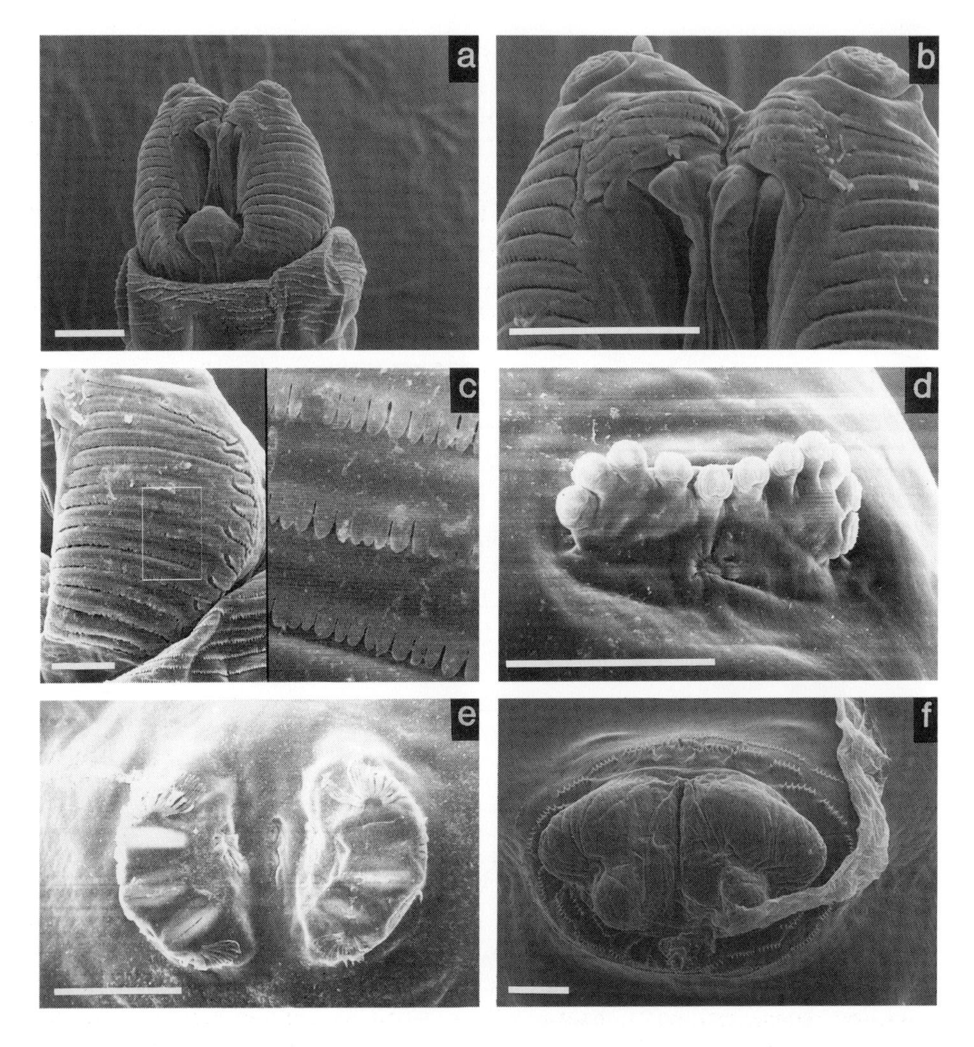

Plate 33. *Dirioxa pornia*, SEMs of 3rd instar larva; a, head and T1, ventral view (scale=0.1mm); b, preoral lobes, ventral view (scale=0.1mm); c, oral ridges, lateral view (scale=50μm); d, anterior spiracle (scale=0.1mm); e, posterior spiracles (scale=0.1mm); f, anal lobes (scale=0.1mm).

Description of third instar larva

Larvae medium-sized, length 5.5-8.5mm; width 1.0-1.5mm. *Head*: Stomal sensory organ (pl. 33.b), small, rounded, flat, with a small sensilla, completely surrounded by 5-6 long serrated preoral lobes resembling small oral ridges; oral ridges (pl. 33.a, c) with 13-18 very long rows of long, parallel sided, bluntly rounded teeth; long deeply serrated accessory plates interlocking with outer edges of oral ridges; mouthhooks long, slender, moderately sclerotised (fig. 160). *Thoracic and abdominal segments*: T1 with a broad band of discontinuous rows of stout sharply pointed spinules surrounding segment, with 5-6 rows dorsally, 5-9 rows laterally, expanding to 9-13 rows ventrally; T2 with smaller spinules in 3-4 discontinuous rows ventrally, but absent dorsally and laterally; T3 similar to T2. Dorsal side of A1-A8 without dorsal spinules. Creeping welts on A1-A8 of short, sharply pointed spinules in 7-11 rows with 1 posterior row of larger, stouter spinules. A8 with well developed sensilla, arranged as 1 dorsally, 3 laterally and one ventrally, on each side of body. *Anterior spiracles* (pl. 33.d): Broad, fan-shaped with 9-13 long, well spaced, finger-like tubules which are slightly expanded apically. *Posterior spiracles* (pl. 33.e): Placed just above the midline; spiracular slits slightly radiating, each about 2.5 times as long as broad, with a sclerotised rima. Spiracular hairs in 4 bundles of broad, flat, branched hairs; 10-15 in dorsal and ventral bundles, 6-9 in each lateral bundle. *Anal area* (pl. 33.f): Lobes very large, unequally bilobed and surrounded by 3-5 discontinuous rows of short, stout, sharply pointed spinules concentrating into a small turret just below anal opening. *Source*: Based on specimens from Queensland, Australia (ex *Chrysophyllum cainito* and *Planchonella australis*).

Distribution

Australia: New South Wales, southern Queensland.
South Pacific: There are no confirmed records from outside Australia (A.C. Courtice, pers. comm., 1990) and the following localities listed by Hardy & Foote (1989) all require confirmation: American Samoa, Fiji, French Polynesia, New Caledonia, New Zealand, Vanuatu.

Other references

BEHAVIOUR, Pritchard (1967) and P.H. Smith (1989); LARVA, Gurney (1912).

Other *Dirioxa* Species of Economic Interest

D. confusa (Hardy), from northern Queensland, only differs from *D. pornia* in being somewhat paler coloured and having smaller hyaline spots in wing cells r_{4+5} and br. Like *D. pornia*, *D. confusa* has been reared from wild species of Sapotaceae, plus the following commercial hosts: banana (*Musa x paradisiaca*), carambola (*Averrhoa carambola*), citrange (*x Citroncirus webberi*), common guava (*Psidium guajava*), grapefruit (*Citrus x paradisi*), feijoa (*Feijoa sellowiana*), kumquat (*Fortunella* sp.), macadamia nut (*Macadamia* sp.), mandarin (*C. reticulata*), mango (*Mangifera indica*), pummelo (*C. maxima*), strawberry guava (*P. littorale*), trifoliate orange (*Poncirus trifoliata*), Valencia orange (*C. sinensis*) and white sapote (*Casimiroa edulis*); it has also been found on (not reared) apple (*Malus domestica*) (ANIC, collection data). *D. confusa* is not considered a pest, and like *D. pornia*, it probably only attacks damaged fruit. Identification: Hardy (1951).

GENUS *Epochra* Loew

Epochra canadensis (Loew)

Currant Fruit Fly
Yellow Currant Fly
Currant and Gooseberry Maggot

(figs 26, 44, 231; pl. 34)

Taxonomic notes
This is the only species of *Epochra*, and it has also been known as *Trypeta canadensis* Loew and *T. lunifera* Hering.

Commercial hosts
This used to be a serious pest of currants (*Ribes* spp.) and gooseberry (*R. uva-crispa*) (e.g. Jones, 1937), but we have been unable to trace any reports of it as a pest since 1950. Other recorded hosts are blackcurrant (*R. nigrum*), golden currant (*R. aureum*) and redcurrant (*R. rubrum*) (Wasbauer, 1972).

Wild hosts
Recorded from several other *Ribes* spp. (Wasbauer, 1972).

Adult identification
A red-brown coloured species (abdomen sometimes dark brown) with a characteristic wing pattern (fig. 231).

Description of third instar larva
Larvae medium-sized, length 7.0-8.0mm; width 1.2-1.5mm. *Head*: Stomal sensory organ large, protuberant with small sensilla; preoral lobes absent; oral ridges (pl. 34.a, b) with 15-17 short rows of long, sharply tapered, well spaced teeth, becoming less serrated towards outer edge; accessory plates absent; mouthhooks large and heavily sclerotised. *Thoracic and abdominal segments*: T1 with a broad band of 6-9 discontinuous rows of small sharply pointed spinules surrounding segment; T2-T3 each with 2-4 discontinuous rows of spinules surrounding segment. A1-A8 with dorsal spinules. Creeping welts of up to 13 rows of spinules. *Anterior spiracles* (pl. 34.c): 14-16 short, stout, rounded tubules, in a single row. *Posterior spiracles* (pl. 34.d): Placed well above midline. Spiracular slits about 3.5 times as long as broad, with sclerotised rimae. Spiracular hairs short (less than length of spiracular slit), thin; 3-5 hairs in dorsal and ventral bundles, and 2-4 hairs in each lateral bundle. *Anal area* (pl. 34.e): Lobes large, protuberant and surrounded by 3-6 discontinuous rows of small spinules.*Source*: Based on 2 specimens from Wyoming, USA.

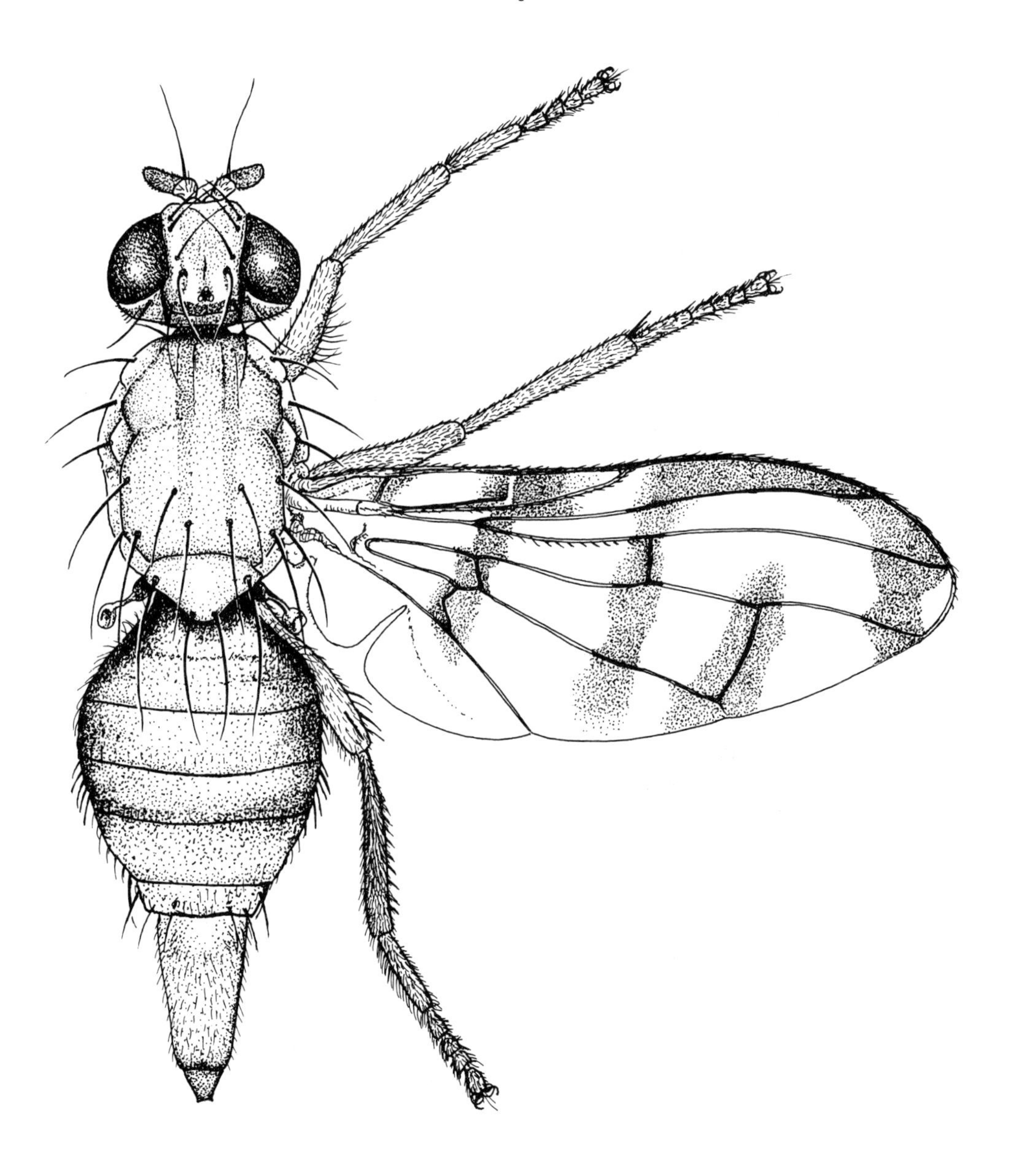

Fig. 231. *Epochra canadensis*, adult female.

Distribution

North America: Southern Canada, northern USA.

Other references

BEHAVIOUR, P.H. Smith (1989); BIOLOGY, Christenson & Foote (1960); LARVA, Phillips (1946).

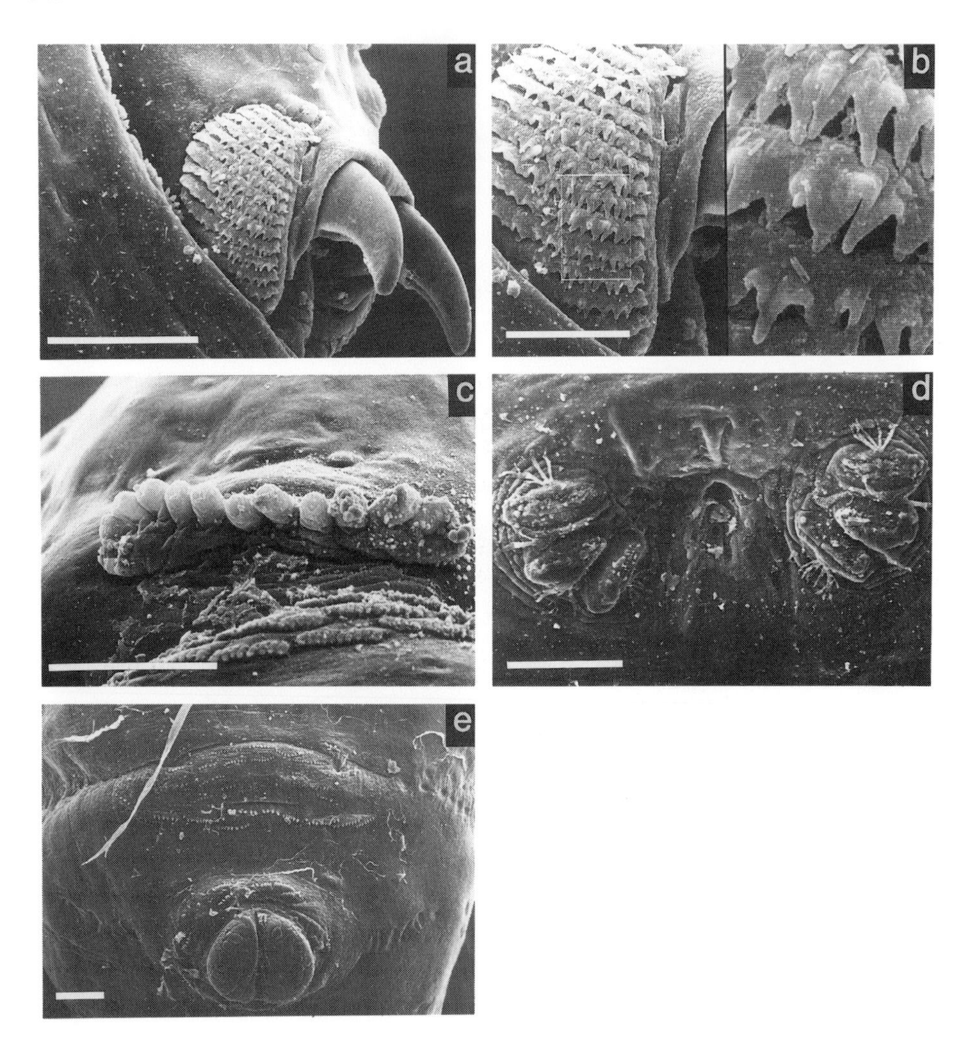

Plate 34. *Epochra canadensis*, SEMs of 3rd instar larva; a, mouthhooks and oral ridges, lateral view (scale=0.1mm); b, oral ridges, lateral view (scale=50μm); c, anterior spiracle (scale=0.1mm); d, posterior spiracles (scale=0.1mm); e, A8, creeping welt and anal lobes (scale=0.1mm).

GENUS *Monacrostichus* Bezzi

Monacrostichus citricola Bezzi

(figs 27, 232)

Taxonomic notes
This is the only species of *Monacrostichus*.

Commercial hosts
Recorded from lemon (*Citrus limon*), lime (*C. limetta*) and pummelo (*C. maxima*) (Hardy, 1974; NHM, collection data). Although very rare, this species has been intercepted in lime fruits imported into Hawaii (Whitney, 1929).

Adult identification
A dark coloured fly with conspicuous yellow markings, long antennae (fig. 27), and characteristic wing markings and venation (fig. 232).

Distribution
Oriental Asia: Malaysia (Peninsular), Philippines (Luzon, Mindanao, Palawan).

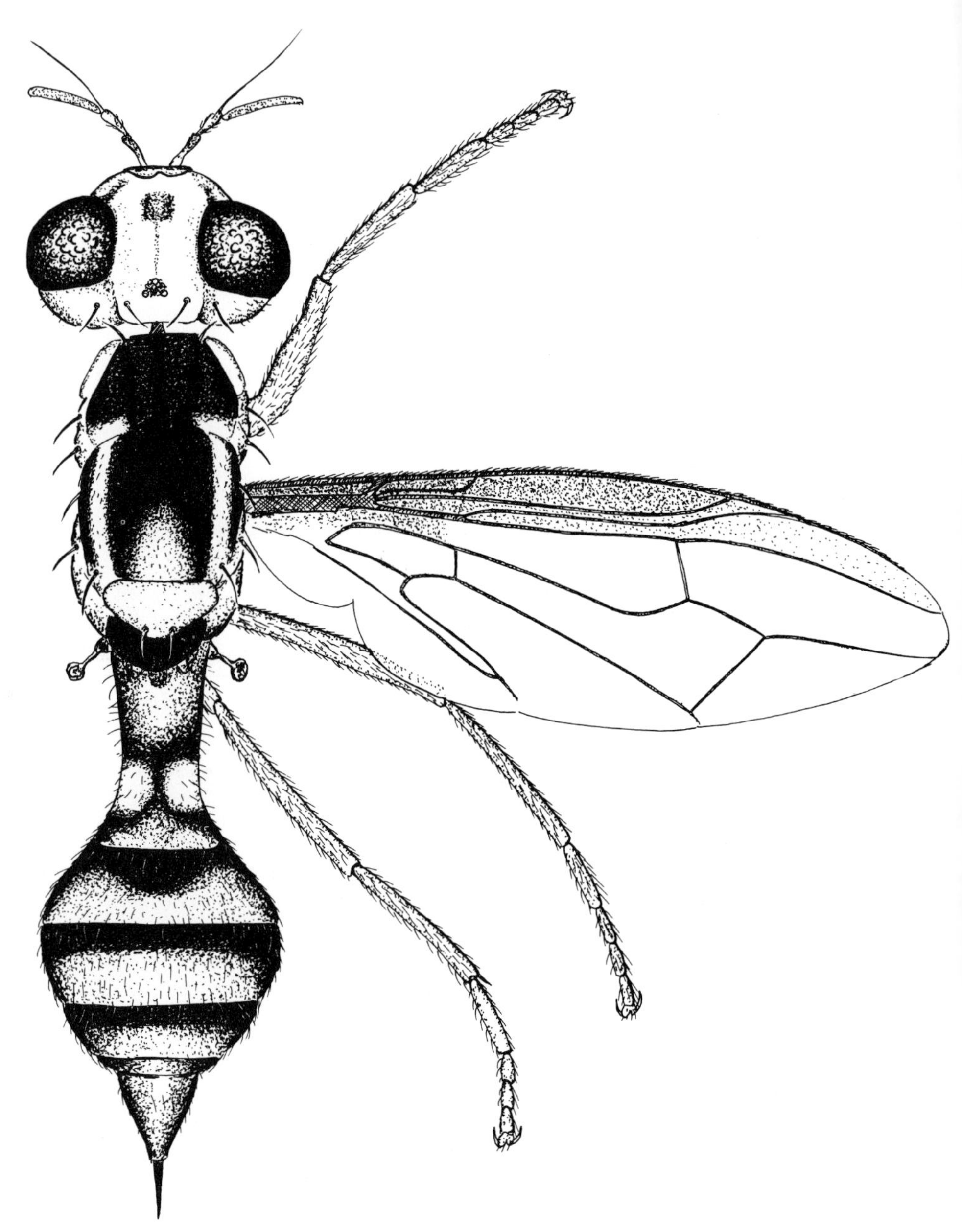

Fig. 232. *Monacrostichus citricola*, adult female.

GENUS *Myiopardalis* Bezzi

A genus of one species which some authors have treated as a subgenus of *Carpomya* Costa.

Myiopardalis pardalina (Bigot)

Baluchistan Melon Fly
Russian Melon Fly

(figs 28, 233)

Taxonomic notes

This species has also been known as *Carpomyia pardalina* Bigot and *Myiopardalis carpalina* Fletcher.

Pest status and commercial hosts

A pest of cucumber (*Cucumis sativus*), melon (*C. melo*) and watermelon (*Citrullus lanatus*); it is also recorded from squirting cucumber (*Ecballium elaterium*) (Kandybina, 1977).

Wild hosts

Recorded from some wild species of Cucurbitaceae (Kandybina, 1977).

Adult identification

An orange-yellow fly with a shiny yellow and black pattern on the scutum and scutellum, and a characteristic pattern of yellow bands on the wing (fig. 233).

Partial description of third instar larva

Larvae medium-sized, length 8.0-10.0mm. *Head*: Antennal sensory organ with a short, broad basal segment and a rounded distal segment; maxillary sensory organ protuberant, with 2 large and 5-7 small sensilla. Stomal sensory organ with small bristle-like sensilla; 3 large preoral teeth at base of stomal sensory organ; oral ridges with 4-7 rows of sharply pointed teeth; mouthhooks large, heavily sclerotised, with a crescent-shaped apical tooth. *Thoracic and abdominal segments*: T1-T3 with several rows of small spinules completely encircling anterior portion of each segment; A1-A6 with rows of small spinules encircling each segment, spinules more numerous ventrally, with 15-20 rows in the creeping welts; A7-A8 without spinules dorsally. *Anterior spiracles*: With a long, irregular row of 26-28 short, closely-spaced tubules. *Posterior spiracles*: Spiracular slits 3-4 times as long as broad. Spiracular hairs short (less than length of a spiracular slit), rarely branched. Dorsal and ventral bundles with 9 hairs, lateral bundles with 4-6 hairs. Area between posterior spiracles with a cellular or

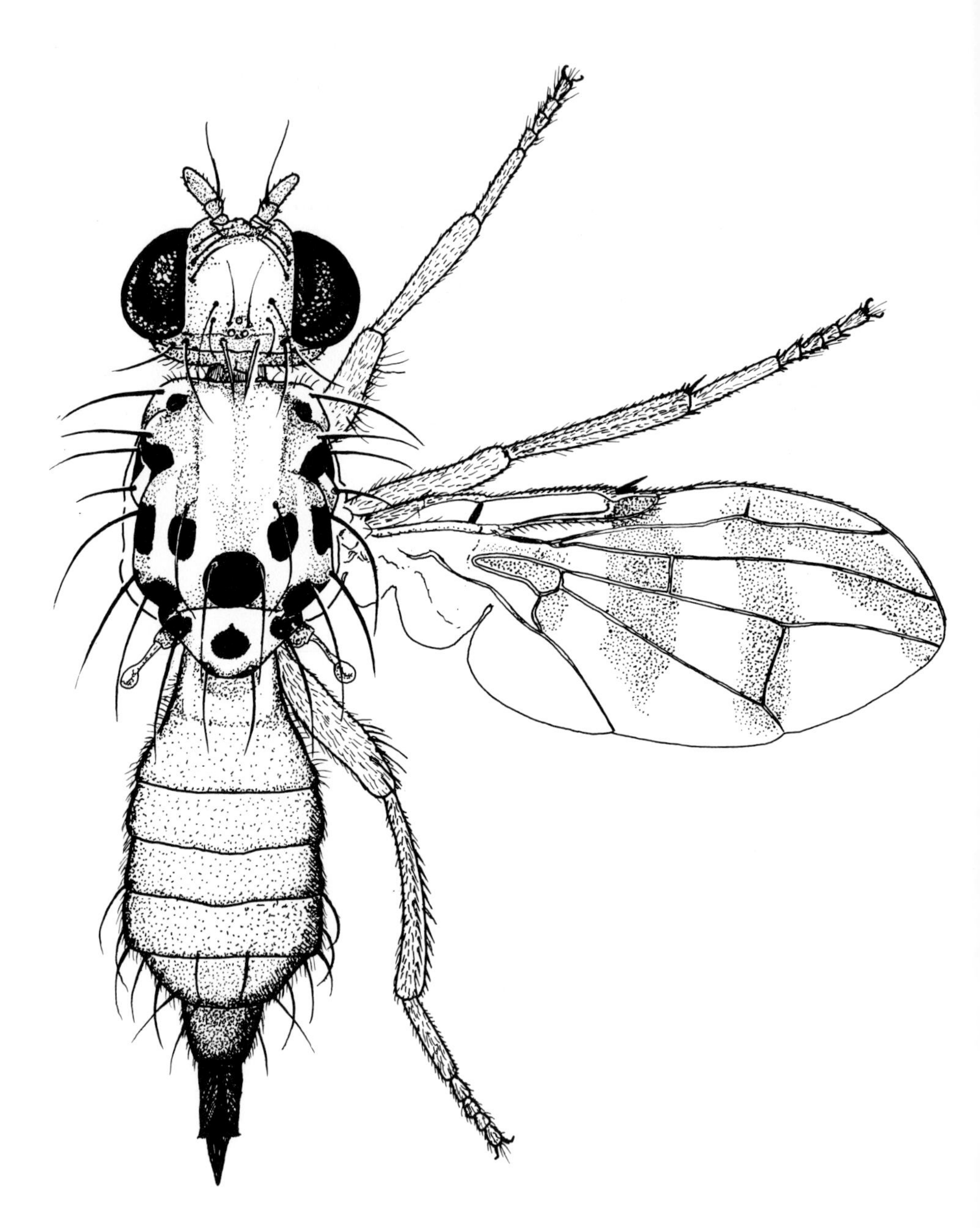

Fig. 233. *Myiopardalis pardalina*, adult female.

rugose appearance. *Anal region*: Anal lobes small, with a few discontinuous rows of small spinules around lobes. *Source*: Based on Kandybina (1965; 1977).

Distribution

Europe: USSR (Caucasus area).
Middle East: Cyprus, Iran, Iraq, Israel, Lebanon, Syria, Turkey.
Oriental Asia: Northern India, Pakistan.
Temperate Asia: Afghanistan.
A distribution map was produced by CIE (1961a); that included a west African record that should be regarded as doubtful.

Other references

AFGHANISTAN, Ullah (1987); BIOLOGY, Christenson & Foote (1960); INDIA, Kapoor (1989).

GENUS *Rhagoletis* Loew

(figs 29, 38, 43)

Taxonomic notes

Some species have formerly been placed in the genera *Megarrhagoletis* Rohdendorf, *Microrrhagoletis* Rohdendorf and *Zonosema* Loew, which are synonyms of *Rhagoletis*.

Distribution and biology

The genus *Rhagoletis* includes about 65 known species, with representatives in the New World, Europe and temperate Asia. Most species of known biology are stenophagous, attacking the fruits of a few closely related plants, typically members of a single genus. Although there are exceptions to that rule, the range of plants attacked by a *Rhagoletis* sp. is typically very narrow compared to the pest species of *Anastrepha*, *Bactrocera*, *Ceratitis* and *Dacus*. The hosts of *Rhagoletis* spp. usually belong to the family Solanaceae in South America, but in the north temperate regions the host families are Berberidaceae, Caprifoliaceae, Cornaceae, Cupressaceae, Elaeagnaceae, Ericaceae, Grossulariaceae, Juglandaceae and Rosaceae. The larvae of *Rhagoletis* spp. develop in the flesh of their host fruit and they are usually univoltine, with pupariation taking place in the soil beneath the host plant, and the adults mate on or near the host; a general review of *Rhagoletis* biology was given by Boller & Prokopy (1976). This genus has been the subject of extensive work on speciation processes and evolutionary genetics, and examples of that extensive literature are Bush (1969) and Berlocher & Bush (1982). Another interesting aspect of *Rhagoletis* research has been the development of a technique for using trace element analysis to determine the larval food of wild caught adults (Diehl & Bush, 1983).

Diagnosis of third instar larva

Antennal sensory organ with a short basal segment and cone-shaped distal segment; maxillary sensory organ flat, with well defined sensilla surrounded by small cuticular folds; stomal sensory organ rounded, with a peg-like sensilla; large, preoral teeth near base of stomal sensory organ; no preoral lobes; oral ridges in 5-13 short, unserrated rows; no accessory plates. Stout spinules forming discontinuous rows on almost all segments. Anterior spiracles with 7-35 stout tubules. Posterior spiracular slits 3-8 times as long as broad, with 3-16 short, branched spiracular hairs. Anal lobes large, protuberant with well defined tubercles and sensilla.

Literature

The following comprehensive regional keys to adults have been produced: North America, Bush (1966); South and Central America, Foote (1981); Europe and temperate Asia, Rohdendorf (1961). The larvae of North American pest species were keyed by Phillips (1946) and the three widespread species of walnut husk maggots were separated by Steyskal (1973); the larvae of European and temperate Asian species were keyed by Kandybina (1961; 1977). Berlocher (1980) produced an electrophoretic key which allowed separation of nine North American *Rhagoletis* spp. as larvae, puparia or adults, including five species in cherry and three sympatric species of the *R. pomonella* complex.

Rhagoletis cerasi (Linnaeus)

European Cherry Fruit Fly

(figs 124, 234; pl. 35)

Taxonomic notes

This species has also been known as *Musca cerasi* Linnaeus, *Rhagoletis cerasorum* (Dufour), *R. liturata* (Robineau-Desvoidy), *R. signata* (Meigen), *Trypeta signata* (Meigen), *Urophora cerasorum* Dufour and *U. liturata* Robineau-Desvoidy. There are also forms and subspecies, which are doubtfully distinct, called *R. cerasi fasciata* Rohdendorf, *R. cerasi nigripes* Rohdendorf and *R. cerasi* form *obsoleta* Hering.

Races

R. cerasi has two races which are referred to as *northern* and *southern*. The *southern race* is found in Italy, Switzerland, southern Germany, south-western France and southern parts of Austria. The *northern race* is found north of those areas from the Atlantic to the Black Sea. There is a unidirectional incompatibility between the races, such that *southern* females and *northern* males are interfertile, but crosses between *southern* males and *northern* females are sterile (Boller, 1989a; Boller *et al.*, 1976). The recently established population in Crete provides an example of how the races can be identified. When *southern* males were crossed with Cretan females the eggs produced were all sterile, but *northern* males and Cretan females produced fertile eggs; the Cretan population therefore belongs to the *northern* race (Neuenschwander *et al.*, 1983). The possibility of using this one way incompatibility as a basis for control (the incompatible insect technique) has been investigated (Boller *et al.*, 1976; Blümel & Russ, 1989).

Pest status and commercial hosts

R. cerasi is the only serious tephritid pest in central and northern Europe and its status in different countries was reviewed by Fischer-Colbrie & Busch-Petersen (1989). Recorded hosts are black cherry (*Prunus serotina*), mahaleb cherry (*P. mahaleb*), sour cherry (*P. cerasus*) and sweet cherry (*P. avium*) (Hendel, 1927; Séguy, 1934; Thiem, 1934). Hendel (1927) also recorded it from berberis (*Berberis vulgaris*), but that was probably based on a misidentification of *R. berberidis* Jermy (p. 388), which is a very similar looking species. Records from Barbary matrimony vine (*Lycium barbarum*) and bilberry (*Vaccinium myrtillus*) (Phillips, 1946) were derived from 19th century data and are likely to have been based on casual observation rather than rearing; Thiem (1934) listed those plants as being free from attack.

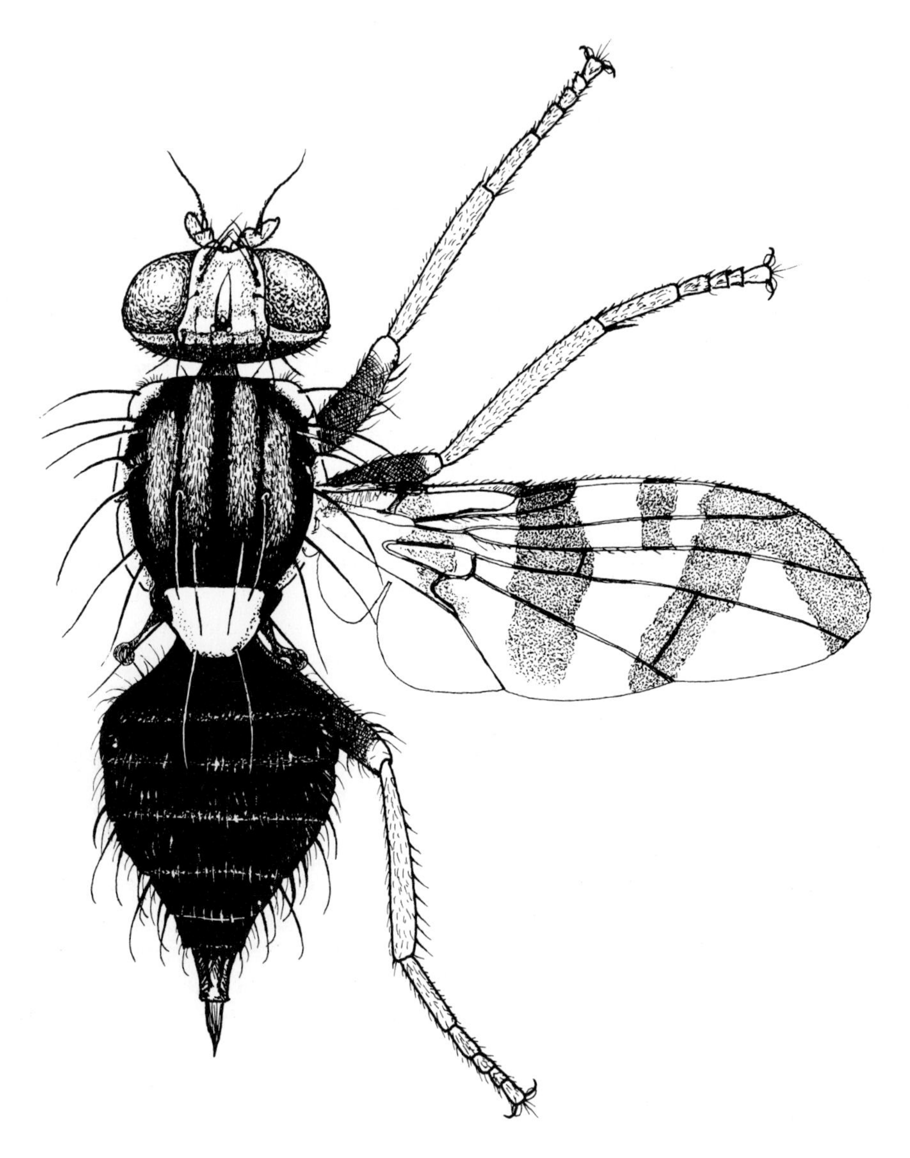

Fig. 234. *Rhagoletis cerasi*, adult female.

Wild hosts

Some wild cherries (*Prunus* spp.) are reservoir hosts for *R. cerasi*. It may also attack wild honeysuckle (*Lonicera* spp.; Caprifoliaceae) (Hendel, 1927). The phenology of *R. cerasi* differs between cherry and honeysuckle associated populations (Haisch & Chwala, 1979) and the honeysuckle population is either a well differentiated host race or possibly a distinct species (G.L. Bush, pers. comm., 1991).

Adult identification
A predominantly black species, with characteristic wing markings (fig. 234) and a scutellum that lacks a black basal mark (fig. 116).

Description of third instar larva
Larvae medium-sized, length 5.0-6.0mm; width 1.2-1.5mm. *Head*: Stomal sensory organs rounded, with 2 small sensilla; preoral teeth large, sclerotised, with 6 strong sharply pointed teeth (pl. 35.a, b); oral ridges not discernible; mouthhooks heavily sclerotised, each with a long, slender curved apical tooth and a much smaller preapical tooth on concave surface. *Thoracic and abdominal segments*: T1 with 3-4 rows of spinules ventrally but none dorsally and laterally; T2 and T3 with 3-5 rows of spinules dorsally and ventrally, but none laterally; A1 with 2-3 rows dorsally and 4-5 rows ventrally; A2-A8 with very few spinules dorsally but 9-12 rows of stout spinules ventrally (pl. 35.c); A8 with intermediate lobes well developed. *Anterior spiracles*: 12-16 tubules. *Posterior spiracles*: Each spiracular slit 4.0-5.0 times as long as broad with a thin, sclerotised rima. Dorsal and ventral hair bundles of 6-7 long (about as long as a spiracular slit) sometimes branched hairs, lateral bundles of 3-5 similar hairs. *Anal area* (pl. 35.c, d): Lobes large, protuberant, with a discontinuous row of small spinules anteriorly and a small concentration of spinules just below anal opening. *Source*: Based on 2 specimens from France (from canned mixed fruit).

Distribution of southern race
Europe: Austria, France (south-western), Germany (southern), Italy (including Sardinia and Sicily), Portugal, Spain, Switzerland.

Distribution of northern race
Europe: Bulgaria, Czechoslovakia, France (except south-west), Germany (except southern Germany), Greece (including Crete), Hungary, Latvia, Lithuania, Netherlands, Norway (not permanently established), Poland, Romania, Sweden, Turkey (north and north-west), USSR (Georgia, Russia, Ukraine), Yugoslavia.
Temperate Asia: Iran, USSR (Kazakhstan, Kirgizia, Russia, Tadzikistan, Turkmenistan).

Several countries impose quarantine controls on cherry importation, including the United Kingdom. A distribution map showing the range of both races was produced by CIE (1989a).

Other references
BEHAVIOUR, Katsoyannos (1989b); BACTERIAL SYMBIONTS, Howard (1989); BIBLIOGRAPHY, Haisch *et al.* (1978) and Zwölfer (1985); ECOLOGY, Debouzie (1989), Meats (1989a) and Ranner (1987; 1988a); CONTROL, Cirio & Vita (1980); CYTOTAXONOMY, Bush & Boller (1977); DEVELOPMENT RATE, Fletcher (1989a); ECOLOGY, Fletcher (1989c); IMPACT, Fimiani (1989); LARVA, Kandybina (1961; 1977), Sabatino (1974) and K.G.V. Smith (1989); PARASITOIDS, Fry (1990), Hennig (1953), Kandybina (1977) and Monaco (1984); PHEROMONES, Averill & Prokopy (1989a), Boller & Hurter (1985), Boller *et al.* (1987), Hurter *et al.* (1987),

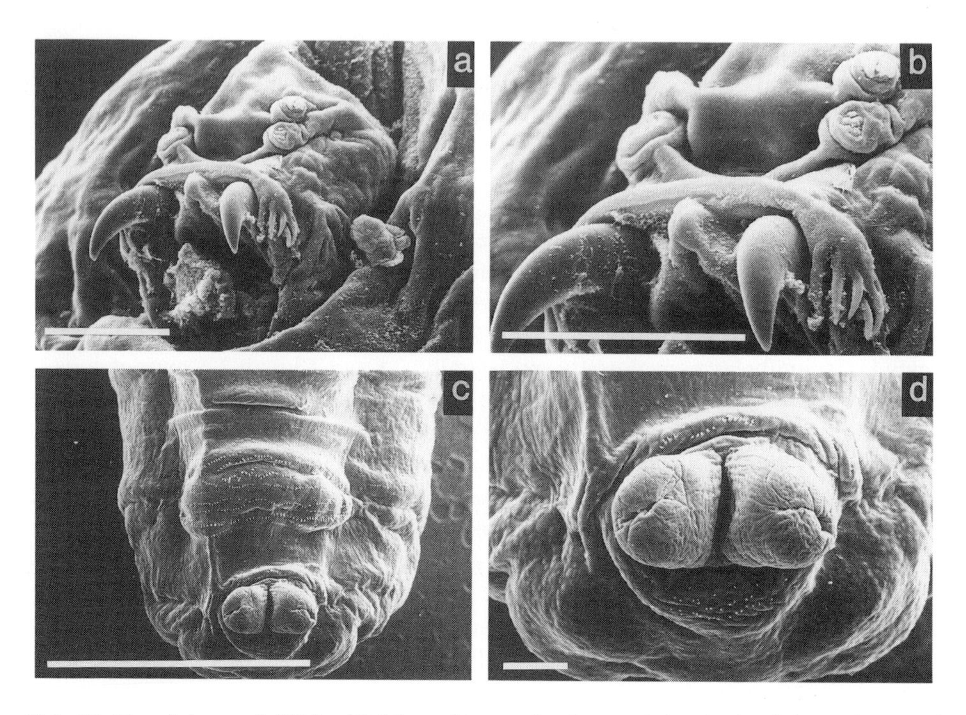

Plate 35. *Rhagoletis cerasi*, SEMs of 3rd instar larva; a, head, ventral view (scale=0.1mm); b, preoral teeth, (scale=0.1mm); c, creeping welt and anal lobes (scale=1.0mm); d, anal lobes (scale=0.1mm).

Katsoyannos (1979; 1982; 1989a), Stadler *et al.* (1987); RACES, Boller & Bush (1974), Faber (1979) and Ranner (1988b); REARING, Boller (1989b) and Tsitsipis (1989); TRAPPING, Economopoulos (1989) and Katsoyannos (1989b).

Rhagoletis cingulata (Loew) Species Complex

North American Cherry Fruit Fly Species Complex

(fig. 235)

Taxonomic notes

This is a complex of four species (Bush, 1966), two of which are serious cherry pests, and two of which attack wild species of Oleaceae. Prior to the work of Bush (1966) both of these cherry pests were regarded as a single nominal species, namely *Rhagoletis cingulata* (Loew), with the western North American form sometimes called *R. cingulata* ssp. *indifferens* Curran. Morphologically the four species are very difficult to separate and full details are beyond the scope of the present work.

Adult identification
A complex of predominantly black species, with a characteristic combination of wing (fig. 235) and scutellar markings (fig. 119).

Rhagoletis cingulata complex: *R. cingulata* (Loew)

Eastern Cherry Fruit Fly

Taxonomic notes
This species has also been known as *Trypeta cingulata* Loew; all pre-1966 references to *Rhagoletis cingulata* in western North America were based on misidentifications of *R. indifferens*.

Pest status and commercial hosts
A pest of sour cherry (*Prunus cerasus*) and sweet cherry (*P. avium*), which is also recorded from black cherry (*P. serotina*) and mahaleb cherry (*P. mahaleb*); it rarely attacks chokecherry (*P. virginiana*) and there is a doubtful record from pin cherry (*P. pensylvanica*) (Bush, 1966).

Adult identification
See Bush (1966) for separation from *R. indifferens*.

Partial description of third instar larva
Very similar to *R. fausta* and *R. indifferens*. *Head*: Stomal sensory organ large, with a single, small, blunt sclerotised preoral tooth; mouthhooks heavily sclerotised, each with a slender curved apical tooth. *Thoracic and abdominal segments*: A8 with intermediate tubercles bifurcated. Anterior spiracles each with an irregular row of 21-31 short tubules. Posterior spiracular slits each about 3-5 times as long as broad (mean 4.2), with a heavily sclerotised rima; spiracular hairs long, about as long as spiracular slit, many branched; dorsal and ventral hair bundles of 12-14 hairs; lateral hair bundles of 4-14 hairs. Anal lobes very large, protuberant. *Source*: Based on Blanc & Keiffer (1955), Kandybina (1977) and Phillips (1946).

Distribution
North America: Eastern USA, Canada (southern Quebec).
A distribution map was produced by CIE (1990a).

Other references
BEHAVIOUR, Katsoyannos (1989b) and Smith (1984); BIBLIOGRAPHY, AliNiazee & Brown (1974); DEVELOPMENT RATE, Fletcher (1989a); ECOLOGY, Debouzie (1989); LARVA, Benjamin (1934); PARASITOIDS, Bush (1966) and Muesebeck (1980); PHEROMONES, Averill & Prokopy (1989a); REVIEW, Blanc & Keifer (1955); TRAPPING, Economopoulos (1989) and Katsoyannos (1989b).

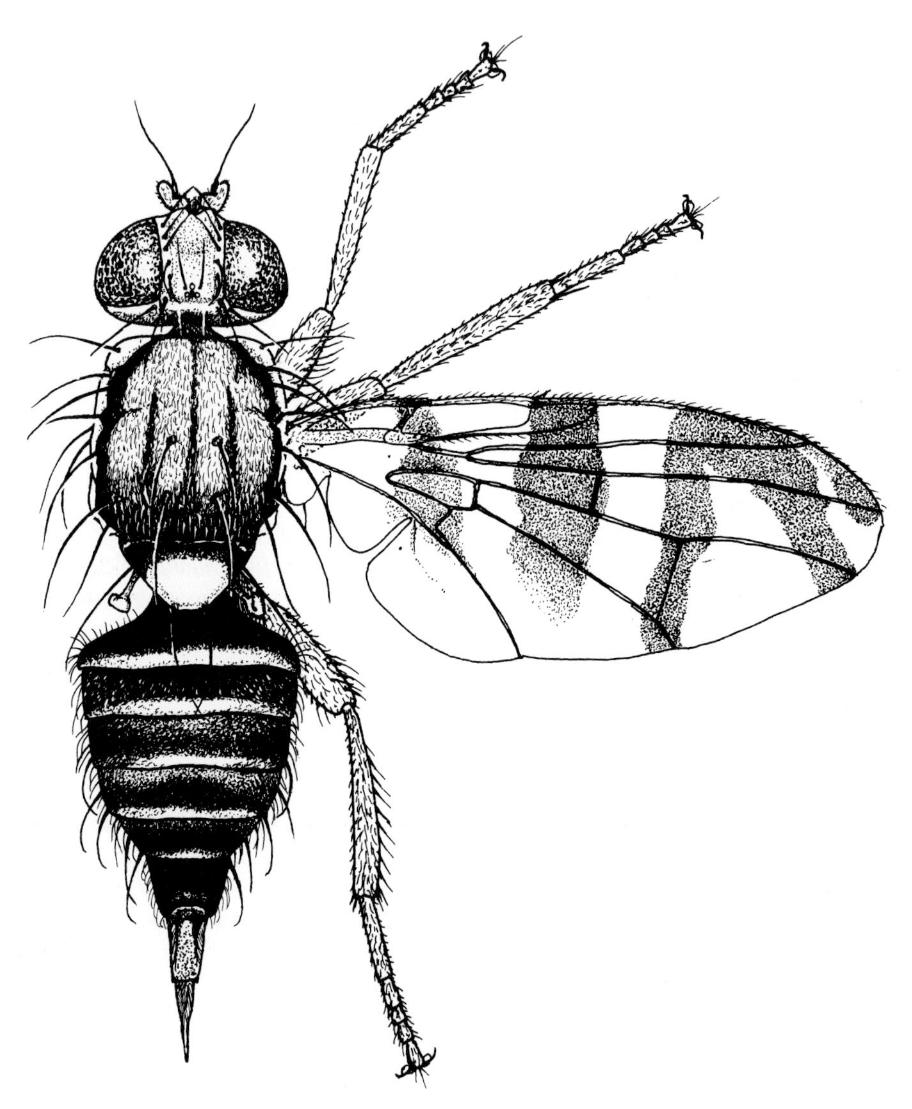

Fig. 235. *Rhagoletis cingulata*, adult female.

Rhagoletis cingulata complex: *R. indifferens* Curran

Western Cherry Fruit Fly

Taxonomic notes

This species has also been known as *Rhagoletis cingulata* ssp. *indifferens* Curran. Most pre-1966 literature does not make any distinction between *R. cingulata* and *R. indifferens*.

Pest status and commercial hosts

This is a pest of sweet cherry (*Prunus avium*), which is also recorded from chokecherry (*P. virginiana*), Japanese plum (*P. salicina*) and Klamath plum (*P. subcordata*) (Bush, 1966). There is also a record from sour cherry (*P. cerasus*) (Foote & Blanc, 1963).

Wild hosts

The main native host of this species is the wild pin cherry (*Prunus emarginata* (Dougl.) D. Dietr.) (Banham, 1971; Bush, 1966).

Adult identification

See Bush (1966) for separation from *R. cingulata*.

Identification of third instar larva

Very similar to *R. cingulata* and *R. fausta*. Blanc & Keiffer (1955) studied long series of both *R. cingulata* and *R. indifferens* and found *R. indifferens* only had 7-19 tubules on each anterior spiracle, and the posterior spiracular slits were usually shorter; mean 3.5 times as long as broad. Phillips (1946) recorded other differences not substantiated by Blanc & Keiffer (1955).

Distribution

North America: Northern California to Montana and Washington State (USA), Colorado (USA), south-eastern British Columbia (Canada).
Europe: Adventive in southern Switzerland (Merz, 1991).
A distribution map was produced by CIE (1990b), but that omitted the Swiss record.

Other references

BIBLIOGRAPHY, AliNiazee & Brown (1974); BIOLOGY, Banham & Arrand (1978); DEVELOPMENT RATE, Fletcher (1989a); ECOLOGY, Fletcher (1989c), Meats (1989a) and Messina (1989b); PARASITOIDS, AliNiazee (1985), Burditt & White (1987) and Bush (1966); PHENOLOGY, AliNiazee (1979); PHEROMONES, Averill & Prokopy (1989a) and Mumtaz & AliNiazee (1983); POST-HARVEST DISINFESTATION, Burditt & Hungate (1988); REARING, Boller (1989b); REVIEW, Blanc & Keifer (1955); TRAPPING, Burditt (1988).

Rhagoletis completa Cresson

Walnut Husk Fly

(figs 125, 236)

Taxonomic notes

This species has also been known as *Rhagoletis suavis* ssp. *completa* Cresson.

Pest status and commercial hosts

A walnut pest whose larvae develop in the husks of black walnut (*Juglans nigra*) throughout its native range. The adventive population in western USA attacks California walnut (*J. californica*), English walnut (*J. regia*) and Hinds' walnut (*J. hindsii*) (Bush, 1966; J. Jenkins, pers. comm., 1991). Peach (*Prunus persica*) is also known to be attacked (Bush, 1966).

Wild hosts

Recorded from some other *Juglans* spp. (Foote, 1981).

Adult identification

A pale coloured species with a characteristic wing pattern (fig. 236).

Partial description of third instar larva

Larvae medium-sized, length 8.0-10.0mm; width 2.0mm. *Head*: Each stomal sensory organ with a single preoral tooth at base; oral ridges in 7 short rows; mouthhooks heavily sclerotised, each with a stout bluntly rounded apical tooth. *Thoracic and abdominal segments*: Spinules confined to creeping welt areas on the ventral half of the body, except for a narrow band on the anterior margin of T1. A8 with intermediate areas well developed, with 1 pair each of intermediate and ventral tubules obvious. Anterior spiracles each with an irregular row of 21 short tubules. Posterior spiracular slits each about 3 times as long as broad, with a heavily sclerotised rima; upper and lower slits at about 60° to each other; spiracular hairs short (about one-third the length of a spiracular slit), mainly branched; dorsal and ventral hair bundles of 8-11 hairs; lateral hair bundles of 4-10 hairs. Anal lobes large, protuberant, surrounded by a few discontinuous rows of small spinules. *Source*: Based on Phillips (1946) and Steyskal (1973).

Distribution

North America: Southern and central USA. Adventive in western USA since the early 1920s (Bush, 1966).
Central America: Mexico (extreme north only).
Europe: Adventive in southern Switzerland (Merz, 1991).

Other references

ANTIMETABOLITES, Tsiropoulos & Hagen (1987); BACTERIAL SYMBIONTS, Tsiropoulos (1976); BAIT SPRAY, Hislop *et al.* (1981); BEHAVIOUR, Katsoyannos (1989b) and P.H. Smith (1989); BIOLOGY, Fletcher (1989b) and Gibson & Kearby (1978); BIOCONTROL, Wharton (1989a); BIONOMICS, Boyce (1934); HOST RESISTANCE, Greany (1989); PARASITOIDS, Bush (1966), Muesebeck (1980) and Wharton (1989a); PHEROMONES, Averill & Prokopy (1989a); INTEGRATED PEST MANAGEMENT, Haley & Baker (1982); TRAPPING & BAITS, Katsoyannos (1989b), Riedl & Hoying (1981) and Riedl & Hislop (1985).

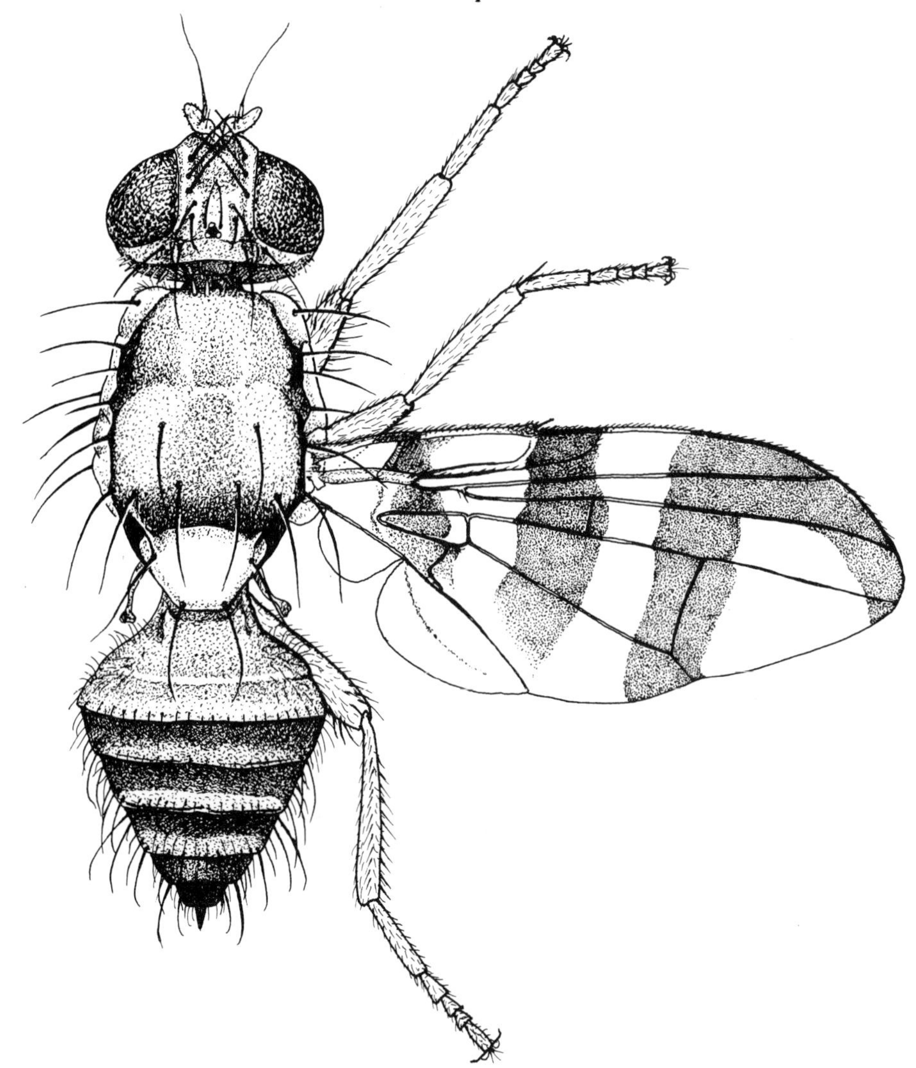

Fig. 236. *Rhagoletis completa*, adult female.

Rhagoletis conversa (Brèthes)

(figs 121, 237)

Taxonomic notes
This species has also been known as *Spilographa conversa* Brèthes.

Commercial hosts
A potential pest of solanaceous crops which has been extensively studied (Frias, 1989; Frias *et al.*, 1987). It was recorded from black nightshade (*Solanum nigrum*) by Frias

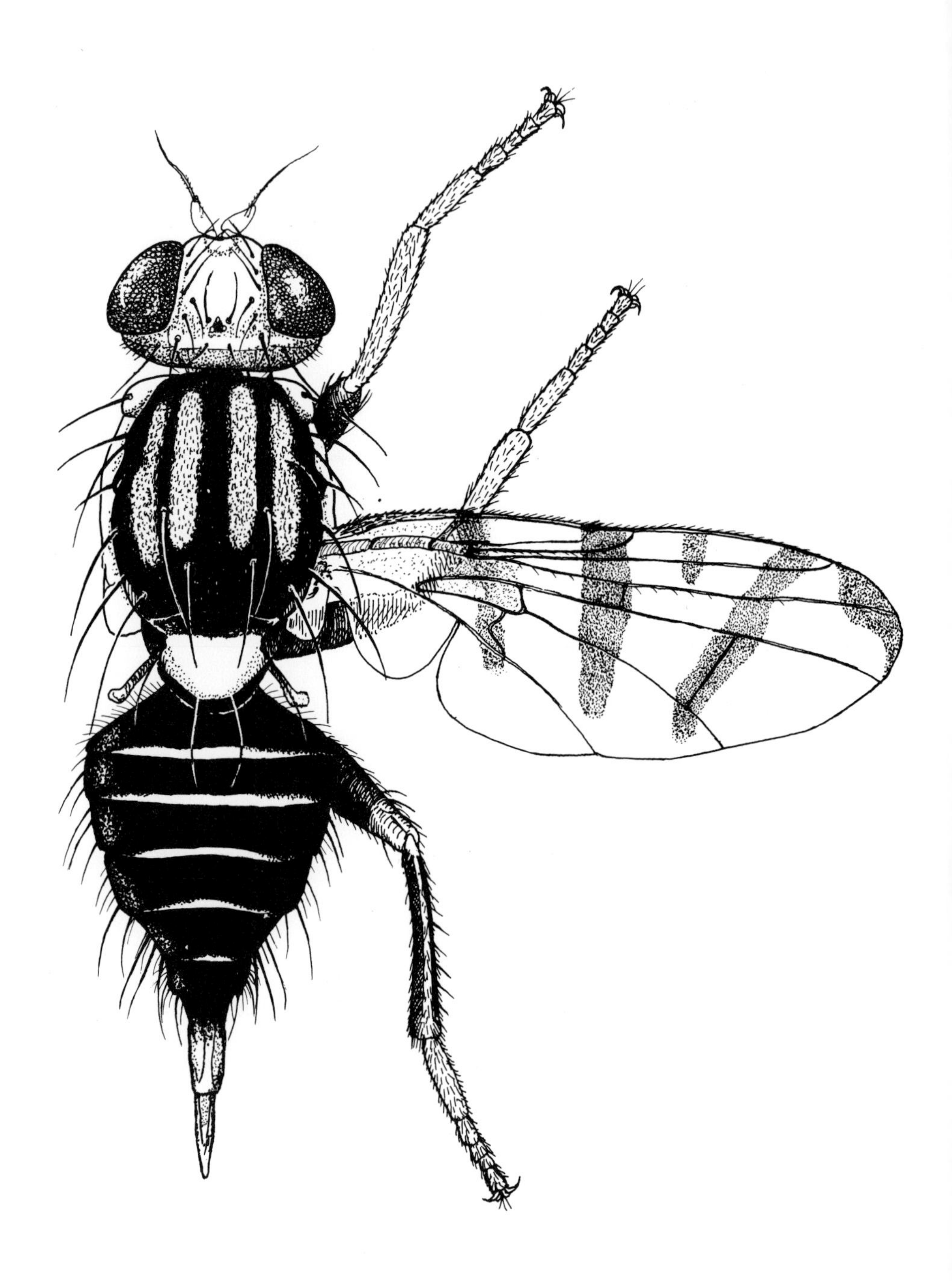

Fig. 237. *Rhagoletis conversa*, adult female.

et al. (1987) and Foote (1981) apparently examined specimens from "husk-tomato" and "tomatillo" both of which are names for *Physalis philadelphica*; the latter host requires further confirmation as "tomatillo" may have been a reference to its wild host which is *S. tomatillo* Phil. f.

Wild hosts
Also recorded from *Solanum tomatillo* (Frias, 1989; Frias *et al.*, 1987).

Adult identification
A predominantly black species, with a characteristic combination of wing (fig. 237) and scutellar markings (fig. 118).

Distribution
South America: Chile.
South Pacific: Easter Island; probably adventive

Other reference
BEHAVIOUR, P.H. Smith (1989).

Rhagoletis fausta **(Osten Sacken)**

Black Cherry Fruit Fly

(fig. 238)

Taxonomic notes
This species has also been known as *Rhagoletis intrudens* Aldrich and *Trypeta (Acidia) fausta* Osten Sacken.

Pest status and commercial hosts
A pest of sour cherry (*Prunus cerasus*) and sweet cherry (*P. avium*), which is also recorded from black cherry (*P. serotina*), chokecherry (*P. virginiana*), mahaleb cherry (*P. mahaleb*) and pin cherry (*P. pensylvanica*) (Bush, 1966).

Wild hosts
Wild plants of pin cherry are the major native host (Bush, 1966).

Adult identification
A predominantly black species, with characteristic wing markings (fig. 238) and a scutellum that lacks a black basal mark (fig. 116). Care should be taken not to confuse trapped specimens of this species with superficially similar members of the leaf mining genus *Euleia* (fig. 252).

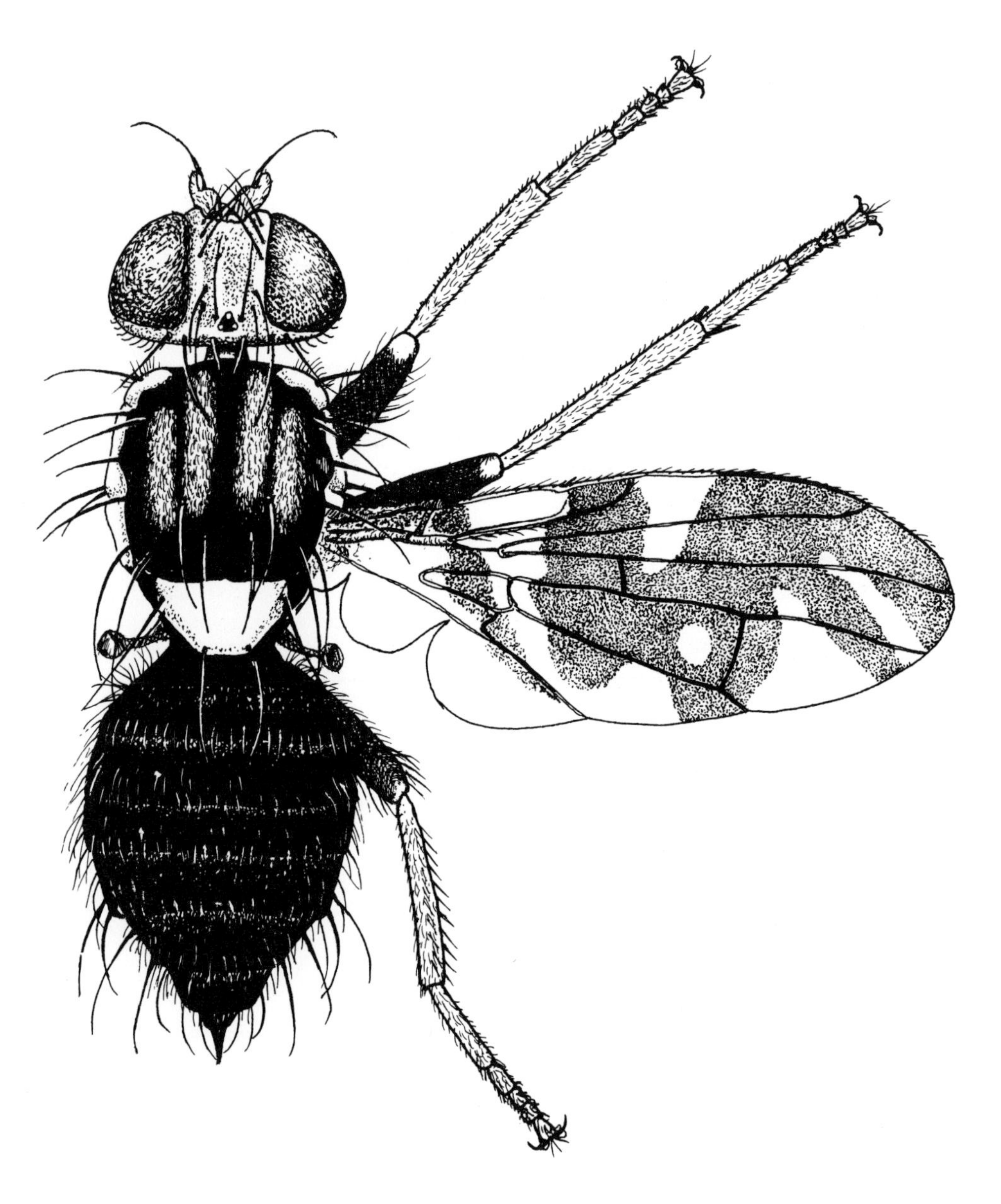

Fig. 238. *Rhagoletis fausta*, adult female.

Partial description of third instar larva
Very similar to *R. cingulata* and *R. indifferens* and further study is needed to substantiate any recorded differences between *R. fausta* and those spp. *Head*: Stomal sensory organ large with a single small, blunt, sclerotised preoral tooth; mouthhooks heavily sclerotised, each with a stout strongly curved apical tooth. *Thoracic and abdominal segments*: A8 with intermediate tubercles single. Anterior spiracles each with a uniform row of 19 short tubules. Posterior spiracular slits each about 4 times as long as broad, with a heavily sclerotised rima; spiracular hairs shorter than length of a spiracular slit, many branched; dorsal and ventral hair bundles of 9-11 hairs, lateral hair bundles of 4-6 hairs. Anal lobes very large, protuberant. *Source*: Based on Phillips (1946).

Distribution
North America: An eastern population from south-central Canada to north-eastern USA, and a western population from central California (USA) to southern Alberta (Canada). A distribution map was produced by CIE (1963a).

Other references
BEHAVIOUR, Katsoyannos (1989b); BIBLIOGRAPHY, AliNiazee & Brown (1974); DEVELOPMENT RATE, Fletcher (1989a); ECOLOGY, Debouzie (1989); PARASITOIDS, Bush (1966) and Muesebeck (1980); PHEROMONES, Averill & Prokopy (1989a); TRAPPING, Economopoulos (1989) and Katsoyannos (1989b).

Rhagoletis juglandis Cresson

(fig. 239)

Commercial hosts
A potential walnut pest whose larvae develop in the husks of English walnut (*Juglans regia*) (Bush, 1966).

Wild hosts
Recorded from *Juglans major* (Torr.) Heller (Bush, 1966).

Adult identification
A pale coloured species with a characteristic wing pattern (fig. 239).

Partial description of third instar larva
Larvae medium-sized, length 8.0mm; width 1.5-2.0mm. *Head*: Oral ridges in 7 rows; mouthhooks heavily sclerotised, each with a long slender strongly curved apical tooth. *Thoracic and abdominal segments*: Spinules small, short, thick, confined to 9 fusiform areas; the 2 anterior areas very small, with only 4-5 short rows of spinules, scarcely extending laterally; other areas wider with 8-10 rows; A8 with surface smooth, unwrinkled, with 4 pairs of tubercles. Anterior spiracles each with a row of 11 large tubules. Posterior spiracular slits long, each about 4 times as long as broad with a

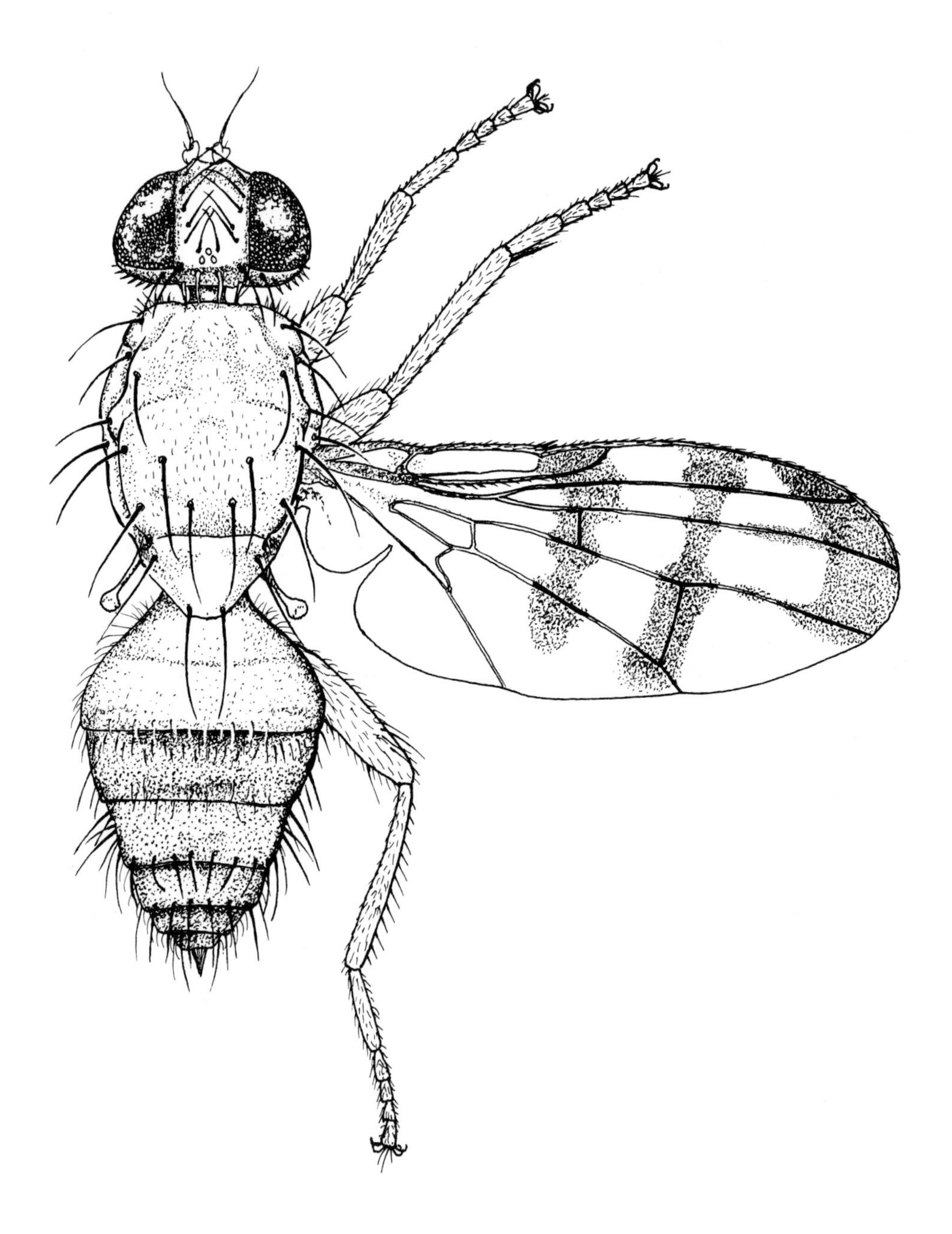

Fig. 239. *Rhagoletis juglandis*, adult female.

heavily sclerotised rima; upper and lower slits at about 90° to each other; spiracular hairs frequently branched; dorsal and ventral hair bundles of 15-17 hairs; lateral hair bundles of 7-10 hairs. Anal lobes large, protuberant. *Source*: Based on Phillips (1946) and Steyskal (1973).

Distribution
Central America: Mexico.
North America: Arizona, New Mexico, Utah (USA).

Other references
PARASITOIDS, Bush (1966) and Muesebeck (1980).

Rhagoletis lycopersella Smyth

(fig. 240)

Taxonomic notes
Some records of the so called *Rhagoletis ochraspis* (Wiedemann) were probably based on misidentifications of this species.

Pest status and commercial hosts
This species is sometimes a pest of tomato (*Lycopersicon esculentum*) in areas where its natural host, the currant tomato (*L. pimpinellifolium* (L.) Mill.) is found (Foote, 1981).

Wild host
Currant tomato (*L. pimpinellifolium*) is an important reservoir host (Foote, 1981).

Adult identification
A predominantly black species, with a characteristic combination of wing (fig. 240) and scutellar markings (figs 118, 119).

Distribution
South America: Peru.

Other references
BIOLOGY, Fletcher (1989b) and Smyth (1960).

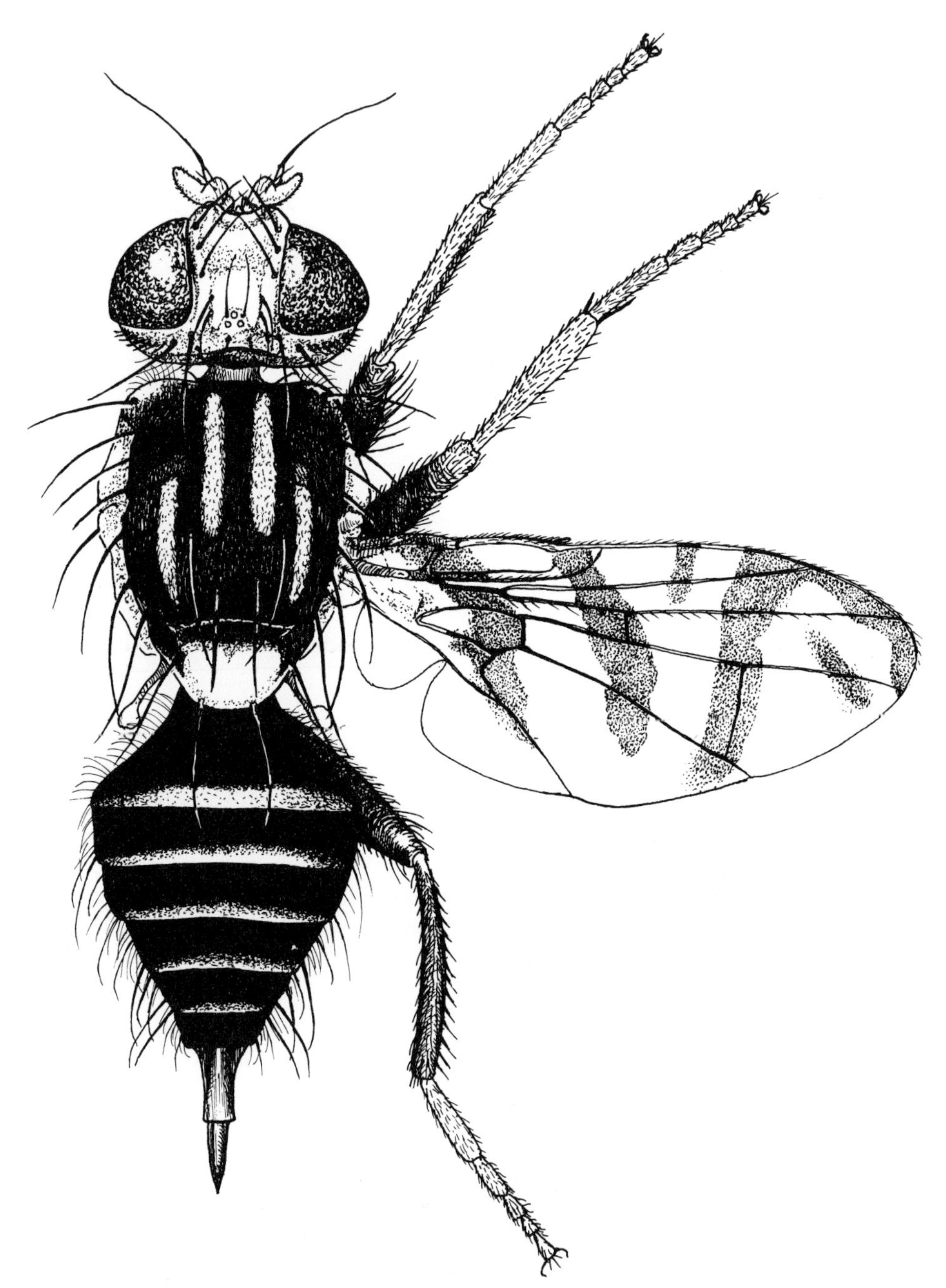

Fig. 240. *Rhagoletis lycopersella*, adult female.

Rhagoletis nova (Schiner)

(figs 122, 241)

Taxonomic notes
This species has also been known as *Spilographa nova* Schiner and some references to the so called *Rhagoletis ochraspis* (Wiedemann) were probably based on misidentifications of this species.

Pest status and commercial hosts
A pest of pepino (*Solanum muricatum*) (Frias, 1986a).

Adult identification
A predominantly black species, with a characteristic combination of wing (fig. 241) and scutellar markings (figs 117, 118).

Partial description of third instar larva
Larva medium-sized, length 5.0-6.0mm. Anterior spiracles with an irregular row of 23 tubules. Posterior spiracular slits very long [according to illustrations], each 6-8 times as long as broad; hair bundles of 8-11 hairs, few branched. *Source*: Based on Frias (1986b).

Distribution
South America: Chile.

Rhagoletis pomonella (Walsh) Species Complex

Apple Maggot Fly Species Complex

(figs 29, 126, 242)

Taxonomic notes
This complex includes four described species, namely *Rhagoletis cornivora* Bush (p. 388), *R. mendax* Curran, *R. pomonella* and *R. zephyria* Snow (p. 389); there are also two unnamed species discovered by Smith (1988a), and S.H. Berlocher and J. Payne (in prep.), which will be formally described by J. Jenkins (in prep.), and there is evidence of further undescribed species (S.H. Berlocher, pers. comm., 1991). Prior to the work of Bush (1966) they were generally regarded as a single nominal species and the genetic basis for recognition of these species has been discussed by a number of authors (Bierbaum & Bush, 1990a; 1990b; Diehl & Prokopy, 1986; Feder *et al.*, 1988; 1989; Feder & Bush, 1989). These species are very difficult to separate morphologically and full details are beyond the scope of the present work.

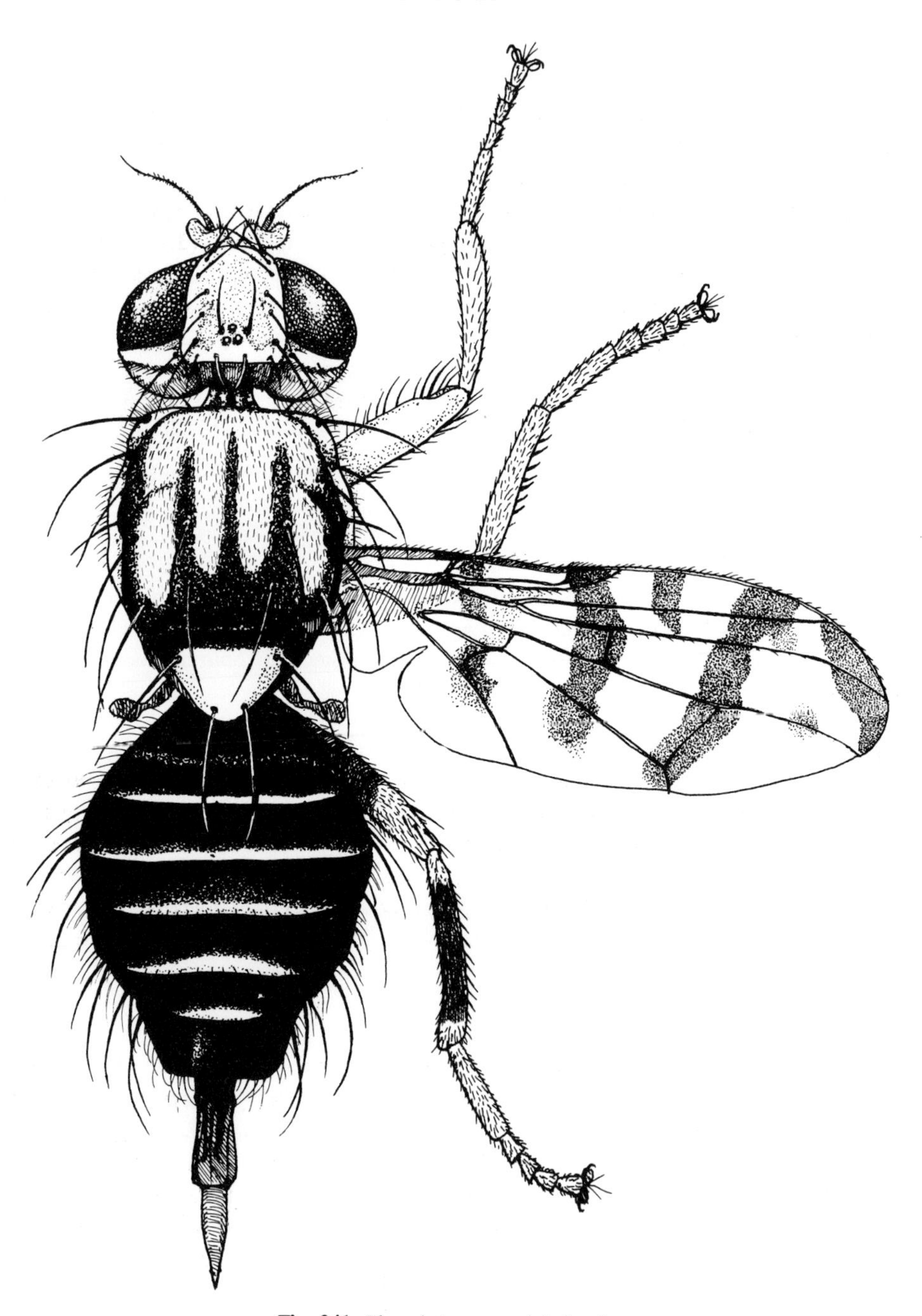

Fig. 241. *Rhagoletis nova*, adult female.

Adult identification

A complex of predominantly black species, with a characteristic combination of wing (fig. 242) and scutellar markings (fig. 119). The four described species can easily be separated according to their host plants, provided that reared specimens are available. If trapped samples have to be identified for a monitoring programme, then the aculeus length measurements given by Bush (1966) should enable the apple race of *R. pomonella* (1.0-1.4mm) and most populations from *Crataegus* (1.1-1.5mm) to be separated from both *R. mendax* and *R. zephyria* (both 0.7-0.9mm). Florida populations of *R. pomonella* from both *Crataegus* and *Prunus* spp. are very small and overlap that separation (aculeus length 0.7-1.0mm); *R. cornivora* also overlaps (aculeus length 0.7-1.1mm). Bush (1966) also gave details of differences in male surstyli form between the four species and Westcott (1982) gave further details of the separation of *R. pomonella* and *R. zephyria*.

Rhagoletis pomonella Complex: *R. mendax* Curran

Blueberry Maggot Fly

(fig. 161; pl. 36)

Taxonomic notes

Records of *Rhagoletis pomonella* (Walsh) from species of Ericaceae were based on misidentifications of *R. mendax*. *R. mendax* has also been confused with at least one other undescribed species which was discovered in recent work by S.H. Berlocher & J. Payne (in prep.). They found that autumn and summer populations attack different hosts, have electrophoretic differences and slight differences in their biology and morphology; the newly recognised species will be formally described by J. Jenkins (in prep.).

Pest status and commercial hosts

A complex of pests which attack cultivated and wild-harvested species of Ericaceae, namely black huckleberry (*Gaylussacia baccata*), dangleberry (*G. frondosa*), dwarf huckleberry (*G. dumosa*), highbush blueberry (*Vaccinium corymbosum*), lowbush blueberry (*V. angustifolium*) and mountain cranberry (*V. vitis-idaea*); cranberry (*V. macrocarpon*) and wintergreen (*Gaultheria procumbens*) are rarely attacked (Bush, 1966); there is also a record from sourtop blueberry (*V. myrtilloides*) (Wasbauer, 1972). Rabbit-eye blueberry (*V. ashei* Reade), which is now being grown extensively in southern USA, is not known to be attacked (G.L. Bush, pers. comm., 1991).

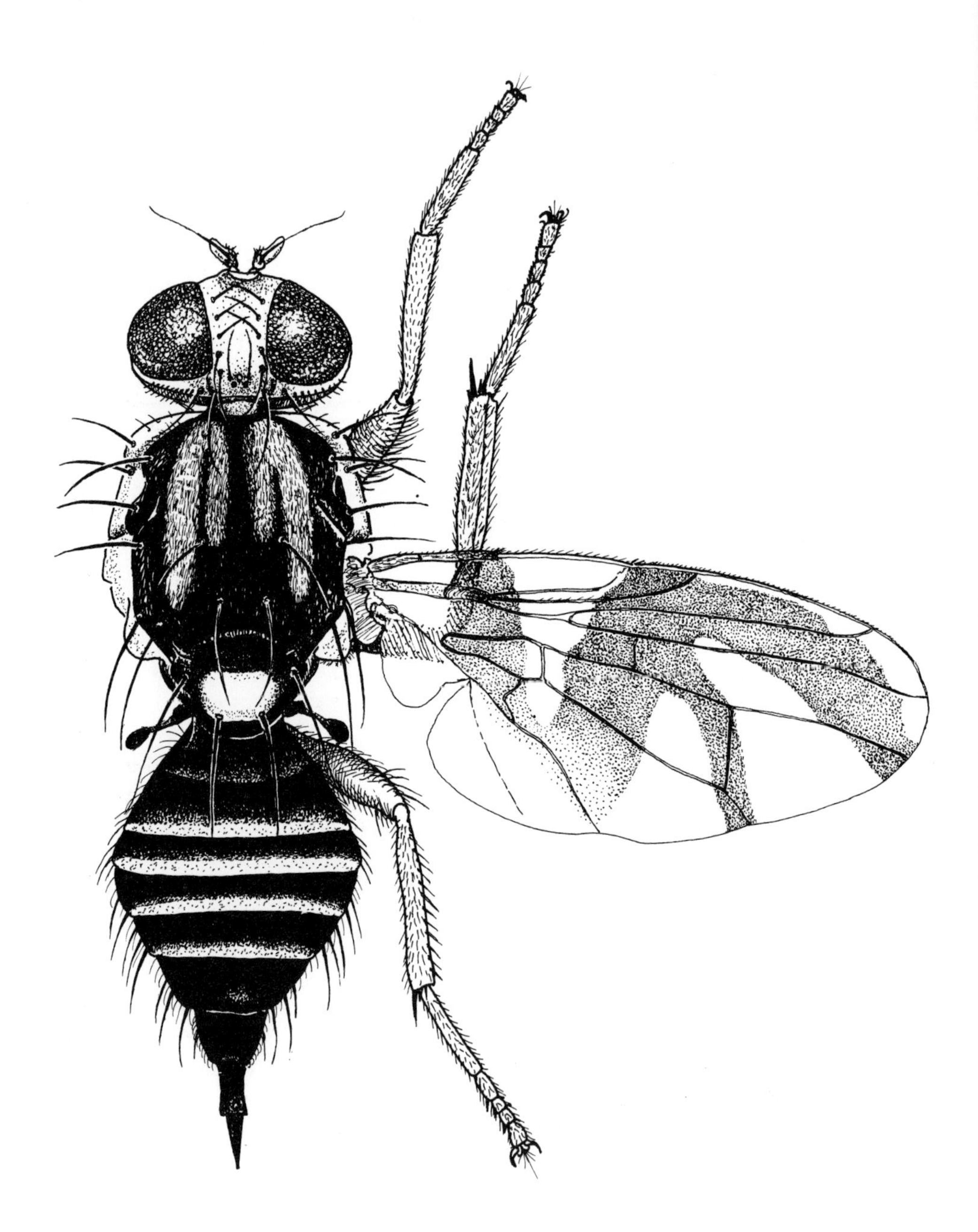

Fig. 242. *Rhagoletis pomonella*, adult female.

Wild hosts

Recorded from some other wild *Gaylussacia* and *Vaccinium* spp. Populations associated with wild plum (*Prunus* sp.) also appear to be *R. mendax* (S.H. Berlocher, pers. comm., 1991).

Adult identification

See Bush (1966) for separation from *R. pomonella*.

Description of third instar larva

Larvae medium to large sized, length 8.0-11.0mm; width 1.5-1.8mm. *Head*: Stomal sensory organ rounded, slightly protuberant with a small peg-like sensillum; preoral lobes absent; preoral teeth (pl. 36.b) large, paired at base of stomal sensory organ; oral ridges (pl. 36.a, b) in 5-7 short, unserrated rows; mouthhooks heavily sclerotised, each with a large, strongly curved apical tooth but lacking a preapical tooth (fig. 161). *Thoracic and abdominal segments*: A distinct anterior band of small spinules, in discontinuous rows, surrounding T2, T3 and A2-A7. T1 and A8 without spinules dorsally. A1-A7 with broad bands of small, stout spinules forming large creeping welts ventrally. A8 with intermediate areas large, well defined, with large tubercles and sensilla; dorsal, lateral and ventral tubercles all very well developed. *Anterior spiracles* (pl. 36.c): A large irregular row of 13-16 stout, finger-like, well spaced tubules. *Posterior spiracles* (pl. 36.d, e): Spiracular slits large, each about 3.5 times as long as wide, with a stout rima. Spiracular hairs branched, relatively short (less than length of a spiracular slit); dorsal and ventral spiracular hair bundles of 9-15 hairs; lateral bundles of 6-10 hairs. *Anal area* (pl. 36.d, f): Lobes very large, protuberant, with 3-5 discontinuous rows of spinules dorsally, extending into lateral rows, but reduced to a small concentration of individual spinules below anal opening. *Source*: Based on specimens from Maine, USA (ex *Vaccinium* sp.).

Distribution

North America: North-eastern USA, south-eastern Canada.

Other references

BEHAVIOUR, Diehl & Prokopy (1986), Katsoyannos (1989b), Sivinski & Burk (1989), P.H. Smith (1989) and Smith & Prokopy (1982); ECOLOGY, Fletcher (1989c); GENETIC SEPARATION FROM *R. POMONELLA*, Feder *et al.* (1989); HOST FRUIT VOLATILES, Frey & Bush (1990) and Lugemwa *et al.* (1989); LARVA, Jones & Kim (1988) and Phillips (1946); PARASITOIDS, Bush (1966) and Geddes *et al.* (1987); PHEROMONES, Averill & Prokopy (1989a); REVIEW, Neilson & Wood (1985); TRAPPING, Economopoulos (1989), Katsoyannos (1989b), Neilson *et al.* (1984), Prokopy & Coll (1978) and Wood *et al.* (1983).

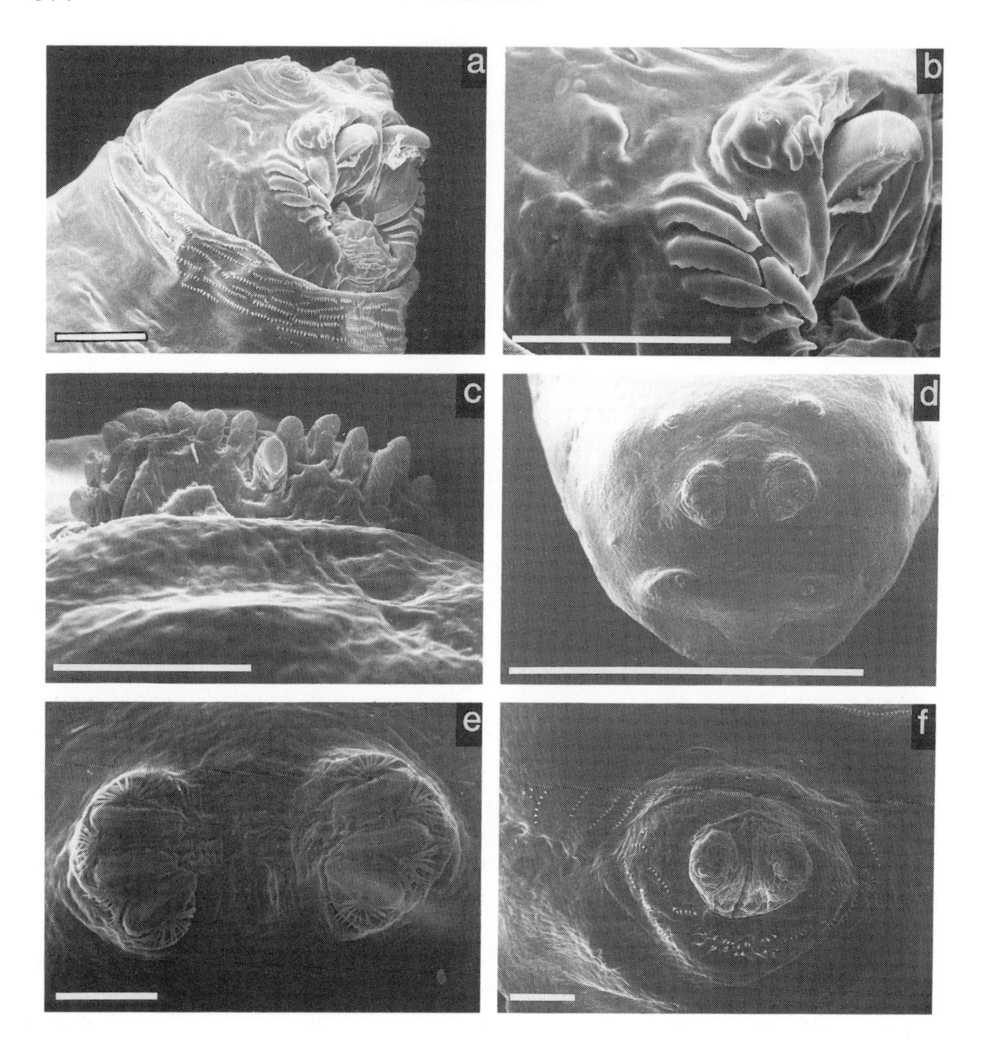

Plate 36. *Rhagoletis mendax*, SEMs of 3rd instar larva; a, head and T1, lateral view (scale=0.1mm); b, stomal sensory organ, preoral teeth and oral ridges, lateral view (scale=0.1mm); c, anterior spiracle (scale=0.1mm); d, A8, posterior view (scale=1.0mm); e, posterior spiracles (scale=0.1mm); f, anal lobes (scale=0.1mm).

Rhagoletis pomonella **Complex:** ***R. pomonella*** **(Walsh)**

Apple Maggot
Apple Maggot Fly

(fig. 162; pl. 37)

Taxonomic notes

This species has also been known as *Trypeta pomonella* Walsh. Many pre-1966 records of *Rhagoletis pomonella* were based on misidentifications of *R. cornivora* (*Cornus* spp.; Cornaceae), *R. mendax* Curran (Ericaceae) and *R. zephyria* (*Symphoricarpos* spp.; Caprifoliaceae); true *R. pomonella* is associated with Rosaceae.

Races

The main commercial host of *R. pomonella* is apple (*Malus domestica*), which was first introduced to eastern North America (New England) some 200 years before the first report of apple being attacked by this fly (Bush, 1966). The original native host of *R. pomonella* was almost certainly hawthorn (*Crataegus* spp.). Apple was introduced to Mexico in 1522, about 100 years earlier than to New England. Mexican and north-eastern United States populations on apple closely match regional varieties of wing patterning found in hawthorn associated populations in those areas, indicating that colonization of apple probably occurred independently in each region (Bush, 1966). Apple fruits mature about one month earlier than those of hawthorn. Correspondingly, the apple race flies emerge about one month earlier than the hawthorn race flies and that severely limits the potential for gene flow between the races (Bush, 1969; Smith, 1988b). Bush (1969) suggested that the populations on apples were initially founded by a few flies that emerged early and that within a few generations a race of early emerging flies with a preference for ovipositing on apples evolved in sympatry with the original hawthorn race. During the relatively short period since the colonization of apple, the apple and hawthorn populations have diverged genetically to form distinct apple and hawthorn races (Bush, 1969; Feder *et al.*, 1989; 1990a; 1990b; McPheron *et al.*, 1988b).

Pest status and commercial hosts

A serious pest of apple (*Malus domestica*), which is also recorded from Chickasaw plum (*Prunus angustifolia*), peach (*P. persica*) and Siberian crabapple (*M. baccata*); larvae have also been found in pear (*Pyrus communis*), but no adults emerged (Bush, 1966). In New England (USA) *R. pomonella* utilizes the hips (fruits) of Japanese rose (*Rosa rugosa*) and *R. carolina* L. (as *R. virginiana*) as alternative hosts (Prokopy & Berlocher, 1980). Recently, *R. pomonella* has adapted to attacking sour cherry (*Prunus cerasus*) in Utah (USA) where it has not been reported from apple (Jorgensen *et al.*, 1986; McPheron *et al.*, 1988a; Messina, 1989a), and there is a record from apricot (*Prunus armeniaca*) in New York (USA) (Lienk, 1970). Ornamental hawthorns (*Crataegus* spp.) may also be attacked by the hawthorn race.

Records from rowan (*Sorbus aucuparia*) and sweet cherry (*P. avium*) (Cresson,

1929; Pickett & Neary, 1940) require confirmation. Records from black huckleberry (*Gaylussacia baccata*), cranberry (*Vaccinium macrocarpon*), "cultivated plum" (presumably *P. domestica*), highbush blueberry (*V. corymbosum*), lowbush blueberry (*V. angustifolium*), mountain cranberry (*V. vitis-idaea*), tomato (*Lycopersicon esculentum*) and wintergreen (*Gaultheria procumbens*) (Pickett, 1937; Pickett & Neary, 1940) were almost certainly based on misidentifications of *R. mendax* and other species.

Wild hosts

The major natural hosts from which the pest populations have evolved are hawthorns (*Crataegus* spp.); *R. pomonella* is also recorded from some *Amelanchier*, *Aronia* and *Cotoneaster* spp. (all Rosaceae) (Bush, 1966).

Adult identification

See Bush (1966) for separation from *R. mendax* and Westcott (1982) for separation from *R. zephyria*.

Description of third instar larva

Larvae medium to large, length 8.0-11.0mm; width 1.5-2.0mm. *Head*: Stomal sensory organ rounded, slightly protuberant, with a small peg like sensillum; preoral lobes absent; preoral teeth (pl. 37.b) large, paired, at base of stomal sensory organ; oral ridges (pl. 37.a, c) in 5-7 short unserrated rows; mouthhooks stout, heavily sclerotised,without preapical teeth (fig. 162). *Thoracic and abdominal segments*: T1 with 3-6 rows of small spinules laterally and ventrally, none dorsally; T2 with 2-3 rows in a similar arrangement to T1; T3 and A1-A7 with broad encircling bands of small, stout spinules forming creeping welts on ventral surfaces; A8 without spinules dorsally but with spinules in discontinuous rows ventrally and laterally. A1-A8 creeping welt spinules with a few rows anteriorly directed but the majority posteriorly directed. A8 intermediate areas large, well defined, with very obvious tubercles and sensilla (pl. 37.e); dorsal, lateral and ventral tubercles all very well developed. *Anterior spiracles* (pl. 37.d): Each with 17-25 stout, well spaced tubules arranged irregularly in 2 rows. *Posterior spiracles* (pl. 37.e): Placed just above the midline; spiracular slits large, well defined, each about 4 times as long as broad with a sclerotised rima. Spiracular hairs, branched, relatively short (less than half the length of spiracular slit); dorsal and ventral bundles of 9-16 hairs, lateral bundles of 6-10 hairs. *Anal area* (pl. 37.f): Lobes very large, protuberant, with 3-5 discontinuous rows of spinules dorsally, extending into lateral rows, but reduced to a small concentration of individual spinules below anal opening. *Source*: Based on specimens from Vermont, USA (ex *Malus domestica*).

Distribution

Central America: Mexico.
North America: Eastern states of USA and Canada; in 1981 it became established in Oregon (western USA), from where it has spread to California, Washington State and

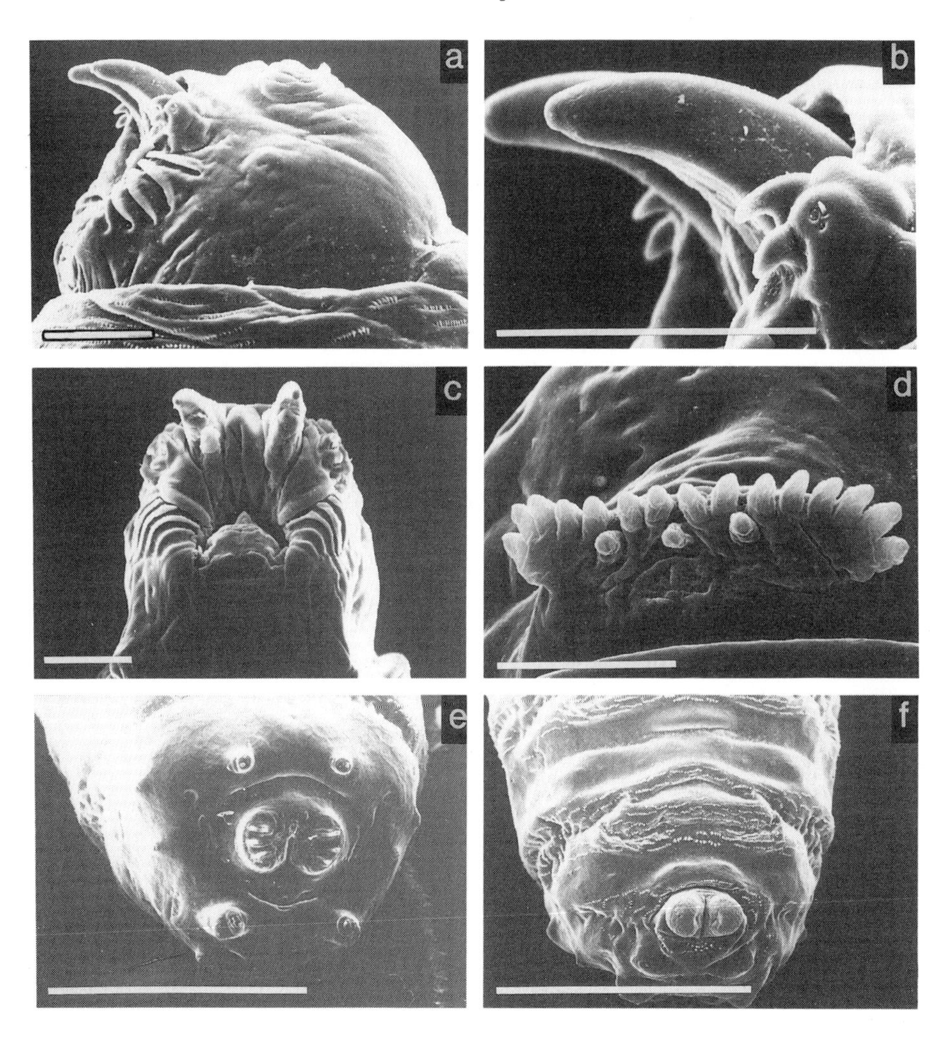

Plate 37. *Rhagoletis pomonella*, SEMs of 3rd instar larva; a, head and T1, lateral view (scale=0.1mm); b, stomal sensory organ, preoral teeth, lateral view (scale=0.1mm); c, head, ventral view (scale=0.1mm); d, anterior spiracle (scale=0.1mm); e, A8, posterior view (scale=1.0mm); f, A8, creeping welt and anal lobes (scale=1.0mm).

Colorado; more recently found in Nebraska (J. Jenkins, pers. comm., 1991) and probably only recently introduced. The recent establishment of *R. pomonella* in western North America was reviewed by AliNiazee & Brunner (1986) and the distribution was mapped by CIE (1989b) (see also AliNiazee & Penrose, 1981; AliNiazee & Westcott, 1986).

Temperate Asia: Recorded from Afghanistan (Ullah, 1988), but that is almost certainly an error.

Other references

BEHAVIOUR, Averill *et al.* (1987), Crnjar *et al.* (1987), Diehl & Prokopy (1986), Katsoyannos (1989b), Mangel & Roitberg (1989), Owens & Prokopy (1986), Papaj & Prokopy (1986), Prokopy (1986), Prokopy & Bush (1973), Prokopy & Roitberg (1989), Prokopy *et al.* (1986; 1989a), Roitberg & Prokopy (1984), Roitberg *et al.* (1982), Sivinski & Burk (1989), P.H. Smith (1989) and Smith & Prokopy (1980); BIBLIOGRAPHY, Rivard (1968) and Zwölfer (1985); BIOLOGY, Fletcher (1989a); CONTROL, Belanger *et al.* (1985) and Prokopy (1988); DATASHEET, Agriculture Canada (1973); DEVELOPMENT RATE, Fletcher (1989a); ECOLOGY, Averill & Prokopy (1989b), Cameron & Morrison (1974b; 1977), Debouzie (1989), Fletcher (1989c) and Meats (1989a); GENETIC SEPARATION FROM *R. MENDAX*, Feder *et al.* (1989); HOST FRUIT VOLATILES, Carle *et al.* (1987) and Frey & Bush (1990); HOST RESISTANCE, Goonewardene & Howard (1989), Greany (1989) and Lamb *et al.* (1988); HOST SELECTION, Prokopy *et al.* (1985); LARVA, Berg (1979), Kandybina (1977) and Snodgrass (1924); PARASITOIDS, AliNiazee (1985), Bush (1966), Cameron & Morrison (1974a), Glas & Vet (1983), Kandybina (1977), Maier (1981), Monteith (1978) and Muesebeck (1980); PATHOGENS, Poinar *et al.* (1978); PHENOLOGY, Opp & Prokopy (1987); PHEROMONES, Averill & Prokopy (1987; 1989a), Katsoyannos (1989a), Prokopy & Webster (1978), Prokopy *et al.* (1988) and Roitberg *et al.* (1984); PREDATORS, Monteith (1977); REARING, Boller (1989b) and Tsitsipis (1989); SEROLOGY, Simon (1969); SYMBIONTS, Howard (1989), Howard & Bush (1989) and Rossiter *et al.* (1983); TRAPPING, Drummond *et al.* (1984), Economopoulos (1989), Johnson (1983), Katsoyannos (1989b) and Reissig *et al.* (1985).

Rhagoletis ribicola Doane

Dark Currant Fly

(figs 127, 243)

Commercial hosts

A potential pest of *Ribes* spp. which has been recorded from European gooseberry (*R. uva-crispa*), golden currant (*R. aureum*) and redcurrant (*R. rubrum*) (Bush, 1966).

Adult identification

A predominantly black species, with a characteristic combination of wing (fig. 243) and scutellar markings (figs 119, 120).

Partial description of third instar larva

Larvae medium-sized, length 8.0mm; width 1.4mm. *Head*: Stomal sensory organs with 6 preoral teeth at base; oral ridges in 12-13 rows; mouthhooks heavily sclerotised, each with a short stout apical tooth. *Thoracic and abdominal segments*: Spinules conspicuous, arranged in 6-8 short discontinuous rows on most segments; T1 with a wide band dorsally, with fewer spinules laterally and ventrally; T2-T3, A1-A2 with

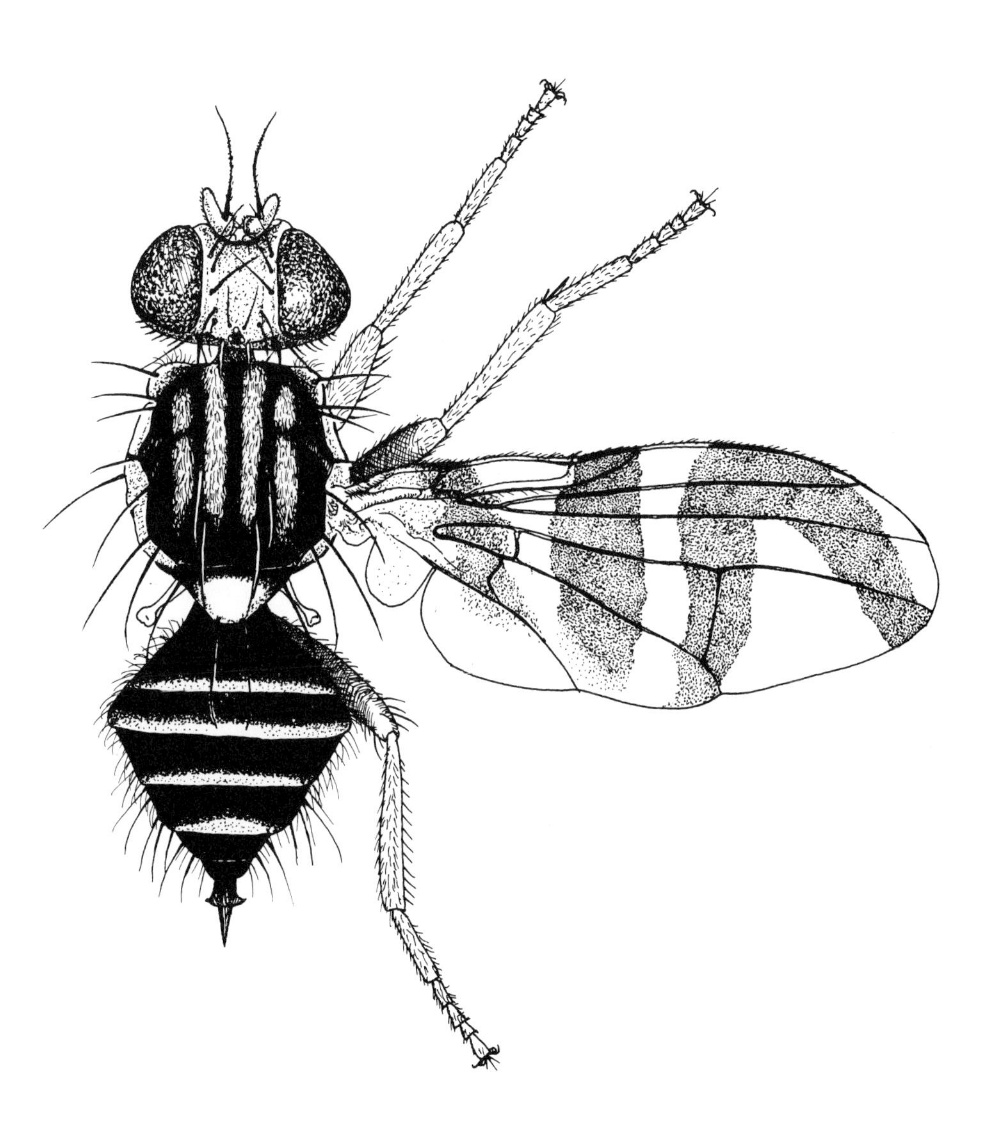

Fig. 243. *Rhagoletis ribicola*, adult female.

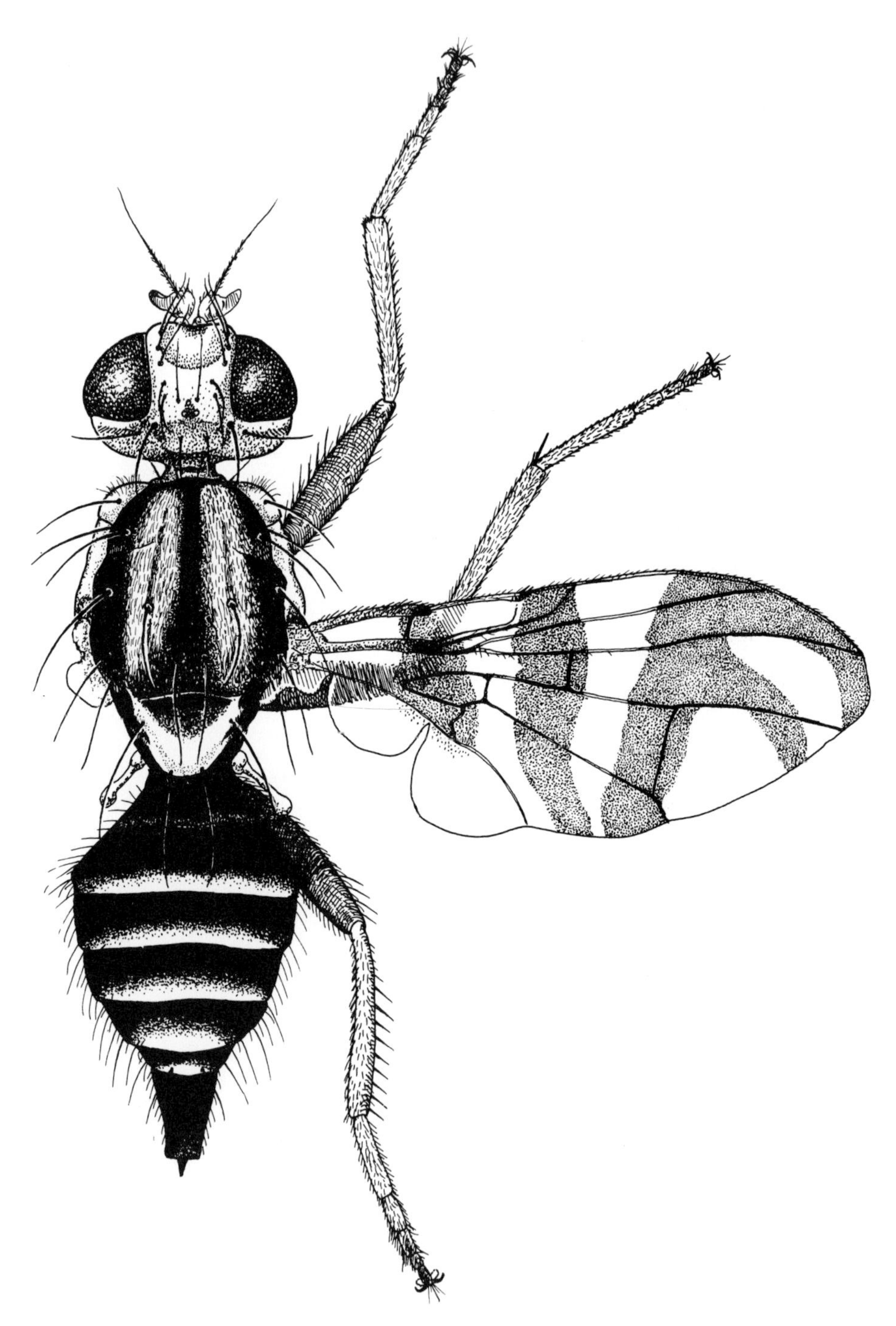

Fig. 244. *Rhagoletis striatella*, adult female.

narrower bands, sometimes only 1 dorsally; A3-A8 with larger creeping welts with spinules extending laterally; A8 with surface smooth, rounded, all depressions very shallow, tubercles small but obvious. Anterior spiracles each with a row of 16 tubules. Posterior spiracular slits each about 3 times as long as broad, with a moderately sclerotised rima; spiracular hairs about half the length of the spiracular slit, with few branches; dorsal and ventral hair bundles of 3-5 hairs, lateral hair bundles of 3 hairs. Anal lobes large and protuberant. *Source*: Based on Phillips (1946).

Distribution

North America: British Columbia (Canada) and western USA.

Other references

NEW MEXICO, Dodson (1987); TRAPPING, Madsen (1970).

Rhagoletis striatella Wulp

(figs 123, 244)

Pest status and commercial hosts

A pest of husk tomato (*Physalis philadelphica*) (Foote, 1981) and a potential pest of other cultivated Solanaceae.

Wild hosts

Recorded from *Physalis lanceolata* Michx. (Norrbom, 1982; as *P. longifolia*).

Adult identification

A predominantly black species, with a characteristic combination of wing (fig. 244) and scutellar markings (fig. 117).

Partial description of third instar larva

Larvae medium-sized, length 8.5-10.0mm. *Head*: Stomal sensory organ large, rounded, with 2 large preoral teeth at base; oral ridges not pronounced. *Thoracic and abdominal segments*: A3-A4 with small encircling rows of spinules and no lateral or ventrolateral tubercles [according to Berg (1979) but not Kandybina (1977); further study needed to confirm the state of this character]. Anterior spiracles each with a row of 33-35 short tubules. Posterior spiracular hairs branched and about half the length of a spiracular slit, with 6-10 hairs per bundle. Anal lobes large, surrounded by several rows of spinules, and bifid [according to Berg (1979)]. *Source*: Based on Berg (1979) and Kandybina (1977).

Distribution

Central America: Mexico.
North America: Southern and central USA; Canada (Ontario).

Rhagoletis suavis (Loew)

(fig. 245)

Taxonomic notes

This species has also been known as *Trypeta suavis* Loew.

Commercial hosts

A potential walnut pest whose larvae develop in the husks of black walnut (*Juglans nigra*), butternut (*J. cinerea*), English walnut (*J. regia*) and Japanese walnut (*J. ailantifolia*) (Bush, 1966); peach (*Prunus persica*) is also an occasional host (Dean, 1969).

Adult identification

A pale coloured species with a characteristic wing pattern (fig. 245).

Partial description of third instar larva

Larvae medium-sized, length 8.0-10.0mm; width 2.0mm. *Head*: Stomal sensory organs and preoral teeth similar to *R. pomonella*; oral ridges in 9-10 rows; mouthhooks heavily sclerotised, each with a large, slender, strongly curved apical tooth. *Thoracic and abdominal segments*: Spinules and spinulose areas similar to *R. pomonella*; A8 with intermediate tubercles directed downwards and very close together, forming a continuous transverse ridge rather than a bifid tubercle. Anterior spiracles each with a uniform row of 25 small tubules. Posterior spiracular slits each 4.0-4.5 times as long as broad, with a heavily sclerotised rima; upper and lower slits at about 90° to each other; spiracular hairs slightly shorter than length of spiracular slit, some branched; dorsal and ventral hair bundles of 5-9 hairs, lateral hair bundles of 5-8 hairs. Anal lobes large, protuberant. *Source*: Based on Phillips (1946) and Steyskal (1973).

Distribution

North America: Eastern half of the USA.

Other references

BEHAVIOUR AND TRAPPING, Katsoyannos (1989b); BIOLOGY, Fletcher (1989b) and Gibson & Kearby (1978); PARASITOIDS, Muesebeck (1980); SYMBIONTS, Howard & Bush (1989).

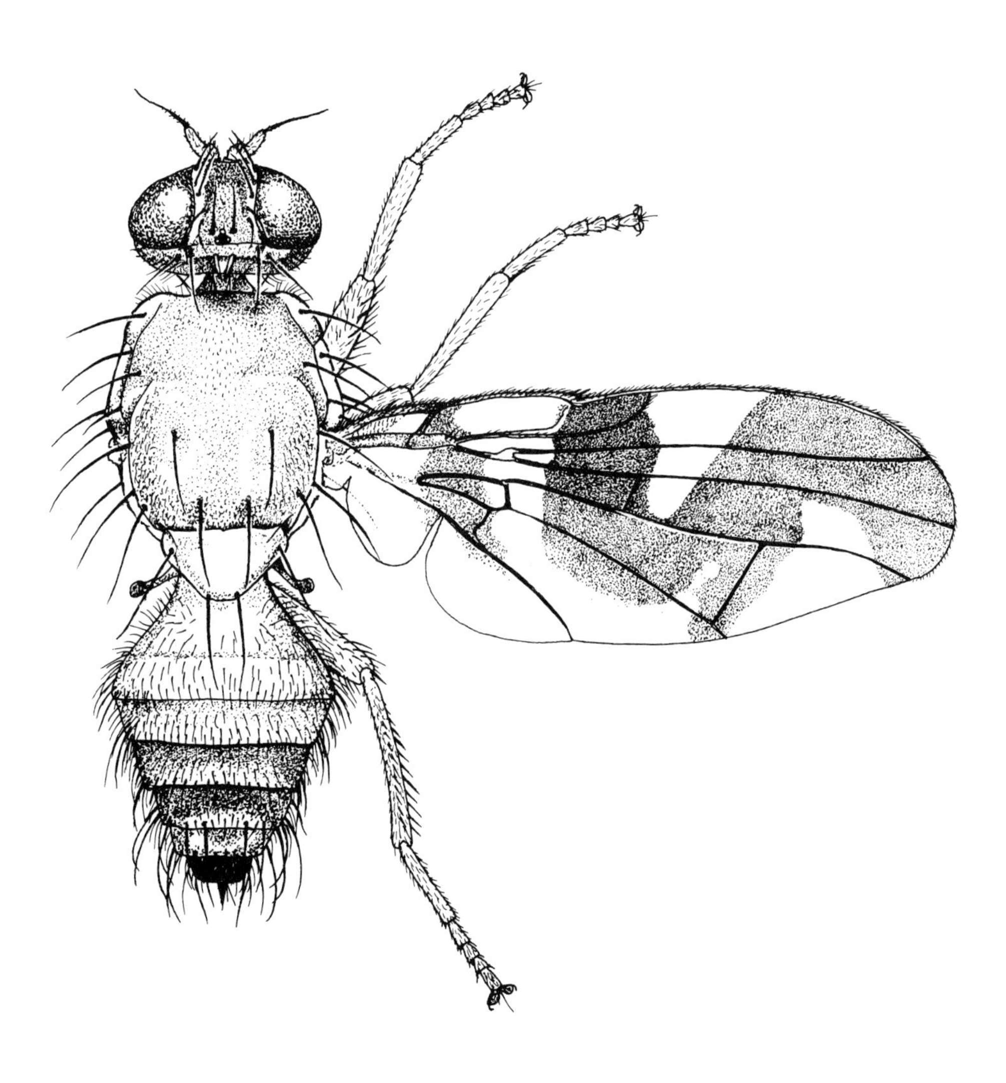

Fig. 245. *Rhagoletis suavis*, adult female.

Rhagoletis tabellaria (Fitch) Species Complex

(fig. 246)

Taxonomic notes
This is a complex of five known species, including *Rhagoletis juniperina* Marcovitch (p. 388), *R. tabellaria* and *R. electromorpha* Berlocher (p. 388). *R. tabellaria* and *R. electromorpha* can only be reliably separated using gel electrophoresis, but some morphological differences were also tabulated by Berlocher (1984). At the time of Berlocher's (1984) work, no overlap of hosts had been found between the eastern *Cornus* attacking population of *R. tabellaria* and *R. electromorpha*; host plants may therefore be used for tentative identification.

Adult identification
A complex of predominantly black species, with a characteristic combination of wing (fig. 246) and scutellar markings (fig. 120).

Rhagoletis tabellaria complex: *R. tabellaria* (Fitch)

Taxonomic notes
This species has also been known as *Tephritis tabellaria* Fitch.

Commercial hosts
A potential pest of *Vaccinium* spp., although only its western North American population attacks *Vaccinium* spp. (Bush, 1966), including cranberry (*V. macrocarpon*) (Wasbauer, 1972). The western North American population is known to attack red-osier dogwood (*Cornus stolonifera*) in Colorado (J. Jenkins, pers. comm., 1991) and the eastern North American population only attacks the fruits of dogwoods, including red-osier dogwood (Berlocher, 1984; Bush, 1966).

Wild hosts
Recorded from silky dogwood (*C. amomum* Mill.) in eastern North America (Bush, 1966).

Adult identification
See Berlocher (1984) for separation from *R. electromorpha*.

Distribution
North America: Eastern race in the northern part of the eastern half of the USA and adjoining Canada; western race from northern California and Colorado to southern British Columbia (Canada).

Other references
BEHAVIOUR, Smith (1985); PHEROMONES, Averill & Prokopy (1989a).

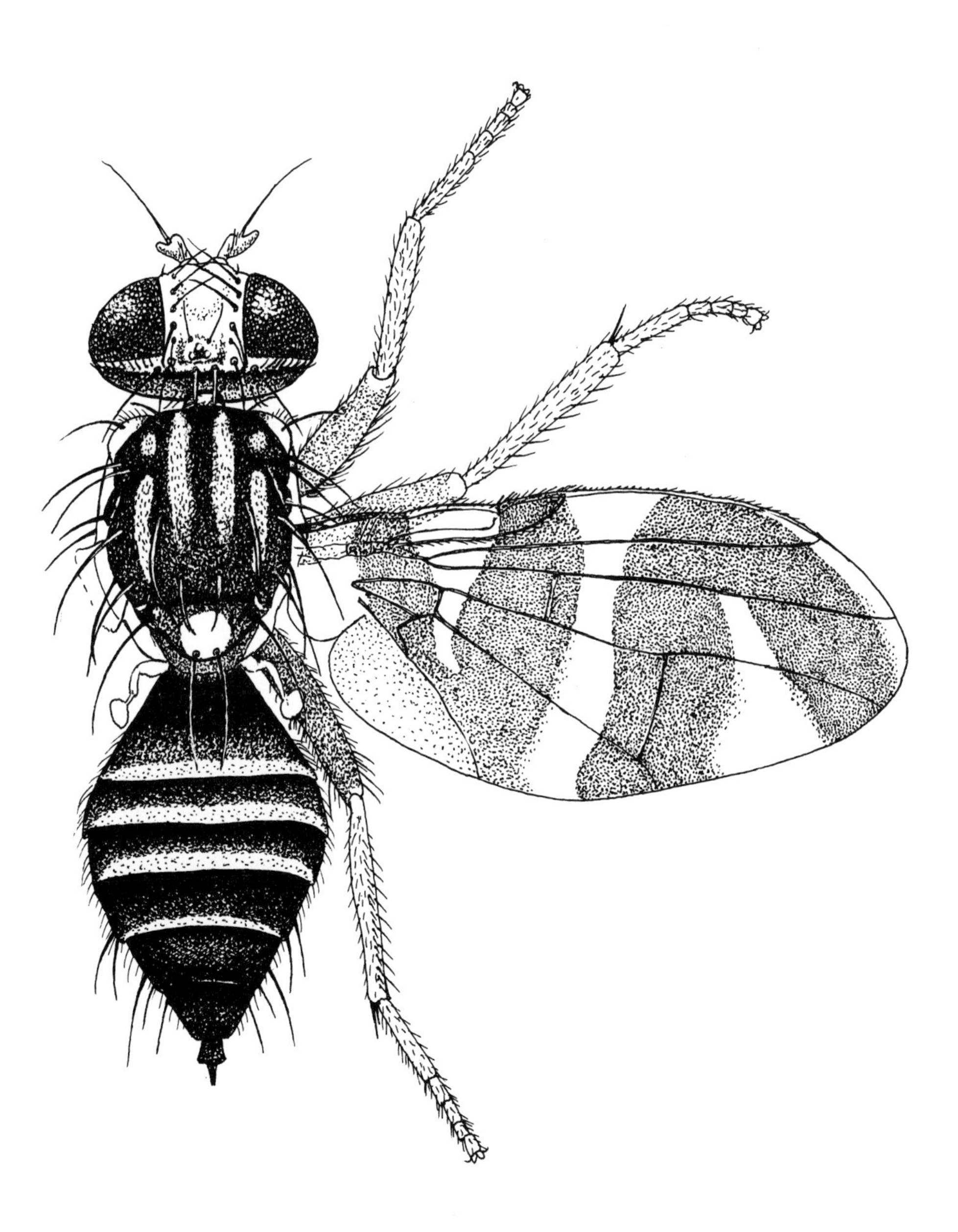

Fig. 246. *Rhagoletis tabellaria*, adult female.

Rhagoletis tomatis Foote

(fig. 247)

Taxonomic notes

Some records of the so called *Rhagoletis ochraspis* (Wiedemann) were probably based on misidentifications of this species.

Commercial hosts

Attacks tomato (*Lycopersicon esculentum*) (Foote, 1981) and it is a potential pest of that crop.

Wild hosts

None known, but probably has native hosts that sustain reservoir populations, similar to *R. lycopersella* (Foote, 1981).

Adult identification

A predominantly black species, with a characteristic combination of wing (fig. 247) and scutellar markings (figs 117, 118).

Distribution

South America: Chile, southern Peru.

Other *Rhagoletis* Species of Economic Interest

R. alternata (Fallén), from Europe and temperate Asia, has been recorded from crops of Japanese rose (*Rosa rugosa*) in Norway (Rygg, 1979) and Poland (Lipa *et al.*, 1976). It is also recorded from some ornamental roses, namely apple rose (*R. villosa*) and dog rose (*R. canina*) (Hendel, 1927). Identification: Rohdendorf (1961); White (1988).

Rhagoletis basiola (Osten Sacken) is a North American species which is very similar to *R. alternata*. It has been recorded from 19 species of rose (*Rosa* spp.) (Bush, 1966), including some ornamental species (Curran, 1924), namely Arkansas rose (*R. arkansana*), dog rose (*R. canina*), French rose (*R. gallica*), Japanese rose (*R. rugosa*) and prairie rose (*R. setigera*). However, in 1981 *Rhagoletis basiola* was reared from apples (*Malus domestica*) in a locality where an infested patch of wild rose had recently been destroyed; small numbers of *R. basiola* were also reared from apples in 1982 and 1983 (Westcott *et al.*, 1985). Identification: Bush (1966).

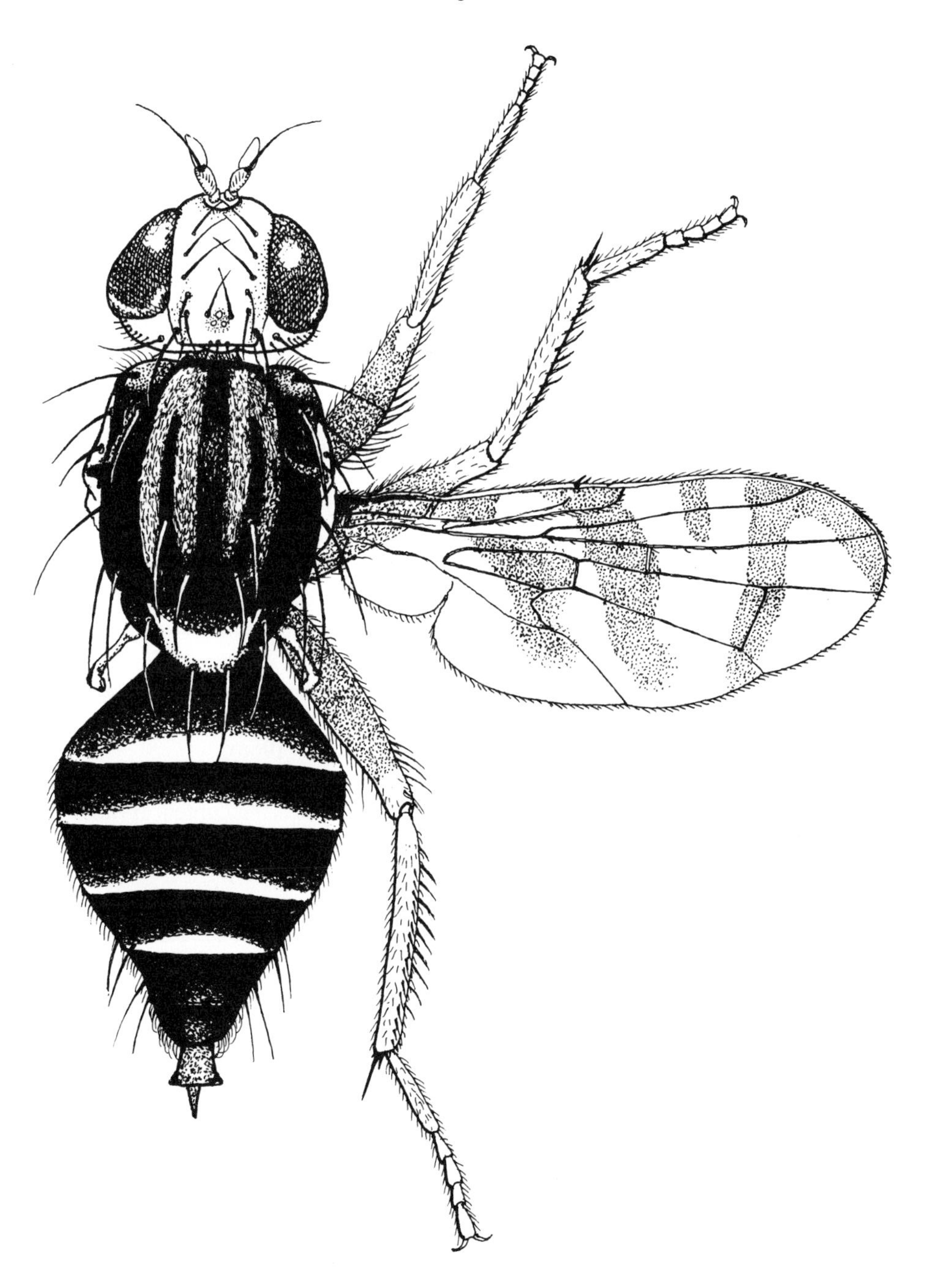

Fig. 247. *Rhagoletis tomatis*, adult female.

R. batava Hering, from Europe and temperate Asia, attacks the fruits of commercially harvested sea buckthorn (*Hippophae rhamnoides*) in the USSR (Andrashchuk, 1981; 1982). Identification: Rohdendorf (1961).

R. berberidis Jermy is found in the mountain areas from Switzerland to Armenian SSR and its larvae develop in the fruits of berberis (*Berberis vulgaris*) (Kandybina, 1977). Identification: Jermy (1961).

R. berberis Curran, from North America, attacks Oregon grape (*Mahonia aquifolium*) (Bush, 1966). Identification: Bush (1966).

R. blanchardi Aczél is an Argentinian species. Foote (1981) examined a female labelled as reared from citrus ("criado de frutas de citrus"), but that is doubtfully correct and he suggested that its normal host is more likely to be a wild *Solanum* sp. (Solanaceae). Identification: Foote (1981).

R. boycei Cresson is a North and Central American species whose larvae develop in the husks of walnuts (*Juglans regia* and some wild *Juglans* spp.). This species has a very restricted distribution, being known only from south-western New Mexico, and adjacent areas of Arizona and Mexico. It differs from other walnut husk flies by having a black body colour. Identification: Bush (1966); Foote (1981).

R. cornivora Bush is a member of the *R. pomonella* complex. It is an eastern North American species whose larvae develop in the fruits of bunchberry (*Cornus canadensis*) and red-osier dogwood (*C. stolonifera*); silky dogwood (*C. amomum*) is an important wild host (Bush, 1966). It is also recorded from flowering dogwood (*C. florida*) (Bush, 1966) but recent evidence indicates that populations on that host are another distinct species (Smith, 1988a; S.H. Berlocher, pers. comm., 1991) which will be formally described by J. Jenkins (in prep.). Identification: Bush (1966).

R. electromorpha Berlocher is a member of the *R. tabellaria* complex. It is so far only known from Illinois, in eastern USA, where it is recorded from roughleaf dogwood (*Cornus drummondii*) and a wild *Cornus* sp. (Berlocher, 1984). Identification: Berlocher (1984).

R. ferruginea Hendel is found in Brazil, Argentina and Uruguay, where it has been recorded from "joá", which is presumed to be a wild *Solanum* sp. (Solanaceae). It has also been recorded from citrus (*Citrus* sp.), but the validity of that record is in doubt (Foote, 1981). Identification: Foote (1981).

R. flavigenualis Hering is a Turkish species recorded from savin (*Juniperus sabina*) (Kandybina, 1977). Identification: Rohdendorf (1961).

R. juniperina Marcovitch is a member of the *R. tabellaria* species complex. It is a North American species whose larvae develop in the fruits of eastern red cedar (*J. virginiana*) (Bush, 1966). There is also a record, which requires confirmation, from one-seed juniper (*J. monosperma*) (Kandybina, 1977). Identification: Bush (1966).

R. meigenii (Loew) is a European species whose larvae develop in the fruits of berberis (*Berberis vulgaris*) (White, 1988). This species is also adventive in North America (A.L. Norrbom, pers. comm., 1990). Identification: Rohdendorf (1961); White (1988).

R. mongolica Kandybina is a Mongolian species recorded from savin (*J. sabina*) (Kandybina, 1977). Identification: Kandybina (1972).

R. psalida Hendel, from Peru, has been reared from the fruits of potato (*Solanum tuberosum*) (Foote, 1981). Identification: Foote (1981).

R. ramosae Hernandez-Ortiz is only known from Mexico, where it has been reared from the husks of a wild species of walnut (*Juglans major*) (Hernandez-Ortiz, 1985). Identification: Hernandez-Ortiz (1985).

R. turanica (Rohdendorf), from Soviet Central Asia, is recorded from dog rose (*Rosa canina*) (Kandybina, 1977). Identification: Rohdendorf (1961).

R. zephyria Snow is a member of the *R. pomonella* complex. The larvae of *R. zephyria* develop in the fruits of snowberries (*Symphoricarpos* spp.), which are sometimes grown as ornamental shrubs. *R. zephyria* is known from most areas where snowberries are grown (G.L. Bush, pers. comm., 1991), and that includes the western half of the USA, northern USA and southern Canada. In the eastern parts of its range it is sympatric with long established *R. pomonella* populations. *R. zephyria* is also sympatric with the newly established populations of *R. pomonella* in Oregon and adjacent areas of the western USA. In those areas it is essential to try to monitor the spread of *R. pomonella* and that is being done by using sticky traps. Unfortunately, *R. zephyria* is often common in the trapping areas and the inability to distinguish accurately between trapped *R. zephyria* and *R. pomonella* without time consuming dissection and detailed measurement, is hindering the monitoring programme (F.L. Banham, pers. comm., 1988). The technique for separating those species described by Bush (1966) was revised by Westcott (1982), who illustrated clear differences in the male surstyli. Some hybridization is occurring between introduced *R. pomonella* and *R. zephyria* (McPheron, 1990).

R. zoqui Bush is a Central American species whose larvae develop in the husks of some wild species of walnut (*Juglans* spp.). This species is only known from the Hidalgo area of Mexico. Identification: Bush (1966); Foote (1981).

GENUS *Toxotrypana* Gerstaecker

A genus of seven South American species, one of which is also found as far north as southern USA, and has previously been referred to the genus *Mikimyia* Bigot, which is a synonym of *Toxotrypana*. *T. australis* Blanchard attacks the pods of a pest species of Asclepiadaceae (p. 423).

Toxotrypana curvicauda Gerstaecker

Papaya Fruit Fly

(figs 30, 248)

Taxonomic notes
This species has also been known as *Mikimyia furcifera* Bigot.

Pest status and commercial hosts
A pest of papaya (*Carica papaya*) which has also been recorded from mango (*Mangifera indica*) (Heppner, 1986). However, the latter record may have been based on traps placed in a mango tree; Aluja *et al.* (1987a) collected *T. curvicauda* in traps placed in mango trees but during their survey it was only reared from papaya. Enkerlin *et al.* (1989) cast doubt on the importance of this species as a pest, although it is an important pest in Trinidad (L.E. McComie, pers. comm., 1991). The larvae feed, at least in part, on the seeds of papaya (Norrbom & Kim, 1988b).

Wild hosts
There is a record of *T. curvicauda* developing in the seed pods of a species of Apocynaceae (Wasbauer, 1972), but that was probably based on a misidentification of another species.

Adult identification
A large ichneumon-fly mimic with a brown and yellow patterned body, and characteristic wing pattern and venation (fig. 248); the female has a very long curved ovipositor.

Partial description of third instar larva
Larvae large, length 13.0-15.0mm; width 4.0mm. *Head*: Antennal sensory organ with a broad stout basal segment and smaller, globular apical segment; maxillary sensory organ broad, protuberant, heavily sclerotised, with 7-8 sensilla; stomal sensory organ and preoral teeth not obvious; oral ridges in 13-15 prominent rows; mouthhooks heavily sclerotised, each with a slender gently curved apical tooth. *Thoracic and abdominal segments*: Spinules small, very sharply pointed; T1-T3 with a few or even

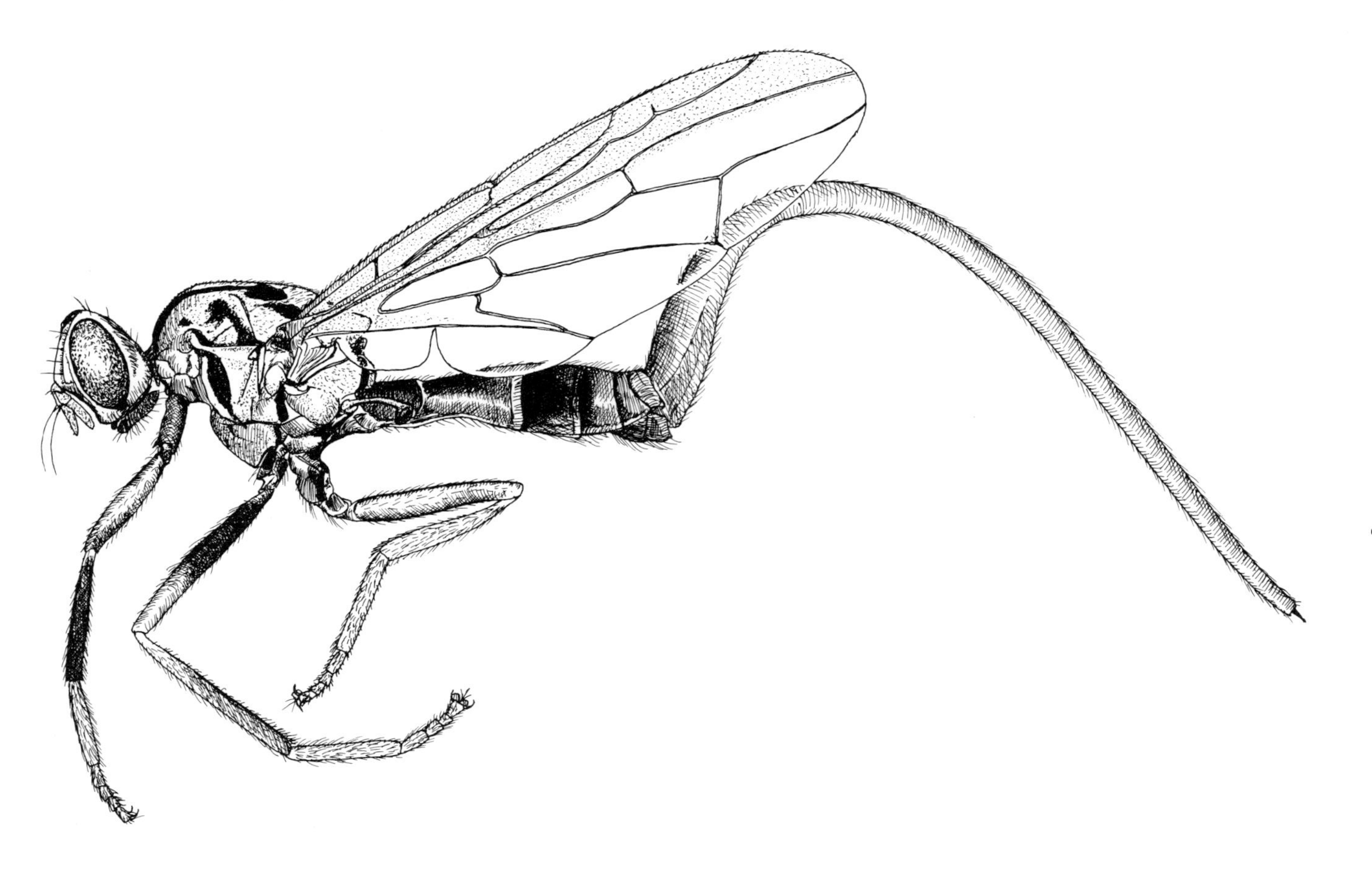

Fig. 248. *Toxotrypana curvicauda*, adult female, lateral view.

a single line of indistinct spinules encircling each segment; A1-A8 with spinules restricted to creeping welts, with 12-15 rows per welt; A8 surface appears smooth with only a few depressions and 6 pairs of small tubercles. Anterior spiracles large, fan shaped, constricted in middle of outer margin, with 22-28 small straight sided tubules, and usually with some in a partial secondary row. Posterior spiracular slits large, each about 4-5 times as long as broad, with a moderately sclerotised rima; spiracular hairs short (less than half length of a spiracular slit), some branched in apical third; dorsal and ventral hair bundles with about 5 hairs, lateral hair bundles with about 4 hairs. Anal lobes large, protuberant. *Source*: Based on Heppner (1986) and Phillips (1946).

Distribution

Central America: Costa Rica, Guatemala, Mexico, Panama.
North America: USA (Florida, Southern Texas).
South America: Brazil, Colombia.
West Indies and nearby islands: Bahamas, Cuba, Dutch Antilles.

Other references

BEHAVIOUR, Landolt (1985a), Landolt & Hendrichs (1983), Sivinski & Burk (1989), Sharp & Landolt (1984), Sivinski & Webb (1985b) and P.H. Smith (1989); BIOLOGY, Landolt (1984); IMPACT, Landolt (1985b); LARVA, Berg (1979); PARASITOIDS, Aluja *et al.* (1990), Narayanan & Chawla (1962), Wharton (1983) and Wharton *et al.* (1981); PHEROMONES, Chuman *et al.* (1987) and Landolt *et al.* (1985).

GENUS *Trirhithromyia* Hendel

A genus of five African species which was originally described as a subgenus of *Ceratitis* but is regarded as a distinct genus by Hancock (1984).

Trirhithromyia cyanescens (Bezzi)

Tomato Fruit Fly

(figs 111, 249)

Taxonomic notes
This species has also been known as *Pardalaspis cyanescens* Bezzi and *Perilampsis bourbonica* Munro, and it was placed in *Trirhithromyia* by Hancock (1984).

Pest status and commercial hosts
A pest of tomato (*Lycopersicon esculentum*) (Orian & Moutia, 1960), which is also recorded from some other species of Solanaceae, namely bell pepper (*Capsicum annuum*), black nightshade (*Solanum nigrum*), eggplant (*S. melongena*) and tabasco pepper (*C. frutescens*) (Étienne, 1972; Hancock, 1984).

Wild hosts
Recorded from some wild *Solanum* spp. (Hancock, 1984).

Adult identification
Easily recognised by its characteristic pattern of brown bands on the wing, and the scutellum being entirely black in the apical half (fig. 249).

Distribution
Indian Ocean: Madagascar; adventive in Mauritius (since 1958) and Réunion (since 1951).
A distribution map was produced by CIE (1962).

Other references
PARASITOIDS, Anon. (1979) and Étienne (1973b); REARING, Étienne (1973a).

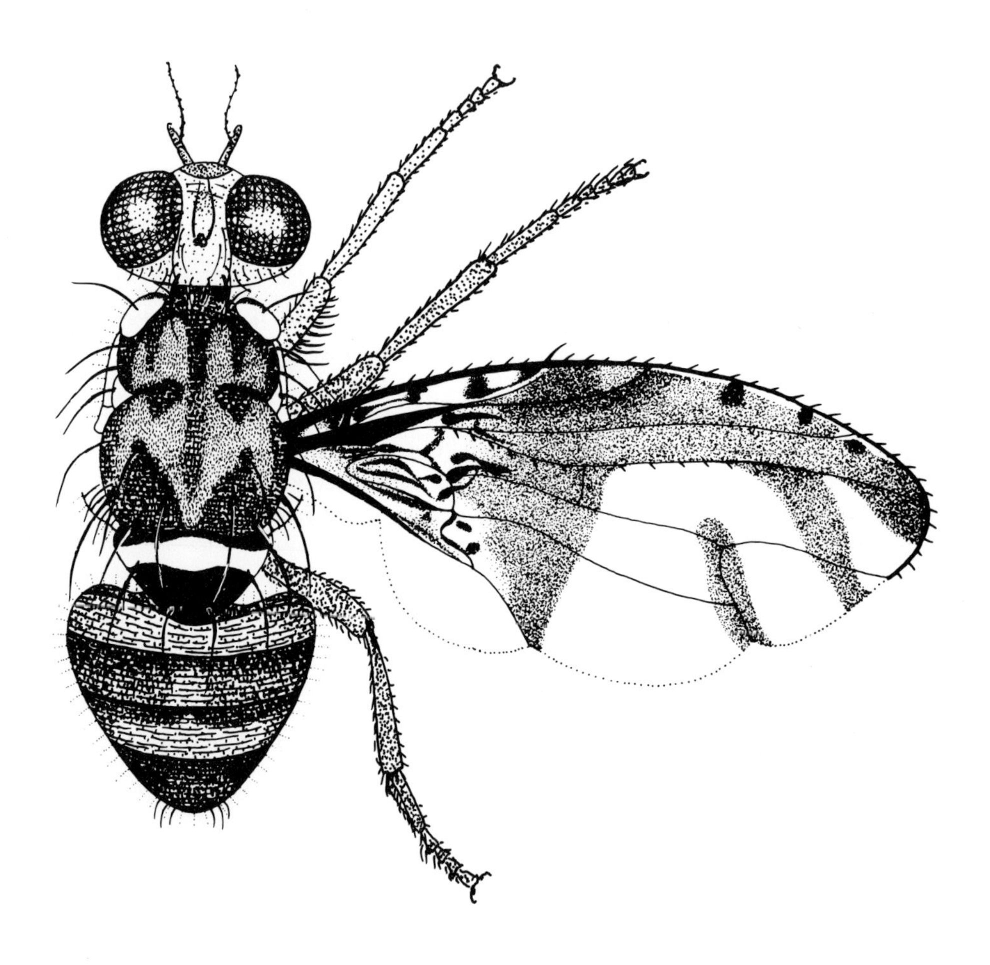

Fig. 249. *Trirhithromyia cyanescens*, adult female.

GENUS *Trirhithrum* Bezzi

(fig. 31)

An African genus of about 40 species, a few of which have been reared from coffee. Most species can be identified using keys by Hancock (1987) and Munro (1934).

Trirhithrum coffeae Bezzi

(figs 31, 250, 128)

Taxonomic notes
This species has also been known as *Trirhithrum nigerrimum* var. *coffeae* Bezzi.

Pest status and commercial hosts
A pest of robusta coffee (*Coffea canephora*), which also attacks arabica coffee (*C. arabica*), but in lower numbers (Greathead, 1972).

Adult identification
A small dark brown fly with a white face and dark sexually dimorphic wing markings (figs 250, 128).

Distribution
Africa: Cameroun, Côte d'Ivoire, Ethiopia, Ghana, Kenya, Tanzania, Togo, Uganda.

Other reference
PARASITOIDS, Steck *et al.* (1986).

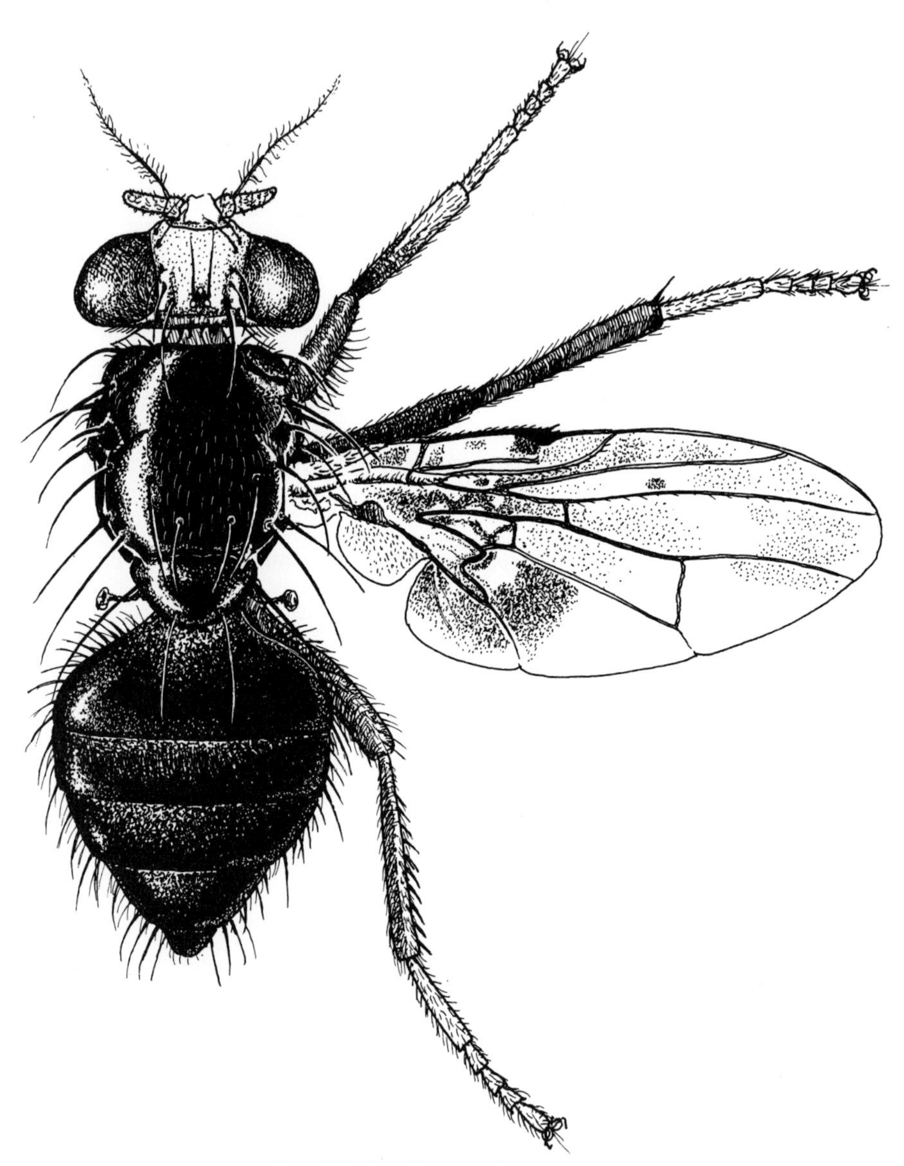

Fig. 250. *Trirhithrum coffeae*, adult male.

Trirhithrum nigerrimum (Bezzi)

(figs 129, 130)

Taxonomic notes
This species has also been known as *Ceratitis nigra* var. *nigerrimum* Bezzi and *C. patagiata* Enderlein, and some records of *Trirhithrum occipitale* Bezzi were based on misidentifications of this species (Munro, 1929b; Hancock, 1989).

Commercial hosts
A potential pest of coffee which has been recorded from arabica coffee (*Coffea arabica*) (Munro, 1935b) and robusta coffee (*C. canephora*) (Steck *et al.*, 1986). There are old records from coca (*Erythroxylum coca*) and Surinam cherry (*Eugenia uniflora*) (Munro, 1934; 1935b; Silvestri, 1913), but those may have been be based on misidentifications or casual observation of adults resting on plants.

Adult identification
A small dark brown fly with a white face and dark sexually dimorphic wing markings (figs 129, 130); general appearance similar to *T. coffeae* (fig. 250).

Partial description of third instar larva
Larvae medium-sized, length 6.0mm; width 1.0mm. *Head*: Maxillary sensory organ large and protuberant; 4 preoral lobes, 4-5 oral ridges, mouthhooks heavily sclerotised, without a preapical tooth. *Thoracic and abdominal segments*: Small spinules dorsally on T2, T3 and A1. Anterior spiracles with 7 stout tubules each with a basal constriction. Posterior spiracles with spiracular slits 3.0 times as long as broad. *Source*: Based on Silvestri (1913).

Distribution
Africa: Cameroun, Ghana, Niger, Nigeria, Rwanda, South Africa, Tanzania, Uganda, Zaire.
Indian Ocean: Comoro Islands.

Other *Trirhithrum* Species of Economic Interest

T. albomaculatum Röder, from southern Africa, has been recorded from kei apple (*Dovyalis caffra*) (Munro, 1925). Identification: Hancock (1987); Munro (1934).

T. basale Bezzi has been recorded as a pest of arabica coffee (*Coffea arabica*) in Uganda (Hancock, 1926). The only confirmed records of this species are from Malawi (Cogan & Munro, 1980; NHM, collection data) and that host record was almost certainly based upon a misidentification of either *T. coffeae* or *T. nigerrimum*. Identification: Hancock (1987); Munro (1934).

T. inscriptum (Graham) has been recorded as a coffee pest in Zaire (Bredo, 1934; Stolp, 1960), and it transmits a bacterium that causes 'potato flavour' in arabica coffee

(*C. arabica*) (Ingram, 1965; Le Pelley, 1968). Identification: Munro (1934).

T. manganum Munro, from Madagascar, has been recorded from coffee (*Coffea* sp.) (Hancock, 1984). It is easily separated from the two coffee pest (or potential pest) species by its bright metallic blue body colour. Identification: Hancock (1984).

T. nigrum Graham has been recorded as a pest of cocoa (*Theobroma cacao*) in Ghana by Munro (1925), but he gave no details. However, Lamborn (1914) said it was trapped in great numbers in cocoa plantations in Nigeria, but was not found attacking the pods. The record given by Munro (1925) may therefore be based on observations that were not confirmed by rearing. Identification: Munro (1934).

T. occipitale Bezzi was recorded from coffee (*Coffea* sp.) in South Africa by Annecke & Moran (1982), but they did not give the source of that record and it is likely to have been based on a misidentification of *T. nigerrimum* which was discussed by Munro (1934). The true hosts of *T. occipitale* are wild grapes (*Cissus* spp.; Vitaceae) (Munro, 1934). Identification: Hancock (1987); Munro (1934).

GENUS *Zonosemata* Benjamin

A New World genus of six species which can be identified using a key by Bush (1965). The two species of known biology attack the fruits of Solanaceae (Bush, 1965).

Zonosemata electa (Say)

(figs 32, 251)

Taxonomic notes

This species has also been known as *Tephritis flavonotata* Macquart and *Trypeta electa* Say.

Pest status and commercial hosts

A pest of bell pepper (*Capsicum annuum*) and eggplant (*Solanum melongena*), which can also attack tomato (*Lycopersicon esculentum*). It also attacks horsenettle (*S. carolinense*) (Bush, 1965) and there is a record from tabasco pepper (*C. frutescens*) (Wasbauer, 1972).

Wild hosts

Wild growing horsenettle is its main natural host and it is also recorded from *S. aculeatissimum* Jacq. (Bush, 1965); there is also an old record, requiring confirmation, from silverleaf nightshade (*S. elaeagnifolium* Cav.) (Benjamin, 1934).

Adult identification

A yellow-orange species with characteristic body and wing markings (fig. 251).

Partial description of third instar larva

Larvae large, length 10.0-12.0mm; width 3.4mm. *Head*: Antennal sensory organ 2 segmented, with a broad basal segment and a smaller globular apical segment; maxillary sensory organ protuberant, with 5-6 sensilla; stomal sensory organ and preoral teeth not obvious; oral ridges of 6 rows; mouthhooks heavily sclerotised, each as deep as long, with a very strong notch in ventral surface setting the distal area off from the broad base. *Thoracic and abdominal segments*: Spinules large, in broad, anterior, discontinuous rows encircling segments; T1 and T2 with 12 and 8 rows respectively; A1-A8 with a broad band of about 20 rows ventrally; A8 with many punctate depressions scattered over area between intermediate and ventral tubercles [Steyskal (1975) refers to these as punctiform, but sometimes short-linear]; tubercles very conspicuous and sometimes bifid. Anterior spiracles fan shaped; irregular row of 27-32 small tubules. Posterior spiracular slits each long and narrow, about 6 times as long as broad, with a heavily sclerotised rima; spiracular hairs short (less than half the

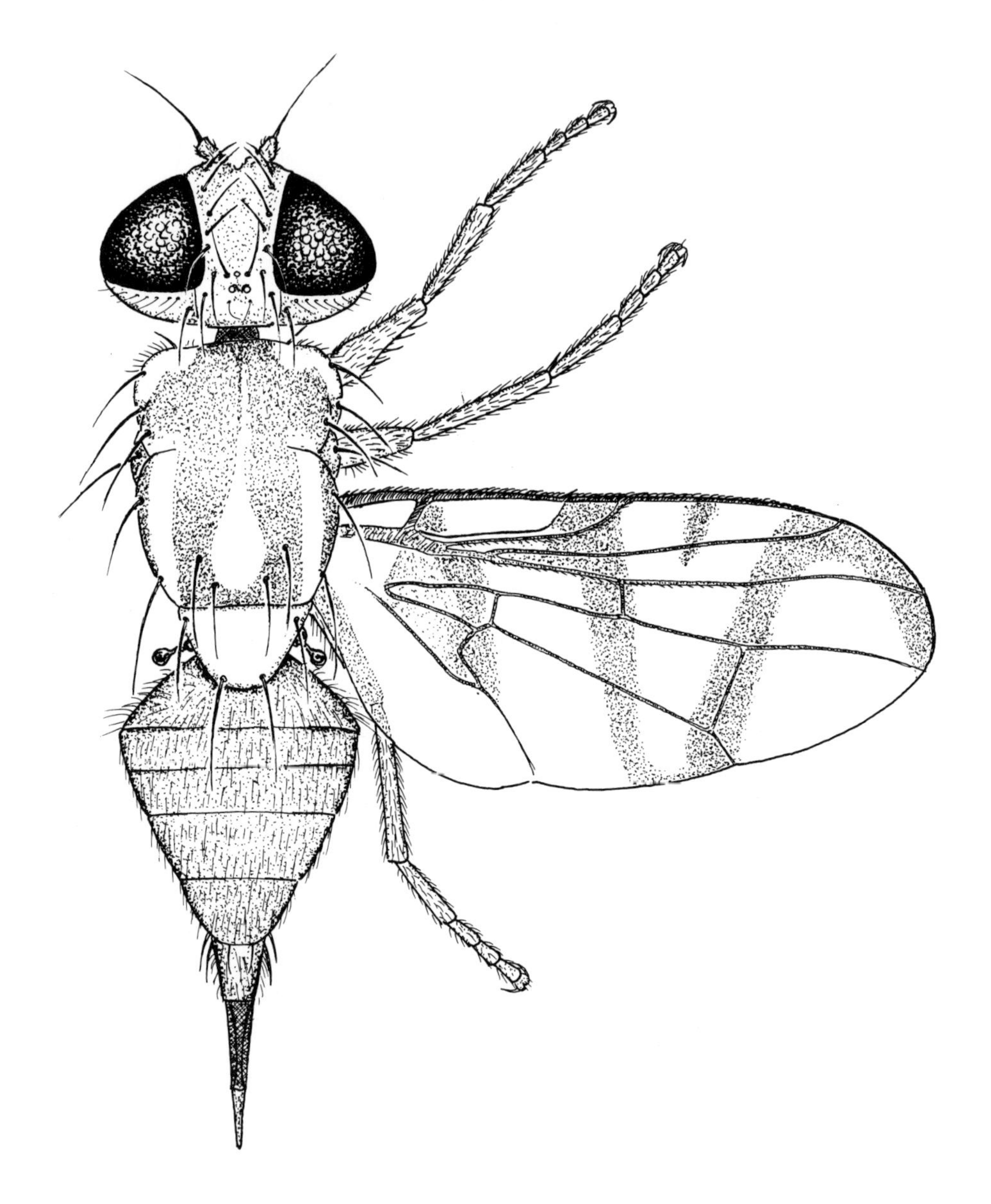

Fig. 251. *Zonosemata electa*, adult female.

length of a spiracular slit), some branched; dorsal and ventral hair bundles with 10-13 hairs, lateral hair bundles with 8-11 hairs. *Source*: Based on Phillips (1946) and Steyskal (1975).

Distribution
North America: Eastern half of USA and Ontario (Canada).

Other reference
BEHAVIOUR, P.H. Smith (1989).

Other Species Associated with Cultivated Fruit

Species belonging to genera containing species of more obvious economic importance are discussed following the species accounts for those genera, for example, species of *Anastrepha* associated with cassava fruits (p. 162).

Acidiella angustifascia (Hering), from China and the far east of the USSR, attacks the fruits of Siberian ginseng (*Eleutherococcus senticosus*) (Kandybina, 1977). Identification: Hering (1936).

Adrama apicalis Shiraki, *A. austeni* Hendel and *A. determinata* (Walker) larvae develop in the seeds of tea (*Camellia sinensis*) in South-east Asia (Hardy, 1986b; Hennig, 1953). Bernard (1917a) showed that *A. determinata* could bore into the husk of a healthy ungerminated seed and it was regarded as a pest in some areas (Bernard, 1917b; Hall, 1917; Stuart, 1921). *A. austeni* has been recorded as causing damage to exposed seeds in germination beds in Sri Lanka (King, 1935) and India (Rao, 1940). However, King (1936) suggested that *A. austeni* only attacked unhealthy tea seeds, to which it was attracted by the products of decomposition. The biology of *A. determinata* was described by Leefmans (1915). Identification: Hardy (1986b).

Anomoia purmunda (Harris), from Europe, temperate Asia and Bhutan, attacks the fruit of some garden shrubs and trees, namely berberis (*Berberis vulgaris*), cotoneaster (*Cotoneaster* spp.) and ornamental hawthorn (*Crataegus* spp.) (White, 1988). There is also a record for rowan (*Sorbus aucuparia*) (Kandybina, 1977), but that requires confirmation. Male attractant: Males known to congregate on fresh paint and lady's bedstraw (*Galium verum*) (Smith, in press), although no lure has yet been isolated. Identification: White (1988).

Callistomyia flavilabris Hering, from Indonesia and Malaysia, has been "reared" from langsat (*Lansium domesticum*), pummelo (*Citrus maxima*) and rambai (*Baccaurea motleyana*) according to Hardy (1988a); however, Hardy (1973) implied that those may not be rearing records and they therefore require confirmation. The related *Callistomyia horni* Hendel has been reared from some species of Rutaceae in Australia which gives some credence to the pummelo record. Identification: Hardy (1973; 1988a).

Several *Euphranta* spp. have been recorded from plants of some economic value, but none appear to be pests. *E. canangae* Hardy, from the Philippines, has been reared from the fruit of ylang-ylang (*Cananga odorata*) (Hardy, 1983b). *E. camelliae* (Ito), from Japan, has been reared from common camellia (*Camellia japonica*) and Japanese chestnut (*Castanea crenata*) (Hardy, 1983b), and *E. mikado* (Matsumura) attacks the fruits of *Camellia* sp. in Japan (Motooka, 1938). *E. cassiae* (Munro), from India, has been reared from the pods of Indian laburnum (*Cassia fistula*) (Hardy, 1983b). *E. connexa* (Fabricius), from Europe, attacks the seed pods of white swallow-wort

(*Vincetoxicum hirudinaria*) (Janzon, 1982). *E. japonica* (Ito) attacks sweet cherry (*Prunus avium*) in Japan (Hardy, 1983b; Ito, 1983-5) and it has been regarded as a pest (Watanabe, 1939); however, Koyama (1989a) did not mention it in his review of Japanese pest species and it is presumably no longer regarded as a pest. *E. lemniscata* (Enderlein), from tropical Asia and the South Pacific, has been recorded from papaya (*Carica papaya*) (Hardy & Adachi, 1956), but they did not say what part of the plant was supposed to be attacked. Some Australian *Euphranta* spp. are reported to develop in the fruits of mangroves growing in salt water (D.K.McAlpine, in Hardy, 1983b). Identification: Hardy (1983b); Hendel (1927) for *E. connexa*; Ito (1983-5) for Japanese species; Munro (1938c) for *E. cassiae*.

Munromyia nudiseta Bezzi, a South African species known as the olive-seed fruit-fly, has only been recorded from bastard ironwood (*Chionanthus foveolata*; Oleaceae) (Munro, 1924) and there are no known records from cultivated olive. Identification: Munro (1924).

Myoleja limata (Coquillett), from North America, has been reared from the fruit of some *Ilex* spp. whose leaves are used to produce a tea (Benjamin, 1934). Some other *Myoleja* spp. mine petioles and twigs (p. 410) and the biology of most species is unknown. *M. nigroscutellata* (Hering), from tropical Asia and the Pacific, has been recorded from unspecified parts of papaya (*Carica papaya*) (Hardy & Adachi, 1956; as *Hendelina bisecta* Hardy & Adachi). Identification: Hardy (1987) for Asian species; Steyskal (1972) for North American species.

Neoceratitis asiatica (Becker), from China, attacks the fruits of *Lycium turcomanicum*, which are used medicinally (Woo *et al.*, 1963). Identification: Zia (1937).

Nitrariomyia lukjanovitshi Rohdendorf, from central Asia and Mongolia, attacks *Nitraria* spp. (Kandybina, 1977), some of which are edible; however, it was not ascertained if the edible species are amongst those attacked. Identification: Kandybina (1972); Richter (1970).

Oedicarena latifrons (Wulp), from Mexico and southern USA, has been reared from "frutos de papa silvestre", which means fruits of wild potato, and may have meant *Solanum tuberosum* or another *Solanum* sp. (Norrbom *et al.*, 1988). Identification: Norrbom *et al.* (1988).

Rhagoletotrypeta uniformis Steyskal, from Mexico, attacks the edible fruits of lowland hackberry (*Celtis laevigata* Willd) (Steyskal, 1981). Identification: Steyskal (1981).

Taomyia marshalli Bezzi, from Africa, has been recorded from the fruits of some species of Agavaceae (*Dracaena* sp. and *Sansevieria* sp.) (Hancock, in press, a). Identification: Bezzi (1924a).

Tomoplagia Coquillett spp. larvae induce galls on species of Asteraceae in the Americas (Freidberg, 1984), and the following records from fruit should be disregarded: *T. cressoni* Aczél from peach (*Prunus persica*) (Wasbauer, 1972) and *T. biseriata* (Loew) from Surinam cherry (*Eugenia uniflora*) (Ihering, 1912).

Species Accounts;
2 - Leaf, Stem and Root Pests

An estimated 600 (16%) of the described species of Tephritidae probably have larvae which mine the leaves, stems or roots of their host plants, although the biology of most of the species which appear to fall into these categories remains unknown. This section includes the Trypetinae whose larvae mine the leaves of crop plants, plus a few stem mining Tephritinae, and those Dacinae (Ceratitini: Gastrozonina) whose larvae develop in bamboo shoots. The occurrence of *Anastrepha manihoti* in the stems of cassava (*Manihot esculenta*) was discussed in the fruit pest section of this work (p. 162).

GENUS *Euleia* Walker

A genus of 11 species found in Africa, temperate Asia, Europe, Madagascar and North America. Species of known biology are leaf-miners in Apiaceae (=Umbelliferae). The adults mate in trees, rather than on the host (Leroi, 1975a), and the larvae form blotch mines in the leaf, with several larvae per mine (Phillips, 1946; Tauber & Toschi, 1965); pupariation may take place either in the plant or in soil (Phillips, 1946; White, 1988).

Euleia fratria (Loew)

Parsnip Leaf Miner

Taxonomic notes
This species has also been known as *Acidia fratria* (Loew), *Trypeta fratria* Loew and *T. liogaster* Thomson.

Pest status and commercial hosts

A pest of parsnip (*Pastinaca sativa*), which has also been recorded from honewort (*Cryptotaenia canadensis*) (Wasbauer, 1972). Pupariation takes place on the soil surface (Tauber & Toschi, 1965); however, Phillips (1946) recorded pupariation in the mine, or sometimes on the leaf surface.

Wild hosts

Recorded from several genera of Apiaceae (Wasbauer, 1972).

Adult identification

An orange-brown species with a characteristic wing pattern (similar to fig. 252, but the hyaline spot in cell br is much wider). Wing length 4.2-5.0mm. Care should be taken not to confuse trapped specimens of this species with superficially similar *Rhagoletis fausta* (fig. 238).

Partial description of third instar larva

Larvae medium-sized, length 8.0mm; width 2.0mm. *Head*: Stomal sensory organs present; mouthhooks heavily sclerotised, each with a small slender apical tooth; hypopharyngeal sclerite 7 times as long as broad. *Thoracic and abdominal segments*: Spinules very small; T1 with a dorsal anterior band; near T2/T3 junction there is a wide encircling band [Phillips (1946) did not indicate which segment this band is on]; A1-A3 with similar arrangement; A4-A8 with fewer bands of spinules; A8 with a raised dorsocentral area with 4 tubercles, 1 at each corner. Anterior spiracles fan shaped with 14-17 uniform tubules. Posterior spiracular slits each about 3.5 times as long as broad, with a heavily sclerotised rima; spiracular hairs short (less than half length of a spiracular slit), unbranched, with 2 lanceolate hairs in each hair bundle. Anal lobes well developed, protuberant. *Source*: Based on Phillips (1946).

Distribution

North America: Eastern Canada and throughout the USA.

Other reference

BEHAVIOUR, P.H. Smith (1989).

Euleia heraclei (Linnaeus)

Celery Fly

(fig. 252)

Taxonomic notes

This species has also been known as *Acidia heraclei* (Linnaeus), *Musca centauriae* Fabricius, *M. onopordinis* Fabricius, *Trupanea berberidis* Schrank and *T. onopordi* Schrank. It was originally called *M. heraclii* Linnaeus and almost always spelt *heraclei*;

it was sometimes misspelt as *heracleii*. The well known spelling of *heraclei* was recently validated (ICZN case 2719, 1989; ICZN opinion 1645, 1991).

Pest status and commercial hosts

A pest of celery (*Apium graveolens*) and sometimes carrot (*Daucus carota*), lovage (*Levisticum officinale*), parsnip (*Pastinaca sativa*) and Russian cow parsnip (*Heracleum sosnowskyi*) (Bevan, 1966; Isart, 1979; Kabysh, 1979; Spitzer, 1964). It is also recorded from alexanders (*Smyrnium olusatrum*), chervil (*Anthriscus cerifolium*) and parsley (*Petroselinum crispum*); there are unconfirmed records from garden angelica (*Angelica archangelica*) (White, 1988). Pupariation normally takes place in the soil, rarely in the mine (Phillips, 1946; White, 1988).

Wild hosts

Recorded from a wide range of Apiaceae (White, 1988).

Adult identification

A red-brown to orange-brown species in its first generation, and dark orange to black in the second generation; with a characteristic wing pattern (fig. 252). Wing length 4.3-5.8mm.

Partial description of third instar larva

Larvae medium-sized, length 10.0mm. *Head*: Antennal sensory organ 1 segmented; maxillary sensory organ with 6 sensilla; mouthhooks heavily sclerotised, each with a large preapical tooth and stout preapical teeth. *Thoracic and abdominal segments*: T1-T3 and A1-A2, each anteriorly encircled by a band of spinules; A3-A8 with spinules confined to anterior parts of ventral and ventrolateral areas. Anterior spiracles with 15-19 tubules. Posterior spiracular slits each about 2 times as long as broad, with a sclerotised rima. *Source*: Keilin & Tate (1943) and White (1988).

Distribution

Europe: Throughout.
Middle East: Israel
Temperate Asia: Afghanistan, USSR (as far east as Kazakhstan).
North Africa: Probably in all countries north of the Sahara.

Other references

BIOLOGY, Labeyrie (1957) and Leroi (1972; 1974; 1975a; 1975b; 1977); LARVA, K.G.V. Smith (1989).

Other *Euleia* Species of Economic Interest

Euleia separata (Becker), from the Canary Islands, has been recorded as a leaf miner of celery (*Apium graveolens*) (Hering, 1927). Identification: Hendel (1927).

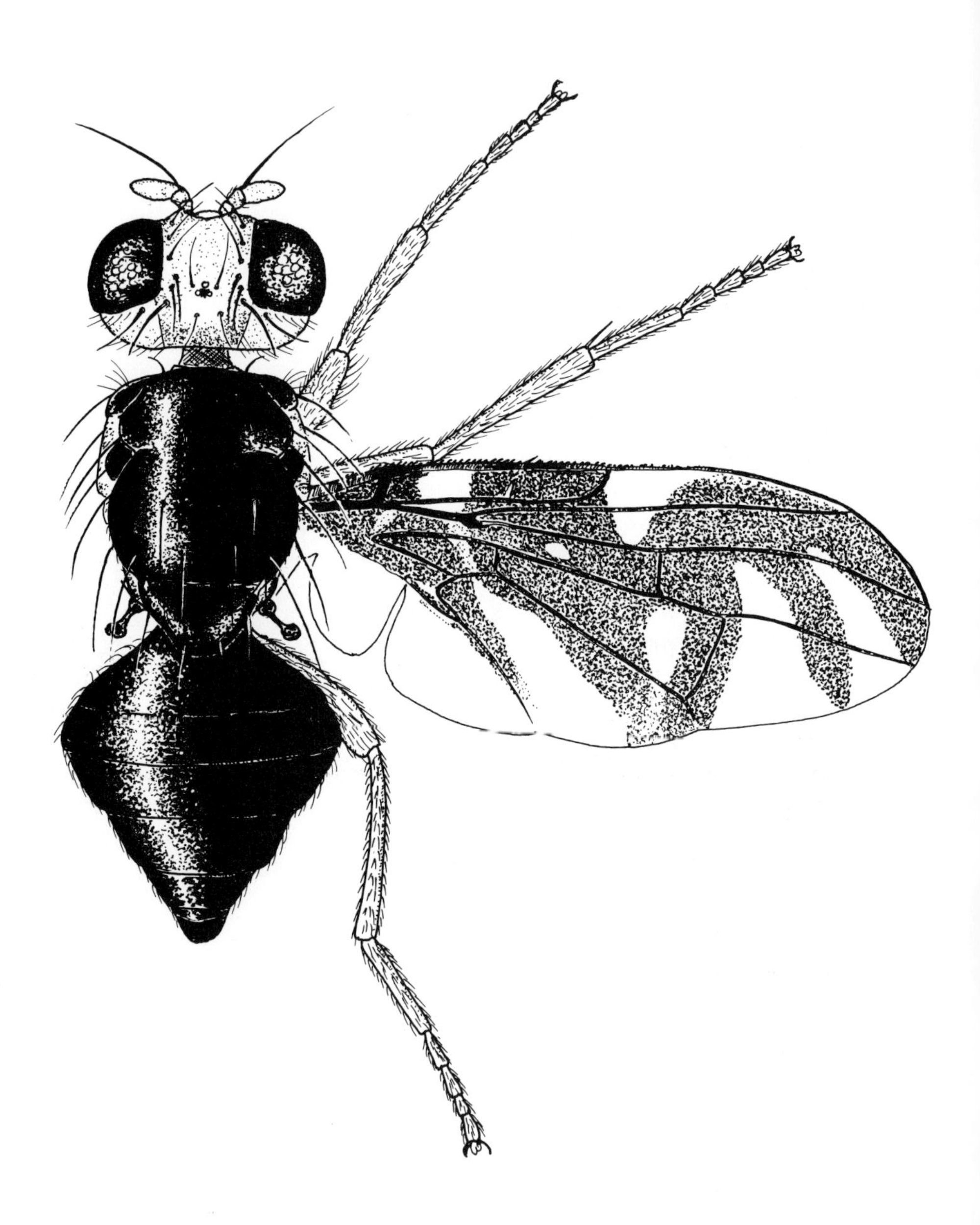

Fig. 252. *Euleia heraclei*, adult female.

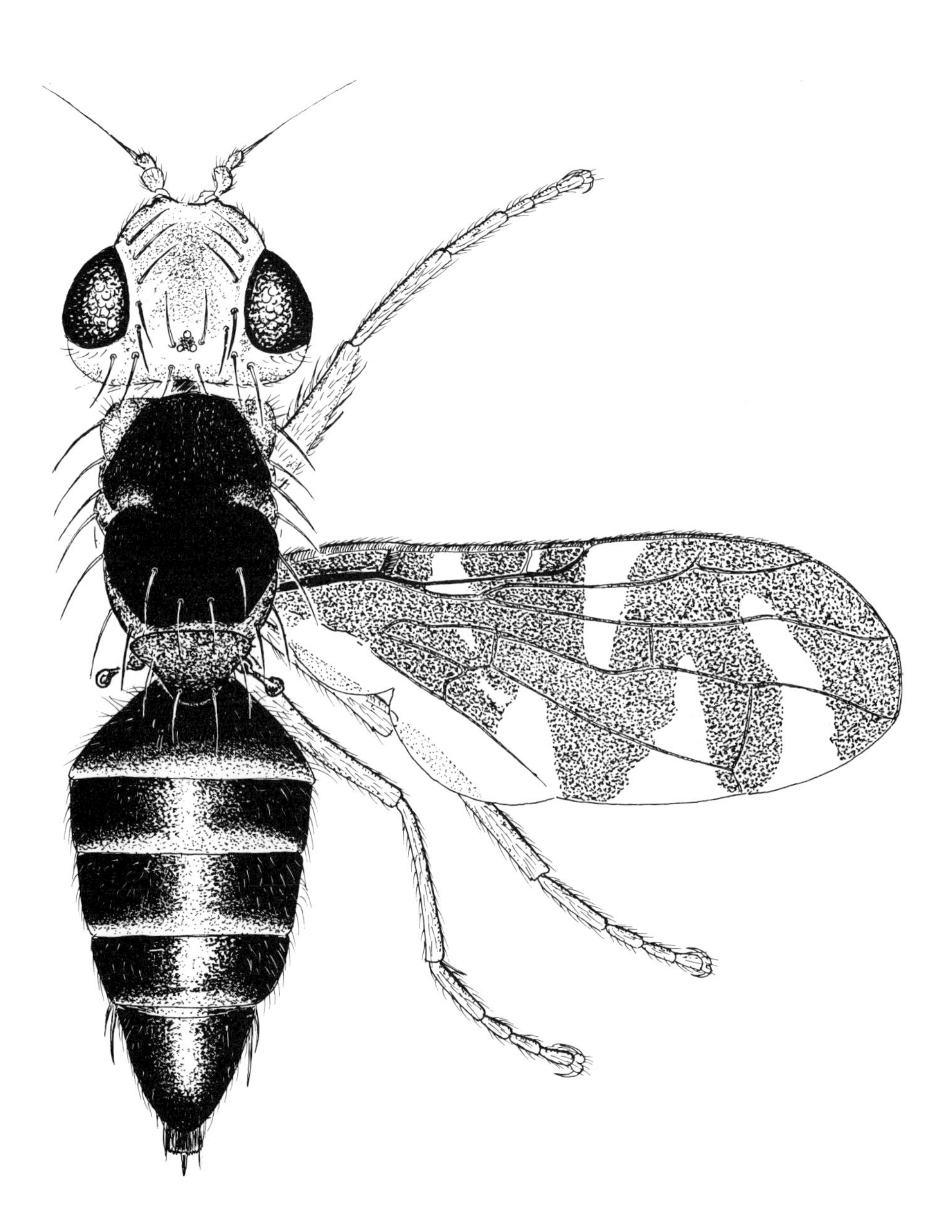

Fig. 253. *Pliorecepta poeciloptera*, adult female.

GENUS *Pliorecepta* Korneyev

Pliorecepta poeciloptera (Schrank)

Asparagus Fly

(fig. 253)

Taxonomic notes
This is the only member of the genus, and it has also been known as *Musca poeciloptera* Schrank, *Ortalis fulminans* Meigen, *Platyparea poeciloptera* (Schrank) and *Poeciloptera poeciloptera* (Schrank).

Pest status and commercial hosts
A pest of the shoots and stems of asparagus (*Asparagus officinalis*) (Dingler, 1934). Fischer-Colbrie & Busch-Petersen (1989) gave details of attack levels of up to 49% of crop area during outbreaks.

Adult identification
A predominantly dark orange to brown species with a black scutum that is covered by a pale tomentum; wing pattern characteristic (figs 253). Wing length 4.3-5.3mm.

Diagnosis of third instar larva
This is the only large dipterous larva likely to be found mining European asparagus. White (1988) examined puparia and noted that the larva had about 36 tubules in each anterior spiracle, arranged in multiple rows; all other species described in the present work have only a single row of tubules. Inadequate descriptions were presented by Dingler (1934) and others (listed by White, 1988).

Distribution
Europe: Reported as a pest in Austria, Germany, Hungary and Sweden (Fischer-Colbrie & Busch-Petersen, 1989); it also occurs in other areas of western Europe, but sometimes only as temporary outbreaks, e.g. it was a pest in some areas of southern England in 1936-7, but has not been found there since 1947 (White, 1988).

Other reference
TRAPPING, Economopoulos (1989).

Tephritidae Associated with Bamboo

Members of several genera of Gastrozonina (Dacinae: Ceratitini) have been reared from bamboos belonging to the genera *Bambusa*, *Dendrocalamus* and *Phyllostachys* (Poaceae). Hardy (1988a) noted that *Acroceratitis*, *Acrotaeniostola*, *Gastrozona* and

Taeniostola spp. have been reared from bamboo, and that *Paraxarnuta*, *Phaeospilodes*, *Pseudacrotoxa* and *Xanthorrachis* spp. have been found in bamboo thickets, suggesting that they may also be bamboo associated. R.A.I. Drew (pers. comm., 1988) observed gastrozonine flies swarming around bamboo shoots being sold at roadside vegetable stalls in Malaysia. Similarly, *Clinotaenia magniceps* (Bezzi), from East Africa, has been reared from guineagrass (*Panicum maximum*) (Le Pelley, 1959; NHM, collection data). There are also some species of Acanthonevrini and Euphrantini associated with bamboo (Hancock, in press, a). The following species of Gastrozonina have been recorded from bamboos of commercial value:

Acroceratitis striata (Froggatt) from giant bamboo (*Dendrocalamus giganteus*) and male bamboo (*D. strictus*) in Sri Lanka (Fletcher, 1920; NHM, collection data). Identification: Hardy (1973).

A. plumosa Hendel has been recorded as a bamboo pest in Taiwan (Issiki *et al.*, 1928; Maki & Rin, 1918). Shiraki (1933) recorded it from *D. latiflorus* and *Phyllostachys pubescens*. Identification: Hardy (1973).

Gastrozona fasciventris (Macquart) has been recorded as a bamboo pest in Taiwan (Issiki *et al.*, 1928; Kurata, 1925). Shiraki (1933) recorded it from puntingpole bamboo (*B. tuldoides*), *D. latiflorus* and *P. pubescens* (as *G. macquarti* Hendel). Identification: Hardy (1988a).

Other Tephritidae Associated with Leaves and Stems

The larvae of *Afrocneros excellens* (Loew), *A. mundus* (Loew), *Ocnerioxa bigemmata* (Bezzi) and *O. sinuata* (Loew), from Africa, have been reared from the parenchyma of umbrella tree (*Cussonia* spp.), some of which are cultivated, and the infestations have been known to kill the trees (Munro, 1967). Identification: Munro (1967).

Diarrhegmoides araucariae (Tryon) larvae, from Australia, develop under lifted bark of hoop pine (*Araucaria cunninghamii*) thinnings, causing a wet rot (Brimblecomb, 1945). The adults are attracted to the resin of the tree (Tryon, 1927). Identification: Tryon (1927).

Chaetostomella vibrissata Coquillett, from temperate Asia, induces galls on the stems of burdock (*Arctium lappa*), which is grown as a vegetable crop in Japan (Shinji, 1939). In Europe *Arctium* spp. have a minor use in soft drink manufacture but tephritids have not been reported as pests. A detailed study of the competitive interactions between *Tephritis bardanae* (Schrank) and *Terellia tussilaginis* (Fabricius), which attack the capitula of *Arctium* spp. in Europe, was carried out by Straw (1989a). Identification: Ito (1983-5) for Asian spp.; White (1988) for European spp.

Euphranta toxoneura (Loew), from Europe, oviposits in young leaf galls of *Pontania* spp. (Hymenoptera, Tenthredinidae) on *Salix* spp. When the fly larva hatches it immediately searches for the *Pontania* larva, which it slits open and sucks dry. There is also a weevil (Coleoptera, Cuculionidae) which develops in *Pontania* galls in a similar manner, and the weevil larvae may also be attacked by *E. toxoneura*. This remarkable biology was described more fully by Kopelke (1984; 1985). Identification: White (1988).

Hemilea araliae Malloch, from the Bismarck Archipelago, has been reared from the leaves of an *Aralia* sp. but it is not known if it was a species with edible leaves (Malloch, 1939b; NHM, collection data). Identification: Malloch (1939b).

The genus *Myoleja* includes both frugivorous species (p. 402), and petiole and twig mining species (Hancock, in press, a). The latter group includes *M. caesio* (Harris) which mines the petioles of nettles (*Urtica* spp.; Urticaceae) in Europe (Beiger, 1968). *M. alboscutellata* (Wulp), from Indonesia, has been recorded as a pest of arabica coffee (*Coffea arabica*) (Leefmans, 1930), and its larvae develop in the green twigs (Hancock, in press, a; de Meijere, 1911). Identification: Hardy (1987) for Asian spp.; White (1988) for European spp.

Orellia falcata (Scopoli) larvae develop in the stem base and roots of meadow salsify (*Tragopogon pratensis*) (White, 1988), but there are no records of it attacking cultivated salsify (*T. porrifolius* L.). Identification: White (1988).

Members of the North American genus *Strauzia* Robineau-Desvoidy attack wild and cultivated sunflowers (*Helianthus* spp.). The larvae of *S. longipennis* (Wiedemann) sometimes develop in the stems of cultivated sunflower (cultivars of *H. annuus*) and Jerusalem artichoke (*H. tuberosus*), reducing the yield of seeds and physically weakening the stems so that the plants becomes prone to wind damage (Steyskal, 1986; Stoltzfus, 1988; Wasbauer, 1972). Identification: Steyskal, 1986; Stoltzfus, 1988.

Members of the genus *Trypeta* Meigen are leaf-miners, and some species have been known to damage cultivated chrysanthemums (*Chrysanthemum* spp.), namely *T. artemisiae* (Fabricius) and *T. zoe* Meigen in western Europe (Andrews, 1941; Fjelddalen, 1953; Hodson & Jary, 1939), and *T. trifasciata* Shiraki in Japan (Koizumi, 1957). In addition, *T. artemisiae* has been recorded from tarragon (*Artemisia dracunculus*) (Hendel, 1927) and *T. zoe* has been recorded from Jerusalem artichoke (*H. tuberosus*) (Séguy, 1934), but those records require confirmation. Identification: Ito (1983-5) for Japanese spp.; White (1988) for European spp.

The larvae of *Zacerata asparagi* Coquillett, from southern Africa, bore in the shoots of wild *Asparagus* spp. and sometimes in cultivated asparagus (*Asparagus officinalis*) (Annecke & Moran, 1982). Identification: Hancock (1986b).

Species Accounts; 3 - Flower Pests

About 1800 (46%) of the described species of Tephritidae are probably flower associated, although only one merits inclusion here as a pest. Almost all Tephritinae of known biology are flower associated, but this section also includes a few Trypetinae whose larvae develop in cultivated flowers or their buds (*Coelotrypes*, *Acidoxantha* and *Macrotrypeta* spp.). There are a few flower and bud associated Dacinae, but they were included in the fruit pest section of this work (p. 128), because there is no clear taxonomic division between flower, fruit and bud associated species in that group, and some cucurbit associated *Bactrocera* spp. may attack either flowers or fruit.

GENUS *Acanthiophilus* Becker

This is a genus of twelve species, found in Africa, Asia and Europe.

Acanthiophilus helianthi (Rossi)

Capsule Fly

(fig. 254)

Taxonomic notes
This species has also been known as *Musca helianthi* Rossi, *Trypeta eluta* Meigen and *Urellia eluta* (Meigen).

Pest status and commercial hosts
Heavy attack by *A. helianthi* prevented the development of safflower (*Carthamus tinctorius*) as a crop in Europe (Féron & Vidaud, 1960). It is also a pest of safflower

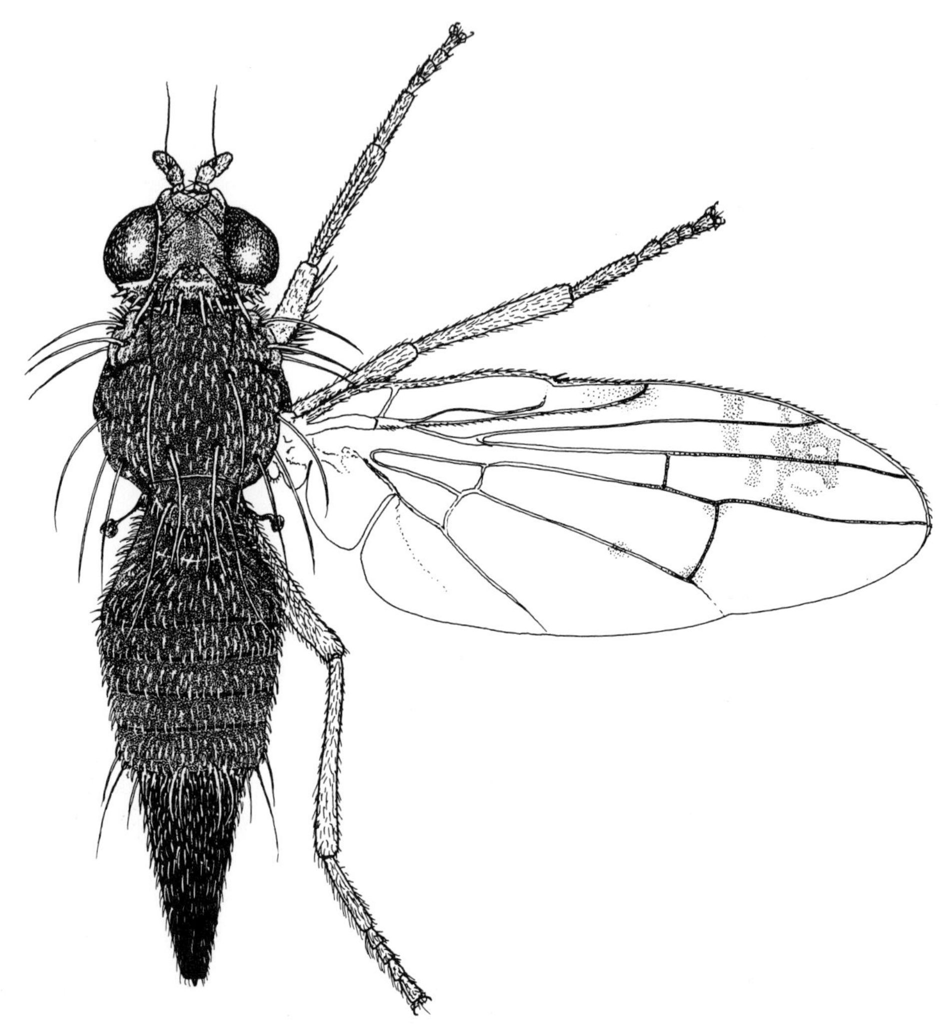

Fig. 254. *Acanthiophilus helianthi*, adult female.

in Egypt (Swailem, 1974), India (Jakhmola & Yadav, 1980; Verma *et al.*, 1974), Iraq (Ali *et al.*, 1977), Israel (Avidov, 1966), Romania (Manolache, 1940), Russia (Rusanova, 1926) and Turkey (Giray, 1979). The value of safflower varieties which are resistant to *A. helianthi* attack was discussed by Jakhmola & Yadav (1980).

A. helianthi (Rossi) has also been reported as a horticultural pest, damaging the commercial production of cornflower (*Centaurea cyanus*) and sweet sultan (*C. moschata*) seed in Hungary (Martinovich, 1966). In India it is recorded as a pest of cornflower (Agarwal & Kapoor, 1982) and it is known to damage *C. americana*, a native of North America (Menon *et al.*, 1969).

In Egypt, globe artichokes (*Cynara scolymus*) have also been attacked by *A. helianthi* (Rossi) (Saddik & Miniawi, 1978), but this is not a host in some other areas (White *et al.*, 1990).

A. helianthi has been reported from sunflower (*Helianthus annuus*) in Italy (Belcari, 1985), but that association is atypical as its other recorded hosts are members of another tribe of Asteraceae (the Cardueae).

Wild hosts
Recorded from a wide range of species belonging to the tribe Cardueae (Asteraceae), some of which were listed by White (1988).

Adult identification
Black body colour obscured by a dense grey or blue-grey tomentum and wing with a characteristic diffuse pattern (fig. 254). Wing length 3.9-5.1mm.

Distribution
Africa: Ethiopia, Kenya, Sudan.
Asia: India, Pakistan, Thailand.
Atlantic Islands: Madeira.
Europe: All southern countries, plus the United Kingdom.

Other references
BIOLOGY, Ali *et al.* (1977); LARVA, White (1988) and Efflatoun (1927) [although he confused his samples of *A. helianthi* and *Chaetorellia conjuncta* (Becker)]; PARASITOIDS, Agarwal & Kapoor (1989) and Narayanan & Chawla (1962).

Other Tephritidae Associated with Oil Seed Crops

Safflower (*Carthamus tinctorius*) may also be attacked by *Chaetorellia carthami* Stackelberg in the Middle East (Ali, 1976; Ali *et al.*, 1979; White & Marquardt, 1989), and by *Terellia luteola* (Wiedemann) in the Mediterranean area (White *et al.*, 1990). Identification: White & Marquardt (1989).

In North America the wild form of the cultivated sunflower (*Helianthus annuus*) is a host of *Eutreta sparsa* (Wiedemann), *Gymnocarena diffusa* (Snow), *Neotephritis finalis* (Loew), *Paracantha cultaris* (Coquillett), *Strauzia longipennis* (Wiedemann), *Trupanea bisetosa* (Coquillett) and *T. nigricornis* (Coquillett). Of those, only *S. longipennis*, *N. finalis*, *P. cultaris* and *T. nigricornis* have been recorded from cultivated sunflower. *S. longipennis* larvae feed in the stem (p. 410), but the larvae of *P. cultaris* eat the bases of immature florets and usually deform the capitulum (Cavender & Goeden, 1984). *N. finalis* and *T. nigricornis* attack the flower heads of a wide range of Asteraceae and sunflower is not their usual host (Cavender & Goeden, 1982; Goeden *et al.*, 1987; Kamali & Schulz, 1974; Schulz & Lipp, 1969), although *N. finalis* has been recorded as a pest of cultivated sunflower (Arthur & Mason, 1989). Identification: Foote & Blanc (1963) for most species.

Dioxyna sororcula (Wiedemann) is a pest of Niger seed (*Guizotia abyssinica*) in India (Jakhmola, 1983). In Ethiopia, an unidentified species of Schistopterini has been reared in large numbers from Niger seed (IIE, unpublished data, 1991). Identification: Freidberg & Kugler (1989).

Tephritidae Associated with Flowers of Food Crops

Globe artichoke (*Cynara scolymus*) is attacked by *Terellia fuscicornis* (Loew) in southern Europe and it has been reported as a pest in Greece (Stavraki & Stavrakis, 1968) and Turkey (Giray, 1979). In North America the crowns of globe artichoke shoots have been attacked by *Paracantha cultaris* (Coquillett), a species normally associated with wild thistles (*Cirsium* spp.) (Lange, 1941). Identification: Freidberg & Kugler (1989) for *T. fuscicornis*; Foote & Blanc (1963) for *P. cultaris*.

The flowers of lettuce (*Lactuca sativa*) are sometimes attacked by *Trupanea amoena* (Frauenfeld) in Europe, and that has been known to cause serious damage to the seeds (presumably the lettuce were for seed production) (Aczél, 1937; Giray, 1979; Zangheri, 1962). Similarly, tall lettuce (*L. canadensis*) flowers are sometimes attacked by *Jamesomyia geminata* (Loew) in eastern North America (Wasbauer, 1972), and in the Philippines *Dioxyna sororcula* (Wiedemann) has been reared from lettuce (Hardy, 1974). Identification: Foote & Steyskal (1987) for *J. geminata*; Freidberg & Kugler (1989) for other species.

Aciura coryli (Rossi) larvae develop in the inflorescence of many species of Lamiaceae (=Labiatae) in the Mediterranean area, and it has been recorded as a pest of Alibotush tea (*Sideritis scardica*) in Bulgaria (Drensky, 1931). Identification: Freidberg & Kugler (1989).

Coelotrypes vittatus Bezzi has been collected on (not reared) sweet potato (*Ipomoea batatas*) in Tanzania (NHM, collection data), and that is likely to be a true host because it has been reared from the buds of other *Ipomoea* spp. (Hancock, 1986b; Munro, 1953). *C. vittatus* has also been trapped in large numbers in a citrus growing area using ammonium phosphate bait (Saraiva, 1965). Identification: Hancock (1986b).

Tephritidae Associated with Medicinal Plants

Many species of Tephritidae are associated with plants that have some actual or historical use in medicine. However, *Tephritis arnicae* (Linnaeus) is one of the few species reported as a pest of a medicinal plant. *T. arnicae* attacks arnica (*Arnica montana*) in many areas of northern, central and eastern Europe, and it is a pest of arnica in Italy (Scaltriti, 1985). Identification: Hendel (1927); Richter (1970).

Tephritidae Associated with Rubber Producing Plants

A few species of Tephritidae are associated with species of Asteraceae which have been used as minor sources of rubber. In the Ukraine *Scorzonera tausaghyz* has been damaged by *Ensina sonchi* (Linnaeus) and *Paroxyna producta* (Loew), and *Taraxacum koksaghyz* by *E. sonchi*, *P. producta* and *Trupanea stellata* (Fuessly) (Dirsh, 1933; Gilyarov & Luk'yanovichi, 1938). Similarly, the North American rubber rabbitbrush (*Chrysothamnus nauseosus*) has been used as a source of chewing gum and is a potential source of rubber, which is attacked by the following stem gall inducing species: *Aciurina bigeloviae* (Cockerell), *A. ferruginea* (Doane), *A. maculata* (Cole), three unidentified *Procecidochares* spp., *Valentibulla californica* (Coquillett) and *V. steyskali* Foote (Wangberg, 1977; 1978; 1980; 1981). Identification: White (1988) for the European spp.; Foote & Blanc (1963) for most of the North American spp.

Tephritidae Associated with Ornamental Flowers

Acidoxantha bombacis de Meijere larvae, from Indonesia, develop in the flowers of red silk-cotton tree (*Bombax ceiba*) (de Meijere, 1938) and other large flowers such as *Bauhinia* and *Hibiscus* spp. (Hardy, 1973; 1974). Similarly, *A. hibisci* Hardy has been reared from the flower heads of sea hibiscus (*H. tiliaceus*) (Hardy, 1974). Identification: Hardy (1974).

Campiglossa hirayamae (Matsumura) has been recorded from cultivated chrysanthemums (*Chrysanthemum* spp.) in Korea (Kwon, 1985). Identification: Ito (1983-5).

Campiglossa misella (Loew) has been known to disfigure glasshouse chrysanthemums in Britain (Andrews, 1941). Identification: White (1988).

Chaetostomella completa (Kapoor, Malla & Ghosh) is a pest of cornflower (*Centaurea cyanus*) in India (Agarwal & Kapoor, 1982). Identification: Kapoor *et al.* (1979).

Craspedoxantha marginalis (Wiedemann) attacks a wide range of Asteraceae and it is reported as a pest of Barberton daisy (*Gerbera jamesonii*), dahlia (*Dahlia* sp.) and zinnia (*Zinnia elegans*) in South Africa (Le Pelley, 1959; Munro, 1964a). *C. octopunctata* Bezzi has been recorded from *Centaurea americana* in India (Menon *et al.*, 1969). Identification: Freidberg (1985).

Dioxyna picciola (Bigot), from North America, sometimes attacks species of Asteraceae of horticultural interest, for example marigold (*Tagetes erecta*) and plains coreopsis (*Coreopsis tinctoria*) (Wasbauer, 1972). Identification: Novak (1974).

Macrotrypeta ortalidina Portschinsky is a pest which attacks the buds of garden peony (*Paeonia chinensis*) in south-eastern Siberia, USSR (Ermolaev *et al.*, 1980). Identification: Ermolaev *et al.* (1980).

Paroxyna murina (Doane), a North American species, has been found amongst seeds of *Aster foliaceus* imported into the United Kingdom (Wilson, 1931). Identification: Novak (1974).

Spathulina acroleuca (Schiner), which is found in most tropical areas of the Old World, has been reared from Mexican sunflower (*Tithonia diversifolia*) in the Philippines (Hardy, 1974). Identification: Hardy (1988b).

Tephritis neesii (Meigen), from Europe, attacks the capitula of *Leucanthemum* spp. (White, 1988), including garden varieties of shasta daisy (*L. maximum*) (I.M. White, unpublished data, 1991).

Trupanea amoena (Frauenfeld) and *T. stellata* (Fuessly) have been recorded as pests of commercially grown pot marigold flowers (*Calendula officinalis*) in India (Nirula, 1942; Trehan, 1947). Identification: White (1988).

Trupanea superdecora Bezzi, from Africa, has been recorded in dahlia (*Dahlia* sp.) (Munro, 1964b). Identification: Munro (1964b).

Xanthaciura tetraspina (Phillips), from the Americas, develops in the flowers of ageratum (*Ageratum houstonianum*) (Wasbauer, 1972). Identification: Aczél (1952).

Species Accounts; 4 - Beneficial Species

About twenty species of Tephritidae, have been used as biological control agents of adventive weeds. Most of the tephritids involved are members of the subfamily Tephritinae, particularly gall inducing Myopitini, and almost all of the weeds are species of Asteraceae (=Compositae). P. Harris (1989) reviewed the use of tephritids for weed biocontrol, and of the 18 species he listed as having been released 10 are well established, two are established at low density, and the remaining six failed to establish in any of the release areas. Julien (1987) catalogued the released biocontrol agents and their target weeds, and White (1991b) discussed the contribution of taxonomy to some of those biocontrol programmes.

Acinia picturata (Snow)

This is a Central American species which was released in the Hawaiian island of Oahu in 1959 for the control of sour bush (*Pluchea odorata* (L.) Cassini; Asteraceae) and it has been accidentally introduced into Johnston Atoll and Wake Island (mid Pacific). Although it has become well established in Oahu it has not proved effective as a biocontrol agent (Goeden, 1978; Julien, 1987). Identification: Hardy & Delfinado (1980).

Ceratitella tomentosa de Meijere

This frugivorous species was introduced into Trinidad from Pakistan in 1978 for the control of bird vine (*Phthirusa adunca* (G.F.W. Mey) Maguire; Loranthaceae), but it failed to become established (Julien, 1987; as *C. asiatica* Hardy). Identification: Hardy (1967; 1987).

GENUS *Chaetorellia* Hendel

This genus includes nine species, some of which are potential biocontrol agents of Palaearctic knapweeds and starthistles (*Centaurea* spp.; Asteraceae) which have become adventive noxious weeds in North America. *Chaetorellia* larvae consume the developing seeds of the host capitulum; the genus was revised by White & Marquardt (1989).

Chaetorellia acrolophi White & Marquardt is native to Central Europe where it attacks spotted knapweed (*Centaurea maculosa* Lam.), which is a serious weed pest in western Canada. Identification: White & Marquardt (1989).

C. australis Hering is native to the Balkan area of Europe where it attacks cornflower (*Centaurea cyanus*) and yellow starthistle (*C. solstitialis* L.), both of which occur in North America, and yellow starthistle is a serious rangeland weed in western USA. Identification: White & Marquardt (1989).

C. succinea (Costa) is very closely related to *C. carthami* (p. 413), which attacks cultivated safflower (*Carthamus tinctorius*). *Chaetorellia succinea* attacks yellow starthistle (*Centaurea solstitialis*) in the Mediterranean area and was at one time considered a potential biocontrol agent of that weed. Identification: White & Marquardt (1989).

GENUS *Euaresta* Loew

This is a New World and temperate Asian genus of 14 species, two of which have been introduced to other areas, either by accident or for weed biocontrol.

Euaresta aequalis Loew is a North American species which was released in Australia in 1932 and Fiji in 1951, for the control of common cocklebur (*Xanthium strumarium* L.; Asteraceae). In Fiji it did not become established, and in Australia it became established but was ineffective for biocontrol (Goeden, 1978; Julien, 1987). The larvae are seed feeders. Identification: Foote & Blanc (1963); Quisenberry (1950).

E. bella Loew is a North American species which was introduced into the USSR in 1969 and 1977, for the control of common ragweed (*Ambrosia artemisiifolia* L.; Asteraceae), but it failed to become established (Julien, 1987). Several other species of *Euaresta* and *Euarestoides* are associated with ragweeds and they have been well studied because of their potential value as biocontrol agents (Batra, 1979; B.A. Foote, 1984; Vitolo & Stiles, 1987; Piper, 1976). Common ragweed is also attacked by *Callachna gibba* (Loew), a North American species which induces a gall (Wasbauer, 1972), and *Procecidochares stonei* Blanc & Foote (Silverman & Goeden, 1980). Identification: Foote & Blanc (1963) for *Euarestoides* spp. and *P. stonei*; Foote & Steyskal (1987) for *C. gibba*; Quisenberry (1950) for *Euaresta* spp.

E. bullans Wiedemann is a South American species which was introduced into Australia (pre-1928) for the control of spiny cocklebur (*X. spinosum* L.), where it is not generally effective, but it may apparently destroy some local stands of the weed. *E. bullans* has also become accidentally established in southern Africa and in some areas of the Palaearctic where adventive populations of its host plant are found (Julien, 1987). Identification: Foote & Blanc (1963); Quisenberry (1950).

Eutreta xanthochaeta Aldrich

This is a native of Central America which induces stem galls on lantana (*Lantana camara* L.; Verbenaceae) (Stoltzfus, 1977). It was introduced into Hawaii in 1902 and together with other biocontrol agents it has achieved partial to substantial control in dry areas. In 1914, 1971 and 1977 it was introduced into Australia, but it failed to become established (Goeden, 1978; Julien, 1987). Identification: Hardy & Delfinado (1980).

GENUS *Procecidochares* Hendel

This is a New World genus of 14 species which induce stem galls, two of which have been introduced into other areas for the biocontrol of noxious weeds.

P. alini Steyskal is a native of Central America. It was introduced into the Hawaiian Islands of Hawaii and Maui in 1974-5 for the control of Hamakua pamakani (*Ageratina riparia* (Regel) R. King & H. Robinson; Asteraceae), and in some areas it has achieved substantial or complete control (Julien, 1987). The biology of this species was described by Hapai & Chang (1986). Identification: Hardy & Delfinado (1980).

P. utilis Stone is also a native of Central America. It was released in the Hawaiian Island of Maui in 1945 for the control of crofton weed or Maui pamakani (*A. adenophora* (Sprengel) R. King & H. Robinson). Crofton weed has been controlled in the dryer areas, but in wet areas the weed populations have not been reduced. *P. utilis* has also been released in Australia (1953), India (1963), Madeira (pre-1971) and New Zealand (1958), but it has not achieved effective control in any of those areas (Goeden, 1978; Julien, 1987); recently it has been released in China (Zhang *et al.*, 1988). Although it has not been an effective means of *Ageratina* control in India, the fly has become sufficiently well established as to have spread to Nepal (NHM, collection data). Identification: Hardy & Delfinado (1980).

Tephritis dilacerata (Loew)

This is a European species which attacks the flower heads of perennial sowthistle (*Sonchus arvensis* L.; Asteraceae). It has been released in Canada on several occasions since 1979 (Julien, 1987) but populations so far tested have been unable to survive the harsh Canadian winter. The larvae feed on the developing ovaries and the receptacle of the host capitulum, which is swollen into a soft gall (Berube 1978a; 1978b; Janzon, 1983; Peschken, 1980; Shorthouse, 1980). *T. formosa* (Loew) has also been studied with regard to the biological control of *Sonchus* spp. in Canada (Berube, 1978). Identification: Hendel (1927); Richter (1970).

GENUS *Terellia* Robineau-Desvoidy

This is a predominantly European genus of 27 species, most of which attack the flower heads of thistles and knapweeds. Two *Terellia* spp. are potential biocontrol agents of European *Centaurea* spp. (Asteraceae) which are adventive in western North America. *T. virens* Loew attacks spotted knapweed (*C. maculosa*), and *T. uncinata* White attacks yellow starthistle (*C. solstitialis*) (White, 1989a). Identification: White (1989a).

Tetreuaresta obscuriventris (Loew)

This is a South American species which attacks the flowers of elephant's foot (*Elephantopus mollis* Humboldt, Bonplaud & Kunth; Asteraceae). In 1957 *T. obscuriventris* was introduced into Fiji for control of *E. mollis*, where it became established but did not control its host. In 1961 it was introduced into the Hawaiian Islands of Kauai and Hawaii but it has not had any effect on its target weed (Goeden, 1978; Julien, 1987). Identification: Hardy & Delfinado (1980).

GENUS *Urophora* Robineau-Desvoidy

This genus includes almost 100 species and it is represented in Europe, temperate Asia, Africa and the New World. Most of the European species attack the flower heads of thistles and knapweeds (Asteraceae), many of which are adventive noxious weeds in other areas. Consequently, several *Urophora* spp. are now being used, or considered for use, as biocontrol agents of noxious weeds. White & Clement (1987) provided a key for the species relevant to North American weed biocontrol programmes and White & Korneyev (1989) revised the European species; those two papers should also be consulted for further references on biology and the use of *Urophora* spp. as biocontrol agents.

U. affinis (Frauenfeld) is a native of central Europe where it attacks diffuse knapweed (*Centaurea diffusa* Lam.) and spotted knapweed (*C. maculosa*), both of which are serious weed pests in western North America. In western Canada *U. affinis* has been successfully established on both spotted and diffuse knapweed (Harris, 1984; Harris & Myers, 1984). Following the use of *U. affinis*, in combination with *U. quadrifasciata*, seed production of those knapweeds in Canada has been reduced, and in certain cases the total plant biomass has declined. *U. affinis* induces lignified unilocular galls of receptacle tissue and the larvae hatch as first instars. Identification: White & Clement (1987); White & Korneyev (1989).

U. cardui (Linnaeus) is a native of cooler areas of Europe, and Asia as far east as Kazakhstan (USSR), where it attacks creeping or Canadian thistle (*Cirsium arvense* (L.) Scop.). *U. cardui* has been released in both western and eastern Canada; it died out in the west, and although established in the east it failed to reduce the areas of its host (Harris, 1984; Peschken, 1984). *U. cardui* induces a lignified multilocular stem gall and the larvae hatch as second instars; this is the only known *Urophora* sp. which does not attack the flower head. Identification: White & Clement (1987); White & Korneyev (1989).

U. jaculata Rondani occurs in Italy and Greece where it induces lignified unilocular galls of receptacle tissue in yellow starthistle (*Centaurea solstitialis*). Misidentified as *U. sirunaseva* (Hering), it was introduced into California in 1969, 1976 and 1977, but it failed to become established. Subsequent experiments in Italy have shown that *U. jaculata* does not attack the variety of yellow starthistle found in western North America (White & Clement, 1987). Identification: White & Clement (1987); White & Korneyev (1989).

U. quadrifasciata (Meigen) is a native of western, central and southern Europe, the Middle East and north Africa. *U. quadrifasciata* attacks the capitula of a wide range of knapweeds and starthistles (*Centaurea* spp.); it induces a gall from the ovary wall and that gall is not lignified. In Canada, *U. quadrifasciata* has been successfully established on the same plants as *U. affinis* (Harris & Myers, 1984) and it has recently extended its range into the USA; it has also become accidentally established in Australia. *U. quadrifasciata* is usually bivoltine, unlike other central European species of known biology, and its larvae hatch as first instars. Identification: White & Clement (1987); White & Korneyev (1989).

U. sirunaseva (Hering) occurs in Moldavia (USSR), Turkey and north-eastern Greece on yellow starthistle (*C. solstitialis*). It induces lignified unilocular galls of receptacle tissue and it is a potential biocontrol agent of yellow starthistle in North America. Identification: White & Clement (1987); White & Korneyev (1989).

U. solstitialis (Linnaeus) occurs in Europe and temperate Asia, as far east as Kazakhstan (USSR); its confirmed hosts are all thistles (*Carduus* spp.), and it induces lignified multilocular galls of receptacle tissue. *U. solstitialis* is a potential biocontrol agent for use against nodding and plumeless thistles (*C. nutans* L. and *C. acanthoides* L.) in North America. Identification: White & Clement (1987); White & Korneyev (1989).

U. stylata (Fabricius) occurs throughout Europe and temperate Asia where it attacks thistles (*Cirsium* and *Carduus* spp.). In British Columbia, *U. stylata* has substantially reduced seed production by bull thistle (*Cirsium vulgare* (Savi) Tenore); it was also released in Quebec, but it died out after the release site was mowed (Harris, 1984; Harris & Wilkinson, 1984); it has also become accidentally established in Australia. *U. stylata* induces a lignified multilocular gall of receptacle tissue and its larvae hatch as second instars. Identification: White & Clement (1987); White & Korneyev (1989).

Xanthaciura connexionis Benjamin

This is a Central American species which was released in Hawaii in 1955 for the control of pamakani weeds (*Ageratina* spp.), but it failed to become established (Julien, 1987). Identification: Hardy & Delfinado (1980).

Other Tephritidae Considered for Use in Weed Biocontrol

Carpomya schineri and *Rhagoletis alternata* have been considered as potential biocontrol agents of sweet briar (*Rosa rubiginosa*), which is an adventive weed in New Zealand, although they are regarded as minor pests of other *Rosa* spp. in some parts of Europe (p. 285, 386) (Eichhorn, 1967). Identification: Freidberg & Kugler (1989) for *C. schineri*; White (1988) for *R. alternata*.

Eurosta solidaginis (Fitch) induces a stem gall (Gilbert & Kurczewski, 1986; McCrea & Abrahamson, 1985; Walton, 1985; 1988) on *Solidago canadensis* L. (Asteraceae), a native of North America, which is an adventive weed in Europe. *Eutreta sparsa* (Wiedemann) and *Procecidochares atra* (Loew) are also recorded from *S. canadensis* in North America (Wasbauer, 1972; as *S. altissima*). Identification: Steyskal & Foote (1977) for *Eurosta* spp.; Stoltzfus (1977) for *Eutreta* spp.

Oxyna parietina (Linnaeus) is a stem miner which has been considered as a potential biocontrol agent of *Artemisia absinthium* L. (Asteraceae), a Canadian weed of Europe origin (Schroeder, 1979). Identification: White (1988).

Species Assisting in the Natural Regulation of Some Weeds

Any tephritid associated with a plant pest is likely to be involved in the natural regulation of the biomass of that pest to a small degree, and examples of potentially useful species follow.

Dacus (Leptoxyda) longistylus Wiedemann has been reported as helping in the natural control of Sodom apple (*Calotropis procera*) which is growing as a weed in the Sudan (Venkatraman & Khidir, 1967); sodom apple is also cultivated in some areas (p. 338). Identification: Munro (1984).

Dioxyna Frey is a genus of 10 species represented in all regions, and although some of the species have been reared from a wide variety of Asteraceae, the species of known biology all seem to have a preference for beggarticks (*Bidens* spp.). The most widespread species in the Old World is *D. sororcula* (Wiedemann) which has an unbroken range from the Mediterranean, through Africa and tropical Asia, to Australia, and is adventive in the Hawaiian Islands. A very similar species, *D. picciola* (Bigot), is found throughout most of the New World. Identification: Freidberg & Kugler (1989) for *D. sororcula*; Novak (1974) for *D. picciola*.

Ensina sonchi (Linnaeus) and *Sphenella marginata* (Fallén) are worth a mention because they are amongst the most widespread Asteraceae associated members of the family. *E. sonchi* is a highly polyphagous native of Europe and temperate Asia, which has been spread to tropical Asia, Ethiopia, Peru and the Hawaiian Islands where it attacks the flower heads of *Sonchus oleraceus* L., which is also a native of Europe. *Sphenella marginata* attacks the flower heads of *Senecio* spp. and it is found in Europe, temperate Asia and Africa. In Australia native and introduced European *Senecio* spp. are attacked by *Sphenella ruficeps* (Macquart), but that species is very similar and doubtfully distinct from *S. marginata*. Jacobean or tansy ragwort (*Senecio jacobaea* L.) is a European species adventive in North America where it is attacked by *Eutreta angusta* Banks, *Paroxyna genalis* Thomson and *Trypeta angustigena* Foote (Wasbauer, 1972). Identification: Foote (1960) for *Trypeta* spp.; Novak (1974) for *Paroxyna* spp.; Stoltzfus (1977) for *Eutreta* spp.; White (1988) for European spp.

Euphranta apicalis Hendel attacks the stems of *Aeginetia indica* L., a species of broomrape (Orobanchaceae) which is a pest of sugar cane (*Saccharum officinarum* L.; Poaceae or Gramineae) (Takano, 1934). *E. apicalis* has been reared from that host in Taiwan, and has also been found in Myanmar, the Philippines and Vietnam (Hardy, 1974; Shiraki, 1933). Identification: Hardy (1974).

Terellia ruficauda (Fabricius) is native to Europe and temperate Asia where it attacks the capitula of some thistles (*Cirsium* spp.). It has become accidentally established in North America where it attacks an adventive European species of thistle (*C. arvense*). Identification: White (1988).

Toxotrypana australis Blanchard, from Argentina, attacks the pods of strangler vine (*Morrenia odorata* (Hook. & Arn.) Lindley; Asclepiadaceae) (Silveira-Guido & Habeck, 1978). Identification: Blanchard (1959).

Trupanea Schrank is the largest genus of Tephritinae, with about 200 species represented in all regions. An example of the genus is *T. glauca* (Thomson), from the Philippines and Australia, which has been reared from the flowers of whiteweed (*Ageratum conyzoides* L.) (Hardy, 1974). Identification: Hardy (1974).

Urophora jaceana (Hering) is a native of western and central Europe where it attacks black knapweed (*Centaurea nigra* L.), and in eastern Canada it has become accidentally established in areas where black knapweed is an adventive weed. *U. jaceana* induces a lignified multilocular gall of receptacle tissue and the larvae hatch as second instars (Varley, 1937; 1947). Identification: White & Clement (1987); White & Korneyev (1989).

Zonosemata vittigera (Coquillett) attacks the fruits of silverleaf nightshade (*Solanum elaeagnifolium*) in Mexico and southern North America, but it has little impact on the weed (Goeden & Ricker, 1971). Identification: Bush (1965).

Distribution of Fruit Pest Tephritidae

The following table lists the fruit associated species discussed in this book and their distribution with respect to the 14 geographic areas used for describing distribution in the species accounts. The column titles are abbreviated as follows, grouped by the regions used in the Diptera catalogues (some catalogue regions are combined in the table to simplify column layout):

Afrotropical (Cogan & Munro, 1980):
- AF Africa;
- AT Atlantic Ocean Islands;
- IO Indian Ocean Islands (including Madagascar).

Australian and Oceanian (Hardy & Foote, 1989):
- AU Australia;
- NG New Guinea area;
- PO Pacific Ocean Islands.

Nearctic (Foote, 1965):
- NA North America.

Neotropical (Foote, 1967):
- CA Central America;
- SA South America;
- WI West Indies.

Oriental (Hardy, 1977):
- OR Tropical Asia;

Palaearctic (R.H. Foote, 1984):
- EU Europe;
- ME Middle East;
- TA Temperate Asia (Asian Palaearctic).

Distribution data uses the following abbreviations:
- A presumed to be adventive;
- E eradicated;
- P present;
- ? possibly present (including adventive).

	Africa			Asia-Pacific				Eurasia			Americas			
Name	AF	AT	IO	AU	NG	PO	OR	EU	ME	TA	CA	NA	SA	WI
Acidiella														
angustifascia										P				
Adrama														
apicalis							P							
austeni							P							
determinata							P							
Anastrepha														
acris											P		P	P
antunesi											P		P	P
bahiensis											P		P	P
bezzii											P		P	
bistrigata													P	
consobrina													P	
daciformis													P	
distincta											P	P	P	
ethalea													P	P
fraterculus											P	P	P	P
grandis											P		P	
leptozona											P		P	
limae											P	P	P	
ludens											P	P		
macrura													P	
manihoti											P		P	
margarita													P	
mburucuyae													P	
minensis													P	
montei											P		P	
nigrifascia												P		
obliqua											P	P	P	P
ocresia												P		P
ornata													P	
pallens											P	P		
pallidipennis											P		P	
panamensis											P			
parishi													P	
passiflorae											P			
perdita													P	

	Africa			Asia-Pacific				Eurasia			Americas			
Name	AF	AT	IO	AU	NG	PO	OR	EU	ME	TA	CA	NA	SA	WI
pickeli											P		P	
pseudoparallela													P	
punctata													P	
rheediae											P		P	P
sagittata											P	P		
schultzi													P	
serpentina											P	P	P	P
sororcula													P	
steyskali													P	
striata											P	P	P	P
suspensa												P		P
turicai													P	
zenildae													P	
Anomoia														
purmunda							P	P		P				
Bactrocera														
albistrigata										P				
aquilonis				P										
arecae							P							
atrisetosa					P									
biguttula	P													
breviaculeus				P										
bryoniae				P	P									
cacuminata				P										
calophylli				P		P	P							
caryeae							P							
caudata							P							
cilifer							P							
correcta							P							
cucumis				P										
cucurbitae	A		A		P	A	P			A		E		
curvipennis						P								
decipiens					P									
depressa							P			P				
distincta						P								
diversa							P							
dorsalis						A	P					E		

	Africa			Asia-Pacific				Eurasia			Americas			
Name	AF	AT	IO	AU	NG	PO	OR	EU	ME	TA	CA	NA	SA	WI
dorsalis (A)							P						A	
dorsalis (B)							P							
dorsalis (C)							P							
dorsalis (D)							P							
duplicata							P							
endiandrae				P	P									
expandens				P			P							
facialis						P								
frauenfeldi				P	P	P								
froggatti						P								
halfordiae				P										
incisa							P							
jarvisi				P										
kirki						P								
kraussi				P										
latifrons						A	P							
limbifera							P							
mayi				P										
mcgregori							P							
melanoscutata					P									
melanota						P								
melas				P										
minax							P							
moluccensis					P		P							
munda							P							
musae				P	P		?							
mutabilis				P										
neohumeralis				P	P									
nigra				P										
nigrotibialis							P							
nigrovittata					P									
obliqua					P									
occipitalis							P							
ochrosiae						P								
oleae	P						P	P	P					
opiliae				P										
pallida				P										

Name	Africa			Asia-Pacific				Eurasia			Americas			
	AF	AT	IO	AU	NG	PO	OR	EU	ME	TA	CA	NA	SA	WI
parvula							P							
passiflorae						P								
psidii						P								
scutellaris							P							
scutellata							P							
simulata					P									
tau							P							
trilineata							P							
trilineola						P								
trimaculata							P							
trivialis				P	P		P							
tryoni				P	?	A								
tsuneonis							P			P				
tuberculata							P							
umbrosa					P	P	P							
versicolor							P							
xanthodes						P								
zonata			A				P					E		
Callistomyia														
flavilabris							P							
Capparimyia														
savastani	P						P	P	P					
Carpomya														
incompleta	P							P	P					
schineri								P		P				
vesuviana			A				P	P		P				
zizyphae							P							
Ceratitis														
aliena	P													
anonae	P													
capitata	P	A	A	A		A		A	A		A	?	A	A
catoirii			P											
colae	P													
cosyra	P													
discussa	P													
flexuosa	P													
giffardi	P													

Name	Africa AF	AT	IO	Asia-Pacific AU	NG	PO	OR	Eurasia EU	ME	TA	Americas CA	NA	SA	WI
malgassa			P											?
morstatti	P													
pedestris	P		P											
penicillata	P													
punctata	P													
quinaria	P								P					
rosa	P		A											
rubivora	P													
silvestrii	P													
turneri	P													
Dacus														
aequalis				P										
africanus	P													
annulatus	P													
arcuatus	P													
armatus	P													
aspilus	P													
axanus				P	P									
bistrigatus	P													
bivittatus	P													
ciliatus	P	?	A				A		A					
demmerezi			P											
disjunctus	P													
eminus	P													
eumenoides							P							
frontalis	P	A							A					
fuscatus	P													
hamatus	P													
humeralis	P													
hyalobasis	P													
inflatus	P													
inornatus	P													
longistylus	P													
lounsburyii	P													
maynei	P													
momordicae	P													
obesus	P													

Name	Africa			Asia-Pacific				Eurasia			Americas			
	AF	AT	IO	AU	NG	PO	OR	EU	ME	TA	CA	NA	SA	WI
pallidilatus	P													
persicus							P							
petioliforma				P										
punctatifrons	P		A						A					
retextus	P													
rufoscutellatus	P													
smieroides							P							
solomonensis					P									
sphaeroidalis							P							
telfaireae	P													
vansomereni	P													
vertebratus	P		A						A					
yangambinus	P													
Dirioxa														
confusa				P										
pornia				P		?								
Epochra														
canadensis												P		
Euphranta														
camelliae										P				
canangae							P							
cassiae							P							
connexa								P						
japonica										P				
lemniscata						P	P							
mikado										P				
Monacrostichus														
citricola							P							
Munromyia														
nudiseta	P													
Myiopardalis														
pardalina							P	P	P	P				
Myoleja														
limata												P		
nigroscutellata						P	P							
Neoceratitis														
asiatica										P				

Name	Africa AF	AT	IO	Asia-Pacific AU	NG	PO	OR	Eurasia EU	ME	TA	Americas CA	NA	SA	WI
Nitrariomyia														
lukjanovitshi										P				
Oedicarena														
latifrons											P	P		
Rhagoletis														
alternata								P		P				
basiola												P		
batava								P		P				
berberidis								P						
berberis												P		
blanchardi													P	
boycei											P	P		
cerasi								P		P				
cingulata												P		
completa								A			P	P		
conversa						A							P	
cornivora												P		
electromorpha												P		
fausta												P		
ferruginea													P	
flavigenualis								P						
indifferens								A				P		
juglandis											P	P		
juniperina												P		
lycopersella													P	
meigenii								P				A		
mendax												P		
mongolica										P				
nova													P	
pomonella											P	P		
psalida													P	
ramosae											P			
ribicola												P		
striatella											P	P		
suavis												P		
tabellaria												P		
tomatis													P	

	Africa			Asia-Pacific				Eurasia			Americas			
Name	AF	AT	IO	AU	NG	PO	OR	EU	ME	TA	CA	NA	SA	WI
turanica										P				
zephyria												P		
zoqui											P			
Rhagoletotrypeta														
uniformis											P			
Taomyia														
marshalli	P													
Toxotrypana														
curvicauda											P	P	P	P
Trirhithromyia														
cyanescens			P											
Trirhithrum														
albomaculatum	P													
basale	P													
coffeae	P													
inscriptum	P													
manganum			P											
nigerrimum	P		P											
nigrum	P													
occipitale	P													
Zonosemata														
electa												P		

Useful Plants and Their Associated Tephritids

The plants in the following list are all of some importance, either as crops, or as harvested wild plants. Some plants of medicinal and ornamental value were also included, but plants which are pests were not. Most of the listed plants were regarded as economically important by Terrell *et al.* (1986) and most of the common or vernacular names were also from that source. Plant names and authors were checked against Mabberley (1987) and Terrell *et al.* (1986), but when they disagreed on synonymy, the opinion of Mabberley (1987) was followed. The source of records, and sometimes other details, were given in the species accounts earlier in this work. Only selected common names were included in this list; to look up a plant by common name a cross reference list has been provided (p. 486). The following symbols and abbreviations are used in the list:

- ! Unusual host.
- ? Possible or likely host, but only known from old records; not confirmed by any known recent survey or authoritative data source.
- ?? Doubtful host; these records are only included to indicate previously published records that should be dismissed as being in error, for example, records which are known, or likely to be based on casual associations of adult flies resting on a plant, rather than reared. Doubtful hosts of *Anastrepha* spp. are not included as they were listed by Norrbom & Kim (1988b).

- B Larvae develop in the bud of the host; not used for the specialized flowers of Asteraceae (=Compositae) (see C for capitulum).
- C Larvae develop in the capitulum (composite flower) of a host belonging to the family Asteraceae (=Compositae).
- F Larvae develop in the fruit (fruit flesh, pod or seed).
- FD Larvae develop in fruit that was already damaged when attacked.
- F?D Larvae develop in fruit, but probably only when already damaged.
- FO Larvae develop in fruit that is already very ripe or over-ripe when attacked.

G Larvae develop in the gall of another insect.
I Larvae develop in the host inflorescence (flower); not used for the specialized flowers of Asteraceae (=Compositae) (see C for capitulum).
L Larvae develop in leaf mines in the host.
R Larvae develop in the root of the host.
S Larvae develop in the stem, trunk or petiole of the host.
SG Larvae induce a stem gall on their host.
X Larval host known, but part of plant attacked not recorded.

Abelmoschus Medikus (Malvaceae)

Abelmoschus esculentus (L.) Moench — okra
- *Bactrocera (Bactrocera) zonata* (Saunders) — F
- *Bactrocera (Zeugodacus) cucurbitae* (Coquillett) ?
- *Dacus (Didacus) ciliatus* Loew ! — F

[*Achras sapota* from New World - see *Pouteria sapota*]
[*Achras sapota* from Old World assumed to be *Manilkara zapota*]
[*Achras zapota* - see *Manilkara zapota*]

Actinidia Lindley

Actinidia chinensis Planchon — kiwi fruit
- *Ceratitis (Ceratitis) capitata* (Wiedemann) ?

Aegle Corr. Serr. (Rutaceae)

Aegle marmelos (L.) Corr.Serr. — Indian bael
- *Bactrocera (Bactrocera) correcta* (Bezzi) ?
- *Bactrocera (Bactrocera) zonata* (Saunders) — F

Ageratum L. (Asteraceae=Compositae)

Ageratum houstonianum Miller — ageratum
- *Xanthaciura tetraspina* (Phillips) — C

Allium L. (Liliaceae)

Allium cepa L. — onion
- *Bactrocera (Zeugodacus) cucurbitae* (Coquillett) ??

Anacardium L. (Anacardiaceae)

Anacardium occidentale L. — cashew
- *Anastrepha fraterculus* (Wiedemann) — F
- *Anastrepha ludens* (Loew) — F
- *Anastrepha obliqua* (Macquart) — F
- *Bactrocera (Bactrocera) aquilonis* (May) — F
- *Bactrocera (Bactrocera)* sp. near *B. dorsalis* (A) — F
- *Bactrocera (Bactrocera) facialis* (Coquillett) — F
- *Bactrocera (Bactrocera) passiflorae* (Froggatt) — F
- *Bactrocera (Bactrocera) tryoni* (Froggatt) — F
- *Ceratitis (Ceratitis) capitata* (Wiedemann) — F

Ananas Miller (Bromeliaceae)

Ananas comosus (L.) Merr.	pineapple	
Bactrocera (Bactrocera) facialis (Coquillett) ?		
Bactrocera (Bactrocera) kirki (Froggatt)		F
Bactrocera (Notodacus) xanthodes (Broun)		F
Ceratitis (Ceratitis) capitata (Wiedemann) ??		

[*Andropogon sorghum* - presumed to be *Sorghum bicolor*]

Angelica L. (Apiaceae=Umbelliferae)

Angelica archangelica L.	garden angelica	
Euleia heraclei (Linnaeus) ?		

Annona L. (Annonaceae)

Annona cherimola Miller	cherimoya	
Anastrepha fraterculus (Wiedemann)		F
Anastrepha ludens (Loew)		F
Ceratitis (Ceratalaspis) giffardi Bezzi ?		
Ceratitis (Ceratitis) capitata (Wiedemann) ?		
Annona glabra L.	pond-apple	
Anastrepha serpentina (Wiedemann)		F
Anastrepha suspensa (Loew)		F
Bactrocera (Bactrocera) sp. near *B. dorsalis* (B)		F
Annona muricata L.	soursop	
Anastrepha fraterculus (Wiedemann)		F
Anastrepha ludens (Loew) ?		
Anastrepha striata Schiner		F
Bactrocera (Afrodacus) jarvisi (Tryon)		F
Bactrocera (Bactrocera) aquilonis (May)		F
Bactrocera (Bactrocera) dorsalis (Hendel)		F
Bactrocera (Bactrocera) sp. near *B. dorsalis* (B)		F
Bactrocera (Zeugodacus) cucurbitae (Coquillett) ?		
Ceratitis (Ceratitis) capitata (Wiedemann)		F
Ceratitis (Pterandrus) anonae Graham ?		
Ceratitis (Pterandrus) rosa Karsch ?		
Annona reticulata L.	custard apple	
Anastrepha ludens (Loew)		F
Anastrepha suspensa (Loew)		F
Bactrocera (Bactrocera) aquilonis (May)		F
Bactrocera (Bactrocera) dorsalis (Hendel)		F
Bactrocera (Bactrocera) tryoni (Froggatt)		F
Bactrocera (Bactrocera) sp. near *B. zonata* (Saunders) ?		
Bactrocera (Zeugodacus) cucurbitae (Coquillett) ?		
Ceratitis (Ceratitis) capitata (Wiedemann)		F
Ceratitis (Ceratitis) catoirii Guérin-Méneville		F
Ceratitis (Pterandrus) rosa Karsch		F
Annona senegalensis Pers.	wild custard-apple	
Ceratitis (Ceratalaspis) cosyra (Walker)		F

Annona squamosa L. sugar-apple
- *Anastrepha fraterculus* (Wiedemann) F
- *Anastrepha ludens* (Loew) F
- *Anastrepha suspensa* (Loew) F
- *Bactrocera (Bactrocera) aquilonis* (May) F
- *Bactrocera (Bactrocera) dorsalis* (Hendel) F
- *Bactrocera (Bactrocera) neohumeralis* (Hardy) F
- *Bactrocera (Bactrocera) tryoni* (Froggatt) F
- *Bactrocera (Bactrocera) zonata* (Saunders) F
- *Bactrocera (Zeugodacus) cucurbitae* (Coquillett) ?
- *Ceratitis (Ceratitis) capitata* (Wiedemann) ?

Anthriscus Pers. (Apiaceae=Umbelliferae)

Anthriscus cerifolium (L.) Hoffm. chervil
- *Euleia heraclei* (Linnaeus) L

Apium L. (Apiaceae=Umbelliferae)

Apium graveolens L. celery
- *Euleia heraclei* (Linnaeus) L

Aralia L. (Araliaceae)

Aralia sp.
- *Hemilea araliae* Malloch L

Araucaria Juss. (Araucariaceae)

Araucaria cunninghamii Aiton ex D. Don hoop pine
- *Diarrhegmoides araucariae* (Tryon) S

Arbutus L. (Ericaceae)

Arbutus unedo L. strawberry tree
- *Ceratitis (Ceratitis) capitata* (Wiedemann) ?

Arctium L. (Asteraceae=Compositae)

Arctium lappa L. burdock
- *Chaetostomella vibrissata* Coquillett SG
- [Plus some other spp. which are only recorded from wild plants; p. 409]

Areca L. (Arecaceae=Palmae)

Areca catechu L. betel nut
- *Bactrocera (Bactrocera) arecae* (Hardy & Adachi) F

Arenga Labill. (Arecaceae)

Arenga pinnata (Wurmb) Merr. sugar palm
- *Bactrocera (Bactrocera)* sp. near *B. dorsalis* (A) F
- *Bactrocera (Bactrocera)* sp. near *B. dorsalis* (B) F

Arnica L. (Asteraceae=Compositae)

Arnica montana L. arnica
- *Tephritis arnicae* (Linnaeus) C

Artemisia L. (Asteraceae=Compositae)

Artemisia dracunculus L. tarragon
Trypeta artemisiae (Fabricius) ?

Artocarpus Forster & Forster f. (Moraceae)

Artocarpus altilis (Parkins.) Fosb. breadfruit
Bactrocera (Bactrocera) distincta (Malloch) F
Bactrocera (Bactrocera) sp. near *B. dorsalis* (A) F
Bactrocera (Bactrocera) sp. near *B. dorsalis* (C) F
Bactrocera (Bactrocera) facialis (Coquillett) F
Bactrocera (Bactrocera) frauenfeldi (Schiner) F
Bactrocera (Bactrocera) passiflorae (Froggatt) F
Bactrocera (Bactrocera) trilineola Drew ?
Bactrocera (Bactrocera) umbrosa (Fabricius) F
Bactrocera (Notodacus) xanthodes (Broun) F
Ceratitis (Ceratitis) capitata (Wiedemann) F
[*Artocarpus champeden* - see *A. integer*]
[*Artocarpus communis* - see *A. altilis*]
Artocarpus elasticus Reinw. ex Blume
Bactrocera (Bactrocera) sp. near *B. dorsalis* (A) F
Artocarpus heterophyllus Lam. jackfruit
Bactrocera (Bactrocera) caryeae (Kapoor) ?
Bactrocera (Bactrocera) sp. near *B. dorsalis* (A) F
Bactrocera (Bactrocera) sp. near *B. dorsalis* (B) F
Bactrocera (Bactrocera) incisa (Walker) ??
Bactrocera (Bactrocera) umbrosa (Fabricius) F
Bactrocera (Zeugodacus) cucurbitae (Coquillett) ! F
Artocarpus integer (Thunb.) Merr. chempedak
Bactrocera (Bactrocera) caryeae (Kapoor) ?
Bactrocera (Bactrocera) umbrosa (Fabricius) F
[*Artocarpus integrifolia* in Fletcher (1920) - see *A. heterophyllus*]
Artocarpus rigidus Blume monkey-jack
Bactrocera (Bactrocera) sp. near *B. dorsalis* (B) F
Artocarpus sp.
Bactrocera (Zeugodacus) tau (Walker) ?

Asclepias L. (Asclepiadaceae)

Asclepias curasavica L. bloodflower milkweed
Dacus (Leptoxyda) annulatus (Becker) S

Asparagus L. (Liliaceae)

Asparagus officinalis L. garden asparagus
Ceratitis (Ceratitis) capitata (Wiedemann) ?
Pliorecepta poeciloptera (Schrank) S
Zacerata asparagi Coquillett ! S

Aster L. (Asteraceae=Compositae)

Aster foliaceus Lindl.		
Paroxyna murina (Doane)		C

Atropa L. (Solanaceae)

Atropa bella-donna L.	belladonna	
Ceratitis (Ceratitis) capitata (Wiedemann) ?		

Averrhoa L. (Oxalidaceae=Averrhoaceae)

Averrhoa bilimbi L.	bilimbi	
Bactrocera (Afrodacus) jarvisi (Tryon)		F
Bactrocera (Bactrocera) sp. near *B. dorsalis* (A)		F
Averrhoa carambola L.	carambola, starfruit	
Anastrepha fraterculus (Wiedemann)		F
Anastrepha obliqua (Macquart)		F
Anastrepha suspensa (Loew)		F

Berberis L. (Berberidaceae)

Berberis vulgaris L.	berberis	
Anomoia purmunda (Harris)		F
Rhagoletis berberidis Jermy		F
Rhagoletis cerasi (Linnaeus) ??		
Rhagoletis meigenii (Loew)		F

Blighia König (Sapindaceae)

Blighia sapida König	akee	
Anastrepha suspensa (Loew)		F
Bactrocera (Bactrocera) aquilonis (May)		F
Bactrocera (Bactrocera) dorsalis (Hendel)		F
Ceratitis (Ceratitis) capitata (Wiedemann)		F

Bombax L. (Bombacaceae)

Bombax ceiba L.	red silk-cotton tree	
Acidoxantha bombacis de Meijere		I
[*Bombax malabaricum* - see *B. ceiba*]		

Brassica L. (Brassicaceae=Cruciferae)

[*Brassica caulorapa* - see *B. oleracea* var. *gongylodes*]		
Brassica juncea (L.) Czernj. & Cosson	Indian mustard	
Bactrocera (Zeugodacus) cucurbitae (Coquillett) !		S?
Brassica oleracea L.	broccoli, cabbage	
Bactrocera (Zeugodacus) cucurbitae (Coquillett) ?		

Brosimum Sw. (Moraceae)

Brosimum alicastrum Sw.	ramón, breadnut	
Anastrepha obliqua (Macquart)		F

Bumelia Sw. (Sapotaceae)

Bumelia languinosa (Michaux) Pers.	gum bumelia	
Anastrepha pallens Coquillett		F

[*Butyrospermum parkii* - see *Vitellaria paradoxa*]

Byrsonima Rich. ex Kunth (Malpighiaceae)

Byrsonima crassifolia (L.) Kunth	nance	
Anastrepha serpentina (Wiedemann)		F

Cajanus DC. (Fabaceae=Leguminosae)

Cajanus cajan (L.) Huth	pigeon pea	
Bactrocera (Zeugodacus) cucurbitae (Coquillett) ?		

Calendula L. (Asteraceae=Compositae)

Calendula officinalis L.	pot marigold	
Trupanea amoena (Frauenfeld)		C
Trupanea stellata (Fuessly)		C

Calophyllum L. (Clusiaceae=Guttiferae)

Calophyllum inophyllum L.	Indian laurel	
Bactrocera (Bactrocera) endiandrae (Perkins & May)		F
Bactrocera (Gymnodacus) calophylli (Perkins & May)		F
Bactrocera (Zeugodacus) cucurbitae (Coquillett) ?		
Ceratitis (Ceratitis) capitata (Wiedemann)		F

Calotropis R.Br. (Asclepiadaceae)

Calotropis procera (Aiton) Dryander ex Aiton f.	Sodom apple	
Dacus (Didacus) aspilus Bezzi		F
Dacus (Didacus) ciliatus Loew ??		
Dacus (Didacus) fuscatus Wiedemann		F
Dacus (Leptoxyda) longistylus Wiedemann		F
Dacus (Leptoxyda) obesus Munro		F
Dacus (Leptoxyda) persicus Hendel		F

Camellia L. (Theaceae)

Camellia japonica L.	common camellia	
Euphranta camelliae (Ito)		F
Camellia sinensis (L.) Kuntze	tea	
Adrama apicalis Shiraki		FD
Adrama austeni Hendel		FD
Adrama determinata (Walker)		FD
Camellia sp.		
Euphranta mikado (Matsumura)		F

Cananga Hook. f. & Thomson (Annonaceae)

Cananga odorata (Lam.) Hook. f. & Thomson	ylang-ylang	
Bactrocera (Bactrocera) arecae (Hardy) ??		
Bactrocera (Bactrocera) endiandrae (Perkins & May)		F
Bactrocera (Bactrocera) kraussi (Hardy) ?		
Bactrocera (Bactrocera) neohumeralis (Hardy)		F
Bactrocera (Bactrocera) tryoni (Froggatt)		F
Euphranta canangae Hardy		F

Canella P. Browne (Canellaceae)

Canella winteriana (L.) Gaertner — wild cinnamon
- *Anastrepha suspensa* (Loew) — F

Capparis L. (Capparidaceae=Capparaceae)

Capparis spinosa L. — caper
- *Capparimyia savastani* (Martelli) — B

Capsicum L. (Solanaceae)

Capsicum annuum L. — bell pepper
- *Anastrepha suspensa* (Loew) — F
- *Bactrocera (Bactrocera) aquilonis* (May) — F
- *Bactrocera (Bactrocera) dorsalis* (Hendel) — F
- *Bactrocera (Bactrocera)* sp. near *B. dorsalis* (A) — F
- *Bactrocera (Bactrocera)* sp. near *B. dorsalis* (B) — F
- *Bactrocera (Bactrocera) facialis* (Coquillett) — F
- *Bactrocera (Bactrocera) kirki* (Froggatt) — F
- *Bactrocera (Bactrocera) latifrons* (Hendel) — F
- *Bactrocera (Bactrocera) simulata* (Malloch) — F
- *Bactrocera (Notodacus) xanthodes* (Broun) — F
- *Bactrocera (Zeugodacus) caudata* (Fabricius) ??
- *Bactrocera (Zeugodacus) cucurbitae* (Coquillett) ?
- *Ceratitis (Ceratitis) capitata* (Wiedemann) — F
- *Ceratitis (Ceratitis) catoirii* Guérin-Méneville — F
- *Trirhithromyia cyanescens* (Bezzi) — F
- *Zonosemata electa* (Say) — F

Capsicum frutescens L. — tabasco pepper
- *Bactrocera (Bactrocera) bryoniae* (Tryon) ??
- *Bactrocera (Bactrocera) cacuminata* (Hering) ??
- *Bactrocera (Bactrocera) dorsalis* (Hendel) ?
- *Bactrocera (Bactrocera) facialis* (Coquillett) — F
- *Bactrocera (Bactrocera) kirki* (Froggatt) — F
- *Bactrocera (Bactrocera) trivialis* (Drew) — F
- *Bactrocera (Bactrocera) tryoni* (Froggatt) — F
- *Bactrocera (Zeugodacus) cucurbitae* (Coquillett) ?
- *Ceratitis (Ceratitis) capitata* (Wiedemann) — F
- *Ceratitis (Pterandrus) rosa* Karsch — F
- *Trirhithromyia cyanescens* (Bezzi) — F
- *Zonosemata electa* (Say) — F

[Doubtfully a separate species from *C. annuum* and therefore the records of these two host species are probably confused].

Capsicum sp.
- *Dacus (Didacus) ciliatus* Loew ??

Carica L. (Caricaceae)

Carica papaya L. — papaya
- *Anastrepha ludens* (Loew) — F
- *Anastrepha suspensa* (Loew) — F
- *Bactrocera (Afrodacus) jarvisi* (Tryon) — F
- *Bactrocera (Austrodacus) cucumis* (French) — F

Bactrocera (Bactrocera) dorsalis (Hendel)		F
Bactrocera (Bactrocera) sp. near *B. dorsalis* (B)		F
Bactrocera (Bactrocera) frauenfeldi (Schiner) ?		
Bactrocera (Bactrocera) musae (Tryon) !		F
Bactrocera (Bactrocera) passiflorae (Froggatt)		F
Bactrocera (Bactrocera) trilineola Drew ?		
Bactrocera (Bactrocera) tryoni (Froggatt)		F
Bactrocera (Bactrocera) zonata (Saunders)		F
Bactrocera (Notodacus) xanthodes (Broun)		F
Bactrocera (Zeugodacus) cucurbitae (Coquillett) !		F
Ceratitis (Ceratitis) capitata (Wiedemann)		FD
Ceratitis (Ceratitis) catoirii Guérin-Méneville		FO
Ceratitis (Pterandrus) rosa Karsch		F
Dacus (Dacus) bivittatus (Bigot) ??		
Euphranta lemniscata (Enderlein) ?		
Myoleja nigroscutellata (Hering)		X
Toxotrypana curvicauda Gerstaecker		F
Carica quercifolia (A. St. Hil.) Hieron	dwarf papaya	
Ceratitis (Ceratitis) capitata (Wiedemann) ?		

Carissa L. (Apocynaceae)

Carissa carandas L.	karanda	
Bactrocera (Bactrocera) correcta (Bezzi) ?		
Carissa edulis Vahl	Egyptian carissa	
Ceratitis (Ceratitis) capitata (Wiedemann)		F
[*Carissa grandiflora* - see *C. macrocarpa*]		
Carissa macrocarpa (Ecklon) A.DC.	Natal plum	
Anastrepha suspensa (Loew)		F
Ceratitis (Ceratitis) capitata (Wiedemann) ?		
Ceratitis (Pterandrus) rosa Karsch		F

Carthamus L. (Asteraceae=Compositae)

Carthamus tinctorius L.	safflower	
Acanthiophilus helianthi (Rossi)		C
Chaetorellia carthami Stackelberg		C
Terellia luteola (Wiedemann)		C

Casimiroa Llave & Lex. (Rutaceae)

Casimiroa edulis Llave & Lex.	white sapote	
Anastrepha ludens (Loew)		F
Anastrepha suspensa (Loew)		F
Bactrocera (Bactrocera) neohumeralis (Hardy)		F
Bactrocera (Bactrocera) tryoni (Froggatt)		F
Bactrocera (Zeugodacus) cucurbitae (Coquillett) ?		
Ceratitis (Ceratitis) capitata (Wiedemann)		F
Dirioxa confusa (Hardy)		F?D
Casirmiroa tetrameria Millsp.	matasano	
Anastrepha ludens (Loew)		F

Cassia L. (Fabaceae=Leguminosae)

Cassia fistula L. — Indian laburnum
- *Euphranta cassiae* (Munro) — F

Castanea Miller (Fagaceae)

Castanea crenata Siebold & Zucc. — Japanese chestnut
- *Euphranta camelliae* (Ito) — F

Celtis L. (Ulmaceae)

Celtis laevigata Willd. — lowland hackberry
- *Rhagoletotrypeta uniformis* Steyskal — F

Centaurea L. (Asteraceae=Compositae)

Centaurea americana Nutt.
- *Acanthiophilus helianthi* (Rossi) — C
- *Craspedoxantha octopunctata* Bezzi — C

Centaurea cyanus L. — cornflower
- *Acanthiophilus helianthi* (Rossi) — C
- *Chaetostomella completa* (Kapoor, Malla & Ghosh) — C

Centaurea moschata L. — sweet sultan
- *Acanthiophilus helianthi* (Rossi) — C

Chrysanthemum L. (Asteraceae=Compositae)

Chrysanthemum sp. — chrysanthemum
- *Campiglossa hirayamae* (Matsumura) — C
- *Campiglossa misella* (Loew) — C
- *Trypeta artemisiae* (Fabricius) — L
- *Trypeta trifasciata* Shiraki — L
- *Trypeta zoe* Meigen — L

Chrysobalanus L. (Chrysobalanaceae)

Chrysobalanus icaco L. — cocoplum
- *Anastrepha suspensa* (Loew) — F
- *Ceratitis (Ceratitis) capitata* (Wiedemann) ?

Chrysophyllum L. (Sapotaceae)

Chrysophyllum albidum G.Don — white star-apple
- *Ceratitis (Pardalaspis) punctata* (Wiedemann) ?
- *Ceratitis (Pterandrus) rosa* Karsch ?

Chrysophyllum cainito L. — star-apple
- *Anastrepha distincta* Greene ! — F
- *Anastrepha leptozona* Hendel — F
- *Anastrepha macrura* Hendel — F
- *Anastrepha margarita* Caraballo — F
- *Anastrepha panamensis* Greene — F
- *Anastrepha serpentina* (Wiedemann) — F
- *Anastrepha steyskali* Korytkowski — F
- *Anastrepha striata* Schiner — F
- *Anastrepha suspensa* (Loew) — F
- *Bactrocera (Bactrocera) aquilonis* (May) — F
- *Bactrocera (Bactrocera) distincta* (Malloch) — F

Bactrocera (Bactrocera) sp. near *B. dorsalis* (A)		FD
Bactrocera (Zeugodacus) cucurbitae (Coquillett) ?		
Ceratitis (Ceratitis) capitata (Wiedemann)		F
Ceratitis (Pterandrus) rubivora (Coquillett) ??		
Dirioxa pornia (Walker)		F?D
Chrysophyllum oliviforme L.	satinleaf	
Anastrepha suspensa (Loew)		F
Bactrocera (Zeugodacus) cucurbitae (Coquillett) ?		
Ceratitis (Ceratitis) capitata (Wiedemann)		F

Chrysothamnus Nutt. (Asteraceae=Compositae)

Chrysothamnus nauseosus (Pal. ex Pursh) Britton	rubber rabbitbrush	
Aciurina bigeloviae (Cockerell)		SG
Aciurina ferruginea (Doane)		SG
Aciurina maculata (Cole)		SG
Procecidochares [3 unidentified spp.]		SG
Valentibulla californica (Coquillett)		SG
Valentibulla steyskali Foote		SG

***x Citrofortunella* J. Ingram & H. Moore (Rutaceae)**

[*Citrus x Fortunella*]

x Citrofortunella x mitis (Blanco) J. Ingram & H. Moore	Panama orange	
Anastrepha serpentina (Wiedemann)		F
Anastrepha suspensa (Loew)		F
Ceratitis (Ceratitis) capitata (Wiedemann) ?		

***x Citroncirus* J. Ingram & H. Moore (Rutaceae)**

[*Citrus sinensis x Poncirus trifoliata*]

x Citroncirus webberi J. Ingram & H. Moore	citrange	
Dirioxa confusa (Hardy)		F?D
Dirioxa pornia (Walker)		F?D

Citrullus Schrader ex Ecklon & Zeyher (Cucurbitaceae)

Citrullus colocynthis (L.) Schrader	colocynth	
Bactrocera (Zeugodacus) cucurbitae (Coquillett)		F
Bactrocera (Zeugodacus) tau (Walker) ?		
Dacus (Didacus) ciliatus Loew		F
Dacus (Didacus) frontalis Becker		F
Citrullus lanatus (Thunb.) Matsum. & Nakai	watermelon	
Anastrepha grandis (Macquart)		F
Bactrocera (Bactrocera) dorsalis (Hendel)		F
Bactrocera (Bactrocera) zonata (Saunders) ?		
Bactrocera (Notodacus) xanthodes (Broun)		F
Bactrocera (Paradacus) depressa (Shiraki) ?		
Bactrocera (Paratridacus) atrisetosa (Perkins)		F
Bactrocera (Zeugodacus) cucurbitae (Coquillett)		F
Bactrocera (Zeugodacus) tau (Walker) ?		
Dacus (Dacus) bivittatus (Bigot)		F
Dacus (Dacus) demmerezi (Bezzi)		F
Dacus (Didacus) ciliatus Loew		F

Dacus (Didacus) frontalis Becker		F
Dacus (Didacus) lounsburyii Coquillett ?		
Dacus (Didacus) vertebratus Bezzi		F
Myiopardalis pardalina (Bigot)		F
[*Citrullus vulgaris* - see *C. lanatus*]		

Citrus L. (Rutaceae)

Citrus aurantiifolia (Christm.) Swingle	lime	
Anastrepha ludens (Loew)		F
Anastrepha suspensa (Loew)		FO
Bactrocera (Bactrocera) kirki (Froggatt)		F
Bactrocera (Bactrocera) passiflorae (Froggatt)		F
Ceratitis (Ceratitis) capitata (Wiedemann)		F
Citrus aurantium L.	sour orange	
Anastrepha fraterculus (Wiedemann)		F
Anastrepha ludens (Loew)		F
Anastrepha obliqua (Macquart)		F
Anastrepha serpentina (Wiedemann)		F
Anastrepha suspensa (Loew)		FO
Bactrocera (Bactrocera) cilifer (Hendel) ?		
Bactrocera (Bactrocera) melanota (Coquillett) ?		
Bactrocera (Bactrocera) tryoni (Froggatt) ?		
Bactrocera (Bactrocera) umbrosa (Fabricius) ?		
Bactrocera (Hemigymnodacus) diversa (Coquillett) ??		
Bactrocera (Tetradacus) minax (Enderlein)		F
Bactrocera (Tetradacus) tsuneonis (Miyake)		F
Bactrocera (Zeugodacus) caudata (Fabricius) ??		
Bactrocera (Zeugodacus) cucurbitae (Coquillett) ?		
Ceratitis (Ceratalaspis) cosyra (Walker)		F
Ceratitis (Ceratitis) capitata (Wiedemann)		F
Ceratitis (Pterandrus) rosa Karsch		F
Dacus (Didacus) aequalis Coquillett ??		
[*Citrus grandis* - see *C. maxima*]		
Citrus limetta Risso	sweet lime	
Anastrepha fraterculus (Wiedemann)		F
Anastrepha ludens (Loew)		F
Anastrepha obliqua (Macquart)		F
Anastrepha suspensa (Loew)		FO
Ceratitis (Ceratitis) capitata (Wiedemann)		F
Monacrostichus citricola Bezzi		F
Citrus limon (L.) Burman f.	lemon	
Bactrocera (Bactrocera) aquilonis (May)		F
Bactrocera (Bactrocera) dorsalis (Hendel)		F
Bactrocera (Bactrocera) facialis (Coquillett)		F
Bactrocera (Bactrocera) latifrons (Hendel) ??		
Bactrocera (Bactrocera) neohumeralis (Hardy)		F
Bactrocera (Bactrocera) tryoni (Froggatt)		F
Bactrocera (Tetradacus) minax (Enderlein)		F
Bactrocera (Zeugodacus) cucurbitae (Coquillett) ?		

Ceratitis (Ceratitis) capitata (Wiedemann)		FD
Dirioxa pornia (Walker)		FD
Monacrostichus citricola Bezzi		F
Citrus x limonia Osbeck	lemandarin	
Anastrepha suspensa (Loew)		FO
Ceratitis (Ceratitis) capitata (Wiedemann)		F
[*Citrus limonum* - see *C. limon*]		
Citrus maxima (Burman) Merr.	pummelo, pomelo	
Anastrepha fraterculus (Wiedemann)		F
Anastrepha ludens (Loew)		F
Anastrepha serpentina (Wiedemann)		F
Anastrepha suspensa (Loew)		FO
Bactrocera (Bactrocera) aquilonis (May)		F
Bactrocera (Bactrocera) dorsalis (Hendel)		F
Bactrocera (Bactrocera) facialis (Coquillett)		F
Bactrocera (Bactrocera) neohumeralis (Hardy)		F
Bactrocera (Bactrocera) trilineola Drew ?		
Bactrocera (Bactrocera) tryoni (Froggatt)		F
Bactrocera (Bactrocera) umbrosa (Fabricius) ?		
Bactrocera (Tetradacus) minax (Enderlein)		F
Bactrocera (Zeugodacus) caudata (Fabricius) ??		
Bactrocera (Zeugodacus) cucurbitae (Coquillett) !		F
Bactrocera (Zeugodacus) tau (Walker) ?		
Callistomyia flavilabris Hering ?		
Ceratitis (Ceratitis) capitata (Wiedemann)		F
Ceratitis (Ceratitis) catoirii Guérin-Méneville ??		
Dirioxa confusa (Hardy)		F?D
Dirioxa pornia (Walker) ?		
Monacrostichus citricola Bezzi		F
Citrus medica L.	citron	
Anastrepha fraterculus (Wiedemann) ?		
Anastrepha ludens (Loew)		F
Bactrocera (Bactrocera) tryoni (Froggatt)		F
Bactrocera (Tetradacus) minax (Enderlein)		F
Ceratitis (Ceratitis) capitata (Wiedemann)		F
Citrus x nobilis Lour.	tangor	
Ceratitis (Ceratitis) capitata (Wiedemann)		F
Citrus x paradisi Macfad.	grapefruit	
Anastrepha fraterculus (Wiedemann)		F
Anastrepha ludens (Loew)		F
Anastrepha obliqua (Macquart)		F
Anastrepha ocresia (Walker) ?		
Anastrepha serpentina (Wiedemann)		F
Anastrepha suspensa (Loew)		FO
Bactrocera (Afrodacus) jarvisi (Tryon)		F
Bactrocera (Bactrocera) aquilonis (May)		F
Bactrocera (Bactrocera) sp. near *B. dorsalis* (A)		FD
Bactrocera (Bactrocera) sp. near *B. dorsalis* (D) ?		
Bactrocera (Bactrocera) facialis (Coquillett)		F

Bactrocera (Bactrocera) halfordiae (Tryon) ??
Bactrocera (Bactrocera) melas (Perkins & May) ?
Bactrocera (Bactrocera) neohumeralis (Hardy) F
Bactrocera (Bactrocera) trivialis (Drew) F
Bactrocera (Bactrocera) tryoni (Froggatt) F
Bactrocera (Zeugodacus) cucurbitae (Coquillett) ?
Ceratitis (Ceratitis) capitata (Wiedemann) F
Ceratitis (Pterandrus) rosa Karsch ?
Dirioxa confusa (Hardy) F?D

Citrus reticulata Blanco — tangerine
Anastrepha fraterculus (Wiedemann) F
Anastrepha ludens (Loew) F
Anastrepha serpentina (Wiedemann) F
Anastrepha suspensa (Loew) FO
Bactrocera (Bactrocera) aquilonis (May) F
Bactrocera (Bactrocera) caryeae (Kapoor) F
Bactrocera (Bactrocera) cilifer (Hendel) ?
Bactrocera (Bactrocera) curvipennis (Froggatt) F
Bactrocera (Bactrocera) dorsalis (Hendel) F
Bactrocera (Bactrocera) sp. near *B. dorsalis* (A) FD
Bactrocera (Bactrocera) facialis (Coquillett) F
Bactrocera (Bactrocera) halfordiae (Tryon) ??
Bactrocera (Bactrocera) kirki (Froggatt) F
Bactrocera (Bactrocera) neohumeralis (Hardy) F
Bactrocera (Bactrocera) passiflorae (Froggatt) F
Bactrocera (Bactrocera) tryoni (Froggatt) F
Bactrocera (Notodacus) xanthodes (Broun) F
Bactrocera (Tetradacus) minax (Enderlein) F
Bactrocera (Tetradacus) tsuneonis (Miyake) F
Bactrocera (Zeugodacus) cucurbitae (Coquillett) ?
Ceratitis (Ceratitis) capitata (Wiedemann) F
Ceratitis (Ceratitis) catoirii Guérin-Méneville F
Ceratitis (Ceratitis) malgassa Munro F
Ceratitis (Pterandrus) rosa Karsch F
Dirioxa confusa (Hardy) F?D
Dirioxa pornia (Walker) FD
[Some of these *Anastrepha* records may be referable to *C. x nobilis* as Norrbom & Kim (1988b) did not separate these two *Citrus* species/hybrids.]

Citrus sinensis (L.) Osbeck — sweet orange
Anastrepha fraterculus (Wiedemann) F
Anastrepha ludens (Loew) F
Anastrepha obliqua (Macquart) F
Anastrepha serpentina (Wiedemann) F
Anastrepha striata Schiner F
Anastrepha suspensa (Loew) FO
Bactrocera (Afrodacus) jarvisi (Tryon) F
Bactrocera (Bactrocera) dorsalis (Hendel) F
Bactrocera (Bactrocera) sp. near *B. dorsalis* (A) FD
Bactrocera (Bactrocera) sp. near *B. dorsalis* (B) F

Bactrocera (Bactrocera) facialis (Coquillett)		F
Bactrocera (Bactrocera) halfordiae (Tryon) ??		
Bactrocera (Bactrocera) kirki (Froggatt)		F
Bactrocera (Bactrocera) latifrons (Hendel) ??		
Bactrocera (Bactrocera) melas (Perkins & May) ?		
Bactrocera (Bactrocera) neohumeralis (Hardy)		F
Bactrocera (Bactrocera) tryoni (Froggatt)		F
Bactrocera (Bactrocera) zonata (Saunders)		F
Bactrocera (Tetradacus) minax (Enderlein)		F
Bactrocera (Tetradacus) tsuneonis (Miyake)		F
Bactrocera (Zeugodacus) cucurbitae (Coquillett) !		F
Ceratitis (Ceratitis) capitata (Wiedemann)		F
Ceratitis (Pterandrus) rosa Karsch		F
Dirioxa confusa (Hardy)		F?D
Citrus x tangelo J. Ingram & H. Moore	tangelo	
Anastrepha ludens (Loew)		F
Anastrepha suspensa (Loew)		FO
Ceratitis (Ceratitis) capitata (Wiedemann)		F
Citrus tangerina Hort. ex Tanaka		
Bactrocera (Tetradacus) minax (Enderlein)		F
Bactrocera (Tetradacus) tsuneonis (Miyake)		F
Citrus sp.		
Bactrocera (Bactrocera) correcta (Bezzi) ?		
Bactrocera (Bactrocera) melanota (Coquillett)		F
Bactrocera (Bactrocera) psidii (Froggatt)		F
Bactrocera (Bactrocera) umbrosa (Fabricius)		F
Ceratitis (Ceratalaspis) discussa (Munro)		F
Ceratitis (Ceratalaspis) quinaria (Bezzi)		F
Rhagoletis blanchardi Aczél ??		
Rhagoletis ferruginea Hendel ??		

Clausena Burman f. (Rutaceae)

Clausena lansium (Lour.) Skeels	wampi	
Anastrepha suspensa (Loew)		F
Bactrocera (Bactrocera) dorsalis (Hendel)		F
Bactrocera (Zeugodacus) cucurbitae (Coquillett) ?		

Coccinia Wight & Arn. (Cucurbitaceae)

Coccinia grandis (L.) Voigt	ivy gourd	
Bactrocera (Javadacus) trilineata Hardy ?		
Bactrocera (Zeugodacus) cucurbitae (Coquillett)		G
Dacus (Didacus) ciliatus Loew		G
[*Coccinia cordifolia* - see *C. grandis*]		
[*Coccinia indica* - see *C. grandis*]		
Coccinia sp.		
Dacus (Didacus) frontalis Becker		F

Coccoloba P. Browne (Polygonaceae)

Coccoloba uvifera (L.) L.	seagrape	
Anastrepha suspensa (Loew)		F

Coffea L. (Rubiaceae)

Coffea arabica L.	arabica coffee	
Anastrepha bahiensis Lima		F
Anastrepha fraterculus (Wiedemann)		F
Anastrepha ludens (Loew)		F
Anastrepha obliqua (Macquart)		F
Anastrepha sororcula Zucchi		F
Bactrocera (Bactrocera) neohumeralis (Hardy)		F
Bactrocera (Bactrocera) tryoni (Froggatt)		F
Ceratitis (Ceratitis) capitata (Wiedemann)		F
Ceratitis (Pardalaspis) punctata (Wiedemann) ?		
Ceratitis (Pterandrus) rosa Karsch		F
Myoleja alboscutellata Wulp		S
Trirhithrum basale Bezzi ??		
Trirhithrum coffeae Bezzi !		F
Trirhithrum inscriptum (Graham)		F
Trirhithrum nigerrimum (Bezzi)		F
Coffea canephora Pierre ex Froehner	robusta coffee	
Bactrocera (Bactrocera) caryeae (Kapoor)		F
Bactrocera (Bactrocera) correcta (Bezzi) ?		
Bactrocera (Bactrocera) nigrotibialis (Perkins)		F
Ceratitis (Pterandrus) anonae Graham		F
Trirhithrum coffeae Bezzi		F
Trirhithrum nigerrimum (Bezzi)		F
Coffea liberica W. Bull ex Hiern	Liberica coffee	
Anastrepha fraterculus (Wiedemann)		F
Ceratitis (Ceratitis) capitata (Wiedemann)		F
[*Coffea robusta* - see *C. canephora*]		
Coffea sp.		
Bactrocera (Bactrocera) dorsalis (Hendel)		F
Bactrocera (Bactrocera) latifrons (Hendel) ??		
Bactrocera (Bactrocera) passiflorae (Froggatt) ?		
Dacus (Dacus) bivittatus (Bigot) ??		
Trirhithrum manganum Munro		F
Trirhithrum occipitale Bezzi ??		

Cola Schott & Endl. (Sterculiaceae)

Cola acuminata (P. Beauv.) Schott & Endl.	abata cola	
Ceratitis (Ceratalaspis) morstatti Bezzi ?		
Ceratitis (Pterandrus) colae Silvestri		F
Ceratitis (Pterandrus) penicillata Bigot ?		

[*Coleus barbatus* - see *Solenostemon barbatus*]

Coreopsis L. (Asteraceae=Compositae)

Coreopsis tinctoria Nutt.	plains coreopsis	
Dioxyna picciola (Bigot)		C

Cornus L. (Cornaceae)

Cornus canadensis L.	bunchberry	
Rhagoletis cornivora Bush		F
Cornus drummondii C. Meyer	roughleaf dogwood	
Rhagoletis electromorpha Berlocher		F
Cornus florida L.	flowering dogwood	
Rhagoletis sp. near *R. cornivora* Bush		F
Cornus stolonifera Michaux	red-osier dogwood	
Rhagoletis cornivora Bush		F
Rhagoletis tabellaria (Fitch)		F

Cotoneaster Medikus (Rosaceae)

Cotoneaster sp.	cotoneaster	
Anomoia purmunda (Harris)		F

Crataegus L. (Rosaceae)

Crataegus sp.	hawthorn	
Anomoia purmunda (Harris)		F
Rhagoletis pomonella (Walsh)		F

Crateva L. (Capparaceae=Capparidaceae)

Crateva religiosa Forster f.	sacred garlic-pear	
Bactrocera (Bactrocera) neohumeralis (Hardy)		F

Cryptotaenia DC. (Apiaceae=Umbelliferae)

Cryptotaenia canadensis (L.) DC.	honewort	
Euleia fratria (Loew)		L

Cucumeropsis Naudin (Cucurbitaceae)

[*Cucumeropsis edulis* - see *C. mannii*]		
Cucumeropsis mannii Naudin	white egusi	
Dacus (Dacus) bivittatus (Bigot)		F
Dacus (Didacus) vertebratus Bezzi		F

Cucumis L. (Cucurbitaceae)

Cucumis melo L.	melon	
Bactrocera (Austrodacus) cucumis (French)		F
Bactrocera (Bactrocera) zonata (Saunders) ?		
Bactrocera (Zeugodacus) cucurbitae (Coquillett)		FI
Bactrocera (Zeugodacus) tau (Walker) ?		
Ceratitis (Pardalaspis) punctata (Wiedemann) ?		
Dacus (Dacus) armatus Fabricius ??		
Dacus (Dacus) bivittatus (Bigot)		F
Dacus (Dacus) demmerezi (Bezzi)		F
Dacus (Didacus) ciliatus Loew		F
Dacus (Didacus) frontalis Becker		F
Dacus (Didacus) lounsburyii Coquillett ?		
Dacus (Didacus) vertebratus Bezzi		F

Dacus (Leptoxyda) longistylus Wiedemann ??		
Myiopardalis pardalina (Bigot)		F
[The *Cucumis cocomon* of Shiraki (1933) was probably an error for *C. melo* var. *conomon* Makino]		
Cucumis metuliferus E. Meyer ex Naudin	African horned cucumber	
Bactrocera (Paradacus) depressa (Shiraki) ?		
Bactrocera (Zeugodacus) cucurbitae (Coquillett) ?		
Dacus (Dacus) bivittatus (Bigot)		F
Dacus (Didacus) ciliatus Loew		F
[*Cucumis pubescens* - see *C. melo*]		
Cucumis sativus L.	cucumber	
Anastrepha grandis (Macquart)		F
Bactrocera (Austrodacus) cucumis (French)		F
Bactrocera (Bactrocera) latifrons (Hendel) ??		
Bactrocera (Paradacus) depressa (Shiraki) ?		
Bactrocera (Paratridacus) atrisetosa (Perkins)		F
Bactrocera (Zeugodacus) caudata (Fabricius) ?		
Bactrocera (Zeugodacus) cucurbitae (Coquillett)		F+S
Bactrocera (Zeugodacus) tau (Walker)		F
Ceratitis (Ceratitis) capitata (Wiedemann) ?		
Dacus (Callantra) eumenoides (Bezzi) ?		
Dacus (Callantra) solomonensis Malloch		F
Dacus (Dacus) armatus Fabricius ??		
Dacus (Dacus) bivittatus (Bigot)		FO
Dacus (Dacus) demmerezi (Bezzi)		F
Dacus (Dacus) punctatifrons Karsch		F
Dacus (Didacus) ciliatus Loew		FO
Dacus (Didacus) frontalis Becker		F
Dacus (Didacus) vertebratus Bezzi		FO
Dacus (Leptoxyda) longistylus Wiedemann ??		
Myiopardalis pardalina (Bigot)		F
[*Cucumis utilissimus* - see *C. melo*]		

Cucurbita L. (Cucurbitaceae)

Cucurbita maxima Duchesne	pumpkin	
Anastrepha grandis (Macquart)		F
Bactrocera (Hemigymnodacus) diversa (Coquillett)		I
Bactrocera (Zeugodacus) cucurbitae (Coquillett)		F
Bactrocera (Zeugodacus) munda (Bezzi)		F or I ?
Bactrocera (Zeugodacus) scutellaris (Bezzi)		I
Bactrocera (Zeugodacus) tau (Walker)		F
Dacus (Dacus) bivittatus (Bigot)		F
Dacus (Dacus) demmerezi (Bezzi)		F
Dacus (Didacus) ciliatus Loew		F
Dacus (Didacus) vertebratus Bezzi		F
Dacus (Dacus) momordicae Bezzi !		F
Dacus (Didacus) ciliatus Loew ?		

Cucurbita moschata (Duchesne) Duchesne ex Poiret	pumpkin	
Anastrepha grandis (Macquart)		F
Bactrocera (Paradacus) depressa (Shiraki)		F
Bactrocera (Paratridacus) expandens (Walker) ?		
Bactrocera (Zeugodacus) cucurbitae (Coquillett) ?		
Bactrocera (Zeugodacus) tau (Walker) ?		
Cucurbita pepo L.	pumpkin	
Anastrepha grandis (Macquart)		F
Bactrocera (Austrodacus) cucumis (French)		F
Bactrocera (Hemigymnodacus) diversa (Coquillett)		I
Bactrocera (Paradacus) decipiens (Drew)		F
Bactrocera (Paratridacus) atrisetosa (Perkins)		F
Bactrocera (Zeugodacus) caudata (Fabricius) ?		
Bactrocera (Zeugodacus) cucurbitae (Coquillett)		F+S
Bactrocera (Zeugodacus) tau (Walker) ?		
Dacus (Callantra) solomonensis Malloch		F
Dacus (Dacus) bivittatus (Bigot)		F
Dacus (Dacus) demmerezi (Bezzi)		F
Dacus (Dacus) punctatifrons Karsch		F
Dacus (Didacus) ciliatus Loew		F
Dacus (Didacus) frontalis Becker		F
Dacus (Didacus) lounsburyii Coquillett ?		
Cucurbita sp.		
Bactrocera (Zeugodacus) caudata (Fabricius)		I (male)
Bactrocera (Zeugodacus) scutellata (Hendel)		B (male)
Ceratitis (Ceratitis) capitata (Wiedemann) ?		

Cussonia Thunb. (Araliaceae)

Cussonia sp.	umbrella tree	
Afrocneros excellens (Loew)		S
Afrocneros mundus (Loew)		S
Ocnerioxa bigemmata (Bezzi)		S
Ocnerioxa sinuata (Loew)		S

Cycas L. (Cycadaceae)

Cycas circinalis L.	queen sago	
Cycasia oculata Malloch ?		
Cycas sp.		
Cycasia oculata Malloch		X

Cydonia Miller (Rosaceae)

Cydonia oblonga Miller	quince	
Anastrepha fraterculus (Wiedemann)		F
Anastrepha ludens (Loew)		F
Anastrepha serpentina (Wiedemann)		F
Bactrocera (Afrodacus) jarvisi (Tryon)		F
Bactrocera (Bactrocera) tryoni (Froggatt)		F
Bactrocera (Bactrocera) zonata (Saunders)		F
Bactrocera (Zeugodacus) cucurbitae (Coquillett) !		F
Ceratitis (Ceratitis) capitata (Wiedemann)		F

Ceratitis (Pterandrus) rosa Karsch F
Dirioxa pornia (Walker) FD
[*Cydonia vulgaris* - see *C. oblonga*]

[*Cynanchum vincetoxicum* - see *Vincetoxicum hirudinaria*]

Cynara L. (Asteraceae=Compositae)

Cynara scolymus L. artichoke
Terellia fuscicornis (Loew) C
Acanthiophilus helianthi (Rossi) ! C

Cyphomandra C. Martius ex Sendtner (Solanaceae)

Cyphomandra crassicaulis (Ortega) Kuntze tree tomato
Bactrocera (Bactrocera) neohumeralis (Hardy) F
Bactrocera (Zeugodacus) cucurbitae (Coquillett) ! F
Ceratitis (Ceratitis) capitata (Wiedemann) F
[*Cyphomandra betacea* - see *C. crassicaulis*]

Dahlia Cav. (Asteraceae=Compositae)

Dahlia sp. dahlia
Craspedoxantha marginalis (Wiedemann) C
Trupanea superdecora Bezzi C

Daucus L. (Apiaceae=Umbelliferae)

Daucus carota L. carrot
Euleia heraclei (Linnaeus) L

Dendrocalamus Nees (Poaceae=Gramineae)

Dendrocalamus giganteus Munro giant bamboo
Acroceratitis striata (Froggatt) S
Dendrocalamus latiflorus Munro
Gastrozona fasciventris (Macquart) S
Acroceratitis plumosa Hendel S
Dendrocalamus strictus (Roxb.) Nees male bamboo
Acroceratitis striata (Froggatt) S

Dimocarpus Lour. (Sapindaceae)

Dimocarpus longan Lour. longan
Bactrocera (Zeugodacus) cucurbitae (Coquillett) ?
Ceratitis (Ceratitis) capitata (Wiedemann) ?

Diospyros L. (Ebenaceae)

Diospyros blancoi A. DC. velvet apple
Anastrepha suspensa (Loew) F
Bactrocera (Bactrocera) sp. near *B. dorsalis* (B) F
Diospyros digyna Jacq. black sapote
Anastrepha obliqua (Macquart) F
Anastrepha serpentina (Wiedemann) F
Anastrepha striata Schiner F
Ceratitis (Ceratitis) capitata (Wiedemann) ?
[*Diospyros discolor* - see *D. blancoi*]

Diospyros kaki L. f.	Japanese persimmon	
Anastrepha fraterculus (Wiedemann)		F
Anastrepha ludens (Loew)		F
Anastrepha suspensa (Loew)		F
Bactrocera (Afrodacus) jarvisi (Tryon)		F
Bactrocera (Bactrocera) tryoni (Froggatt)		F
Ceratitis (Ceratitis) capitata (Wiedemann)		F
Diospyros lotus L.	date-plum	
Bactrocera (Bactrocera) dorsalis (Hendel)		F
Diospyros texana Scheele	Texas persimmon	
Anastrepha ludens (Loew) ?		
Diospyros virginiana L.	common persimmon	
Anastrepha suspensa (Loew)		F
Ceratitis (Pterandrus) rosa Karsch ?		
Diospyros sp.		
Dirioxa pornia (Walker) ?		

[*Dolichos lablab* - see *Lablab purpureus*]

Dovyalis E. Meyer ex Arn. (Flacourtiaceae)

Dovyalis caffra (Hook. f. & Harv.) Warb.	kei apple	
Anastrepha suspensa (Loew)		F
Bactrocera (Bactrocera) tryoni (Froggatt)		F
Ceratitis (Ceratitis) capitata (Wiedemann)		F
Ceratitis (Pterandrus) rosa Karsch ?		
Trirhithrum albomaculatum Röder		F
Dovyalis hebecarpa (Gardner) Warb.	ketembilla	
Anastrepha antunesi Lima		F
Anastrepha fraterculus (Wiedemann)		F
Anastrepha obliqua (Macquart)		F
Anastrepha serpentina (Wiedemann)		F
Anastrepha suspensa (Loew)		F
Ceratitis (Ceratitis) capitata (Wiedemann) ?		

Dracaena Vand. ex L. (Agavaceae)

Dracaena sp.		
Taomyia marshalli Bezzi		F

Dracontomelon Blume (Anacardiaceae)

Dracontomelon dao (Blanco) Merr. & Rolfe	Argus pheasant tree	
Bactrocera (Bactrocera) limbifera (Bezzi)		F
Bactrocera (Zeugodacus) cucurbitae (Coquillett) ?		
Dracontomelon sp.		
Bactrocera (Zeugodacus) tau (Walker) ?		

Durio Adanson (Bombacaceae)

Durio zibethinus Murray	durian	
Bactrocera (Bactrocera) dorsalis (Hendel) ??		

Ecballium A. Rich. (Cucurbitaceae)

Ecballium elaterium (L.) A. Rich. squirting cucumber
Myiopardalis pardalina (Bigot) F

Eleutherococcus Maxim. (Araliaceae)

Eleutherococcus senticosus (Rupr. & Maxim.) Maxim. Siberian ginseng
Acidiella angustifascia Hering F

Eremocitrus Swingle (Rutaceae)

Eremocitrus glauca (Lindley) Swingle Australian desert lime
Bactrocera (Bactrocera) tryoni (Froggatt) F

Eriobotrya Lindley (Rosaceae)

Eriobotrya japonica (Thunb.) Lindley loquat
Anastrepha minensis Lima F
Anastrepha fraterculus (Wiedemann) F
Anastrepha obliqua (Macquart) F
Anastrepha suspensa (Loew) F
Bactrocera (Bactrocera) aquilonis (May) F
Bactrocera (Bactrocera) dorsalis (Hendel) F
Bactrocera (Bactrocera) halfordiae (Tryon) ??
Bactrocera (Bactrocera) melas (Perkins & May) ?
Bactrocera (Bactrocera) neohumeralis (Hardy) F
Bactrocera (Bactrocera) tryoni (Froggatt) F
Bactrocera (Zeugodacus) cucurbitae (Coquillett) ?
Ceratitis (Ceratitis) capitata (Wiedemann) F
Ceratitis (Ceratitis) catoirii Guérin-Méneville F
Ceratitis (Pterandrus) rosa Karsch F

Erythroxylum P. Browne (Erythroxylaceae)

Erythroxylum coca Lam. coca
Trirhithrum nigerrimum (Bezzi) ??

Eugenia L. (Myrtaceae)

Eugenia aggregata (Vell.) Kiaersk. Rio Grande cherry
Anastrepha suspensa (Loew) F
[*Eugenia aqueum* - see *Syzygium aqueum*]
Eugenia brasiliensis Lam. Brazil cherry
Anastrepha fraterculus (Wiedemann) F
Anastrepha suspensa (Loew) F
Ceratitis (Ceratitis) capitata (Wiedemann) F
[*Eugenia cumini* - see *Syzygium cumini*]
[*Eugenia dombeyi* - see *E. brasiliensis*]
[*Eugenia jambolana* - see *Syzygium cumini*]
[*Eugenia jambos* - see *Syzygium jambos*]
[*Eugenia javanica* - see *Syzygium samarangense*]
[*Eugenia malaccensis* - see *Syzygium malaccense*]
Eugenia uniflora L. Surinam cherry
Anastrepha fraterculus (Wiedemann) F
Anastrepha sororcula Zucchi F

Anastrepha suspensa (Loew)		F
Bactrocera (Bactrocera) aquilonis (May)		F
Bactrocera (Bactrocera) correcta (Bezzi) ?		
Bactrocera (Bactrocera) dorsalis (Hendel)		F
Bactrocera (Bactrocera) sp. near *B. dorsalis* (A)		F
Bactrocera (Bactrocera) sp. near *B. dorsalis* (B)		F
Bactrocera (Bactrocera) facialis (Coquillett)		F
Bactrocera (Bactrocera) halfordiae (Tryon) ??		
Bactrocera (Bactrocera) kirki (Froggatt)		F
Bactrocera (Bactrocera) neohumeralis (Hardy)		F
Bactrocera (Bactrocera) tryoni (Froggatt)		F
Ceratitis (Ceratitis) capitata (Wiedemann)		F
Ceratitis (Ceratitis) catoirii Guérin-Méneville		F
Ceratitis (Pterandrus) rosa Karsch		F
Tomoplagia biseriata (Loew) ??		
Trirhithrum nigerrimum (Bezzi) ??		

[*Euphoria longan* - see *Dimocarpus longan*]

Feijoa O. Berg (Myrtaceae)

Feijoa sellowiana (O. Berg) O. Berg	feijoa	
Anastrepha fraterculus (Wiedemann)		F
Anastrepha ludens (Loew)		F
Bactrocera (Bactrocera) dorsalis (Hendel)		F
Bactrocera (Bactrocera) halfordiae (Tryon) ??		
Bactrocera (Bactrocera) melas (Perkins & May)		F
Bactrocera (Bactrocera) neohumeralis (Hardy)		F
Bactrocera (Bactrocera) tryoni (Froggatt)		F
Bactrocera (Bactrocera) zonata (Saunders) ?		
Bactrocera (Zeugodacus) cucurbitae (Coquillett) !		F
Ceratitis (Ceratitis) capitata (Wiedemann)		F
Ceratitis (Pterandrus) rosa Karsch ?		
Dirioxa confusa (Hardy)		F?D
Dirioxa pornia (Walker)		F?D

Ficus L. (Moraceae)

Ficus benjamini L.	weeping fig	
Bactrocera (Bactrocera) neohumeralis (Hardy)		F
Bactrocera (Bactrocera) tryoni (Froggatt)		F
Ficus carica L.	common fig	
Anastrepha fraterculus (Wiedemann)		F
Anastrepha suspensa (Loew)		F
Bactrocera (Bactrocera) melas (Perkins & May) ?		
Bactrocera (Bactrocera) tryoni (Froggatt)		F
Bactrocera (Bactrocera) zonata (Saunders) ??		
Bactrocera (Zeugodacus) cucurbitae (Coquillett) !		F
Ceratitis (Ceratitis) capitata (Wiedemann)		F
Ceratitis (Pterandrus) rosa Karsch		F

Ficus racemosa L.	cluster fig	
Bactrocera (Bactrocera) halfordiae (Tryon) ??		
Bactrocera (Bactrocera) tryoni (Froggatt)		F
Ficus sp.	fig	
Bactrocera (Bactrocera) caryeae (Kapoor)		F
Bactrocera (Bactrocera) dorsalis (Hendel)		F
Ceratitis (Ceratalaspis) quinaria (Bezzi) ?		

Flacourtia Comm. ex L'Hér. (Flacourtiaceae)

Flacourtia indica (Burman f.) Merr.	governor's plum	
Anastrepha suspensa (Loew)		F
Ceratitis (Ceratitis) catoirii Guérin-Méneville ??		
Flacourtia jangomas (Lour.) Räusch.	Indian plum	
Bactrocera (Bactrocera) aquilonis (May)		F
Bactrocera (Bactrocera) tryoni (Froggatt)		F
Flacourtia rukam Zoll. & Moritzi	Indian prune	
Bactrocera (Bactrocera) aquilonis (May)		F

***Fortunella* Swingle (Rutaceae)**

Fortunella x crassifolia Swingle	meiwa kumquat	
Anastrepha suspensa (Loew)		F
Bactrocera (Bactrocera) aquilonis (May)		F
Bactrocera (Bactrocera) minax (Enderlein)		F
Bactrocera (Tetradacus) tsuneonis (Miyake)		F
Fortunella japonica (Thunb.) Swingle	round kumquat	
Anastrepha fraterculus (Wiedemann)		F
Bactrocera (Bactrocera) halfordiae (Tryon) ??		
Bactrocera (Bactrocera) melas (Perkins & May) ?		
Bactrocera (Bactrocera) mutabilis (May) ?		
Bactrocera (Bactrocera) neohumeralis (Hardy)		F
Bactrocera (Bactrocera) tryoni (Froggatt)		F
Bactrocera (Tetradacus) tsuneonis (Miyake)		F
Ceratitis (Ceratitis) capitata (Wiedemann)		F
Fortunella margarita (Lour.) Swingle	oval kumquat	
Anastrepha suspensa (Loew)		F
Bactrocera (Tetradacus) tsuneonis (Miyake)		F
Fortunella sp.		
Dirioxa confusa (Hardy)		F?D

***Fragaria* L. (Rosaceae)**

Fragaria chiloensis (L.) Duchesne	Chilean strawberry	
Bactrocera (Zeugodacus) cucurbitae (Coquillett) ?		
Fragaria vesca L.	strawberry	
Anastrepha fraterculus (Wiedemann)		F
Fragaria sp.		
Bactrocera (Bactrocera) neohumeralis (Hardy)		F

Garcinia L. (Clusiaceae=Guttiferae)

Garcinia dulcis (Roxb.) Kurz	mundur	
Bactrocera (Paratridacus) expandens (Walker)		F
Garcinia livingstonei T. Anderson	imbe	
Anastrepha suspensa (Loew)		F
Ceratitis (Ceratitis) capitata (Wiedemann)		F
Ceratitis (Pterandrus) rosa Karsch ?		
Garcinia mangostana L.	mangosteen	
Ceratitis (Ceratitis) capitata (Wiedemann)		F
Ceratitis (Pterandrus) rosa Karsch		F
Garcinia xanthochymus Hook. f. ex T. And.	egg tree	
Bactrocera (Hemisurstylus) melanoscutata Drew		F
Bactrocera (Paratridacus) expandens (Walker)		F

Gaultheria L. (Ericaceae)

Gaultheria procumbens L.	wintergreen	
Rhagoletis mendax Curran !		F
Rhagoletis pomonella (Walsh) ??		

Gaylussacia Kunth (Ericaceae)

Gaylussacia baccata (Wangenh.) K. Koch	black huckleberry	
Rhagoletis mendax Curran		F
Rhagoletis pomonella (Walsh) ??		
Gaylussacia dumosa (Andrews) Torrey & A. Gray	dwarf huckleberry	
Rhagoletis mendax Curran		F
Gaylussacia frondosa (L.) Torrey & A. Gray ex Torrey	dangleberry	
Rhagoletis mendax Curran		F

Genipa L. (Rubiaceae)

Genipa americana L.	marmelade-box	
Anastrepha antunesi Lima		F

Gerbera L. (Asteraceae=Compositae)

Gerbera jamesonii Bolus ex Hook. f.	Barberton daisy	
Craspedoxantha marginalis (Wiedemann)		C

Gloriosa L. (Liliaceae)

Gloriosa sp.		
Dacus (Dacus) punctatifrons Karsch ??		

Gnetum L. (Gnetaceae)

Gnetum gnemon L.	buko	
Bactrocera (Bactrocera) mcgregori (Bezzi)		F

Gossypium L. (Malvaceae)

Gossypium sp.		
Ceratitis (Ceratitis) capitata (Wiedemann) ??		
Dacus (Didacus) ciliatus Loew !		F

Grewia L. (Tiliaceae)

Grewia asiatica L. phalsa
Bactrocera (Bactrocera) zonata (Saunders) F

Guizotia Cass. (Asteraceae=Compositae)

Guizotia abyssinica (L. f.) Cass. Niger seed
Dioxyna sororcula (Wiedemann) C
unidentified Schistopterini C

Gymnema R. Br. (Asclepiadaceae)

Gymnema sylvestre (Retz.) Schultes miracle fruit
Dacus (Didacus) eminus Munro S

Harpephyllum Bernh. ex K. Krausse (Anacardiaceae)

Harpephyllum caffrum Bernh. ex K. Krausse kaffir plum
Ceratitis (Ceratitis) capitata (Wiedemann) F

Helianthus L. (Asteraceae=Compositae)

Helianthus annuus L. sunflower
Acanthiophilus helianthi (Rossi) ! C
Neotephritis finalis (Loew) ! C
Paracantha cultaris (Coquillett) C
Strauzia longipennis (Wiedemann) S

Trupanea nigricornis (Coquillett) ! C
[Plus some other spp. which are only recorded from wild plants; p. 413.]
Helianthus tuberosus L. Jerusalem artichoke
Strauzia longipennis (Wiedemann) S
Trypeta zoe Meigen ?

Heracleum L. (Apiaceae=Umbelliferae)

Heracleum sosnowskyi Maneden Russian cow parsnip
Euleia heraclei (Linnaeus) L

Hevea Aublet (Euphorbiaceae)

Hevea brasiliensis (Willd. ex Adr. Juss.) Muell. Arg. rubber tree
Rioxa sexmaculata (Wulp) ?

Hibiscus L. (Malvaceae)

[*Hibiscus esculentus* - see *Abelmoschus esculentus*]
Hibiscus tiliaceus L. sea hibiscus
Bactrocera (Bactrocera) distincta (Malloch) ! F
Bactrocera (Bactrocera) facialis (Coquillett) ! F
Bactrocera (Bactrocera) kirki (Froggatt) ! F
Bactrocera (Bactrocera) pallida (Perkins & May) F
Bactrocera (Notodacus) xanthodes (Broun) ! F
Acidoxantha hibisci Hardy I
Hibiscus sp.
Acidoxantha bombacis de Meijere I

***Hippomane* L. (Euphorbiaceae)**

Hippomane mancinella L.	manchineel	
Anastrepha acris Stone		F

***Hippophae* L. (Elaeagnaceae)**

Hippophae rhamnoides L.	sea buckthorn	
Rhagoletis batava Hering		F

***Ilex* L. (Aquifoliaceae)**

Ilex cassine L.	dahoon	
Myoleja limata (Coquillett)		F
Ilex vomitoria Aiton	yaupon	
Myoleja limata (Coquillett)		F

***Inga* Miller (Fabaceae=Leguminosae)**

Inga edulis C. Martius	ice cream bean	
Anastrepha distincta Greene		F
Anastrepha fraterculus (Wiedemann)		F
Inga laurina Willd.	caspirol	
Anastrepha distincta Greene		F
Inga paterna Harms	paterna	
Anastrepha distincta Greene		F
[*Inga ruiziana* - see *I. laurina*]		
Inga spectabilis Willd.	guavo real	
Anastrepha distincta Greene		F

***Inocarpus* Forster & Forster f. (Fabaceae=Leguminosae)**

[*Inocarpus edulis* - see *I. fagifer*]		
Inocarpus fagifer (Parkinson) Fosb.	Tahiti chestnut	
Bactrocera (Bactrocera) facialis (Coquillett)		F
Bactrocera (Bactrocera) kirki (Froggatt)		F
Bactrocera (Bactrocera) moluccensis (Perkins)		F
Bactrocera (Bactrocera) psidii (Froggatt) ??		

***Ipomoea* L. (Convolvulaceae)**

Ipomoea batatas (L.) Lam.	sweet potato	
Coelotrypes vittatus Bezzi ?		

***Juglans* L. (Juglandaceae)**

Juglans ailantifolia Carrière	Japanese walnut	
Rhagoletis suavis (Loew)		F
Juglans californica S. Watson	California walnut	
Rhagoletis completa Cresson		F
Juglans cinerea L.	butternut	
Rhagoletis suavis (Loew)		F
Juglans hindsii (Jepson) Jepson ex R.E. Smith	Hinds' walnut	
Bactrocera (Bactrocera) dorsalis (Hendel)		F
Bactrocera (Zeugodacus) cucurbitae (Coquillett) !		F
Rhagoletis completa Cresson		F
Juglans neotropica Diels	Andean walnut	
Anastrepha bahiensis Lima		F
Anastrepha fraterculus (Wiedemann)		F

Juglans nigra L.	black walnut	
Rhagoletis completa Cresson		F
Rhagoletis suavis (Loew)		F
Juglans regia L.	English walnut	
Anastrepha bahiensis Lima		F
Anastrepha fraterculus (Wiedemann)		F
Bactrocera (Bactrocera) tryoni (Froggatt)		F
Ceratitis (Ceratitis) capitata (Wiedemann)		F
Rhagoletis boycei Cresson		F
Rhagoletis completa Cresson		F
Rhagoletis juglandis Cresson		F
Rhagoletis suavis (Loew)		F

Juniperus L. (Cupressaceae)

Juniperus monosperma (Engelm.) Sarg.	one-seed juniper	
Rhagoletis juniperina Marcovitch ?		
Juniperus sabina L.	savin	
Rhagoletis flavigenualis Hering		F
Rhagoletis mongolica Kandybina		F
Juniperus virginiana L.	eastern red cedar	
Rhagoletis juniperina Marcovitch		F

Lablab Adanson (Fabaceae=Leguminosae)

Lablab purpureus (L.) Sweet	hyacinth bean	
Bactrocera (Zeugodacus) cucurbitae (Coquillett) ?		

Lactuca L. (Asteraceae=Compositae)

Lactuca canadensis L.	tall lettuce	
Jamesomyia geminata (Loew)		C
Lactuca sativa L.	garden lettuce	
Dioxyna sororcula (Wiedemann)		C
Trupanea amoena (Frauenfeld)		C

Lagenaria Ser. (Cucurbitaceae)

Lagenaria siceraria (Molina) Standley	white-flowered gourd	
Anastrepha grandis (Macquart)		F
Bactrocera (Bactrocera) parvula Hendel ??		
Bactrocera (Bactrocera) zonata (Saunders) ?		
Bactrocera (Hemigymnodacus) diversa (Coquillett)		I
Bactrocera (Paradacus) depressa (Shiraki) ?		
Bactrocera (Zeugodacus) caudata (Fabricius) ?		
Bactrocera (Zeugodacus) cucurbitae (Coquillett)		F
Bactrocera (Zeugodacus) scutellaris (Bezzi)		I
Bactrocera (Zeugodacus) tau (Walker) ?		
Dacus (Dacus) bivittatus (Bigot)		F
Dacus (Dacus) demmerezi (Bezzi)		F
Dacus (Didacus) ciliatus Loew		F
[*Lagenaria vulgaris* - see *L. siceraria*]		
Lagenaria sp.		
Dacus (Dacus) telfaireae (Bezzi) ??		

Lansium **Correa (Meliaceae)**

Lansium domesticum Correa	langsat	
Bactrocera (Bactrocera) sp. near *B. dorsalis* (B)		F
Callistomyia flavilabris Hering ?		

Leucanthemum **Miller (Asteraceae=Compositae)**

Leucanthemum maximum (Ramond) DC.	shasta daisy	
Tephritis neesii (Meigen)		C

Levisticum **Hill (Apiaceae=Umbelliferae)**

Levisticum officinale Koch	lovage	
Euleia heraclei (Linnaeus)		L

Litchi **Sonn. (Sapindaceae)**

Litchi chinensis Sonn.	lychee	
Anastrepha suspensa (Loew) !		F
Bactrocera (Bactrocera) latifrons ??		
Ceratitis (Ceratitis) capitata (Wiedemann)		F
Ceratitis (Pterandrus) rosa Karsch		F

[*Lucuma salicifolia* - see *Pouteria campechiana*]

Luffa **Miller (Cucurbitaceae)**

Luffa acutangula (L.) Roxb.	angled luffa	
Bactrocera (Hemigymnodacus) diversa (Coquillett)		I
Bactrocera (Zeugodacus) caudata (Fabricius) ?		
Bactrocera (Zeugodacus) cucurbitae (Coquillett)		F
Bactrocera (Zeugodacus) tau (Walker)		F
Dacus (Callantra) smieroides (Walker)		F
Dacus (Dacus) demmerezi (Bezzi)		F
Dacus (Didacus) ciliatus Loew		F
Luffa aegyptiaca Miller	luffa	
Bactrocera (Austrodacus) cucumis (French)		F
Bactrocera (Hemigymnodacus) diversa (Coquillett)		I
Bactrocera (Zeugodacus) caudata (Fabricius) ?		
Bactrocera (Zeugodacus) cucurbitae (Coquillett)		F
Bactrocera (Zeugodacus) tau (Walker) ?		
Dacus (Callantra) axanus (Hering)		F
Dacus (Dacus) bivittatus (Bigot)		F
Dacus (Didacus) ciliatus Loew		F
[*Luffa cylindrica* - see *L. aegyptiaca*]		
Luffa sp.		
Bactrocera (Bactrocera) zonata (Saunders) ?		

Lycium **L. (Solanaceae)**

Lycium barbarum L.	Barbary matrimony vine	
Rhagoletis cerasi (Linnaeus) ??		
Lycium turcomanicum Turcz. ex Miers		
Neoceratitis asiatica (Becker)		F

Lycopersicon Miller (Solanaceae)

Lycopersicon esculentum Miller	tomato	
Anastrepha suspensa (Loew)		F
Bactrocera (Austrodacus) cucumis (French)		F
Bactrocera (Bactrocera) aquilonis (May)		F
Bactrocera (Bactrocera) cacuminata (Hering) ??		
Bactrocera (Bactrocera) dorsalis (Hendel)		F
Bactrocera (Bactrocera) sp. near *B. dorsalis* (A)		F
Bactrocera (Bactrocera) facialis (Coquillett)		F
Bactrocera (Bactrocera) latifrons (Hendel)		F
Bactrocera (Bactrocera) neohumeralis (Hardy)		F
Bactrocera (Bactrocera) tryoni (Froggatt)		F
Bactrocera (Bactrocera) zonata (Saunders) ?		
Bactrocera (Notodacus) xanthodes (Broun)		F
Bactrocera (Paradacus) depressa (Shiraki) ??		
Bactrocera (Paratridacus) atrisetosa (Perkins)		F
Bactrocera (Zeugodacus) caudata (Fabricius) ??		
Bactrocera (Zeugodacus) cucurbitae (Coquillett) !		F+S
Bactrocera (Zeugodacus) tau (Walker) ?		
Ceratitis (Ceratitis) capitata (Wiedemann) ?		
Ceratitis (Ceratitis) catoirii Guérin-Méneville		F
Ceratitis (Pterandrus) pedestris (Bezzi)		F
Ceratitis (Pterandrus) rosa Karsch		F
Dacus (Dacus) bivittatus (Bigot) ??		
Dacus (Didacus) ciliatus Loew !		F
Rhagoletis lycopersella Smyth		F
Rhagoletis pomonella (Walsh) ??		
Rhagoletis tomatis Foote		F
Trirhithromyia cyanescens (Bezzi)		F
Zonosemata electa (Say) !		F
[*Lycopersicon lycopersicon* - see *L. esculentum*]		
Lycopersicon pimpinellifolium (L.) Miller	currant tomato	
Bactrocera (Bactrocera) latifrons (Hendel)		F
Rhagoletis lycopersella Smyth		F

Macadamia F.Muell. (Proteaceae)

Macadamia sp.	macadamia nut	
Dirioxa confusa (Hardy)		F?D

Mahonia Nutt. (Berberidaceae)

Mahonia aquifolium (Pursh) Nutt.	mahonia	
Rhagoletis berberis Curran		F

Malpighia L. (Malpighiaceae)

Malpighia glabra L.	huesito	
Anastrepha obliqua (Macquart)		F
Anastrepha suspensa (Loew)		F
Bactrocera (Bactrocera) aquilonis (May)		F
Bactrocera (Bactrocera) dorsalis (Hendel)		F
Bactrocera (Bactrocera) sp. near *B. dorsalis* (A)		F

Bactrocera (Bactrocera) ochrosiae (Malloch) ?		
Ceratitis (Ceratitis) capitata (Wiedemann)		F
[*Malpighia punicifolia* - see *M. glabra*]		

Malus Miller (Rosaceae)

Malus baccata (L.) Borkh.	Siberian crabapple	
Rhagoletis pomonella (Walsh) !		F
Malus domestica Borkh.	apple	
Anastrepha fraterculus (Wiedemann)		F
Anastrepha ludens (Loew)		F
Anastrepha obliqua (Macquart)		F
Anastrepha serpentina (Wiedemann)		F
Anastrepha suspensa (Loew)		F
Bactrocera (Bactrocera) aquilonis (May)		F
Bactrocera (Bactrocera) dorsalis (Hendel)		F
Bactrocera (Bactrocera) latifrons (Hendel) !		F
Bactrocera (Bactrocera) melas (Perkins & May) ?		
Bactrocera (Bactrocera) neohumeralis (Hardy)		F
Bactrocera (Bactrocera) tryoni (Froggatt)		F
Bactrocera (Bactrocera) zonata (Saunders)		F
Bactrocera (Zeugodacus) cucurbitae (Coquillett) ?		
Ceratitis (Ceratitis) capitata (Wiedemann)		F
Ceratitis (Pterandrus) rosa Karsch		F
Dirioxa confusa (Hardy) ?		
Dirioxa pornia (Walker)		FD
Rhagoletis basiola (Osten Sacken) !		F
Rhagoletis pomonella (Walsh)		F
Malus pumila Miller	paradise apple	
Anastrepha ludens (Loew)		F
Bactrocera (Bactrocera) dorsalis (Hendel)		F
Bactrocera (Bactrocera) zonata (Saunders)		F
Ceratitis (Ceratitis) capitata (Wiedemann)		F
Malus sylvestris Miller	crabapple	

[References to this species by Drew (1989a), Freidberg & Kugler (1989), Munro (1925), Norrbom & Kim (1988b), Smith *et al.* (1988) and Syed (1971) were assumed to be errors for *M. domestica*.]

Mammea L. (Clusiaceae=Guttiferae)

Mammea americana L.	mammy-apple	
Anastrepha ludens (Loew)		F
Anastrepha serpentina (Wiedemann)		F
Bactrocera (Notodacus) xanthodes (Broun) ?		
Ceratitis (Ceratitis) capitata (Wiedemann) ?		

Mangifera L. (Anacardiaceae)

Mangifera foetida Lour.	bachang	
Bactrocera (Zeugodacus) tau (Walker)		F
Mangifera indica L.	mango	
Anastrepha distincta Greene !		F
Anastrepha fraterculus (Wiedemann)		F

Anastrepha ludens (Loew)	F
Anastrepha obliqua (Macquart)	F
Anastrepha pseudoparallela (Loew) !	F
Anastrepha serpentina (Wiedemann)	F
Anastrepha striata Schiner	F
Anastrepha suspensa (Loew)	F
Bactrocera (Afrodacus) jarvisi (Tryon)	F
Bactrocera (Bactrocera) aquilonis (May)	F
Bactrocera (Bactrocera) caryeae (Kapoor)	F
Bactrocera (Bactrocera) correcta (Bezzi)	F
Bactrocera (Bactrocera) dorsalis (Hendel)	F
Bactrocera (Bactrocera) sp. near *B. dorsalis* (A)	F
Bactrocera (Bactrocera) sp. near *B. dorsalis* (B)	F
Bactrocera (Bactrocera) sp. near *B. dorsalis* (C)	F
Bactrocera (Bactrocera) sp. near *B. dorsalis* (D)	F
Bactrocera (Bactrocera) facialis (Coquillett)	F
Bactrocera (Bactrocera) frauenfeldi (Schiner)	F
Bactrocera (Bactrocera) froggatti (Bezzi) ?	
Bactrocera (Bactrocera) incisa (Walker) ??	
Bactrocera (Bactrocera) kirki (Froggatt)	F
Bactrocera (Bactrocera) latifrons (Hendel) ??	
Bactrocera (Bactrocera) melanota (Coquillett)	F
Bactrocera (Bactrocera) neohumeralis (Hardy)	F
Bactrocera (Bactrocera) occipitalis (Bezzi)	F
Bactrocera (Bactrocera) opiliae (Drew & Hardy) !	F
Bactrocera (Bactrocera) passiflorae (Froggatt)	F
Bactrocera (Bactrocera) psidii (Froggatt)	F
Bactrocera (Bactrocera) trilineola Drew ?	
Bactrocera (Bactrocera) tryoni (Froggatt)	F
Bactrocera (Bactrocera) tuberculata (Bezzi)	F
Bactrocera (Bactrocera) versicolor (Bezzi) ??	
Bactrocera (Bactrocera) zonata (Saunders)	F
Bactrocera (Hemigymnodacus) diversa (Coquillett) ??	
Bactrocera (Notodacus) xanthodes (Broun) ?	
Bactrocera (Zeugodacus) cucurbitae (Coquillett) !	F
Bactrocera (Zeugodacus) tau (Walker) ?	
Ceratitis (Ceratalaspis) cosyra (Walker)	F
Ceratitis (Ceratitis) capitata (Wiedemann)	F
Ceratitis (Ceratitis) catoirii Guérin-Méneville	F
Ceratitis (Pardalaspis) punctata (Wiedemann) ?	
Ceratitis (Pterandrus) anonae Graham	F
Ceratitis (Pterandrus) flexuosa (Walker) ?	
Ceratitis (Pterandrus) rosa Karsch	F
Dirioxa confusa (Hardy)	F?D
Dirioxa pornia (Walker) ?	
Toxotrypana curvicauda Gerstaecker ??	

Manihot Miller (Euphorbiaceae)

Manihot esculenta Crantz	cassava, manioc	
Anastrepha manihoti Lima		S
Anastrepha montei Lima		F
Anastrepha pickeli Lima		F
Anastrepha striata Schiner		F

Manilkara Adanson (Sapotaceae)

[*Manilkara achras* - see *M. zapota*]		
Manilkara kauki (L.) Dubard	sauh	
Bactrocera (Bactrocera) frauenfeldi (Schiner)		F
Manilkara zapota (L.) P. Royen	sapodilla, chicle	
Anastrepha antunesi Lima		F
Anastrepha fraterculus (Wiedemann)		F
Anastrepha nigrifascia Stone		F
Anastrepha obliqua (Macquart)		F
Anastrepha ocresia (Walker)		F
Anastrepha serpentina (Wiedemann)		F
Anastrepha suspensa (Loew)		F
Bactrocera (Bactrocera) aquilonis (May)		F
Bactrocera (Bactrocera) correcta (Bezzi)		F
Bactrocera (Bactrocera) sp. near *B. dorsalis* (A)		F
Bactrocera (Bactrocera) sp. near *B. dorsalis* (B)		F
Bactrocera (Bactrocera) versicolor (Bezzi)		F
Bactrocera (Bactrocera) zonata (Saunders) ??		
Bactrocera (Zeugodacus) caudata (Fabricius) ??		
Bactrocera (Zeugodacus) tau (Walker) ?		
Ceratitis (Ceratitis) capitata (Wiedemann)		F
Ceratitis (Ceratitis) catoirii Guérin-Méneville ??		
Ceratitis (Pterandrus) rosa Karsch		F
[Records may be confused with *Pouteria sapota*]		

Mespilus L. (Rosaceae)

Mespilus germanica L.	medlar	
Ceratitis (Ceratitis) capitata (Wiedemann)		F

Mimusops L. (Sapotaceae)

Mimusops elengi L.	Spanish cherry	
Bactrocera (Bactrocera) tryoni (Froggatt)		F
Ceratitis (Ceratitis) capitata (Wiedemann) ?		

Momordica L. (Cucurbitaceae)

Momordica balsamina L.	balsam-apple	
Anastrepha suspensa (Loew)		F
Bactrocera (Zeugodacus) cucurbitae (Coquillett)		F
Dacus (Didacus) ciliatus Loew		F
Momordica charantia L.	bitter gourd	
Anastrepha suspensa (Loew)		F
Bactrocera (Austrodacus) cucumis (French)		F
Bactrocera (Bactrocera) sp. near *B. dorsalis* (B)		F
Bactrocera (Bactrocera) umbrosa (Fabricius)		F

Bactrocera (Bactrocera) zonata (Saunders) F
Bactrocera (Zeugodacus) caudata (Fabricius) ?
Bactrocera (Zeugodacus) cucurbitae (Coquillett) F
Bactrocera (Zeugodacus) tau (Walker) F
Bactrocera (Zeugodacus) trimaculata (Hardy & Adachi) F ?
Dacus (Dacus) bivittatus (Bigot) F
Dacus (Dacus) demmerezi (Bezzi) F
Dacus (Dacus) disjunctus (Bezzi) ??
Dacus (Dacus) punctatifrons Karsch F
Dacus (Dacus) yangambinus Munro F
Dacus (Didacus) ciliatus Loew F

Momordica cochinchinensis (Lour.) Sprengel — teruah
Bactrocera (Bactrocera) cilifer (Hendel) ?
Bactrocera (Zeugodacus) cucurbitae (Coquillett) F
Bactrocera (Zeugodacus) trimaculata (Hardy & Adachi) F ?

Morus L. (Moraceae)

Morus alba L. — white mulberry
Bactrocera (Bactrocera) tryoni (Froggatt) F
Ceratitis (Ceratitis) capitata (Wiedemann) ?

Morus nigra L. — black mulberry.
Bactrocera (Bactrocera) neohumeralis (Hardy) F
Bactrocera (Bactrocera) tryoni (Froggatt) F

Morus sp.
Bactrocera (Zeugodacus) tau (Walker) ?
Dirioxa pornia (Walker) ?

Muntingia L. (Elaeocarpaceae)

Muntingia calabura L. — calabur
Anastrepha suspensa (Loew) F
Bactrocera (Bactrocera) dorsalis (Hendel) F
Bactrocera (Zeugodacus) tau (Walker) F
Ceratitis (Ceratitis) capitata (Wiedemann) F

Murraya J. König ex L. (Rutaceae)

Murraya paniculata (L.) Jack — orange jessamine
Anastrepha suspensa (Loew) F
Bactrocera (Bactrocera) dorsalis (Hendel) F
Ceratitis (Ceratitis) capitata (Wiedemann) F

Musa L. (Musaceae)

Musa acuminata Colla — dwarf banana
Bactrocera (Afrodacus) jarvisi (Tryon) F
Bactrocera (Bactrocera) aquilonis (May) F
Bactrocera (Bactrocera) musae (Tryon) F
Bactrocera (Bactrocera) tryoni (Froggatt) F

Musa nana Lour.
Ceratitis (Pterandrus) rosa Karsch ?

Musa x *paradisiaca* L. — banana
Bactrocera (Bactrocera) caryeae (Kapoor) !?
Bactrocera (Bactrocera) curvipennis (Froggatt) ??

Bactrocera (Bactrocera) dorsalis (Hendel) F
Bactrocera (Bactrocera) sp. near *B. dorsalis* (A) F
Bactrocera (Bactrocera) sp. near *B. dorsalis* (B) F
Bactrocera (Bactrocera) frauenfeldi (Schiner) ?
Bactrocera (Bactrocera) latifrons (Hendel) ??
Bactrocera (Bactrocera) musae (Tryon) F
Bactrocera (Bactrocera) trilineola Drew ?
Bactrocera (Bactrocera) tryoni (Froggatt) ?
Bactrocera (Hemigymnodacus) diversa (Coquillett) ?
Bactrocera (Zeugodacus) cucurbitae (Coquillett) ?
Ceratitis (Pterandrus) rosa Karsch ?
Dirioxa confusa (Hardy) F?D
Dirioxa pornia (Walker) F?D

[*Musa x sapientum* - see *M. x paradisiaca*]

Musa sp.
Bactrocera (Bactrocera) bryoniae (Tryon) ??
Ceratitis (Ceratitis) capitata (Wiedemann) FD
Ceratitis (Pterandrus) rosa Karsch F?D

Myrciaria O. Berg (Myrtaceae)

Myrciaria cauliflora (C. Martius) O. Berg — jaboticaba
Anastrepha suspensa (Loew) F
Ceratitis (Ceratitis) capitata (Wiedemann) F

Myristica Gronov. (Myristicaceae)

Myristica fragrans Houtt. — nutmeg, mace
Ceratitis (Ceratitis) malgassa Munro F

Myristica sp.
Bactrocera (Hemigymnodacus) diversa (Coquillett) ??

Nephelium L. (Sapindaceae)

Nephelium lappaceum L. — rambutan
Bactrocera (Bactrocera) sp. near *B. dorsalis* (B) F

[*Nephelium litchi* - see *Litchi chinensis*]

Nephelium sp.
Bactrocera (Bactrocera) psidii ??

[*Nephetium* - assumed to be error for *Nephelium*]

Nitraria L. (Zygophyllaceae=Nitrariaceae)

Nitraria schoberi L. — nitre bush
Nitrariomyia lukjanovitshi Rohdendorf F

Nitraria sibirica Pall.
Nitrariomyia lukjanovitshi Rohdendorf F

Olea L. (Oleaceae)

Olea europaea L. — olive
Bactrocera (Bactrocera) tryoni (Froggatt) F
Bactrocera (Daculus) oleae (Gmelin) F
Ceratitis (Ceratitis) capitata (Wiedemann) ?
[See notes on *Bactrocera (Afrodacus) biguttula* (Bezzi), *B. (Melanodacus) nigra* (Tryon) (p. 274) and *Munromyia nudiseta* Bezzi (p. 402), which attack wild *Olea* spp. and related genera.]

Opuntia Miller (Cactaceae)

Opuntia ficus-indica (L.) Miller	Indian fig prickly pear	
Bactrocera (Bactrocera) tryoni (Froggatt)		F
Ceratitis (Ceratitis) capitata (Wiedemann) ?		
Ceratitis (Pterandrus) rosa Karsch ?		
Opuntia tuna (L.) Miller	elephant-ear prickly pear	
Ceratitis (Ceratitis) capitata (Wiedemann) ?		
Ceratitis (Pterandrus) rosa Karsch ?		
Opuntia vulgaris Miller	prickly pear, tuna	
Ceratitis (Ceratitis) capitata (Wiedemann)		F

Paeonia L. (Paeoniaceae)

Paeonia chinensis Hort.	garden peony	
Macrotrypeta ortalidina Portschinsky		I

Panicum L. (Poaceae=Gramineae)

Panicum maximum Jacq.	guineagrass	
Clinotaenia magniceps (Bezzi)		S?

Parmentiera DC. (Bignoniaceae)

Parmentiera aculeata (Kunth) Seemann	cuachilote	
Ceratitis (Ceratitis) capitata (Wiedemann)		F

Passiflora L. (Passifloraceae)

Passiflora alata Dryander	maracuja grande	
Anastrepha pseudoparallela (Loew)		F
Passiflora caerulea L.	blue passion fruit	
Anastrepha mburucuyae Blanchard		F
Anastrepha pseudoparallela (Loew)		F
Ceratitis (Ceratitis) capitata (Wiedemann)		F
Passiflora edulis Sims	purple granadilla	
Anastrepha ludens (Loew)		F
Anastrepha pseudoparallela (Loew)		F
Bactrocera (Austrodacus) cucumis (French) !		F
Bactrocera (Bactrocera) sp. near *B. dorsalis* (B)		F
Bactrocera (Bactrocera) kirki (Froggatt)		F
Bactrocera (Bactrocera) passiflorae (Froggatt)		F
Bactrocera (Bactrocera) tryoni (Froggatt)		F
Bactrocera (Zeugodacus) cucurbitae (Coquillett) ?		
Ceratitis (Ceratitis) capitata (Wiedemann) ?		
Passiflora foetida L.	wild waterlemon	
Bactrocera (Bactrocera) bryoniae (Tryon) ?		
Bactrocera (Zeugodacus) cucurbitae (Coquillett) ?		
Ceratitis (Ceratitis) capitata (Wiedemann) ?		
Dacus (Dacus) punctatifrons Karsch !		F
Passiflora laurifolia L.	yellow granadilla	
Anastrepha ethalea (Walker)		F
Bactrocera (Zeugodacus) cucurbitae (Coquillett) ?		
Passiflora ligularis A.L. Juss.	sweet granadilla	
Anastrepha pseudoparallela (Loew)		F
Passiflora mollisima (Kunth) L.H. Bailey	banana passion fruit	

Bactrocera (Bactrocera) dorsalis (Hendel)		F
Passiflora quadrangularis L.	giant granadilla	
Anastrepha consobrina (Loew)		F
Anastrepha ethalea (Walker)		F
Anastrepha limae Stone		F
Anastrepha obliqua (Macquart)		F
Anastrepha pallidipennis Greene		F
Anastrepha pseudoparallela (Loew)		F
Bactrocera (Bactrocera) facialis (Coquillett)		F
Bactrocera (Bactrocera) passiflorae (Froggatt)		F
Bactrocera (Bactrocera) psidii (Froggatt)		F
Bactrocera (Bactrocera) tryoni (Froggatt)		F
Bactrocera (Bactrocera) umbrosa (Fabricius) ?		
Bactrocera (Notodacus) xanthodes (Broun)		F
Bactrocera (Zeugodacus) cucurbitae (Coquillett) ?		
Ceratitis (Ceratitis) capitata (Wiedemann) ?		
Dacus (Dacus) bivittatus (Bigot) ??		
Passiflora suberosa L.		
Bactrocera (Bactrocera) bryoniae (Tryon) !		F
Bactrocera (Bactrocera) neohumeralis (Hardy)		F
Passiflora vitifolia Kunth		
Anastrepha passiflorae Greene		F
Passiflora sp.		
Bactrocera (Zeugodacus) cucurbitae (Coquillett) !		F
Ceratitis (Pterandrus) punctata (Wiedemann) ?		
Dacus (Didacus) vertebratus Bezzi !		F
Dirioxa pornia (Walker) !		F

Pastinaca L. (Apiaceae=Umbelliferae)

Pastinaca sativa L.	parsnip	
Euleia fratria (Loew)		L
Euleia heraclei (Linnaeus)		L

Pereskia Miller (Cactaceae)

Pereskia aculeata Miller	Barbados gooseberry	
Ceratitis (Ceratitis) capitata (Wiedemann) ?		

Pergularia L. (Asclepiadaceae)

Pergularia daemia (Forssk.) Chiov.		
Dacus (Didacus) arcuatus Munro		F

Persea Miller (Lauraceae)

Persea americana Miller	avocado	
Anastrepha fraterculus (Wiedemann)		F
Anastrepha ludens (Loew)		F
Anastrepha serpentina (Wiedemann)		F
Anastrepha striata Schiner		F
Anastrepha suspensa (Loew)		F
Bactrocera (Bactrocera) aquilonis (May)		F
Bactrocera (Bactrocera) dorsalis (Hendel)		F
Bactrocera (Bactrocera) facialis (Coquillett)		F

Bactrocera (Bactrocera) passiflorae (Froggatt)		F
Bactrocera (Bactrocera) trilineola Drew ?		
Bactrocera (Bactrocera) tryoni (Froggatt)		F
Bactrocera (Zeugodacus) cucurbitae (Coquillett) !		F
Ceratitis (Ceratalaspis) cosyra (Walker)		F
Ceratitis (Ceratitis) capitata (Wiedemann)		FD
Ceratitis (Ceratitis) catoirii Guérin-Méneville		F
Ceratitis (Pterandrus) anonae Graham ?		
Ceratitis (Pterandrus) rosa Karsch		F
Persea gratissima Gaertner f.		
Ceratitis (Pterandrus) rosa Karsch ?		

[*Persica vulgaris* - see *Prunus persica*]

Petroselinum Hill (Apiaceae=Umbelliferae)

Petroselinum crispum (Miller) Nyman ex A.W. Hill	parsley	
Euleia heraclei (Linnaeus)		L

Phaseolus L. (Fabaceae=Leguminosae)

[*Phaseolus limensis* - see *P. lunatus*]

Phaseolus lunatus L.	Lima bean	
Bactrocera (Zeugodacus) cucurbitae (Coquillett) ?		

[*Phaseolus radiatus* - see *Vigna radiata*]

Phaseolus vulgaris L.	garden bean	
Bactrocera (Zeugodacus) cucurbitae (Coquillett) !		F
Ceratitis (Ceratitis) capitata (Wiedemann) ??		
Phaseolus sp.		
Dacus (Didacus) ciliatus Loew !		F

Phoenix L. (Arecaceae=Palmae)

Phoenix dactylifera L.	date palm	
Anastrepha suspensa (Loew)		F
Bactrocera (Bactrocera) melas (Perkins & May) ?		
Bactrocera (Bactrocera) neohumeralis (Hardy)		F
Bactrocera (Bactrocera) tryoni (Froggatt)		F
Bactrocera (Bactrocera) zonata (Saunders)		F
Bactrocera (Zeugodacus) cucurbitae (Coquillett) ?		
Ceratitis (Ceratitis) capitata (Wiedemann)		F

Phyllanthus L. (Euphorbiaceae)

Phyllanthus acidus (L.) Skeels	Otaheite gooseberry	
Bactrocera (Bactrocera) aquilonis (May)		F

Phyllostachys Siebold & Zucc. (Poaceae=Gramineae)

Phyllostachys pubescens Mazel ex Houz.		
Acroceratitis plumosa Hendel		S
Gastrozona fasciventris (Macquart)		S

Physalis L. (Solanaceae)

[*Physalis ixocarpa* - see *P. philadelphica*]

Physalis peruviana L.	Cape gooseberry	
Bactrocera (Bactrocera) dorsalis (Hendel)		F
Bactrocera (Bactrocera) tryoni (Froggatt)		F
Ceratitis (Ceratitis) capitata (Wiedemann)		F
Physalis philadelphica Lam.	tomatillo	
Rhagoletis conversa (Brèthes) ?		
Rhagoletis striatella Wulp		F

Pimenta Lindley (Myrtaceae)

Pimenta dioica (L.) Merr.	allspice, pimento	
Anastrepha suspensa (Loew)		F
Ceratitis (Ceratitis) capitata (Wiedemann) ?		

Pometia Forster & Forster f. (Sapindaceae)

Pometia pinnata Forst. & Forst. f.	Pacific lychee	
Bactrocera (Bactrocera) distincta (Malloch) ?		
Bactrocera (Bactrocera) facialis (Coquillett)		F
Bactrocera (Bactrocera) passiflorae (Froggatt) ?		
Bactrocera (Bactrocera) simulata (Malloch) ?		
Bactrocera (Notodacus) xanthodes (Broun) ?		

Poncirus Raf. (Rutaceae)

Poncirus trifoliata (L.) Raf.	trifoliate orange	
Dirioxa confusa (Hardy)		F?D
Dirioxa pornia (Walker) ?		

[*Poupartia caffra* - see *Sclerocarya birrea*]

Pouteria Aublet (Sapotaceae)

Pouteria caimito (Ruiz & Pavon) Radlk.	abiu	
Anastrepha leptozona Hendel		F
Anastrepha serpentina (Wiedemann)		F
Pouteria campechiana (Kunth) Baehni	egg-fruit tree	
Anastrepha leptozona Hendel		F
Anastrepha sagittata Stone)		F
Anastrepha serpentina (Wiedemann)		F
Anastrepha suspensa (Loew)		F
Ceratitis (Ceratitis) capitata (Wiedemann) ?		
Pouteria obovata Kunth	lucmo	
Anastrepha fraterculus (Wiedemann)		F
Anastrepha serpentina (Wiedemann)		F
Pouteria sapota (Jaqu.) H. Moore & Stearn	sapote	
Anastrepha ludens (Loew)		F
Anastrepha obliqua (Macquart)		F
Anastrepha serpentina (Wiedemann)		F
Bactrocera (Bactrocera) dorsalis (Hendel)		F
Ceratitis (Ceratitis) capitata (Wiedemann)		F
[Records may be confused with *Manilkara zapota*]		

Pouteria viridis (Pittier) Cronq.	green sapote	
Anastrepha obliqua (Macquart)		F
Anastrepha serpentina (Wiedemann)		F
Anastrepha striata Schiner		F
Ceratitis (Ceratitis) capitata (Wiedemann)		F

Prunus L. (Rosaceae)

Prunus angustifolia Marshall	Chickasaw plum	
Rhagoletis pomonella (Walsh) !		F
Prunus armeniaca L.	apricot	
Anastrepha fraterculus (Wiedemann)		F
Bactrocera (Afrodacus) jarvisi (Tryon) !		F
Bactrocera (Bactrocera) dorsalis (Hendel)		F
Bactrocera (Bactrocera) mayi (Hardy) ??		
Bactrocera (Bactrocera) neohumeralis (Hardy)		F
Bactrocera (Bactrocera) tryoni (Froggatt)		F
Bactrocera (Zeugodacus) cucurbitae (Coquillett) ?		
Ceratitis (Ceratalaspis) quinaria (Bezzi)		F
Ceratitis (Ceratitis) capitata (Wiedemann)		F
Ceratitis (Pterandrus) rosa Karsch		F
Rhagoletis pomonella (Walsh) !		F
Prunus avium (L.) L.	sweet cherry	
Bactrocera (Bactrocera) tryoni (Froggatt)		F
Euphranta japonica (Ito)		F
Rhagoletis cerasi (Linnaeus)		F
Rhagoletis cingulata (Loew)		F
Rhagoletis fausta (Osten Sacken)		F
Rhagoletis indifferens Curran		F
Rhagoletis pomonella (Walsh) ?		
Prunus cerasus L.	sour cherry	
Rhagoletis cerasi (Linnaeus)		F
Rhagoletis cingulata (Loew)		F
Rhagoletis fausta (Osten Sacken)		F
Rhagoletis indifferens Curran		F
Rhagoletis pomonella (Walsh) !		F
Ceratitis (Ceratitis) capitata (Wiedemann) ?		
Prunus cerasifera Ehrh.	myrobalan plum	
Bactrocera (Bactrocera) tryoni (Froggatt)		F
Prunus domestica L.	plum	
Anastrepha fraterculus (Wiedemann) ?		
Anastrepha ludens (Loew) ?		
Bactrocera (Bactrocera) caryeae (Kapoor) ?		
Bactrocera (Bactrocera) dorsalis (Hendel)		F
Bactrocera (Bactrocera) melas (Perkins & May) ?		
Bactrocera (Bactrocera) neohumeralis (Hardy)		F
Bactrocera (Bactrocera) tryoni (Froggatt)		F
Ceratitis (Ceratitis) capitata (Wiedemann)		F
Ceratitis (Pterandrus) rosa Karsch		F
Dirioxa pornia (Walker) ?		

Rhagoletis pomonella (Walsh) ??		
Prunus dulcis (Miller) D.A. Webb	almond	
Anastrepha obliqua (Macquart)		F
Prunus mahaleb L.	mahaleb cherry	
Rhagoletis cerasi (Linnaeus)		F
Rhagoletis cingulata (Loew)		F
Rhagoletis fausta (Osten Sacken)		F
Prunus pensylvanica L. f.	pin cherry	
Rhagoletis cingulata (Loew) ??		
Rhagoletis fausta (Osten Sacken)		F
Prunus persica (L.) Batsch	peach	
Anastrepha daciformis Bezzi		F
Anastrepha fraterculus (Wiedemann)		F
Anastrepha ludens (Loew)		F
Anastrepha minensis Lima		F
Anastrepha serpentina (Wiedemann)		F
Anastrepha striata Schiner		F
Anastrepha suspensa (Loew)		F
Anastrepha turicai Blanchard		F
Bactrocera (Afrodacus) jarvisi (Tryon)		F
Bactrocera (Bactrocera) aquilonis (May)		F
Bactrocera (Bactrocera) correcta (Bezzi)		F
Bactrocera (Bactrocera) dorsalis (Hendel)		F
Bactrocera (Bactrocera) facialis (Coquillett)		F
Bactrocera (Bactrocera) kirki (Froggatt)		F
Bactrocera (Bactrocera) melas (Perkins & May) ?		
Bactrocera (Bactrocera) neohumeralis (Hardy)		F
Bactrocera (Bactrocera) sp. [Bhutan]		F
Bactrocera (Bactrocera) trivialis (Drew)		F
Bactrocera (Bactrocera) tryoni (Froggatt)		F
Bactrocera (Bactrocera) tuberculata (Bezzi)		F
Bactrocera (Bactrocera) zonata (Saunders)		F
Bactrocera (Zeugodacus) caudata (Fabricius) ??		
Bactrocera (Zeugodacus) cucurbitae (Coquillett) !		F
Bactrocera (Zeugodacus) duplicata (Bezzi) ?		
Ceratitis (Ceratalaspis) cosyra (Walker)		F
Ceratitis (Ceratalaspis) quinaria (Bezzi)		F
Ceratitis (Ceratitis) capitata (Wiedemann)		F
Ceratitis (Ceratitis) catoirii Guérin-Méneville		F
Ceratitis (Pterandrus) rosa Karsch		F
Dirioxa pornia (Walker)		FD
Rhagoletis completa Cresson !		F
Rhagoletis pomonella (Walsh) !		F
Rhagoletis suavis (Loew) !		F
Tomoplagia cressoni Aczél ??		
Prunus salicina Lindley	Japanese plum	
Anastrepha obliqua (Macquart)		F
Rhagoletis indifferens Curran		F

Prunus serotina Ehrh.	black cherry	
Rhagoletis cerasi (Linnaeus)		F
Rhagoletis cingulata (Loew)		F
Rhagoletis fausta (Osten Sacken)		F
Prunus subcordata Benth.	Klamath plum	
Rhagoletis indifferens Curran		F
Prunus virginiana L.	chokecherry	
Rhagoletis cingulata (Loew) !		F
Rhagoletis fausta (Osten Sacken)		F
Rhagoletis indifferens Curran		F

Psidium L. (Myrtaceae)

[*Psidium cattleianum* - see *P. littorale*]		
Psidium friedrichsthalianum (O. Berg) Niedenzu	wild guava	
Anastrepha striata Schiner		F
Anastrepha suspensa (Loew)		F
Psidium guajava L.	common guava	
Anastrepha antunesi Lima		F
Anastrepha bahiensis Lima ?		
Anastrepha bistrigata Bezzi		F
Anastrepha fraterculus (Wiedemann)		F
Anastrepha grandis (Macquart) !		F
Anastrepha ludens (Loew)		F
Anastrepha minensis Lima		F
Anastrepha obliqua (Macquart)		F
Anastrepha ocresia (Walker)		F
Anastrepha ornata Aldrich		F
Anastrepha parishi Stone		F
Anastrepha punctata Hendel		F
Anastrepha schultzi Blanchard		F
Anastrepha serpentina (Wiedemann)		F
Anastrepha sororcula Zucchi		F
Anastrepha striata Schiner		F
Anastrepha suspensa (Loew)		F
Anastrepha zenildae Zucchi		F
Bactrocera (Afrodacus) jarvisi (Tryon)		F
Bactrocera (Bactrocera) aquilonis (May)		F
Bactrocera (Bactrocera) breviaculeus (Hardy) ??		
Bactrocera (Bactrocera) caryeae (Kapoor)		F
Bactrocera (Bactrocera) correcta (Bezzi)		F
Bactrocera (Bactrocera) dorsalis (Hendel)		F
Bactrocera (Bactrocera) sp. near *B. dorsalis* (A)		F
Bactrocera (Bactrocera) sp. near *B. dorsalis* (B)		F
Bactrocera (Bactrocera) facialis (Coquillett)		F
Bactrocera (Bactrocera) frauenfeldi (Schiner)		F
Bactrocera (Bactrocera) incisa (Walker) ??		
Bactrocera (Bactrocera) kirki (Froggatt)		F
Bactrocera (Bactrocera) latifrons (Hendel) ??		
Bactrocera (Bactrocera) melanota (Coquillett)		F

Bactrocera (Bactrocera) melas (Perkins & May) ?		
Bactrocera (Bactrocera) musae (Tryon) !		F
Bactrocera (Bactrocera) neohumeralis (Hardy)		F
Bactrocera (Bactrocera) passiflorae (Froggatt)		F
Bactrocera (Bactrocera) psidii (Froggatt)		F
Bactrocera (Bactrocera) trilineola Drew ?		
Bactrocera (Bactrocera) trivialis (Drew)		F
Bactrocera (Bactrocera) tryoni (Froggatt)		F
Bactrocera (Bactrocera) versicolor (Bezzi) ?		
Bactrocera (Bactrocera) zonata (Saunders)		F
Bactrocera (Hemigymnodacus) diversa (Coquillett) ??		
Bactrocera (Notodacus) xanthodes (Broun)		F
Bactrocera (Zeugodacus) caudata (Fabricius) ??		
Bactrocera (Zeugodacus) cucurbitae (Coquillett) ?		
Bactrocera (Zeugodacus) tau (Walker) ?		
Carpomya vesuviana Costa ??		
Ceratitis (Ceratalaspis) cosyra (Walker)		F
Ceratitis (Ceratalaspis) quinaria (Bezzi)		F
Ceratitis (Ceratitis) capitata (Wiedemann)		F
Ceratitis (Ceratitis) catoirii Guérin-Méneville		F
Ceratitis (Ceratitis) malgassa Munro		F
Ceratitis (Pardalaspis) punctata (Wiedemann) ?		
Ceratitis (Pterandrus) anonae Graham ?		
Ceratitis (Pterandrus) rosa Karsch		F
Dirioxa confusa (Hardy)		F?D
Psidium guineense Sw.	Brazilian guava	
Anastrepha bistrigata Bezzi		F
Anastrepha fraterculus (Wiedemann)		F
Anastrepha ludens (Loew) ?		
Anastrepha striata Schiner		F
Anastrepha suspensa (Loew)		F
Psidium littorale Raddi	strawberry guava	
Anastrepha fraterculus (Wiedemann)		F
Anastrepha ludens (Loew)		F
Anastrepha obliqua (Macquart)		F
Anastrepha striata Schiner		F
Anastrepha suspensa (Loew)		F
Bactrocera (Afrodacus) jarvisi (Tryon)		F
Bactrocera (Bactrocera) aquilonis (May)		F
Bactrocera (Bactrocera) dorsalis (Hendel)		F
Bactrocera (Bactrocera) neohumeralis (Hardy)		F
Bactrocera (Bactrocera) passiflorae (Froggatt) ?		
Bactrocera (Bactrocera) psidii (Froggatt)		F
Bactrocera (Bactrocera) tryoni (Froggatt)		F
Bactrocera (Zeugodacus) cucurbitae (Coquillett) ?		
Ceratitis (Ceratitis) capitata (Wiedemann)		F
Ceratitis (Ceratitis) catoirii Guérin-Méneville		F
Ceratitis (Pterandrus) anonae Graham ?		
Ceratitis (Pterandrus) rosa Karsch		F

Dirioxa confusa (Hardy)		F?D
Psidium sp.		
Anastrepha daciformis Bezzi		F

Punica L. (Punicaceae)

Punica granatum L.	pomegranate	
Anastrepha fraterculus (Wiedemann)		F
Anastrepha ludens (Loew)		F
Anastrepha suspensa (Loew)		F
Bactrocera (Afrodacus) jarvisi (Tryon)		F
Bactrocera (Bactrocera) tryoni (Froggatt)		F
Bactrocera (Bactrocera) zonata (Saunders)		F
Ceratitis (Ceratitis) capitata (Wiedemann)		FD
Ceratitis (Ceratitis) catoirii Guérin-Méneville		F

Pyrus L. (Rosaceae)

Pyrus communis L.	pear	
Anastrepha fraterculus (Wiedemann)		F
Anastrepha ludens (Loew)		F
Anastrepha obliqua (Macquart)		F
Anastrepha ornata Aldrich		F
Anastrepha serpentina (Wiedemann)		F
Anastrepha suspensa (Loew)		F
Bactrocera (Afrodacus) jarvisi (Tryon)		F
Bactrocera (Bactrocera) dorsalis (Hendel)		F
Bactrocera (Bactrocera) melas (Perkins & May) ?		
Bactrocera (Bactrocera) neohumeralis (Hardy)		F
Bactrocera (Bactrocera) tryoni (Froggatt)		F
Bactrocera (Zeugodacus) cucurbitae (Coquillett) ?		
Bactrocera (Zeugodacus) scutellata (Hendel) ??		
Ceratitis (Ceratitis) capitata (Wiedemann)		F
Ceratitis (Pterandrus) rosa Karsch		F
Dirioxa pornia (Walker)		FD
Rhagoletis pomonella (Walsh) ?		
[*Pyrus malus* - see *Malus domestica*]		
Pyrus pyrifolia (Burman f.) Nakai	sand pear	
Anastrepha suspensa (Loew)		F
Dirioxa pornia (Walker) ?		

Quassia L. (Simaroubaceae)

Quassia simarouba L. f.	paradise tree	
Bactrocera (Bactrocera) dorsalis (Hendel)		F

Raphanus L. (Brassicaceae=Cruciferae)

Raphanus sp.	radish	
Bactrocera (Hemigymnodacus) diversa (Coquillett) ??		

Rheedia L. (Clusiaceae=Guttiferae)

Rheedia acuminata (Ruiz & Pavon) Planchon & Triana	madrono	
Anastrepha rheediae Stone		F
Rheedia braziliensis Planchon & Triana	bakupari	
Anastrepha rheediae Stone		F

Ribes L. (Grossulariaceae)

Ribes aureum Pursh	golden currant	
Epochra canadensis (Loew)		F
Rhagoletis ribicola Doane		F
[*Ribes grossularia* - see *R. uva-crispa*]		
Ribes nigrum L.	blackcurrant	
Epochra canadensis (Loew)		F
Ribes rubrum L.	redcurrant	
Epochra canadensis (Loew)		F
Rhagoletis ribicola Doane		F
Ribes uva-crispa L.	European gooseberry	
Epochra canadensis (Loew)		F
Rhagoletis ribicola Doane		F
[*Ribes vulgare* - see *R. rubrum*]		

Ricinus L. (Euphorbiaceae)

Ricinus communis L.	castorbean	
Bactrocera (Bactrocera) correcta (Bezzi) ?		

Rollinia A. St. Hil. (Annonaceae)

[*Rollinia deliciosa* - see *R. pulchrinervis*]		
Rollinia mucosa (Jacq.) Baillon	wild sweetsop	
Bactrocera (Bactrocera) aquilonis (May)		F
Rollinia pulchrinervis A. DC.	biribá	
Bactrocera (Bactrocera) aquilonis (May)		F

Rosa L. (Rosaceae)

Rosa arkansana Porter	Arkansas rose	
Rhagoletis basiola (Osten Sacken)		F
Rosa canina L.	dog rose	
Carpomya schineri (Loew)		F
Rhagoletis alternata (Fallén)		F
Rhagoletis basiola (Osten Sacken)		F
Rhagoletis turanica (Rohdendorf)		F
Rosa damascena Miller	damask rose	
Carpomya schineri (Loew)		F
Rosa gallica L.	French rose	
Carpomya schineri (Loew)		F
Rhagoletis basiola (Osten Sacken)		F
Rosa rugosa Thunb.	Japanese rose	
Carpomya schineri (Loew)		F
Rhagoletis alternata (Fallén)		F
Rhagoletis basiola (Osten Sacken)		F
Rhagoletis pomonella (Walsh) !		F

Rosa setigera Michaux	prairie rose	
Rhagoletis basiola (Osten Sacken)		F
Rosa villosa L.	apple rose	
Carpomya schineri (Loew)		F
Rhagoletis alternata (Fallén)		F

Rubus L. (Rosaceae)

Rubus flagellaris Willd. x *R. loganobaccus*	youngberry	
Ceratitis (Ceratitis) capitata (Wiedemann)		F
Ceratitis (Pterandrus) rubivora (Coquillett)		F
Rubus fruticosa L.	blackberry	
Bactrocera (Bactrocera) tryoni (Froggatt)		F
Ceratitis (Pterandrus) rosa Karsch ?		
Ceratitis (Pterandrus) rubivora (Coquillett)		F
Rubus glaucus Benth.	Andes berry	
Anastrepha fraterculus (Wiedemann)		F
Rubus idaeus L.	raspberry	
Ceratitis (Pterandrus) rubivora (Coquillett)		F
Rubus loganobaccus L.H. Bailey	loganberry	
Ceratitis (Pterandrus) rubivora (Coquillett)		F
Rubus rosifolius Smith	Mauritius raspberry	
Bactrocera (Bactrocera) neohumeralis (Hardy)		F
Rubus ursinus Cham. & Schldl.	California berry	
Bactrocera (Bactrocera) tryoni (Froggatt)		F

Salix L. (Salicaceae)

Salix alba L.	white willow	
Euphranta toxoneura (Loew)		G
Salix caprea L.	goat willow	
Euphranta toxoneura (Loew)		G
Salix fragilis L.	crack willow	
Euphranta toxoneura (Loew)		G

Sansevieria Thunb. (Agavaceae)

Sansevieria sp.		
Taomyia marshalli Bezzi		F

Santalum L. (Santalaceae)

Santalum album L.	sandalwood	
Bactrocera (Bactrocera) correcta (Bezzi) ?		
Ceratitis (Ceratitis) capitata (Wiedemann)		F

Sclerocarya Hochst. (Anacardiacae)

Sclerocarya birrea (A. Rich.) Hochst.	maroola plum	
Ceratitis (Ceratalaspis) cosyra (Walker)		F
Ceratitis (Ceratalaspis) giffardi Bezzi		F
Ceratitis (Ceratitis) malgassa Munro		F

[*Sclerocarya caffra* - see *S. birrea*]

Scorzonera **L. (Asteraceae=Compositae)**

Scorzonera tausaghyz Lipsch. & Bosse
- *Ensina sonchi* (Linnaeus) — C
- *Paroxyna producta* (Loew) — C

Sechium **P. Browne (Cucurbitaceae)**

Sechium edule (Jacq.) Sw. — chayote
- *Bactrocera (Zeugodacus) cucurbitae* (Coquillett) — F
- *Dacus (Dacus) bivittatus* (Bigot) — F
- *Dacus (Dacus) demmerezi* (Bezzi) — F
- *Dacus (Dacus) punctatifrons* Karsch ?
- *Dacus (Didacus) ciliatus* Loew — F

Sesbania **Scop. (Fabaceae=Leguminosae)**

Sesbania grandiflora (L.) Poiret
- *Bactrocera (Bactrocera) dorsalis* (Hendel) ! — I
- *Bactrocera (Zeugodacus) cucurbitae* (Coquillett) ! — I

Sideritis **L. (Lamiaceae=Labiatae)**

Sideritis scardica Griseb. — Alibotush tea
- *Aciura coryli* (Rossi) — I

[*Simarouba glauca* - *Quassia simarouba*]

Smyrnium **L. (Apiaceae=Umbelliferae)**

Smyrnium olusatrum L. — alexanders
- *Euleia heraclei* (Linnaeus) — L

Solanum **L. (Solanaceae)**

Solanum carolinense L. — horsenettle
- *Zonosemata electa* (Say) — F

Solanum ferox L.
- *Bactrocera (Bactrocera)* sp. near *B. dorsalis* (A) — F
- *Bactrocera (Bactrocera)* sp. near *B. dorsalis* (B) — F

Solanum incanum L.
- *Bactrocera (Bactrocera) latifrons* (Hendel) — F
- *Ceratitis (Ceratitis) capitata* (Wiedemann) — F

Solanum laciniatum Aiton — kangaroo apple
- *Bactrocera (Bactrocera) neohumeralis* (Hardy) — F
- *Bactrocera (Bactrocera) tryoni* (Froggatt) — F

[*Solanum lycopersicon* - see *Lycopersicon esculentum*]

Solanum melongena L. — eggplant, aubergine
- *Bactrocera (Bactrocera)* sp. near *B. dorsalis* (D) ?
- *Bactrocera (Bactrocera) latifrons* (Hendel) — F
- *Bactrocera (Bactrocera) passiflorae* (Froggatt) ! — F
- *Bactrocera (Bactrocera) zonata* (Saunders) ?
- *Bactrocera (Zeugodacus) cucurbitae* (Coquillett) ?
- *Ceratitis (Ceratitis) capitata* (Wiedemann) ?
- *Trirhithromyia cyanescens* (Bezzi) — F
- *Zonosemata electa* (Say) — F

Solanum muricatum Aiton	pepino	
Rhagoletis nova (Schiner)		F
Solanum nigrum L.	black nightshade	
Bactrocera (Bactrocera) latifrons (Hendel)		F
Ceratitis (Ceratalaspis) aliena (Bezzi)		F
Ceratitis (Ceratitis) capitata (Wiedemann)		F
Rhagoletis conversa (Brèthes)		F
Trirhithromyia cyanescens (Bezzi)		F
Solanum nodiflorum Jacq.		
Ceratitis (Ceratalaspis) turneri Munro		F
Solanum pseudocapsicum L.	Jerusalem cherry	
Ceratitis (Ceratitis) capitata (Wiedemann)		F
Solanum quitoense Lam.	naranjilla, lulo	
Anastrepha fraterculus (Wiedemann)		F
Solanum seaforthianum Andrews		
Bactrocera (Bactrocera) dorsalis (Hendel)		F
Bactrocera (Bactrocera) neohumeralis (Hardy)		F
Bactrocera (Bactrocera) tryoni (Froggatt)		F
Solanum sisymbriifolium Lam.		
Bactrocera (Bactrocera) latifrons (Hendel)		F
Solanum torvum Sw.	terongan	
Bactrocera (Bactrocera) sp. near *B. dorsalis* (B)		F
Bactrocera (Bactrocera) latifrons (Hendel)		F
Bactrocera (Bactrocera) nigrovittata Drew		F
Solanum tuberosum L.	potato	
Oedicarena latifrons (Wulp) ?		
Rhagoletis psalida Hendel		F

Sorbus L. (Rosaceae)

Sorbus aucuparia L.	rowan	
Anomoia purmunda (Harris) ?		
Rhagoletis pomonella (Walsh) ?		

Sorghum Moench (Poaceae=Gramineae)

Sorghum bicolor (L.) Moench.	grain sorghum	
Bactrocera (Zeugodacus) caudata (Fabricius) ??		

Spondias L. (Anacardiaceae)

Spondias cytherea Sonn.	jew plum	
Anastrepha fraterculus (Wiedemann)		F
Anastrepha obliqua (Macquart)		F
Anastrepha suspensa (Loew)		F
Bactrocera (Afrodacus) jarvisi (Tryon)		F
Bactrocera (Bactrocera) aquilonis (May)		F
Bactrocera (Bactrocera) sp. near *B. dorsalis* (B)		F
Bactrocera (Bactrocera) neohumeralis (Hardy)		F
Bactrocera (Bactrocera) tryoni (Froggatt)		F
Ceratitis (Ceratitis) capitata (Wiedemann)		F
Ceratitis (Ceratitis) catoirii Guérin-Méneville ??		
[*Spondias dulcis* - see *S. cytherea*]		

Spondias mombin L.	hog-plum	
Anastrepha antunesi Lima		F
Anastrepha bahiensis Lima ?		
Anastrepha fraterculus (Wiedemann)		F
Anastrepha obliqua (Macquart)		F
Anastrepha perdita Stone		F
Anastrepha serpentina (Wiedemann)		F
Anastrepha striata Schiner		F
Anastrepha suspensa (Loew)		F
Bactrocera (Bactrocera) dorsalis (Hendel)		F
Ceratitis (Ceratitis) capitata (Wiedemann) ?		
Spondias purpurea L.	red mombin	
Anastrepha antunesi Lima		F
Anastrepha fraterculus (Wiedemann)		F
Anastrepha ludens (Loew)		F
Anastrepha obliqua (Macquart)		F
Anastrepha serpentina (Wiedemann)		F
Anastrepha striata Schiner		F
Bactrocera (Bactrocera) dorsalis (Hendel)		F
Ceratitis (Ceratitis) capitata (Wiedemann)		F

Sterculia L. (Sterculiaceae)

Sterculia apetala (Jacq.) Karsten	Panama tree	
Anastrepha bezzii Lima		F

Strychnos L. (Loganiaceae)

Strychnos nux-vomica L.	strychnine tree	
Bactrocera (Zeugodacus) cucurbitae (Coquillett) ?		

Symphoricarpos Duhamel (Caprifoliaceae)

Symphoricarpos alba (L.) S.F. Blake	snowberry	
Rhagoletis zephyria Snow		F
Symphoricarpos racemosa Michx.	snowberry	
Rhagoletis zephyria Snow		F

Synsepalum (A. DC.) Daniell (Sapotaceae)

Synsepalum dulcificum (Schumacher & Thonn.) Daniell	miraculous berry	
Anastrepha suspensa (Loew)		F

Syzygium Gaertner (Myrtaceae)

Syzygium aqueum (Burman f.) Alston	watery rose-apple	
Bactrocera (Bactrocera) albistrigata (de Meijere)		F
Bactrocera (Bactrocera) aquilonis (May)		F
Bactrocera (Bactrocera) dorsalis (Hendel)		F
Bactrocera (Bactrocera) sp. near *B. dorsalis* (A)		F
Bactrocera (Bactrocera) sp. near *B. dorsalis* (B)		F
Bactrocera (Bactrocera) neohumeralis (Hardy)		F
Bactrocera (Bactrocera) tryoni (Froggatt)		F
Bactrocera (Zeugodacus) cucurbitae (Coquillett) ?		
Ceratitis (Ceratitis) catoirii Guérin-Méneville		F

Ceratitis (Pterandrus) rosa Karsch		F
Syzygium cumini (L.) Skeels	Java plum	
Anastrepha suspensa (Loew)		F
Bactrocera (Hemigymnodacus) diversa (Coquillett) ??		
Ceratitis (Ceratitis) capitata (Wiedemann)		F
Ceratitis (Ceratitis) catoirii Guérin-Méneville ??		
Ceratitis (Pterandrus) rosa Karsch		F
Syzygium jambos (L.) Alston	rose-apple	
Anastrepha fraterculus (Wiedemann)		F
Anastrepha ludens (Loew)		F
Anastrepha obliqua (Macquart)		F
Anastrepha striata Schiner		F
Anastrepha suspensa (Loew)		F
Bactrocera (Bactrocera) aquilonis (May)		F
Bactrocera (Bactrocera) correcta (Bezzi)		F
Bactrocera (Bactrocera) dorsalis (Hendel)		F
Bactrocera (Bactrocera) sp. near *B. dorsalis* (A)		F
Bactrocera (Bactrocera) sp. near *B. dorsalis* (B)		F
Bactrocera (Bactrocera) facialis (Coquillett)		F
Bactrocera (Bactrocera) frauenfeldi (Schiner) ?		
Bactrocera (Bactrocera) kirki (Froggatt)		F
Bactrocera (Bactrocera) ochrosiae (Malloch) ?		
Bactrocera (Bactrocera) tryoni (Froggatt)		F
Ceratitis (Ceratitis) capitata (Wiedemann)		F
Ceratitis (Pterandrus) flexuosa (Walker) ?		
Ceratitis (Pterandrus) rosa Karsch		F
Syzygium malaccense (L.) Merr. & Perry	Malay-apple	
Anastrepha fraterculus (Wiedemann)		F
Anastrepha obliqua (Macquart)		F
Anastrepha striata Schiner		F
Anastrepha suspensa (Loew)		F
Bactrocera (Afrodacus) jarvisi (Tryon)		F
Bactrocera (Bactrocera) albistrigata (de Meijere)		F
Bactrocera (Bactrocera) aquilonis (May)		F
Bactrocera (Bactrocera) dorsalis (Hendel)		F
Bactrocera (Bactrocera) sp. near *B. dorsalis* (A)		F
Bactrocera (Bactrocera) sp. near *B. dorsalis* (B)		F
Bactrocera (Bactrocera) sp. near *B. dorsalis* (C)		F
Bactrocera (Bactrocera) facialis (Coquillett) ?		
Bactrocera (Bactrocera) frauenfeldi (Schiner)		F
Bactrocera (Bactrocera) kirki (Froggatt)		F
Bactrocera (Bactrocera) obliqua (Malloch)		F
Bactrocera (Bactrocera) passiflorae (Froggatt) ?		
Bactrocera (Bactrocera) trilineola Drew ?		
Bactrocera (Zeugodacus) tau (Walker)		F
Ceratitis (Ceratitis) capitata (Wiedemann)		F
Ceratitis (Pterandrus) rosa Karsch		F

Syzygium samarangense (Blume) Merr. & Perry — water apple
- *Anastrepha suspensa* (Loew) — F
- *Bactrocera (Bactrocera) albistrigata* (de Meijere) — F
- *Bactrocera (Bactrocera) dorsalis* (Hendel) — F
- *Bactrocera (Bactrocera)* sp. near *B. dorsalis* (A) — F
- *Bactrocera (Zeugodacus) caudata* (Fabricius) ??
- *Bactrocera (Zeugodacus) cucurbitae* (Coquillett) ! — F
- *Bactrocera (Zeugodacus) tau* (Walker) ?
- *Ceratitis (Ceratitis) capitata* (Wiedemann) — F

Tagetes L. (Asteraceae=Compositae)

Tagetes erecta L. — marigold
- *Dioxyna picciola* (Bigot) — C

Taraxacum Wigg. (Asteraceae=Compositae)

Taraxacum koksaghyz Rodin — Russian dandelion
- *Ensina sonchi* (Linnaeus) — C
- *Paroxyna producta* (Loew) — C
- *Trupanea stellata* (Fuessly) — C

Telfairea Hook. (Cucurbitaceae)

Telfairea pedata (Smith & Sims) Hook. — oysternut
- *Dacus (Dacus) bivittatus* (Bigot) ! — F
- *Dacus (Dacus) telfaireae* (Bezzi) — F

Telosma Cov. (Asclepiadaceae)

Telosma cordata (Burm. f.) Merr. — Chinese violet
- *Dacus (Callantra) sphaeroidalis* (Bezzi) — F

Terminalia L. (Combretaceae)

Terminalia catappa L. — tropical almond
- *Anastrepha bezzii* Lima ?
- *Anastrepha fraterculus* (Wiedemann) — F
- *Anastrepha striata* Schiner — F
- *Anastrepha suspensa* (Loew) — F
- *Bactrocera (Afrodacus) jarvisi* (Tryon) — F
- *Bactrocera (Bactrocera) albistrigata* (de Meijere) — F
- *Bactrocera (Bactrocera) aquilonis* (May) — F
- *Bactrocera (Bactrocera) correcta* (Bezzi) — F
- *Bactrocera (Bactrocera) dorsalis* (Hendel) — F
- *Bactrocera (Bactrocera)* sp. near *B. dorsalis* (A) — F
- *Bactrocera (Bactrocera) facialis* (Coquillett) — F
- *Bactrocera (Bactrocera) frauenfeldi* (Schiner) — F
- *Bactrocera (Bactrocera) kirki* (Froggatt) — F
- *Bactrocera (Bactrocera) neohumeralis* (Hardy) — F
- *Bactrocera (Bactrocera) ochrosiae* (Malloch) ?
- *Bactrocera (Bactrocera) tryoni* (Froggatt) — F
- *Bactrocera (Bactrocera) zonata* (Saunders) — F
- *Ceratitis (Pterandrus) anonae* Graham — F
- *Ceratitis (Ceratitis) capitata* (Wiedemann) — F

Ceratitis (Ceratitis) catoirii Guérin-Méneville		F
Ceratitis (Pterandrus) rosa Karsch		F
Terminalia chebula (Gaertner) Roxb.	myrobalan nut	
Ceratitis (Ceratitis) capitata (Wiedemann) ?		

[*Thea sinensis* - see *Camellia sinensis*]

Theobroma L. (Sterculiaceae)

Theobroma cacao L.	cocoa	
Anastrepha fraterculus (Wiedemann)		F
Bactrocera (Bactrocera) passiflorae (Froggatt)		F
Ceratitis (Ceratitis) capitata (Wiedemann) ?		
Ceratitis (Pardalaspis) punctata (Wiedemann)		F
Ceratitis (Pterandrus) anonae Graham ??		
Ceratitis (Pterandrus) rosa Karsch		F
Trirhithrum nigrum Graham ??		

Thevetia L. (Apocynaceae)

Thevetia peruviana (Pers.) Schumann	lucky nut	
Bactrocera (Bactrocera) kraussi (Hardy) ?		
Ceratitis (Ceratitis) capitata (Wiedemann)		F

Tithonia Desf. ex A.L. Juss. (Asteraceae=Compositae)

Tithonia diversifolia (Hemsley) A.Gray	Mexican sunflower	
Spathulina acroleuca (Schiner)		C

Tragopogon L. (Asteraceae=Compositae)

Tragopogon pratensis L.	meadow salsify	
Orellia falcata (Scopoli)		R/S

Trichosanthes L. (Cucurbitaceae)

[*Trichosanthes anguina* - see *T. cucumerina*]

Trichosanthes cucumerina L.	snakegourd	
Bactrocera (Austrodacus) cucumis (French)		F
Bactrocera (Bactrocera) latifrons (Hendel) ??		
Bactrocera (Zeugodacus) caudata (Fabricius) ?		
Bactrocera (Zeugodacus) cucurbitae (Coquillett)		F
Bactrocera (Zeugodacus) tau (Walker) ?		
Dacus (Callantra) axanus (Hering)		F
Dacus (Callantra) eumenoides (Bezzi)		F
Dacus (Callantra) petioliforma (May)		F
Dacus (Dacus) demmerezi (Bezzi)		F
Dacus (Didacus) ciliatus Loew		F
Trichosanthes cucumeroides Maxim.		
Bactrocera (Zeugodacus) scutellata (Hendel)		X
Trichosanthes dioica Roxb.	pointed gourd	
Bactrocera (Zeugodacus) cucurbitae (Coquillett) ?		

Triphasia Lour. (Rutaceae)

Triphasia trifolia (Burman f.) P. Wilson — limeberry
- *Anastrepha suspensa* (Loew) — F
- *Bactrocera (Bactrocera) dorsalis* (Hendel) — F
- *Bactrocera (Zeugodacus) cucurbitae* (Coquillett) ! — F

Vaccinium L. (Ericaceae)

Vaccinium angustifolium Aiton — lowbush blueberry
- *Rhagoletis mendax* Curran — F
- *Rhagoletis pomonella* ??

Vaccinium corymbosum L. — highbush blueberry
- *Rhagoletis mendax* Curran — F
- *Rhagoletis pomonella* (Walsh) ??

Vaccinium macrocarpon Aiton — cranberry
- *Rhagoletis mendax* Curran ! — F
- *Rhagoletis pomonella* (Walsh) ??
- *Rhagoletis tabellaria* (Fitch) — F

Vaccinium myrtilloides Michaux — sourtop blueberry
- *Rhagoletis mendax* Curran — F

Vaccinium myrtillus L. — bilberry, whortleberry
- *Rhagoletis cerasi* (Linnaeus) ??

Vaccinium vitis-idaea L. — mountain cranberry
- *Rhagoletis mendax* Curran — F
- *Rhagoletis pomonella* (Walsh) ??

Vanda Jones ex R. Br. (Orchidaceae)

Vanda teres Lind. x *V. hookerana* Reichb. f. — Miss Joaquim orchid
- *Bactrocera (Bactrocera) dorsalis* (Hendel) ! — I

Vigna Savi (Fabaceae=Leguminosae)

Vigna radiata (L.) R. Wilczek — mung bean
- *Bactrocera (Zeugodacus) cucurbitae* (Coquillett) ?

Vigna unguiculata (L.) Walp. — southern pea
- *Bactrocera (Zeugodacus) cucurbitae* (Coquillett) ! — F

Vincetoxicum Wolf (Asclepiadaceae)

Vincetoxicum hirudinaria Medikus — white swallow-wort
- *Euphranta connexa* (Fabricius) — F

Vitellaria Gaertner f. (Sapotaceae)

Vitellaria paradoxa Gaertner f. — shea butter
- *Ceratitis (Ceratalaspis) silvestrii* Bezzi ?

Vitis L. (Vitaceae)

Vitis labrusca L. — fox grape
- *Bactrocera (Bactrocera) neohumeralis* (Hardy) — F
- *Bactrocera (Bactrocera) tryoni* (Froggatt) — F

Vitis vinifera L. — wine grape
- *Anastrepha fraterculus* (Wiedemann) — F
- *Bactrocera (Bactrocera) tryoni* (Froggatt) — F
- *Ceratitis (Ceratitis) capitata* (Wiedemann) — F
- *Ceratitis (Pterandrus) rosa* Karsch — F

Zinnia **L. (Asteraceae=Compositae)**

Zinnia elegans Jacq.	zinnia	
Craspedoxantha marginalis (Wiedemann)		C

Ziziphus **Miller (Rhamnaceae)**

Ziziphus jujuba Miller	common jujube	
Bactrocera (Bactrocera) correcta (Bezzi)		F
Bactrocera (Bactrocera) sp. near *B. dorsalis* (A)		F
Bactrocera (Bactrocera) zonata (Saunders) ?		
Carpomya incompleta (Becker)		F
Carpomya vesuviana Costa		F
Ceratitis (Ceratitis) catoirii Guérin-Méneville		F
Ceratitis (Pterandrus) rosa Karsch		F
Ziziphus lotus (L.) Lam.	lotus [of ancients]	
Carpomya incompleta (Becker)		F
Ziziphus mauritiana Lam.	Indian jujube	
Bactrocera (Bactrocera) aquilonis (May)		F
Bactrocera (Bactrocera) sp. near *B. dorsalis* (B)		F
Bactrocera (Bactrocera) tryoni (Froggatt)		F
Carpomya vesuviana Costa		F
Ceratitis (Ceratitis) capitata (Wiedemann) ?		
Ziziphus nummularia (Burm. f.) Wight & Arn.		
Carpomya vesuviana Costa		F
Ziziphus sp.		
Carpomya zizyphae Agarwal & Kapoor		F

[*Zizyphus* - see *Ziziphus*]

Common Names of Host Plants

The following list of common or vernacular plant names is intended to permit cross reference with the list of "Useful Plants and Their Associated Tephritids" (p. 433). Major sources of common names were Barlow et al. (1991), Hedrick (1972), Litsinger *et al.* (1991), Norrbom & Kim (1988b), Mabberley (1987) and Terrell *et al.* (1986).

Aak - *Calotropis procera*
aba - *Brassica juncea*
abacate - *Persea americana*
abata cola - *Cola acuminata*
abieiro - *Pouteria caimito*
abio - *P. campechiana*
abio or abiu - *P. caimito*
abobora - *Cucurbita pepo*
abrico-do-Para - *Mammea americana*
abricot - *Prunus armeniaca*
acerola - *Malpighia glabra*
ache - *Apium graveolens*
agatti-keerai - *Sesbania grandiflora*
ageratum - *Ageratum houstonianum*
aguacate or ahuacate - *Persea americana*
akee - *Blighia sapida*
alexanders - *Smyrnium olusatrum*
alicastrum snakewood - *Brosimum alicastrum*
alkekengi - *Physalis peruviana*
allspice - *Pimenta dioica*
almendra or almendron - *Terminalia catappa*
almond - *Prunus dulcis*
almond, country, Indian or tropical - *Terminalia catappa*
amanaku-maram - *Ricinus communis*
amatingula or amatungula - *Carissa macrocarpa*
amba - *Mangifera indica*
ambarella - *Spondias cytherea*
ameixa - *Prunus domestica*
ameixa Japonesa - *Eriobotrya japonica*
anakoya-pallam - *Persea americana*
angelica - *Angelica archangelica*
anggur - *Vitis vinifera*
annasi - *Ananas comosus*
annona - *Annona reticulata*
anoda - *A. reticulata* or *A. squamosa*
anon - *A. reticulata* or *A. squamosa*
anon de puerco - *A. glabra*
anona - *A. squamosa*
anone - *A. reticulata*
apple - *Malus domestica*
apple, alligator - *Annona glabra*
apple, balsam - *Momordica balsamina* or *M. charantia*
apple, beef - *Manilkara zapota*
apple, bell - *Passiflora laurifolia*
apple, bitter - *Citrullus colocynthis*
apple, cane - *Arbutus unedo*
apple, cashew - *Anacardium occidentale*
apple, common - *Malus domestica*
apple, Curaçao - *Syzygium samarangense*
apple, custard - *Annona cherimola*, *A. reticulata* or *A. squamosa*
apple, gold - *Lycopersicon esculentum*
apple, golden - *Spondias cytherea*
apple, golden - *Aegle marmelos*
apple, Jamaica - *Annona cherimola*
apple, Java - *Syzygium samarangense*
apple, Jew's - *Solanum melongena*

apple, kangaroo - *S. laciniatum*
apple, kei - *Dovyalis caffra*
apple, love - *Lycopersicon esculentum*
apple, mad - *Solanum melongena*
apple, Malay - *Syzygium malaccense*
apple, mammee or mammy - *Mammea americana*
apple, Mexican - *Casimiroa edulis*
apple, mountain or Otaheite - *Syzygium malaccense*
apple, paradise - *Malus pumila*
apple, pond - *Annona glabra*
apple, prickly custard - *A. muricata*
apple, rose - *Syzygium malaccense* or *S. jambos*
apple, Sodom - *Calotropis procera*
apple, star - *Chrysophyllum cainito*
apple, sugar - *A. reticulata* or *A. squamosa*
apple, velvet - *Diospyros blancoi*
apple, water - *Syzygium samarangense*
apple, watery rose - *S. aqueum*
apple, wax - *S. samarangense*
apple, white star - *Chrysophyllum albidum*
apple, wild - *Malus sylvestris*
apple, wild custard - *Annona senegalensis*
apple, wild Transvaal - *A. senegalensis*
apple, wonder - *Momordica balsamina*
apricot - *Prunus armeniaca*
apricot, St Domingo or South American - *Mammea americana*
apu - *Anacardium occidentale*
ara - *Ficus carica*
araca or araxa - *Psidium guineense*
areca nut - *Areca catechu*
argus pheasant tree - *Dracontomelon dao*
arbute - *Arbutus unedo*
artichoke, globe - *Cynara scolymus*
artichoke, Jerusalem - *Helianthus tuberosus*
asam kandis - *Garcinia xanthochymus*
ash, mountain - *Sorbus aucuparia*
asparagus, garden - *Asparagus officinalis*
ata - *Annona squamosa*
ati - *Calophyllum inophyllum*
aubergine - *Solanum melongena*
auyama - *Cucurbita moschata*
avocado, avocat or avoka - *Persea americana*

Ba-tala - *Ipomoea batatas*
bacang or bachang - *Mangifera foetida*
bachelor's-buttons - *Centaurea cyanus*
badamier - *Terminalia catappa*
badinjan - *Solanum melongena*
bael or baeli - *Aegle marmelos*
bagas de cafe - *Coffea arabica*
bakupari - *Rheedia braziliensis*
ball tree - *Aegle marmelos*
bambalinas - *Citrus maxima*
bamboo, giant - *Dendrocalamus giganteus*
bamboo, male - *D. strictus*
bamboo, puntingpole - *Bambusa tuldoides*
banana, dwarf - *Musa acuminata*
banana or banane - *M. x paradisiaca*
bandak-kai, bandakai or bandakka - *Abelmoschus esculentus*
barberry - *Berberis vulgaris*
barmi-ah - *Abelmoschus esculentus*
bean, asparagus - *Vigna unguiculata*
bean, baked - *Phaseolus vulgaris*
bean, bonavist - *Lablab purpureus*
bean, Burma, butter, duffin, Lima or Rangoon - *Phaseolus lunatus*
bean, dwarf, flageolet, French, garden, green, haricot - *P. vulgaris*
bean, guada - *Trichosanthes cucumerina*
bean, hyacinth or lablab - *Lablab purpureus*
bean, ice cream - *Inga edulis*
bean, mung - *Vigna radiata*
bean, yard-long - *V. unguiculata*
bedara - *Ziziphus mauritiana*
bela tree - *Aegle marmelos*
belimbing manis - *Averrhoa carambola*
belladonna - *Atropa bella-donna*
ber - *Ziziphus mauritiana*
berberis - *Berberis vulgaris*
berry, Andes - *Rubus glaucus*
berry, California - *R. ursinus*
berry, miraculous - *Synsepalum dulcificum*
berry, Pacific - *Rubus ursinus*
berry, Panama - *Muntingia calabura*
berry, partridge or tea - *Gaultheria procumbens*
betel nut - *Areca catechu*
betik - *Carica papaya*
bibace - *Eriobotrya japonica*
bidara - *Ziziphus mauritiana*
bigarade - *Citrus aurantium*
bilak - *Aegle marmelos*
bilberry - *Vaccinium myrtillus*
bilimbi, bilimbing, bilimbikal or biling - *Averrhoa bilimbi*
biribá - *Rollinia pulchrinervis*
bissy nut - *Cola acuminata*
black-til - *Guizotia abyssinica*
blackberry - *Rubus fruticosa*
blackberry, western - *R. ursinus*
bladderseed - *Levisticum officinale*
blaeberry - *Vaccinium myrtillus*
blimbing - *Averrhoa bilimbi*

blood flower - *Asclepias curasavica*
blueberry, high or highbush - *Vaccinium corymbosum*
blueberry, late sweet or lowbush - *V. angustifolium*
blueberry, sourtop - *V. myrtilloides*
blueberry, swamp - *V. corymbosum*
blueberry, sweet - *V. angustifolium*
blueberry, velvetleaf - *V. myrtilloides*
bluebottle - *Centaurea cyanus*
boysenberry - *Rubus loganobaccus*
box, China - *Murraya paniculata*
box, marmelade - *Genipa americana*
bramble - *Rubus fruticosa*
breadfruit *Artocarpus altilis*
breadnut - *Brosimum alicastrum*
brède martin - *Solanum nigrum*
bringellier - *S. auriculatum*
broccoli - *Brassica oleracea*
broomcorn - *Sorghum bicolor*
buah mentega - *Diospyros blancoi* or *Persea americana*
buah perian - *Artocarpus rigidus*
buah susu - *Passiflora laurifolia*
buap-kom - *Luffa aegyptiaca*
buap liem - *L. acutangula*
buap ugu - *Trichosanthes cucumerina*
buckthorn, false - *Bumelia languinosa*
buckthorn, sea - *Hippophae rhamnoides*
buko - *Gnetum gnemon*
bullock's heart - *Annona reticulata*
bully tree - *Manilkara zapota*
bumblekites - *Rubus fruticosa*
bumelia, gum - *Bumelia languinosa*
bunchberry - *Cornus canadensis*
burdock - *Arctium lappa*
bush pais - *Myristica fragrans*
butter, shea - *Vitellaria paradoxa*
butternut - *Juglans cinerea*
butterseed or buttertree - *Vitellaria paradoxa*

Cabai-acong - *Capsicum annuum*
cabbage tree - *Cussonia* sp.
cabbage - *Brassica oleracea*
cacaoyer - *Theobroma cacao*
caco - *Diospyros kaki*
cactus, leafy - *Pereskia aculeata*
cactus, spineless - *Opuntia ficus-indica*
café - *Coffea arabica*
caimito - *Chrysophyllum cainito*
caja - *Spondias cytherea* or *S. purpurea*
caju or cajugaha - *Anacardium occidentale*
calabash - *Lagenaria siceraria*
calabaza - *Cucurbita pepo*
calabur or calabura - *Muntingia calabura*
calamondin - *x Citrofortunella x mitis*
calebasse - *Lagenaria siceraria*
camansi - *Artocarpus altilis*
camasa - *Lagenaria siceraria*
camellia, common - *Camellia japonica*
camias - *Averrhoa bilimbi*
camoruco - *Sterculia apetala*
canistel - *Pouteria campechiana*
cantaloupe - *Cucumis melo*
caper or caperbush - *Capparis spinosa*
capulin - *Muntingia calabura* or *Prunus serotina*
caqui - *Diospyros kaki*
caramba or carambola - *Averrhoa carambola*
caranda - *Carissa carandas*
caraunda - *C. macrocarpa*
carilla fruit - *Momordica charantia*
carissa, Egyptian - *Carissa edulis*
carnel, dwarf - *Cornus canadensis*
carrot - *Daucus carota*
cas - *Psidium friedrichsthalianum*
cashew - *Anacardium occidentale*
caspirol - *Inga ruiziana*
cassava - *Manihot esculenta*
cassia, purging - *Cassia fistula*
cassina - *Ilex cassine*
castor oil plant or castorbean - *Ricinus communis*
catechu - *Areca catechu*
cat's eyes - *Dimocarpus longan*
catjang - *Vigna unguiculata*
catoche - *Annona muricata*
cauliflower - *Brassica oleracea*
cedar - *Juniperus virginiana*
ceku - *Manilkara zapota*
celery - *Apium graveolens*
celtuce - *Lactuca sativa*
cempedak - *Artocarpus integer*
cerezo - *Malpighia glabra*
cerise - *Eugenia uniflora*
cermai - *Phyllanthus acidus*
cermai belanda - *Eugenia uniflora*
cha-e or chai - *Camellia sinensis*
cham-pa-da - *Artocarpus integer*
chapote - *Diospyros texana*
chayote - *Sechium edule*
checkerberry - *Gaultheria procumbens*
chempedak - *Artocarpus integer*
cherimalla, cherimola, cherimoya or cherimuyu - *Annona cherimola*
chermai - *Phyllanthus acidus*
cherry, amarelle - *Prunus cerasus*
cherry, American - *P. serotina*

cherry, Barbados - *Malpighia glabra*
cherry, bird - *Prunus avium* or *P. pensylvanica*
cherry, black - *P. serotina*
cherry, Brazil - *Eugenia brasiliensis* or *E. uniflora*
cherry, Cayenne - *E. uniflora*
cherry, fire - *Prunus pensylvanica*
cherry, Jamaica - *Muntingia calabura*
cherry, Jamaican - *Malpighia glabra*
cherry, Jerusalem - *Solanum pseudocapsicum*
cherry, mahaleb - *Prunus mahaleb*
cherry, maraschino or morello - *P. cerasus*
cherry, Peruvian - *Physalis peruviana*
cherry, pie - *Prunus cerasus*
cherry, pin - *P. pensylvanica*
cherry, Puerto Rican - *Malpighia glabra*
cherry, purple winter - *Physalis philadelphica*
cherry, Rio Grande - *Eugenia aggregata*
cherry, rum - *Prunus serotina*
cherry, sour - *P. cerasus*
cherry, Spanish - *Eugenia brasiliensis* or *Mimusops elengi*
cherry, Surinam - *E. uniflora*
cherry, sweet - *Prunus avium*
cherry, West Indian - *Malpighia glabra*
cherry, wild - *Prunus avium*
cherry, wild red - *P. pensylvanica*
cherry, winter - *Physalis peruviana* or *Solanum pseudocapsicum*
chervil - *Anthriscus cerifolium*
chestnut, Japanese - *Castanea crenata*
chestnut, O'taheite or Tahiti - *Inocarpus fagifer*
chicle, chico, chico-zapote, chiku or chikku - *Manilkara zapota*
Chinese lime - *Triphasia trifolia*
cho-cho - *Sechium edule*
chocolath-gas - *Theobroma cacao*
chokecherry - *Prunus virginiana*
cholam - *Sorghum bicolor*
chom-phu-pa - *Syzygium aqueum*
chom-phu-sa-raek - *S. malaccense*
chom-phy-khao - *S. samarangense*
chom pu nam - *S. jambos*
chouchou or christophine - *Sechium edule*
chrysanthemum - *Chrysanthemum* sp.
cidran - *Citrus medica*
ciku - *Manilkara zapota*
cinnamon, wild - *Canella winteriana*
ciruela - *Prunus salicina*
ciruela - *Spondias purpurea*
ciruela del gobernador - *Dovyalis hebecarpa*
ciruelo - *Prunus domestica*
citrange x *Citroncirus webberi*
citron - *Citrus medica*
citrouille - *Cucurbita pepo*
cluster tree - *Ficus racemosa*
coca - *Erythroxylum coca*
cochin-goraka - *Garcinia xanthochymus*
coco-maram or cocoa - *Theobroma cacao*
cocoplum - *Chrysobalanus icaco*
coffee, arabica - *Coffea arabica*
coffee, Congo - *C. canephora*
coffee, Liberian or Liberica - *C. liberica*
coffee, robusta - *C. canephora*
coing de Chine - *Diospyros kaki*
cola nut or tree - *Cola acuminata*
colocynth - *Citrullus colocynthis*
concombre - *Cucumis sativus*
corazon - *Annona reticulata*
coreopsis, plains - *Coreopsis tinctoria*
corn, Guinea- or kaffir - *Sorghum bicolor*
cornflower - *Centaurea cyanus*
corossol - *Annona muricata* or *A. reticulata*
cotoneaster - *Cotoneaster* sp.
cotton, upland - *Gossypium hirsutum*
cowberry - *Vaccinium vitis-idaea*
cowpea - *Vigna unguiculata*
crabapple - *Malus sylvestris*
crabapple, Siberian - *M. baccata*
cranberry - *Vaccinium vitis-idaea*
cranberry, American or large - *V. macrocarpon*
cranberry, mountain or rock - *V. vitis-idaea*
crookneck, Canada or winter - *Cucurbita moschata*
crown-flower, small - *Calotropis procera*
cuachilote - *Parmentiera aculeata*
cucumber - *Cucumis sativus*
cucumber, African horned - *C. metuliferus*
cucumber, bitter - *Momordica charantia*
cucumber, serpent - *Trichosanthes cucumerina*
cucumber, squirting - *Ecballium elaterium*
cucumber, sweet - *Solanum muricatum*
cucumber tree - *Averrhoa bilimbi*
currant, buffalo, golden or Missouri - *Ribes aureum*
currant, white - *R. rubrum*
cushaw - *Cucurbita moschata*

Dahlia - *Dahlia* sp.
daisy, Barberton - *Gerbera jamesonii*
daisy, dog or ox-eye - *Leucanthemum vulgare*
damson - *Prunus domestica*
dandelion, Russian - *Taraxicum koksaghyz*
dangleberry - *Gaylussacia frondosa*
dara-vetakola - *Luffa acutangula*
date, Chinese - *Ziziphus jujuba*
dawa - *Pometia pinnata*

dehi - *Citrus aurantiifolia*
delan - *Punica granatum*
delima - *P. granatum*
dewberry - *Rubus ursinus*
dhaisi-kai - *Citrus aurantiifolia*
diya-labu - *Lagenaria siceraria*
dogwood, flowering - *Cornus florida*
dogwood, red-osier - *C. stolonifera*
dogwood, roughleaf - *C. drummondii*
domba or dommakottai - *Calophyllum inophyllum*
duku - *Lansium domesticum*
duraznero or durazno - *Prunus persica*
durian - *Durio zibethinus*
durian belanda - *Annona muricata*
durra - *Sorghum bicolor*

Egg tree - *Garcinia xanthochymus*
egg-fruit tree - *Pouteria campechiana*
eggplant - *Solanum melongena*
egusi - *Cucumeropsis mannii*
emi - *Vitellaria paradoxa*
endaru-tel - *Ricinus communis*
erapillakai - *Artocarpus altilis*
eschalot - *Allium cepa*
et-pera - *Persea americana*

Fak kao - *Momordica cochinchinensis*
fak neo - *Cucurbita pepo*
fak tawng - *C. moschata*
farang - *Psidium guajava*
fao - *Ochrosia oppositifolia*
fau - *Hibiscus tiliaceus*
feaberry - *Ribes uva-crispa*
feijoa - *Feijoa sellowiana*
fekika - *Syzygium malaccense*
fekika papalangi - *S. jambos*
feterita - *Sorghum bicolor*
fiana - *Ananas comosus*
fig, Barbary - *Opuntia ficus-indica*
fig, Barberry - *O. vulgaris*
fig, cluster - *Ficus racemosa*
fig, common - *F. carica*
fig, Indian - *Opuntia ficus-indica* or *O. tuna*
fig, keg- *Diospyros kaki*
fig, weeping - *Ficus benjamini*
five corner - *Averrhoa carambola*
forbidden fruit - *Citrus maxima*
foxberry - *Vaccinium vitis-idaea*
frijoles - *Phaseolus vulgaris*

Gada-kaduru - *Strychnos nux-vomica*
gajas - *Anacardium occidentale*
gallito - *Sesbania grandiflora*
gari - *Manihot esculenta*
garnetberry - *Ribes rubrum*
gas-takkali - *Cyphomandra crassicaulis*
gean - *Prunus avium*
genip, genipap or genipapeiro - *Genipa americana*
gherkin - *Cucumis sativus*
ginseng, Siberian - *Eleutherococcus senticosus*
girasole - *Helianthus tuberosus*
gnemom tree - *Gnetum gnemon*
goatsbeard - *Tragopogon pratensis*
gobo, gobba or gobbo - *Abelmoschus esculentus*
god tree - *Bombax ceiba*
goiaba - *Psidium guajava*
golden fruit of the Andes - *Solanum quitoense*
golden-shower - *Cassia fistula*
gombo - *Abelmoschus esculentus*
goora or gooroo nut - *Cola acuminata*
gooseberry, Barbados - *Pereskia aculeata* or *Physalis peruviana*
gooseberry, Cape - *P. peruviana*
gooseberry, Ceylon - *Dovyalis hebecarpa*
gooseberry, Chinese - *Actinidia chinensis*
gooseberry, country - *Averrhoa carambola*
gooseberry, European - *Ribes uva-crispa*
gooseberry, Indian, Malay or star - *Phyllanthus acidus*
gooseberry tree, Otaheite - *P. acidus*
goraka-jambu - *Eugenia uniflora*
gourd - *Cucurbita pepo*
gourd, bitter - *Citrullus colocynthis* or *Momordica charantia*
gourd, bonnet - *Luffa aegyptiaca*
gourd, bottle or calabash - *Lagenaria siceraria*
gourd, Chinese fuzzy - *Benincasa hispida*
gourd, club - *Trichosanthes cucumerina*
gourd, dish-cloth - *Luffa aegyptiaca*
gourd, ivy - *Coccinia grandis*
gourd, leprosy - *Momordica charantia*
gourd, pointed - *Trichosanthes dioica*
gourd, scarlet-fruited - *Coccinia grandis*
gourd, sponge - *Luffa aegyptiaca*
gourd, trumpet - *Lagenaria siceraria*
gourd, wax or white - *Benincasa hispida*
gourd, white-flowered - *Lagenaria siceraria*
gourd, viper's - *Trichosanthes cucumerina*
goyave - *Psidium guajava*
goyavier - *P. littorale*
gram - *Vigna radiata*
granada - *Punica granatum*
granadilla - *Passiflora ligularis* or *P. quadrangularis*
granadilla, giant - *P. quadrangularis*

granadilla, purple - *P. edulis*
granadilla, sweet - *P. ligularis*
granadilla, yellow - *P. laurifolia*
grape - *Vitis vinifera*
grape, fox - *V. labrusca*
grape, isabella or scunk - *V. labrusca*
grape, seaside - *Coccoloba uvifera*
grape tree, Brazilian - *Myrciaria cauliflora*
grape, wild - *Nitraria schoberi*
grape, wine - *Vitis vinifera*
grapefruit - *Citrus x paradisi*
graviola - *Annona muricata*
greens, mustard - *Brassica juncea*
groundcherry, Peruvian - *Physalis peruviana*
groundcherry, purple or tomatill - *P. philadelphica*
grumichama or grumixama - *Eugenia brasiliensis*
guanábana - *Annona muricata*
guava, apple - *Psidium guajava*
guava, Brazilian - *P. guineense*
guava, Calcutta, cattley, cherry or China - *P. littorale*
guava, common - *P. guajava*
guava, Costa Rican - *P. friedrichsthalianum*
guava, Guinea - *P. guineense*
guava, pear - *P. guajava*
guava, pineapple - *Feijoa sellowiana* or *Psidium littorale*
guava, purple or strawberry - *P. littorale*
guava, wild - *P. friedrichsthalianum*
guava, yellow - *P. guajava*
guavo real - *Inga spectabilis*
guayaba fresa - *Psidium littorale*
guayabo de agua - *P. friedrichsthalianum*
guayabo peruano or guayabo, falso - *Feijoa sellowiana*
guayabo sabanero - *Psidium littorale*
guineagrass - *Panicum maximum*
guisaro - *Psidium guineense*
gumbo - *Abelmoschus esculentus*
guyaba - *Psidium guajava*
guyaba de sabana - *P. guineense*
guyabo - *P. guajava*

Hackberry - *Celtis laevigata*
hagberry - *Prunus avium*
hawthorn - *Crataegus* sp.
hibiscus, sea - *Hibiscus tiliaceus*
holly, dahoon - *Ilex cassine*
honewort - *Cryptotaenia canadensis*
honeysuckle, Jamaica - *Passiflora laurifolia*
horsenettle - *Solanum carolinense*
hu kwang - *Terminalia catappa*
huckleberry, black - *Gaylussacia baccata*
huckleberry, dwarf - *G. dumosa*
huesito - *Malpighia glabra*
huevo vegetal - *Pouteria campechiana*
humming-bird, vegetable - *Sesbania grandiflora*

Ifi - *Inocarpus fagifer*
ilantai - *Ziziphus mauritiana*
imbe - *Garcinia livingstonei*
imbu - *Spondias tuberosa*
Indian root - *Asclepias curasavica*
inga or ingazeira - *Inga edulis*
inga-til - *Guizotia abyssinica*
injerto - *Pouteria viridis*
ipecac, false or ipecacuanha, bastard - *Asclepias curasavica*

Jaboticaba - *Myrciaria cauliflora*
jack or jackfruit - *Artocarpus heterophyllus*
jacote - *Spondias purpurea*
jak - *Artocarpus heterophyllus*
jam fruit - *Muntingia calabura*
jamalac - *Syzygium malaccense*
jamberry - *Physalis philadelphica*
jamblica - *Citrus maxima*
jamblong or jambolan - *Syzygium cumini*
jambola - *Citrus maxima*
jambool or jambu - *Syzygium jambos*
jambos - *S. jambos* or *S. malaccense*
jambosa - *S. samarangense*
jambu air - *S. aqueum*
jambu air rhio - *S. samarangense*
jambu batu - *Psidium guajava*
jambu bol - *Syzygium malaccense*
jambu mawar or jamrosat - *S. jambos*
jambu merah - *S. malaccense*
jambu - *S. malaccense*
jasmine or jessamine, orange - *Murraya paniculata*
jivat - *Syzygium cumini*
jobo - *Spondias mombin*
jobo de India - *S. cytherea*
Johnny-go-to-bed-at-noon - *Tragopogon pratensis*
juar - *Sorghum bicolor*
jujube, Chinese, common, or French - *Ziziphus jujuba*
jujube, Indian - *Z. mauritiana*
jumrool - *Syzygium samarangense*
juniper, one-seed - *Juniperus monosperma*
jushte - *Brosimum alicastrum*

Kacang bendi or kachieb - *Abelmoschus esculentus*
kaduga - *Brassica juncea*
kaki- *Diospyros kaki*
kalyana pushini - *Cucurbita moschata*
kam-foi - *Carthamus tinctorius*
kamani, false or winged - *Terminalia catappa*
kamani, ball - *Calophyllum inophyllum*
kamaranga - *Averrhoa carambola*
kamoti - *Chrysobalanus icaco*
kanchurai - *Strychnos nux-vomica*
kap-pe - *Coffea arabica*
karal-iringu - *Sorghum bicolor*
karanda - *Carissa carandas*
karawila - *Momordica charantia*
karri-kochika - *Capsicum annuum*
kating - *Calophyllum inophyllum*
katrikai - *Solanum melongena*
katu-anoda - *Annona muricata*
katuru-marunga or ke - *Sesbania grandiflora*
kedondong - *Spondias cytherea*
kehel - *Musa x paradisiaca*
kelengkeng - *Litchi chinensis*
kerenda - *Carissa carandas*
kerkup - *Flacourtia jangomas*
kerkup besar - *F. jangomas*
kerkup kecil, kerkup kocil or lesser - *F. indica*
kerkup siam - *Muntingia calabura*
kesumba - *Carthamus tinctorius*
ketapang - *Terminalia catappa*
ketembilla - *Dovyalis hebecarpa*
ketola - *Luffa acutangula*
kha-pun - *Artocarpus heterophyllus*
khrop-dong - *Flacourtia rukam*
kidanar-attankai - *Citrus limon*
kidney, navy or string bean - *Phaseolus vulgaris*
kino, Jamaican - *Coccoloba uvifera*
kitembilla - *Dovyalis hebecarpa*
kiwi - *Actinidia chinensis*
kluai - *Musa x paradisiaca*
kohlrabi - *Brassica oleracea*
koiya-pallam - *Psidium guajava*
kok-saghyz - *Taraxicum koksaghyz*
koko - *Theobroma cacao*
kola [Tonga] - *Citrus aurantium*
kola or kolanut - *Cola acuminata*
komadu - *Citrullus lanatus*
kopai or kopi - *Coffea arabica*
kos - *Artocarpus heterophyllus*
kotamba or kottai - *Terminalia catappa*
kuaka - *Psidium guajava*
kuchla - *Strychnos nux-vomica*
kumara - *Ipomoea batatas*
kumquat - *Fortunella margarita*
kumquat, meiwa - *F. x crassifolia*
kumquat, oval - *F. margarita*
kumquat, round - *F. japonica*
kurdee or kusuma - *Carthamus tinctorius*
kweme - *Telfairea pedata*

Labu air - *Lagenaria siceraria* or *Cucurbita pepo*
labu merah or labu parang - *C. moschata*
laburnum, Indian - *Cassia fistula*
lady's-fingers - *Abelmoschus esculentus*
laimi - *Citrus aurantiifolia*
lamut - *Mangifera foetida*
langsat or lansek - *Lansium domesticum*
laranja azeda - *Citrus aurantium*
laranje doce - *C. sinensis*
lasa - *Lansium domesticum*
laurel, Indian or laurelwood - *Calophyllum inophyllum*
lemandarin - *Citrus x limonia*
lemani or lemon - *C. limon*
lemon, yellow water - *Passiflora laurifolia*
lessi - *Carica papaya*
letchi - *Litchi chinensis*
lettuce, Canada wild - *Lactuca canadensis*
lettuce, garden - *L. sativa*
lettuce, tall - *L. canadensis*
lima chica or lima grande - *Citrus aurantiifolia*
lima Palestina - *C. limetta*
limao cravo - *C. x limonia*
limau asam - *C. aurantiifolia*
limau besar - *C. maxima*
limau kikir - *Triphasia trifolia*
limau kupas - *Citrus reticulata*
limau manis - *C. sinensis*
limau susu - *C. medica*
lime - *C. aurantiifolia*
lime, Australian desert - *Eremocitrus glauca*
lime, Mandarin - *Citrus x limonia*
lime, myrtle - *Triphasia trifolia*
lime, Rangpur - *Citrus x limonia*
lime, sweet - *C. limetta*
limeberry - *Triphasia trifolia*
limon - *Citrus aurantiifolia*
limoncillo - *Hippomane mancinella*
lin chi - *Litchi chinensis*
lingberry or lingen - *Vaccinium vitis-idaea*
linmangkon - *Passiflora edulis*
litchi - *Litchi chinensis*
loganberry - *Rubus loganobaccus*
longan or longyen - *Dimocarpus longan*
loof or loofah - *Luffa aegyptiaca*
loquat - *Eriobotrya japonica*

lote-tree, false - *Diospyros lotus*
lotus - *Ziziphus lotus*
lovage - *Levisticum officinale*
lucky nut - *Thevetia peruviana*
lucmo - *Pouteria obovata*
luffa, angled - *Luffa acutangula*
lulo - *Solanum quitoense*
lungan - *Dimocarpus longan*
lychee - *Litchi chinensis*
lychee, Pacific - *Pometia pinnata*

Ma-da-luang - *Garcinia xanthochymus*
ma-fai farang - *Baccaurea motleyana*
ma-fu'ang - *Averrhoa carambola*
ma-ha - *Syzygium cumini*
ma-khanat - *Ananas comosus*
ma-kok farang - *Spondias cytherea*
ma-la-ke - *Carica papaya*
ma-muang-himma-phan - *Anacardium occidentale*
ma-muang - *Mangifera indica*
ma-phut - *Garcinia dulcis*
ma-pun - *Psidium guajava*
ma-rit - *Diospyros blancoi*
ma-tan - *Ziziphus mauritiana*
ma-tum - *Aegle marmelos*
ma-yom-farang - *Eugenia uniflora*
ma-yom - *Phyllanthus acidus*
mabola or mabolo - *Diospyros blancoi*
maca - *Malus domestica*
macadamia nut - *Macadamia integrifolia*
macanchi - *Triphasia trifolia*
mace - *Myristica fragrans*
madalan-kai - *Punica granatum*
Madeira nut - *Juglans regia*
madrono or madruno - *Rheedia acuminata*
maha-dan - *Syzygium cumini*
maha-karamba - *Carissa carandas*
mahoe - *Hibiscus tiliaceus*
mahonia - *Mahonia aquifolium*
mahung - *Ricinus communis*
majagua - *Hibiscus tiliaceus*
mak - *Areca catechu*
malu-miris - *Capsicum annuum*
mamey - *Mammea americana* or *Pouteria sapota*
mamey de tierre - *P. sapota*
mamo tet - *Triphasia trifolia*
mamoncillo - *Genipa americana*
man sam parang - *Manihot esculenta*
man thet - *Ipomoea batatas*
manchineel - *Hippomane mancinella*
Mandarin - *Citrus reticulata*
mandioca - *Manihot esculenta*
mang-khut - *Garcinia mangostana*
manga or mangga - *Mangifera indica*
manggis - *Garcinia mangostana*
mango - *Mangifera indica*
mango, horse - *M. foetida*
mangosteen - *Garcinia mangostana*
mangosteen, African - *G. livingstonei*
mangue - *Mangifera indica*
mangus or mangus-kai - *Garcinia mangostana*
manioc or manyokka - *Manihot esculenta*
manzana rosa - *Syzygium jambos*
manzano - *Malus domestica*
manzinilla de playa - *Hippomane mancinella*
mara - *Momordica charantia*
maracuja - *Passiflora edulis* or *P. quadrangularis*
maracuja grande - *P. alata*
maracuja redondo - *P. edulis*
maracuya - *P. caerulea*
maranon - *Anacardium occidentale*
marapa - *Spondias mombin*
marasco - *Prunus cerasus*
maravali-kelangu - *Manihot esculenta*
margose - *Momordica charantia*
marguerite - *Leucanthemum vulgare*
marigold - *Tagetes erecta*
marigold, pot - *Calendula officinalis*
markesa - *Passiflora quadrangularis*
marmelade fruit - *Pouteria sapota*
marrow - *Cucurbita maxima* or *C. pepo*
marrow, Madeira - *Sechium edule*
masan - *Ziziphus mauritiana*
mata kucing - *Dimocarpus longan*
matac - *Asclepias curasavica*
matasano - *Casimiroa edulis* or *C. tetrameria*
mazzagua - *Sorghum bicolor*
mburucuya - *Passiflora caerulea*
medlar - *Mespilus germanica* or *Mimusops elengi*
medlar, Japanese - *Eriobotrya japonica*
mei - *Artocarpus altilis*
melancia - *Citrullus lanatus*
melindjo or melinjau - *Gnetum gnemon*
melocotonero - *Prunus persica*
melon - *Cucumis melo*
melon, Chinese winter - *Benincasa hispida*
melon tree - *Carica papaya*
membrillo - *Cydonia oblonga*
meninjau - *Gnetum gnemon*
mentegn, bush - *Diospyros blancoi*
menteng negeri - *Baccaurea motleyana*
merey - *Anacardium occidentale*
mexerica - *Citrus reticulata*
mikan - *C. reticulata*

milkweed, bloodflower - *Asclepias curasavica*
millet, great or Turkish, or milo - *Sorghum bicolor*
miracle fruit - *Gymnema sylvestre* or *Synsepalum dulcificum*
mkweme - *Telfairea pedata*
moli kai melie - *Citrus sinensis*
moli kalepi - *C. x paradisi*
moli peli - *C. reticulata*
moli Tonga - *C. maxima*
mombin - *Spondias purpurea*
mombin, red - *S. purpurea*
mombin, yellow - *S. mombin*
monkey-jack - *Artocarpus rigidus*
mora - *Rubus glaucus* or *Dimocarpus longan*
morango - *Fragaria vesca*
moz - *Musa x paradisiaca*
mukalai - *Mimusops elengi*
mulberry, black or common - *Morus nigra*
mulberry, white - *M. alba*
mundirimaram - *Anacardium occidentale*
mundu - *Garcinia xanthochymus*
mundur - *G. dulcis*
muskmelon - *Cucumis melo*
mustard - *Brassica juncea*
myrobalan nut - *Terminalia chebula*

Na-val - *Syzygium cumini*
nagka - *Artocarpus heterophyllus*
nam prom - *Carissa carandas*
nam tao - *Lagenaria siceraria*
nanas - *Ananas comosus*
nance - *Byrsonima crassifolia*
nani - *Ananas comosus*
naran-kai - *Citrus sinensis* or *C. reticulata*
naranja agria - *C. aurantium*
naranja dulce - *C. sinensis*
naranjilla - *Solanum quitoense*
naranjo dulce - *Citrus sinensis*
naseberry - *Manilkara zapota*
natran - *Citrus limon*
ngo - *Nephelium lappaceum*
Niger seed - *Guizotia abyssinica*
nightshade, black or common - *Solanum nigrum*
nightshade, deadly - *Atropa bella-donna*
nispero - *Eriobotrya japonica* or *Manilkara zapota*
nispero del Japon - *E. japonica*
nitre bush - *Nitraria schoberi*
niyan-vetakola - *Luffa aegyptiaca*
nogal - *Juglans neotropica* or *J. regia*
noi-nong - *Annona squamosa*
noinang - *A. reticulata*
nona - *A. glabra*
nona kapri - *A. reticulata*
nona seri kaya or srikaya - *A. squamosa*
noog - *Guizotia abyssinica*
nuez - *Juglans neotropica* or *Juglans regia*
nug - *Guizotia abyssinica*
nurai - *Dimocarpus longan*
nut, cashew - *Anacardium occidentale*
nutmeg - *Myristica fragrans*
nux-vomica tree - *Strychnos nux-vomica*

Ocra or ochro - *Abelmoschus esculentus*
ohia - *Syzygium malaccense*
okra - *Abelmoschus esculentus*
oleander, yellow - *Thevetia peruviana*
olive - *Olea europaea*
onion - *Allium cepa*
orange, bergamot or bitter - *Citrus aurantium*
orange, king - *C. x nobilis*
orange, Mandarin - *C. reticulata*
orange, marmelade - *C. aurantium*
orange, mock [of Hawaii] - *Murraya paniculata*
orange, navel - *Citrus sinensis*
orange, Otaheite - *C. x limonia*
orange, Panama - *x Citrofortunella x mitis*
orange, Satsuma - *Citrus reticulata*
orange, Seville or sour - *C. aurantium*
orange, sweet - *C. sinensis*
orange, temple - *C. x nobilis*
orange, trifoliate - *Poncirus trifoliata*
orange, Valencia - *Citrus sinensis*
orchid, Miss Joaquim - *Vanda teres x hookerana*
oysternut - *Telfairea pedata*

Paingani - *Solanum melongena*
pak-ku - *Areca catechu*
pakal - *Momordica charantia*
palm, African date - *Ziziphus lotus*
palm, date - *Phoenix dactylifera*
palm, sago - *Cycas circinalis*
palm, betel - *Areca catechu*
palma-christi - *Ricinus communis*
palta - *Persea americana*
pamplemousse - *Citrus maxima*
pan - *Areca catechu*
Panama tree - *Sterculia apetala*
paniala - *Flacourtia jangomas*
papengaye - *Luffa acutangula*
papaw or papaya - *Carica papaya*
papaya, dwarf - *C. quercifolia*
papeta or pappali - *C. papaya*
paprika - *Capsicum annuum*

parengi-kai - *Cucurbita pepo*
parsley - *Petroselinum crispum*
parsley, horse - *Smyrnium olusatrum*
parsnip - *Pastinaca sativa*
parsnip, Russian cow - *Heracleum sosnowskyi*
pasionaria - *Passiflora caerulea*
pasione - *P. quadrangularis*
passion fruit - *P. edulis*
passion fruit, banana - *P. mollisima*
passion fruit, blue - *P. caerulea*
paterna - *Inga paterna*
patola - *Trichosanthes cucumerina*
pavakai - *Momordica charantia*
pawpaw - *Carica papaya*
pea, black-eyed - *Vigna unguiculata*
pea, Congo or pigeon - *Cajanus cajan*
pea, crowder or southern - *Vigna unguiculata*
peach - *Prunus persica*
pear - *Pyrus communis*
pear, alligator - *Persea americana*
pear, balsam - *Momordica charantia*
pear, Chinese or Japanese - *Pyrus pyrifolia*
pear, melon - *Solanum muricatum*
pear, oriental - *Pyrus pyrifolia*
pear, sacred garlic - *Crateva religiosa*
pear, sand - *Pyrus pyrifolia*
pear, vegetable - *Sechium edule*
pecego or pêche - *Prunus persica*
pekan-kai - *Luffa acutangula*
penaga laut - *Calophyllum inophyllum*
peni dodan - *Citrus sinensis*
peni-komadu - *Citrullus lanatus*
peony, garden - *Paeonia chinensis*
pepino - *Cucumis sativus* or *Solanum muricatum*
pepino dulce - *S. muricatum*
pepol - *Carica papaya*
pepper, bell - *Capsicum annuum*
pepper, cayenne - *C. frutescens*
pepper, chili - *C. annuum* or *C. frutescens*
pepper, green, mango or sweet - *C. annuum*
pepper, tabasco - *C. frutescens*
pepper, Jamaica - *Pimenta dioica*
pera - *Pyrus communis*
pera de agua - *Syzygium malaccense*
pera - *Psidium guajava*
peria jambu - *Syzygium malaccense*
peria - *Momordica charantia*
perita - *Syzygium samarangense*
persimmon, American - *Diospyros virginiana*
persimmon, black - *D. texana*
persimmon, common - *D. virginiana*
persimmon, Japanese or kaki - *D. kaki*
persimmon, lotus - *D. lotus*
persimmon, Mexican - *D. texana*
persimmon, Oriental - *D. kaki*
persimmon, Texas - *D. texana*
perunkila - *Carissa carandas*
petoh buntal or petola manis - *Luffa aegyptiaca*
petola sanding - *L. acutangula*
petola ular - *Trichosanthes cucumerina*
pey-pichuleku - *Luffa acutangula*
phalsa - *Grewia asiatica*
phi-kun - *Mimusops elengi*
phruan - *Nephelium lappaceum*
pichukku - *Luffa aegyptiaca*
pickle fruit - *Averrhoa bilimbi*
pikku - *Luffa aegyptiaca*
pilla-kai - *Artocarpus heterophyllus*
piment - *Capsicum frutescens*
pimento - *C. annuum* or *Pimenta dioica*
pimiento - *C. annuum*
pinang - *Areca catechu*
pine, hoop - *Araucaria cunninghamii*
pineapple - *Ananas comosus*
pini jambu - *Syzygium samarangense*
pipangaille - *Luffa acutangula*
pipingha - *Cucumis sativus*
pisang - *Musa x paradisiaca*
pitanga - *Eugenia uniflora*
pitti-kekiri - *Cucumis melo*
plantain - *Musa x paradisiaca*
plum, beach - *Chrysobalanum icaco*
plum - *Prunus domestica*
plum, black - *Syzygium cumini*
plum, bullace - *Prunus domestica*
plum, chickasaw - *P. angustifolia*
plum, cocoa - *Chrysobalanus icaco*
plum, cultivated - *Prunus domestica*
plum, damson - *Chrysophyllum oliviforme*
plum, date - *Diospyros lotus* or *D. kaki*
plum, European - *Prunus domestica*
plum, garden - *P. domestica*
plum, governor's - *Flacourtia indica*
plum, hog - *Spondias mombin* or *S. cytherea*
plum, Indian - *Flacourtia jangomas*
plum, Jambolan - *Syzygium cumini*
plum, Japanese - *Eriobotrya japonica* or *Prunus salicina*
plum, Java - *Syzygium cumini*
plum, Jew - *Spondias cytherea*
plum, kaffir - *Harpephyllum caffrum*
plum, Klamath - *Prunus subcordata*
plum, Madagascar - *Flacourtia indica*
plum, Malabar - *Syzygium jambos*
plum, marmelade - *Pouteria sapota*
plum, maroola - *Sclerocarya birrea*
plum, myrobalan - *Prunus cerasus*

plum, Natal - *Carissa macrocarpa*
plum, Pacific - *Prunus subcordata*
plum, red coat - *Spondias purpurea*
plum, sapodilla - *Manilkara zapota*
plum, Spanish - *Spondias purpurea* or *S. mombin*
podalangai or podivilangu - *Trichosanthes cucumerina*
poire - *Pyrus communis*
poivron - *Capsicum annuum*
pokok tanjong - *Mimusops elengi*
polo or polo palangi - *Capsicum annuum*
polong wattaka - *Cucurbita moschata*
pomagas - *Syzygium malaccense*
pomarrosa - *S. jambos*
pomegranate - *Punica granatum*
pomelo - *Citrus maxima*
pomerac - *Syzygium malaccense*
pomme - *Malus domestica*
pomme d'Or - *Passiflora laurifolia*
pompelmous - *Citrus maxima*
Poonay-oil plant - *Calophyllum inophyllum*
potato - *Solanum tuberosum*
potato, sweet - *Ipomoea batatas*
prickly pear - *Opuntia vulgaris*
prickly pear, elephant-ear - *O. tuna*
prickly pear, Indian fig - *O. ficus-indica*
priek-yai - *Capsicum annuum*
prune - *Prunus domestica*
prune, Indian - *Flacourtia rukam*
pudding-pipe tree - *Cassia fistula*
pummelo - *Citrus maxima*
pumpkin - *Cucurbita maxima*, *C. moschata* or *C. pepo*
punnai nut - *Calophyllum inophyllum*
puwak - *Areca catechu*

Quelghen - *Fragaria chiloensis*
quetembila - *Dovyalis hebecarpa*
quince - *Cydonia oblonga*
quince, Bengal - *Aegle marmelos*

Rabbitbrush, rubber - *Chrysothamnus nauseosus*
radish - *Raphanus* sp.
rai - *Brassica juncea*
rambai - *Baccaurea motleyana*
rambutan - *Nephelium lappaceum*
rameaux - *Murraya paniculata*
ramón - *Brosimum alicastrum*
ramontchi - *Flacourtia indica*
ramphala - *Annona reticulata*
rampostan - *Nephelium lappaceum*
ramput benggala - *Panicum maximum*
ramsita - *Annona reticulata* or *A. squamosa*
ramtil - *Guizotia abyssinica*
ramtum - *Nephelium lappaceum*
rasa-mora - *Dimocarpus longan*
raspberry or raspberry, red European - *Rubus idaeus*
raspberry, Mauritius - *R. rosifolius*
rata-del - *Artocarpus altilis*
rata-goraka - *Garcinia xanthochymus*
rata-karapincha - *Clausena lansium*
rata-labu - *Cucurbita maxima*
rata-lauvulu - *Chrysobalanus icaco*
rata-mi - *Manilkara zapota*
rata-nelli - *Phyllanthus acidus*
rata-puhul - *Passiflora quadrangularis*
rata-tana - *Panicum maximum*
rata-uguressa - *Flacourtia jangomas*
rokam mania - *F. rukam*
rose, Arkansas - *Rosa arkansana*
rose, apple - *R. villosa*
rose, brier - *R. canina*
rose, damask - *R. damascena*
rose, dog - *R. canina*
rose, French - *R. gallica*
rose, Japanese - *R. rugosa*
rose, prairie - *R. setigera*
rose, Turkestan - *R. rugosa*
rowan - *Sorbus aucuparia*
rubber tree - *Hevea brasiliensis*
ruddles - *Calendula officinalis*
rukam - *Flacourtia rukam*
rukam manis - *F. rukam*

Sa-ke - *Artocarpus altilis*
sa-wa-rot - *Passiflora laurifolia*
safflower or saffron, false - *Carthamus tinctorius*
sago, queen - *Cycas circinalis*
sallow thorn - *Hippophae rhamnoides*
salmonberry - *Rubus ursinus*
salsify, meadow - *Tragopogon pratensis*
sandalwood - *Santalum album*
sapodilla - *Manilkara zapota*
sapote - *Pouteria sapota*
sapote, black - *Diospyros digyna*
sapote domingo - *Mammea americana*
sapote, green - *Pouteria viridis*
sapote, grosse or mammee - *P. sapota*
sapote, white - *Casimiroa edulis*
sapote, yellow - *Pouteria campechiana*
sapoti or sapotiseiro - *Manilkara zapota*
sappa-rot - *Ananas comosus*
satinleaf - *Chrysophyllum oliviforme*
Satsuma - *Citrus reticulata*

sauh - *Manilkara kauki*
savin - *Juniperus sabina*
sawi-sawi - *Brassica juncea*
sawo - *Manilkara zapota*
scaldberry - *Rubus fruticosa*
seagrape - *Coccoloba uvifera*
seemai-goraka - *Garcinia xanthochymus*
seemaipala-pallum - *Chrysobalanus icaco*
seemaisora-kai - *Passiflora quadrangularis*
seeni-jambu - *Syzygium jambos*
seetha - *Annona muricata*
semeruco - *Malpighia glabra*
semolok - *Diospyros blancoi*
sengkuang - *Dracontomelon dao*
shaddock - *Citrus maxima*
shallot - *Allium cepa*
shimai-eluppai - *Manilkara zapota*
silk-cotton tree, red - *Bombax ceiba*
sineguelas - *Spondias purpurea*
siri-nelli - *Phyllanthus acidus*
sitaphal - *Annona squamosa*
smallage - *Apium graveolens*
snakegourd - *Trichosanthes cucumerina*
snowberry - *Symphoricarpos alba* or *S. racemosa*
soldier's butter - *Persea americana*
som-mai-fai - *Clausena lansium*
som-mal-mao - *Citrus aurantiifolia*
som-mu - *C. medica*
som o - *C. maxima*
sorakai - *Lagenaria siceraria*
sorghum, grain - *Sorghum bicolor*
soursop - *Annona muricata*
spadic - *Erythroxylum coca*
sponge, vegetable - *Luffa aegyptiaca*
squash - *Cucurbita maxima*, *C. moschata* or *C. pepo*
squash, autumn or summer - *C. pepo*
squash, turban - *C. maxima*
squash, winter - *C. maxima* or *C. moschata*
starfruit - *Averrhoa carambola*
strawberry - *Fragaria vesca*
strawberry, alpine, perpetual or wood - *F. vesca*
strawberry, Chilean or pine - *F. chiloensis*
strawberry tree - *Arbutus unedo*
strychnine tree - *Strychnos nux-vomica*
su-khatha-rot - *Passiflora quadrangularis*
sugarberry - *Celtis laevigata*
sukun - *Artocarpus altilis*
sunchoke - *Helianthus tuberosus*
sunflower - *H. annuus*
sunflower, Mexican - *Tithonia diversifolia*
swallow-wort - *Asclepias curasavica*
swallow-wort, white - *Vincetoxicum hirudinaria*
sweet cup - *Passiflora edulis*
sweet sultan - *Centaurea moschata*
sweetsop - *Annona squamosa*
sweetsop, wild - *Rollinia mucosa*

Ta-khop-pa - *Flacourtia indica*
ta-khop-thai - *F. jangomas*
ta-ling-pring - *Averrhoa bilimbi*
taeng-ma - *Citrullus lanatus*
takop farang - *Muntingia calabura*
tamanu - *Calophyllum inophyllum*
tamarilla - *Cyphomandra crassicaulis*
tamarta - *Averrhoa carambola*
tangelo - *Citrus x tangelo*
tangerine - *C. reticulata*
tangor - *C. x nobilis*
tanjong tree - *Mimusops elengi*
tapanima - *Averrhoa carambola*
tapereba - *Spondias mombin*
tapioca - *Manihot esculenta*
tarragon - *Artemisia dracunculus*
taun - *Pometia pinnata*
taxa - *P. pinnata*
tea - *Camellia sinensis*
tea, Alibotush - *Sideritis scardica*
teh - *Camellia sinensis*
telfairea nut - *Telfairea pedata*
Teli huhu'a - *Anacardium occidentale*
telie - *Terminalia catappa*
temata - *Lycopersicon esculentum*
tembikai - *Citrullus lanatus*
temponek - *Artocarpus rigidus*
teng kai - *Cucumis melo*
teng ran - *C. sativus*
terongan - *Solanum torvum*
teruah - *Momordica cochinchinensis*
terung - *Solanum melongena*
tesa - *Pouteria campechiana*
tey-ile - *Camellia sinensis*
thap-thim - *Punica granatum*
thay-gas or thay-kola - *Camellia sinensis*
thorn, Jew - *Ziziphus lotus*
thurian-khaek - *Annona muricata*
Ti-Es - *Pouteria campechiana*
timun belanda - *Passiflora quadrangularis*
timun - *Cucumis sativus*
tippari - *Physalis peruviana*
tolok - *Inocarpus fagifer*
tomate - *Lycopersicon esculentum*
tomatillo - *Physalis philadelphica*
tomato - *Lycopersicon esculentum*
tomato, cherry - *Physalis peruviana*

tomato, currant - *Lycopersicon pimpinellifolium*
tomato, husk or purple strawberry - *Physalis philadelphica*
tomato, tree - *Cyphomandra crassicaulis*
toranja - *Citrus maxima*
toronja - *C. x paradisi*
towelgourd, singkwa - *Luffa acutangula*
tuna - *Opuntia tuna* or *O. vulgaris*
turi - *Sesbania grandiflora*
turkeyberry - *Solanum torvum*

Ubi kayu - *Manihot esculenta*
ubi keledek - *Ipomoea batatas*
umbrella tree - *Cussonia* sp

Vaine - *Passiflora edulis*
vala - *Musa x paradisiaca*
vel-kelengu - *Ipomoea batatas*
veli jambu - *Syzygium jambos*
vellari-kai - *Cucumis sativus*
veta-kola - *Luffa acutangula*
vilvam - *Aegle marmelos*
vine, bulso - *Gnetum gnemon*
vine, grape - *Vitis vinifera*
vine, lemon - *Pereskia aculeata*
vine of Sodom - *Citrullus colocynthis*
vine, strainer - *Luffa acutangula*
vine, Barbary matrimony - *Lycium barbarum*
violet, Chinese - *Telosma cordata*

Wa - *Syzygium cumini*
walnut - *Juglans regia*
walnut, Andean - *J. neotropica*
walnut, black - *J. nigra*
walnut, California or California black - *J. californica*
walnut, English - *J. regia*
walnut, Hinds' or Hinds' black - *J. hindsii*
walnut, Japanese - *J. ailantifolia*
walnut, Persian - *J. regia*
wambotu - *Solanum melongena*
waccmpi - *Clausena lansium*
waterlemon - *Passiflora laurifolia*
waterlemon, wild - *P. foetida*
watermelon - *Citrullus lanatus*
wattaka - *Cucurbita maxima*
wax jambu - *Syzygium samarangense*
whinberry - *Vaccinium myrtillus*
whitewood, Bahama - *Canella winteriana*
whortleberry - *Vaccinium myrtillus*
wi - *Spondias cytherea*
willow, crack - *Salix fragilis*
willow, goat - *S. caprea*
willow, white - *S. alba*
wintergreen - *Gaultheria procumbens*

Yam [of USA] - *Ipomoea batatas*
yash-tel - *Pouteria viridis*
yaupon - *Ilex vomitoria*
yawa - *Vigna unguiculata*
ylang-ylang - *Cananga odorata*
youngberry - *Rubus flagellaris* x *R. loganobaccus*
yuca - *Manihot esculenta*
yucca - *M. esculenta*

Zapote amarillo - *Pouteria campechiana*
zapote, black - *Diospyros digyna*
zapote blanco - *Casimiroa edulis*
zapote colorado - *Pouteria sapota*
zapote de Santo Domingo - *Mammea americana*
zapote - *Pouteria sapota*
zinnia - *Zinnia elegans*

Glossary

Simple terms that can be found in any general introduction to entomology and terms that are easily explained by figs 5-10 are omitted. Equivalent terms used in other key works are given in *italics* and cross-referenced to the preferred term.

1st FLAGELLOMERE - This is the 3rd segment of the antenna (fig. 5). It tends to be apically pointed in *Rhagoletis* spp. (fig. 29), long in *Bactrocera* and *Dacus* spp. (fig. 19), and very long in *D. (Callantra)* spp. (fig. 24).

A (LARVAL TERM) - Abbreviation for abdominal segment.

ACCESSORY PLATES (LARVAL TERM) - Small plates, often toothed, interlocking with the outer ends of the oral ridges (fig. 11; pl. 20.c).

ACCESSORY COSTAL CROSSBAND - This is a short crossband between the discal and preapical crossbands of some *Rhagoletis* spp. (fig. 38).

ACROSTICHALS - see prescutellar acrostichals.

ACULEUS - This is the piercing part of the female ovipositor which is normally retracted into the oviscape (fig. 10). To examine the aculeus, it must be dissected from the rest of the abdomen, slide mounted, and then examined with a compound microscope (p. 27). Some authors call this structure the *gynium*, *ovipositor piercer* or *ovipositor*. The ultrastructure of the aculeus was discussed by Stoffolano (1989).

ADVENTIVE - Species which have been introduced to a new geographic region as a result of human activity are said to be adventive in that region.

AEDEAGAL GLANS - This is the apical part of the male aedeagus or penis. In the tephritids this is a very complex structure which was first used for species diagnosis by Munro (1947), but use of this character has been avoided in the present work. It is also known as the *distiphallus* and a method for its dissection was given by White (1988).

ALLOTYPE - see type specimen.

ANAL ELEVATION (LARVAL TERM) - The area surrounding the anal opening (fig. 16).

ANAL LOBES (LARVAL TERM) - The pair of lobes flanking the anal opening in fruit feeding larvae (fig. 16; pls 4.e, 23.d).

ANAL STREAK - This is a diagonal marking that covers cell cu*p* and part of cell cua_1 in most *Bactrocera* and *Dacus* spp. (fig. 36). It is sometimes called the *anal stripe*.

ANAL STRIPE - see anal streak.

ANATERGITE - This sclerite is just below the scutellum and just above the haltere (fig. 8). The Adramini and Euphrantini differ from other tephritids by having long pale coloured hairs on the anatergite (fig. 44). In most *Bactrocera* and *Dacus* spp. the anatergite is bright yellow or orange, but it is a different colour in a few species; that colour difference is used in the identification of some *Dacus* spp. The anatergite and the katatergite together form the laterotergite, which is the *pleurotergite* of many authors.

ANEPISTERNUM - A large pleural sclerite of the thorax (fig. 8). Many tephritids have a vertical suture just anterior to the series of setae along the posterior edge of the anepisternum; that suture is usually well developed in the Dacinae and Trypetinae, but obscured by tomentum in most Tephritinae. The anepisternum is the *mesopleuron* of many authors.

ANTENNAL SENSORY ORGANS (LARVAL TERM) - Small paired sensory papillae consisting of 1-3 segments, placed anteriorly on head (fig. 11; pl. 5.b).

ANTERIOR SPIRACLES (LARVAL TERM) - Project laterally from the first thoracic segment (T1) with a variable number of tubules along the outer edge (fig. 15; pl. 4.c).

ANTERIOR SCLERITE (LARVAL TERM) - Occurs on either side of the pharyngeal sclerite projecting anteriorly from just below the dorsal bridge (fig. 14).

APICAL CROSSBANDS - see crossbands.

APPARENT GENITAL OPENING - An apparent hole near the apex of the aculeus. It is a useful point of reference when describing structures near the aculeus apex in *Anastrepha* spp. (figs 50-63).

ARISTA - Style or seta-like part of the antenna attached near the base of the 1st flagellomere (fig. 5). Most tephritids have a micropubescent arista, i.e. it is covered in a microscopic downy pile. However, some tephritids, such as the Gastrozonina and many Acanthonevrini and Euphrantini, have a plumose or pectinate arista (fig. 25).

BASICOSTAL BAND - This is a coloured band along the costal edge of the wing from the wing base to cell sc. It is here used for *Anastrepha* spp. (fig. 35) because the term *costal band* used by other workers could be confused with the costal band of *Bactrocera* and *Dacus* spp. which usually only runs from cell sc to the wing apex.

BASIPHALLUS - This is the long tubular basal part of the male aedeagus or penis.

BIVOLTINE - Having two generations per year.

BRISTLES - Called setae in the present work.

CAPITULA (singular: CAPITULUM) (BOTANICAL TERM) - The flower bud, flower or seed head of a plant belonging to the family Asteraceae (=Compositae).

CAUDAL SEGMENT (LARVAL TERM) - Abdominal segment VIII (A8) (fig. 15).

CELL bm - A basal wing cell bounded anteriorly by vein M, apically by crossvein bm-cu and posteriorly by vein CuA_2 (fig. 6).

CELL cu*p* - A basal wing cell bounded anteriorly and apically by vein CuA_2 and posteriorly by vein A_1 (fig. 6). In the Myopitini, and in a few genera of other groups, this cell is closed by a convex vein CuA_2. In other groups vein CuA_2 is concave or oblique across the apical side of cell cu*p*, forming an acute extension which may be small, e.g. *Rhagoletis* spp. (fig. 43), long, e.g. in *Ceratitis* spp. (fig. 42), or very long, e.g. *Bactrocera* spp. (fig. 40). In most Ceratitina the anterior margin of the cu*p* extension is curved forwards (fig. 42), rather than straight.

CELL sc - The cell distal to the faint part of vein Sc, bounded anteriorly by C and enclosed by R_1 both posteriorly and distally (fig. 6). This is the *stigma* of many authors.

CEROMAE - see ceromata

CEROMATA - A pair of slightly depressed shiny areas on tergite 5 of *Bactrocera* and *Dacus* spp. They were called *ceromae* by Munro (1984) and *shining spots* by Drew (1982a). These areas are covered in wax glands (Munro, 1984).

CHORION (EGG TERM) - The outer surface of the egg which may appear smooth or reticulate (pl. 3.b).

CILIATE - see pecten.

COMPLEX - see species complex.

CONVERGENT SETAE - Any setae which lean towards the midline of the fly. The Terelliini have the posterior pair of orbital setae converging in this manner.

COSTAL EDGE - The leading (anterior) wing edge along which vein C runs (fig. 6).

COSTAL BAND - A characteristic feature of most *Bactrocera* and *Dacus* spp. is the coloured band along the anterior edge of the wing (fig. 36). In most species this runs from cell sc to the wing apex, but in a few species it is reduced to an apical spot (fig. 196), and in others it is extended back to the wing base (fig. 193). The depth of the band is an important feature of some species, and in some species it is apically expanded into a large spot (fig. 206).

CREEPING WELT (LARVAL TERM) - A locomotory structure consisting of numerous rows of small spinules or rounded projections on the ventral surface of an abdominal segment (fig. 15; pls 4.f, 15.e).

CROSSBANDS - The system of named wing crossbands used here is based on that of Steyskal (1979); that was devised for use with *Urophora* spp. but is here applied to other genera when appropriate. *Urophora* spp., and many other banded winged species, have up to 4 crossbands, namely: subbasal, through the basal cells bm and cu*p*; discal, through cell sc, the r-m crossvein and the middle of cell dm; preapical, through the dm-cu crossvein; apical, at the wing apex. Many *Rhagoletis* spp. also have a short *accessory costal crossband* and a *posterior apical crossband* (fig. 38). *Anastrepha* spp. have *S* and *V bands* (fig. 35).

CROSSVEIN DM-CU - This links veins M and CuA_1 about two-thirds the way along the wing (fig. 6). It may be covered by the preapical crossband (figs 35, 38).

CROSSVEIN R-M - This links veins R_{4+5} and M, joining M part way along cell dm (fig. 6). In some *Bactrocera* and *Dacus* spp. this is covered by a short crossband (fig. 180).

DENTAL SCLERITES (LARVAL TERM) - A pair of small sclerites lying close to the ventral margin of the mouthhooks (fig. 14). They are common in the Dacini but absent or inconspicuous in other groups.

DISCAL CROSSBAND - see crossbands.

DISTIPHALLUS - see aedeagal glans.

DORSAL AREA (LARVAL TERM) - The area above the posterior spiracles on A8 (fig. 16).

DORSAL CORNUA (LARVAL TERM) - The two dorsal, wing-like portions of the pharyngeal sclerite (fig. 14). They are frequently sclerotised and often cleft on the outer margins.

DORSAL BRIDGE (LARVAL TERM) - Anteriorly joins the dorsal cornua (fig. 14).

DORSOCENTRAL SETAE - Tephritids have at most 1 pair of postsutural dorsocentral setae; a term here contracted to dorsocentral setae (fig. 7). The relative position of these setae to the anterior supra-alar setae is of some use in the higher classification of tephritids; in Tephritini the dorsocentrals are usually placed in front of an imaginary line between the anterior supra-alar setae (right half of fig. 7); in the Trypetinae and Terelliini they are usually on or behind that line (left half of fig. 7). The Dacini lack dorsocentral setae. The dorsocentrals should not be confused with the prescutellar acrostichal setae. A few tephritids also have presutural dorsocentral setae, e.g. *Chaetorellia* spp.

ECDYSIAL SCAR (LARVAL TERM) - A mark on the posterior spiracle which marks the area occupied by the spiracle of the previous instar (fig. 17; pl. 5.d).

EPISTOME - see lower facial margin.

EVERSIBLE OVIPOSITOR SHEATH - This is a membraneous tube between the apex of the oviscape and the base of the aculeus, which allows the aculeus to slide part way out of the oviscape for oviposition or copulation. The sheath is covered in scales and the basal scales of the sheath are greatly enlarged in *Anastrepha* and *Toxotrypana* spp. (figs 45-46).

FACE MASK (LARVAL TERM) - A term used by Kandybina (1977) for the area on the head surrounding the antennal and maxillary sensory organs and part of the mouth (fig. 13).

FLAGELLOMERE - see 1st flagellomere.

FRONS - The anterodorsal area of the head, bounded laterally by the eyes, posteriorly by the ocellar triangle, and anteriorly by the lunule (fig. 5).

FRONTAL SETAE - The row of setae next to each eye in the lower part of the frons (fig. 5). Most tephritids have between 1 and 3 pairs, i.e. 1-3 setae next to each eye. They are usually incurved and many authors call them *inferior* or *lower fronto-orbitals*.

FRUGIVOROUS - A species whose larvae feed on fruit.

GYNIUM - see aculeus.

HOLOTYPE - see type specimen.

HOMONYM - If the same name has been used for more than one taxon (e.g. species, genus or family) it is a homonym. For example, *Dacus humeralis*, from Africa, was described in 1915 by M. Bezzi; *Chaetodacus humeralis*, from Australia, was described in 1934 by F.A. Perkins. When transferred to *Dacus* the Australian species became a junior homonym of the more senior African name. Consequently, the Australian species had to be given a new name; Hardy (1951) gave it the name *D. neohumeralis* which is still the valid name even though it is now called *Bactrocera neohumeralis*.

HYALINE WING - A wing with no markings.

HYPOPHARYNGEAL SCLERITE (LARVAL TERM) - Consists of 2 elongate, sclerotised, posteriorly directed plates connected by a small crossbar and giving an H-shaped appearance; it articulates anteriorly with the mouthhooks (fig. 14).

INFERIOR ORBITAL SETAE - see frontal setae.

INTERMEDIATE AREAS (LARVAL TERM) - Areas on A8, each side of the midline, bounded by the posterior spiracles and the ventral area; often protuberant and in some species, e.g. *Bactrocera cucurbitae*, almost linked by a pigmented transverse band.

INTRA-ALAR SETAE - A series of setae between the dorsocentral and supra-alar series; tephritids only have one pair of intra-alars, placed level with the prescutellar acrostichal setae (fig. 7).

KATEPISTERNUM - The triangular sclerite between the coxae of the fore- and midlegs (fig. 8). Most species have a well developed seta near the posterior corner, but that is absent in the Adramini, Dacini and some leaf and stem mining Trypetini. It is the *sternopleuron* of many authors.

LABIAL SCLERITES (LARVAL TERM) - 2 sclerites forming a V-shape in the floor of the mouth between the hypopharyngeal sclerites and the mouthhooks (fig. 14).

LABIUM (LARVAL TERM) - A large triangular fleshy lobe on the ventral margin of the mouth (fig. 11; pl. 4.a).

LATERAL AREA (LARVAL TERM) - The area lateral to the posterior spiracles on A8 (fig. 16).

LECTOTYPE - see type specimen.

LOWER ORBITAL SETAE - see frontal setae.

LOWER FACIAL MARGIN - The lower anterior part of the head, below the face and above or in front of the mouth opening. The lower facial margin is the *epistome* of many authors.

LUNULE - The semicircular plate above the antennal bases and below the ptilinal fissure (fig. 5).

MAXILLARY SENSORY ORGANS (LARVAL TERM) - Small paired sensory papillae, usually consisting of two well defined groups of sensilla surrounded by cuticular folds, on the head, posterior to the antennal sensory organs (fig. 11; pl. 5.b).

MEDIOTERGITE - This is the swollen sclerite below the scutellum and subscutellum (fig. 8).

MESONOTUM - see scutum.

MESOPLEURON - see anepisternum.

MICROPYLE (EGG TERM) - A small, often nipple-like, structure at the anterior end of the egg (pl. 3.b).

MICROTRICHIA - Minute tooth-like structures that cover most of the wing surface.

MONOPHAGOUS - A species which only attacks one species of host plant.

MONOTYPY - see type species.

MORPHOSPECIES - see species complex.

MOUTHHOOKS (LARVAL TERM) - Paired, curved sclerotised hooks which articulate posteriorly with the hypopharyngeal sclerite and may have additional small teeth along the ventral margin (fig. 14; pl. 28.d).

MULTILOCULAR - Describes a gall which has many chambers; not to be confused with capitula containing several unilocular galls that have fused.

MULTIVOLTINE - Having more than one generation per year.

NOMINAL SPECIES - Any named species is a nominal species even if it is a synonym of another biological species.

NOTOPLEURAL SETAE - The notopleuron (fig. 8) is a lateral thoracic sclerite, which has an anterior and a posterior setal position. The colour of the notopleura is sometimes used as a character in *Bactrocera* spp.

OCELLAR SETAE - This is the pair of setae that are based one in front of each posterior ocellus (fig. 5). They are absent or very reduced in *Bactrocera*, *Dacus* and some other genera.

OCELLAR TRIANGLE - The subtriangular area that encloses the 3 ocelli, which are themselves arranged as a triangle.

OLIGOPHAGOUS - A species which attacks a limited range of closely related host plants, e.g. all of its hosts are members of a single family (but broader than *stenophagous*).

ORAL RIDGES (LARVAL TERM) - Each side of the mouth opening there are several rows of ridges (fig. 11) which may be entire (unserrated) (pl. 7.b) or toothed (pl. 9.a) on their lower (posterior) edge.

ORBITAL SETAE - The setae in the upper part of the frons which usually form a series of 1-3 reclinate setae near the upper part of each eye (fig. 5). Most species have 2 pairs, i.e. 2 reclinate setae next to the upper corner of each eye, but they are absent in some genera. Strictly, these setae are "reclinate orbital setae" and they are the *superior* or *upper fronto-orbitals* of many authors. The prefix "reclinate" is superfluous because tephritids lack proclinate orbitals. Furthermore, the Terelliini have the posterior pair of orbitals convergent rather than reclinate.

ORIGINAL DESIGNATION - see type species.

OVIPOSITOR - This term is used here to refer to the entire female terminalia; oviscape, eversible sheath and aculeus. Some authors use this term to mean only the aculeus; see Norrbom & Kim (1988a) for discussion.

OVIPOSITOR PIERCER - see aculeus.

OVISCAPE - This is the tubular basal segment of the ovipositor which is formed from the fusion of tergite 7 and sternite 7. Strictly, it should be called *syntergosternite 7*; see Norrbom & Kim (1988a) for discussion.

PARALECTOTYPE - see type specimen.

PARASTOMAL BARS (LARVAL TERM) - Two long rod-shaped sclerites lying dorsally, parallel to the hypopharyngeal sclerite (fig. 14).

PARATYPE - see type specimen.

PECTEN - The rows of setae on each side of the posterior margin of tergite 3 of the males of most *Bactrocera* and *Dacus* spp. (fig. 9). The pecten is part of the stridulatory apparatus (see Kanmiya, 1988). Some authors call this a *comb* and tergite 3 is said to be *ciliate* by some other authors.

PEDICEL - The second segment of the antenna.

PHARYNGEAL SCLERITE (LARVAL TERM) - The combination of the dorsal and ventral cornua (fig. 14).

PLEURA (singular: PLEURON) - The lateral sclerites of the thorax (fig. 8).

PLEUROTERGITE - see anatergite.

POLYPHAGOUS - A species with a broad range of host plants, often belonging to unrelated groups, for example several plant families.

POSTERIOR APICAL CROSSBAND - This is a short band that runs between wing cells r_{4+5} and m, crossing the apical section of vein M (fig. 38). It is a feature of some *Anastrepha* (fig. 165), *Ceratitis* (fig. 97) and *Rhagoletis* (fig. 240) spp. In most *Anastrepha* spp. it forms part of the *V band* (figs 35, 163) and in *Rhagoletis* spp. it is often joined to the preapical crossband (fig. 244).

POSTERIOR SPIRACLES (LARVAL TERM) - Positioned on the caudal segment from the midline to higher up towards the dorsal edge, consisting of 2-3 paired *spiracular slits*, frequently arranged almost parallel to each other (figs 16, 17; pl. 4.d).

POSTOCELLAR SETAE - This is a pair of setae placed behind the ocellar triangle.

POSTOCULAR SETAE - The row of small setae behind each eye (fig. 5). These are usually thin and black, but in the Tephritini at least some of these setae are white and scale-like.

PREAPICAL TOOTH (LARVAL TERM) - An additional tooth on the ventral surface of the mouthhook (fig. 156; pl. 28.d). More than 1 tooth may be present in some species.

PREAPICAL CROSSBANDS - see crossbands.

PREORAL LOBES (LARVAL TERM) - A small series of lobes or ridges, with entire or serrated edges, associated with the stomal sensory organ (fig. 11; pl. 23.c).

PREORAL TEETH (LARVAL TERM) - Small teeth or finger-like projections at the base of the stomal sensory organ (fig. 12). They only occur in a small number of genera e.g. *Rhagoletis* (pl. 36.b) and *Carpomya*.

PRESCUTELLAR ACROSTICHAL SETAE - The acrostichal setae are the setae nearest to the midline of the scutum; tephritids have at most one pair and they are placed just in front of the scutellum (fig. 7). Consequently, some authors call them the *prescutellar setae*. They are lacking in all *Dacus* spp. and some subgenera of *Bactrocera*.

PRESCUTELLAR SETAE - see prescutellar acrostichal setae.

PROCLINATE SETAE - Any setae which lean forwards.

PTILINAL FISSURE - The inverted U- or V-shaped slit which runs over the antennal bases, ending in the genal grooves (fig. 5). It marks the edge of the facial sclerite that is pushed forwards when the *ptilinum* is expanded.

PTILINUM - A sack-like structure which is inflated by the adult as a mechanism for bursting the puparium.

PUPARIATION - The formation of the *puparium*.

PUPARIUM - The hardened skin of the last larval instar within which pupation takes place (p. 34).

RECEPTACLE (BOTANICAL TERM) - The solid, sometimes fleshy, basal part of the *capitulum*, beneath where the seeds form.

RECLINATE SETAE - Any setae which lean back.

RIMA (LARVAL TERM) - The marginal supporting sclerotisation of each spiracular opening or slit (fig. 17).

S BAND - This is a coloured band that most *Anastrepha* spp. have; it runs from the area of cell cu*p*, diagonally across the r-m crossvein to join the costa in the apical part of cell r_1, and then follows the edge of the wing to the wing apex (fig. 35). It may be joined to the basicostal band, that runs from the wing base to cell sc, and in some species it is joined to the V band.

SCUTUM - The scutum and scutellum form the dorsal part of the thorax. The scutum includes both the pre- and postsutural areas (fig. 7).

SENSILLA (singular: SENSILLUM) - Simple sense organs, e.g. on the larva (pl. 22.c) or at the aculeus apex (fig. 81).

SHINING SPOTS - see ceromata.

SIBLING SPECIES - see species complex.

SPECIES COMPLEX - A group of species which cannot be distinguished using the criteria normally used to identify members of the genus (or other wider group). For example, most *Rhagoletis* spp. can be identified using simple combinations of colour characters, but some groups of species that are genetically distinct are difficult or impossible to distinguish morphologically. Some authors call these groups *morphospecies* and the individual species within them *sibling species*. Sibling species is a convenient term but it unfortunately implies that they are not 'real' species. The term species complex is used by some authors in a wider sense, to refer to any group of species (e.g. Drew, 1989a). Conversely, Bush (1966) used the term *species group* to mean species complex.

SPERMATHECAE - Female internal organs used to store sperm until needed to fertilize eggs; these can be found in segment 6 when an aculeus is being dissected. Most Tephritinae have 2 spermathecae; Dacinae and most Trypetinae have 3.

SPINULE (LARVAL TERM) - A small spine-like projection (pls 2.d, 23.b).

SPIRACULAR SLITS (LARVAL TERM) - The external openings of the spiracular chamber (fig. 17; pl. 4.d). Slits are variable in shape and length.

SPIRACULAR HAIRS (LARVAL TERM) - Translucent hairs arranged in four groups or bundles around the outer edge of the posterior spiracles (fig. 17; pl. 4.d).

STENOPHAGOUS - A species which attacks a very narrow range of closely related host plants, e.g. all members of a single genus.

STERNITES - The shape of sternite 5 of the male is sometimes a useful character. The subgeneric classification of *Bactrocera* (see Drew, 1989a), makes much use of this character. Its use has been avoided in the present work because it is difficult to interpret in badly preserved specimens, especially reared specimens that have not been fed for a few days before being killed. Use of that character is unavoidable when the full range of species is being studied.

STERNOPLEURON - see katepisternum.

STIGMA - see cell sc.

STOMAL SENSORY ORGAN (LARVAL TERM) - A small rounded sensory organ at each anterolateral corner of the mouth (fig. 11; pl. 23.c).

SUBBASAL CROSSBAND - see crossbands.

SUBSEQUENT DESIGNATION - see type species.

SUPERIOR ORBITAL SETAE - see orbital setae.

SUPRA-ALAR SETAE - Tephritids have up to 3 pairs of supra-alar setae, as follows: presutural; anterior postsutural, a term usually reduced to anterior; posterior postsutural, a term usually reduced to posterior (fig. 7). It is usually only the anterior supra-alar setae that are referred to in the keys, and the presence or absence of these setae is important in the identification of *Bactrocera* and *Dacus* spp.

SURSTYLI (singular: SURSTYLUS) - Part of the external male terminalia; see White (1988) for details.

SYMPATRY - Found in the same place. Two species which evolved from one, in one area, and with the potential to interbreed, are said to have evolved sympatrically (see Bush, 1966; 1969).

SYNONYM - When a taxon (genus, species, etc.) has been described by more than one author only the first author's name is valid; subsequent names are synonyms. An exception occurs when the senior (oldest) name is a junior *homonym*. In that case the next to oldest name is used (if not also a junior homonym) or a new name must be published.

SYNTERGOSTERNITE 7 - see oviscape.

SYNTYPE - see type specimen.

T (LARVAL TERM) - Abbreviation for thoracic segment.

TENERAL - A freshly emerged adult with a soft pale coloured body and poorly formed wing markings, and often with the *ptilinum* exposed. Reared specimens should always be kept alive for a few days to allow their bodies to harden and colours to develop before being killed.

TERGITES - In the higher Diptera the first visible dorsal abdominal sclerite is formed by the fusion of tergites 1 and 2, sometimes called syntergite 1+2 (fig. 9). In *Dacus* spp. all the tergites are fused, but shiny lines or depressions marking the tergite borders can still be seen in most species. Males have 5 tergites (fig. 9) and females have 6 (fig. 10). However, tergite 6 is very short in *Anastrepha* spp. and in the Dacinae; in the Dacini it is hidden under tergite 5 so that it is not visible from above.

TOMENTOSE/TOMENTOSITY/TOMENTUM - A covering of ultramicroscopic pubescence on any part of the body, so that the body surface is matt; *pollinose* (*pollinosity*) of many authors. Note that the tomentum can only be seen in dry mounted specimens; specimens preserved in alcohol, or dampened in any way before dry mounting, may have a damaged tomentum.

TRANSVERSE SUTURE - Calyptrate Diptera have a distinct suture across the scutum, between the notopleura. In the Acalyptratae, e.g. the Tephritidae, this suture is absent across the central part of the scutum and is only represented by vestiges of the suture next to each notopleuron (fig. 7). Even so, this vestigial suture marks the boundary between the presutural and postsutural scutum.

TUBERCLE (LARVAL TERM) - A small raised area often forming a base for a sensillum on A8 (pl. 22.c).

TUBULE (LARVAL TERM) - A small tubular or finger-like process on the outer edge of the anterior spiracles (pl. 4.c).

TYPE SPECIES - Every genus has one species which is called its 'type'; its function is to identify which species rightfully hold an existing generic name when generic (or subgeneric) limits are revised. In a genus originally described for a single species, that species is the type by *monotypy*. A genus originally described for several species must have a type chosen from amongst those originally included species and this is called *subsequent designation*. Present day taxonomists are obliged to choose a type species at the time of description and this is called type by *original designation*. Through any changes of generic limits the name of the genus always stays with the type species, even if that means placing it in synonymy with another genus or retaining the name of a large well known genus for a single species. For example, almost all Dacini used to be called *Dacus*, but when it was decided that the group was really two distinct genera (Drew, 1989a), only the minority of species could retain the name *Dacus* because the type species (*D. armatus*) belonged to the smaller of the two groups; most species, including most pests, were transferred to *Bactrocera*.

TYPE SPECIMEN - When a new (previously unknown) species is described the taxonomist is obliged to designate a type specimen; its function is to identify which species holds the name if there is need for further revision. For example, when Drew (1991) discovered that the Oriental fruit fly was a complex of several species that had previously all been known as *B. dorsalis* (Hendel), they examined Hendel's original specimens from Taiwan to establish which of the many members of the complex rightfully held that name. Modern taxonomists select a single specimen called a *holotype* and name any other specimens that they use for describing the species as *paratypes*; only the holotype matters if the subsequent identity of the species is disputed. Authors such as F. Hendel, working in the early part of this century, called all the specimens they used to describe a species 'types'. Modern authors call the specimens in a series which lack a single holotype *syntypes*, and to guard against future dispute they may select a single specimen as a *lectotype* (a sort of type by *subsequent designation*); all remaining specimens in the series are *paralectotypes*. The term *allotype* was used by authors such as E.M. Hering to mean a specimen of the opposite sex (normally female) to the holotype, but that term has no present day validity.

UNIVOLTINE - Having one generation per year.

UPPER ORBITAL SETAE - see orbital setae.

V BAND - This is an inverted V-shaped coloured band that is present on the wings of most *Anastrepha* spp. (fig. 35). It is a fusion of two other crossbands; the preapical crossband covers the dm-cu crossvein; it then curves round in cell r_{4+5} to join the posterior apical crossband. In many species the V band is reduced, typically to just the preapical crossband covering the dm-cu crossvein. In some species the V band may be joined to the S band. Other authors have used the terms *proximal* and *distal* to refer to the two parts of the V band.

VEIN M - The position of this vein is shown in fig. 6. In *Anastrepha* spp. it is curved forwards at its apex (fig. 35).

VEIN R_{4+5} - Most Trypetini, e.g. *Zonosemata* (fig. 251), and a few genera of Tephritini have a series of setulae (small hairs) along the dorsal side of R_{4+5}. Some Tephritini also have setulae along the ventral side of this vein, for example most *Tephritis* spp.

VENTRAL CORNUA (LARVAL TERM) - The two ventral wing-like portions of the pharyngeal sclerite (fig. 14). They may have a clear or unsclerotised area called a window.

VENTRAL AREA (LARVAL TERM) - The area immediately dorsal to the anal elevation on A8 (fig. 16).

VERTEX - The uppermost part of the head, between the eyes and around the ocellar triangle (fig. 5).

VERTICAL SETAE - There are 2 pairs of verticals on the vertex, placed between the eyes and the ocelli (fig. 5). The pair nearest the ocelli are the inner verticals which are usually long and black. The pair nearest the eyes are the outer verticals. In many species these are well developed black setae; in the Tephritini the outer verticals are white and often scale-like, and sometimes they are difficult to distinguish from the postocular setae.

VITTAE (singular: VITTA) - This is a term for any stripes or other regions of colour that contrast sharply with the adjacent ground colour. The term is used extensively in other author's keys to *Bactrocera* and *Dacus* spp.; to avoid confusion we have used the term 'stripe'.

WINDOW (LARVAL TERM) - see ventral cornua.

XANTHINE - Used by Munro (1984) to mean a 'stripe'; see also *vittae*.

References

References to abstracts in *Review of Applied Entomology* (Series A: Agricultural) [vols **1**-**77**] and its successor, *Review of Agricultural Entomology* [vol. **78** onwards], are given after many of the references [RAE]. Most references that did not include taxonomic or systematic information such as host plant data, were only seen in abstract form, notably those references listed in the "Other references" part of each species account.

Aczél, M.L. (1937) Trypetiden Studien. *Allattani Közlemények*, **34**, 80-82. [RAE **25**:538]

Aczél, M.L. (1952) Further revision of the genus *Xanthaciura* Hendel (Trypetidae, Dipt.). *Acta Zoologica Lilloana*, **10**, 199-243.

Agarwal, M.L. & Kapoor, V.C. (1982) *Acanthiophilus helianthi* (Rossi) and *Chetostoma completum* Kapoor et al. (Diptera: Tephritidae), serious pests of *Centaurea cyanus* Linnaeus (Compositae) in India. *Journal of Entomological Research*, **6**, 102-104. [RAE **72**:4457]

Agarwal, M.L. & Kapoor, V.C. (1985) On a collection of Trypetinae (Diptera: Tephritidae) from northern India. *Annals of Entomology*, **3**(2), 59-64.

Agarwal, M.L. & Kapoor, V.C. (1989; dated 1986) New records of some hymenopterous parasites of fruit flies (Diptera: Tephritidae) from India. *Bulletin of Entomology*, **27**, 193. [RAE **78**:2692]

Agarwal, M.L., Sharma, D.D. & Rahman, O. (1987) Melon fruit-fly and its control. *Indian Horticulture*, **32**(2), 10-11. [RAE **77**:411]

Agriculture Canada (1973) The apple maggot, *Rhagoletis pomonella* (Walsh). *Agriculture Canada Insect Identification Sheet*, (1), 1-2.

Ali, A.M., Awadallah, A., Khalil, F.M. & Abd el-Hamid, M.A. (1975) Ecological studies on the zizyphus fruit fly *Carpomyia incompleta*, Becker. *Agricultural Research Review*, **53**, 115-121. [RAE **65**:5575]

Ali, A.S. Al- (1976) Outbreaks and new records. *Plant Protection Bulletin, Food and Agriculture Organization*, **24**, 133-137. [RAE **65**:5033]

Ali, A.S. Al-, Neamy, I.K. Al-, Abbas, S.A., Abdul-Masih, A.M.E. (1977) On the life-history of the safflower fly *Acanthiophilus helianthi* Rossi (Dipt., Tephritidae) in Iraq. *Zeitschrift für Angewandte Entomologie*, **83**, 216-223. [RAE **66**:824]

Ali, A.S. Al-, Abbas, S.A., Neamy, I.K. Al- & Abdul-Masih, A.M.E. (1979) On the biology of the yellow safflower-fly *Chaetorellia carthami* Stack. (Dipt., Tephritidae) in Iraq. *Zeitschrift für Angewandte Entomologie*, **87**, 439-445. [RAE **68**:477]

AliNiazee, M.T. (1979) A computerized phenology model for predicting biological events of *Rhagoletis indifferens* (Diptera: Tephritidae). *Canadian Entomologist*, **111**, 1101-1109. [RAE **68**:1984]

AliNiazee, M.T. (1985) Opiine parasitoids (Hymenoptera: Braconidae) of *Rhagoletis pomonella* and *R. zephyria* (Diptera: Tephritidae) in the Willamette Valley, Oregon. *Canadian Entomologist*, **117**, 163-166. [RAE **73**:4293]

AliNiazee, M.T. & Brown, R.D. (1974) A bibliography of North American cherry fruit flies (Diptera: Tephritidae). *Bulletin of the Entomological Society of America*, **20**, 93-101. [RAE **63**:905]

AliNiazee, M.T. & Brunner, J.F. (1986) Apple maggot in the western United States: A review of its establishment and current approaches to management. *Journal of the Entomological Society of British Columbia*, **83**, 49-53. [RAE **77**:5530]

AliNiazee, M.T. & Penrose, R.L. (1981) Apple maggot in Oregon: a possible threat to the Northwest apple industry. *Bulletin of the Entomological Society of America*, **27**, 245-246. [RAE **70**:6599]

AliNiazee, M.T. & Westcott, R.L. (1986) Distribution of the apple maggot, *Rhagoletis pomonella* (Diptera: Tephritidae). *Journal of the Entomological Society of British Columbia*, **83**, 54-56. [RAE **77**:5531]

Allman, S.L. (1941) Observations of various species of fruit flies. *Journal of the Australian Institute of Agricultural Science*, **7**, 155-156. [RAE **30**:499]

Aluja, M., Cabrera, M., Rios, E., Guillen, J., Celedonio, H., Hendrichs, J. & Liedo, P. (1987a) A survey of the economically important fruit flies (Diptera: Tephritidae) present in Chiapas and a few other fruit growing regions in Mexico. *Florida Entomologist*, **70**, 320-329. [RAE **76**:8838]

Aluja, M., Guillen, J., Rosa, G. de la, Cabrera, M., Celedonio, H., Liedo, P. & Hendrichs, J. (1987b) Natural host plant survey of the economically important fruit flies (Diptera: Tephritidae) of Chiapas, Mexico. *Florida Entomologist*, **70**, 329-330. [RAE **76**:8839]

Aluja, M., Guillen, J., Liedo, P., Cabrera, M. Rios, E., Rosa, G. de la, Celdonio, H. & Mota, D. (1990) Fruit infesting tephritids (Dipt.: Tephritidae) and associated parasitoids in Chiapas, Mexico. *Entomophaga*, **35**, 39-48. [RAE **78**:8636]

Amador, J.P.Ros (1988) La mosca mediterranea de la fruta, *Ceratitis capitata* Wied. biologia y metodos de control. *Hojas Divulgadores, Ministerio de Agricultura, Pesca y Alimentacion Spain*, (8), 1-28. [RAE **77**:3920]

Anderson, D.T. (1963) The larval development of *Dacus tryoni* (Frogg.) (Diptera: Trypetidae). I. Larval instars, imaginal discs, and haemocytes. *Australian Journal of Zoology*, **11**, 202-218.

Anderson, S.S., McCrea, K.D., Abrahamson, W.G. & Hartzel, L.M. (1989) Host genotype choice by the ball gallmaker *Eurosta solidaginis* (Diptera: Tephritidae). *Ecology*, **70**, 1048-1054. [RAE **78**:3616]

Andrashchuk, V.V. (1981) Susceptibility of the adult sea-buckthorn fly to entomopathogenic microorganisms. *Izvestiya Sibirskogo Otdeleniya Akademii Nauk SSSR, Seriya Biologicheskikh Nauk*, **1981**(15, 3), 119-125. [in Russian]. [RAE **70**:6249]

Andrashchuk, V.V. (1982) Susceptibility of the immature stages of the sea-buckthorn fly to entomopathogenic microorganisms. *Izvestiya Sibirskogo Otdeleniya Akademii Nauk SSSR, Seriya Biologicheskikh Nauk*, **1982**(5, 1), 95-103. [in Russian]. [RAE **70**:6250]

Andrews, H.W. (1941) *Paroxyna misella* Lw. and *Oxyna parietina* L. (Diptera: Trypetidae). A record of a failure and a success. *Proceedings of the South London Entomological and Natural History Society*, **1940-41**, 36-38.

Angood, S.A.S. Ba- (1977) Control of the melon fruit fly, *Dacus frontalis* Becker (Diptera: Trypetidae), on cucurbits. *Journal of Horticultural Science*, **52**, 545-547. [RAE **66**:2114]

Annecke, D.P. & Moran, V.C. (1982) *Insects and Mites of Cultivated Plants in South Africa*. Butterworths, Durban.

Anon. (1975) Outbreaks and new records. *Plant Protection Bulletin, Food and Agriculture Organization*, **23**, 26-28. [RAE **64**:2667]

Anon. (1979) Rapport annuel 1978. *Rapport, Institut de Recherches Agronomiques Tropicales et des Cultures Vivrières*. [RAE **68**:694]

Anon. (1986) Report of the expert consultation on progress and problems in controlling fruit fly infestation, Bangkok, 1986. *RAPA Publication*, **1986**(28), 1-18. Food and Agriculture Organisation, Regional Office for Asia and the Pacific, Bangkok.

Anon. (1987) Outbreaks and new records. USA. Eradication of Oriental fruit fly. *Plant Protection Bulletin, Food and Agriculture Organization*, **35**, 166. [RAE **78**:490]

Anon. (1991) Huge spraying operation in the Riverland; $170m fruit fly threat. *The Advertiser, Adelaide*. **133**(41295), 1-2.

Arakaki, N., Kuba, H. & Soemori, H. (1984) Mating behaviour of the Oriental fruit fly, *Dacus dorsalis* Hendel (Diptera: Tephritidae). *Applied Entomology and Zoology*, **19**, 42-51. [RAE **72**:6233]

Arambourg, Y. & Onillon, J. (1971; dated 1970) Élevage d'*Opius longicaudatus* Ash. *taiensis* Full., Hym. Braconidae, parasite de Trypetidae. *Annales de Zoologie Écologie Animale*, **2**, 663-665. [RAE **61**:1587]

Araoz, D. Fernandez de, & Nasca, A.J. (1985; dated 1984) Especies de Braconidae (Hymenoptera: Ichneumonoidea) parasitoides de moscas de los frutos (Diptera: Tephritidae) colectados en la provincia de Tucuman (Argentina). *Revista de Investigacion CIRPON*, **2**, 37-46. [RAE **73**:7525]

Arita, L.H. & Kaneshiro, K.Y. (1986) Structure and function of the rectal epithelium and anal glands during mating behavior in the mediterranean fruit fly male. *Proceedings of the Hawaiian Entomological Society*, **26**, 27-30. [RAE **75**:5176]

Arita, L.H. & Kaneshiro, K.Y. (1989) Sexual selection and lek behaviour in the Mediterranean fruit fly, *Ceratitis capitata* (Diptera: Tephritidae). *Pacific Science*, **43,** 135-143. [RAE **78:**766]

Armstrong, J.W. (1983) Infestation biology of three fruit fly (Diptera: Tephritidae) species on 'Brazilian', 'Valery' and 'William's' cultivars of banana in Hawaii. *Journal of Economic Entomology*, **76,** 539-543. [RAE **72:**730]

Armstrong, J.W. & Couey, H.M. (1989) Control; fruit disinfestation; fumigation, heat and cold, In: Robinson, A.S. & Hooper, G. (eds), Fruit flies; their biology, natural enemies and control. *World Crop Pests*, **3**(B), 411-424. Elsevier, Amsterdam.

Armstrong, J.W. & Vargas, R.I. (1982) Resistance of pineapple variety '59-656' to field populations of Oriental fruit flies and melon flies (Diptera: Tephritidae). *Journal of Economic Entomology*, **75**, 781-782. [RAE **71:**1257]

Armstrong, J.W., Mitchell, W.C. & Farias, G.J. (1983) Resistance of 'Sharwil' avocados at harvest maturity to infestation of three fruit fly species (Diptera: Tephritidae) in Hawaii. *Journal of Economic Entomology*, **76,** 119-121. [RAE **71:**7505]

Armstrong, J.W., Vriesenga, J.D. & Lee, C.Y.L. (1979) Resistance of pineapple varieties D-10 and D-20 to field populations of Oriental fruit flies and melon flies. *Journal of Economic Entomology*, **72,** 6-7. [RAE **67:**3878]

Armstrong, J.W., Schneider, E.L., Garcia, D.L. & Couey, H.M. (1984) Methyl bromide quarantine fumigation for strawberries infested with Mediterranean fruit fly (Diptera: Tephritidae). *Journal of Economic Entomology*, **77,** 680-682. [RAE **73:**469]

Armstrong, J.W., Harvey, J.M., Garcia, D.L., Menezes, T.D. & Brown, S.A. (1988) Methyl bromide fumigation for control of Oriental fruit fly (Diptera: Tephritidae) in California. *Journal of Economic Entomology*, **81,** 1120-1123. [RAE **77:**8571]

Arthur, A.P. & Mason, P.G. (1989) Description of the immature stages and notes on the biology of *Neotephritis finalis* (Loew.) (Diptera: Tephritidae), a pest of sunflowers in Saskatchewan, Canada. *Canadian Entomologist*, **121,** 729-735. [RAE **78:**8767]

Ashley, T.R. & Chambers, D.L. (1979) Effects of parasite density and host availability on progeny production by *Biosteres (Opius) longicaudatus*, (Hym., Braconidae), a parasite of *Anastrepha suspensa* (Dip.: Tephritidae). *Entomophaga*, **24,** 363-369.

Attenborough, D. (1990) *The Trials of Life*. Collins and BBC Books, London.

Averill, A.L. & Prokopy, R.J. (1987) Residual activity of oviposition-deterring pheromone in *Rhagoletis pomonella* (Diptera: Tephritidae) and female response to infested fruit. *Journal of Chemical Ecology*, **13,** 167-177. [RAE **75:**5612]

Averill, A.L. & Prokopy, R.J. (1989a) Biology and physiology; host-marking pheromones, In: Robinson, A.S. & Hooper, G. (eds), Fruit flies; their biology, natural enemies and control. *World Crop Pests*, **3**(A), 207-219. Elsevier, Amsterdam.

Averill, A.L. & Prokopy, R.J. (1989b) Distribution patterns of *Rhagoletis pomonella* (Diptera: Tephritidae) eggs in hawthorn. *Annals of the Entomological Society of America*, **82,** 38-44.

Averill, A.L., Bowdan, E.S. & Prokopy, R.J. (1987) Acid rain affects egg-laying behaviour of apple maggot flies. *Experimentia*, **43**, 939-942. [RAE **76**:2080]

Avidov, Z. (1966) Studies in agricultural entomology and plant pathology. *Scripta Hierosolymitana*, **18**, 1-208. [RAE **54**:636]

Avilla, J. & Albajes, R. (1984) Estudio de la relacion huesped-parasitoide *Ceratitis capitata* Wied. - *Opius concolor* Szepl. en conditiones de laboratorio: I. - Mortalidad de los estados postembrionarios del parasitoide. *Anales del Instituto Nacional de Investigaciones Agrarias, Agricola*, **27**, 65-77. [RAE **73**:2161]

Awadallah, A., Selim, O.F., Fouda, S. & Hashem, A.G. (1975) The chemical control of the zizyphus fruit fly, *Carpomyia incompleta*, Becker, in the Assiut area. *Agricultural Research Review*, **53**, 123-125. [RAE **65**:5576]

Ayatollahi, M. (1971) Importance of the study of Diptera and their role in biological control. *Entomologie et Phytopathologie Appliquées*, **31**, 20-28. [in Persian]. [RAE **61**:2778]

Azab, A.K., Nahal, A.K.M. El-, & Swailem, S.M. (1971; dated 1970) The immature stages of the melon fruit fly, *Dacus ciliatus* Loew (Diptera: Trypanaeidae). *Bulletin de la Société Entomologique d'Egypt*, **54**, 243-247. [RAE **62**:1913]

Back, E.A. & Pemberton, C.E. (1914) Life history of the melon fly. *Journal of Agricultural Research*, **3**, 269-274. [RAE **3**:163]

Baker, P.S., Hendrichs, J. & Liedo, P. (1988) Improvement of attractant dispensing systems for the Mediterranean fruit fly (Diptera: Tephritidae) sterile release program in Chiapas, Mexico. *Journal of Economic Entomology*, **81**, 1068-1072. [RAE **77**:9781]

Baker, R.T. & Cowley, J.M. (1991) A New Zealand view of quarantine security with special reference to fruit flies, In: Vijaysegaran, S. & Ibrahim, A.G. (eds), *First International Symposium on Fruit Flies in the Tropics*, Kuala Lumpur, 1988. Malaysian Agricultural Research and Development Institute, Kuala Lumpur. pp. 396-408.

Bala, A. (1987) Effect of nutritionally different diets on preoviposition period of the fruit fly, *Dacus tau* (Walker). *Annals of Agricultural Research*, **8**, 258-260. [RAE **77**:7290]

Banham, F.L. (1971) Native hosts of western cherry fruit fly (Diptera: Tephritidae) in the Okanagan Valley of British Columbia. *Journal of the Entomological Society of British Columbia*, **68**, 29-32. [RAE **61**:4958]

Banham, F.L. & Arrand, J.C. (1978) Biology and control of cherry fruit flies in British Columbia. *Publications of the British Columbia Ministry of Agriculture*, (78-13), 1-6.

Barlow, H.S., Enoch, I. & Russell, R.A. (1991) *H.F. Macmillan's Tropical Planting and Gardening*. 6th ed. Malayan Nature Society, Kuala Lumpur.

Barnes, B.N. (1976) Mass rearing the Natal fruit fly *Pterandrus rosa* (Ksh.) (Diptera: Trypetidae). *Journal of the Entomological Society of Southern Africa*, **39**, 121-124. [RAE **65**:681]

Barnes, B.N. (1979) Alternative larval media for mass rearing the Natal and Mediterranean fruit flies. *Journal of the Entomological Society of Southern Africa*, **42**, 55-56. [RAE **68**:177]

Barros, M.D., Novaes, M. & Malavasi, A. (1983) Estudos do comportamento de oviposicao de *Anastrepha fraterculus* (Wiedemann, 1830) (Diptera, Tephritidae) em condicoes naturais e de laboratorio. *Anais de Sociedade Entomologica do Brasil*, **12,** 243-247. [RAE **72:**6881]

Bateman, M.A. (1972) The ecology of fruit flies. *Annual Review of Entomology*, **17,** 493-518. [RAE **60:**1564]

Bateman, M.A. (1982) III. Chemical methods for suppression or eradication of fruit fly populations, In: Drew, R.A.I., Hooper, G.H.S. & Bateman, M.A. (eds), *Economic Fruit Flies of the South Pacific Region*. 2nd ed. Brisbane. pp. 115-128.

Batra, H.N. (1964) The population, behaviour, host specificity and development potential of fruit flies bred at constant temperature in winter. *Indian Journal of Entomology*, **26,** 195-206. [RAE **54:**218]

Batra, H.N. (1968) Biology and bionomics of *Dacus (Zeugodacus) hageni* de Meijere (=*D. caudatus* Fabr.). *Indian Journal of Agricultural Science*, **38,** 1015-1020.

Batra, S.W.T. (1979) Reproductive behaviour of *Euaresta bella* and *E. festiva* (Diptera: Tephritidae), potential agents for the biological control of adventive North American ragweeds (*Ambrosia* spp.) in Eurasia. *Journal of the New York Entomological Society*, **87,** 118-125. [RAE **68:**3351]

Beiger, M. (1968) Notatki o polskich muchówkach z rodziny Trypetidae (Diptera). *Fragmenta Faunistica*, **15,** 45-49.

Belanger, A., Bostanian, N.J. & Rivard, I. (1985) Apple maggot (Diptera: Trypetidae) control with insecticides and their residues in and on apples. *Journal of Economic Entomology*, **78,** 463-466. [RAE **74:**843]

Belcari, A. (1985) Presenza di *Acanthiophilus helianthi* su girasole in Toscana. *Informatore Fitopatologico*, **35,** 23-26. [RAE **74:**2969]

Bellas, T.E. & Fletcher, B.S. (1979) Identification of the major components in the secretion from the rectal pheromone glands of the Queensland fruit flies *Dacus tryoni* and *Dacus neohumeralis* (Diptera: Tephritidae). *Journal of Chemical Ecology*, **5,** 795-803. [RAE **68:**3983]

Ben-Hamouds, M.H. & Ben-Salah, H. (1984) Incidence parasitaire d'*Opius concolor* (Spzl.) (Hymenoptera, Braconidae) sur un hôte de remplacement *Ceratitis capitata* Wied. (Diptera, Trypetidae). *Archives de l'Institut Pasteur de Tunis*, **61,** 153-164. [RAE **73:**8041]

Ben-Salah, H., Cheikh, M. & Ben-Hamouda, M.H. (1989) Serodiagnostic of the host-parasite relationships of some fruit flies (Diptera, Tephritidae) and *Opius concolor* Sz. (Hymenoptera, Braconidae), In: Cavalloro, R. (ed.), *Fruit Flies of Economic Importance 87. Proceedings of the CEC/IOBC International Symposium*, Rome, 1987. Balkema, Rotterdam. pp. 295-310.

Benjamin, F.H. (1934) Descriptions of some native trypetid flies with notes on their habits. *Technical Bulletin, United States Department of Agriculture*, **401,** 1-95. [RAE **22:**220]

Berdyeva, N.G. (1978) The dynamics of emergence of the jujube fly in Turkmenia. *Izvestiya Akademii Nauk Turkmenskoi SSR, Biologicheskikh Nauk*, **2,** 91-93. [RAE **67:**1110]

Berg, G.H. (1979) *Pictorial Key to Fruit Fly Larvae of the Family Tephritidae.* Organismo International Regional de Sanidad Agropecuaria, San Salvador. [RAE **69**:3411]

Berlocher, S.H. (1980) An electrophoretic key for distinguishing species of the genus *Rhagoletis* (Diptera: Tephritidae) as larvae, pupae or adults. *Annals of the Entomological Society of America*, **73**, 131-137. [RAE **68**:5597]

Berlocher, S.H. (1984) A new North American species of *Rhagoletis* (Diptera: Tephritidae), with records of host plants of *Cornus*-infesting *Rhagoletis*. *Journal of the Kansas Entomological Society*, **57**, 237-242.

Berlocher, S.H. & Bush, G.L. (1982) An electrophoretic analysis of *Rhagoletis* (Diptera: Tephritidae) phylogeny. *Systematic Zoology*, **31**, 136-155.

Bernard, C. (1917a) Verslag van het proefstation voor thee over het jaar 1916. *Mededeelingen van het Proefstation voor Thee*, **53**, 1-12. [RAE **5**:416]

Bernard, C. (1917b) Over eenige ziekten en plagen van de thee op de oostkust van Sumatra. *Mededeelingen van het Proefstation voor Thee*, **54**, 1-21. [RAE **6**:37]

Berrigan, D.A., Carey, J.R., Guillen, J. & Celedonio, H. (1988) Age and host effects on clutch size in the Mexican fruit fly, *Anastrepha ludens*. *Entomologia Experimentalis et Applicata*, **47**, 73-80. [RAE **76**:7785]

Berube, D.E. (1978) The basis for host plant specificity in *Tephritis dilacerata* and *T. formosa* (Dipt.: Tephritidae). *Entomophaga*, **23**, 331-337. [RAE **67**:3748]

Bevan, W.J. (1966) Control of carrot fly on celery, with notes on other pests. *Plant Pathology*, **15**, 101-108. [RAE **57**:1399]

Bezzi, M. (1912) Intorno ad alcune *Ceratitis* raccolte nel'Africa occidentale. *Bollettino del Laboratorio di Zoologia Generale e Agraria della Facoltà Agraria in Portici*, **7**, 3-16. [RAE **1**:91]

Bezzi, M. (1916) On the fruit flies of the genus *Dacus* occuring in India, Burma and Ceylon. *Bulletin of Entomological Research*, **7**, 99-121. [RAE **5**:3]

Bezzi, M. (1924a) Further notes on the Ethiopian fruit-flies, with keys to all the known genera and species. *Bulletin of Entomological Research*, **15**, 73-118. [RAE **13**:15]

Bezzi, M. (1924b) Further notes on the Ethiopian fruit-flies, with keys to all the known genera and species -(cont.). *Bulletin of Entomological Research*, **15**, 121-155. [RAE **13**:15]

Bezzi, M. (1924c) 9. South African trypaneid Diptera in the collection of the South African Museum. *Annals of the South African Museum*, **19**, 449-577.

Bezzi, M. (1928) *Diptera Brachycera and Athericera of the Fiji Islands Based on Material in the British Museum (Natural History)*. British Museum (Natural History), London.

Bhatia, S.K. & Mahto, Y. (1968) Notes on breeding of fruit flies *Dacus ciliatus* Loew and *D. cucurbitae* Coquillett in stem galls of *Coccinia indica* W. & A. *Indian Journal of Entomology*, **30**, 244-245. [RAE **59**:1365]

Bierbaum, T.J. & Bush, G.L. (1990a) Genetic differentiation in the viability of sibling species of *Rhagoletis* fruit flies on host plants, and the influence of reduced hybrid viability on reproductive isolation. *Entomologia Experimentalis et Applicata*, **55**, 105-118. [RAE **78**:11616]

Bierbaum, T.J. & Bush, G.L. (1990b) Host fruit chemical stimuli eliciting distinct ovipositional responses from sibling species of *Rhagoletis* fruit flies. *Entomologia Experimentalis et Applicata*, **56**, 165-177. [RAE **79**:521]

Bigler, F. & Delucchi, V. (1981) Wichtigste Mortalitätsfaktoren während der präpupalen Entwicklung der Olivenfliege, *Dacus oleae* Gmel. (Dipt., Tephritidae) auf Oleastern und kultivierten Oliven in Westkreta, Griechenland. *Zeitschrift für Angewandte Entomologie*, **92**, 343-363. [RAE **70**:1558]

Bigler, F., Neuenschwander, P., Delucchi, V. & Michelakis, S. (1986) Natural enemies of preimaginal stages of *Dacus oleae* Gmel. (Dipt., Tephritidae) in western Crete. II. Impact on olive fly populations. *Bolletino del Laboratorio di Entomologia Agraria 'Filippo Silvestri'*, **43**, 79-96. [RAE **77**:4746]

Blanc, F.L. & Keifer, H.H. (1955) The cherry fruit fly in North America. *Bulletin of the California Department of Agriculture*, **44**, 77-88.

Blanchard, E.E. (1959) El genero "*Toxotrypana*" en la Republica Argentine. *Acta Zoologià Lilloana*, **17**, 33-44.

Blümel, S. & Russ, K. (1989) Control; manipulation of races, In: Robinson, A.S. & Hooper, G. (eds), Fruit flies; their biology, natural enemies and control. *World Crop Pests*, **3**(B), 387-389. Elsevier, Amsterdam.

Boller, E.F. (1989a) Genetics; cytoplasmic incompatibility in *Rhagoletis cerasi*, In: Robinson, A.S. & Hooper, G. (eds), Fruit flies; their biology, natural enemies and control. *World Crop Pests*, **3**(B), 69-74. Elsevier, Amsterdam.

Boller, E.F. (1989b) Rearing; small-scale rearing; *Rhagoletis* spp., In: Robinson, A.S. & Hooper, G. (eds), Fruit flies; their biology, natural enemies and control. *World Crop Pests*, **3**(B), 119-127. Elsevier, Amsterdam.

Boller, E.F. & Bush, G.L. (1974) Evidence for genetic variation in populations of the European cherry fruit fly, *Rhagoletis cerasi* (Diptera: Tephritidae) based on physiological parameters and hybridization experiments. *Entomologia Experimentalis et Applicata*, **17**, 279-293. [RAE **64**:73]

Boller, E.F. & Hurter, J. (1985) Oviposition deterring pheromone in *Rhagoletis cerasi*: behavioural laboratory test to measure pheromone activity. *Entomologia Experimentalis et Applicata*, **39**, 163-169. [RAE **74**:602]

Boller, E.F. & Prokopy, R.J. (1976) Bionomics and management of *Rhagoletis*. *Annual Review of Entomology*, **21**, 223-246. [RAE **64**:5971]

Boller, E.F., Russ, K., Vallo, V. & Bush, G.L. (1976) Incompatible races of European cherry fruit fly, *Rhagoletis cerasi* (Diptera: Tephritidae), their origin and potential use in biological control. *Entomologia Experimentalis et Applicata*, **20**, 237-247. [RAE **65**:3627]

Boller, E.F., Schoni, R. & Bush, G.L. (1987) Oviposition deterring pheromone in *Rhagoletis cerasi*: biological activity of a pure single compound verified in semi-field test. *Entomologia Experimentalis et Applicata*, **45**, 17-22. [RAE **76**:3336]

Bose, P.C. & Mehrotra, K.N. (1986) Thrust of the ovipositor of fruit fly, *Dacus dorsalis* Hendel. *Current Science, India*, **55**, 1004-1005. [RAE **76**:640]

Boyce, A.M. (1934) Bionomics of the walnut husk fly, *Rhagoletis completa*. *Hilgardia*, **8**, 363-579. [RAE **23**:413]

Braga, M.A. & Zucoloto, F.S. (1981) Estudos sobre a melhor concentaçao de aminoácidos para moscas adultas de *Anastrepha obliqua* (Diptera, Tephritidae). *Revista Brasileira de Biologia*, **41,** 75-79. [RAE **70:**1327]

Bredo, H.J. (1934) Catalogue des principaux insectes et nématodes parasites des caféiers dans les Uelés. *Bulletin Agricole du Congo Belge*, **25,** 494-514. [RAE **23:**378]

Brimblecomb, A.R. (1945) The biology, economic importance and control of the pine bark weevil, *Aesiotes notabilis* Pasc. *Queensland Journal of Agricultural Science*, **2,** 1-88. [RAE **34:**258]

Bueno, A. Montiel & Mata, M.A. Simon (1985) La interrupcion de la comunicacion sexual de la mosca del olivo (*Dacus oleae* Gmel.) como estrategia de lucha integrada en olivar. *Boletin del Servicio de Defensa contra Plagas e Inspeccion Fitopatologica*, **11,** 11-23. [RAE **74:**1382]

Burditt, A.K. (1982) *Anastrepha suspensa* (Loew) (Diptera: Tephritidae) McPhail traps for survey and detection. *Florida Entomologist*, **65,** 367-373. [RAE **71:**3541]

Burditt, A.K. (1988) Western cherry fruit fly (Diptera: Tephritidae): Efficacy of homemade and commercial traps. *Journal of the Entomological Society of British Columbia*, **85,** 53-57. [RAE **77:**8576]

Burditt, A.K. & Balock, J.W. (1985) Refrigeration as a quarantine treatment for fruits and vegetables infested with eggs and larvae of *Dacus dorsalis* and *Dacus cucurbitae* (Diptera: Tephritidae). *Journal of Economic Entomology*, **78,** 885-887. [RAE **74:**1539]

Burditt, A.K. & Hungate, F.P. (1988) Gamma irradiation as a quarantine treatment for cherries infested by western cherry fruit fly (Diptera: Tephritidae). *Journal of Economic Entomology*, **81,** 859-862. [RAE **77:**5916]

Burditt, A.K. & White, L.D. (1987) Parasitization of western cherry fruit fly by *Pachycrepoideus vindemiae*. *Florida Entomologist*, **70,** 405-406. [RAE **76:**8878]

Burk, T. (1983) Behavioral ecology of mating in the Caribbean fruit fly, *Anastrepha suspensa* (Loew) (Diptera: Tephritidae). *Florida Entomologist*, **66,** 330-334. [RAE **72:**2143]

Burk, T. (1984) Male-male interactions in Caribbean fruit flies, *Anastrepha suspensa* (Loew) (Diptera: Tephritidae): territorial fights and signalling stimulation. *Florida Entomologist*, **67,** 542-547. [RAE **73:**5025]

Burk, T. & Calkins, C.O. (1983) Medfly mating behavior and control strategies. *Florida Entomologist*, **66,** 3-18. [RAE **71:**6759]

Burk, T. & Webb, J.C. (1983) Effect of male size on calling propensity, song parameters, and mating succcess in Caribbean fruit flies, *Anastrepha suspensa* (Loew) (Diptera: Tephritidae). *Annals of the Entomological Society of America*, **76,** 678-682. [RAE **72:**571]

Bush, G.L. (1965) The genus *Zonosemata* with notes on the cytology of two species. *Psyche*, **72,** 307-323.

Bush, G.L. (1966) The taxonomy, cytology and evolution of the genus *Rhagoletis* in North America (Diptera: Tephritidae). *Bulletin of the Museum of Comparative Zoology, Harvard*, **134,** 431-526.

Bush, G.L. (1969) Sympatric host formation and speciation in frugivorous flies of the genus *Rhagoletis*. *Evolution*, **23**, 237-251.

Bush, G.L. & Boller, E. (1977) The chromosome morphology of the *Rhagoletis cerasi* species complex (Diptera, Tephritidae). *Annals of the Entomological Society of America*, **70**, 316-318. [RAE **65**:6422]

Butani, D.K. (1976) Insect pests of fruit crops and their control - custard apple. *Pesticides*, **10**, 27-28. [RAE **66**:2728]

Butani, D.K. (1978) Insect pests of fruit crops and their control: 25 - Mulberry. *Pesticides*, **12**, 53-59. [RAE **68**:378]

Butt, F.H. (1937) The posterior stigmatic apparatus of trypetid larvae. *Annals of the Entomological Society of America*, **30**, 487-491.

Calkins, C.O. (1989) Rearing; quality control, In: Robinson, A.S. & Hooper, G. (eds), Fruit flies; their biology, natural enemies and control. *World Crop Pests*, **3**(B), 153-165. Elsevier, Amsterdam.

Calkins, C.O. & Webb, J.C. (1988) Temporal and seasonal differences in movement of the Caribbean fruit fly larvae in grapefruit and the relationship to detection by acoustics. *Florida Entomologist*, **71**, 409-416. [RAE **78**:3818]

Calkins, C.O., Schroeder, W.J. & Chambers, D.L. (1984) Probability of detecting Caribbean fruit fly, *Anastrepha suspensa* (Loew) (Diptera: Tephritidae), populations with McPhail traps. *Journal of Economic Entomology*, **77**, 198-201. [RAE **72**:6488]

Cals-Usciati, J. (1972) Les relations hôte-parasite dans le couple *Ceratitis capitata* Wiedemann (Diptera, Trypetidae) et *Opius concolor* Szepligetti (Hymenoptera, Braconidae). I. Morphologie et organogenèse de *Ceratitis capitata* (développement embryonnaire et larvaire). *Annales de Zoologie, Écologie Animale*, **4**, 427-481. [RAE **62**:4071]

Cals-Usciati, J., Cals, P. & Pralavorio, R. (1985) Adaptations fonctionnelles des structures liées à la prise alimentaire chez la larve primaire de *Trybliographa daci* Weld (Hymenoptera Cynipoidea), endoparasitoide de la mouche de fruits *Ceratitis capitata*. *Comptes Rendus Hebdomadaires des Séances de l'Academie des Sciences, III (Sciences de la Vie)*, **300**, 103-108. [RAE **74**:3911]

Cameron, P.J. & Morrison, F.O. (1974a) *Psilus* sp. (Hymenoptera: Diapriidae), a parasite of the pupal stage of the apple maggot, *Rhagoletis pomonella* (Diptera: Tephritidae) in southwestern Quebec. *Phytoprotection*, **55**, 13-16. [RAE **63**:2531]

Cameron, P.J. & Morrison, F.O. (1974b) Sampling methods for estimating the abundance and distribution of all life stages of the apple maggot, *Rhagoletis pomonella* (Diptera: Tephritidae). *Canadian Entomologist*, **106**, 1025-1034. [RAE **63**:3464]

Cameron, P.J. & Morrison, F.O. (1977) Analysis of mortality in the apple maggot, *Rhagoletis pomonella* (Diptera: Tephritidae), in Quebec. *Canadian Entomologist*, **109**, 769-788. [RAE **66**:758]

Caraballo de Valdivieso, J. (1985) Nuevas especies del genero *Anastrepha* Schiner, 1868 (Diptera: Tephritidae) de Venezuela. *Boletín de Entomologia Venezolana*, **4**, 25-32.

Carey, J.R. (1989) Ecology; demographic analysis of fruit flies, In: Robinson, A.S. & Hooper, G. (eds), Fruit flies; their biology, natural enemies and control. *World Crop Pests*, **3(B)**, 253-265. Elsevier, Amsterdam.

Carey, J.R. & Dowell, R.V. (1989) Exotic fruit pests and California agriculture. *California Agriculture*, **43**(5/6), 38-40. [RAE **78**:3753]

Carle, S.A., Averill, A.L., Rule, G.S., Reissig, W.H. & Roelofs, W.L. (1987) Variation in host fruit volatiles attractive to apple maggot fly, *Rhagoletis pomonella*. *Journal of Chemical Ecology*, **13**, 795-805. [RAE **75**:6446]

Carlson, D.A. & Yocom, S.R. (1986) Cuticular hydrocarbons from six species of tephritid fruit flies. *Archives of Insect Biochemistry and Physiology*, **3**, 397-412.

Carroll, L.E. (1986) Larval morphology of Tephritidae, In: Darvas, B. & Papp, L. (eds), *1st International Congress of Dipterology, Abstract Volume*. p. 34.

Carroll, L.E. & Wharton, R.A. (1989) Morphology of the immature stages of *Anastrepha ludens* (Diptera: Tephritidae). *Annals of the Entomological Society of America*, **82**, 201-214.

Carter, W. (1950) The Oriental fruit fly: progress on research. *Journal of Economic Entomology*, **43**, 677-683. [RAE **39**:175]

Cavalloro, R. (ed.) (1983) *Fruit Flies of Economic Importance. Proceedings of the CEC/IOBC International Symposium*, Athens, 1982. Balkema, Rotterdam. [RAE **71**:6672]

Cavalloro, R. (ed.) (1986) *Fruit Flies of Economic Importance 84. Proceedings of the CEC/IOBC 'ad-hoc meeting'*, Hamburg, 1984. Balkema, Rotterdam. [RAE **74**:4900]

Cavalloro, R. (ed.) (1989) *Fruit Flies of Economic Importance 87. Proceedings of the CEC/IOBC International Symposium*, Rome, 1987. Balkema, Rotterdam. [RAE **78**:4240]

Cavender, G.L. & Goeden, R.D. (1982) Life history of *Trupanea bisetosa* (Diptera: Tephritidae) on wild sunflower in southern California. *Annals of the Entomological Society of America*, **75**, 400-406. [RAE **71**:467]

Cavender, G.L. & Goeden, R.D. (1984) The life history of *Paracantha cultaris* (Coquillett) on wild sunflower, *Helianthus annuus* L. ssp. *lenticularis* (Douglas) Cockerell, in southern California (Diptera: Tephritidae). *Pan-Pacific Entomologist*, **60**, 213-218. [RAE **73**:3563]

Celedonio-Hurtado, H., Liedo, P., Aluja, M., Guillen, J., Berrigan, D. & Carey, J. (1988) Demography of *Anastrepha ludens, A. obliqua* and *A. serpentina* (Diptera: Tephritidae) in Mexico. *Florida Entomologist*, **71**, 111-120. [RAE **77**:6161]

Chao, Y.-s. (1987) The two species of fruit flies on oranges in China. *Technical Bulletin of Plant Quarantine Research*, **1987**(5), 1-10.

Chen, S.H. & Zia, Y. (1955) Taxonomic notes on the Chinese citrus fly, *Tetradacus citri* (Chen). *Acta Entomologica Sinica*, **5**, 123-126.

Cheng, C.C. & Lee, W.Y. (1991) Fruit flies in Taiwan, In: Vijaysegaran, S. & Ibrahim, A.G. (eds), *First International Symposium on Fruit Flies in the Tropics*, Kuala Lumpur, 1988. Malaysian Agricultural Research and Development Institute, Kuala Lumpur. pp. 152-160.

Christenson, L.D. & Foote, R.H. (1960) Biology of fruit flies. *Annual Review of Entomology*, **5**, 171-192. [RAE **48**:184]

Chughtai, G.H., Khan, S. & Baloch, U.K. (1985) A new record of infestation of melon fruits by an anthomyiid fly in Indus River beach areas of D.I. Khan. *Pakistan Journal of Zoology*, **17**, 165-168. [RAE **75**:1238]

Chuman, T., Landolt, P.J., Heath, R.R. & Tumlinson, J.H. (1987) Isolation, identification, and synthesis of male-produced sex pheromone of papaya fruit fly, *Toxotrypana curvicauda* Gerstaecker (Diptera: Tephritidae). *Journal of Chemical Ecology*, **13**, 1979-1992. [RAE **76**:6937]

CIE (1957) *Dacus oleae* (Gmel.) (the olive fly). *Commonwealth Institute of Entomology, Distribution Maps of Insect Pests*, series A, (74), 1-2. Commonwealth Institute of Entomology, London. [RAE **45**:290]

CIE (1958a) *Anastrepha fraterculus* (Wied.) (Dipt., Trypetidae). *Commonwealth Institute of Entomology, Distribution Maps of Insect Pests*, series A, (88), 1-2. Commonwealth Institute of Entomology, London. [RAE **46**:240]

CIE (1958b) *Anastrepha ludens* (Lw.) (Dipt., Trypetidae) (Mexican fruit-fly). *Commonwealth Institute of Entomology, Distribution Maps of Insect Pests*, series A, (89), 1-2. Commonwealth Institute of Entomology, London. [RAE **46**:240]

CIE (1961a) *Myiopardalis pardalina* (Big.) (Dipt., Trypetidae) (Baluchistan melon fly). *Commonwealth Institute of Entomology, Distribution Maps of Pests*, series A (Agricultural), (124), 1-2. Commonwealth Institute of Entomology, London. [RAE **49**:344]

CIE (1961b) *Dacus zonatus* Saund. (Dipt., Trypetidae) (peach fruit-fly (India)). *Commonwealth Institute of Entomology, Distribution Maps of Pests*, series A (Agricultural), (125), 1-2. Commonwealth Institute of Entomology, London. [RAE **49**:344]

CIE (1962) *Pardalaspis cyanescens* Bez. (Dipt., Trypetidae). *Commonwealth Institute of Entomology, Distribution Maps of Pests*, series A (Agricultural), (140), 1-2. Commonwealth Agricultural Bureaux, London. [RAE **50**:357]

CIE (1963a) *Rhagoletis fausta* (O.-S.) (Dipt., Trypetidae) (black cherry fruit-fly, of North America). *Commonwealth Institute of Entomology, Distribution Maps of Pests*, series A (Agricultural), (160), 1-2. Commonwealth Agricultural Bureaux, London. [RAE **51**:389]

CIE (1963b) *Pardalaspis quinaria* Bez. (Dipt., Trypetidae) (Rhodesian fruit-fly). *Commonwealth Institute of Entomology, Distribution Maps of Pests*, series A (Agricultural), (161), 1-2. Commonwealth Agricultural Bureaux, London. [RAE **51**:389]

CIE (1966) *Ceratitis catoirii* Guérin-Méneville (Dipt., Tephritidae). *Commonwealth Institute of Entomology, Distribution Maps of Pests*, series A (Agricultural), (226), 1-2. Commonwealth Agricultural Bureaux, London. [RAE **55**:648]

CIE (1974) *Dacus ciliatus* Lw. (*D. brevistylus* Bez.) (Dipt., Tephritidae) (lesser pumpkin fly). *Commonwealth Institute of Entomology, Distribution Maps of Pests*, series A (Agricultural), (323), 1-2. Commonwealth Agricultural Bureaux, London. [RAE **62**:3544]

CIE (1978) *Dacus cucurbitae* Coq. (Dipt., Tephritidae) (melon fruit-fly). *Commonwealth Institute of Entomology, Distribution Maps of Pests*, series A (Agricultural), (64), revised, 1-2. Commonwealth Agricultural Bureaux, London. [RAE **66**:5841]

CIE (1986a; dated 1985) *Pterandrus rosa* (Karsch) [Diptera: Tephritidae] Natal fruit fly. *Commonwealth Institute of Entomology, Distribution Maps of Pests*, series A (Agricultural), (153), 1-2. Commonwealth Agricultural Bureaux, London. [RAE **74**:2650]

CIE (1986b) *Dacus dorsalis* Hendel [Diptera: Tephritidae] Oriental fruit fly. *CAB International Institute of Entomology, Distribution Maps of Pests*, series A (Agricultural), (109), revised, 1-2. CAB International, London.

CIE (1988a) *Ceratitis capitata* (Wiedemann) [Diptera: Tephritidae] Mediterranean fruit fly, medfly. *CAB International Institute of Entomology, Distribution Maps of Pests*, series A (Agricultural), (1), revised, 1-3. CAB International, London.

CIE (1988b) *Anastrepha obliqua* (Macquart) (=*A. mombinpraeoptans* Sein) [Diptera: Tephritidae] West Indian fruit-fly. *CAB International Institute of Entomology, Distribution Maps of Pests*, series A (Agricultural), (90), revised, 1-2. CAB International, London.

CIE (1989a) *Rhagoletis cerasi* (Linnaeus) [Diptera: Tephritidae] cherry fruit fly. *CAB International Institute of Entomology, Distribution Maps of Pests*, series A (Agricultural), (65), revised, 1-3. CAB International, London. [RAE **79**:196]

CIE (1989b) *Rhagoletis pomonella* (Walsh) [Diptera: Tephritidae] apple maggot, apple fruit fly. *CAB International Institute of Entomology, Distribution Maps of Pests*, series A (Agricultural), (48), revised, 1-3. CAB International, London.

CIE (1990a) *Rhagoletis cingulata* (Loew) [Diptera: Tephritidae] eastern cherry fruit fly, North American cherry fruit fly. *CAB International Institute of Entomology, Distribution Maps of Pests*, series A (Agricultural), (159), revised, 1-2. CAB International, London. [RAE **79**:2209]

CIE (1990b) *Rhagoletis indifferens* Curran [Diptera: Tephritidae] western cherry fruit fly, North American cherry fruit fly. *CAB International Institute of Entomology, Distribution Maps of Pests*, series A (Agricultural), (513), 1-2. CAB International, London.

Cirio, U. & Vita, G. (1980) Fruit fly control by chemical attractants and repellents. *Bollettino del Laboratorio di Entomologia Agraria 'Filippo Silvestri', Portici*, **37**, 127-139. [RAE **69**:5937]

Clausen, C.P. (1978) Tephritidae (Trypetidae, Trupaneidae), In: Clausen, C.P. (ed.), Introduced parasites and predators of arthropod pests and weeds: a world review. *Agricultural Handbook, United States Department of Agriculture*, **480**, 320-335.

Clausen, C.P., Clancy, D.W. & Chock, Q.C. (1965) Biological control of the Oriental fruit fly (*Dacus dorsalis* Hendel) and other fruit flies in Hawaii. *Technical Bulletin, United States Department of Agriculture*, **1322**, 1-102.

Cochereau, P. (1970) Les mouches des fruits et leurs parasites dans la zone Indo-Australo-Pacifique et particulièrement en Nouvelle-Calédonie. *Cahiers ORSTOM, Série Biologie*, **12**, 15-50. [RAE **63**:838]

Cogan, B.H. & Munro, H.K. (1980) 40. Family Tephritidae, In: Crosskey, R.W. (ed.), *Catalogue of the Diptera of the Afrotropical Region*. British Museum (Natural History), London. pp. 518-554.

Cooley, S.S., Prokopy, R.J., McDonald, P.T. & Wong, T.T.Y. (1986) Learning in oviposition site selection by *Ceratitis capitata* flies. *Entomologia Experimentalis et Applicata*, **40,** 47-51. [RAE **74:**1854]

Cordero-Jenkins, L.E., Jiron, L.F. & Lezama, H.J. (1990) Notes on the biology and ecology of *Ptecticus testaceus* (Diptera: Stratiomyidae) a soldier fly associated with mango fruit (*M. indica*) in Costa Rica. *Tropical Pest Management*, **36,** 285-286. [RAE **79:**2661]

Couey, H.M. & Hayes, C.F. (1986) Quarantine procedure for Hawaiian papaya using fruit selection and a two-stage hot-water immersion. *Journal of Economic Entomology*, **79,** 1307-1314. [RAE **75:**2538]

Couey, H.M., Armstrong, J.W., Hylin, J.W., Thornburg, W., Nakamura, A.N., Linse, E.S., Ogata, J. & Vetro, R. (1985) Quarantine procedure for Hawaii papaya, using a hot-water treatment and high temperature, low-dose ethylene dibromide fumigation. *Journal of Economic Entomology*, **78,** 879-884. [RAE **74:**1553]

Cowley, J.M., Page, F.D., Nimmo, P.R. & Cowley, D.R. (1990) Comparison of effectiveness of two traps for *Bactrocera tryoni* (Diptera: Tephritidae) and implications for quarantine surveillance systems. *Journal of the Australian Entomological Society*, **29,** 171-176. [RAE **79:**5668]

Cresson, E.T. (1929) A revision of the North American species of fruit-flies of the genus *Rhagoletis* (Diptera: Trypetidae). *Transactions of the American Entomological Society*, **55,** 401-414. [RAE **18:**207]

Crnjar, R., Angioy, A.M., Stoffolano, J.G., Barbarossa, I.T. & Pietra, P. (1987) Taste and olfactory responses of ovipositor chemosensilla in *Rhagoletis pomonella*. *Chemical Senses*, **12,** 208. [RAE **76:**6909]

Cunningham, R.T. (1989a) Biology and physiology; parapheromones, In: Robinson, A.S. & Hooper, G. (eds), Fruit flies; their biology, natural enemies and control. *World Crop Pests*, **3**(A), 221-230. Elsevier, Amsterdam.

Cunningham, R.T. (1989b) Population detection and assessment; population detection, In: Robinson, A.S. & Hooper, G. (eds), Fruit flies; their biology, natural enemies and control. *World Crop Pests*, **3**(B), 169-173. Elsevier, Amsterdam.

Cunningham, R.T. (1989c) Control; insecticides; male annihilation, In: Robinson, A.S. & Hooper, G. (eds), Fruit flies; their biology, natural enemies and control. *World Crop Pests*, **3**(B), 345-351. Elsevier, Amsterdam.

Cunningham, R.T. & Suda, D.Y. (1985) Male annihilation of the Oriental fruit fly, *Dacus dorsalis* Hendel (Diptera: Tephritidae): a new thickener and extender for methyl eugenol formulations. *Journal of Economic Entomology*, **78,** 503-504. [RAE **74:**1143]

Curran, C.H. (1924) *Rhagoletis pomonella* and two allied species (Trypaneidae, Diptera). *54th Annual Report of the Entomological Society of Ontario*. **1923,** 56-57. [RAE **12:**532]

Dadour, I.R., Yeates, D.K. & Postle, A.C. (in press) Two rapid diagnostic techniques for distinguishing Mediterranean fruit fly from Queensland fruit fly (Diptera: Tephritidae). *Journal of Economic Entomology.*

Daramola, A.M. & Ivbijaro, M.F. (1975) The ecology and distribution of kola weevils in Nigeria. *Nigerian Journal of Plant Protection*, **1**, 5-9. [RAE **64**:2003]

Daxl, R. (1978) Mediterranean fruit fly ecology in Nicaragua, and a proposal for integrated control. *Plant Protection Bulletin, Food and Agriculture Organization*, **26**, 150-157. [RAE **67**:4204]

Dean, R.W. (1969) Infestation of peaches by *Rhagoletis suavis*. *Journal of Economic Entomology*, **62**, 940-941. [RAE **58**:124]

Debouzie, D. (1989) Ecology; biotic mortality factors in tephritid populations, In: Robinson, A.S. & Hooper, G. (eds), Fruit flies; their biology, natural enemies and control. *World Crop Pests*, **3**(B), 221-227. Elsevier, Amsterdam.

Delrio, G. (1979) *Dacus oleae* Gmel.: bibliographia 1966-1979. *CNR Programma Finalizzato 'Fitofarmaci e Fitoregolatori'*, Firenze.

Delrio, G. & Ortu, S. (1988) Attraction of *Ceratitis capitata* to sex pheromones, trimedlure, ammonium and protein. *Bulletin, Section Regionale Ouest Palearctique, Organisation Internationale de Lutte Biologique*, **11**(6), 20-25. [RAE **77**:9580]

Delrio, G. & Prota, R. (1977; dated 1975/6) Osservazioni eco-etologiche sul *Dacus oleae* Gmelin nella Sardegna nord-occidentale. *Bollettino di Zoologia Agraria e di Bachicoltura*, **13**, 49-118. [RAE **66**:773]

Dickens, J.C., Hart, W.G., Light, D.M. & Jang, E.B. (1988) Tephritid olfaction: morphology of the antennae of four tropical species of economic importance (Diptera: Tephritidae). *Annals of the Entomological Society of America*, **81**, 325-331. [RAE **76**:7373]

Diehl, S.R. & Bush, G.L. (1983) The use of trace element analysis for determining the larval host plant of adult *Rhagoletis* (Diptera: Tephritidae), In: Cavalloro, R. (ed.), *Fruit Flies of Economic Importance. Proceedings of the CEC/IOBC International Symposium*, Athens, 1982. Balkema, Rotterdam. pp. 276-284. [RAE **71**:6322]

Diehl, S.R. & Prokopy, R.J. (1986) Host-selection behavior differences between the fruit fly sibling species *Rhagoletis pomonella* and *R. mendax* (Diptera: Tephritidae). *Annals of the Entomological Society of America*, **79**, 266-271. [RAE **74**:5151]

Dingler, M. (1934) Die Spargelfliege (*Platyparea poeciloptera* Schrank). *Arbeiten über Physiologische und Angewandte Entomologie*, **1**, 131-217. [RAE **22**:725]

Dirlbek, J. (1973) Supplement to the key to the determination of Paleoarctic species of the genus *Myopites* containing recently discovered species. *Annotationes Zoologicae et Botanicae, Slovenské Národné Múzeum v Bratislava*, **73**, 1-3.

Dirlbek, J. (1974) Contribution to the knowledge of Trypetidae in Cyprus (Diptera). *Sborník Faunistickych Prací Entomologického Oddeleni Národního Musea v Praze*, **15**, 69-78.

Dirsh, V. (1933) Pests of rubber-producing plants in the Ukraine. *Zhurnal Heleoho-Heohrafichnoho Tsyklu*, **4**, 41-58. [in Russian]. [RAE **23**:339]

Dodson, G. (1982) Mating and territoriality in wild *Anastrepha suspensa* (Diptera: Tephritidae) in field cages. *Journal of the Georgia Entomological Society*, **17,** 189-200. [RAE **70:**5375]

Dodson, G. (1987) Host-plant records and life history notes on New Mexico Tephritidae (Diptera). *Proceedings of the Entomological Society of Washington*, **89,** 607-615. [RAE **76:**6743]

Dodson, G. & Daniels, G. (1988) Diptera reared from *Dysoxylum gaudichaudianum* (Juss.) Miq. at Iron Range, northern Queensland. *Australian Entomological Magazine*, **15,** 77-79. [RAE **78:**1875]

Doss, M.C. (1989) Mediterranean fruit fly. *Pest Data Sheet*, (32), 1-6. Asean Plant Quarantine Centre and Training Institute, Serdang.

Dowell, R.V. & Wange, L.K. (1986) Process analysis and failure avoidance in fruit fly programs, In: Mangel, M., Carey, J.R. & Plant, R.E. (eds), Pest control: operations and systems analysis in fruit fly management. *NATO Advanced Science Institutes Series G: Ecological Sciences*, **11,** 43-65. Springer Verlag, Berlin.

Drensky, P. (1931) Pests of Alibotush tea - *Siderites scardica. Izvestiya na Bulgarskoto Entomologichno Druzhestvo*, **6,** 139-141. [in Bulgarian]. [RAE **19:**426]

Drew, R.A.I. (1972a) The generic and subgeneric classification of Dacini (Diptera: Tephritidae) from the South Pacific area. *Journal of the Australian Entomological Society*, **11,** 1-22. [RAE **61:**1860]

Drew, R.A.I. (1972b) Additions to the species of Dacini (Diptera: Tephritidae) from the South Pacific area with keys to species. *Journal of the Australian Entomological Society*, **11,** 185-231. [RAE **61:**3162]

Drew, R.A.I. (1973) Revised descriptions of species of Dacini (Diptera: Tephritidae) from the South Pacific area. I. Genus *Callantra* and the *Dacus* group of subgenera of the genus *Dacus*. *Queensland Division of Plant Industry Bulletin*, **652,** 1-39.

Drew, R.A.I. (1974) The responses of fruit fly species (Diptera: Tephritidae) in the South Pacific area to male attractants. *Journal of the Australian Entomological Society*, **13,** 267-270. [RAE **64:**3344]

Drew, R.A.I. (1982a) I. Taxonomy, In: Drew, R.A.I., Hooper, G.H.S. & Bateman, M.A. (eds), *Economic Fruit Flies of the South Pacific Region*. 2nd ed. Brisbane. pp. 1-97.

Drew, R.A.I. (1982b) IV. Fruit fly collecting, In: Drew, R.A.I., Hooper, G.H.S. & Bateman, M.A. (eds), *Economic Fruit Flies of the South Pacific Region*. 2nd ed. Brisbane. pp. 129-139.

Drew, R.A.I. (1987a) Behavioural strategies of fruit flies of the genus *Dacus* (Diptera: Tephritidae) significant in mating and host-plant relationships. *Bulletin of Entomological Research*, **77,** 73-81. [RAE **75:**2871]

Drew, R.A.I. (1987b) Reduction in fruit fly (Tephritidae: Dacinae) populations in their endemic rainforest habitat by frugivorous vertebrates. *Australian Journal of Zoology*, **35,** 283-288.

Drew, R.A.I. (1989a) The tropical fruit flies (Diptera: Tephritidae: Dacinae) of the Australasian and Oceanian regions. *Memoirs of the Queensland Museum*, **26,** 1-521. [RAE **78:**13]

Drew, R.A.I. (1989b) Taxonomy and zoogeography; taxonomic characters used in identifying Tephritidae, In: Robinson, A.S. & Hooper, G. (eds), Fruit flies; their biology, natural enemies and control. *World Crop Pests*, **3**(A), 3-7. Elsevier, Amsterdam.

Drew, R.A.I. (1989c) Taxonomy and zoogeography; the taxonomy and distribution of tropical and subtropical Dacinae (Diptera: Tephritidae), In: Robinson, A.S. & Hooper, G. (eds), Fruit flies; their biology, natural enemies and control. *World Crop Pests*, **3**(A), 9-14. Elsevier, Amsterdam.

Drew, R.A.I. (1991) Taxonomic studies on Oriental fruit fly, In: Vijaysegaran, S. & Ibrahim, A.G. (eds), *First International Symposium on Fruit Flies in the Tropics*, Kuala Lumpur, 1988. Malaysian Agricultural Research and Development Institute, Kuala Lumpur. pp. 63-66.

Drew, R.A.I. & Allwood, A.J. (1985) A new family of Strepsiptera parasitizing fruit flies (Tephritidae) in Australia. *Systematic Entomology*, **10,** 129-134. [RAE **73**:4869]

Drew, R.A.I. & Fay, H.A.C. (1988) Comparison of the roles of ammonia and bacteria in the attraction of *Dacus tryoni* (Froggatt) (Queensland fruit fly) to proteinaceous suspensions. *Journal of Plant Protection in the Tropics*, **5,** 127-130.

Drew, R.A.I. & Hardy, D.E. (1981) *Dacus (Bactrocera) opiliae*, a new sibling species of the *dorsalis* complex of fruit flies from Northern Australia (Diptera: Tephritidae). *Journal of the Australian Entomological Society*, **20,** 131-137. [RAE **70**:646]

Drew, R.A.I. & Hooper, G.H.S. (1981) The responses of fruit fly species (Diptera: Tephritidae) in Australia to various attractants. *Journal of the Australian Entomological Society*, **20,** 201-205. [RAE **70**:2416]

Drew, R.A.I. & Lambert, D.M. (1986) On the specific status of *Dacus (Bactrocera) aquilonis* and *D. (Bactrocera) tryoni* (Diptera: Tephritidae). *Annals of the Entomological Society of America*, **79,** 870-878. [RAE **75**:5166]

Drew, R.A.I. & Lloyd, A.C. (1989) Biology and physiology; nutrition; bacteria associated with fruit flies and their host plants, In: Robinson, A.S. & Hooper, G. (eds), Fruit flies; their biology, natural enemies and control. *World Crop Pests*, **3**(A), 131-140. Elsevier, Amsterdam.

Drew, R.A.I., Courtice, A.C. & Teakle, D.S. (1983) Bacteria as a natural source of food for adult fruit flies (Diptera: Tephritidae). *Oecologia*, **60,** 279-284.

Drew, R.A.I., Hooper, G.H.S. & Bateman, M.A. (eds) (1982) *Economic fruit flies of the South Pacific Region*. 2nd ed. Brisbane.

Drummond, F., Groden, E. & Prokopy, R.J. (1984) Comparitive efficacy and optimal positioning of traps for monitoring apple maggot flies (Diptera: Tephritidae). *Environmental Entomology*, **13,** 232-235. [RAE **72**:5921]

Dubois, J. (1965a) Possibilité de lutte de la région des Hauts-Plateaux à Madagascar. *Congrès de la Protection des Cultures Tropicales, Compte Rendu des Travaux, Marseilles, 1965*. 465-468. [RAE **55**:445]

Dubois, J. (1965b) La mouche des fruits malgache (*Ceratitis malagassa* Munro) et autres insectes des agrumes, pêchers et pruniers à Madagascar. *Fruits*, **20,** 435-460. [RAE **56**:302]

Economopoulos, A.P. (1987) *Fruit flies. Proceedings of the Second International Symposium*, Crete, 1986. Elsevier Science Press, Amsterdam. [RAE **76**:1419]

Economopoulos, A.P. (1989) Control; use of traps based on color and/or shape, In: Robinson, A.S. & Hooper, G. (eds), Fruit flies; their biology, natural enemies and control. *World Crop Pests*, **3**(B), 315-327. Elsevier, Amsterdam.

Economopoulos, A.P., Raptis, A., Stavropoulou-Delivoria, A. & Papadopoulos, A. (1986) Control of *Dacus oleae* by yellow sticky traps combined with ammonium acetate slow-release dispensers. *Entomologia Experimentalis et Applicata*, **41**, 11-16. [RAE **74**:5224]

Efflatoun, H.C. (1927) On the morphology of some Egyptian trypaneid larvae (Diptera), with descriptions of some hitherto unknown forms. *Bulletin, Société Entomologique d'Egypte, (N.S.),* **11**, 18-50. [RAE **16**:241]

Eichhorn, O. (1967) Insects attacking rose hips in Europe. *Technical Bulletin of the Commonwealth Institute of Biological Control*, **8**, 83-102. [RAE **56**:1735]

Eisemann, C.H. & Rice, M.J. (1989) Behavioural evidence for hygro- and mechanoreception by ovipositor sensilla of *Dacus tryoni* (Diptera: Tephritidae). *Physiological Entomology*, **14**, 273-277. [RAE **78**:1108]

Elson-Harris, M.M. (1988) Morphology of the immature stages of *Dacus tryoni* (Froggatt) (Diptera: Tephritidae). *Journal of the Australian Entomological Society*, **27**, 91-98. [RAE **76**:7368]

Elson-Harris, M.M. (1991) Studies in larval taxonomy of tropical fruit flies (Tephritidae), In: Vijaysegaran, S. & Ibrahim, A.G. (eds), *First International Symposium on Fruit Flies in the Tropics*, Kuala Lumpur, 1988. Malaysian Agricultural Research and Development Institute, Kuala Lumpur. pp. 67-70.

Enkerlin, D., Garcia, R. Laura & Lopez, M. Fidel (1989) Pest status; Mexico, Central and South America, In: Robinson, A.S. & Hooper, G. (eds), Fruit flies; their biology, natural enemies and control. *World Crop Pests*, **3**(A), 83-90. Elsevier, Amsterdam.

Entwistle, P.F. (1972) *Pests of Cocoa*. Longman, London. [RAE **62**:651]

Ermolaev, V.P., Kandybina, M.N. & Tobias, V.I. (1980) The paeony fly *Macrotrypeta ortalidina* Portsch. (Diptera, Tephritidae) and its parasite *Opius (Biosteres) paeoniae* Tobias, sp.n. (Hymenoptera, Braconidae). *Entomologicheskoe Obozrenie*, **59**, 895-903. [English translation, *Entomological Review, Washington*, **59**(4), 143-150.] [RAE **69**:4732]

Eskafi, F.M. & Cunningham, R.T. (1987) Host plants of fruit flies (Diptera: Tephritidae) of economic importance in Guatemala. *Florida Entomologist*, **70**, 116-123. [RAE **76**:6867]

Espinosa, W. Olarte (1980) *Dinámica Poblacional del Complejo Constituido por las Moscas de las Frutas* Anastrepha striata *Schiner y* Anastrepha fraterculus *Wiedemann en el Medio Ecológico del sur de Santander.* Universidad Industrial de Santander, Colombia. [RAE **70**:5372]

Eta, C.R. (1986) Review - Eradication of the melonfly from Shortland Islands, Western Province, Solomon Islands. *Solomon Islands, Agriculture Quarantine Service, 1985 annual report.* Solomon Islands Agricultural Quarantine Service, Honiara. 14-23. [RAE **74**:5248]

Étienne, J. (1972) Les principales Trypétides nuisibles d l'ile de la Réunion. *Annales de la Societé Entomologique de France* (N.S.), **8**, 485-491. [RAE **61**:4829]

Étienne, J. (1973a) Élevage permanent de *Pardalaspis cyanescens* (Dipt. Trypetidae) sur hôte végétal de remplacement. *Annales de la Société Entomologique de France*, **9**, 853-860. [RAE **62**:4664]

Étienne, J. (1973b) Lutte biologique et aperçu sur les études entomologiques diverses effectuées ces dernières années à La Réunion. *Agronomie Tropical*, **28**, 683-687. [RAE **62**:5231]

Exley, E.M. (1955) Comparative morphological studies of the larvae of some Queensland Dacinae (Trypetidae, Diptera). *Queensland Journal of Agricultural Science*, **12**, 119-150.

Faber, B. (1979) Investigations on the distribution of the unidirectional incompatibility of the European cherry fruit fly in Austria. *Bulletin, Section Regionale Ouest Palearctique, Organisation International de Lutte Biologique*, **2**, 156. [RAE **69**:3587]

Fang, M.N. (1989) A nonpesticide method for the control of melon fly. *Special Publication of the Taichung District Agricultural Improvement Station*, **16**, 193-205. [RAE **79**:2698]

Fang, M.N. & Chang, C.P. (1984) The injury and seasonal occurrence of melon fly, *Dacus cucurbitae* Coquillett, in central Taiwan (Trypetidae, Diptera). *Plant Protection Bulletin, Taiwan*, **26**, 241-248. [RAE **73**:868]

Fay, H.A.C. (1989) Rearing; small-scale rearing; multi-host species of fruit fly, In: Robinson, A.S. & Hooper, G. (eds), Fruit flies; their biology, natural enemies and control. *World Crop Pests*, **3**(B), 129-140. Elsevier, Amsterdam.

Feder, J.L. & Bush, G.L. (1989) A field test of differential host-plant usage between two sibling species of *Rhagoletis pomonella* fruit flies (Diptera: Tephritidae) and its consequences for sympatric models of speciation. *Evolution*, **43**, 1813-1819. [RAE **78**:4751]

Feder, J.L., Chilcote, C.A. & Bush, G.L. (1988) Genetic differentiation between sympatric host races of the apple maggot fly *Rhagoletis pomonella*. *Nature, UK*, **336**, 61-64. [RAE **77**:8585]

Feder, J.L., Chilcote, C.A. & Bush, G.L. (1989) Are the apple maggot, *Rhagoletis pomonella*, and blueberry maggot, *R. mendax*, distinct species? Implications for sympatric speciation. *Entomologia Experimentalis et Applicata*, **51**, 113-123. [RAE **78**:1538]

Feder, J.L., Chilcote, C.A. & Bush, G.L. (1990a) The geographic pattern of genetic differentiation between host associated populations of *Rhagoletis pomonella* (Diptera: Tephritidae) in the eastern United States and Canada. *Evolution*, **44**, 570-594. [RAE **79**:562]

Feder, J.L., Chilcote, C.A. & Bush, G.L. (1990b) Regional, local and microgeographic allele frequency variation between apple and hawthorn populations of *Rhagoletis pomonella* in western Michigan. *Evolution*, **44**, 595-608. [RAE **79**:563]

Féron, M. & Vidaud, J. (1960) La mouche du carthame *Acanthiophilus helianthi* Rossi (Dipt. Trypetidae) en France. *Revue de Pathologie Végétale et d'Entomologie Agricole de France*, **39** (fasc. 1), 1-12. [RAE **49**:656]

Ferrar, P. (1987) A guide to the breeding habits and immature stages of Diptera Cyclorrhapha. *Entomonograph*, **8**(1), 1-478; **8**(2), 479-907.

Filho, N. Suplicy, Sampaio, A.S. & Myazaki, I. (1979) Flutuacao populacional das 'moscas das frutas' (*Anastrepha* spp. e *Ceratitis capitata* (Wied., 1824) em citros na fazenda Guanabara, Barretos, SP. *Biologico*, **44,** 279-284. [RAE **67**:4545]

Fimiani, P. (1989) Pest status; Mediterranean region, In: Robinson, A.S. & Hooper, G. (eds), Fruit flies; their biology, natural enemies and control. *World Crop Pésts*, **3**(A), 37-50. Elsevier, Amsterdam.

Fischer-Colbrie, P. & Busch-Petersen, E. (1989) Pest status; temperate Europe and west Asia, In: Robinson, A.S. & Hooper, G. (eds), Fruit flies; their biology, natural enemies and control. *World Crop Pests*, **3**(A), 91-99. Elsevier, Amsterdam.

Fitt, G.P. (1981) Inter- and intraspecific responses to sex pheromones in laboratory bioassays by females of three species of tephritid fruit flies from northern Australia. *Entomologia Experimentalis et Applicata*, **30,** 40-44. [RAE **70**:1271]

Fitt, G.P. (1984) Oviposition behaviour of two tephritid fruit flies, *Dacus tryoni* and *Dacus jarvisi*, as influenced by the presence of larvae in the host fruit. *Oecologia*, **62,** 37-46. [RAE **73**:2051]

Fitt, G.P. (1986) The influence of a shortage of hosts on the specificity of oviposition behaviour in species of *Dacus* (Diptera, Tephritidae). *Physiological Entomology*, **11,** 133-143. [RAE **74**:4547]

Fitt, G.P. (1989) Ecology; the role of interspecific interactions in the dynamics of tephritid populations, In: Robinson, A.S. & Hooper, G. (eds), Fruit flies; their biology, natural enemies and control. *World Crop Pests*, **3**(B), 281-300. Elsevier, Amsterdam.

Fitt, G.P. & O'Brien, R.W. (1985) Bacteria associated with four species of *Dacus* (Diptera: Tephritidae) and their role in the nutrition of the larvae. *Oecologia*, **67,** 447-454. [RAE **74**:3455]

Fjelddalen, J. (1953) Systemiske midler mot skadedyr på frukttraer, baervekster og prydplanter. *Meldinger fra Statens Plantevern*, **8,** 1-40. [RAE **42**:237]

Fletcher, B.S. (1987) The biology of dacine fruit flies. *Annual Review of Entomology*, **32,** 115-144. [RAE **75**:3189]

Fletcher, B.S. (1989a) Biology and physiology; temperature - development rate relationships of the immature stages and adults of tephritid fruit flies, In: Robinson, A.S. & Hooper, G. (eds), Fruit flies; their biology, natural enemies and control. *World Crop Pests*, **3**(A), 273-289. Elsevier, Amsterdam.

Fletcher, B.S. (1989b) Ecology; life history strategies of tephritid fruit flies, In: Robinson, A.S. & Hooper, G. (eds), Fruit flies; their biology, natural enemies and control. *World Crop Pests*, **3**(B), 195-208. Elsevier, Amsterdam.

Fletcher, B.S. (1989c) Ecology; movements of tephritid fruit flies, In: Robinson, A.S. & Hooper, G. (eds), Fruit flies; their biology, natural enemies and control. *World Crop Pests*, **3**(B), 209-219. Elsevier, Amsterdam.

Fletcher, B.S. & Giannakakis, A. (1973) Factors limiting the response of females of the Queensland fruit fly, *Dacus tryoni*, to the sex pheromone of the male. *Journal of Insect Physiology*, **19,** 1147-1155. [RAE **62:**2995]

Fletcher, T.B. (1919) Second hundred notes on Indian insects. *Bulletin of the Agricultural Research Institute, Pusa*, **89,** 1-102. [RAE **8:**99]

Fletcher, T.B. (1920) 2.-Annotated list of Indian crop pests. *Proceedings of the Third Entomological Meeting at Pusa*, 33-314. [RAE **9:**68]

Flitters, N.E. (1951) Vanda 'Miss Agnes Joaquim', a host of *Dacus dorsalis*. *Journal of Economic Entomology*, **44,** 799-802. [RAE **40:**89]

Foote, B.A. (1984) Host plant records for North American ragweed flies (Diptera: Tephritidae). *Entomological News*, **95,** 51-54. [RAE **72:**7753]

Foote, R.H. (1960) The genus *Trypeta* Meigen in America north of Mexico (Diptera, Tephritidae). *Annals of the Entomological Society of America*, **53,** 253-260.

Foote, R.H. (1965) Family Tephritidae (Trypetidae, Trupaneidae), In: Stone, A. (principal ed.), A catalog of the Diptera of America north of Mexico. *Agricultural Handbook, United States Department of Agriculture*, **276,** 658-678.

Foote, R.H. (1967) Family Tephritidae, In: Vanzolini, E.P. & Papavero, N. (eds), *A Catalog of the Diptera of the Americas South of the United States*, **57,** 1-91. Sao Paulo.

Foote, R.H. (1980) Fruit fly genera south of the United States (Diptera: Tephritidae). *Technical Bulletin, United States Department of Agriculture*, **1600,** iv + 1-79. [RAE **69:**4125]

Foote, R.H. (1981) The genus *Rhagoletis* Loew south of the United States (Diptera: Tephritidae). *Technical Bulletin, United States Department of Agriculture*, **1607,** iv + 1-75. [RAE **70:**4288]

Foote, R.H. (1984) Tephritidae (Trypetidae), In: Soos, A. & Papp, L. (eds) *Catalogue of Palaearctic Diptera*. **9,** 66-149. Akadémiai Kiadó, Budapest.

Foote, R.H. & Blanc, F.L. (1963) The fruit flies or Tephritidae of California. *Bulletin of the California Insect Survey*, **7,** 1-117.

Foote, R.H. & Steyskal, G.C. (1987) 66. Tephritidae, In: McAlpine, J.F. (ed.), Manual of Nearctic Diptera-2. *Monograph of the Biosystematic Research Centre*, **28,** 817-831. Agriculture Canada, Ottawa.

Freeman, P. & Lane, R.P. (1985) Bibionid and scatopsid flies (Diptera: Bibionidae & Scatopsidae). *Handbooks for the Identification of British Insects*, **9**(7), 1-74.

Freeman, R. & Carey, J.R. (1990) Interaction of host stimuli in the ovipositional response of the Mediterranean fruit fly (Diptera: Tephritidae). *Environmental Entomology*, **19,** 1075-1080. [RAE **79:**1250]

Freidberg, A. (1980) On the taxonomy and biology of the genus *Myopites* (Diptera: Tephritidae). *Israel Journal of Entomology*, **13,** 13-26. [RAE **69:**7]

Freidberg, A. (1981) Mating behaviour of *Schistopterum moebiusi* Becker (Diptera: Tephritidae). *Israel Journal of Entomology*, **15,** 89-95.

Freidberg, A. (1982) *Urophora neuenschwanderi* and *Terellia sabroskyi* (Diptera: Tephritidae), two new species reared from *Ptilostemon gnaphaloides* in Crete. *Memoirs of the Entomological Society of Washington*, **10,** 56-64.

Freidberg, A. (1984) 6. Gall Tephritidae (Diptera), In: Ananthakrishnan, T.N. (ed.), *Biology of gall insects*. Edward Arnold, London. pp. 129-167. [RAE **73**:5403]

Freidberg, A. (1985) The genus *Craspedoxantha* Bezzi (Diptera: Tephritidae: Terelliinae). *Annals of Natal Museum*, **27**, 183-206.

Freidberg, A. (1991) A new species of *Ceratitis (Ceratitis)* (Diptera: Tephritidae), key to species of subgenera *Ceratitis* and *Pterandrus*, and record of *Pterandrus* fossil. *Bishop Museum Occasional Papers*, **31**, 166-173.

Freidberg, A. & Kugler, J. (1989) Diptera: Tephritidae. *Fauna Palaestina, Insecta*, **4**, 1-212. [RAE **78**:10241]

Freidberg, A. & Mathis, W.N. (1986) Studies of Terelliinae (Diptera: Tephritidae): A revision of the genus *Neaspilota* Osten Sacken. *Smithsonian Contributions to Zoology (Entomology)*, **439**, 1-75.

Frey, J.E. & Bush, G.L. (1990) *Rhagoletis* sibling species and host races differ in host odor recognition. *Entomologia Experimentalis et Applicata*, **57**, 123-131. [RAE **79**:3561]

Frias, D. (1986a) Biologia poblacional de *Rhagoletis nova* (Schiner) (Diptera: Tephritidae). *Revista Chilena de Entomologia*, **13**, 75-84. [RAE **76**:9005]

Frias, D. (1986b) Algunas consideraciones sobre la taxonomia de *Rhagoletis nova* (Schiner) (Diptera: Tephritidae). *Revista Chilena de Entomologia*, **13**, 59-73. [RAE **76**:9004]

Frias, D. (1989) Diferenciacion ecologica y reproductiva de dos razas huespedes de *Rhagoletis conversa* (Brethes) (Diptera: Tephritidae). *Acta Entomologica Chileana*, **15**, 163-170. [RAE **78**:4882]

Frias, D., Ibarra, M. & Llanca, B.A.M. (1987) Un nuevo diseno alar en *Rhagoletis conversa* (Brethes) (Diptera: Tephritidae). *Revista Chilena de Entomologia*, **15**, 21-26. [RAE **76**:4414]

Froggatt, W.W. (1899) Notes of fruit maggot flies. *Agricultural Gazette of New South Wales*, **10**, 497-504.

Froggatt, W.W. (1909) *Official Report on Fruit Fly and other Pests in Various Countries, 1907-8*. New South Wales Department of Agriculture, Sydney. [Reprinted in *Farmer's Bulletin, New South Wales Department of Agriculture*, **24**, 1-56.]

Fry, J.M. (1990) *Natural Enemy Databank, 1987; a Catalogue of the Natural Enemies of Arthropods Derived from Records in the CIBC Natural Enemy Databank*. CAB International, Wallingford.

Gariboldi, P., Jommi, G., Rossi, R. & Vita, G. (1982) Studies on the chemical constitution and sex pheromone activity of volatile substances emitted by *Dacus oleae*. *Experimentia*, **38**, 441-444. [RAE **71**:383]

Geddes, P.S., Blanc, J.P.R. le & Yule, W.N. (1987) The blueberry maggot, *Rhagoletis mendax* (Diptera: Tephritidae) in eastern North America. *Revue d'Entomologie du Quebec*, **32**, 16-24. [RAE **78**:3764]

Gellatley, J.G. & Turpin, J.W. (1987) Queensland fruit fly. *Agfacts*, (No. AE 50), 1-7. [RAE **76**:6007]

Giannakakis, A. & Fletcher, B.S. (1985) Morphology and distribution of antennal sensilla of *Dacus tryoni* (Froggatt) (Diptera: Tephritidae). *Journal of the Australian Entomological Society*, **24**, 31-35. [RAE **73**:5335]

Giannakakis, A., Fletcher, B.S., Bartell, R.J. & Shorey, H.H. (1978) An improved bioassay technique for the sex pheromone of male *Dacus tryoni* (Diptera: Tephritidae). *Canadian Entomologist*, **110**, 125-129. [RAE **66**:3455]

Gibson, K.E. & Kearby, W.H. (1978) Seasonal life history of the walnut husk fly and husk maggot in Missouri. *Environmental Entomology*, **7**, 81-87. [RAE **66**:4428]

Gilbert, G.S. & Kurczewski, F.E. (1986) Soil nutrient effects on goldenrod galls formed by *Eurosta solidaginis* (Diptera: Tephritidae). *Entomological News*, **97**, 28-32. [RAE **74**:2739]

Gilmore, J.E. (1989) Control; sterile insect technique (SIT); overview, In: Robinson, A.S. & Hooper, G. (eds), Fruit flies; their biology, natural enemies and control. *World Crop Pests*, **3**(B), 353-363. Elsevier, Amsterdam.

Gilyarov, M.S. & Luk'yanovichi, L.M. (1938) The insect pests of the seeds of *Taraxacum kok-saghyz* Rod. and other rubber-producing plants. *Pests & Diseases of Rubber Producing Plants*, (Ser. Art. 2), 22-48. [in Russian] [RAE **26**:618]

Gingrich, R.E. (1987) Demonstration of *Bacillus thuringiensis* as a potential control agent for the adult Mediterranean fruit fly, *Ceratitis capitata* (Wied.). *Journal of Applied Entomology*, **104**, 378-385. [RAE **76**:9206]

Giray, H. (1979) Turkiye Trypetidae (Diptera) faunasina ait ilk liste. *Turkiye Bitki Koruma Dergisi*, **3**, 35-46. [RAE **68**:1311]

Girolami, V. (1983) Fruit fly symbiosis and adult survival: general aspects, In: Cavalloro, R. (ed.), *Fruit Flies of Economic Importance. Proceedings of the CEC/IOBC International Symposium*, Athens, 1982. Balkema, Rotterdam. pp. 74-76. [RAE **71**:6262]

Girolami, V. (1986) Mediterranean fruit fly associated bacteria: transmission and larval survival, In: Mangel, M., Carey, J.R. & Plant, R.E. (eds), Pest Control: operations and systems analysis in fruit fly management. *NATO Advanced Science Institutes Series G: Ecological Sciences*, **11**, 135-146. Springer-Verlag, Berlin. [RAE **75**:2347]

Glas, P.C.G. & Vet, L.E.M. (1983) Host-habitat location and host location by *Diachasma alloeum* Muesebeck (Hym.; Braconidae), a parasitoid of *Rhagoletis pomonella* Walsh (Dipt.; Tephritidae). *Netherland Journal of Zoology*, **33**, 41-54. [RAE **72**:1725]

Glenn, H. & Baranowski, R.M. (1987) Establishment of *Trybliographa daci* (Eucoilidae, Hymenoptera) in Florida. *Florida Entomologist*, **70**, 183. [RAE **76**:6930]

Goeden, R.D. (1978) II: Biological control of weeds, In: Clausen, C.P. (ed.), Introduced parasites and predators of arthropod pests and weeds: a world review. *Agricultural Handbook, United States Department of Agriculture*, **480**, 357-414.

Goeden, R.D. & Ricker, D.W. (1971) Biology of *Zonosemata vittigera* relative to silverleaf nightshade. *Journal of Economic Entomology*, **64**, 417-421. [RAE **59**:2771]

Goeden, R.D., Cadatal, T.D. & Cavender, G.A. (1987) Life history of *Neotephritis finalis* (Loew) on native Asteraceae in southern California (Diptera: Tephritidae). *Proceedings of the Entomological Society of Washington*, **89**, 552-558. [RAE **76**:7064]

Gonzàlez, R.H. (1978) Introduction and spread of agricultural pests in Latin America: analysis and prospects. *Plant Protection Bulletin, Food and Agriculture Organization*, **26**, 41-52. [RAE **67**:2616]

Goonewardene, H.F. & Howard, P.H. (1989) E7-47, E7-54, E29-56, and E31-10 apple germplasm with multiple pest resistance. *HortScience*, **24**, 167-169. [RAE **78**:1564]

Gowda, G. & Ramaiah, E. (1979) Incidence of *Dacus dorsalis* Hendel (Diptera: Tephritidae) on cashew (*Anacardium occidentale* L.). *Current Research*, **8**, 98-99. [RAE **68**:2431]

Greany, P.D. (1989) Behaviour; host plant resistance to tephritids: an under-exploited control strategy, In: Robinson, A.S. & Hooper, G. (eds), Fruit flies; their biology, natural enemies and control. *World Crop Pests*, **3**(A), 353-362. Elsevier, Amsterdam.

Greathead, D.J. (1972) Notes on coffee fruit-flies and their parasites at Kawanda (Uganda). *Technical Bulletin of the Commonwealth Institute of Biological Control*, **15**, 11-18. [RAE **64**:1496]

Greene, C.T. (1929) Characters of the larvae and pupae of certain fruit flies. *Journal of Agricultural Research*, **38**, 489-498. [RAE **17**:468]

Grewal, J.S. & Kapoor, V.C. (1989) Karyotypes of some fruit fly species (Tephritidae) of India, In: Cavalloro, R. (ed.), *Fruit Flies of Economic Importance 87. Proceedings of the CEC/IOBC International Symposium*, Rome, 1987. Balkema, Rotterdam. pp. 237-244.

Grewal, J.S. & Malhi, C.S. (1987) *Prunus persica* Batsch damage by birds and fruit fly pests in Ludhiana (Punjab). *Journal of Entomological Research*, **11**, 119-120. [RAE **76**:8879]

Gurney, W.B. (1910) Fruit flies and other insects attacking cultivated and wild fruits in New South Wales. *Agricultural Gazette of New South Wales*, **21**, 423-433.

Gurney, W.B. (1912) Fruit-flies and other insects attacking cultivated and wild fruits in New South Wales. Part III. *Agricultural Gazette of New South Wales*, **23**, 75-80.

Habu, N., Iga, M. & Numazawa, K. (1984) An eradication program of the oriental fruit fly *Dacus dorsalis* Hendel (Diptera: Tephritidae), in the Ogasawara (Bonin) Islands. I. Eradication field test using sterile fly release method on small islets. *Applied Entomology and Zoology*, **19**, 1-7. [RAE **72**:6441]

Haisch, A. & Chwala, D. (1979) Host influences on the diapause of *Rhagoletis cerasi* Linnaeus 1758. *Bulletin, Section Regionale Ouest Palearctique, Organisation Internationale de Lutte Biologique*, **2**, 153-155. [RAE **69**:3586]

Haisch, A., Boller, E., Russ, K., Vallo, V. & Fimiani, P. (1978) Bibliography of *Rhagoletis cerasi* L. (1971-1976). *Bulletin, Section Regionale Ouest Palearctique, Organisation Internationale de Lutte Biologique*, **1**(3), 1-43.

Haley, M.J. & Baker, L. (eds) (1982) *Integrated Pest Management for Walnuts*. University of California, Berkeley. [RAE **71**:341]

Hall, C.J.J. van (1917) Ziekten en plagen der cultuurgewassen in Nederlandsch-Indië in 1916. *Mededelingen van de Laboratorium voor Plantenzeikten*, **29**, 1-37. [RAE **6**:447]

Hancock, D.L. (1981) Some economic Zimbabwean fruit flies (Diptera: Tephritidae). *Hortus (Zimbabwe)*, **27**, 11-15.

Hancock, D.L. (1984) Ceratitinae (Diptera: Tephritidae) from the Malagasy subregion. *Journal of the Entomological Society Southern Africa*, **47**, 277-301. [RAE **73**:4250]

Hancock, D.L. (1985a) Two new species of African Ceratitinae (Diptera: Tephritidae). *Arnoldia, Zimbabwe*, **9**, 291-297. [RAE **74**:1600]

Hancock, D.L. (1985b) New species and records of African Dacinae (Diptera: Tephritidae). *Arnoldia, Zimbabwe*, **9**, 299-314. [RAE **74**:1601]

Hancock, D.L. (1985c) A key to *Clinotaenia* Bezzi and related genera (Diptera: Tephritidae) with description of a new species. *Transactions of the Zimbabwe Scientific Association*, **62**, 56-65.

Hancock, D.L. (1986a) New genera and species of African Tephritinae (Diptera: Tephritidae), with comments on some currently unplaced or misplaced taxa and on classification. *Transactions of the Zimbabwe Scientific Association*, **63**, 16-34. [RAE **75**:518]

Hancock, D.L. (1986b) Classification of the Trypetinae (Diptera: Tephritidae), with a discussion of the Afrotropical fauna. *Journal of the Entomological Society of Southern Africa*, **49**, 275-305. [RAE **76**:459]

Hancock, D.L. (1987) Notes on some African Ceratitinae (Diptera: Tephritidae), with special reference to the Zimbabwean fauna. *Transactions of the Zimbabwe Scientific Association*, **63**, 47-57.

Hancock, D.L. (1989) Pest status; southern Africa, In: Robinson, A.S. & Hooper, G. (eds), Fruit flies; their biology, natural enemies and control. *World Crop Pests*, **3**(A), 51-58. Elsevier, Amsterdam.

Hancock, D.L. (1990) Notes on the Tephrellini-Aciurini (Diptera: Tephritidae), with a checklist of the Zimbabwe species. *Transactions of the Zimbabwe Scientific Association*, **64**, 41-48.

Hancock, D.L. (in press, a) Revised tribal classification of various genera of Trypetinae and Ceratitinae, and the description of a new species of *Taomyia* Bezzi (Diptera: Tephritidae). *Journal of the Entomological Society of Southern Africa*.

Hancock, D.L. (in press, b) Tephrellini (Diptera: Tephritidae: Tephritinae) from Madagascar. *Journal of the Entomological Society of Southern Africa*.

Hancock, G.L.R. (1926) Annual report of the assistant entomologist. *Annual Report of the Department of Agriculture, Uganda*, **1925**, 25-28. [RAE **14**:553]

Haniotakis, G., Francke, W., Mori, K., Redlich, H. & Schurig, V. (1986) Sex specific activity of (R)-(-)-and (S)-(+)-1, 7-dioxaspiro [5, 5] undecane, the major pheromone of *Dacus oleae*. *Journal of Chemical Ecology*, **12**, 1559-1568. [RAE **74**:5799]

Haniotakis, G.E. (1977) Male olive fly attraction to virgin females in the field. *Annales de Zoologie. Ecologie Animale*, **9**, 273-276. [RAE **66**:1810]

Haniotakis, G.E. (1987) Experiments towards disrupting pheromonal communication in *Dacus oleae*. *Bulletin, Section Regionale Ouest Palearctique, Organisation Internationale de Lutte Biologique*, **10,** 55-56. [RAE **77**:2530]

Haniotakis, G.E. & Vassilio-Waite, A. (1987) Effect of combining food and sex attractants on the capture of *Dacus oleae* flies. *Entomologica Helenica*, **5,** 27-33. [RAE **77**:1862]

Haniotakis, G.E., Mavraganis, V.G. & Ragoussis, V. (1989) 1, 5, 7-trioxaspiro (5.5) undecane, a pheromone analog with high biological activity for the olive fruit fly, *Dacus oleae*. *Journal of Chemical Ecology*, **15,** 1057-1065. [RAE **78**:577]

Hansen, J.D., Webb, J.C., Armstrong, J.W. & Brown, S.A. (1988) Acoustical detection of Oriental fruit fly (Diptera: Tephritidae) larvae in papaya. *Journal of Economic Entomology*, **81,** 963-965. [RAE **77**:5573]

Hapai, M.N. & Chang, F. (1986) The induction of gall formation in *Ageratina riparia* by *Procecidochares alini* (Diptera: Tephritidae). I. Gall histology and gross morphology of the third instar. *Proceedings of the Hawaiian Entomological Society*, **26,** 59-64. [RAE **75**:5331]

Haramoto, F.H. & Bess, H.A. (1970) Recent studies of the Oriental and Mediterranean fruit flies and the status of their parasites. *Proceedings of the Hawaiian Entomological Society*, **20,** 551-566.

Hardy, D.E. (1948) New host of melon fly, In: Notes and exhibitions (November 9, 1948). *Proceedings of the Hawaiian Entomological Society*, **13,** 339.

Hardy, D.E. (1949) Studies in Hawaiian fruit flies. *Proceedings of the Entomological Society of Washington*, **51,** 181-205. [RAE **39**:444]

Hardy, D.E. (1951) The Krauss collection of Australian fruit flies (Tephritidae-Diptera). *Pacific Science*, **5,** 115-189.

Hardy, D.E. (1955) The *Dacus (Afrodacus)* Bezzi of the world (Tephritidae, Diptera). *Journal of the Kansas Entomological Society*, **28,** 3-15.

Hardy, D.E. (1967) Studies of fruitflies associated with mistletoe in Australia and Pakistan with notes and descriptions on genera related to *Perilampsis* Bezzi. *Beiträge zur Entomologie*, **17,** 127-149.

Hardy, D.E. (1969) Taxonomy and distribution of the Oriental fruit fly and related species (Tephritidae-Diptera). *Proceedings of the Hawaiian Entomological Society*, **20,** 395-428. [RAE **61**:4270]

Hardy, D.E. (1973) The fruit flies (Tephritidae - Diptera) of Thailand and bordering countries. *Pacific Insects Monograph*, **31,** 1-353. [RAE **62**:2962]

Hardy, D.E. (1974) The fruit flies of the Philippines (Diptera: Tephritidae). *Pacific Insects Monograph*, **32,** 1-266. [RAE **63**:780]

Hardy, D.E. (1977) Tephritidae (Trypetidae, Trupaneidae), In: Delfinado, M.D. & Hardy, D.E. (eds), *A Catalog of the Diptera of the Oriental Region*, **3,** 44-134. University of Hawaii, Honolulu.

Hardy, D.E. (1982a) The *Epacrocerus* complex of genera in New Guinea (Diptera: Tephritidae: Acanthonevrini). *Memoirs of the Entomological Society of Washington*, **10,** 78-92.

Hardy, D.E. (1982b) The Dacini of Sulawesi (Diptera: Tephritidae). *Treubia*, **28,** 173-241.

Hardy, D.E. (1983a) The fruit flies of the genus *Dacus* Fabricius of Java, Sumatra and Lombok, Indonesia (Diptera: Tephritidae). *Treubia*, **29**, 1-45.

Hardy, D.E. (1983b) The fruit flies of the tribe Euphrantini of Indonesia, New Guinea, and adjacent islands (Tephritidae: Diptera). *International Journal of Entomology*, **25**, 152-205. [RAE **72**:2815]

Hardy, D.E. (1985) The Schistopterinae of Indonesia and New Guinea (Tephritidae: Diptera). *Proceedings of the Hawaiian Entomological Society*, **25**, 59-74.

Hardy, D.E. (1986a) Fruit flies of the subtribe Acanthonevrina of Indonesia, New Guinea, and the Bismarck and Solomon Islands (Diptera: Tephritidae: Trypetinae: Acanthonevrini). *Pacific Insects Monograph*, **42**, 1-191. [RAE **77**:8130]

Hardy, D.E. (1986b) The Adramini of Indonesia, New Guinea and adjacent islands (Diptera: Tephritidae: Trypetinae). *Proceedings of the Hawaiian Entomological Society*, **27**, 53-78. [RAE **76**:789]

Hardy, D.E. (1987) The Trypetini, Aciurini and Ceratitini of Indonesia, New Guinea and adjacent islands of the Bismarcks and Solomons (Diptera: Tephritidae: Trypetinae). *Entomography*, **5**, 247-373. [RAE **76**:7340]

Hardy, D.E. (1988a) Fruit flies of the subtribe Gastrozonina of Indonesia, New Guinea and the Bismarck and Solomon Islands (Diptera, Tephritidae, Trypetinae, Acanthonevrini). *Zoologica Scripta*, **17**, 77-121.

Hardy, D.E. (1988b) The Tephritinae of Indonesia, New Guinea, the Bismarck and Solomon Islands (Diptera: Tephritidae). *Bishop Museum Bulletins in Entomology*, **1**, vii+1-92. [RAE **79**:1118]

Hardy, D.E. (1991) Contribution of taxonomic studies to the integrated pest management of fruit flies with emphasis on the Asia-Pacific region, In: Vijaysegaran, S. & Ibrahim, A.G. (eds), *First International Symposium on Fruit Flies in the Tropics*, Kuala Lumpur, 1988. Malaysian Agricultural Research and Development Institute, Kuala Lumpur. pp. 44-48.

Hardy, D.E. & Adachi, M.S. (1954) Studies in the fruit flies of the Philippine Islands, Indonesia and Malaya, Part I. Dacini (Tephritidae-Diptera). *Pacific Science*, **8**, 147-204.

Hardy, D.E. & Adachi, M.S. (1956) Diptera: Tephritidae. *Insects of Micronesia*, **14**(1), 1-28.

Hardy, D.E. & Delfinado, M.D. (1980) Diptera: Cyclorrhapha III, Series Schizophora, Section Acalypterae, exclusive of family Drosophilidae. *Insects of Hawaii*, **13**, vii+1-451.

Hardy, D.E. & Foote, R.H. (1989) 66. Family Tephritidae, In: Evenhuis, N.L. (ed.), *Catalog of the Diptera of the Australasian and Oceanian Regions*. Bishop Museum Press, Honolulu. pp. 502-531.

Hargreaves, H. (1924) Annual report of the Government Entomologist. *Annual Report of the Department of Agriculture, Uganda*. **1923**, 15-21. [RAE **12**:470]

Hargreaves, J.R., Murray, D.A.H. & Cooper, L.P. (1986) Studies of the stinging of passion fruit by Queensland fruit fly, *Dacus tryoni* and its control by bait and cover sprays. *Queensland Journal of Agricultural and Animal Sciences*, **43**, 33-40. [RAE **75**:3424]

Harris, E.J. (1989) Pest status; Hawaiian Islands and North America, In: Robinson, A.S. & Hooper, G. (eds), Fruit flies; their biology, natural enemies and control. *World Crop Pests*, **3**(A), 73-81. Elsevier, Amsterdam.

Harris, E.J. & Lee, C.Y.L. (1986) Seasonal and annual occurrence of Mediterranean fruit flies (Diptera: Tephritidae) in Makaha and Waianae Valleys, Oahu, Hawaii. *Environmental Entomology*, **15,** 507-512. [RAE **74:**5157]

Harris, E.J., Takara, J.M. & Nishida, T. (1986b) Distribution of melon fly, *Dacus cucurbitae* (Diptera: Tephritidae), and host plants on Kauai, Hawaiian Islands. *Environmental Entomology*, **15,** 488-493. [RAE **74:**5229]

Harris, E.J., Cunningham, R.T., Tanaka, N., Ohinata, K. & Schroeder, W.J. (1986a) Development of the sterile-insect technique on the island of Lanai, Hawaii, for suppression of the Mediterranean fruit fly. *Proceedings of the Hawaiian Entomological Society*, **26,** 77-88. [RAE **75:**5478]

Harris, P. (1984) 26. Current approaches to biological control of weeds, In: Kelleher, J.S. & Hulme, M.A. (eds), *Biological Control Programmes Against Insects and Weeds in Canada 1969-1980.* Commonwealth Agricultural Bureaux, Slough. pp. 95-104.

Harris, P. (1989) The use of Tephritidae for the biological control of weeds. *Biocontrol News and Information*, **10,** 7-16. [RAE **77:**9376]

Harris, P. & Myers, J.H. (1984) 31. *Centaurea diffusa* Lam. and *C. maculosa* Lam. s.lat., diffuse and spotted knapweed (Compositae), In: Kelleher, J.S. & Hulme, M.A. (eds), *Biological Control Programmes Against Insects and Weeds in Canada 1969-1980.* Commonwealth Agricultural Bureaux, Slough. pp. 127-137.

Harris, P. & Wilkinson, A.T.S. (1984) 33. *Cirsium vulgare* (Savi) Ten., bull thistle (Compositae), In: Kelleher, J.S. & Hulme, M.A. (eds), *Biological Control Programmes Against Insects and Weeds in Canada 1969-1980.* Commonwealth Agricultural Bureaux, Slough. pp. 147-154.

Harten, A. van (1987) Biologische Schädlingsbekämpfung auf den Kapverdischen Inseln. *Courier Forschungsinstituts Senckenberg*, **95,** 29-39. [RAE **78:**907]

Hashem, A.G., Saafan, M.H. & Harris, E.J. (1987) Population ecology of the Mediterranean fruit fly in the reclaimed area in the western desert of Egypt (South Tahrir sector). *Annals of Agricultural Science, Ain Shams University*, **32,** 1803-1811. [RAE **77:**1540]

Hayes, C.F., Chingon, H.T.G., Nitta, F.A. & Wang, W.J. (1984) Temperature control as an alternative to ethylene dibromide fumigation for the control of fruit flies (Diptera: Tephritidae) in papaya. *Journal of Economic Entomology*, **77,** 683-686. [RAE **73:**470]

Heather, N.W. (1985) Alternatives to EDB fumigation as post-harvest treatment for fruit and vegetables. *Queensland Agricultural Journal*, **111,** 321-323. [RAE **75:**906]

Heather, N.W. (1989) Control; fruit disinfestation; insecticidal dipping, In: Robinson, A.S. & Hooper, G. (eds), Fruit flies; their biology, natural enemies and control. *World Crop Pests*, **3**(B), 435-440. Elsevier, Amsterdam.

Hedrick, U.P. (ed.) (1972) *Sturtevant's Edible Plants of the World*. Dover, New York. [Facsimile reprint of Sturtevant (1919) Sturtevant's notes on edible plants. *27th Annual Report of New York Agricultural Experiment Station*, **2**(2), 1-686.]

Hedstrom, I. & Jimenez, J. (1988) Evaluacion de campo de sustancias atrayentes en la captura de *Anastrepha* spp. (Diptera, Tephritidae), plaga de frutales en America tropical. II. Acetato de amonio y torula boratada. *Revista Brasileira de Entomologia*, **32**, 319-322. [RAE **77**:7719]

Hedstrom, I. & Jiron, L.F. (1985) Evaluacion de campo de sustancias atrayentes en la captura de *Anastrepha* spp. (Diptera, Tephritidae), plaga de frutales en America tropical. I. Melaza y torula. *Revista Brasileira de Entomologia*, **29**, 515-520. [RAE **75**:2917]

Hendel, F. (1927) 49. Trypetidae, In: Lindner, E. (ed.), *Die Fliegen der Palaearktischen Region*, **5**(1), 1-221. Stuttgart.

Hennig, W. (1953) Diptera, Zweiflügler, In: Sorauer, P.C.M. *Handbuch der Pflanzenkrankheiten*. **5**(5), Auflage (Lieferung 1), 1-166. Berlin.

Henshaw, D.J. de Courcy (1980) Observations on the preparation of Berlese's Fluid. *Entomologist's Monthly Magazine*, **116**, 206.

Henshaw, D.J. de Courcy & Howse, D. (1982) A micro clearing and bleaching technique. *Entomologist's Monthly Magazine*, **118**, 144.

Heppner, J.B. (1984) Larvae of fruit flies. I. *Anastrepha ludens* (Mexican fruit fly) and *Anastrepha suspensa* (Caribbean fruit fly) (Diptera: Tephritidae). *Entomology Circular, Division of Plant Industry, Florida Department of Agriculture and Consumer Services*, (260), 1-4. [RAE **72**:7129]

Heppner, J.B. (1985) Larvae of fruit flies. II. *Ceratitis capitata* (Mediterranean fruit fly) (Diptera: Tephritidae). *Entomology Circular, Division of Plant Industry, Florida Department of Agriculture and Consumer Services*, (273), 1-2. [RAE **74**:860]

Heppner, J.B. (1986) Larvae of fruit flies. III. *Toxotrypana curvicauda* (papaya fruit fly) (Diptera: Tephritidae). *Entomology Circular, Division of Plant Industry, Florida Department of Agriculture and Consumer Services*, (282), 1-2. [RAE **75**:4820]

Heppner, J.B. (1988) Larvae of fruit flies. IV. *Dacus dorsalis* (Oriental fruit fly) (Diptera: Tephritidae). *Entomology Circular, Division of Plant Industry, Florida Department of Agricultural and Consumer Services*, (303), 1-2. [RAE **77**:4693]

Heppner, J.B. (1989) Larvae of fruit flies. V. *Dacus cucurbitae* (melon fly) (Diptera: Tephritidae). *Entomology Circular, Division of Plant Industry, Florida Department of Agriculture and Consumer Services*, (315), 1-2.

Hering, [E.] M. (1927) Die Minenfauna der Canarischen Inseln. *Zoologische Jahrbücher. Abteilung für Systematik, Ökologie und Geographie der Tiere*, **53**, 405-486.

Hering, [E.] M. (1935) Neue Bohrfliegen aus Africa. *Konowia*, **14**, 154-158.

Hering, [E.] M. (1936) Bohrfliegen aus der Mandschurei. *Konowia*, **15**, 180-189.

Hernandez, A.G. & Tejada, L.D. (1979) Fluctuación de la población de *Anastrepha ludens* (Loew) y de sus enemigos naturales en *Sargentia greggii* S. Watts. *Folia Entomologica Mexicana*, **41**, 49-60. [RAE **68**:3356]

Hernandez, A.G. & Tejada, L.D. (1980) Especies de *Anastrepha* (Diptera: Tephritidae) en el Estado de Nuevo Leon, Mexico. *Folia Entomologica Mexicana*, **44,** 121-128. [RAE **70:**112]

Hernandez-Ortiz, V. (1985) Descripcion de una nueva especie Mexicana del genero *Rhagoletis* Loew (Diptera: Tephritidae). *Folia Entomologica Mexicana*, **64,** 73-79.

Herrera, A.J.M., & Vinas, V.L.E. (1977) 'Moscas de la fruta' (Dipt.: Tephritidae) en mangos de Chulucanas, Piura. *Revista Peruana de Entomologia*, **20,** 107-114. [RAE **67:**1123]

Herting, B. & Simmonds, F.J. (eds) (1978) *A Catalogue of Parasites and Predators of Terrestrial Arthropods; Section A, Host or Prey/Enemy*, volume **V** (Neuroptera, Diptera, Siphonaptera). Commonwealth Institute of Biological Control, Commonwealth Agricultural Bureaux, Slough.

Hilburn, D.J. & Dow, R.L. (1990) Mediterranean fruit fly, *Ceratitis capitata*, eradicated from Bermuda. *Florida Entomologist*, **73,** 342-343. [RAE **79:**979]

Hill, A.R. (1986a) Reduction in trap captures of female fruit flies (Diptera: Tephritidae) when synthetic male lures are added. *Journal of the Australian Entomological Society*, **25,** 211-214. [RAE **75:**2875]

Hill, A.R. (1986b) Choice of insecticides in Steiner traps affects the capture rate of fruit flies (Diptera: Tephritidae). *Journal of Economic Entomology*, **79,** 533-536. [RAE **74:**5215]

Hill, A.R. (1987) Comparison between trimedlure and capilure-attractants for male *Ceratitis capitata* (Wiedemann) (Diptera: Tephritidae). *Journal of the Australian Entomological Society*, **26,** 35-36. [RAE **75:**5947]

Hill, A.R. & Hooper, G.H.S. (1984) Attractiveness of various colours to Australian tephritid fruit flies in the field. *Entomologia Experimentalis et Applicata*, **35,** 119-128. [RAE **72:**6888]

Hill, A.R., Rigney, C.J. & Sproul, A.N. (1988) Cold storage of oranges as a disinfestation treatment against the fruit flies *Dacus tryoni* (Froggatt) and *Ceratitis capitata* (Wiedemann) (Diptera: Tephritidae). *Journal of Economic Entomology*, **81,** 257-260. [RAE **76:**7201]

Hill, G.F. (1921) Notes on some Diptera found in association with termites. *Proceedings of the Linnean Society of New South Wales*, **46,** 216-220. [RAE **10:**59]

Hislop, R.G., Riedl, H. & Joos, J.L. (1981) Control of walnut husk fly with pyrethroids and bait. *California Agriculture*, **35**(9/10), 23-25. [RAE **70:**2114]

Hodson, W.E.H. & Jary, S.G. (1939) A new insect pest of chrysanthemum. *Agriculture, London*, **46,** 54-56. [RAE **27:**519]

Hong, T.K. & Serit, M. (1988) Movements and population density comparisons of native male adult *Dacus dorsalis* and *Dacus umbrosus* (Diptera: Tephritidae) among three ecosystems. *Journal of Plant Protection in the Tropics*, **5,** 17-21. [RAE **77:**3282]

Hooper, G.H.S. (1978a) Rearing larvae of the Queensland fruit fly, *Dacus tryoni* (Froggatt) (Diptera: Tephritidae), on a bran-based medium. *Journal of the Australian Entomological Society*, **17,** 143-144. [RAE **67:**525]

Hooper, G.H.S. (1978b) Effect of combining methyl eugenol and cuelure on the capture of male tephritid fruit flies. *Journal of the Australian Entomological Society*, **17**, 189-190. [RAE **67**:526]

Hooper, G.H.S. & Drew, R.A.I. (1979) Effect of height of trap on capture of tephritid fruit flies with cuelure and methyl eugenol in different environments. *Environmental Entomology*, **8**, 786-788. [RAE **68**:3701]

Hooper, G.H.S. & Drew, R.A.I. (1989) Pest status; Australia and South Pacific islands, In: Robinson, A.S. & Hooper, G. (eds), Fruit flies; their biology, natural enemies and control. *World Crop Pests*, **3**(A), 67-72. Elsevier, Amsterdam.

Howard, D.J. (1989) Biology and physiology; nutrition; the symbionts of *Rhagoletis*, In: Robinson, A.S. & Hooper, G. (eds), Fruit flies; their biology, natural enemies and control. *World Crop Pests*, **3**(A), 121-129. Elsevier, Amsterdam.

Howard, D.J. & Bush, G.L. (1989) Influence of bacteria on larval survival and development in *Rhagoletis* (Diptera: Tephritidae). *Annals of the Entomological Society of America*, **82**, 633-640. [RAE **78**:9284]

Howse, P.E. & Foda, M.E. (1986) Pheromone communication in the Mediterranean fruit fly (*Ceratitis capitata* Wied.), In: Mangel, M., Carey, J.R. & Plant, R.E. (eds), Pest control: operations and systems analysis in fruit fly management. *NATO Advanced Science Institutes Series G: Ecological Sciences*, **11**, 189. Springer-Verlag, Berlin. [RAE **75**:2385]

Huffaker, C.B. & Caltagirone, L.E. (1986) The impact of biological control on the development of the Pacific. *Agriculture, Ecosystems and Environment*, **15**, 95-107. [RAE **74**:4262]

Hurter, J., Boller, E.F., Stadler, E., Blattmann, B., Buser, H.R., Bosshard, N.U., Damm, L., Kozlowski, M.W., Schoni, R., Raschdorf, F., Dahinden, R., Schlumpf, E., Fritz, H., Richter, W.J. & Schreiber, J. (1987) Oviposition-deterring pheromone in *Rhagoletis cerasi* L.: purification and determination of the chemical composition. *Experientia*, **43**, 157-164. [RAE **76**:629]

IAEA (1990) *Genetic Sexing of the Mediterranean Fruit Fly; Proceedings of the Final Research Co-ordination Meeting*, Crete, 1988. International Atomic Energy Agency, Vienna.

Ibrahim, A.G. & Ibrahim, R. (1990) *Handbook on Identification of Fruit Flies in the Tropics*. Universiti Pertanian Malaysia, Serdang.

Ibrahim, A.G., Singh, G. & King, H.S. (1979) Trapping of the fruit-flies, *Dacus* spp. (Diptera: Tephritidae) with methyl eugenol in orchards. *Pertanika*, **2**, 58-61. [RAE **69**:1128]

Ibrahim, Y. & Mohamad, R. (1978) Pupal distribution of *Dacus dorsalis* Hendel in relation to host plants and its pupation depth. *Pertanika*, **1**, 66-69. [RAE **67**:1948]

Ihering, R. von (1912) *As Moscas das Fructas e sua Destruiçao*, 2ª ediç. Sao Paulo.

IIE (1991a) *Bactrocera tryoni* (Froggatt) (=*Dacus tryoni* (Froggatt)), Diptera: Tephritidae, Queensland fruit-fly. *International Institute of Entomology, Distribution Maps of Pests*, Series A (Agricultural), (110), revised, 1-2. CAB International, London.

IIE (1991b) *Bactrocera tsuneonis* (Miyake) (=*Tetradacus tsuneonis* (Miyake)), Diptera: Tephritidae, Japanese orange fruit-fly, Japanese orange fly. *International Institute of Entomology, Distribution Maps of Pests*, Series A (Agricultural), (410), revised, 1-2. CAB International, London.

IIE (1991c) *Bactrocera minax* (Enderlein) (= *Tetradacus citri* (Chen); *Callantra minax* (Enderlein)), Diptera: Tephritidae, Chinese citrus fly. *International Institute of Entomology, Distribution Maps of Pests*, Series A (Agricultural), (526), 1-2. CAB International, London.

Ingram, W.R. (1965) An evaluation of several insecticides against berry borer and fruit fly in Uganda robusta coffee. *East African Agricultural and Forestry Journal*, **30**, 259-262. [RAE **53**:576]

Irwin, A.G. (1978) 1. Collecting and recording; curating, In: Stubbs, A. & Chandler, P. (eds), A dipterist's handbook. *Amateur Entomologist*, **15**, 7-17.

Isart, J. (1979; dated 1977) Observaciones sobre *Euleia heraclei* (Linneo, 1758) en España (Dipt. Tephritidae). *Graellsia*, **33**, 261-278. [RAE **68**:2463]

Issiki, S., Sonan, J. & Takahashi, R. (1928) Studies on bamboo trypetids, I. *Bulletin of the Department of Agriculture, Research Institute, Formosa*, **61**, 1-16. [in Japanese]. [RAE **16**:481]

Ito, P.J., Kunimoto, R. & Ko, W.H. (1979) Transmission of mucor rot of guava fruits by three species of fruit flies. *Tropical Agriculture*, **56**, 49-52. [RAE **67**:2848]

Ito, S. (1983-5) *Die Japanischen Bohrfliegen*. Osaka. [Privately published in 7 parts.]

Iwahashi, O. (1977) Eradication of the melon fly, *Dacus cucurbitae*, from Kume Island, Okinawa with the sterile insect release method. *Researches on Population Ecology*, **19**, 87-98. [RAE **66**:2411]

Iwahashi, O. & Majima, T. (1986) Lek formation and male-male competition in the melon fly, *Dacus cucurbitae* Coquillett (Diptera: Tephritidae). *Applied Entomology and Zoology*, **21**, 70-75. [RAE **74**:4113]

Jack, R.W. (1943) Report of the Division of Entomology [Southern Rhodesia] for the year 1941. *Review of Applied Entomology, Series A: Agricultural*, **31**, 86-87. [The original report was a 19 page typescript; the abstract in *Review of Applied Entomology* is presumed to be the only published version].

Jakhmola, S.S. (1983) Niger grain fly, *Diozina sororcula* (Wiedemann), a serious pest of niger in central India. *Journal of the Bombay Natural History Society*, **80**, 439-440. [RAE **73**:329]

Jakhmola, S.S. & Yadav, H.S. (1980) Incidence of and losses caused by capsule fly *Acanthiophilus helianthi* Rossi in different varieties of safflower. *Indian Journal of Entomology*, **42**, 48-53. [RAE **69**:5409]

Jang, E.B. (1986) Kinetic of thermal death in eggs and first instars of three species of fruit flies (Diptera: Tephritidae). *Journal of Economic Entomology*, **79**, 700-705. [RAE **74**:6057]

Jang, E.B., Light, D.M., Flath, R.A., Nagata, J. & Mon, T.R. (1989) Electroantennogram responses of Mediterranean fruit fly, *Ceratitis capitata* to identified volatile constituents from calling males. *Entomologia Experimentalis et Applicata*, **50**, 7-19. [RAE **77**:6180]

Janzon, L.-Å. (1982) Description of the egg and larva of *Euphranta connexa* (Fabricius) (Diptera: Tephritidae) and of the egg of its parasitoid *Scambus brevicornis* (Gravenhorst) (Hymenoptera: Ichneumonidae). *Entomologica Scandinavica*, **13,** 313-316.

Janzon, L.-Å. (1983) *Pteromalus sonchi* n.sp. (Hymenoptera: Chalcidoidea), a parasitoid of *Tephritis dilacerata* (Loew) (Diptera: Tephritidae), living in flower-heads of *Sonchus arvensis* L. (Asteraceae) in Sweden. *Entomologica Scandinavica*, **14,** 309-315. [RAE **72:**380]

Javaid, I. (1986) Causes of damage to some wild mango fruit trees in Zambia. *International Pest Control*, **28,** 98-99. [RAE **74:**5804]

Jepson, F.P. (1935) Report on the work of the Entomological Division. *Ceylon Administration Reports, Agriculture*, **1934,** D132-D147. [RAE **24:**102]

Jermy, T. (1961) Eine neue *Rhagoletis*-Art (Diptera: Trypetidae) aus den Fruchten von *Berberis vulgaris* L. *Acta Zoologica Hungarica*, **7,** 133-137.

Jirón, L.F. & Mexzon, R.G. (1989) Parasitoid hymenopterans of Costa Rica: geographical distribution of the species associated with fruit flies (Diptera: Tephritidae). *Entomophaga*, **34,** 53-60. [RAE **78:**4479]

Jirón, L.F. & Zeledón, R. (1979) El género *Anastrepha* (Diptera; Tephritidae) en las principales frutas de Costa Rica y su relación con pseudomiasis humana. *Revista de Biología Tropical*, **27,** 155-161. [RAE **69:**620]

Jirón, L.F., Soto-Manitiu, J. & Norrbom, A.L. (1988) A preliminary list of the fruit flies of the genus *Anastrepha* (Diptera: Tephritidae) in Costa Rica. *Florida Entomologist*, **71,** 130-137. [RAE **77:**6280]

Johnson, P.C. (1983) Response of adult apple maggot (Diptera: Tephritidae) to pherocon A.M. traps and red spheres in a non-orchard habitat. *Journal of Economic Entomology*, **76,** 1279-1284. [RAE **72:**3824]

Jones, O.T. (1989) Biology and physiology; mating pheromones; *Ceratitis capitata*, In: Robinson, A.S. & Hooper, G. (eds), Fruit flies; their biology, natural enemies and control. *World Crop Pests*, **3**(A), 179-183. Elsevier, Amsterdam.

Jones, O.T., Lisk, J.C., Longhurst, C., Howse, P.E., Ramos, P. & Campos, M. (1983) Development of a monitoring trap for the olive fly, *Dacus oleae* (Gmelin) (Diptera: Tephritidae), using a component of its sex pheromone as lure. *Bulletin of Entomological Research*, **73,** 97-106. [RAE **71:**3826]

Jones, S.C. (1937) The currant and gooseberry maggot or yellow currant fly (*Epochra canadensis*) Loew. *Circular of the Oregon Agricultural Experimental Station*, (121), 1-11. [RAE **26:**60]

Jones, S.R. & Kim, K.C. (1988) Ultrastructure of the posterior spiracles of larval *Rhagoletis mendax* Curran (Diptera: Tephritidae). *Annals of the Entomological Society of America*, **81,** 511-515. [RAE **77:**2936]

Jorgensen, C.D., Allred, D.B. & Westcott, R.L. (1986) Apple maggot (*Rhagoletis pomonella*) adaptation for cherries in Utah. *Great Basin Naturalist*, **46,** 173-174. [RAE **75:**737]

Julien, M.H. (1987) *Biological Control of Weeds: A World Catalogue of Agents and their Target Weeds*. 2nd Ed. CAB International, Slough. [RAE **76:**4752]

Kabysh, T.A. (1979) Varying the dates of harvest of cow parsnip for the control of a trypetid. *Zashchita Rastenii*, **5,** 41. [in Russian.] [RAE **67:**3880]

Kamali, K. & Schulz, J.T. (1974) Biology and ecology of *Gymnocarena diffusa* (Diptera: Tephritidae) on sunflower in North Dakota. *Annals of the Entomological Society of America*, **67**, 695-699. [RAE **63**:2134]

Kamasaki, H., Sutton, R., Lopez, F. & Selhime, A. (1970) Laboratory culture of the Caribbean fruit fly, *Anastrepha suspensa* in Florida. *Annals of the Entomological Society of America*, **63**, 639-642. [RAE **59**:361]

Kamikado, T., Chisaki, N., Kamiwada, H. & Tanaka, A. (1987) Mass rearing of the melon fly, *Dacus cucurbitae* Coquillett, by the sterile insect release method. I. Changes in the amount of eggs laid and the longevity of mass reared adults. *Proceedings of the Association for Plant Protection of Kyushu*, **33**, 164-166. [in Japanese]. [RAE **77**:7235]

Kandybina, M.N. (1961) On the diagnostics of the larvae of fruit flies of the family Trypetidae (Diptera). *Entomologicheskoe Obozrenie*, **40**, 202-213. [in Russian: English translation in *Entomological Review, Washington*, **40**, 103-110].

Kandybina, M.N. (1965) The larvae of fruit flies flies of the genus *Carpomyia* A. Costa (Diptera, Trypetidae). *Entomologicheskoe Obozrenie*, **44**, 665-672. [in Russian: English translation in *Entomological Review, Washington*, **44**, 390-394].

Kandybina, M.N. (1972) Contribution to the study of the Tephritidae (Dipt.) of the Mongolian People's Republic. *Entomologicheskoe Obozrenie*, **51**, 909-918. [in Russian: English translation in *Entomological Review, Washington*, **51**, 540-545].

Kandybina, M.N. (1977) The larvae of fruit-flies (Diptera, Tephritidae). *Opredeliteli po Faune SSR, Izdavaemye Zoologicheskim Muzeem Akademii Nauk*, **114**, 1-212. [in Russian: unpublished English translation, 1987, produced by National Agricultural Library, Beltsville, Maryland, USA]. [RAE **66**:2961]

Kanmiya, K. (1988) Acoustic studies on the mechanism of sound production in the mating songs of the melon fly, *Dacus cucurbitae* Coquillett (Diptera: Tephritidae). *Journal of Ethology*, **6**, 143-151. [RAE **78**:2313]

Kanmiya, K., Tanaka, A., Kamiwada, H. Nakagawa, K. & Nishioka, T. (1987) Time-domain analysis of the male coutship songs produced by wild, mass-reared, and by irradiated melon flies, *Dacus cucurbitae* Coquillett (Diptera: Tephritidae). *Applied Entomology and Zoology*, **22**, 181-194. [RAE **76**:25]

Kapatos, E.T. (1989) Control; integrated pest management systems of *Dacus oleae*, In: Robinson, A.S. & Hooper, G. (eds), Fruit flies; their biology, natural enemies and control. *World Crop Pests*, **3**(B), 391-398. Elsevier, Amsterdam.

Kapatos, E.T. & Fletcher, B.S. (1984) The phenology of the olive fly, *Dacus oleae* (Gmel.) (Diptera, Tephritidae), in Corfu. *Zeitschrift für Angewandte Entomologie*, **97**, 360-370. [RAE **72**:5416]

Kapatos, E., Fletcher, B.S., Pappas, S. & Laudeho, Y. (1977a) The release of *Opius concolor* and *O. concolor* var. *siculus* (Hym.: Braconidae) against the spring generation of *Dacus oleae* (Dipt.: Trypetidae) on Corfu. *Entomophaga*, **22**, 265-270. [RAE **66**:1366]

Kapatos, E., McFadden, M.W. & Pappas, S. (1977b) Sampling techniques and preparation of partial life tables for the olive fly, *Dacus oleae* (Diptera: Trypetidae). *Ecological Entomology*, **2**, 193-196. [RAE **66**:1367]

Kapatos, E., Pappas, S. & McFadden, M.W. (1977c) Ecological studies on the olive fly *Dacus oleae* Gmel. in Corfu. II. Mortality of the immature stages in the fruit. *Bollettino del Laboratorio di Entomologia Agraria 'Filippo Silvestri', Portici*, **34,** 74-79. [RAE **67:**1118]

Kapoor, V.C. (1970) Indian Tephritidae with their recorded hosts. *Oriental Insects*, **4,** 207-251.

Kapoor, V.C. (1971) Four new species of fruitflies (Tephritidae) from India. *Oriental Insects*, **5,** 477-482.

Kapoor, V.C. (1989) Pest status; Indian sub-continent, In: Robinson, A.S. & Hooper, G. (eds), Fruit flies; their biology, natural enemies and control. *World Crop Pests*, **3**(A), 59-62. Elsevier, Amsterdam.

Kapoor, V.C. & Agarwal, M.L. (1983) Fruit flies and their increasing host plants in India, In: Cavalloro, R. (ed.), *Fruit Flies of Economic Importance. Proceedings of the CEC/IOBC International Symposium*, Athens, 1982. Balkema, Rotterdam. 252-257. [RAE **71:**6277]

Kapoor, V.C., Hardy, D.E., Agarwal, M.L. & Grewal, J.S. (1980) *Fruit fly (Diptera: Tephritidae) Systematics of the Indian Subcontinent.* Export India Publications, Jullunder.

Kapoor, V.C., Malla, Y.K. & Ghosh, K. (1979) On a collection of fruit flies from Kathmandu Valley. *Oriental Insects*, **13,** 81-85. [RAE **68:**4299]

Karpati, J.F. (1983) *The Mediterranean Fruit Fly (its Importance, Detection and Control).* Food and Agriculture Organization, Rome.

Katsoyannos, B. (1979) Das Markierungspheromon der Kirschenfliege: biologische Bedeutung und praktische Anwendung. *Mitteilungen der Schweizerischen Entomologischen Gesellschaft*, **52,** 444. [RAE **69:**1558]

Katsoyannos, B.I. (1982) Male sex pheromones of *Rhagoletis cerasi* L. (Diptera, Tephritidae): factors affecting release and response and its role in mating behavior. *Zeitschrift für Angewandte Entomologie*, **94,** 187-198. [RAE **70:**6840]

Katsoyannos, B.I. (1987) Effect of color properties of spheres on their attractiveness for *Ceratitis capitata* (Wiedemann) flies in the field. *Journal of Applied Entomology*, **104,** 79-85. [RAE **76:**7453]

Katsoyannos, B.I. (1989a) Biology and physiology; mating pheromones; *Rhagoletis* spp., In: Robinson, A.S. & Hooper, G. (eds), Fruit flies; their biology, natural enemies and control. *World Crop Pests*, **3**(A), 185-188. Elsevier, Amsterdam.

Katsoyannos, B.I. (1989b) Behaviour; response to shape size and color, In: Robinson, A.S. & Hooper, G. (eds), Fruit flies; their biology, natural enemies and control. *World Crop Pests*, **3**(A), 307-324. Elsevier, Amsterdam.

Katsoyannos, B.I., Panagiotidou, K. & Kechagia, I. (1986) Effect of color properties on the selection of oviposition site by *Ceratitis capitata. Entomologia Experimentalis et Applicata*, **42,** 187-193. [RAE **75:**3410]

Keilin, D. & Tate, P. (1943) The larval stages of the celery fly (*Acidia heraclei* L.) and of the braconid *Adelura apii* (Curtis), with notes upon an associated parasitic yeast-like fungus. *Parasitology, Cambridge*, **35,** 27-36.

Keiser, I. (1989) Control; insecticides; insecticide resistance status, In: Robinson, A.S. & Hooper, G. (eds), Fruit flies; their biology, natural enemies and control. *World Crop Pests*, **3**(B), 337-344. Elsevier, Amsterdam.

Khan, R. Jabbar & Khan, M.A. Jabbar (1987) A comparative morphological study on third instar larvae of some *Dacus* species (Tephritidae: Diptera) in Pakistan. *Pakistan Journal of Scientific and Industrial Research*, **30**, 534-538.

King, C.B.R. (1935) Report of the Entomologist for the year 1934. *Bulletin of the Tea Research Institute of Ceylon*, **12**, 26-31. [RAE **23**:638]

King, C.B.R. (1936) Report of the Entomologist for the year 1935. *Bulletin of the Tea Research Institute of Ceylon*, **13**, 35-40. [RAE **24**:633]

Kitching, W., Lewis, J.A., Perkins, M.V., Drew, R.A.I., Moore, C.J., Schurig, V., Konig, W.A. & Francke, W. (1989) Chemistry of fruit flies. Composition of rectal gland secretion of (male) *Dacus cucumis* (cucumber fly) and *Dacus halfordiae*, characterization of (Z, Z)-2, 8-dimethyl-1, 7-dioxaspiro[5.5] undecane. *Journal of Organic Chemistry*, **54**, 3893-3902. [RAE **78**:4307]

Kitto, G.B. (1983) An immunological approach to the phylogeny of the Tephritidae, In: Cavalloro, R. (ed.), *Fruit Flies of Economic Importance. Proceedings of the CEC/IOBC International Symposium*, Athens, 1982. Balkema, Rotterdam. pp. 203-211. [RAE **71**:6028]

Koizumi, K. (1957) Notes on some dipterous pests of economic plants in Japan. *Botyu-Kagaku*, **22**, 223-227. [in Japanese]. [RAE **48**:51]

Kolbe, M.E. & Eskafi, F.M. (1990) Method to rank host plants infested with Mediterranean fruit fly, *Ceratitis capitata* in multiple host situations in Guatemala. *Florida Entomologist*, **73**, 708-711. [RAE **79**:523]

Kopelke, J.P. (1984) Der erste Nachweis eines Brutparasiten unter den Bohrfliegen. *Natur und Museum*, **114**, 1-28.

Kopelke, J.P. (1985) Biologie und Parasiten der gallenbilden Blattwespe *Pontania proxima* (Lepeletier, 1823) (Insecta: Hymenoptera: Tenthredinidae). *Senckenbergiana Biologica*, **65**, 215-239. [RAE **73**:5830]

Korneyev, V. (1985) Fruit flies of the tribe Terelliini (Diptera, Tephritidae) in the USSR. *Entomologicheskoe Obozrenie*, **64**, 626-644. [in Russian: English translation in *Entomological Review, Washington*, **65**(1), 35-55].

Koyama, J. (1989a) Pest status; south-east Asia and Japan, In: Robinson, A.S. & Hooper, G. (eds), Fruit flies; their biology, natural enemies and control. *World Crop Pests*, **3**(A), 63-66. Elsevier, Amsterdam.

Koyama, J. (1989b) Biology and physiology; mating pheromones; tropical dacines, In: Robinson, A.S. & Hooper, G. (eds), Fruit flies; their biology, natural enemies and control. *World Crop Pests*, **3**(A), 165-168. Elsevier, Amsterdam.

Krainacker, D.A., Carey, J.R. & Vargas, R.I. (1987) Effect of larval host on life history traits of the Mediterranean fruit fly, *Ceratitis capitata*. *Oecologia*, **73**, 583-590. [RAE **76**:8834]

Kuba, H. & Koyama, J. (1985) Mating behavior of wild melon flies, *Dacus cucurbitae* Coquillett (Diptera: Tephritidae) in a field cage: courtship behavior. *Applied Entomology and Zoology*, **20**, 365-372. [RAE **74**:1696]

Kuba, H. & Sokei, Y. (1988) The production of pheromone clouds by spraying in the melon fly, *Dacus cucurbitae* Coquillett (Diptera: Tephritidae). *Journal of Ethology*, **6**, 105-110. [RAE **78**:2217]

Kuba, H., Koyama, J. & Prokopy, R.J. (1984) Mating behavior of wild melon flies, *Dacus cucurbitae* Coquillett (Diptera: Tephritidae) in a field cage: distribution and behavior of flies. *Applied Entomology and Zoology*, **19**, 367-373. [RAE **73**:1205]

Kurata, U. (1925) A dipterous pest found in the young shoots of bamboo imported into Formosa. *Journal of Plant Protection*, **12**, 674-677 [in Japanese]. [RAE **14**:93]

Kwon, Y.J. (1985) Classification of the fruitfly-pests from Korea. *Insecta Koreana* (Series 5), **1985**, 49-112.

Labeyrie, V. (1957) Observations sur le comportement de ponte de la mouche du céleri (*Philophylla heraclei* L.). *Annales des Epiphyties (et de Phytogénétique)*, **8**, 171-183. [RAE **47**:213]

Lakra, R.K. & Singh, Z. (1983) Oviposition behaviour of ber fruit fly, *Carpomyia vesuviana* Costa and relationship between its incidence and ruggedness in fruits in Haryana. *Indian Journal of Entomology*, **45**, 48-59. [RAE **71**:7950]

Lakra, R.K. & Singh, Z. (1984) Calender of losses due to ber fruit fly, *Carpomyia vesuviana* Costa (Diptera: Tephritidae) in different *Zizyphus* spp. in Haryana. *Indian Journal of Entomology*, **46**, 261-269. [RAE **75**:2379]

Lakra, R.K. & Singh, Z. (1985) Seasonal fluctuations in incidence of ber fruitfly *Carpomyia vesuviana* Costa (Diptera: Tephritidae) under agro-climatic conditions of Hisar. *Haryana Agricultural University Journal of Research*, **15**, 42-50. [RAE **73**:8244]

Lakra, R.K. & Singh, Z. (1989; dated 1986) Bionomics of *Zizyphus* fruitfly, *Carpomyia vesuviana* Costa (Diptera: Tephritidae) in Haryana. *Bulletin of Entomology*, **27**, 13-27. [RAE **78**:3013]

Lamb, R.C., Brown, S.K. & Reissig, W.H. (1988) Breeding for arthropod resistance in apple. *Acta Horticulturae*, **224**, 123-131. [RAE **78**:1563]

Lamborn, W.A. (1914) The agricultural pests of the southern provinces, Nigeria. *Bulletin of Entomological Research*, **5**, 197-214. [RAE **3**:164]

Landolt, P.J. (1984) Reproductive maturation and premating period of the papaya fruit fly, *Toxotrypana curvicauda* (Diptera: Tephritidae). *Florida Entomologist*, **67**, 240-244. [RAE **72**:7943]

Landolt, P.J. (1985a; dated 1984) Behavior of the papaya fruit fly *Toxotrypana curvicauda* Gerstaecker (Diptera: Tephritidae), in relation to its host plant, *Carica papaya* L. *Folia Entomologica Mexicana*, **61**, 215-224. [RAE **74**:877]

Landolt, P.J. (1985b) Papaya fruit fly eggs and larvae (Diptera: Tephritidae) in field-collected papaya fruit. *Florida Entomologist*, **68**, 354-356. [RAE **74**:2919]

Landolt, P.J. & Hendrichs, J. (1983) Reproductive behavior of the papaya fruit fly, *Toxotrypana curvicauda* Gerstaecker (Diptera: Tephritidae). *Annals of the Entomological Society of America*, **76**, 413-417. [RAE **72**:736]

Landolt, P.J., Heath, R.R. & King, J.R. (1985) Behavioral responses of female papaya fruit flies, *Toxotrypana curvicauda* (Diptera: Tephritidae), to male-produced sex pheromone. *Annals of the Entomological Society of America*, **78**, 751-755. [RAE **74**:1343]

Lange, W.H. (1941) The artichoke plume moth and other pests injurious to the globe artichoke. *Bulletin of the California Agricultural Experiment Station*, **653**, 1-71. [RAE **30**:554]

Lawrence, P.O. (1979) Immature stages of the Caribbean fruit fly, *Anastrepha suspensa*. *Florida Entomologist*, **62**, 214-219. [RAE **68**:2188]

Lawrence, P.O. (1986) The role of 20-hydroxyecdysone in the moulting of *Biosteres longicaudatus*, a parasite of the Caribbean fruit fly, *Anastrepha suspensa*. *Journal of Insect Physiology*, **32**, 329-337. [RAE **74**:3904]

Lawrence, P.O. (1988) Intraspecific competition among first instars of the parasitic wasp *Biosteres longicaudatus*. *Oecologia*, **74**, 607-611. [RAE **77**:5263]

Le Pelley, R.H. (1959) *Agricultural Insects of East Africa*. East Africa High Commission, Nairobi. [RAE **48**:94]

Le Pelley, R.H. (1968) *Pests of Coffee*. Longmans, London. [RAE **57**:391]

Lee, H.S. (1972) A study on the ecology of melon fly. *Plant Protection Bulletin, Taiwan*, **14**, 175-182.[in Chinese] [RAE **63**:932]

Leefmans, S. (1915) De theezaadvlieg. *Mededeelingen van het Proefstation voor Thee*, **15**, 1-15. [RAE **3**:434]

Leefmans, S. (1930) Ziekten en plagen der cultuurgewassen in Nederlandsch Oost-Indië in 1929. *Medeedelingen van het Instituut voor Plantziektenkunde*, **79**, 1-100. [RAE **19**:397]

Leonhardt, B.A., Rice, R.E, Harte, E.M. & Cunningham, R.T. (1984) Evaluation of dispensers containing trimedlure, the attractant for the Mediterranean fruit fly (Diptera: Tephritidae). *Journal of Economic Entomology*, **77**, 744-749. [RAE **73**:216]

Leonhardt, B.A., Cunnigham, R.T., Rice, R.E., Harte, E.M. & McGovern, T.P. (1987) Performance of controlled-release formulations of trimedlure to attract the Mediterranean fruit fly, *Ceratitis capitata*. *Entomologia Experimentalis et Applicata*, **44**, 45-51. [RAE **75**:5712]

Leppla, N.C. (1989) Rearing; laboratory colonization of fruit flies, In: Robinson, A.S. & Hooper, G. (eds), Fruit flies; their biology, natural enemies and control. *World Crop Pests*, **3**(B), 91-103. Elsevier, Amsterdam.

Leroi, B. (1972) Données expérimentales sur les changements de galerie des larves mineuses de *Philophylla heraclei* (Diptera, Tephritidae). *Entomologia Experimentalis et Applicata*, **15**, 351-359. [RAE **62**:596]

Leroi, B. (1974) A study of natural populations of the celery leaf-miner, *Philophylla heraclei* L. (Diptera, Tephritidae). II. Importance of changes of mines for larval populations. *Researches on Population Ecology*, **15**, 163-182. [RAE **63**:3209]

Leroi, B. (1975a) Importance des arbres pour les populations d'adultes de la mouche du céleri, *Philophylla heraclei* L. (Diptère, Tephritidae). *Comptes Rendus Hebdomadaires des Séances de l'Académie des Sciences, Paris*, D, **281**, 289-292. [RAE **65**:809]

Leroi, B. (1975b) Influence d'une plant-hôte des larves (*Apium graveolens* L.) sur la stimulation de la ponte et de la production ovarienne de *Philophylla heraclei* L. (Diptère, Tephritidae). *Comptes Rendus Hebdomadaires des Séances de l'Académie des Sciences, Paris*, D, **281**, 1015-1018. [RAE **65**:609]

Leroi, B. (1977) Relations biocoenotiques de la mouche du celeri, *Philophylla heraclei* L. (Diptère, Tephritidae): Necessité de végetaux complementaires pour les populations vivant sur céleri. *Colloques Internationaux du Centre National de la Recherche Scientifique*, **265**, 443-454. [RAE **66**:1159]

Levinson, H.Z., Levinson, A.R. & Schafer, K. (1987) Pheromone biology of the Mediterranean fruit fly (*Ceratitis capitata* Wied.) with emphasis on the functional anatomy of the pheromone glands and antennae as well as mating behaviour. *Journal of Applied Entomology*, **104**, 448-461. [RAE **76**:8401]

Levinson, H.Z., Levinson, A.R. & Schäfer, K. (1989) New aspects of the pheromone biology of the Mediterranean fruit fly, In: Cavalloro, R. (ed.), *Fruit Flies of Economic Importance 87. Proceedings of the CEC/IOBC International Symposium*, Rome, 1987. Balkema, Rotterdam. pp. 113-128.

Lewis, J.A., Moore, C.J., Fletcher, M.T., Drew, R.A.I. & Kitching, W. (1988) Volatile compounds from the flowers of *Spathiphyllum cannaefolium*. *Phytochemistry*, **27**, 2755-2757. [RAE **77**:3840]

Leyva-Vazquez, J.L. (1988) Temperatura umbral y unidades calor requeridas por los estados inmaduros de *Anastrepha ludens* (Loew) (Diptera: Tephritidae). *Folia Entomologica Mexicana*, **74**, 189-196. [RAE **77**:6237]

Liaropoulos, C., Louskas, C., Canard, M. & Laudeho, Y. (1977) Releases of *Opius concolor* (Hym.: Braconidae) among the spring population of *Dacus oleae* (Dipt.: Trypetidae). An experiment in continental Greece. *Entomophaga*, **22**, 259-264. [RAE **66**:1365]

Lienk, S.E. (1970) Apple maggot infesting apricot. *Journal of Economic Entomology*, **63**, 1684. [RAE **59**:667]

Lifschitz, E. & Cladera, J.L. (1989) Genetics; *Ceratitis capitata*; cytogenetics and sex determination, In: Robinson, A.S. & Hooper, G. (eds), Fruit flies; their biology, natural enemies and control. *World Crop Pests*, **3**(B), 3-11. Elsevier, Amsterdam.

Lipa, J.J., Borusiewicz, K. & Balazy, S. (1976) Noxiousness of the rose fruit fly (*Rhagoletis alternata* Meigen) and infection of its puparia by a fungus *Scopulariopsis brevicaulis* (Sacc.) Bainier. *Bulletin de l'Academie Polonaise de Sciences, Sciences Biologiques*, **24**, 451-456. [RAE **66**:1940]

Liquido, N.J. & Cunningham, R.T. (1991) Ecological considerations in eradicating exotic fruit fly introductions, In: Vijaysegaran, S. & Ibrahim, A.G. (eds), *First International Symposium on Fruit Flies in the Tropics*, Kuala Lumpur, 1988. Malaysian Agricultural Research and Development Institute, Kuala Lumpur. pp. 235-241.

Litsinger, J.A., Fakalata, O.K., Faluku, T.L., Crooker, P.S. & Keyserlingk, N. von (1991) A study of fruit fly species (Tephritidae) occurring in the Kindom of Tonga, In: Vijaysegaran, S. & Ibrahim, A.G. (eds), *First International Symposium on Fruit Flies in the Tropics*, Kuala Lumpur, 1988. Malaysian Agricultural Research and Development Institute, Kuala Lumpur. pp. 71-80.

Lloyd, A.C. (1991) The distribution of alimentary tract bacteria in the host tree by *Dacus tryoni*, In: Vijaysegaran, S. & Ibrahim, A.G. (eds), *First International Symposium on Fruit Flies in the Tropics*, Kuala Lumpur, 1988. Malaysian Agricultural Research and Development Institute, Kuala Lumpur. pp. 289-295.

Loiacono, M.S. (1981) Notas sobre Diapriinae Neotropicales (Hymenoptera, Diapriidae). *Revista de la Sociedad Entomológica Argentina*, **40,** 237-241. [RAE **71:**1851]

Longo, S. & Siscaro, G. (1989) Notes on behaviour of *Capparimyia savastanoi* (Martelli) (Diptera, Tephritidae) in Sicily, In: Cavalloro, R. (ed.), *Fruit flies of Economic Importance 87. Proceedings of the CEC/IOBC International Symposium*, Rome, 1987. Balkema, Rotterdam. pp. 81-89.

Lorraine, H. & Chambers, D.L. (1989) Control; eradication of exotic species; recent experiences in California, In: Robinson, A.S. & Hooper, G. (eds), Fruit flies; their biology, natural enemies and control. *World Crop Pests*, **3**(B), 399-410. Elsevier, Amsterdam.

Louis, C., Lopez-Ferber, M., Kuhl, G. & Arnoux, M. (1989) Importance des drosophiles comme 'mouches des fruits', In: Cavalloro, R. (ed.), *Fruit Flies of Economic Importance 87. Proceedings of the CEC/IOBC International Symposium*, Rome, 1987. Balkema, Rotterdam. pp. 595-602.

Louskas, C., Liaropoulos, C., Canard, M. & Laudého, Y. (1980) Infestation estivale précoce des olives par *Dacus oleae* (Gmel.) (Diptera, Trypetidae) et rôle limitant du parasite *Eupelmus urozonus* Dalm. (Hymenoptera, Eupelmidae) dans une oliveraie grecque. *Zeitschrift für Angewandte Entomologie*, **90,** 473-481. [RAE **69:**3764]

Lugemwa, F.N., Lwande, W., Bentley, M.D., Mendel, M.J. & Alford, A. Randall (1989) Volatiles of wild blueberry, *Vaccinium angustifolium*: possible attractants for the blueberry maggot fly, *Rhagoletis mendax*. *Journal of Agricultural and Food Chemistry*, **37,** 232-233. [RAE **78:**508]

Mabberley, D.J. (1987) *The Plant-Book; a Portable Dictionary of the Higher Plants*. Cambridge University Press, Cambridge.

McAlpine, D.K. & Schneider, M.A. (1978) A systematic study of *Phytalmia* (Diptera, Tephritidae) with description of a new genus. *Systematic Entomology*, **3,** 159-175.

McAlpine, J.F. (1981) 2. Morphology and terminology-adults, In: McAlpine, J.F., Peterson, B.V., Shewell, G.E., Teskey, H.J., Vockeroth, J.R. & Wood, D.M. (eds), Manual of Nearctic Diptera-1. *Monograph of the Biosystematics Research Institute*, **27,** 9-63. Agriculture Canada, Ottawa.

McBride, O.C. & Tanada, Y. (1949) A revised list of host plants of the melon fly in Hawaii. *Proceedings of the Hawaiian Entomological Society*, **13,** 411-421.

McCrea, K.D. & Abrahamson, W.G. (1985) Evolutionary impacts of the goldenrod ball gallmaker on *Solidago altissima* clones. *Oecologia*, **68,** 20-22. [RAE **76:**7644]

McDonald, P.T. (1987) Intragroup stimulation of pheromone release by male Mediterranean fruit flies (Diptera: Tephritidae). *Annals of the Entomological Society of America*, **80,** 17-20. [RAE **75:**6094]

McFadden, M.W., Kapatos, E., Pappas, S. & Carvounis, G. (1977) Ecological studies on the olive fly *Dacus oleae* Gmel. in Corfu. I. The yearly life cycle. *Bollettino del Laboratorio di Entomologia Agraria 'Filippo Silvestri', Portici*, **34,** 43-50. [RAE **67:**1117]

McGovern, T.P. & Cunningham, R.T. (1988) Attraction of Mediterranean fruit fly (Diptera: Tephritidae) to analogs of selected trimedlure isomers. *Journal of Economic Entomology*, **81,** 1052-1056. [RAE **77:**8994]

McGovern, T.P., Cunningham, R.T. & Leonhardt, B.A. (1987) Attractiveness of trans-trimedlure and its four isomers in field tests with the Mediterranean fruit fly (Diptera: Tephritidae). *Journal of Economic Entomology*, **80,** 617-620 [RAE **76:**622]

McKechnie, S.W. (1975) Enzyme polymorphism and species discrimination in fruit flies of the genus *Dacus* (Tephritidae). *Australian Journal of Biological Sciences*, **28,** 405-411. [RAE **64:**4595]

McPheron, B.A. (1990) Genetic structure of apple maggot fly (Diptera: Tephritidae) populations. *Annals of the Entomological Society of America*, **83,** 568-577. [RAE **79:**4714]

McPheron, B.A., Jorgensen, C.D. & Berlocher, S.H. (1988a) Low genetic variability in a Utah cherry-infesting population of the apple maggot, *Rhagoletis pomonella*. *Entomologia Experimentalis et Applicata*, **46,** 155-160. [RAE **76:**3338]

McPheron, B.A., Smith, D.C. & Berlocher, S.H. (1988b) Genetic differences between host races of *Rhagoletis pomonella*. *Nature, UK*, **336,** 64-67. [RAE **77:**8586]

Madsen, H.F. (1970) Observations on *Rhagoletis indifferens* and related species in the Okanagan Valley of British Columbia. *Journal of the Entomological Society of British Columbia*, **67,** 13-16. [RAE **61:**3922]

Maehler, K.L. (1948) The Oriental fruit fly in Guam. *Journal of Economic Entomology*, **41,** 991-992. [RAE **38:**75]

Maehler, K.L. (1951) Notes and exhibitions (March 13, 1950). *Proceedings of the Hawaiian Entomological Society*, **14,** 205-207.

Maier, C.T. (1981) Parasitoids emerging from puparia of *Rhagoletis pomonella* (Diptera: Tephritidae) infesting hawthorn and apple in Connecticut. *Canadian Entomologist*, **113,** 867-870. [RAE **70:**2741]

Maki, S. & Rin, G. (1918) Takenoko no gaichu ni tsuite. *Ringyo Shikenyo Hokoku, Formosan Government Industry Bureau, Taipe*. **104**(5), 85-100. [RAE **6:**402]

Malan, E.M. & Giliomee, J.H. (1969) Morphology and descriptions of the larvae of three species of Dacinae (Diptera: Trypetidae). *Journal of the Entomological Society of South Africa*, **32,** 259-271.

Malavasi, A., Duarte, A.L. & Cabrini, G. (1990) Field evaluation of three baits for South American cucurbit fruit fly (Diptera: Tephritidae) using McPhail traps. *Florida Entomologist*, **73,** 510-512. [RAE **79:**1311]

Malavasi, A. & Morgante, J.S. (1980) Biologia de 'moscas-das-frutas' (Diptera, Tephritidae). II: indices de infestacao em diferentes hospedeiros e localidades. *Revista Brasileira de Biologia*, **40,** 17-24. [RAE **69:**1291]

Malavasi, A., Morgante, J.S. & Prokopy, R.J. (1983) Distribution and activities of *Anastrepha fraterculus* (Diptera: Tephritidae) flies on host and nonhost trees. *Annals of the Entomological Society of America*, **76,** 286-292. [RAE **72:**573]

Malavasi, A., Morgante, J.S. & Zucchi, R.A. (1980) Biologia de 'moscas-das-frutas' (Diptera, Tephritidae). I: lista de hospedeiros e ocorrencia. *Revista Brasileira de Biologia*, **40,** 9-16. [RAE **69:**1290]

Malio, E. (1979) Observations on the mango fruit fly *Ceratitis cosyra* in the Coast Province, Kenya. *Kenya Entomologist's Newsletter*, **10,** 7. [RAE **68:**5644]

Malloch, J.R. (1938) Trypetidae of the Mangarevan Expedition. *Occasional Papers of the Bernice Pauahi Bishop Museum*, **14,** 111-116. [RAE **27:**294]

Malloch, J.R. (1939a) The Diptera of the Territory of New Guinea. XI. Family Trypetidae. *Proceedings of the Linnean Society of New South Wales*, **64,** 409-465.

Malloch, J.R. (1939b) Solomon Islands Trypetidae. *Annals and Magazine of Natural History*, (11) **4,** 228-278.

Malloch, J.R. (1942) Trypetidae, Otitidae, Heleomyzidae and Clusiidae of Guam (Diptera). *Bulletin of the Bernice Pauahi Bishop Museum*, **172,** 201-210.

Malo, E., Baker, P.S. & Valenzuela, J. (1987) The abundance of species of *Anastrepha* (Diptera: Tephritidae) in the coffee producing area of coastal Chiapas, southern Mexico. *Folia Entomologica Mexicana*, **73,** 125-140. [RAE **77:**3964]

Mangel, M., Carey, J.R. & Plant, R.E. (eds) (1986) Pest control: operations and systems analysis in fruit fly management. *NATO Advanced Science Institutes Series G: Ecological Sciences*, **11,** xii+1-465. Springer Verlag, Berlin.

Mangel, M. & Roitberg, B.D. (1989) Dynamic information and host acceptance by a tephritid fruit fly. *Ecological Entomology*, **14,** 181-189. [RAE **78:**3009]

Manolache, C. (1940) *Acanthiophilus helianthi* Rossi. *Viata Agricola*, **31,** 65. [RAE **31:**227]

Manousis, T. & Ellar, D.J. (1988) *Dacus oleae* microbial symbionts. *Microbiological Sciences*, **5,** 149-152. [RAE **76:**7379]

Manousis, T. & Moore, N.F. (1987) Cricket paralysis virus, a potential control agent for the olive fruit fly, *Dacus oleae* Gmel. *Applied and Environmental Microbiology*, **53,** 142-148. [RAE **76:**5220]

Margaritis, L.H. (1985) Comparative study of the eggshell of the fruit flies *Dacus oleae* and *Ceratitis capitata* (Diptera: Trypetidae). *Canadian Journal of Zoology*, **63,** 2194-2206. [RAE **74:**808]

Martinovich, V. (1966) Pórsáfrány-légy (*Acanthiophilus helianthi* Rossi) a *Centaurea* magtermesztés kártevoje Magyarországon (Dipt. Trypetidae). *Folia Entomologica Hungarica*, **19,** 375-402. [RAE **55:**1522]

Martins, D. dos Santos & Alves, F. de Lima (1988) Ocorrencia da mosca-das-frutas *Ceratitis capitata* (Wiedemann, 1824) (Diptera: Tephritidae) na cultura do mamoeiro (*Carica papaya* L.) no norte do estado do Espirito Santo. *Anais da Sociedade Entomologica do Brasil*, **17,** 227-229. [RAE **77:**5572]

Matanmi, B.A. (1975) The biology of tephritid fruitflies (Diptera, Tephritidae) attacking cucurbits at Ile-Ife, Nigeria. *Nigerian Journal of Entomology*, **1,** 153-159. [RAE **68:**422]

Matanmi, J. (1972; dated 1970) Observations on fruit flies at Ile-Ife. *Nigerian Entomologists' Magazine*, **2,** 59-61. [RAE **61:**1661]

Mather, M.H. & Roitberg, B.D. (1987) A sheep in wolf's clothing: tephritid flies mimic spider predators. *Science, USA*, **236**(4799), 308-310. [RAE **75:**5255]

May, A.W.S. (1953) Queensland host records for the Dacinae (fam. Trypetidae). *Queensland Journal of Agricultural Science*, **10,** 36-79. [RAE **42:**344]

Mayo, I., Anderson, M., Burguete, J. & Chillida, E.M. Robles (1987) Structure of superficial chemoreceptive sensilla on the third antennal segment of *Ceratitis capitata* (Wiedemann) (Diptera: Tephritidae). *International Journal of Insect Morphology & Embryology*, **16**, 131-141. [RAE **76**:18]

Mazomenos, B.E. (1989) Biology and physiology; mating pheromones; *Dacus oleae*, In: Robinson, A.S. & Hooper, G. (eds), Fruit flies; their biology, natural enemies and control. *World Crop Pests*, **3**(A), 169-178. Elsevier, Amsterdam.

Mazor, M., Gothilf, S. & Galun, R. (1987) The role of ammonia in the attraction of females of the Mediterranean fruit fly to protein hydrolysate baits. *Entomologia Experimentalis et Applicata*, **43**, 25-29. [RAE **75**:4324]

Meats, A. (1989a) Ecology; abiotic mortality factors - temperature, In: Robinson, A.S. & Hooper, G. (eds), Fruit flies; their biology, natural enemies and control. *World Crop Pests*, **3**(B), 229-239. Elsevier, Amsterdam.

Meats, A. (1989b) Ecology; bioclimatic potential, In: Robinson, A.S. & Hooper, G. (eds), Fruit flies; their biology, natural enemies and control. *World Crop Pests*, **3**(B), 241-252. Elsevier, Amsterdam.

Meijere, J.C.H. de (1911) Studien über südostasiatische Dipteren, VI. *Tijdschrift voor Entomologie*, **54**, 258-342.

Meijere, J.C.H. de (1938) *Acidoxantha bombacis* n.sp. *Tijdschrift voor Entomologie*, **81**, 122-123.

Menon, M.G. Ramdas, Kapoor, V.C. & Mahto, Y. (1969) *Centaurea americana* as a new host plant record for the fruit flies, *Acanthiophilus helianthi* Rossi and *Craspedoxantha octopunctata* Bezzi in India. *Indian Journal of Entomology*, **30**, 316. [RAE **59**:2522]

Menon, M.G. Ramdas, Mahto, Y., Kapoor, V.C. & Bhatia, S.K. (1968) Identifies of the immature stages of three species of Indian fruit-flies *Dacus cucurbitae* Coquillett, *D. diversus* Coquillett, and *D. ciliatus* Loew (Diptera, Trypetidae). *Bulletin of Entomology. Entomological Society of India*, **9**, 87-94.

Merz, B. (1991) *Rhagoletis completa* Cresson und *Rhagoletis indifferens* Curran, zwei wirtschaftlich bedeutende nordamerikanische Fruchtfliegen, neu für Europa (Diptera: Tephritidae). *Mitteilungen der Schweizerischen Entomologischen Gesellschaft*, **64**, 55-57.

Message, C.M. & Zucoloto, F.S. (1980) Valor nutritivo do levedo de cerveja para *Anastrepha obliqua* (Diptera, Tephritidae). *Ciência e Cultura*, **32**, 1091-1094. [RAE **69**:4334]

Messina, F.J. (1989a) Host preferences of cherry- and hawthorn-infesting populations of *Rhagoletis pomonella* in Utah. *Entomologia Experimentalis et Applicata*, **53**, 89-95. [RAE **78**:7738]

Messina, F.J. (1989b) Host-plant variables influencing the spatial distribution of a frugivorous fly, *Rhagoletis indifferens*. *Entomologia Experimentalis et Applicata*, **50**, 287-294. [RAE **77**:8597]

Milani, R., Gasperi, G. & Malacrida, A. (1989) Genetics; *Ceratitis capitata*; biochemical genetics, In: Robinson, A.S. & Hooper, G. (eds), Fruit flies; their biology, natural enemies and control. *World Crop Pests*, **3**(B), 33-56. Elsevier, Amsterdam.

Mitchell, W.C. & Saul, S.H. (1990) Current control methods for the Mediterranean fruit fly, *Ceratitis capitata*, and their application in the USA. *Review of Agricultural Entomology*, **78,** 923-930. [RAE **79:**968]

Mitchell, W.C., Metcalf, R.L., Metcalf, E.R. & Mitchell, S. (1985) Candidate substitutes for methyl eugenol as attractants for the area-wide monitoring and control of the Oriental fruitfly, *Dacus dorsalis* Hendel (Diptera: Tephritidae). *Environmental Entomology*, **14,** 176-181. [RAE **74:**554]

Miyake, T. (1919) Studies on the fruit-flies of Japan. I. Japanese orange fly. *Bulletin of the Imperial Central Agricultural Experiment Station of Japan*, **2,** 85-165. [RAE **7:**238]

Monaco, R. (1978) Note sui parassiti del *Dacus oleae* Gmel. (Dipt.-Tephritidae) in Sud-Africa. *Atti XI Congresso Nazionale Italiano di Entomologia*, 303-310. [RAE **67:**977]

Monaco, R. (1984) *Opius magnus* Fischer (Braconidae), parassita di *Rhagoletis cerasi* L. su *Prunus mahaleb*. *Entomologica*, **19,** 75-80. [RAE **75:**1822]

Monteith, L.G. (1977; dated 1976) Field studies of potential predators of the apple maggot *Rhagoletis pomonella* (Diptera: Tephritidae) in Ontario. *Proceedings of the Entomological Society of Ontario*, **107,** 23-30. [RAE **66:**5482]

Monteith, L.G. (1978; dated 1977) Additional records and the role of the parasites of the apple maggot *Rhagoletis pomonella* (Diptera: Tephritidae) in Ontario. *Proceedings of the Entomological Society of Ontario*, **108,** 3-6. [RAE **67:**4526]

Monty, J. (1973) Rearing of the Natal fruit-fly *Ceratitis (Pterandrus) rosa* Karsch (Diptera, Trypetidae) in the laboratory. *Revue Agricole et Sucrière de l'Ile Maurice*, **52,** 133-135. [RAE **63:**1592]

Morgante, J.S., Malavasi, A & Prokopy, R.J. (1983) Mating behavior of wild *Anastrepha fraterculus* (Diptera: Tephritidae) on a caged host tree. *Florida Entomologist*, **66,** 234-241. [RAE **72:**731]

Moshonas, M.G. & Shaw, P.E. (1984) Effects of low-dose-irradiation on grapefruit products. *Journal of Agricultural and Food Chemistry*, **32,** 1098-1101. [RAE **74:**5384]

Motooka, K. (1938) On insect pests of *Camellia* L. *Journal of Plant Protection*, **25,** 181-185. [in Japanese]. [RAE **26:**311]

Muesebeck, C.F.W. (1980) The Nearctic parasitic wasps of the genera *Psilus* Panzer and *Coptera* Say (Hymenoptera, Proctotrupoidea, Diapriidae). *Technical Bulletin, United States Department of Agriculture*, **1617,** iv+1-71. [RAE **69:**5814]

Mumtaz, M.M. & AliNiazee, M.T. (1983) The oviposition-deterring pheromone in the western cherry fruit fly, *Rhagoletis indifferens* Curran (Dipt., Tephritidae). 1. Biological properties. *Zeitschrift für Angewandte Entomologie*, **96,** 83-93. [RAE **71:**7040]

Munro, H.K. (1924) Fruit flies of wild olives. *Entomology Memoirs, Department of Agriculture, Union of South Africa*, [1] (2), 5-17. [RAE **13:**281]

Munro, H.K. (1925) Biological notes on the South African Trypaneidae (fruit-flies) I. *Entomology Memoirs, Department of Agriculture, Union of South Africa*, [1] (3), 40-67. [RAE **13:**525]

Munro, H.K. (1926) Biological notes on the South African Trypaneidae (Trypetidae: fruit-flies) II. *Entomology Memoirs, Department of Agriculture, Union of South Africa*, [1] (5), 17-40. [RAE **16**:303]

Munro, H.K. (1929a) Additional trypetid material in the collection of the South African Museum (Trypetidae, Diptera). *Annals of the South African Museum*, **29**, 1-39. [RAE **17**:664]

Munro, H.K. (1929b) Biological notes on the South African Trypetidae (fruit-flies. Diptera) III. *Entomology Memoirs, Department of Agriculture, Union of South Africa*, [1] (6), 9-17. [RAE **18**:420]

Munro, H.K. (1932) Notes on *Dacus ciliatus* Lw., and certain related species (Dipt. Trypetidae). *Stylops*, **1**, 151-158. [RAE **20**:499]

Munro, H.K. (1934) A review of the species of the subgenus *Trirhithrum*, Bezzi (Trypetidae, Diptera). *Bulletin of Entomological Research*, **25**, 473-489. [RAE **23**:152]

Munro, H.K. (1935a) Some new species of the subgenus *Pardalaspis* Bez. (Trypetidae, Diptera). *Annals and Magazine of Natural History*, (10) **15**, 301-313.

Munro, H.K. (1935b) Biological and systematic notes and records of South African Trypetidae (fruit-flies, Diptera) with descriptions of new species. *Entomology Memoirs, Department of Agriculture, Union of South Africa*, [1] (9), 18-59. [RAE **23**:689]

Munro, H.K. (1935c) Observations and comments on the Trypetidae (Dipt.) of Formosa. *Arbeiten über Physiologische und Angewandte Entomologie aus Berlin-Dahlem*, **2**, 195-203, 253-271. [RAE **24**:199]

Munro, H.K. (1937) Some new Trypetidae from Kenya Colony. *Journal of the East Africa and Uganda Natural History Society*, **5** (supplement), 1-13.

Munro, H.K. (1938a) New Trypetidae from Kenya Colony II. *Journal of the East Africa and Uganda Natural History Society*, **13**, 159-167.

Munro, H.K. (1938b) Quelques diptères trypétides du Congo Belge avec descriptions d'espèces nouvelles. *Revue de Zoologie et de Botanique Africaines*, **31**, 163-173.

Munro, H.K. (1938c) Studies on Indian Trypetidae. *Record of the Indian Museum*, **40**, 21-37. [RAE **26**:602]

Munro, H.K. (1947) African Trypetidae (Diptera); a review of the transition genera between Tephritinae and Trypetinae, with a preliminary study of the male terminalia. *Memoirs of the Entomological Society of Southern Africa*, **1**, 1-284.

Munro, H.K. (1953) Records of some Trypetidae (Diptera) collected on the Bernard Carp Expedition to Barotseland, 1952, with a new species from Kenya. *Journal of the Entomological Society of Southern Africa*, **16**, 217-226.

Munro, H.K. (1964a) *Some Fruitflies of Economic Importance in South Africa*. Department of Agricultural Technical Services, Pretoria.

Munro, H.K. (1964b) The genus *Trupanea* in Africa; an analytical study in bio-taxonomy. *Entomology Memoirs, Department of Agricultural Technical Services, Republic of South Africa*, **8**, 1-101.

Munro, H.K. (1967) Fruitflies allied to species of *Afrocneros* and *Ocnerioxa* that infest *Cussonia*, the umbrella tree or kiepersol (Araliaceae) (Diptera: Trypetidae). *Annals of the Natal Museum*, **18**, 571-594.

Munro, H.K. (1984) A taxonomic treatise on the Dacidae (Tephritoidea, Diptera) of Africa. *Entomology Memoirs, Department of Agriculture and Water Supply, Republic of South Africa*, **61,** ix + 1-313. [RAE **73:**4969]

Mustafa, T.M. & Zaghal, K. Al- (1987) Frequency of *Dacus oleae* (Gmelin) immature stages and their parasites in seven olive varieties in Jordan. *Insect Science and its Application*, **8,** 165-169. [RAE **76:**8892]

Nahal, A.K.M. El-, Azab, A.K. & Swailem, S.M. (1971; dated 1970) Studies on the biology of the melon fruit fly, *Dacus ciliatus* Loew (Diptera: Trypanaeidae). *Bulletin de la Société Entomologique d'Egypt*, **54,** 231-241. [RAE **62:**1912]

Nakagawa, S. & Yamada, T. (1965) Two varieties of *Sesbania grandiflora* as fruit fly hosts. *Journal of Economic Entomology*, **58,** 796. [RAE **53:**564]

Nakagawa, S., Farias, G.J. & Urago, T. (1968) Newly recognized hosts of the Oriental fruit fly, melon fly, and Mediterranean fruit fly. *Journal of Economic Entomology*, **61,** 339-340. [RAE **56:**1593]

Narayanan, E.S. (1953) Seasonal pests of crops: fruit fly pests of orchards and kitchen gardens. *Indian Farming*, **3**(4), 8-11, 29-31.

Narayanan, E.S. & Chawla, S.S. (1962) Parasites of fruit fly pests of the world with brief notes on their bionomics, habits and distribution. *Beiträge zur Entomologie*, **12,** 437-476. [RAE **51:**516]

Nascimento, A.S. & Zucchi, R.A. (1981) Dinâmica populacional das moscas-das-frutas do genêro *Anastrepha* (Dip., Tephritidae) no Recôncavo Baiano. I. Levantamento das espécies. *Pesquisa Agropecuária Brasileira*, **16,** 763-767. [RAE **70:**6042]

Nath, D.K. (1973; dated 1972) *Callantra minax* (Enderlein) (Tephritidae: Diptera), a new record of a ceratitinid fruitfly on orange fruits (*Citrus reticulata* Blanco) in India. *Indian Journal of Entomology*, **34,** 246. [RAE **64:**2031]

Nation, J.L. (1981) Sex-specific glands in tephritid fruit flies of the genera *Anastrepha, Ceratitis, Dacus* and *Rhagoletis* (Diptera: Tephritidae). *International Journal of Insect Morphology and Embryology*, **10,** 121-129. [RAE **70:**2486]

Nation, J.L. (1989a) Biology and physiology; mating pheromones; the role of pheromones in the mating system of *Anastrepha* fruit flies, In: Robinson, A.S. & Hooper, G. (eds), Fruit flies; their biology, natural enemies and control. *World Crop Pests*, **3**(A), 189-205. Elsevier, Amsterdam.

Nation, J.L. (1989b) Biology of pheromone release by male Caribbean fruit flies, *Anastrepha suspensa* (Diptera: Tephritidae). *Journal of Chemical Ecology*, **16,** 553-572. [RAE **78:**7350]

Neilson, W.T.A. & Wood, G.W. (1985) The blueberry maggot: distribution, economic importance, and management practices. *Acta Horticulturae*, **165,** 171-175. [RAE **74:**5742]

Neilson, W.T.A., Knowlton, A.D. & Fuller, M. (1984) Capture of blueberry maggot adults, *Rhagoletis mendax* (Diptera: Tephritidae), on Pherocon AM traps and on tartar red sticky spheres in lowbush blueberry fields. *Canadian Entomologist*, **116,** 113-118. [RAE **72:**2953]

Neuenschwander, P. (1982) Searching parasitoids of *Dacus oleae* (Gmel.) (Dipt., Tephritidae) in South Africa. *Zeitschrift für Angewandte Entomologie*, **94,** 509-522. [RAE **71:**2759]

Neuenschwander, P. (1984) Observations on the biology of two species of fruit flies (Diptera: Tephritidae), and their competition with a moth larva. *Israel Journal of Entomology*, **18,** 95-97.

Neuenschwander, P. & Michelakis, S. (1978) The infestation of *Dacus oleae* (Gmel.) (Diptera, Tephritidae) at harvest time and its influence on yield and quality of olive oil in Crete. *Zeitschrift für Angewandte Entomologie*, **86,** 420-433. [RAE **67**:3327]

Neuenschwander, P., Bigler, F., Delucchi, V. & Michelakis, S. (1983) Natural enemies of preimaginal stages of *Dacus oleae* Gmel. (Dipt., Tephritidae) in western Crete. I. Bionomics and phenologies. *Bollettino del Laboratorio di Entomologia Agraria 'Filippo Silvestri'*, **40,** 3-32. [RAE **72:**7195]

Neuenschwander, P., Michelakis, S., Holloway, P. & Berchtold, W. (1985) Factors affecting the susceptibility of fruits of different olive varieties to attack by *Dacus oleae* (Gmel.) (Dipt., Tephritidae). *Zeitschrift für Angewandte Entomologie*, **100,** 174-188. [RAE **74:**165]

Nirula, K.K. (1942) *Trypanea stellata* Fuessly a new pest of some Compositae flowers. *Indian Journal of Entomology*, **4,** 90. [RAE **32:**408]

Nishida, T. (1980; dated 1977) Food system of tephritid flies in Hawaii. *Proceedings of the Hawaiian Entomological Society*, **23,** 245-254. [RAE **69:**565]

Nishida, T., Harris, E.J., Vargas, R.I. & Wong, T.T.Y. (1985) Distributional loci and host fruit utilization patterns of the Mediterranean fruit fly, *Ceratitis capitata* (Diptera: Tephritidae), in Hawaii. *Environmental Entomology*, **14,** 602-606. [RAE **74:**805]

Nishida, R., Tan, K.H., Serit, M., Lajis, N.H., Sukari, A.M., Takahasi, S. & Fukami, H. (1988) Accumulation of phenylpropanoids in the rectal glands of males of the Oriental fruit fly, *Dacus dorsalis*. *Experientia*, **44,** 534-536. [RAE **78:**2215]

Nishijima, K.A., Couey, H.M. & Alvarez, A.M. (1987) Internal yellowing, a bacterial disease of papaya fruits caused by *Enterobacter cloacae*. *Plant Disease*, **71,** 1029-1034. [RAE **76:**6938]

Norrbom, A.L. (1982) A new host for *Rhagoletis striatella* (Diptera: Tephritidae). *Melsheimer Entomological Series*, **32,** 11. [RAE **71:**7045]

Norrbom, A.L. (1987) A revision of the Neotropical genus *Polionota* Wulp (Diptera: Tephritidae). *Folia Entomológica Mexicana*, **73,** 101-123.

Norrbom, A.L. (1989) The status of *Urophora acuticornis* and *U. sabroskyi* (Diptera: Tephritidae). *Entomological News*, **100,** 59-66. [RAE **77:**9069]

Norrbom, A.L. (1991) The species of *Anastrepha* (Diptera: Tephritidae) with a *grandis*-type wing pattern. *Proceedings of the Entomological Society of Washington*, **93,** 101-124.

Norrbom, A.L. & Foote, R.H. (1989) Taxonomy and zoogeography; the taxonomy and zoogeography of the genus *Anastrepha* (Diptera: Tephritidae), In: Robinson, A.S. & Hooper, G. (eds), Fruit flies; their biology, natural enemies and control. *World Crop Pests*, **3**(A), 15-26. Elsevier, Amsterdam.

Norrbom, A.L. & Kim, K.C. (1988a) Revision of the *schausi* group of *Anastrepha* Schiner (Diptera: Tephritidae), with a discussion of the terminology of the female terminalia in the Tephritoidea. *Annals of the Entomological Society of America*, **81,** 164-173. [RAE **76:**7356]

Norrbom, A.L. & Kim, K.C. (1988b) *A List of the Reported Host Plants of the Species of Anastrepha (Diptera: Tephritidae)*. United States Department of Agriculture (APHIS 81-52), Washington.

Norrbom, A.L., Ming, Y. & Hernandez-Ortiz, V. (1988) A revision of the genus *Oedicarena* Loew (Diptera: Tephritidae). *Folia Entomológica Mexicana*, **75,** 93-117.

Novak, J.A. (1974) A taxonomic revision of *Dioxyna* and *Paroxyna* (Diptera: Tephritidae) for America north of Mexico. *Melanderia*, **16,** 1-53.

Ogbalu, O.N. (1989) The susceptibility of pepper fruits (*Capsicum* species) to oviposition by the pepper fruitfly, *Atherigona orientalis* (Schiner) in Port Harcourt, Nigeria. *Tropical Pest Management*, **35,** 392-393. [RAE **79:**4863]

Okinawa Prefecture (1987) *Melon Fly Eradication Project in the Okinawa Prefecture*. Okinawa Prefectural Fruit Fly Eradication Project Office, Naha.

O'Loughlin, G.T., East, R.W., Kenna, G.B. & Harding, E. (1983) Comparison of the effectiveness of traps for the Queensland fruit fly, *Dacus tryoni* (Froggatt) (Diptera: Tephritidae). *General and Applied Entomology*, **15,** 3-6. [RAE **72:**570]

Ooi, C.S. (1991) Genetic variation in populations of two sympatric taxa in the *Dacus dorsalis* complex and their relative infestation levels in various fruit hosts, In: Vijaysegaran, S. & Ibrahim, A.G. (eds), *First International Symposium on Fruit Flies in the Tropics*, Kuala Lumpur, 1988. Malaysian Agricultural Research and Development Institute, Kuala Lumpur. pp. 71-80.

Opp, S.B. & Prokopy, R.J. (1987) Seasonal changes in resightings of marked, wild *Rhagoletis pomonella* (Diptera: Tephritidae) flies in nature. *Florida Entomologist*, **70,** 449-457. [RAE **76:**8876]

Orian, A.J.E. (1962) Pest control recommendations made by the Division of Entomology of the Department of Agriculture, Mauritius. *Revue Agricole et Sucrière de l'Ile Maurice*, **41,** 87-116.

Orian, A.J.E. & Moutia, L.A. (1960) Fruit flies (Trypetidae) of economic importance in Mauritius. *Revue Agricole et Sucrière de l'Ile Maurice*, **39,** 142-150. [RAE **49:**667]

Owens, E.D. & Prokopy, R.J. (1986) Relationship between reflectance spectra of host plant surfaces and visual detection of host fruit by *Rhagoletis pomonella* flies. *Physiological Entomology*, **11,** 297-307. [RAE **75:**1819]

Owusu-Manu, E. & Bonku, M.E. (1987) Kola entomology. *Annual Report, Cocoa Research Institute, Ghana, 1985/86*. 45-46. [RAE **77:**5736]

Papaj, D.R. & Prokopy, R.J. (1986) Phytochemical basis of learning in *Rhagoletis pomonella* and other herbivorous insects. *Journal of Chemical Ecology*, **12,** 1125-1143. [RAE **74:**5540]

Papaj, D.R., Roitberg, B.D., Opp, S.B., Aluja, M., Prokopy, R.J. & Wong, T.T.Y. (1990) Effect of marking pheromone on clutch size in the Mediterranean fruit fly. *Physiological Entomology*, **15,** 463-468. [RAE **79:**1600]

Papaj, D.R., Katsoyannos, B.I. & Hendrichs, J. (1989a) Use of fruit wounds in oviposition by Mediterranean fruit flies. *Entomologia Experimentalis et Applicata*, **53,** 203-209. [RAE **79:**2649]

Papaj, D.R., Opp, S.B., Prokopy, R.J. & Wong, T.T.Y. (1989b) Cross-induction of fruit acceptance by the medfly *Ceratitis capitata*: the role of fruit size and chemistry. *Journal of Insect Behaviour*, **2,** 241-254. [RAE **78**:494]

Pappas, S., Kapatos, E. & McFadden, M.W. (1977) Ecological studies on the olive fly *Dacus oleae* Gmel. in Corfu. III. The action of hymenopterous parasites. *Bollettino del Laboratorio di Entomologia Agraria 'Filippo Silvestri', Portici*, **34,** 80-86. [RAE **67**:1119]

Parihar, D.R. (1984) Breeding biology of aak fruitfly, *Dacus (Leptoxyda) longistylus* Wied. (Diptera: Tephritidae) on *Calotropis procera* plantations in Indian desert. *Indian Journal of Forestry*, **7,** 213-216. [RAE **74**:2971]

Paripurna, K.A. & Srivastava, B.G. (1988; dated 1987) Effect of different quantities of water in the chemically defined diet on the growth and development of *Dacus cucurbitae* (Coquillett) maggots under aseptic condition. *Indian Journal of Entomology*, **49,** 259-262. [RAE **78**:169]

Parlati, M.V., Petruccioli, G. & Turco, D. (1986; dated 1981/3) Effetti dell'attacco del '*Dacus*' sulla qualita dell'olio. *Annali dell'Instituto Sperimentale per l'Olivicoltura*, **7,** 21-29. [RAE **75**:4816]

Paulian, R. (1953) Recherches sur les insectes d'importance biologique à Madagascar; XII. - les mouches des fruits. *Mémoires de l'Institut Scientifique de Madagascar*, Series E, **3,** 1-7.

Pemberton, C.E. (1946) A new fruit fly in Hawaii. *Hawaiian Planters' Record*, **50,** 53-55. [RAE **36**:249]

Perkins, F.A. (1938) Studies in Oriental and Australian Trypaneidae - part 2. *Proceedings of the Royal Society of Queensland*, **49,** 120-144.

Perkins, F.A. (1939) Studies in Oriental and Australian Trypetidae - part 3. *Papers from the Department of Biology, University of Queensland*, **1,** 1-35.

Perkins, M.V., Fletcher, M.T., Kitching, W., Drew, R.A.I. & Moore, C.J. (1990) Chemichal studies of rectal gland secretions of some species of *Bactrocera dorsalis* complex of fruit flies (Diptera: Tephritidae). *Journal of Chemical Ecology*, **16,** 2475-2487. [RAE **79**:65]

Peschken, D.P. (1980; dated 1979). Host specificity and suitability of *Tephritis dilacerata* (Dip.: Tephritidae): a candidate for the biological control of perennial sow-thistle (*Sonchus arvensis*) (Compositae) in Canada. *Entomophaga*, **24,** 455-461. [RAE **69**:168]

Peschken, D.P. (1984). *Cirsium arvense* (L.) Scop., Canada thistle (Compositae), In: Kelleher, J.S. & Hulme, M.A. (eds), *Biological Control Programmes Against Insects and Weeds in Canada 1969-1980*. Commonwealth Agricultural Bureaux, Slough. pp. 139-146.

Philip, A. (1950) Description of one new species of *Strumeta* Walker (Trypetidae: Diptera) from Burma and a record of one far-eastern species of the genus from India. *Indian Journal of Entomology*, **10,** 31-32.

Phillips, V.T. (1946) The biology and identification of trypetid larvae. *Memoirs of the American Entomological Society*, **12,** 1-161. [RAE **35**:376]

Pickett, A.D. (1937) Studies on the genus *Rhagoletis* (Trypetidae) with special reference to *Rhagoletis pomonella* (Walsh). *Canadian Journal of Research*, Series D, **15**, 53-75. [RAE **25**:471]

Pickett, A.D & Neary, M.E. (1940) Further studies on *Rhagoletis pomonella* (Walsh). *Scientific Agriculture*, **20**, 551-556. [RAE **29**:226]

Piper, G.L. (1976) Bionomics of *Euarestoides acutangulus* (Diptera: Tephritidae). *Annals of the Entomological Society of America*, **69**, 381-386. [RAE **64**:6659]

Pittara, I.S. & Katsoyannos, B.I. (1990) Evidence for a host-marking pheromone in *Chaetorellia australis*. *Entomologia Experimentalis et Applicata*, **54**, 287-295. [RAE **78**:10508]

Plant, R.E. (1986) The sterile insect technique: a theoretical perspective, In: Mangel, M., Carey, J.R. & Plant, R.E. (eds), Pest control: operations and systems analysis in fruit fly management. *NATO Advanced Science Institutes Series G: Ecological Sciences*, **11**, 361-386. Springer-Verlag, Berlin. [RAE **75**:2353]

Podoler, H. & Mazor, M. (1981a) *Dirhinus giffardii* Silvestri (Hym.: Chalcididae) as a parasite of the Mediterranean fruit fly, *Ceratitis capitata* (Wiedemann) (Dip.: Tephritidae). 1. Some biological studies. *Acta Oecologia, Oecologia Applicata*, **2**, 255-265. [RAE **70**:1352]

Podoler, H. & Mazor, M. (1981b) *Dirhinus giffardii* Silvestri (Hym.: Chalcididae) as a parasite of the Mediterranean fruit fly, *Ceratitis capitata* (Wiedemann) (Dip.: Tephritidae). 2. Analysis of parasite responses. *Acta Oecologia, Oecologia Applicata*, **2**, 299-309. [RAE **70**:1373]

Podoler, H. & Mendel, Z. (1977) Analysis of solitariness in a parasite-host system (*Muscidifurax raptor*, Hymenoptera: Pteromalidae - *Ceratitis capitata*, Diptera: Tephritidae). *Ecological Entomology*, **2**, 153-160. [RAE **65**:6056]

Podoler, H. & Mendel, Z. (1979) Analysis of a host-parasite (*Ceratitis-Muscidifurax*) relationship under laboratory conditions. *Ecological Entomology*, **4**, 45-59. [RAE **67**:3743]

Poinar, G.D., Thomas, G. & Prokopy, R.J. (1978; dated 1977) Microorganisms associated with *Rhagoletis pomonella* (Diptera: Tephritidae) in Massachusetts. *Proceedings of the Entomological Society of Ontario*, **108**, 19-22. [RAE **67**:4161]

Pritchard, G. (1967) Laboratory observations on the mating behaviour of the island fruit fly *Rioxa pornia* (Diptera: Tephritidae). *Journal of the Australian Entomological Society*, **6**, 127-132. [RAE **58**:3617]

Prokopy, R.J. (1977) Stimuli influencing trophic relations in Tephritidae. *Colloques Internationaux du Centre National de la Recherche Scientifique*, **265**, 305-336. [RAE **66**:1150]

Prokopy, R.J. (1986) Alightment of apple maggot flies on fruit mimics in relation to contrast against background. *Florida Entomologist*, **69**, 716-721. [RAE **76**:627]

Prokopy, R.J. (1988) Apple IPM in Massachusetts: a progress report. *IPM Practitioner*, **10**(5), 7-8. [RAE **77**:384]

Prokopy, R.J. & Berlocher, S.H. (1980) Establishment of *Rhagoletis pomonella* (Diptera: Tephritidae) on rose hips in southern New England. *Canadian Entomologist*, **112**, 1319-1320. [RAE **69**:5332]

Prokopy, R.J. & Bush, G.L. (1973) Mating behaviour of *Rhagoletis pomonella* (Diptera: Tephritidae). IV. Courtship. *Canadian Entomologist*, **105**, 873-891. [RAE **62**:1321]

Prokopy, R.J. & Coll, W.M. (1978) Selective traps for monitoring *Rhagoletis mendax* flies. *Protection Ecology*, **1**, 45-53. [RAE **66**:6080]

Prokopy, R.J. & Koyama, J. (1982) Oviposition site partitioning in *Dacus cucurbitae*. *Entomologia Experimentalis et Applicata*, **31**, 428-432. [RAE **70**:5809]

Prokopy, R.J. & Roitberg, B.D. (1989) Behaviour; fruit fly foraging behavior, In: Robinson, A.S. & Hooper, G. (eds), Fruit flies; their biology, natural enemies and control. *World Crop Pests*, **3**(A), 293-306. Elsevier, Amsterdam.

Prokopy, R.J. & Webster, R.P. (1978) Oviposition-deterring pheromone of *Rhagoletis pomonella*: a kairomone for its parasitoid *Opius lectus*. *Journal of Chemical Ecology*, **4**, 481-494. [RAE **67**:64]

Prokopy, R.J., Cooley, S.S. & Opp, S.B. (1989a) Prior experience influences the fruit residence of male apple maggot flies, *Rhagoletis pomonella*. *Journal of Insect Behavior*, **2**, 39-48. [RAE **77**:9555]

Prokopy, R.J., Green, T.A. & Wong, T.T.Y. (1989b) Learning to find fruit in *Ceratitis capitata* flies. *Entomologia Experimentalis et Applicata*, **53**, 65-72. [RAE **78**:7897]

Prokopy, R.J., Kallet, C. & Cooley, S.S. (1985) Fruit-acceptance pattern of *Rhagoletis pomonella* (Diptera: Tephritidae) flies from different geographic regions. *Annals of the Entomological Society of America*, **78**, 799-803. [RAE **74**:1359]

Prokopy, R.J., Ziegler, J.R. & Wong, T.T.Y. (1978) Deterrence of repeated oviposition by fruit-marking pheromone in *Ceratitis capitata* (Diptera: Tephritidae). *Journal of Chemical Ecology*, **4**, 55-63. [RAE **66**:4709]

Prokopy, R.J., Papaj, D.R., Cooley, S.S. & Kallet, C. (1986) On the nature of learning in oviposition site acceptance by apple maggot flies. *Animal Behaviour*, **34**, 98-107. [RAE **74**:4570]

Prokopy, R.J., Powers, P.J., Heath, R.R., Deuben, B.D. & Tumlinson, J.H. (1988) Comparative laboratory methods for assaying behavioural responses of *Rhagoletis pomonella* flies to host marking pheromones. *Journal of Applied Entomology*, **106**, 437-443. [RAE **78**:548]

Quisenberry, B.F. (1950) The genus *Euaresta* in the United States. *Journal of the New York Entomological Society*, **58**, 9-38.

Qureshi, Z.A., Ashraf, M. & Bughio, A.R. (1974a) Relative abundance of *Dacus cucurbitae* and *Dacus ciliatus* in common hosts. *Pakistan Journal of Scientific and Industrial Research*, **17**, 123-124. [RAE **64**:2922]

Qureshi, Z.A., Ashraf, M., Bughio, A.R. & Hussain, S. (1974b) Rearing, reproductive behaviour and gamma sterilization of fruit fly, *Dacus zonatus* (Diptera: Tephritidae). *Entomologia Experimentalis et Applicata*, **17**, 504-510. [RAE **64**:1785]

Qureshi, Z.A., Hussain, T. & Siddiqui, Q.H. (1987) Interspecific competition of *Dacus cucurbitae* Coq. and *Dacus ciliatus* Loew in mixed infestations of cucurbits. *Journal of Applied Entomology*, **104**, 429-432. [RAE **76**:8941]

Ramsamy, M.P., Rawanansham, T. & Joomaye, A. (1987) Studies on the control of *Dacus cucurbitae* Coquillet and *Dacus d'emmerezi* Bezzi (Diptera: Tephritidae) by male annihilation. *Revue Agricole et Sucrière de l'Ile Maurice*, **66,** 105-114.

Ranaldi, F. & Santoni, M. (1987) I parasiti della mosca olearia *Dacus oleae* (Gmel.). *Informatore Fitopatologico*, **37**(11), 15-18. [RAE **77**:8613]

Ranner, H. (1987) Untersuchungen zur Biologie und Bekämpfung der Kirschfruchtfliege, *Rhagoletis cerasi* L. (Diptera, Trypetidae) - II. Statistischer Verleich von Eiern und Puparien der Kirschfruchtfliege aus verschiedenen Wirtspflanzen, von verschiedenen Sammelorten und Jahren. *Pflanzenschutzberichte*, **48,** 27-43. [RAE **76**:6913]

Ranner, H. (1988a) Untersuchungen zur Biologie und Bekämpfung der Kirschfruchtfliege, *Rhagoletis cerasi* L. (Diptera, Trypetidae) - III. Statistischer Vergleich der Schlupfperioden und Schlupfraten der Kirschfliege. *Pflanzenschutzberichte*, **49,** 17-26. [RAE **77**:2516]

Ranner, H. (1988b) Untersuchungen zur Biologie und Bekämpfung der Kirschfruchtfliege, *Rhagoletis cerasi* L. (Diptera, Trypetidae) - IV. Statistischer Auswertung von Krezenversuchen mit Kirschfliegen verschiedenen Alters and Puppengewichts, verschiedener Wirtsplanzenherkunft und Rassenzugehorigkeit. *Pflanzenschutzberichte*, **49,** 74-86. [RAE **78**:3801]

Rao, S. Ananda (1940) Report of the Entomologist 1939-40. *Report of the Tea Science Department, U.P.A.S.I.* **1939-40,** 12-20. [RAE **30**:336]

Rao, Y.R. (1956) *Dacus dorsalis* Hend. in banana fruits at Bangalore. *Indian Journal of Entomology*, **18,** 471-472. [RAE **46**:475]

Ravindranath, K. & Pillai, K.S. (1986) Control of fruitfly of bitter gourd using synthetic pyrethroids. *Entomon*, **11,** 269-272. [RAE **76**:8940]

Reissig, W.H., Stanley, B.H., Roelofs, W.L. & Schwarz, M.R. (1985) Tests of synthetic apple volatile in traps as attractants for apple maggot flies (Diptera: Tephritidae) in commercial apple orchards. *Environmental Entomology*, **14,** 55-59. [RAE **74**:422]

Rejesus, R.S., Baltazar, C.R. & Manoto, E.C. (1991) Fruit flies in the Philippines: current status and future prospects, In: Vijaysegaran, S. & Ibrahim, A.G. (eds), *First International Symposium on Fruit Flies in the Tropics*, Kuala Lumpur, 1988. Malaysian Agricultural Research and Development Institute, Kuala Lumpur. pp. 108-124.

Rhode, R.H. & Sanchez, R.M. (1982) Field evaluation of McPhail and gallon plastic tub traps against the Mexican fruit fly. *Southwestern Entomologist*, **7,** 98-100. [RAE **71**:2075]

Rice, R.E, Cunningham, R.T. & Leonhardt, B.A. (1984) Weathering and efficacy of trimedlure dispensers for attraction of Mediterranean fruit flies (Diptera: Tephritidae). *Journal of Economic Entomology*, **77,** 750-756. [RAE **73**:217]

Richter, V.A. (1970) 62. Tephritidae (Trypetidae), In: Bienko, G. Ya. Bei (ed.), Keys to the insects of the European part of the USSR. *Opredeliteli po Faune SSSR*, **103**(5), 132-172. [in Russian: English translation published 1988, Smithsonian Institution Libraries, Washington, **5**(2), 21-276].

Riedl, H. & Hislop, R. (1985) Visual attraction of the walnut husk fly (Diptera: Tephritidae) to color rectangles and spheres. *Environmental Entomology*, **14**, 810-814. [RAE **74**:1871]

Riedl, H. & Hoying, S.A. (1981) Evaluation of trap designs and attractants for monitoring the walnut husk fly, *Rhagoletis completa* Cresson (Diptera, Tephritidae). *Zeitschrift für Angewandte Entomologie*, **91**, 510-520. [RAE **70**:321]

Rigney, C.J. (1989) Control; fruit disinfestation; radiation-disinfestation of fresh fruit, In: Robinson, A.S. & Hooper, G. (eds), Fruit flies; their biology, natural enemies and control. *World Crop Pests*, **3**(B), 425-434. Elsevier, Amsterdam.

Ripley, L.B. & Hepburn, G.A. (1930) A menace to the fruit industry: "Bug-tree" harbours fruit-fly. *Farming in South Africa*, (72), 1-2. [RAE **19**:167]

Rivard, I. (1968) Synopsis et bibliographie annotée sur la mouche de la pomme, *Rhagoletis pomonella* (Walsh), (Diptères: Tephritidae). *Memoirs of the Entomological Society of Québec*, **2**, 1-158.

Robacker, D.C. & Hart, W.G. (1985a) (Z)-3-nonenol, (Z, Z)-3, 6-nonadienol and (S, S)-(-)-epianastrephin: male produced pheromones of the Mexican fruit fly. *Entomologia Experimentalis et Applicata*, **39**, 103-108. [RAE **74**:809]

Robacker, D.C. & Hart, W.G. (1985b) Courtship and territoriality of laboratory-reared Mexican fruit flies, *Anastrepha ludens* (Diptera: Tephritidae), in cages containing host and nonhost trees. *Annals of the Entomological Society of America*, **78**, 488-494. [RAE **74**:858]

Robacker, D.C. & Hart, W.G. (1986) Behavioural responses of male and female Mexican fruit flies, *Anastrepha ludens*, to male-produced chemicals in laboratory experiments. *Journal of Chemical Ecology*, **12**, 39-47. [RAE **74**:2835]

Robacker, D.C. & Hart, W.G. (1987) Electroantennograms of male and female Caribbean fruit flies (Diptera: Tephritidae) elicited by chemicals produced by males. *Annals of the Entomological Society of America*, **80**, 508-512. [RAE **76**:39]

Robacker, D.C. & Moreno, D.S. (1988) Responses of female Mexican fruit flies at various distances from male-produced pheromone. *Southwestern Entomologist*, **13**, 95-100. [RAE **77**:4427]

Robacker, D.C. & Wolfenbarger, D.A. (1988) Attraction of laboratory-reared, irradiated Mexican fruit flies to male produced pheromone in the field. *Southwestern Entomologist*, **13**, 75-80. [RAE **77**:4741]

Robacker, D.C., Chapa, B.E. & Hart, W.G. (1986) Electroantennagrams of Mexican fruit flies to chemicals produced by males. *Entomologia Experimentalis et Applicata*, **40**, 123-127. [RAE **74**:3209]

Robacker, D.C., Ingle, S.J. & Hart, W.G. (1985) Mating frequency and response to male-produced pheromone by virgin and mated females of the Mexican fruit fly. *Southwestern Entomologist*, **10**, 215-221. [RAE **74**:2185]

Robinson, A.S. (1989) Genetics; *Ceratitis capitata*; genetic sexing methods in the Mediterranean fruit fly, *Ceratitis capitata* (Wiedemann), In: Robinson, A.S. & Hooper, G. (eds), Fruit flies; their biology, natural enemies and control. *World Crop Pests*, **3**(B), 57-67. Elsevier, Amsterdam.

Robinson, A.S. & Hooper, G. (eds) (1989) Fruit flies; their biology, natural enemies and control. *World Crop Pests*, **3**(A), xii+1-372; **3**(B), xv+1-447. Elsevier, Amsterdam. [RAE **78**:4234/5]

Roessler, Y. (1989a) Genetics; *Ceratitis capitata*; genetic maps and markers, In: Robinson, A.S. & Hooper, G. (eds), Fruit flies; their biology, natural enemies and control. *World Crop Pests*, **3**(B), 13-18. Elsevier, Amsterdam.

Roessler, Y. (1989b) Control; insecticides; insecticidal bait and cover sprays, In: Robinson, A.S. & Hooper, G. (eds), Fruit flies; their biology, natural enemies and control. *World Crop Pests*, **3**(B), 329-336. Elsevier, Amsterdam.

Rohani, I. (1987) Identification of larvae of common fruit fly pest species in West Malaysia. *Journal of Plant Protection in the Tropics*, **4,** 135-137. [RAE **77**:8138]

Rohdendorf, B.B. (1961) Palaearctic fruit flies (Diptera, Trypetidae) of the genus *Rhagoletis* Loew and closely related genera. *Entomologicheskoe Obozrenie*, **40,** 176-201. [in Russian: English translation in *Entomological Review, Washington*, **40,** 89-102].

Roitberg, B.D. & Prokopy, R.J. (1984) Host visitation sequence as a determinant of search persistance in fruit parasitic tephritid flies. *Oecologia*, **62,** 7-12. [RAE **73**:2056]

Roitberg, B.D., Cairl, R.S. & Prokopy, R.J. (1984) Oviposition deterring pheromone influences dispersal distance in tephritid fruit flies. *Entomologia Experimentalis et Applicata*, **35,** 217-220. [RAE **72**:6885]

Roitberg, B.D., Lenteren, J.C. van, Alphen, J.M.J. van, Galis, F., & Prokopy, R.J. (1982) Foraging behaviour of *Rhagoletis pomonella*, a parasite of hawthorn (*Crataegus viridis*), in nature. *Journal of Animal Ecology*, **51,** 307-325. [RAE **70**:4564]

Rossiter, M.C., Howard, D.J. & Bush, G.L. (1983) Symbiotic bacteria of *Rhagoletis pomonella*, In: Cavalloro, R. (ed.), *Fruit Flies of Economic Importance. Proceedings of the CEC/IOBC International Symposium*. Athens, 1982. Balkema, Rotterdam. pp. 77-84. [RAE **71**:6333]

Roy, B.D. (1977) Annual report for the year 1975. *Report, Ministry of Agriculture and Natural Resources, Mauritius*, **11,** 1-184. [RAE **66**:3349]

Rusanova, V.N. (1926) *Urellia eluta*, Mgn., as a pest of *Carthamus tinctorius*. *Journal of Experimental Agronomy, South-East*, **3**(1), 1-19. [in Russian]. [RAE **14**:603]

Russell, T.A. (1936) Plant pathological report, 1935. *Agricultural Journal, Department of Science and Agriculture, Barbados*, **1935,** 18-23. [RAE **24**:565]

Rygg, T. (1979) Undersoekelser over nypeflue, *Rhagoletis alternata*, Fall. (Diptera: Trypetidae). *Forskning og Forsoek im Landbruket*, **30,** 269-277. [RAE **68**:5013]

Sabatino, A. (1974) Distinctive morphological characters of the larvae of *Dacus oleae* Gmel., *Ceratitis capitata* Wied., *Rhagoletis cerasi* L. (Dipt., Tephritidae). *Entomologica*, **10,** 109-116. [RAE **64**:4001]

Saddik, A. & Miniawi, S.F. El- (1978) A study on infestation of artichoke *Cynara cardunculus* v. *scolymus* L. by looper caterpillars and other pests. *Proceedings of the IV Conference of Pest Control, 1978, Plant Protection Research Institute, Agricultural research Centre, Dokki, Egypt*. Academy of Scientific Research and Technology and National Research Centre, Cairo. pp. 179-187. [RAE **68**:4350]

Saraiva, A. Coutinho (1965) As moscas dos frutos - Diptera Trupaneidae - no arquipélago de Cabo Verde. *Garcia de Orta*, **13**, 491-505. [RAE **56**:660]

Saravia, G. & Freidberg, A. (1989) Comportamiento de oviposición de *Anastrepha striata* (Diptera, Tephritidae) en Pakitza (Manu-Perú). *Revista Peruana de Entomologia*, **31**, 91-93.

Satoh, I., Yamabe, M., Satoh, S. & Ohki, A. (1985) Study of the frequency of finding of the fruit flies infesting the fruit imported as air baggage. *Research Bulletin of the Plant Protection Service, Japan*, **21**, 71-73. [RAE **74**:3454]

Saul, S.H., Mau, R.F.L. & Oi, D. (1985) Laboratory trials of methoprene-impregnated waxes for disinfesting papayas and peaches of the Mediterranean fruit fly (Diptera: Tephritidae). *Journal of Economic Entomology*, **78**, 652-655. [RAE **74**:1342]

Saul, S.H., Mau, R.F.L., Kobayashi, R.M., Tsuda, D.M. & Nishina, M.S. (1987) Laboratory trials of methoprene-impregnated waxes for preventing survival of adult Oriental fruit flies (Diptera: Tephritidae) from infested papayas. *Journal of Economic Entomology*, **80**, 494-496. [RAE **76**:384]

Scalera, G., Bigiani, A., Crnjar, R. & Pietra, P. (1987) A morpho-functional investigation on the antennal olfactory receptors in the med-fly, *Ceratitis capitata* Wied. *Chemical Senses*, **12**, 213. [RAE **76**:6586]

Scaltriti, G.P. (1985) Gli insetti delle piante officinali. I nota. i'*Arnica montana* L. e due suoi fitofagi specifici. *Redia*, **48**, 355-364. [RAE **75**:3878]

Schmidt, C.T. (1967) A fruit-fly attacking coffee cherries in São Tomé. *Garcia de Orta*, **15**, 329-331.

Schroeder, D. (1979) Investigations on *Euzophera cinerosella* (Zeller) (Lep.: Pyralidae), a possible agent for the biological control of the weed *Artemisia absinthium* L. (Compositae) in Canada. *Mitteilungen der Schweizerischen Entomologischen Gesellschaft*, **52**, 91-101. [RAE **68**:1905]

Schulz, J.T. & Lipp, W.V. (1969) The status of the sunflower insect complex in the Red River Valley of North Dakota. *Proceedings of the North Central Branch, American Association of Economic Entomologists*, **24**, 99-100. [RAE **58**:1535]

Schwarz, A.J., Liedo, J.P. & Hendrichs, J.P. (1989a) Control; sterile insect technique (SIT); current programme in Mexico, In: Robinson, A.S. & Hooper, G. (eds), Fruit flies; their biology, natural enemies and control. *World Crop Pests*, **3**(B), 375-386. Elsevier, Amsterdam.

Schwarz, A.J., Zambada, A., Orozco, D.H.S. & Zavala, J.L. (1985) Mass production of the Mediterranean fruit fly at Metapa, Mexico. *Florida Entomologist*, **68**, 467-477. [RAE **74**:4392]

Schwarz, A.J., Reyes, J., Villaseñor, A., Liedo, P. & Gutierrez, J. (1989b) Fruit fly control in Latin America: Research and training coordination, In: Cavalloro, R. (ed.), *Fruit Flies of Economic Importance 87. Proceedings of the CEC/IOBC International Symposium*, Rome, 1987. Balkema, Rotterdam. pp. 525-532.

Segarra, A.E. (1988) Identity and economic importance of fruit flies attacking mango in Puerto Rico: a two year survey. *Journal of Agriculture of the University of Puerto Rico*, **72**, 325. [RAE **77**:5570]

Séguy, E. (1934) *Faune de France*. **28** [Diptères (Brachycères) (Muscidae Acalypterae et Scatophagidae)]. Paris.

Seitz, A. & Komma, M. (1984) Genetic polymorphism and its ecological background in *Tephritid* populations (Diptera: Tephritidae, In: Wöhrmann, K. & Loeschcke, V. (eds), *Population Biology and Evolution*. Springer-Verlag, Berlin. pp. 143-158.

Sengalevich, G. (1970) New possibilities for the control of the rose-hip fly. *Rastitelna Zashchita*, **18,** 17-20. [in Bulgarian]. [RAE **61:**728]

Shah, A.H. & Patel, R.C. (1976) Role of tulsi plant (*Ocimum sanctum*) in control of mango fruitfly, *Dacus correctus* Bezzi (Tephritidae: Diptera). *Current Science*, **45,** 313-314. [RAE **65:**3546]

Shah, A.H. & Vora, V.J. (1975; dated 1974) Occurence of *Dacus correctus* Bezzi (Tephritidae: Diptera) on mango and chiku in south Gujarat. *Indian Journal of Entomology*, **36,** 76. [RAE **64:**7326]

Shaheen, A.H., Samhan, M. & Elezz, A.A. (1973) Cucurbit pests at Komombo. *Agricultural Research Review*, **51,** 97-101. [RAE **63:**2648]

Sharp, J.L. (1986) Hot-water treatment for control of *Anastrepha suspensa* (Diptera: Tephritidae) in mangos. *Journal of Economic Entomology*, **79,** 706-708. [RAE **74:**6058]

Sharp, J.L. (1987) Laboratory and field experiments to improve enzymatic casein hydrolysate as an arrestant and attractant for Caribbean fruit fly, *Anastrepha suspensa* (Diptera: Tephritidae). *Florida Entomologist*, **70,** 225-233. [RAE **76:**7299]

Sharp, J.L. & Chambers, D.L. (1983) Aggregation response of *Anastrepha suspensa* (Diptera: Tephritidae) to proteins and amino acids. *Environmental Entomology*, **12,** 923-928. [RAE **72:**729]

Sharp, J.L. & Chew, V. (1987) Time/mortality relationships for *Anastrepha suspensa* (Diptera: Tephritidae) eggs and larvae submerged in hot water. *Journal of Economic Entomology*, **80,** 646-649. [RAE **76:**728]

Sharp, J.L. & Landolt, P.J. (1984) Gustatory and olfactory behaviour of the papaya fruit fly, *Toxotrypana curvicauda* Gerstaecker, (Diptera: Tephritidae) in the laboratory with notes on longevity. *Journal of the Georgia Entomological Society*, **19,** 176-182. [RAE **72:**7202]

Sharp, J.L., Ouye, M.T., Thalman, R., Hart, W. Ingle, S. & Chew, V. (1988a) Submersion of 'Francis' mango in hot water as a quarantine treatment for the West Indian fruit fly and the Caribbean fruit fly (Diptera: Tephritidae). *Journal of Economic Entomology*, **81,** 1431-1436. [RAE **77:**7007]

Sharp, J.L., Thalman, R.K., Webb, J.C. & Masuda, S. (1988b) Flexible acoustical device to detect feeding sounds of Caribbean fruit fly (Diptera: Tephritidae) larvae in mango, cultivar Francis. *Journal of Economic Entomology*, **81,** 406-409. [RAE **76:**6935]

Shiga, M. (1989) Control; sterile insect technique (SIT); current programme in Japan, In: Robinson, A.S. & Hooper, G. (eds), Fruit flies; their biology, natural enemies and control. *World Crop Pests*, **3**(B), 365-374. Elsevier, Amsterdam.

Shinji, O. (1939) On the Trypetidae of north-eastern Japan, with the description of new species. *Insect World*, **43,** 288-291. [in Japanese]. [RAE **28:**207]

Shiraki, T. (1933) A systematic study of the Trypetidae of the Japanese Empire. *Memoirs of the Faculty of Science and Agriculture, Taihoku Imperial University*, **2,** 1-509. [RAE **21:**632]

Shiraki, T. (1968) Fruit flies of the Ryukyu Islands. *United States National Museum Bulletin*, **263,** 1-104.

Shorthouse, J.D. (1980) Modification of the flower heads of *Sonchus arvensis* (family Compositae) by the gall former *Tephritis dilacerata* (Order Diptera, Family Tephritidae). *Canadian Journal of Botany*, **58,** 1534-1540. [RAE **69:**4461]

Silva, M.T. da, Polloni, Y.J. & Bressan, S. (1985) Mating behaviour of some fruit flies of the genus *Anastrepha* Schiner, 1868 (Diptera, Tephritidae) in the laboratory. *Revista Brasileira de Entomologia*, **29,** 155-164. [RAE **74:**1697]

Silveira-Guido, A. & Habeck, D.H. (1978) Natural enemies of strangler, *Morrenia odorata*, and two closely related species, *M. brachystephana* and *Araujia hortorum* in Uruguay. *Proceedings of the IVth International Symposium on Biological Control of Weeds, Gainesville.* 128-131. [RAE **66:**5991]

Silverman, J. & Goeden, R.D. (1980) Life history of a fruit fly, *Procecidochares* sp., on the ragweed, *Ambrosia dumosa* (Gray) Payne, in southern California (Diptera: Tephritidae). *Pan-Pacific Entomologist*, **56,** 283-288. [RAE **69:**5836]

Silvestri, F. (1913) Viaggio in Africa per cercare parassiti di mosche dei frutti. *Bolletino del Laboratorio di Zoologia Generale e Agraria della R. Scuola Superiore d'Agricoltura, Portici*, **8,** 1-164. [English version published in 1914 in *Territory of Hawaii, Board of Agriculture and Forestry, Division of Entomology Bulletin*, **3,** 1-146.] [RAE **2:**316]

Simmonds, H.W. (1936) Fruit fly in Fiji. *Agricultural Journal, Department of Agriculture, Fiji*, **8,** 22-23. [RAE **24:**446]

Simon, J.P. (1969) Comparative serology of a complex species-group of food-plant specialists: the *Rhagoletis pomonella* complex (Dipt., Tephritidae). *Systematic Zoology*, **18,** 169-184.

Singh, A. & Premlata (1985) The male terminalia of seven species of genus *Dacus* (Diptera: Tephritidae). *Uttar Pradesh Journal of Zoology*, **5,** 63-68.

Singh, M.P. (1984) Studies on the field resistance of different jujube cultivars to the fruit fly *Carpomyia vesuviana* Costa. *Madras Agricultural Journal*, **71,** 413-415. [RAE **74:**2358]

Singh, M.P. (1989) *Biosteres vandenboschi* Fullaway - a new braconid parasite of *Carpomyia vesuviana* Costa from the Indian desert. *Entomon*, **14,** 169. [RAE **78:**6847]

Singh, M.P. & Vashishta, B.B. (1985; dated 1984) Field screening of some ber cultivars for resistance to ber fruit fly, *Carpomyia vesuviana* Costa. *Indian Journal of Plant Protection*, **12,** 55-56. [RAE **74:**3512]

Singh, P., Leppla, N.C. & Adams, F. (1988) Feeding behavior and dietary substrates for rearing larvae of the Caribbean fruit fly, *Anastrepha suspensa*. *Florida Entomologist*, **71,** 380-384. [RAE **77:**2230]

Sivinski, J. (1984) Effect of sexual experience on male mating success in a lek forming tephritid *Anastrepha suspensa* (Loew). *Florida Entomologist*, **67,** 126-130. [RAE **72:**6232]

Sivinski, J. (1987) Acoustical oviposition cues in the Caribbean fruit fly, *Anastrepha* suspensa (Diptera: Tephritidae). *Florida Entomologist*, **70,** 171-172. [RAE **76:**6928]

Sivinski, J. (1988) What do fruit fly songs mean? *Florida Entomologist*, **71,** 462-466. [RAE **78:**3378]

Sivinski, J. (1989) Lekking and the small-scale distribution of sexes in the Caribbean fruit fly, *Anastrepha suspensa* (Loew). *Journal of Insect Behavior*, **2,** 3-13. [RAE **77:**9591]

Sivinski, J. & Burk, T. (1989) Behaviour; reproductive and mating behaviour, In: Robinson, A.S. & Hooper, G. (eds), Fruit flies; their biology, natural enemies and control. *World Crop Pests*, **3**(A), 343-351. Elsevier, Amsterdam.

Sivinski, J.M. & Calkins, C. (1986) Pheromones and parapheromones in the control of tephritids. *Florida Entomologist*, **69,** 157-168. [RAE **75:**5794]

Sivinski, J. & Webb, J.C. (1985a) Sound production and reception in the caribfly, *Anastrepha suspensa* (Diptera: Tephritidae). *Florida Entomologist*, **68,** 273-278. [RAE **74:**2611]

Sivinski, J. & Webb, J.C. (1985b) The form and function of acoustic courtship signals of the papaya fruit fly, *Toxotrypana curvicauda* (Tephritidae). *Florida Entomologist*, **68,** 634-641. [RAE **75:**2392]

Sivinski, J. & Webb, J.C. (1986) Changes in a Caribbean fruit fly acoustic signal with social situation (Diptera: Tephritidae). *Annals of the Entomological Society of America*, **79,** 146-149. [RAE **74:**4965]

Smith, D. & Nannan, L. (1988) Yeast autolysate sprays for control of Queensland fruit fly on passion fruit in Queensland. *Queensland Journal of Agricultural and Animal Sciences*, **45,** 169-177. [RAE **78:**7757]

Smith, D.C. (1984) Feeding, mating, and oviposition by *Rhagoletis cingulata* (Diptera: Tephritidae) flies in nature. *Annals of the Entomological Society of America*, **77,** 702-704. [RAE **73:**2448]

Smith, D.C. (1985) General activity and reproductive behavior of *Rhagoletis tabellaria* (Diptera: Tephritidae) flies in nature. *Journal of the Kansas Entomological Society*, **58,** 737-739. [RAE **75:**625]

Smith, D.C. (1988a) Reproductive differences between *Rhagoletis* (Diptera: Tephritidae) fruit parasites of *Cornus amomum* and *C. florida* (Cornaceae). *Journal of the New York Entomological Society*, **96,** 327-331. [RAE **77:**8845]

Smith, D.C. (1988b) Heritable divergence of *Rhagoletis pomonella* host races by seasonal asynchrony. *Nature, UK*, **336,** 66-67. [RAE **78:**543]

Smith, D.C. & Prokopy, R.J. (1980) Mating behavior of *Rhagoletis pomonella* (Diptera: Tephritidae). VI. Site of early-season encounters. *Canadian Entomologist*, **112,** 589-590. [RAE **69:**2015]

Smith, D.C. & Prokopy, R.J. (1982) Mating behavior of *Rhagoletis mendax* (Diptera: Tephritidae) flies in nature. *Annals of the Entomological Society of America*, **75,** 388-392. [RAE **71:**329]

Smith, E.S.C. (1977) Studies on the biology and commodity control of the banana fruit fly, *Dacus musae* (Tryon), in Papua New Guinea. *Papua New Guinea Agricultural Journal*, **28**(2/4), 47-56. [RAE **67:**242]

Smith , E.S.C., Chin, D., Allwood, A.J. & Collins, S.G. (1988) A revised host list of fruit flies (Diptera: Tephritidae) from the Northern Territory of Australia. *Queensland Journal of Agricultural and Animal Sciences*, **45,** 19-28. [RAE **77**:6492]

Smith, K.G.V. (1989) An introduction to the immature stages of British flies; Diptera larvae, with notes on eggs, puparia and pupae. *Handbooks for the Identification of British Insects*, **10**(14), 1-280.

Smith, K.G.V. (in press) An aggregation of *Anomoia purmunda* (Harris) (Dipt., Tephritidae) on *Galium verum* L. *Entomologist's Monthly Magazine.*

Smith, P.H. (1979) Genetic manipulation of the circadian clock's timing of sexual behaviour in the Queensland fruit flies, *Dacus tryoni* and *Dacus neohumeralis. Physiological Entomology*, **4,** 71-78. [RAE **67**:4058]

Smith, P.H. (1989) Behaviour; behaviour partitioning of the day and circadian rhythmicity, In: Robinson, A.S. & Hooper, G. (eds), Fruit flies; their biology, natural enemies and control. *World Crop Pests*, **3**(A), 325-341. Elsevier, Amsterdam.

Smyth, E.G. (1960) A new tephritid fly injurious to tomatoes in Peru. *Bulletin of the California Department of Agriculture*, **49,** 16-22. [RAE **49**:614]

Snodgrass, R.E. (1924) Anatomy and metamorphosis of the apple maggot. *Journal of Agricultural Research*, **28,** 1-36. [RAE **12**:150]

Solinas, M. & Nuzzaci, G. (1984) Functional anatomy of *Dacus oleae* Gmel. female genitalia in relation to insemination and fertilization processes. *Entomologica*, **19,** 135-165. [RAE **75**:1568]

Soria, F. & Yana, A. (1962) Contribution à l'étude de la bio-écologie de la mouche des câpres (*Capparimyia savastanii* Mart.) et du parasitisme de cette trypétide par *O. concolor* Szépl. en Tunisie. *Annales de l'Institut National de la Recherche Agronomique Tunisie*, **32,** 125-157. [RAE **52**:570]

Souza, H.M.L. de, Cytrynowicz, M., Morgante, J.S. & Pavan, O.H. de O. (1983) Occurrence of *Anastrepha fraterculus* (Wied.), *Ceratitis capitata* (Wied.) (Diptera, Tephritidae) and *Silba* spp. (Diptera, Lonchaeidae) eggs in oviposition bores on three host fruits. *Revista Brasileira de Entomologia*, **27,** 191-195. [RAE **72**:4626]

Souza, H.M.L. de, Pavan, O.H.O., Silva, I.D. & Matioli, S.R. (1986) New fruit flies bait solutions. *Revista Brasileira de Entomologia*, **30,** 251-255. [RAE **77**:7695]

Spalding, D.H., King, J.R. & Sharp, J.L. (1988) Quality and decay of mangos treated with hot water for quarantine control of fruit fly. *Tropical Science*, **28,** 95-101. [RAE **77**:2718]

Spaugy, L. (1988) Fruit flies. Two more eradication projects over. *Citrograph*, **73,** 168. [RAE **77**:6625]

Spitler, G.H., Armstrong, J.W. & Couey, H.M. (1984) Mediterranean fruit fly (Diptera: Tephritidae) host status of commercial lemon. *Journal of Economic Entomology*, **77,** 1441-1444. [RAE **73**:5229]

Spitzer, K. (1964) Beitrag über Einfluss biotischer Faktoren auf die Population der *Philophylla heraclei* L. in der Tschechoslowakei. *Zoologické Listy*, **13,** 155-160. [RAE **53**:628]

Stadler, E., Schoni, R., Hurter, J. & Boller, E. (1987) Electrophysiological recordings used as bioassay for the isolation and identification of a non-volatile pheromone. *Chemical Senses*, **12,** 192-193. [RAE **76:**6914]

Stavraki, H. & Stavrakis, G. (1968) Trois insectes nuisibles signalés aux capitules d'artichaut en Attique et Péloponnése. *Annales de l'Institute Phytopathologique Benaki (N.S.)*, **8,** 150-152. [RAE **59:**3135]

Stechmann, D.H., Englberger, K. & Langi, T.F. (1988) Estimation of mortality of *Dacus xanthodes* (Broun) maggots in fumigated and non-fumigated watermelons, a fruitfly (Dipt.: Tephritidae) of plant quarantine importance in the Pacific region. *Anzeiger für Schädlingskunde, Pflanzenschutz, Umweltschutz*, **61,** 125-129. [RAE **78:**4085]

Steck, G.J. (1991) Biochemical systematics and population genetic structure of *Anastrepha fraterculus* and related species (Diptera: Tephritidae). *Annals of the Entomological Society of America*, **84,** 10-28.

Steck, G.J. & Malavasi, A. (1988) Description of the immature stages of *Anastrepha bistrigata* (Diptera: Tephritidae). *Annals of the Entomological Society of America*, **81,** 1004-1009. [RAE **78:**2142]

Steck, G.J. & Wharton, R.A. (1988) Description of immature stages of *Anastrepha interrupta, A. limae*, and *A. grandis* (Diptera: Tephritidae). *Annals of the Entomological Society of America*, **81,** 994-1003. [RAE **78:**2141]

Steck, G.J., Carroll, L.E., Celedonio-Hurtado, H, & Guillen-Aguilar, J. (1990) Methods for identification of *Anastrepha* larvae (Diptera: Tephritidae), and key to 13 species. *Proceedings of the Entomological Society of Washington*, **92,** 333-346.

Steck, G.J., Gilstrap, F.E., Wharton, R.A. & Hart, W.G. (1986) Braconid parasitoids of Tephritidae (Diptera) infesting coffee and other fruits in west-central Africa. *Entomophaga*, **31,** 59-67. [RAE **74:**5324]

Steffens, R.J. (1983) Ecology and approach to integrated control of *Dacus frontalis* on the Cape Verde Islands, In: Cavalloro, R. (ed.), *Fruit Flies of Economic Importance. Proceedings of the CEC/IOBC International Symposium*, Athens, 1982. Balkema, Rotterdam. pp. 632-638. [RAE **71:**6407]

Steyskal, G.C. (1972) A new species of *Myoleja* with a key to North American species (Diptera: Tephritidae). *Florida Entomologist*, **55,** 207-211.

Steyskal, G.C. (1973) Distinguishing characters of the walnut husk maggots of the genus *Rhagoletis* (Diptera, Tephritidae). *Cooperative Economic Insect Report*, **23,** 522. [RAE **62:**313]

Steyskal, G.C. (1975) Recognition characters for larvae of the genus *Zonosemata* (Diptera, Tephritidae). *Cooperative Economic Insect Report*, **25,** 231-232. [RAE **63:**3402]

Steyskal, G.C. (1977) *Pictorial Key to the Species of the Genus* Anastrepha *(Diptera: Tephritidae)*. Entomological Society of Washington, Washington.

Steyskal, G.C. (1979) *Taxonomic Studies on Fruit Flies of the Genus* Urophora *(Diptera: Tephritidae)*. Entomological Society of Washington, Washington.

Steyskal, G.C. (1981) A new species of *Rhagoletotrypeta* (Diptera: Tephritidae) from Texas, with a key to the known species. *Proceedings of the Entomological Society of Washington*, **83,** 707-712. [RAE **70:**3325]

Steyskal, G.C. (1982) A second species of *Ceratitis* (Diptera: Tephritidae) adventive in the New World. *Proceedings of the Entomological Society of Washington*, **84,** 165-166. [RAE **70:**6448]

Steyskal, G.C. (1986) Taxonomy of the adults of the genus *Strauzia* Robineau-Desvoidy (Diptera, Tephritidae). *Insecta Mundi*, **1,** 101-117.

Steyskal, G.C. & Foote, R.H. (1977) Revisionary notes on North American Tephritidae (Diptera), with keys and descriptions of new species. *Proceedings of the Entomological Society of Washington*, **79,** 146-155.

Stoffolano, J.G. (1989) Structure and function of the ovipositor of the tephritids, In: Cavalloro, R. (ed.), *Fruit Flies of Economic Importance 87. Proceedings of the CEC/IOBC International Symposium*, Rome, 1987. Balkema, Rotterdam. pp. 141-146.

Stolp, H. (1960) Über das Zusammenwirken von Bakterien und Insekten bei der Entstehung einer Geschmacksbeeinträchtigung des Kivu-Kaffees und die Rolle von Bakteriophagen bei der Aufklärung der Zusammenhäge. *Phytopathologische Zeitschrift*, **39,** 1-15. [RAE **52:**132]

Stoltzfus, W.B. (1977) The taxonomy and biology of *Eutreta* (Diptera: Tephritidae). *Iowa State Journal of Research*, **51,** 369-438.

Stoltzfus, W.B. (1988) The taxonomy and biology of *Strauzia* (Diptera: Tephritidae). *Journal of Iowa Academy of Science*, **95,** 117-126. [RAE **77:**8720]

Stone, A. (1942) The fruitflies of the genus *Anastrepha*. *Miscellaneous Publications of the United States Department of Agriculture*, **439,** 1-112. [RAE **30:**515]

Straw, N.A. (1989a) The timing of oviposition and larval growth by two tephritid fly species in relation to host-plant development. *Ecological Entomology*, **14,** 443-454. [RAE **78:**6418]

Straw, N.A. (1989b) Evidence for an oviposition-deterring pheromone in *Tephritis bardanae* (Schrank) (Diptera: Tephritidae). *Oecologia*, **78,** 121-130. [RAE **78:**6666]

Stuart, C.P. Cohen (1921) De theezaadtuinen van Java en Sumatra. *Mededeelingen van het Proefstation voor Thee*, **75,** 1-32. [RAE **9:**535]

Su, C.Y. (1986) Seasonal population fluctuation of *Dacus cucurbitae* in southern Taiwan. *Plant Protection Bulletin, Taiwan*, **28,** 171-178. [in Chinese]. [RAE **76:**1155]

Sugimoto, S., Kaneda, M., Tanaka, K. & Tao, M. (1988) Some biological notes on *Dacus scutellatus* (Hendel), (Diptera: Tephritidae). *Research Bulletin of the Plant Protection Service, Japan*, **24,** 49-51. [in Japanese]. [RAE **77:**6755]

Sun, C.-y., Du, I.-l. & Liao, Y.-m. (1958) The preliminary studies on Chinese citrus fly, *Tetradacus citri* Chen (Trypetidae, Diptera). *Acta Oeconomico-Entomologica Sinica*, **1,** 175-187. [in Chinese; English summary]. [RAE **49:**16]

Sungawa, K., Kume, K., Ishikawa, A., Sugimoto, T. & Tanabe, K. (1988) Efficacy of vapour heat treatment for bitter momordica fruit infested with melon fly, *Dacus cucurbitae* (Coquillett) (Diptera: Tephritidae). *Research Bulletin of the Plant Protection Service, Japan*, **24,** 1-5. [in Japanese]. [RAE **77:**6759]

Swailem, S.M. (1974; dated 1973) On the binomics of *Acanthiophilus helianthi* Rossi (Diptera: Tephritidae). *Bulletin de la Societé Entomologique d'Egypt*, **57**, 165-173. [RAE **63**:3097]

Syed, R.A. (1970) Studies on trypetids and their natural enemies in West Pakistan. *Dacus* species of lesser importance. *Pakistan Journal of Zoology*, **2**, 17-24. [RAE **63**:90]

Syed, R.A. (1971) Studies on trypetids and their natural enemies in West Pakistan. V. *Dacus (Strumeta) cucurbitae* Coquillett. *Technical Bulletin of the Commonwealth Institute of Biological Control*, **14**, 63-75. [RAE **62**:4787]

Syed, R.A., Ghani, M.A. & Murtaza, M. (1970a) Studies on the trypetids and their natural enemies in West Pakistan. III. *Dacus (Strumeta) zonatus* (Saunders). *Technical Bulletin of the Commonwealth Institute of Biological Control*, **13**, 1-16. [RAE **62**:3180]

Syed, R.A., Ghani, M.A. & Murtaza, M. (1970b) Studies on the trypetids and their natural enemies in West Pakistan. IV. *Dacus (Strumeta) dorsalis* Hendel. *Technical Bulletin of the Commonwealth Institute of Biological Control*, **13**, 17-30. [RAE **62**:3181]

Takano, S. (1934) Insects attacking *Aeginetia indica* Roxb. *Journal of the Formosan Sugar Planters Association*, **12**, 150. [in Japanese]. [RAE **23**:43]

Tan, K.H. (1984) Description of a new attractant trap and the effect of placement height on catches of two *Dacus* species (Diptera: Tephritidae). *Journal of Plant Protection in the Tropics*, **1**, 117-120. [RAE **73**:8003]

Tan, K.H. (1985) Estimation of native populations of male *Dacus* spp. by Jolly's stochastic method using a new designed attractant trap in a village ecosystem. *Journal of Plant Protection in the Tropics*, **2**, 87-95. [RAE **76**:6869]

Tan, K.H. & Lee, S.L. (1982) Species diversity and abundance of *Dacus* (Diptera: Tephritidae) in five ecosystems of Penang, West Malaysia. *Bulletin of Entomological Research*, **72**, 709-716. [RAE **71**:3313]

Tanaka, K. (1936) On *Zeugodacus bezzii* Miyake. *Nojikairyo-shiryo*, **106**, 42-46. [in Japanese]. [RAE **24**:693]

Tanev, I. (1967) Studies on the bionomics of the rose Tephritid (*Carpomyia schineri* Loew.) and the possibilities of control. *Rastenievudni Nauki*, **4**, 59-68. [in Bulgarian]. [RAE **57**:2657]

Tauber, M.J. & Toschi, C.A. (1965) Bionomics of *Euleia fratria* (Loew) (Diptera: Tephritidae). I. Life history and mating behaviour. *Canadian Journal of Zoology*, **43**, 369-379. [RAE **54**:159]

Teles, M. Da Costa & Polloni, Y.J. (1989) Structure and development of specific sex glands in males of some Brasilian fruit flies of the genus *Anastrepha* Schiner, 1868 (Diptera, Tephritidae), In: Cavalloro, R. (ed.), *Fruit Flies of Economic Importance 87. Proceedings of the CEC/IOBC International Symposium*, Rome, 1987. Balkema, Rotterdam. pp. 179-189.

Terrell, E.T., Hill, S.R., Wiersema, J.H. & Rice, W.E. (1986) A checklist of names for 3,000 vascular plants of economic importance (revised ed.). *Agricultural Handbook, United States Department of Agriculture*, **505**, 1-241.

Teskey, H.J. (1981) 3. Morphology and terminology-larvae, In: McAlpine, J.F., Peterson, B.V., Shewell, G.E., Teskey, H.J., Vockeroth, J.R. & Wood, D.M. (eds), Manual of Nearctic Diptera-1. *Monograph of the Biosystematics Research Institute*, **27**, 65-88. Agriculture Canada, Ottawa.

Thakur, J.N. & Kumar, A. (1986) Effects of thiotepa concentrations on the reproductive biology of juvenile and matured fruit fly, *Dacus dorsalis* Hendel (Diptera: Tephritidae). *Journal of Advanced Zoology*, **7**, 75-78. [RAE **76**:8438]

Thakur, J.N. & Kumar, A. (1988) Effect of ethylmethane sulphonate (EMS) on the reproductive potential of fruit fly, *Dacus dorsalis* Hendel (Diptera: Tephritidae). *Entomon*, **13**, 69-73. [RAE **76**:8439]

Thiem, von H. (1934) Beträge zur Epidemiologie und Bekämpfung der Kirschfruchtfliege (*Rhagoletis cerasi* L.). *Arbeiten über Physiologische und Angewandte Entomologie aus Berlin-Dahlem*, **1**, 7-79. [RAE **22**:247]

Thompson, W.R. (ed.) (1943) *A Catalogue of the Parasites and Predators of Insect Pests; Section 1, Parasite Host Catalogue; Part 2, Parasites of the Dermaptera and Diptera*. Imperial Agricultural Bureaux, Slough. [RAE **32**:106]

Trehan, K.N. (1947) Biological observations on *Trypanea amoena* Frfld. *Indian Journal of Entomology*, **8**, 107-109. [RAE **39**:28]

Tryon, H. (1927) Queensland fruit flies (Trypetidae), series 1. *Proceedings of the Royal Society of Queensland*, **38**, 176-223. [RAE **15**:317]

Tsiropoulos, G.J. (1976) Bacteria associated with the walnut husk fly, *Rhagoletis completa*. *Environmental Entomology*, **5**, 83-86. [RAE **64**:7295]

Tsiropoulos, G.J. & Hagen, K.S. (1987) Effect of nutritional deficiencies, produced by antimetabolites, on the reproduction of *Rhagoletis completa* Cresson (Dipt., Tephritidae). *Journal of Applied Entomology*, **103**, 351-354. [RAE **75**:6034]

Tsitsipis, J.A. (1977) An improved method for the mass rearing of the olive fruit fly, *Dacus oleae* (Gmel.) (Diptera, Tephritidae). *Zeitschrift für Angewandte Entomologie*, **83**, 419-426. [RAE **66**:1848]

Tsitsipis, J.A. (1989) Biology and physiology; nutrition; requirements, In: Robinson, A.S. & Hooper, G. (eds), Fruit flies; their biology, natural enemies and control. *World Crop Pests*, **3**(A), 103-119. Elsevier, Amsterdam.

Tsubaki, Y. & Sokei, Y. (1988) Prolonged mating in the melon fly, *Dacus cucurbitae* (Diptera: Tephritidae): competition for fertilization by sperm-loading. *Researches on Population Ecology*, **30**, 343-352. [RAE **79**:633]

Tychsen, P.H. & Bateman, M.A. (1977) Mating behaviour of the Queensland fruit fly, *Dacus tryoni* (Diptera: Tephritidae), in field cages. *Journal of the Australian Entomological Society*, **16**, 459-465. [RAE **66**:4452]

Tzanakakis, M.E. (1989) Rearing; small-scale rearing; *Dacus oleae*, In: Robinson, A.S. & Hooper, G. (eds), Fruit flies; their biology, natural enemies and control. *World Crop Pests*, **3**(B), 105-118. Elsevier, Amsterdam.

Tzanakakis, M.E. & Economopoulos, A.P. (1967) Two efficient larval diets for continuous rearing of the olive fruit fly. *Journal of Economic Entomology*, **60**, 660-663. [RAE **55**:2069]

Ullah, M. (1987) Economic insect pests and phytophagous mites associated with melon crops in Afghanistan. *Tropical Pest Management*, **33,** 29-31, 101, 105. [RAE **75:**3448]

Ullah, M. Mohammad (1988) Major insect pests and phytophagous mites associated with deciduous orchards in Afghanistan. *Tropical Pest Management*, **34,** 215-217. [RAE **77:**4712]

Vargas, R.I. (1989) Rearing; mass production of tephritid fruit flies, In: Robinson, A.S. & Hooper, G. (eds), Fruit flies; their biology, natural enemies and control. *World Crop Pests*, **3**(B), 141-151. Elsevier, Amsterdam.

Vargas, R.I. & Mitchell, S. (1987) Two artificial larval diets for rearing *Dacus latifrons* (Diptera: Tephritidae). *Journal of Economic Entomology*, **80,** 1337-1339. [RAE **76:**6661]

Vargas, R.I. & Nishida, T. (1985a) Life history and other demographic parameters of *Dacus latifrons* (Diptera: Tephritidae). *Journal of Economic Entomology*, **78,** 1242-1244. [RAE **74:**1986]

Vargas, R.I. & Nishida, T. (1985b) Survey for *Dacus latifrons* (Diptera: Tephritidae). *Journal of Economic Entomology*, **78,** 1311-1314. [RAE **74:**1981]

Vargas, R.I. & Nishida, T. (1989) Distribution and abundance patterns for Mediterranean fruit fly in Hawaii: development of eradication strategies for Kauai, In: Cavalloro, R. (ed.), *Fruit Flies of Economic Importance 87. Proceedings ofthe CEC/IOBC International Symposium*, Rome, 1987. Balkema, Rotterdam. pp. 41-48.

Vargas, R.I., Chang, H. & Williamson, D.L. (1983a) Evaluation of a sugarcane bagasse larval diet for mass production of the Mediterranean fruit fly (Diptera: Tephritidae) in Hawaii. *Journal of Economic Entomology*, **76,** 1360-1362. [RAE **72:**3559]

Vargas, R.I., Harris, E.J. & Nishida, T. (1983b) Distribution and seasonal occurrence of *Ceratitis capitata* (Wiedemann) (Diptera: Tephritidae) on the Island of Kauai in the Hawaiian Islands. *Environmental Entomology*, **12,** 303-310. [RAE **72:**565]

Vargas, R.I., Nishida, T. & Beardsley, J.W. (1983c) Distribution and abundance of *Dacus dorsalis* (Diptera: Tephritidae) in native and exotic forest areas on Kauai. *Environmental Entomology*, **12,** 1185-1189. [RAE **72:**3867]

Varley, G.C. (1937) The life-history of some trypetid flies with descriptions of the early stages (Diptera). *Proceedings of the Royal Entomological Society of London (A)*, **12,** 109-122.

Varley, G.C. (1947) The natural control of population balance in the knapweed gall-fly (*Urophora jaceana*). *Journal of Animal Ecology*, **16,** 139-187.

Venkatraman, T.V. & Khidir, E. El (1967) Observations on crop pests in the Sudan in 1966/67. *Plant Protection Bulletin, Food and Agriculture Organization*, **15,** 115-116. [RAE **56:**1402]

Verma, A.N., Singh, R. & Mehrotra, N. (1974; dated 1972) *Acanthiophilus helianthi* Rossi a serious pest of safflower in Haryana. *Indian Journal of Entomology*, **34,** 364-365. [RAE **64:**2080]

Vijaysegaran, S. (1984) The occurrence of Oriental fruit fly on starfruit in Serdang and the status of its parasitoids. *Journal of Plant Protection in the Tropics*, **1**, 93-98. [RAE **74**:4099]

Vijaysegaran, S. (1989) *Dacus dorsalis* complex. *Pest Profile* (14), 1. Malaysian Plant Protection Society, Serdang.

Vijaysegaran, S. & Ibrahim, A.G. (eds) (1991) *First International Symposium on Fruit Flies in the Tropics*, Kuala Lumpur, 1988. Malaysian Agricultural Research and Development Institute, Kuala Lumpur.

Vitolo, D.B. & Stiles, E.W. (1987) The effect of density of *Ambrosia trifida* L. on seed predation by *Euaresta festiva* (Loew) (Diptera: Tephritidae). *Journal of the New York Entomological Society*, **95**, 491-494. [RAE **77**:3830]

Vuttanatungum, A. & Hooper, G.H.S. (1974) Biology and chemical sterilization of the fruit fly *Dacus cucumis* French (Diptera: Tephritidae). *Journal of the Australian Entomological Society*, **13**, 169-178. [RAE **63**:1550]

Walker, A.K. & Crosby, T.K. (1979) The preparation and curation of insects. *New Zealand Department of Scientific and Industrial Research Information Series*, **130**, 1-55.

Walton, R. (1985; dated 1984) Density-dependent mortality on galls of the goldenrod gall fly, *Eurosta solidaginis*. *Proceedings of the Indiana Academy of Science*, **94**, 214. [RAE **75**:1683]

Walton, R. (1988) The distribution of risk and density-dependent mortality in the galls of *Eurosta solidaginis*, the goldenrod gall fly. *Ecological Entomology*, **13**, 347-354. [RAE **76**:8663]

Wangberg, J.K. (1977) A new *Tetrastichus* parasitising gall-formers on *Chrysothamnus* in Idaho (Hymenoptera: Eulophidae). *Pan-Pacific Entomologist*, **53**, 237-240. [RAE **66**:2796]

Wangberg, J.K. (1978) Biology of gall-formers of the genus *Valentibulla* (Diptera: Tephritidae) on rabbitbrush in Idaho. *Journal of the Kansas Entomological Society*, **51**, 472-483.

Wangberg, J.K. (1980) Comparative biology of gall-formers in the genus *Procecidochares* (Diptera: Tephritidae) on rabbitbrush in Idaho. *Journal of the Kansas Entomological Society*, **53**, 401-420. [RAE **69**:1412]

Wangberg, J.K. (1981) Gall-forming habits of *Aciurina* species (Diptera: Tephritidae) on rabbitbrush (Compositae: *Chrysothamnus* spp.) in Idaho. *Journal of the Kansas Entomological Society*, **54**, 711-732.

Warthen, J.D. & McInnes, D.O. (1989) Isolation and identification of male medfly attractive components in *Litchi chinensis* stems and *Ficus* spp. stem exudates. *Journal of Chemical Ecology*, **15**, 1931-1946. [RAE **78**:564]

Wasbauer, M.S. (1972) An annotated host catalog of the fruit flies of America north of Mexico (Diptera: Tephritidae). *Occasional Papers, Bureau of Entomology, California Department of Agriculture*, **19**, 1-172. [RAE **61**:1848]

Watanabe, C. (1939) On the parasitic wasps of fruit flies attacking cherry. *Oyo-Dobutsugaku-Zasshi*, **11**, 123-128. [in Japanese] [RAE **27**:620]

Webb, J.C., Burk, T. & Sivinski, J. (1983a) Attraction of female Caribbean fruit flies, *Anastrepha suspensa* (Diptera: Tephritidae), to the presence of males and male-produced stimuli in field cages. *Annals of the Entomological Society of America*, **76**, 996-998. [RAE **72**:2687]

Webb, J.C., Calkins, C.O., Chambers, D.L., Schwiebacher, W. & Russ, K. (1983b) Acoustical aspects of behavior of Mediterranean fruit fly, *Ceratitis capitata*: Analysis and identification of courtship sounds. *Entomologia Experimentalis et Applicata*, **33**, 1-8. [RAE **71**:3154]

Webb, J.C., Sivinski, J. & Litzkow, C. (1984) Acoustical behavior and sexual success in the Caribbean fruit fly, *Anastrepha suspensa* (Loew) (Diptera: Tephritidae). *Environmental Entomology*, **13**, 650-656. [RAE **72**:7633]

Webb, J.C., Slaughter, D.C. & Litzkow, C.A. (1988) Acoustical systems to detect larvae in infested commodities. *Florida Entomologist*, **71**, 492-504. [RAE **78**:4080]

Weems, H.V. (1963) Mexican fruit fly (*Anastrepha ludens* (Loew)) (Diptera: Tephritidae). *Entomology Circular, Division of Plant Industry, Florida Department of Agriculture and Consumer Services*, (16), 1-2.

Weems, H.V. (1964a) Oriental fruit fly (*Dacus dorsalis* Hendel) (Diptera: Tephritidae). *Entomology Circular, Division of Plant Industry, Florida Department of Agriculture and Consumer Services*, (21), 1-2.

Weems, H.V. (1964b) Melon fly (*Dacus cucurbitae* Coquillett) (Diptera: Tephritidae). *Entomology Circular, Division of Plant Industry, Florida Department of Agriculture and Consumer Services*, (29), 1-2.

Weems, H.V. (1965a) Queensland fruit fly (*Dacus tryoni* (Froggatt)) (Diptera: Tephritidae). *Entomology Circular, Division of Plant Industry, Florida Department of Agriculture and Consumer Services*, (34), 1-2.

Weems, H.V. (1965b) *Anastrepha suspensa* (Loew) (Diptera: Tephritidae). *Entomology Circular, Division of Plant Industry, Florida Department of Agriculture and Consumer Services*, (38), 1-4.

Weems, H.V. (1966a) Olive fruit fly (*Dacus oleae* (Gmelin)) (Diptera: Tephritidae). *Entomology Circular, Division of Plant Industry, Florida Department of Agriculture and Consumer Services*, (44), 1-2.

Weems, H.V. (1966b) Natal fruit fly (*Ceratitis rosa* Karsch) (Diptera: Tephritidae). *Entomology Circular, Division of Plant Industry, Florida Department of Agriculture and Consumer Services*, (51), 1-2.

Weems, H.V. (1967) Japanese orange fly (*Dacus tsuneonis* Miyake) (Diptera: Tephritidae). *Entomology Circular, Division of Plant Industry, Florida Department of Agriculture and Consumer Services*, (56), 1-2.

Weems, H.V. (1968) *Anastrepha ocresia* (Wlk.) (Diptera: Tephritidae). *Entomology Circular, Division of Plant Industry, Florida Department of Agriculture and Consumer Services*, (71), 1-2. [RAE **59**:567]

Weems, H.V. (1980) *Anastrepha fraterculus* (Wiedemann) (Diptera: Tephritidae). *Entomology Circular, Division of Plant Industry, Florida Department of Agriculture and Consumer Services*, (217), 1-4. [RAE **69**:2387]

Weems, H.V. (1981) Mediterranean fruit fly, *Ceratitis capitata* (Wiedemann) (Diptera: Tephritidae). *Entomology Circular, Division of Plant Industry, Florida Department of Agriculture and Consumer Services*, (230), 1-12. [RAE **70**:4643]

Weems, H.V. (1982) *Anastrepha striata* Schiner (Diptera: Tephritidae). *Entomology Circular, Division of Plant Industry, Florida Department of Agriculture and Consumer Services*, (245), 1-2. [RAE **71**:3537]

Weems, H.V. (1987) Guava fruit fly, *Dacus (Strumeta) correctus* (Bezzi) (Diptera: Tephritidae). *Entomology Circular, Division of Plant Industry, Florida Department of Agriculture and Consumer Services*, (291), 1-4. [RAE **76**:6866]

Westcott, R.L. (1982) Differentiating adults of apple maggot, *Rhagoletis pomonella* (Walsh) from snowberry maggot *R. zephyria* Snow (Diptera: Tephritidae) in Oregon. *Pan-Pacific Entomologist*, **58,** 25-30. [RAE **73**:802]

Wharton, R.A. (1983) Variation in *Opius hirtus* Fischer and discussion of *Desmiostoma* Foerster (Hymenoptera: Braconidae). *Proceedings of the Entomological Society of Washington*, **85,** 327-330. [RAE **71**:6869]

Wharton, R.H. (1989a) Control; classical biological control of fruit-infesting Tephritidae, In: Robinson, A.S. & Hooper, G. (eds), Fruit flies; their biology, natural enemies and control. *World Crop Pests*, **3**(B), 303-313. Elsevier, Amsterdam.

Wharton, R.A. (1989b) Biological control of fruit-infesting Tephritidae, In: Cavalloro, R. (ed.), *Fruit Flies of Economic Importance 87. Proceedings of the CEC/IOBC International Symposium*, Rome, 1987. Balkema, Rotterdam. pp. 323-332.

Wharton, R.A. & Gilstrap, F.E. (1983) Key to and status of opiine braconid (Hymenoptera) parasitoids used in biological control of *Ceratitis* and *Dacus* s.l. (Diptera: Tephritidae). *Annals of the Entomological Society of America*, **76,** 721-742. [RAE **72**:572]

Wharton, R.A., Gilstrap, F.E., Rhode, R.H., Fischel-M., M. & Hart, W.G. (1981) Hymenopterous egg-pupal and larval-pupal parasitoids of *Ceratitis capitata* and *Anastrepha* spp. (Dip.: Tephritidae) in Costa Rica. *Entomophaga*, **26,** 285-290. [RAE **70**:3184]

Whervin, L.W. van (1974) Some fruitflies (Tephritidae) in Jamaica. *PANS; Pest Articles and News Summaries*, **20,** 11-19. [RAE **62**:3758]

White, I.M. (1988) Tephritid flies (Diptera: Tephritidae). *Handbooks for the Identification of British Insects*, **10**(5a), 1-134. [RAE **78**:7262]

White, I.M. (1989a) A new species of *Terellia* Robineau-Desvoidy associated with *Centaurea solstitialis* and a revision of the *Terellia virens* (Loew) species group (Diptera: Tephritidae). *Entomologist's Monthly Magazine*, 125:53-61.

White, I.M. (1989b) The state of fruit fly taxonomy and future research priorities, In: Cavalloro, R. (ed.), *Fruit Flies of Economic Importance 87. Proceedings of the CEC/IOBC International Symposium*, Rome, 1987. Balkema, Rotterdam. pp. 543-552.

White, I.M. (1991a) The taxonomy of tropical fruit flies, In: Vijaysegaran, S. & Ibrahim, A.G. (eds), *First International Symposium on Fruit Flies in the Tropics*, Kuala Lumpur, 1988. Malaysian Agricultural Research and Development Institute, Kuala Lumpur. pp. 171-176.

White, I.M. (1991b) The application of tephritid taxonomy to problems in plant quarantine and weed biological control, In: Weismann, L., Országh, I. & Pont,

A.C. (eds), *Proceedings of the Second International Congress of Dipterology*, Bratislava, 1990. SPB Academic Publishing, The Hague. pp. 341-349.

White, I.M. & Clement, S.L. (1987) Systematic notes on *Urophora* (Diptera, Tephritidae) species associated with *Centaurea solstitialis* (Asteraceae, Cardueae) and other Palaearctic weeds adventive in North America. *Proceedings of the Entomological Society of Washington*, **89,** 571-580. [RAE **76:**6756]

White, I.M. & Korneyev, V.A. (1989) A revision of the western Palaearctic species of *Urophora* (Robineau-Desvoidy). *Systematic Entomology*, **14,** 327-374. [RAE **79:**1138]

White, I.M. & Marquardt, K. (1989) A revision of the genus *Chaetorellia* Hendel (Diptera: Tephritidae) including a new species associated with spotted knapweed, *Centaurea maculosa* Lam. (Asteraceae). *Bulletin of Entomological Research*, **79,** 453-487. [RAE **78:**6639]

White, I.M., & Wang, X.-j. (in press) Taxonomic notes on some dacine (Diptera: Tephritidae) fruit flies associated with citrus, olives and cucurbits. *Bulletin of Entomological Research.*

White, I.M., Groppe, K. & Sobhian, R. (1990) Tephritids of knapweeds, starthistles and safflower: results of a host choice experiment and the taxonomy of *Terellia luteola* (Wiedemann) (Diptera: Tephritidae). *Bulletin of Entomological Research*, **80,** 107-111. [RAE **78:**6643]

Whitman, D.W., Orsak, L. & Greene, E. (1988) Spider mimicry in fruit flies (Diptera: Tephritidae): further experiments on the deterrence of jumping spiders (Araneae: Salticidae) by *Zonosemata vittigera* (Coquillett). *Annals of the Entomological Society of America*, **81,** 532-536. [RAE **77:**3167]

Whitney, L.A. (1929) Reports of the Associate Plant Inspector, February 1929. *Hawaiian Forestry and Agriculture*, **26,** 81-82. [RAE **17:**708]

Whittle, K. & Norrbom, A.L. (1987) A fruit fly *Anastrepha grandis* (Macquart). *U.S. Department of Agriculture, Pests not known to occur in the United States or of Limited Distribution*, (82), 1-8.

Willers, P. (1979) Suitability of *Harpephyllum caffrum* (kaffir plum) as host for Mediterranean fruit fly and false codling moth. *Citrus and Subtropical Fruit Journal*, **543,** 5-6. [RAE **68:**1466]

Williamson, D.L. (1989) Biology and physiology; oogenesis and spermatogenesis, In: Robinson, A.S. & Hooper, G. (eds), Fruit flies; their biology, natural enemies and control. *World Crop Pests*, **3**(A), 141-151. Elsevier, Amsterdam.

Williamson, D.L. & Hart, W.G. (1989) Current status of Mexican fruit fly research in the Rio Grande Valley of Texas, In: Cavalloro, R. (ed.), *Fruit Flies of Economic Importance 87. Proceedings of the CEC/IOBC International Symposium*, Rome, 1987. Balkema, Rotterdam. pp. 563-569.

Wilson, G.F. (1931) Insects associated with the seeds of garden plants. *Journal of the Royal Horticultural Society*, **61,** 31-47. [RAE **19:**246]

Windeguth, D.L. von, Burditt, A.K. & Spalding, D.H. (1976) Phosphine as a fumigant for grapefruit infected by Caribbean fruit fly larvae. *Florida Entomologist*, **59,** 285-286. [RAE **65:**4612]

Witherell, P.C. (1982) Efficacy of two types of survey traps for Caribbean fruit fly, *Anastrepha suspensa* (Loew). *Florida Entomologist*, **65**, 580-581. [RAE **71**:4207]

Wong, T.T.Y. & Ramadan, M.M. (1987) Parasitization of the Mediterranean and Oriental fruit flies (Diptera: Tephritidae) in the Kula area of Maui, Hawaii. *Journal of Economic Entomology*, **80**, 77-80. [RAE **75**:6435]

Wong, M.A. & Wong, T.T.Y. (1988) Predation of the Mediterranean fruit fly and the Oriental fruit fly (Diptera: Tephritidae) by the fire ant (Hymenoptera: Formicidae) in Hawaii. *Proceedings of the Hawaiian Entomological Society*, **28**, 169-177. [RAE **77**:5569]

Wong, T.T.Y., Kobayashi, R.M. & McInnis, D.O. (1986) Mediterranean fruit fly (Diptera: Tephritidae): methods of assessing the effectiveness of sterile insect releases. *Journal of Economic Entomology*, **79**, 1501-1506. [RAE **75**:3408]

Wong, T.T.Y., McInnes, D.O. & Nishimoto, J.I. (1989) Relationship of sexual maturation rate to response of Oriental fruit fly strains (Diptera: Tephritidae) to methyl eugenol. *Journal of Chemical Ecology*, **15**, 1399-1405. [RAE **78**:582]

Wong, T.T.Y., Mochizuki, N. & Nishimoto, J.I. (1984b) Seasonal abundance of parasitoids of the Mediterranean and Oriental fruit flies (Diptera: Tephritidae) in the Kula area of Maui, Hawaii. *Environmental Entomology*, **13**, 140-145. [RAE **72**:5344]

Wong, T.T.Y., Nishimoto, J.I. & Mochizuki, N. (1983) Infestation pattern of Mediterranean fruit fly and the Oriental fruit fly (Diptera: Tephritidae) in the Kula area of Maui, Hawaii. *Environmental Entomology*, **12**, 1031-1039. [RAE **72**:3816]

Wong, T.T.Y., McInnis, D.O., Nashimoto, J.I., Ota, A.K. & Chang, V.C.S. (1984a) Predation of the Mediterranean fruit fly (Diptera: Tephritidae) by the Argentine ant (Hymenoptera: Formicidae) in Hawaii. *Journal of Economic Entomology*, **77**, 1454-1458. [RAE **73**:4967]

Woo, F.-c., Whang, Y.-c., Mang, C.-c. & Liang, T.-k. (1963) Studies on the life history and the control of *Lycium* fruit fly, *Neoceratitis asiatica* (Becker) (Diptera, Trypetidae). *Acta Phytophylactica Sinica*, **2**, 387-398. [RAE **52**:566]

Wood, G.W., Crozier, L.M. & Neilson, W.T.A. (1983) Monitoring the blueberry maggot, *Rhagoletis mendax* (Diptera: Tephritidae), with Pherocon AM traps. *Canadian Entomologist*, **115**, 219-220. [RAE **71**:4211]

Wood, R.J. & Harris, D.J. (1989) Genetics; *Ceratitis capitata*; artificial and natural selection, In: Robinson, A.S. & Hooper, G. (eds), Fruit flies; their biology, natural enemies and control. *World Crop Pests*, **3**(B), 19-31. Elsevier, Amsterdam.

Yang, P. (1988) *Status of Fruit Fly Research in China*. Research Institute of Entomology, Zhongshan (Sun Yatsen) University, Guangzhou, China.

Yeates, D. (1990) Queensland fruit fly eradication campaign in Perth. *Myrmecia*, **26**(2), 24-27.

Yong, H.S. (1988) Allozyme variation in the *Artocarpus* fruit fly, *Dacus umbrosus* (Insecta: Tephritidae) from Peninsular Malaysia. *Comparative Biochemistry and Physiology, B (Comparative Biochemistry)*, **91**, 85-89. [RAE **77**:8157]

Yong, H.S. (1990a) Flower of *Couroupita guianensis*: a male fruit-fly attractant of the methyl eugenol group. *Nature Malaysiana*, **15**, 92-97.

Yong, H.S. (1990b) Fruit fly of seashore mangosteen (*Garcinia hombroniana*). *Nature Malaysiana*, **15,** 98-99.

Yount, L. (1981) The fly in our fruit. *International Wildlife*, **11**(6), 17-19.

Yukawa, J. (1984) Fruit flies of the genus *Dacus* (Diptera: Tephritidae) on the Krakatau Islands in Indonesia, with special reference to an outbreak of *Dacus albistrigatus* de Meijere. *Japanese Journal of Ecology*, **34,** 281-288.

Yunus, A. & Ho, T.H. (1980) List of economic pests, host plants, parasites and predators in West Malaysia (1920-1978). *Bulletin, Ministry of Agriculture, Malaysia*, **153,** 1-538.

Zaka-ur-Rab, M. (1978a) Skeleto-muscular mechanism of the cephalo-pharyngeal skeleton of the mature larva of the melon-fly, *Dacus (Strumeta) cucurbitae* Coquillett. *Beiträge zur Entomologie*, **28,** 251-255. [RAE **67**:4308]

Zaka-ur-Rab, M. (1978b) Studies on the tracheal system of the mature larva of the melon-fly, *Dacus (Strumeta) cucurbitae* Coquillett. *Beiträge zur Entomologie*, **28,** 257-262. [RAE **67**:4309]

Zangheri, S. (1962) Due ditteri dannosi - *Trypanea amoena* e *Phorbia platura*. *Informe Fitopatologica*, **12,** 380-382. [RAE **53**:271]

Zervas, G.A. (1982) A new long-life trap for olive fruit fly, *Dacus oleae* (Gmelin) (Dipt., Tephritidae) and other Diptera. *Zeitschrift für Angewandte Entomologie*, **94,** 522-529. [RAE **71**:2760]

Zhang, Z.Y., Wei, Y. & He, D.Y. (1988) Biology of the gall fly, *Procecidochares utilis* (Dip.: Tephritidae) and its impact on croftonweed, *Eupatorium adenophorum*. *Chinese Journal of Biological Control*, **4,** 10-13. [RAE **77**:7459]

Zia, Y. (1937) Study on the Trypetidae or fruit flies of China. *Sinensia, Shanghai* **8,** 103-226.

Zia, Y. & Chen, C.H. (1938) Trypetidae of North China. *Sinensia, Shanghai*, **9,** 1-172.

Zoheiry, M.S. el- (1950) *Leptoxyda (Dacus) longistylus* Wiedemann (Diptera: Trypaneidae), a new pest of cucurbitaceous plants in Upper Egypt. *Proceedings of the VIIIth International Congress of Entomology*, Stockholm, 1948, 721-726. [RAE **39**:301]

Zouros, E. & Loukas, M. (1989) Genetics; biochemical and colonization genetics of *Scaus oleae* (Gmelin), In: Robinson, A.S. & Hooper, G. (eds), Fruit flies; their biology, natural enemies and control. *World Crop Pests*, **3**(B), 75-87. Elsevier, Amsterdam.

Zucchi, R.A. (1979) Novas espécies de *Anastrepha* Schiner, 1868 (Diptera, Tephritidae). *Revista Brasileira de Entomologia*, **23,** 35-41. [RAE **69**:2281]

Zwölfer, H. (1974a) Das Treffpunkt-Prinzip als Kommunikationsstrategie und Isolationsmechanismus bei Bohrfliegen (Diptera: Trypetidae). *Entomologica Germanica*, **1,** 11-20. [RAE **64**:5333]

Zwölfer, H. (1974b) Innerartliche Kommunikationssysteme bei Bohrfliegen. *Biologie in Unserer Zeit*, **5,** 146-153.

Zwölfer, H. (1983) Life systems and strategies of resource exploitation in tephritids, In: Cavalloro, R. (ed.), *Fruit Flies of Economic Importance. Proceedings of the*

CEC/IOBC International Symposium, Athens, 1982. Balkema, Rotterdam. pp. 16-30. [RAE **71**:6261]

Zwölfer, H. (ed.) (1985) Bibliography of fruit fly literature (1977-1983); 'a review of recent publications'. *Bulletin, Section Regionale Ouest Palearctique, Organisation Internationale de Lutte Biologique*, **8**(2), 1-35. [RAE **74**:2533]

Indexes

1; Plant Genera

Names in *italics* are synonyms, misidentifications or misspellings.

2; Insects

Principal entries are shown in **bold** type and illustration pages in *italics*; names in *italics* are, synonyms, misidentifications or misspellings.

3; General

Common names of fruit flies and selected references to non-taxonomic subjects.